Food Processing By-Products and their Utilization

Food Processing By-Products and their Utilization

Edited by

Anil Kumar Anal
Asian Institute of Technology, Thailand

WILEY Blackwell

This edition first published 2018

Registered Offices
John Wiley & Sons, Inc., 111 River Street, Hoboken, NJ 07030, USA
John Wiley & Sons Ltd, The Atrium, Southern Gate, Chichester, West Sussex, PO19 8SQ, UK

Editorial Office
The Atrium, Southern Gate, Chichester, West Sussex, PO19 8SQ, UK

For details of our global editorial offices, customer services, and more information about Wiley products visit us at www.wiley.com.

Wiley also publishes its books in a variety of electronic formats and by print-on-demand. Some content that appears in standard print versions of this book may not be available in other formats.

Library of Congress Cataloging-in-Publication Data

Names: Anal, Anil Kumar, editor.
Title: Food processing by-products and their utilization / edited by Dr. Anil Kumar Anal.
Description: Hoboken, NJ : John Wiley & Sons, 2017. | Includes bibliographical references and index. |
Identifiers: LCCN 2017017235 (print) | LCCN 2017037253 (ebook) | ISBN 9781118432938 (pdf) | ISBN 9781118432891 (epub) | ISBN 9781118432884 (cloth)
Subjects: LCSH: Food processing by-products industry. | Food industry and trade–By-products.
Classification: LCC HD9495.A2 (ebook) | LCC HD9495.A2 .A53 2017 (print) | DDC 664/.08–dc23
LC record available at https://lccn.loc.gov/2017017235

Cover Design: Wiley
Cover Images: (Left image) © Allan Baxter/Gettyimages; (Center image) © IP Galanternik D.U./iStockphoto; (Right image) Fuse/Getty Images

Set in 9.5/11.5pt TimesTenLTstd-Roman by SPi-global, Chennai, India.

10 9 8 7 6 5 4 3 2 1

Contents

About the IFST Advances in Food Science Book Series

The Institute of Food Science and Technology (IFST) is the leading qualifying body for food professionals in Europe and the only professional organisation in the UK concerned with all aspects of food science and technology. Its qualifications are internationally recognised as a sign of proficiency and integrity in the industry. Competence, integrity, and serving the public benefit lie at the heart of the IFST philosophy. IFST values the many elements that contribute to the efficient and responsible supply, manufacture and distribution of safe, wholesome, nutritious and affordable foods, with due regard for the environment, animal welfare and the rights of consumers.

IFST Advances in Food Science is a series of books dedicated to the most important and popular topics in food science and technology, highlighting major developments across all sectors of the global food industry. Each volume is a detailed and in-depth edited work, featuring contributions by recognized international experts, and which focuses on new developments in the field. Taken together, the series forms a comprehensive library of the latest food science research and practice, and provides valuable insights into the food processing techniques that are essential to the understanding and development of this rapidly evolving industry.

The IFST Advances series is edited by Dr Brijesh Tiwari, who is Senior Research Officer at Teagasc Food Research Centre in Ireland.

Forthcoming titles in the IFST series

Herbs and Spices: Processing Technology and Health Benefits, edited by Mohammad B. Hossain, Nigel P. Brunton and Dilip K Rai

List of Contributors

Ali Akbar, Department of Microbiology, Faculty of Life Sciences, University of Balochistan Quetta, Pakistan

Imran Ali, Plant Biomass Utilization Research Unit, Department of Botany, Chulalongkorn University, Bangkok, Thailand and Institute of Biochemistry, Faculty of Life Sciences, University of Balochistan Quetta, Pakistan

Anil Kumar Anal, Food Engineering and Bioprocess Technology, Department of Food Agriculture and Bioresources, Asian Institute of Technology, Klong Luang, Pathumthani, Thailand

Manisha Anand, Food Engineering and Bioprocess Technology, Department of Food Agriculture and Bioresources, Asian Institute of Technology, Klong Luang, Pathumthani, Thailand

Gabriel Arome Ataguba, Aquaculture and Aquatic Resources Management (AARM), Department of Food Agriculture and Bioresources, Asian Institute of Technology, Klong Luang, Pathumthani, Thailand and University of Agriculture, Makurdi, Nigeria

Dattatreya Banavara, Global Innovation, Firmenich Inc, Plainsboro, NJ, USA

Deepak Bhopatkar, Global Research and Development, Mead Johnson Nutrition, Evansville, IN, US

Arup Jyoti Das, Department of Food Engineering & Technology, Tezpur University, Napaam, Sonitpur, Assam, India

Avishek Datta, Agricultural Systems and Engineering, Department of Food Agriculture and Bioresources, Asian Institute of Technology, Pathumthani, Thailand

Sankar Chandra Deka, Department of Food Engineering & Technology, Tezpur University, Napaam, Sonitpur, Assam, India

Lavaraj Devkota, Department of Chemical Engineering, Monash University, Clayton, Australia

Damodar Dhakal, Food Engineering and Bioprocess Technology, Department of Food Agriculture and Bioresources, Asian Institute of Technology, Klong Luang, Pathumthani, Thailand

Taslim Ersam, Department of Chemistry, Faculty of Mathematics and Science, Institut Teknologi Sepuluh Nopember, Surabaya, Indonesia

Zannatul Ferdous, Agricultural Systems and Engineering, Department of Food Agriculture and Bioresources, Asian Institute of Technology, Pathumthani, Thailand

Ganjyal Girish, School of Food Science, Washington State University, Pullman, WA, USA

Juan M. Gonzalez, Global Research and Development, PepsiCo. Barrington, IL, USA

Wan Rosli Bin Wan Ishak, School of Health Sciences, Universiti Sains Malaysia Health Campus, Kubang Kerian, Kota Bharu, Kelantan, Malaysia

Surangna Jain, Food Engineering and Bioprocess Technology, Department of Food Agriculture and Bioresources, Asian Institute of Technology, Klong Luang, Pathumthani, Thailand

Manoj Tukaram Kamble, Aquaculture and Aquatic Resources Management (AARM), Department of Food Agriculture and Bioresources, Asian Institute of Technology, Klong Luang, Pathumthani, Thailand

Mandeep Kaur, Amity Institute of Food Technology, Amity University, Noida, India

Prerna Khawas, Department of Food Engineering & Technology, Tezpur University, Napaam, Sonitpur, Assam, India

Maushmi S. Kumar, Department of Pharmaceutical Biotechnology, Shobhaben Pratapbhai Patel School of Pharmacy and Technology Management, SVKM'S NMIMS, Vile Parle West, Mumbai, India

Navneet Kumar, Department of Processing and Food Engineering, College of Agricultural Engineering & Technology, Anand Agricultural University, Godhra (Gujarat), India

Md. Abdul Matin, Farm Machinery and Postharvest Process Engineering Division, Bangladesh Agricultural Research Institute, Gazipur, Bangladesh

Seema Medhe, Food Engineering and Bioprocess Technology, Department of Food Agriculture and Bioresources, Asian Institute of Technology, Klong Luang, Pathumthani, Thailand

Medina-Meza Ilce Gabriela, Department of Biosystems and Agricultural Engineering. Michigan State University, USA

Didier Montet, Food Safety Team Leader, UMR Qualisud, CIRAD, Montpellier, France

Taslima Ayesha Aktar Nasrin, Postharvest Technology Section, Horticulture Research Centre, Bangladesh Agricultural Research Institute, Gazipur, Bangladesh

Ngo Dang Nghia, Institute of Biotechnology and Environment, Nha Trang University, Vietnam

Zjahra Vianita Nugraheni, Department of Chemistry, Faculty of Mathematics and Science, Institut Teknologi Sepuluh Nopember, Surabaya, Indonesia

Muhammad Bilal Sadiq, Food Engineering and Bioprocess Technology, Department of Food Agriculture and Bioresources, Asian Institute of Technology, Klong Luang, Pathumthani, Thailand

Krishna R. Salin, Aquaculture and Aquatic Resources Management (AARM), Department of Food Agriculture and Bioresources, Asian Institute of Technology, Klong Luang, Pathumthani, Thailand

H.K. Sharma, Food Engineering and Technology Department, Sant Longowal Institute of Engineering and Technology, Sangrur, Punjab, India

Sajal Man Shrestha, Food Engineering and Bioprocess Technology, Department of Food Agriculture and Bioresources, Asian Institute of Technology, Klong Luang, Pathumthani, Thailand

Rahul Shrivastava, Maulana Azad National Institute of Technology, Bhopal MP, India

Manisha Singh, Food Engineering and Bioprocess Technology, Department of Food Agriculture and Bioresources, Asian Institute of Technology, Klong Luang, Pathumthani, Thailand

M.K. Tripathi, Agro Produce Processing Division, ICAR-CIAE, Nabi Bagh, Bhopal MP, India

Kittima Triratanasirichai, Food Engineering and Bioprocess Technology, Department of Food Agriculture and Bioresources, Asian Institute of Technology, Klong Luang, Pathumthani, Thailand

Hayat Ullah, Agricultural Systems and Engineering, Department of Food Agriculture and Bioresources, Asian Institute of Technology, Pathumthani, Thailand

Santad Wichienchot, Interdisciplinary Graduate School of Nutraceutical and Functional Food, Prince of Songkla University, Hat Yai, Songkhla, Thailand

1

Food Processing By-Products and their Utilization: Introduction

Anil Kumar Anal
Asian Institute of Technology, Klong Luang, Pathumthani, Thailand

1.1 Introduction

Food industries are growing rapidly to huge numbers due to globalization and population increase and are providing a wider range of food products to satisfy the needs of the consumers. The major food industries of the world include dairy, fruits and vegetables, meat and poultry, seafood and cereal. However, these industries generate huge amounts of by-products and wastes, which consist of high amounts of organic matter leading to problems regarding disposal, environmental pollution and sustainability (Russ and Pittroff, 2004). In addition, there is the loss of biomass and valuable nutrients that can be used for developing value-added products. Food industries are currently focusing on solving the problems of waste management and recycling by valorization, i.e. utilization of the by-products and discarded materials and developing new value-added products from them for commercial applications. Waste valorization is an interesting new concept that offers a range of alternatives for management of waste other than disposal or land-filling. Valorization allows exploration of the possibility of reusing nutrients in the production of main products, and thus highlights the potential gains that can be achieved. Traditional methods of waste utilization include their use as animal feed, fertilizer or disposal (Jayathilakan *et al.*, 2012). However, their use has been limited due to legal restrictions, ecological problems and cost issues. Therefore, efficient, cheap and ecologically sound methods for utilization of wastes are being focused upon, which can minimize the quantities of wastes exposed to the environment and the subsequent health hazards.

Wastes from the food industries generally comprise of dietary fibers, proteins and peptides, lipids, fatty acids and phenolic compounds, depending on the nature of the product produced. For example, the wastes from meat and poultry industries comprise of proteins and lipids, while waste from fruit and vegetable processing industries and cereal industries comprise of phenolic compounds and dietary fibers. The recovery of

Food Processing By-Products and their Utilization, First Edition. Edited by Anil Kumar Anal.

Table 1.1 Different food processing industries and their wastes (Ezejiofor *et al.*, 2014)

Food processing industry	Waste materials generated
Cereal processing	Husks, hull, rice bran
Fruits and vegetable processing	Skin, peels, pulp, seeds, stem, fiber
Poultry processing	Skin, bones, blood, feathers, liver, intestines
Marine products processing	Viscera, heads, backbones, blood and shells
Dairy products processing	Whey, lactose

these bioactive compounds is important for their commercialization, so that they can be utilized as nutraceuticals and pharmaceutical products.

1.2 Food Processing Wastes and By-Products for Industrial Applications

Food-processing wastes and by-products are generated during processing of the various food products by the industries, which have not already been used for other purposes and have not been recycled. Crude raw materials such as cereals, fruits, vegetables and animals are processed to final products with the production of large amounts of materials in the form of wastes (Ezejiofor *et al.*, 2014). These wastes emerging from the food processing industries differ from one another, depending on the type of product being produced and the production technique used. Even the amount and concentrations of wastes differ and do not remain constant. For example, wastes from the fruit and vegetable processing industries comprise of high concentrations of polyphenols and dietary fibers, whereas wastes from meat processing industries comprise of high protein and fat content. The food processing wastes also possess characteristics, such as large amounts of organic materials in the form of lipids, proteins and carbohydrates and high chemical oxygen demand (COD) and biochemical oxygen demand (BOD) (Ezejiofor *et al.*, 2014). Hence, they are harmful and affect the environment and human health. Appropriate technologies that focus on their reuse for creation of valuable products, whose costs exceed the costs of reprocessing, should be considered. The different types of wastes produced by the different food processing industries are listed in Table 1.1.

1.3 By-Products from Cereal Processing Industries

Cereals are the edible seeds derived from plants, which are a good source of carbohydrates. They contribute to 60% of the total world food production (Krishna and Chandrasekaran, 2013), with the main seeds being maize and wheat. Wastes from cereal processing are produced during the harvesting period, post-harvesting and the production period. Presently, these by-products are used as animal feed. However, they need to be utilized more efficiently as they comprise of proteins, dietary fibers and small amounts of unsaturated fatty acids.

Rice bran is an important cereal industry by-product, which is generated during the production of white rice. It is generated during the milling process, where it is separated from the rice to produce white rice. The rice bran production is 60–66 million tonnes annually (Ryan, 2011) and it is mostly used as animal feed or in the production of edible cooking oil. Rice bran is a rich source of nutrients, proteins and peptides, with a wide range of nutritional and functional applications. Defatted rice bran is another by-product, which is produced after oil extraction from the rice bran, also a good source of proteins and dietary fibers (Anal, 2013a). It are currently being utilized in food supplements and in the production of bakery items.

1.4 Fruits and Vegetables By-Products

The world production of fruits and vegetables has increased rapidly. As crop production increases, there is a concomitant increase in the quantity of by-products generated (FAO, 2009). The fruit and vegetable processing by-products are regarded as waste and disposed of in the environment, which causes ecosystem problems as they are prone to microbial degradation. However, fruit and vegetable by-products and wastes are very good sources of bioactive compounds, such as dietary fibers and phenolic compounds with antibacterial, cardio-protective and antitumor activities (Khao and Chen, 2013). Efforts are being made to develop methods to reuse these wastes and by-products by obtaining bioactive compounds for health benefits, profit-making and allowing their environmental-friendly disposal.

The total worldwide production of citrus fruits was reported as 7.78 million tonnes in 2009 (FAO, 2009). These include oranges, lemons, grapefruits and limes They are commonly used forms are as fresh pulps or juice, but following their processing, the by-products such as peels, pulp and seeds remain that make up 50% of the fresh fruit weight (Khao and Chen, 2013). From these wastes, fibers, flavanoids, pectins and limonene can be produced. The major flavanoids found in the citrus peels and seeds include hesperidin, narirutin, naringin and eriocitrin (Mouly *et al.*, 1994). These flavanoids have found to have antioxidant activities (Manthey *et al.*, 2001). Limonin, nimolin and nomilinic acid are major limonoids found mainly in the peels, and demonstrate antibacterial, antiviral and antimicrobial activities (Djilas *et al.*, 2009).

Banana is the largest growing tropical fruit following citrus fruits, contributing to 16% of total fruit production worldwide (Mohapatra *et al.*, 2010). Waste from banana products includes the peels that represent about 40% of the total weight of the fresh bananas (Tchobanoglous *et al.*, 1993). These peels are utilized in animal feed and the preparation of banana chips and banana powder. However, still huge amounts of the peels are being under-utilized and disposed of, resulting in environmental pollution. These banana wastes contain dietary fibers, proteins and different bioactive compounds such as phenolic compounds with reported antioxidant activities (Anal *et al.*, 2014). Hence they need to be recycled so that they can be used for producing various valuable products.

Mango (*Mangifera indica* L., Anacardiaceae) is a common seasonal fruit, which is mainly processed to produce products such as juices, pickles, purees and canned products (Aslam *et al.*, 2014). Recent researches have indicated that mango wastes, which mostly include the peels (7–24%) and the kernels (9–40%), are good sources of bioactive compounds. The mango peels comprise of functional compounds such as polyphenols, carotenoids, vitamins C and E, dietary fibers and natural antioxidants (Ajila *et al.*, 2007), whereas the kernels are sources of essential amino acids like lysine, valine and leucine (Abdalla *et al.*, 2007), phenolic compounds, edible oils and high amounts of unsaturated fatty acids. These wastes show huge potential to be used as valuable ingredients for the purpose of making functional foods.

Mangosteen (*Garcinia mangostana* L) is a popular fruit of several Asian countries. However, the increasing consumption of this fruit has led to the generation of ample abandoned mangosteen pericarps. It has been reported that 10 kg of harvested mangosteens lead to the generation of about 6 kg of pericarps (Mohammad *et al.*, 2014). These pericarps are woody in texture, comprising of bitter substances such as xanthones, tannins and anthocyanins (Lim *et al.*, 2013) that have medicinal properties and are being used as dietary supplements. The therapeutic benefits of these components include hypolipidemia, anti-inflammatory, anti-microbial and anti-carcinogenic properties (Zafra-Stone *et al.*, 2007; Mishra *et al.*, 2016). Another by-product from the processing of mangosteens is their seeds, which contain 21.18% oil (Ajayi *et al.*, 2006) with essential and non-essential fatty acids. They have been reported to be safe for the heart and liver; hence they can be used as edible oils.

The apple processing wastes are termed apple pomace, which makes up 25–35% of the total apple wastes (Dijlas *et al.*, 2009). The apple pomace includes the peels, seeds, stems, core and the soft tissues. They are good sources of polyphenols, which are mainly present in the peels such as catechin, quercetin, hydroxycinnamates, chlorogenic acid and epicatechins (Mamma *et al.*, 2009) and pectins, proteins and vitamins. However, they are mainly utilized in the production of pectins, which can be co-precipitated out from the apple pomace. These pectins demonstrate good gelling properties, even better than citrus pectins.

Tomato is an important vegetable, with a world total production of 141 million tonnes in 2009 (FAO, 2009). The major products produced using tomatoes are soups, ketchup, juice and paste. Along with their high consumption, there is the generation of huge amounts of by-products and wastes accounting for 40% of the total fresh weight of the tomatoes. These include the seeds (33%), peels (27%) and the pulp (40%) (Encinar *et al.*, 2008; Kaur *et al.*, 2008). These wastes are good sources of proteins (35%) and fats (25%) (Anal *et al.*, 2013a). In addition, they contain high amounts of unsaturated fatty acids due to which the tomato seed oil is used as edible oil. Lycopene, an important carotenoid, is also present in large amounts in tomato wastes.

Carrot processing, for the production of carrot juice, generates wastes in the form of peels and pomace (Chantaro *et al.*, 2008). These wastes make up 12% of the fresh carrot weight and comprises of several valuable compounds such as carotenes, uronic acids and sugars, which are generally discarded or used in feeds and fertilizers. These compounds have important beneficial properties and hence can be utilized for value addition. The carrot waste also contains huge amounts of fibers, including cellulose, hemicelluloses, lignin and pectin (Nawirska and Kewasniewska, 2005). Studies are being done to recover these fibers from the carrot waste residues, as they have been reported to have cholesterol-lowering effects that can protect against coronary heart diseases. Also, various attempts are being made to incorporate the valuable compounds from carrot wastes into the production of functional foods and beverages.

The total world onion production in 2009 was reported to be 72 million tonnes (FAO, 2009). During the processing of onions, the major wastes that are generated are the peels and roots. They are a serious threat to environmental pollution, as they are not suitable for fodder because of their aroma or as fertilizers due to the fast development of phytogenetic agents, and also they contribute to toxicity in animals during digestion

(Bello *et al.*, 2013). Hence, new applications need to be found for these wastes, which contain high amounts of polyphenols and dietary fibers.

Cabbage is also a vegetable that has a high production yield; however, since it is consumed either in the raw form or the fresh form, wastes generated are very little. The main wastes are their outer leaves which are disposed off. These leaves can be used mainly for the production of biofuels by the anaerobic digestion process (Liu *et al.*, 2006).

1.5 By-Products from the Meat and Poultry Processing Industries

Meat and poultry processing generates a number of organic by-products like bones, blood, feathers, head etc. (Lasekan *et al.*, 2013). The majority of these by-products are produced during the slaughtering process. The slaughterhouse waste comprises of the portion that cannot be utilized or sold as meat. This includes bones, skin, blood and internal organs (Lasekan, *et al.*, 2013). Currently, these wastes are under-utilized, discarded and disposed of in landfills. However, they must be dealt with efficiently, as the growth of these industries mainly depends on the management of their by-products (Jayathilakan *et al.*, 2012). The disposal of these wastes can also be difficult, due to their high water content, susceptibility to oxidation and changes caused by enzymatic activity that results in serious environmental pollution and hazards. Hence, it is essential to find applications for these wastes, which are becoming a serious environmental issue.

Blood is the first and most inevitable by-product of the meat and the poultry industries, which is a major problem due to its high pollutant load. However, blood comprises of a number of compounds that have potential value and is a good source of proteins which makes it an important edible by-product (Jayathilakan *et al.*, 2012). Blood from a healthy animal is generally sterile. It will be approved for use in food products, if it has been obtained from bleeding a healthy animal. Due to an increasing trend in worldwide protein deficiency, usage of animal blood as a source of protein should be investigated and further extraction of bioactive peptides can be carried out to allow for large-scale utilization of the blood.

A great amount of poultry feathers of about 1.8 million tones are generated every year in the form of wastes (Wang and Cao, 2012). These feathers are an important waste product and are used mainly as animal feed; however, research is being made into their new applications. Feathers are composed of 90% proteins with the main one being keratin (Wang and Cao, 2012). The remainder comprises of 1% lipids and 8% water. Keratins are the major structural proteins found in feathers and are characterized by high amounts of cysteine and hydroxyl amino acids such as serine.

Bones are not usually consumed and have no value for the meat and poultry processing industries; hence they are discarded. Approximately 16–45 million tons of bones are discarded worldwide (Dong *et al.*, 2014). They can also be utilized in feed products, as they comprise of proteins, calcium, essential minerals and lipids, which are useful for bodily function. Therefore, studies about comprehensive utilization of bones are required for developing an effective way to utilize the huge amount of bones as potential protein sources.

Skin is also an important and valuable by-product obtained from animals. Just like bones, the skin also contains huge amounts of proteins such as collagen. Gelatin is also one important protein that can be obtained after the hydrolysis of collagen under controlled conditions (Jayathilakan *et al.*, 2012). Both of these proteins have been reported to have various functional and biological properties.

Another product of the poultry industry, which is largely produced and consumed, is poultry eggs. Their high nutritional value and relative low cost has led to their increased production worldwide. According to the FAO, global egg production in 2012 was reported as 65 million tonnes (FAO, 2012), which includes all types of eggs, including hatching eggs. However, the egg processing industries generate huge amounts of wastes, of about 1.5 million tonnes annually, in the form of shell wastes (Wei *et al.*, 2009), which are discarded and disposed of in landfills. This contributes to environmental pollution and hazards and loss of potential revenues. By-products of the egg-processing industries comprise of the eggshells and the eggshell membrane (ESM) that represents 11% of the total egg weight (Stadelman, 2000). The ESM mainly are a very good source of bioactive compounds such as proteins and polysaccharides, together with high amounts of polypeptides (Zhao and Chi, 2009; Jain and Anal, 2016). Collagen makes up 10% of the total proteins, whereas the rest (70–75%) comprises of the glycoproteins. Due to their high protein content, they can be used for production of proteins, and peptides from them can be used in a wide range of food and nutraceutical applications.

1.6 Seafood Processing By-Products

Marine organisms are an important food source for many countries and contain value-added compounds such as lipids, amino acids, proteins and polysaccharides, which are crucial for human health. Industrial processing of these marine organisms leads to the generation of huge amounts of waste that are either discarded or used as fertilizers and fish meals. By-products from seafood processing include viscera, heads, backbones, skin, tail, blood and shells, which comprise of important bioactive compounds that can be used in pharmaceutical and nutraceutical applications (Anal *et al.*, 2013b). Some of the valuable components that can be obtained from seafood processing are the bioactive peptides, proteins such as collagen, polyunsaturated fatty acids and chitin (Suresh and Prabhu, 2013).

Collagen is a major protein obtained from seafood processing by-products, which are mainly obtained from the skin, bone, tendons etc. (Regenstein and Zhou, 2007). They have a wide range of applications, such as gel formation, water binding, formation of stable emulsions and formation of films (Gomez-Guillen *et al.*, 2011). They are also a good source of bioactive peptides. Gelatin, another protein, can also be derived from collagen, which have many applications in food industries. They can be used as food additives for improving the texture and stability of food products such as meat, bakery goods etc. (Mariod and Fadul, 2013). They are also used in the pharmaceutical industries for making capsules and tablet coatings.

Proteins from seafood by-products can be used to recover protein hydrolysates and peptides, by using various methods such as chemical hydrolysis, enzymatic hydrolysis, microbial fermentation, microwave and ultrasonic irradiation (Anal *et al.*, 2013b). These protein hydrolysates and peptides possess strong biological activities such as antioxidant, antimicrobial and antihypertensive.

Marine fishes are mainly very good sources of polyunsaturated fatty acids such as omega-3 fatty acids. The by-products of fish processing, such as viscera, stomach, liver etc. can be used to recover polyunsaturated fatty acids with nutraceutical and pharmaceutical applications (Analava *et al.*, 2014). Fatty acids such as eicosapentenoic acid (EPA) and docosahexenoic acid (DHA) can be obtained by molecular distillation. These omega-3 fatty acids have remarkable health benefits, such as protective effects against cardiovascular diseases, nerve and brain disorders and anti-inflammatory effects in diseases like Crohn's disease and kidney diseases.

Processing of crustaceans, such as shrimp and crab, generate solid by-products from which chitin can be obtained. Chitin is a linear amino polysaccharide and the most abundant biopolymer (Tharanathan and Kittur, 2003). They can be extracted from the crustacean by-products by enzymatic methods and fermentation by lactic acid bacteria. They also have a wide range of biological applications such as in edible packing, as bio-preservatives, food additives and nutritional and functional ingredients.

1.7 By-Products from the Dairy Processing Industries

The dairy industries are also major food processing industries that generate large amounts of by-products and waste during the manufacturing of various dairy food products and milk processing. These wastes contain high amounts of proteins, lipids, vitamins etc. and their utilization for the purpose of value addition can greatly enhance the profit of the dairy industries. Whey is a major by-product generated during the manufacturing of cheese, cottage cheese etc. which can be subdivided into rennet whey and acid whey. The whey comprises of high amounts of lactose and proteins. The whey proteins are composed of a number of proteins with a very high biological value, more than that of casein and soy proteins (Mandal *et al.*, 2013). Lactose, on the other hand, can be used for the production of organic acids such as citric acid, gluconic acid and lactic acid by the process of microbial fermentation.

1.8 Conclusion

The food processing industries will continue to grow throughout the world, along with the demands of the consumers. This will also result in the generation of huge quantities of by-products and wastes that are currently being under-utilized. However, due to the growing concerns regarding environmental conservation, intensive research needs to be carried out such that the food wastes can be utilized for the purpose of value addition and human consumption. This will lead to maximum benefits to the industries, environment and the consumers.

References

Abdalla, A.E.M., Darwish, S.M., Ayad, E.H.E. and El-Hamahmy, R.M. (2007) Egyptian mango by-product. 1: Compositional quality of mango seed kernel. *Food Chemistry*, **103**: 1134–1140.

Ajayi, I.A., Oderinde, R.A., Ogunkoya, B.O., Egunyomi, A. and Taiwo, V.O. (2006) Chemical analysis and preliminary toxicological evaluation of *Garcinia mangostana* seeds and seed oil. *Journal of Food Chemistry*, **101**: 999–1004.

Ajila, C.M., Naidu, K.A., Bhat, S.G. and Prasada Rao, U.J.S. (2007) Bioactive compounds and antioxidant potential of mango peel extract. *Food Chemistry*, **105**: 982–988.

Anal, A.K. (2013a) Food processing by-products. In: *Handbook of Plant Food Phytochemicals: Sources, stability and extraction*, Chapter 8 (eds B. Tiwari, N.P. Brunton and C.S. Brennan). Wiley Science Publishers, Oxford, pp. 180–192.

Anal, A.K., Noomhorm, A. and Vongsawasdi, P. (2013b) Protein hydrolysates and bioactive peptides from seafood and crustacean waste: Their extraction, bioactive properties and industrial perspectives. In: *Marine Proteins and Peptides: Biolgical activities and applications*, Chapter 36 (ed. S.K. Kim). John Wiley & Sons, Chichester, UK, pp. 709–735.

Anal, A.K., Jaisanti, S. and Noomhorm, A. (2014) Enhanced yield of phenolic extracts from banana peels (*Musa acuminata* Colla AAA) and cinnamon barks (*Cinnamomum varum*) and their antioxidative potentials in fish oil. *Journal of Food Science Technology*, **51**(10): 2632–2639.

Analava, M., Baishakhi, D. and Anindya, M. (2014) Recovery of omega-3 health boosters from fisheries and poultry wastes and their micro-delivery techniques. *International Journal of Drug Delivery Science*, **1**(1): 1–13.

Aslam, H.K.W., Raheem, M.I.U., Ramzan, R., Shakeel, A., Shoaib, M. and Sakandar, H.A. (2014) Utilization of mango waste material (peel, kernel) to enhance dietary fiber content and antioxidant properties of biscuit. *Journal of Global Innovation Agricultural Social Science*, **2**(2): 76–81.

Bello, M.O., Olabanji, I.O., Abdul-Hammad, M. and Okunade, T.D. (2013) Characterization of domestic onion wastes and bulb fatty acids and metal contents. *International Food Research Journal*, **20**(5): 2153–2158.

Chantaro, P., Devhastin, S. and Chiewchan, N. (2008) Production of antioxidant high dietary fiber powder from carrot peels. *LWT – Food Science and Technology*, **41**(10): 1987–1994.

Dijlas, S., Canadanovic, B.J. and Cetkovic, G. (2009) By-products of fruit processing as a source of phytochemicals. *Chemical Industry and Chemical Engineering*, **15**: 191–202.

Dong, X.B., Li, X., Zhang, C.H., Wang, J.Z., Tang, C.H. *et al.* (2014) Development of a novel method for hot-pressure extraction of protein from chicken bone and the effect of enzymatic hydrolysis on the extracts. *Food Chemistry*, **157**: 339–346.

Encinar, J.M., Gonzalez, J.F. and Martinez, G. (2008) Energetic use of the tomato plant waste. *Fuel Processing Technology*, **89**: 1193–1200.

Ezejiofor, T.I.N., Enebaku, U.E. and Ogueke, C. (2014) Waste to wealth-value recovery from agro-food processing wastes using biotechnology: A review. *British Biotechnology Journal*, **4**(4): 418–481.

FAO (2009) *High Level Expert Forum – How to Feed the World in 2050*. FAO, Rome.

Gómez-Guillén, M., Giménez, B., López-Caballero, M.A. and Montero, M. (2011) Functional and bioactive properties of collagen and gelatin from alternative sources: A review. *Food Hydrocolloids*, **25**(8): 1813–1827.

Jain, S. and Anal, A. K. (2016) Optimization of extraction of functional protein hydrolysates from chicken Egg Shell Membrane (ESM) by Ultrasonic Assisted Extraction (UAE) and Enzymatic Hydrolysis. *LWT – Food Science and Technology*, **69**: 295–302.

Jayathilakan, K., Sultana, K., Radhakrishna, K. and Bawa, A.S. (2012) Utilization of by-products and waste materials from meat, poultry and fish processing industries: A review. *Journal of Food Science and Technology*, **49**(3): 278–293.

Kaur, D., Wani, A.A., Oberoi, D.P.S. and Sogi,D.S. (2008) Effect of extraction conditions on lycopene extractions from tomato processing waste skin using response surface methodology. *Food Chemistry*, **108**: 711–718.

Khao, T.H. and Chen, B.H. (2013) Fruits and vegetables. In: *Valorization of Food Processing By-Products*, Chapter 18 (ed. M. Chandrasekaran). Taylor and Francis Group, Boca Raton, FL, pp. 517–557.

Krishna, J.G. and Chandrasekaran, M. (2013) Cereals. In: *Valorization of Food Processing By-Products*, Chapter 12 (ed. M. Chandrasekaran). Taylor and Francis Group, Boca Raton, FL, pp. 303–330.

Lasekan, A., Bakar, F.A. and Hashim, D. (2013) Potential of chicken by-products as sources of useful biological resources. *Waste Management*, **33**: 552–565.

Lim, Y.S., Lee, S.S.H. and Tan, B.C. (2013) Antioxidant capacity and antibacterial activity of different parts of mangosteen (*Garcinia mangostana* Linn) extracts. *Fruits*, **68**(6): 484–489.

Liu, D., Liu, D., Zeng, R.J. and Angelidaki, I. (2006) Hydrogen and methane production from household solid waste in the two-stage fermentation process. *Water Resources*, **40**: 2230–2236.

Mamma, D., Topakas, E., Vafiadi, C. and Christakopoulos, P. (2009) Biotechnological potential of fruit processing industry residues. In: *Biotechnology for Agro-Industrial Residue Utilization*, Chapter 14 (eds P.S. Nigam and A. Pandey). Springer Press, Hiedelberg, pp. 273–291.

Mandal, S., Puniya, M., Sangu, K.P.S., Dagar, S.S., Singh, R. and Puniya, A.K. (2013) Dairy By-Products: Wastes or resources. In: *Valorization of Food Processing By-Products*, Chapter 21 (ed. M. Chandrasekaran). Taylor and Francis Group, Boca Raton, FL, pp. 617–648.

Manthey, J.A., Grohmann, K. and Guthrie, N. (2001) Biological properties of citrus flavanoids pertaining to cancer and inflammation. *Current Medicinal Chemistry*, **8**(2): 135–153.

Mariod, A.A. and Fadul, H. (2013) Review: Gelatin, source, extraction and industrial applications. *ACTA Scientiarum Polonorum Technologia Alimentaria*, **12**(2): 135–147.

Mishra, S., Kumar, M.S., Stanley H.R.C. and Anal, A.K. (2016) Modulation of digestive enzymes and lipoprotein metabolism by alpha mangosteen extracted from mangosteen (*Garcinia Mangostana*) fruit peels. *Journal of Microbiology, Biotechnology and Food Sciences*, **6**(1): 717–721.

Mohammad, M.A., Shitu, A., Tadda, M.A. and Ngabura, M. (2014) Utilization of various agricultural waste materials in the treatment of industrial wastewater containing heavy metals: A review. *International Research Journal of Environmental Sciences*, **3**(3): 62–71.

Mohapatra, D., Mishra, S. and Sutar, N. (2010) Banana and its by-product utilization: An overview. *Journal of Scientific and Industrial Research*, **69**: 323–329.

Mouly, P.P., Arzouyan, C.R., Gaydou, E.M. and Estienne, J.M. (1994) Differentiation of citrus juices by factorial discriminant analysis using liquid chromatography of flavones glycosides. *Journal of Agricultural Food Chemistry*, **42**: 70–79.

Nawirska, A. and Kwasniewska, M. (2005) Dietary fiber fractions from fruit and vegetable processing waste. *Food Chemistry*, **91**: 221–225.

Regenstein, J.M. and Zhou, P. (2007) Collagen and gelatin from marine by-products. In: *Maximising the Value of Marine By-Products*, 1st edition (ed. F Shahidi). Woodhead Publishing Ltd, Cambridge, pp. 279–303.

Russ, W. and Pitroff, R. (2004) Utilizing waste products from the food production and processing industries. *Critical Reviews in Food Science and Nutrition*, **44**(1): 57–62.

Ryan, P.E. (2011) Bioactive food components and health properties of rice bran. *Journal of the American Veterinary Medical Association*, **238**(5): 593–600.

Stadelman, W.J. (2000) Eggs and egg products. In: *Encylopaedia of Food Science and Technology* (ed. F.J. Francis). John Wiley & Sons, New York, pp. 593–599.

Suresh, P.V. and Prabhu, N. (2013). *Valorization of Food Processing By-Products* (ed. M. Chandrasekaran). Taylor and Francis Group, Boca Raton, FL, pp. 687–736.

Tchobanoglous, G., Theisen, H.M. and Vigil, S.A. (1993) *Integrated Solid Waste Management: Engineering principals and management issues*. McGraw Hill, New York, pp. 3–22.

Tharanathan, R.N. and Kittur, F.S. (2003) Chitin – The undisputed biomolecules of great potential. *Critical Reviews in Food Science and Nutrition*, **43**(1): 61–87.

Wang, Y.X. and Cao, X.J. (2012) Extracting keratin from chicken feathers by using a hydrophobic ionic liquid. *Process Biochemistry*: **47**: 896–899.

Wei, Z., Xu, C. and Li, B. (2009) Application of waste eggshell as low-cost solid catalyst for biodiesel production. *Bioresource Technology*, **100**(11): 2883–2885.

Zafra-Stone, S., Yasmin, T., Bagchi, M., Chatterjee, A., Vinson, J.A. and Bagchi, D. (2007) Berry anthocyanins as novel antioxidants in human health and disease prevention. *Molecular Nutrition Food Research*, **51**: 675–583.

Zhao, Y.H. and Chi, Y.J. (2009) Characterization of collagen from eggshell membrane. *Biotechnology*, **8**: 254–258.

2

Fruit Processing By-Products: A Rich Source for Bioactive Compounds and Value Added Products

Medina-Meza Ilce Gabriela[1] and Ganjyal Girish[2]

[1]*Biosystems and Agricultural Engineering, Michigan State University, USA*
[2]*School of Food Science, Washington State University, USA*

2.1 Introduction

The expansion of fruit-processing worldwide has generated huge quantities of fruit wastes (Ayala-Zavala *et al.*, 2011). Fruits are processed into various fruit-based products, such as juices, jams, jellies, concentrates, alcoholic beverages, vinegar etc. Fruit pomace, consisting of peel, seeds, core, stems and exhausted soft tissue, is the left-over solid biomass obtained as a by-product during the processing of fruits. In the tropics, fruits such as mango, pineapple, passion fruit and papaya contribute to higher fruit pomace generation and in the sub-tropics, mainly apples, grapes, oranges and berries generate higher processing by-products as pomace. The chemical composition of fruit pomace varies according to the type of fruit. Fruit pomace possesses a high phytochemical content that can be recovered for secondary food and non-food applications (Djilas *et al.*, 2009). Currently, this is used as a cattle feed and the potential use of various fruit pomace as functional foods has been evaluated in various studies (Nawirska and Kwasniewska, 2005; Sun-Waterhouse, 2011). On the other hand, the negative concerns of the consumer versus synthetic products, has led to the possibility of using such by-products as an alternative source of natural antioxidants, especially considering the higher demand of additives to prevent lipid oxidation and oxidative rancidity in meat products, as well as to retard development of off-flavors, and to improve color stability.

This chapter focuses on the presence of bioactive functional ingredients in fruit-processing by-products, mainly fruit pomaces, and presents an overview of their value-added qualities. In the first sections of this chapter, a brief and comprehensive view of the different bioactive compounds in fruit pomaces is presented. Then, several categories of agro-industry by-products are discussed, with a particular emphasis

Food Processing By-Products and their Utilization, First Edition. Edited by Anil Kumar Anal.

on their potential as sources of bioactive molecules. Finally, comments on future directions in the field are provided.

2.2 Phenolic Compounds as Functional foods

Phenolic compounds, or polyphenols, constitute one of the largest and widely-distributed groups of phytochemicals. More than 8000 phenolic structures are known and among them around 4000 flavonoids have been identified (Tsao, 2010). Polyphenols are secondary metabolites that are derivatives of the pentose phosphate, shikimate and phenylpropanoid pathways in plants (Boudet, 2007; Tsao, 2010). They comprise a wide variety of molecules that have a polyphenol structure (i.e. several hydroxyl groups on aromatic rings), but also molecules with one phenol ring such as phenolic acids and phenolic alcohols. Phenolics have considerable physiological and morphological importance for plants; they play an important role in plant pigmentation, reproductions, UV-light protection, antioxidative and anti-feedant effects, as well as providing protection against pathogens and predators (Treutter, 2006).

Currently, there is an enormous interest in this class of compounds, due to their capacity to improve public health through their intake, where preventative health care can be promoted through a diet rich in fruit and vegetables. Studies have shown that phenolic compounds exhibit an extensive range of physiological properties; they may prevent degenerative diseases, and cardiovascular and neurodegenerative diseases, as well as some types of cancers (Tsao, 2010). Their potent antioxidants properties and their effects in prevention of oxidative stress-associated diseases are also known (Scalbert *et al.*, 2005). Fruits, vegetables, whole grains, tea, chocolate and wine are rich sources of polyphenols and natural antioxidants. The most relevant groups of phenolic compounds for human health are phenolic acids, flavonoids, tannins, stilbenes and lignans. Generally, one or more sugar residues are linked to hydroxyl groups. These sugars can be present as monosaccharides, disaccharides or oligosaccharides (Bravo, 1998), and may also occur as functional derivatives such as esters and methyl esters (Harborne and Baxter, 1999).

2.2.1 Phenolic Acids

Phenolic acids comprise about a one-third part of dietary phenols, which can be present in the plant kingdom as free and bounds forms (Robbins, 2003). Phenolic acids consist of two subgroups, i.e. the hydroxycinnamic acids and the hydroxybenzoic acids. Hydroxycinnamic acids include aromatic compounds with a three-carbon side chain (C_6–C_3), with caffeic, ferullic, *p*-coumaric and sinapic acids being the most representatives. On the other hand, hydroxybenzoic acids are p-hydroxybenzoic acids, protocatechuic acids, and vanillic, syringic and gallic acids, the latter being the most representative of this group, which have in common the C_6–C_1 structure (Bravo, 1998). While fruits and vegetables contain many free phenolic acids, in seeds and grains they are often presents in their bound form, especially in bran or hull (Shi *et al.*, 2003). Bound phenolic acids can be released by alkaline or acid hydrolysis, or even by enzymatic catalysis.

2.2.2 Flavonoids

Flavonoids are the most-common and widely-distributed group of plant phenolics, with exception of the algae and fungi kingdoms. They are low molecular weight compounds, with a C6–C3–C6 general structural backbone in which two C6 units (Ring A and Ring B) are of a phenolic nature (Tsao, 2010). Flavonoids can be further divided into sub-groups such as flavonols (or catequins), flavan-2-ols, flavones, flavanones, isoflavones and anthocyanins (Bravo, 1998). Flavonoids rarely appear in plants as aglycones, as they usually exist as glycosides. Flavonoids are important antioxidants as a result of their high redox potential, which allows them to act as hydrogen doncrs, reducing agents and singlet oxygen quenchers. Beyond that, they also present a metal chelating potential (Tsao, 2010; Tsao *et al.*, 2003).

2.2.2.1 Isoflavones Isoflavones have structural similarities to estrogens, and like estradiol molecules, they have a ring B attached to the C3 position of ring C. These phytochemicals are found in many plants, mostly in the leguminous family. Genistein and daidzein are two main isoflavones found in soy beans, along with glycetein, biochanin A and formononetin; they are also found in red clovers (Tsao, 2010). The open ring chalcones can be found in fruits such as apples, and beverages as beers (Zhao *et al.*, 2005).

2.2.2.2 Flavones, Flavanols, Flavanones, Flavonols and Flavanonols Flavones and their 3-hydroxyl derivatives flavonols, including their glycosides, methoxides and acylated products on all three rings, form the largest subgroup among polyphenols. Flavanones, such as naringenin and hesperidin, are especially abundant in citrus foods and prunes. Taxifolin is a well-known flavanonol from citrus fruits (Grayer *et al.*, 2006). On the other hand, flavanols (also called catechins) differ from other flavonoids, because there are no double bonds between C2 and C3, and no C4 carbonyl in ring C.

Catechin and epicatechin are monomeric flavanols, usually are found in fruits, especially in the skins of grapes, apples and blueberries (Awad *et al.*, 2000; Määttä-Riihinen *et al.*, 2005; Wolfe *et al.*, 2003); their derivatives (i.e. gallocatechins) are the major flavonoids in tea leaves and cacao (Lee *et al.*, 2003). These monomers can form polymers, which are usually referred to as proanthocyanidins, because an acid-catalyzed cleavage of the polymeric chains generates anthocyanidins (Tsao, 2010).

2.2.2.3 Anthocyanins Anthocyanins (from Greek *anthos* = flower and *kianos* = blue) are the most important group of water soluble plant pigments and are responsible for the color of flowers, leaves, stems, roots and fruits. Anthocyanidin is the basic structure of these compounds, and when they are found in their glycoside form (bonded to a sugar moiety), they are called anthocyanins. They may appear pink, red, purple or blue, depending on pH. However, the isolated anthocyanins are highly instable and susceptible to degradation (Giusti and Wrolstad, 2003). Other factors, such as degree of hydroxylation or methylation arrangement of the aromatic rings, storage temperature, light, oxygen, solvents, enzymes, proteins and metallic ions, can also affect their color (Rein, 2005).

Anthocyanins can act as antioxidants by donating hydrogen to highly reactive radicals. Their antioxidant potential is dependent on the number and arrangement of the hydroxyl groups and structural conjugation, as well as the presence of electron-donating/withdrawing substitutions in the ring structure (Lapornik *et al.*, 2005). Anthocyanins are potential substitutes for prohibited food dyes, providing additional health benefits during their intake.

2.2.3 Tannins

Tannins are compounds of intermediate to high molecular weight. Goldstein and Swain (1963) defined plant tannins as water-soluble phenolic compounds having a molecular weight. Tannins have been found in carob pods; they are highly hydroxylated molecules and can form insoluble complexes with carbohydrates and protein. This capacity is responsible for the astringency of tannin-rich food, due to the precipitation of salivary proteins (Bravo, 1998). Within this general character, tannins exhibit a number of various bioactivities, which are often related to their antioxidant activity. Tannins are classified into two major groups on the basis of their structure: the hydrolysable tannins and the condensed tannins. Plants are able to biosynthesize gallotannins, ellagitannins, or a mixture of both types of hydrolysable tannins. While condensed tannins are present in many species of higher plants, the presence of hydrolysable tannins is limited to Angiospermae and Dicotyledons (Koleckar *et al.*, 2008).

2.2.3.1 Hydrolysable Tannins Hydrolysable tannins are compounds containing a central core of glucose or another polyol esterified with gallic acid. As the name indicates, these compounds are easily hydrolyzed by acid alkali, hot water and enzymic action, which produce polyhydric alcohol and phenylcarboxylic acid. According to this, they are also called gallotannins, or with hexahydroxydiphenic acid, are called ellagitannins (Bravo, 1998). Gallotannins consist of a central molecule, such as glucose, surrounded by gallic acid units. Tannic acid is the best know hydrolysable tannin, consisting of a gallotanin with a pentagalloyl glucose molecule.

2.2.3.2 Condensed Tannins Condensed tannins are a major group of oligomeric and polymeric dietary polyphenols made up of flavan-3-ol and flavan-3,4-diols widely-distributed in plant foods, where they affect sensory properties such as astringency, bitterness, aroma and color. They are also named as Proanthocyanidins, because they decompose to anthocyanidins in heated ethanol solutions. The most frequent basic units of condensed tannins are derivatives of flavan-3-ols: (+)-catechin, (-)-epicatechin, (+)-gallocatechin and major polyphenols of green tea: (–)-pigallocatechin and (–)-epigallocatehin gallate (Koleckar *et al.*, 2008). Proanthocyanidins can occur as polymers with 50 degrees of polymerization and greater, having molecular weights of 5000 Da. Proanthocyanidins are classified according to their hydroxylation pattern into several subgroups, i.e. procyanidins and prodelphinidins (Cos *et al.*, 2004).

Procyanidins of the B-type (dimeric) and C-type (trimeric) are characterized by single linked flavanyl units, usually between C-4 of the flavan-3-ol of the upper unit and C-6 or C-8 of the lower unit, while Proanthocyanidins of the A-type possess an additional ether linkage between C-2 of the upper unit and a 7- or 5-OH of the lower unit. (Koleckar *et al.*, 2008). The largest group of proanthocyanidins is formed by procyanidins. Procyanidin B-1 is present in grapefruit, sorghum and cranberries, B-2 in apples, cocoa beans and cherries, B-3 in strawberries and hops, and B-4 in raspberries and blackberries (Xie and Dixon, 2005). Furthermore, red wine, green tea, cocoa and chocolate are well-known sources of dietary protoanthocyanidins (Scalbert *et al.*, 2005).

2.2.4 Stilbenes and Lignans

These groups of compounds are less common in the human diet; however, they are considered important to human health. The most representative stilbene is resveratrol, that can appear in both *cis* and *trans* isomeric forms, mostly glycosylated forms (Delmas *et al.*, 2011). It can be found in grapes and red wine (0.3–7 mg aglycones/L and 15 mg glycosides/L) (Vitrac *et al.*, 2002). Due to its anticarcinogenic effects, shown during screening of medicinal plants, it has been extensively studied.

Lignans are produced by oxidative dimerization of two phenyl propane units; they usually exist in the free form, while their glycoside derivatives are only a minor form. The richest dietary source is linseed, which contains secoisolariciresinol (up to 3.7 g/kg dry wt) and low quantities of matairesinol (Mazur *et al.*, 1998). Other cereals (triticale and wheat), grains (lentils), fruit (pears, prunes) and certain vegetables (garlic, asparagus, carrots) also contain traces of these lignans, but concentrations in linseed are approximately 1000 times as high as concentrations in these other food sources. The interest in lignans has increased due to its potential applications in cancer chemotherapy and other different pharmacological effects (Saleem *et al.*, 2005).

2.3 Fruit By-Products Sources

2.3.1 Agro-Industrial By-Products

The processing of fruits, oilseeds and vegetables produce high amounts of waste materials such as pomace, peels, seeds and oilseed meals. These by-products are generally utilized as animal feed or fertilizer; however, they still contain a huge amount of phenolic compounds and are potential sources of antioxidants. Hence the interest on further exploitation of by-products to produce food additives or supplements with high nutritional value has increased, since their recovery may be economically attractive.

2.3.1.1 Citrus Fruit Citrus is the largest fruit crop in the world. Its worldwide production is over 88 million tons and one-third is processed. Oranges, grapefruit, lemons and mandarins represent up to 98% of the entire industrialized crop. They are processed principally to obtain juice, jam and segments of mandarin for canning industry. The USA and Brazil are the major producers; they harvest and process around 60% and 85% of the world's oranges, respectively (Djilas *et al.*, 2009).

The citrus fruit industry generates large amounts of peel (albedo and flaveldo), pulp and seed after juice extraction, which represents more than 50% of the yield (Lario *et al.*, 2004). Albedo is a white, spongy and cellulosic tissue, which is the main citrus peel component. Due to its high fiber content, albedo is considered as a potential for fiber source (Fernandez-Gines *et al.*, 2004).

The juice pulp from the finishing process and the essences recovered from juice processing, along with peel press liquor, add up to 5% of fresh fruit. The peels especially have been found to contain high quantities of phenolics. Oranges, lemons and grapefruit peels contain 15% more polyphenols than the edible portions (Gorinstein *et al.*, 2001). Similarly, lemon has been reported to have high amounts of antioxidant compounds (Marin *et al.*, 2002). In addition, peels of yellow and white nectarines contain at least twice more phenolic compounds than the flesh (Gil-Izquierdo *et al.*, 2002). Citrus seeds represent 0.1–5% of the whole fruit, and depending on the variety, they can be used for oil extraction and terpenoids recovery (Djilas *et al.*, 2009).

Flavonoids are the largest group of bioactive compounds in citrus fruit. The main flavonoids found are hesperidine, narirutin, naringin and eriocitrin (Schieber *et al.*, 2001). Otherwise, lemon peels contain mainly hesperidin and erocitrin, while naringin and erocitrin predominate from liquid residues (Coll *et al.*, 1998). It has been reported that bergamot peel contains large amounts of naringin (up to 1 mg/g) and neohesperidin (1 mg/g), among others (Mandalari *et al.*, 2007). It is well known that flavonoids from citrus have potential antioxidative, anticancer and anti-inflammatory activities. In addition, ascorbic acid is present in citrus fruit in high amounts. Ascorbic acid content is higher in peel than in flesh, and also is higher in peeled oranges and lemons than in grapefruits and their peels (Gorinstein *et al.*, 2001).

2.3.1.2 Grape The grape is one of the largest fruit crops, with more than 68 million tons produced annually (FAOSTAT, 2012); 80% of the crop is used in the wine industry and around 12 million tons of grape pomace is produced within a few weeks of the harvest period. Grape juices and wine processing generates seeds, skin and pomace as principal by-products. The pomace represents around 20% in weight of fresh fruit. Grape and products such as wine, juice, jams and raisins are economically important for the industry.

Wine industry wastes, which consist mainly of solid by-products, include marcs, pomace and stems, and account on average for almost 30% (w/w) of the grapes used for wine production. Grape by-products have been utilized for animal feed, but the presence of lignin reduces digestibility due to inhibition of cellulolytic and proteolytic enzymes, as well as the rumen microbiota (Kammerer *et al.*, 2004). Furthermore, the final solid residue from ethanol production is generally used as fertilizer, although the high polyphenols content inhibits seed germination (Fontana *et al.*, 2013).

Grape pomace is rich in polyphenols, with catechin, epicatechin, procyanidin B_1, quercetin and kaempferol being the most relevant (Lu and Foo, 1999). Good amounts of anthocyanins are also reported; several glycosylated forms of malvinidin, as well as cyanidin and peonidin were recovered from grape pomace from several Italian varieties (Ruberto *et al.*, 2007). On the other hand, grape seeds form a considerable portion of the grape, ranging from 38–52% of dry weight basis (Schieber *et al.*, 2002). The seed oil is notable for its high unsaturated fatty acid content (especially linoleic acid) and is a great source of phenolic compounds such as gallic acid, catechin and epicatechin; as well as proanthocyanidins. In certain varieties, the content of phenolic acids is 2- or 3-fold higher in seeds than in press residue, whereas flavonoids tend to be concentrated in press residue (Maier *et al.*, 2009). Thus, grape by-products are considered a valuable source of phytochemicals as functional compounds for food industries.

2.3.1.3 Apple Apple is the most selected and a widespread fruit in the world. World apple production is nearly 76 million tons, of which 4 million tons are produced in the USA (FAOSTAT, 2012). Around 70% of apples are consumed as fresh, the rest being processed into value products such as juice, apple cider, jams, jelly and purees, with wine, vermouth and dried apple products too. In large-scale apple processing, by-products can be classified into two categories. The first is the fruit discarded into the sorting belt due to its partially bruised/spoiled nature, usually known as belt rejection. The second is the apple pomace obtained after juice extraction. The belt rejection apples are also discharged along with apple pomace as waste (Shalini and Gupta, 2010). About 65% of processed apples are used for juice concentrate; then apple pomace represents 25–30% of the original fruit and is generated during fruit pressing (Schieber *et al.*, 2003). Apple pomace has high water content and is mainly composed of insoluble carbohydrates such as cellulose, hemicellulose and lignin. Apple pomace contains up to 5–10 g/kg phenolics, being phloridzin, quercetin and epicatechin as the most representative (Lu and Foo, 1999; Schieber *et al.*, 2001); proanthocyanidins, chlorogenic acid and phloretin are also present in considerable amounts.

However, apple pomace composition varies according to the apple variety and the type of process used for juice extraction, mainly on how many times the fruit is pressed (Cetkovic *et al.*, 2008). Traditional apple juice processing results in a juice poor in phenolics, with around 3–10% of the antioxidant activity (van der Sluis *et al.*, 2002). It is well known that the phenolic compounds content is larger in peels than in flesh. Flesh contains catechins, procyanidins, phloridzin, phloretin glycosides, caffeic acid and chlorogenic acid; while the peels contains flavonoids not found in the flesh, such as quercetin and cyaniding glycosides (Wolfe *et al.*, 2003). Hence, the commercial exploitation of apple by-products for phenolic compounds recovery seems promising.

Pectin represents around 10–15% of apple pomace on a dry weight basis (Oreopoulou and Tzia, 2007) and generally is recovered by acid extraction followed by precipitation. In general, apple by-products are a good natural source of bioactive compounds, with different health benefits and applications that should be further explored.

2.3.1.4 Tropical Fruits Tropical fruit consumption has increased in the international markets, because of their remarkable flavor and nutritional properties (Ayala-Zavala *et al.*, 2011). Some tropical fruits are considered exotic fruits when grown outside their country of origin. There is a wide list of tropical fruit products, but the most common are banana, papaya, mango, avocado, pineapple and peach. Those fruits are usually consumed directly by humans in their own country, but storage and processing steps are required for exportation to other locations. Thus, separation of the desired value products from plant tissue involves different processing steps and by-products generation as a consequence. The most common bioactive compounds present in tropical fruits are vitamins C and E, carotenoids, phenolic compounds and dietary fiber (Gonzalez-Aguilar *et al.*, 2008).

Peach (*Prunus persica*) has been used in the food industry for the production of peach slices, syrups, juices and jam. By-products after processing are the kernels and peels. They are mostly used to obtain dietary fibers and pectin. Peach fiber has been incorporated into muffins, improving texture and flavor (Grigelmo-Miguel *et al.*, 2001). Likewise, Bitter apricot (*Prunus armeniaca L., Rosaceae*) seeds are by-products of the apricot processing industry, with the peeled seeds serving as a raw material for the production of persipan (Schieber *et al.*, 2001).

Mango (*Mangifera indica L., Anacardiaceae*) is one of the most important tropical fruits and its consumption is widely increasing in the international market. Major wastes of mango processing are peels and stones, amounting to 35–60% of the total fruit weight (Larrauri *et al.*, 1996). Mango seed kernels are a good source of natural antioxidants. The main antioxidant are phenolic compounds (gallic and ellagic acids, and gallates) and phospholipids (Puravankara *et al.*, 2000). Also, gallotannins and condensed tannin-related polyphenols were reported in mango kernels (Arogba, 2000).

Banana (*Musa paradisiaca* L., Musaceae) is another important crop worldwide. Peels constitute up to 30% of the ripe fruit. About 1000 banana plants are estimated to yield 20–25 tons of pseudostems, providing about 5% edible starch (Anand and Maini, 1997). Furthermore, anthocyanin pigments from banana bracts have been evaluated for their potential application as natural food colorants. It was concluded that the bracts proved to be a good and abundant source of anthocyanins of attractive appearance, as well as being a useful tool in anthocyanin identification, since all six most common anthocyanidins (delphinidin, cyanidin, pelargonidin, peonidin, petunidin and malvidin) are present (Pazmino-Duran *et al.*, 2001).

2.4 Dietary Fibers-Rich By-Products

In recent years, consumers are becoming more interested in a diet rich in fruits and vegetables for a healthy lifestyle. Thus dietary fiber intake has been shown to include several health benefits, including reduced risk of coronary heart disease, obesity and diabetes, as well as high cholesterol in blood (O'Shea *et al.*, 2012). Generally, fibers from cereals have been used in food production; however, novel sources of fiber from fruit and especially fruit by-product sources has been discovered and used in food processing. Cereals are the principal source of cellulose, lignin and hemicelluloses, whereas fruits are the main source of pectin, gums and mucilage (Normand *et al.*, 1987). Moreover, fibers from citrus fruits have an added advantage over traditional dietary fibers sources, because of the presence of great amounts of bioactive compounds and their antioxidant properties.

Dietary fiber (DF) is a group of compounds that includes plant carbohydrate polymers (both oligosaccharides and polysaccharides), i.e. cellulose, hemicellulose, pectin, gums, resistant starch and inulin, that may be associated with lignin (Elleuch *et al.*, 2011). Dietary fibers function as a bulking agent, helping digestion and adsorption in the human small intestine, increasing the intestinal mobility and moisture content of the faeces (Sudha *et al.*, 2007); stimulates colonic fermentation, reduces post-prandial blood glucose (insulin response) and reduces pre-prandial cholesterol.

DF is classified as soluble or insoluble, based on the capacity to mix with water. Soluble dietary fibers include pectins, gums, mucilage and some hemicelluloses, whereas cellulose and lignin are the insoluble fraction (Davidson and McDonald, 1998). For example, citrus by-products have a high content of water (80%), which needs to be removed for further application in the food industry, and it is necessary in their processing to obtain a powder with around 7% water (Fernandez-Lopez *et al.*, 2004). Thus, the development of the process for reused fruit by-products, which can reduce or minimize the losses of bioactive compounds, should be achieved.

In food processing, the soluble fraction has shown a high capacity to increase viscosity, ability to form gels and act as emulsifiers, improve the texture, and is easy to incorporate into the processed food. Furthermore, DF can provide functional properties to foodstuff, i.e. increase water adsorption capacity, oil holding capacity, emulsification and gel formation. The soluble fraction can also improve the viscosity, texture, sensory characteristics, and shelf life of food products (Aguedo *et al.*, 2012).

2.4.1 Hemicelluloses

Hemicelluloses are attractive as biopolymers, which can be utilized in their native and modified forms in various areas, including the food industry, such as films, adhesives, gelling, stabilizing and viscosity enhancing additives. In the literature, various fruit pomaces have been reported to extract hemicelluloses (Rabetafika *et al.*, 2014; Watt *et al.*, 1999). Hemicelluloses were produced from pear pomace using direct alkaline extraction (NaOH and H_2O_2) and two-step extraction with delignification pre-treatment (acidified $NaClO_2$/NaOH) (Rabetafika *et al.*, 2014).

2.4.2 Pectins

Pectin is a family of complex polysaccharides in which the presence of partly methyl-esterified galacturonic acid and, to a lesser extent, rhamnose is a distinctive feature. Industrially, pectins are extracted from fruit pomaces including apples, grapes and citrus peels (Minjares-Fuentes *et al.*, 2014; Wang *et al.*, 2014). Pectin has been widely used in food, cosmetics and pharmaceutical industries as gelling and thickening agents (Atmodjo *et al.*, 2013; Minjares-Fuentes *et al.*, 2014; Voragen *et al.*, 2009). Applications of pectin involving fruit products are jams. Peach dietary fiber has replaced industrial pectin in strawberry jams, and sensory evaluation shows that jams enriched with high peach fruit dietary fiber is as acceptable as the conventional jam (Grigelmo-Miguel and Martin-Belloso, 1999). Moreover, pectins are used for cardiovascular disease therapy, induction of apoptosis, colon-specific drug delivery, anti-inflammation, and even a new raw material for porous materials production (Wang and Lu, 2014).

2.5 Value-Added Products from Fruit By-Products

In the last few years, DF has been incorporated into food products, for example, dairy and bakery products, as well as meat and ready-to-eat products. Fiber-rich by-products can be incorporated into the food products to enhance water and oil retention and improve emulsification properties, as well as non-caloric bulking agents for partial replacement of fat, sugars or flour.

2.5.1 Meat Products

Fiber has been introduced into meat products with the aim to substitute fat, reduce caloric content and improve the texture and stability of meat products. Addition of fruit fibers (1.5%), especially from oranges, to dry fermented sausages, results in a product sensory similar to the conventional high-fat product (Garcia *et al.*, 2002). Also raw (5–10%) and cooked albedo (2.5–7%) from lemons has been added to cooked sausages and dry sausages. The most remarkable result is the reduction of residual nitrite levels. It is well known that nitrite levels are associated with nitrosamines formation, which represents a risk during consumption of meat products.

Furthermore, orange fiber powder (~2%), with a high quantity of phenolic compounds, has been used in the preparation of bologna sausages and chicken burgers. Results showed an effective inhibition of the lipid oxidation, through improving the stability, reducing the residual nitrite levels and prolonging the shelf life of the product (Fernandez-Gines *et al.*, 2004).

2.5.2 Dairy Products

The use of dietary fibers in dairy products is also widespread, especially the addition of inulin, which promoted a mouth feel and bulk in cheese and ice cream; it also reduces syneresis in yogurt and other fermented milk products (Blecker *et al.*, 2001). In ice cream, fiber also provides a homogeneous smooth bulk, resistance to melting and improving handling properties by hindering crystal growth, as temperature fluctuates during storage (Regand and Goff, 2003). The use of dietary fibers, such as apple-derived inulin, has been used as crystallization and recrystallization controllers in frozen dairy products with excellent results (Soukoulis *et al.*, 2009).

2.5.3 Baking Products

Dietary fibers from different fruit pomaces have been used to replace or reduce wheat flour in the preparation of bakery products. Usually, dietary fibers are incorporated into bakery preparations to preserve freshness due to their capacity to retain water. Moreover, dietary fibers can improve the nutritional content while not compromising the loaf volume, firmness and crumb structure of the bread. However, the amount of fiber for incorporation depends on the fiber source.

Several studies using fruit by-products to substitute for wheat in baking formulations has been performed. Apple (Masoodi *et al.*, 2002; Sudha *et al.*, 2007), mango (Ajila *et al.*, 2008) and orange (Fernández-López *et al.*, 2009; Kohajdova *et al.*, 2011) by-products are the most studied. Apple pomace containing around 36.8% has been used in apple pie filling and in oatmeal cookies (Carson *et al.*, 1994). Citrus fibers have been introduced into biscuits (Kohajdova *et al.*, 2011) and bread and frozen dough (Bchir *et al.*, 2014; Ocen and Xu, 2013). Also, 5.5% of orange pomace has been included into gluten-free bread formulations, resulting in an increase from 2.1% to 3.9% of total fiber content in the final product. Thus, orange pomace has a potential application in preserving loaf volume and texture (O'Shea *et al.*, 2015).

2.5.4 Ready-To-Eat Products

Ready-to-eat products have established themselves as one of the most popular foods in the market. They are obtained through a variety of industrial processing; however, extrusion has found the largest application in the ready-to-eat area such as in breakfast cereals, salty or sweet snacks, co-extruded snacks, croutons for soups and salads, pet and fish foods, nutritious pre-cooked food mixtures for infants, and confectionery products (Singh *et al.*, 2007). Extrusion cooking is a high temperature, short-time process in which food material is plasticized and cooked in a tube by combination of pressures, moisture, temperature and mechanical shear (Harper, 1990) that can increase the total dietary fiber content. This process can reduce the molecular weight of pectin and hemicellulose molecules, resulting in increased water solubility of fiber. Several reactions with fiber may occur during extrusion, for example, the formation of retrograded amylose, which is insoluble at room temperature (Vasanthan *et al.*, 2002). Also pressure cooking may reduce cellulose and hemicellulose content by degradation of polysaccharides into simple sugars. Furthermore, the properties of fiber components in orange pulps using extrusion were modified by Larrea *et al.* (2005), who reported that extrusion conditions decreased insoluble dietary fiber and increased soluble fiber. Stojceska *et al.* (2008) reported on the incorporation of cauliflower trimmings into ready-to-eat expanded products of up to the 10% level.

2.6 Future Perspectives

Fruit by-products that are unwanted cast-offs of processing can be used as value-added food supplements, because of their high dietary fiber and bioactive compounds contents. They can enhance water and oil retention, serve as non-caloric bulking agents, and improve emulsification properties and oxidation stability. The exploitation of fruit by-products is promising, not only for environmental issues but also because the innovative sector of functional food is growing and becoming more decisive. In this sense, simplified processes with fewer extraction and purification steps, likewise the use of emerging technologies, will be advanced, allowing an accessible scale-up as well as economical production.

Furthermore, novel and nutritionally products should be developed and validated using dietary fiber and bioactive compounds from fruit by-products. The characterization of extracts from by-products is an important issue, the identity and individual concentration of recovered phenolics is a relevant fact, and this information will support specific applications of extracted bioactive compounds in food industry.

References

Aguedo, M., Kohnen, S., Rabetafika, N., Vanden Bossche, S., Sterckx, J. *et al.* (2012) Composition of by-products from cooked fruit processing and potential use in food products. *Journal of Food Composition and Analysis*, **27**: 61–69.

Ajila, C.M., Leelavathi, K. and Rao, U.J.S.P. (2008) Improvement of dietary fiber content and antioxidant properties in soft dough biscuits with the incorporation of mango peel powder. *Journal of Cereal Science*, **48**: 319–326.

Anand, J. and Maini, S. (1997) Utilisation of fruit and vegetable wastes. *Indian Food Packer*, **51**: 45–63.

Arogba, S.S. (2000) Mango (*Mangifera indica*) kernel: Chromatographic analysis of the tannin, and stability study of the associated polyphenol oxidase activity. *Journal of Food Composition and Analysis*, **13**: 149–156.

Atmodjo, M.A., Hao, Z.Y. and Mohnen, D. (2013) Evolving views of pectin biosynthesis. *Annual Review of Plant Biology*, **64**: 747–79.

Awad, M.A., de Jager, A. and van Westing, L.M. (2000) Flavonoid and chlorogenic acid levels in apple fruit: Characterisation of variation. *Scientia Horticulturae*, **83**: 249–263.

Ayala-Zavala, J.F., Vega-Vega, V., Rosas-Dominguez, C., Palafox-Carlos, H., Villa-Rodriguez, J.A. *et al.* (2011) Agro-industrial potential of exotic fruit by-products as a source of food additives. *Food Research International*, **44**: 1866–1874.

Bchir, B., Rabetafika, H.N., Paquot, M. and Blecker, C. (2014) Effect of pear, apple and date fibres from cooked fruit by-products on dough performance and bread quality. *Food and Bioprocess Technology*, **7**: 1114–1127.

Blecker, C., Chevalier, J.-P., Van Herck, J.-C., Fougnies, C., Deroanne, C. and Paquot, M. (2001) Inulin: Its physicochemical properties and technological functionality. *Recent Research Developments in Agricultural and Food Chemistry*, **5**: 125–131.

Boudet, A.-M. (2007) Evolution and current status of research in phenolic compounds. *Phytochemistry*, **68**: 2722–2735.

Bravo, L. (1998) Polyphenols: Chemistry, dietary sources, metabolism, and nutritional significance. *Nutrition reviews*, **56**, 317–333.

Carson, K.J., Collins, J.L. and Penfield, M.P. (1994) Unrefined, dried apple pomace as a potential food ingredient. *Journal of Food Science*, **59**: 1213–1215.

Cetkovic, G., Canadanovic-Brunet, J., Djilas, S., Savatovic, S., Mandic, A. and Tumbas, V. (2008) Assessment of polyphenolic content and *in vitro* antiradical characteristics of apple pomace. *Food Chemistry*, **109**, 340–347.

Coll, M.D., Coll, L., Laencina, J. and Tomas-Barberan, F.A. (1998) Recovery of flavanones from wastes of industrially processed lemons. *Zeitschrift Fur Lebensmittel-Untersuchung Und-Forschung a-Food Research and Technology*, **206**: 404–407.

Cos, P., De Bruyne, T., Hermans, N., Apers, S., Vanden Berghe, D. and Vlietinck, A.J. (2004) Proanthocyanidins in health care: Current and new trends. *Current Medicinal Chemistry*, **11**: 1345–1359.

Davidson, M.H. and McDonald, A. (1998) Fiber: Forms and functions. *Nutrition Research*, **18**: 617–624.

Delmas, D., Aires, V., Limagne, E., Dutartre, P., Mazue, F. *et al.* (2011) Transport, stability, and biological activity of resveratrol. *Resveratrol and Health*, **1215**: 48–59.

Djilas, S., Canadanovic-Brunet, J. and Cetkovic, G. (2009) By-products of fruits processing as a source of phytochemicals. *Chemical Industry and Chemical Engineering Quarterly*, **15**: 191–202.

Elleuch, M., Bedigian, D., Roiseux, O., Besbes, S., Blecker, C. and Attia, H. (2011) Dietary fibre and fibre-rich by-products of food processing: Characterisation, technological functionality and commercial applications: A review. *Food Chemistry*, **124**: 411–421.

FAOSTAT (2012) *Food and Agricultural Commodities Production*. FAO, Rome.

Fernandez-Gines, J.M., Fernandez-Lopez, J., Sayas-Barbera, E., Sendra, E. and Perez-Alvarez, J.A. (2004) Lemon albedo as a new source of dietary fiber: Application to Bologna sausages. *Meat Science*, **67**: 7–13.

Fernandez-Lopez, J., Fernandez-Gines, J.M., Aleson-Carbonell, L., Sendra, E., Sayas-Barbera, E. and Perez-Alvarez, J.A. (2004) Application of functional citrus by-products to meat products. *Trends in Food Science and Technology*, **15**(3): 176–185.

Fernández-López, J., Sendrao-Nadal, E., Navarro, C., Sayas, E., Viudao-Martos, M. and Alvarez, J.A.P. (2009) Storage stability of a high dietary fibre powder from orange by-products. *International Journal of Food Science and Technology*, **44**: 748–756.

Fontana, A.R., Antoniolli, A. and Bottini, R. (2013) Grape pomace as a sustainable source of bioactive compounds: extraction, characterization, and biotechnological applications of phenolics. *Journal of Agricultural and Food Chemistry*, **61**: 8987–9003.

Garcıa, M.L., Dominguez, R., Galvez, M.D., Casas, C. and Selgas, M.D. (2002) Utilization of cereal and fruit fibres in low fat dry fermented sausages. *Meat Science*, **60**(3): 227–236.

Gil-Izquierdo, A., Gil, M.I. and Ferreres, F. (2002) Effect of processing techniques at industrial scale on orange juice antioxidant and beneficial health compounds. *Journal of Agricultural and Food Chemistry*, **50**: 5107–5114.

Giusti, M.M. and Wrolstad, R.E. (2003) Acylated anthocyanins from edible sources and their applications in food systems. *Biochemical Engineering Journal*, **14**: 217–225.

Goldstein, J.L. and Swain, T. (1963) Changes in tannins in ripening fruits. *Phytochemistry*, **2**: 371–383.

Gonzalez-Aguilar, G.A., Celis, J., Sotelo-Mundo, R.R., De La Rosa, L.A., Rodrigo-Garcia, J. and Alvarez-Parrilla, E. (2008) Physiological and biochemical changes of different fresh-cut mango cultivars stored at 5°C. *International Journal of Food Science and Technology*, **43**: 91–101.

Gorinstein, S., Martin-Belloso, O., Park, Y.S., Haruenkit, R., Lojek, A. *et al.* (2001) Comparison of some biochemical characteristics of different citrus fruits. *Food Chemistry*, **74**: 309–315.

Grayer, R.J. and N.C. Veitch (2006) Flavanones and dihydroflavonols. In: *Flavonoids: Chemistry, Biochemistry and Applications* (eds Ø.M. Andersen and K.R. Markham). CRC Press, Boca Raton, FL, pp. 917–1002.

Grigelmo-Miguel, N. and Martin-Belloso, O. (1999) Influence of fruit dietary fibre addition on physical and sensorial properties of strawberry jams. *Journal of Food Engineering*, **41**: 13–21.

Grigelmo-Miguel, N., Carreras-Boladeras, E. and Martin-Belloso, O. (2001) Influence of the addition of peach dietary fiber in composition, physical properties and acceptability of reduced-fat muffins. *Food Science and Technology International*, **7**: 425–431.

Harborne, J.B. and Baxter, H. (1999) *The Handbook of Natural Flavonoids*, vols 1 and 2. John Wiley & Sons. Chichester, UK.

Harper, J.M. (1990) Extrusion of foods. In: *Biotechnology and Food Process Engineering*, vol. **1**. CRC Press, Boca Roton, FL, pp. 295–308.

Kammerer, D., Claus, A., Carle, R. and Schieber, A. (2004) Polyphenol screening of pomace from red and white grape varieties (*Vitis vinifera* L.) by HPLC-DAD-MS/MS. *Journal of Agricultural and Food Chemistry*, **52**: 4360–4367.

Kohajdova, Z., Karovicova, J., Jurasova, M. and Kukurova, K. (2011) Application of citrus dietary fibre preparations in biscuit production. *Journal of Food and Nutrition Research*, **50**: 182–190.

Koleckar, V., Kubikova, K., Rehakova, Z., Kuca, K., Jun, D. *et al.* (2008) Condensed and hydrolysable tannins as antioxidants influencing the health. *Mini-Reviews in Medicinal Chemistry*, **8**: 436–447.

Lapornik, B., Prosek, M. and Wondra, A.G. (2005) Comparison of extracts prepared from plant by-products using different solvents and extraction time. *Journal of Food Engineering*, **71**: 214–222.

Lario, Y., Sendra, E., Garcí, J., Fuentes, C., Sayas-Barberá, E. *et al.* (2004) Preparation of high dietary fiber powder from lemon juice by-products. *Innovative Food Science and Emerging Technologies*, **5**: 113–117.

Larrauri, J.A., Ruperez, P., Borroto, B. and Saura-Calixto, F. (1996) Mango peels as a new tropical fibre: Preparation and characterization. *Lebensmittel-Wissenschaft and Technologie*, **29**: 729–733.

Larrea, M.A., Chang, Y.K. and Martínez Bustos, F. (2005) Effect of some operational extrusion parameters on the constituents of orange pulp. *Food Chemistry*, **89**: 301–308.

Lee, K.W., Kim, Y.J., Lee, H.J. and Lee, C.Y. (2003) Cocoa has more phenolic phytochemicals and a higher antioxidant capacity than teas and red wine. *Journal of Agricultural and Food Chemistry*, **51**: 7292–7295.

Lu, Y.R. and Foo, L.Y. (1999) The polyphenol constituents of grape pomace. *Food Chemistry*, **65**: 1–8.

Määttä-Riihinen, K.R., Kähkönen, M.P., Törrönen, A.R. and Heinonen, I.M. (2005) Catechins and procyanidins in berries of vaccinium species and their antioxidant activity. *Journal of Agricultural and Food Chemistry*, **53**: 8485–8491.

Maier, T., Schieber, A., Kammerer, D.R. and Carle, R. (2009) Residues of grape (*Vitis vinifera* L.) seed oil production as a valuable source of phenolic antioxidants. *Food Chemistry*, **112**: 551–559.

Mandalari, G., Bennett, R.N., Bisignano, G., Trombetta, D., Saija, A. *et al.* (2007) Antimicrobial activity of flavonoids extracted from bergamot (*Citrus bergamia* Risso) peel, a by-product of the essential oil industry. *Journal of Applied Microbiology*, **103**: 2056–2064.

Marin, F.R., Martinez, M., Uribesalgo, T., Castillo, S. and Frutos, M.J. (2002) Changes in nutraceutical composition of lemon juices according to different industrial extraction systems. *Food Chemistry*, **78**: 319–324.

Masoodi, F.A., Sharma, B. and Chauhan, G.S. (2002) Use of apple pomace as a source of dietary fiber in cakes. *Plant Foods for Human Nutrition*, **57**: 121–128.

Mazur, W.M., Duke, J.A., Wahala, K., Rasku, S. and Adlercreutz, H. (1998) Isoflavonoids and lignans in legumes: Nutritional and health aspects in humans. *Journal of Nutritional Biochemistry*, **9**: 193–200.

Minjares-Fuentes, R., Femenia, A., Garau, M.C., Meza-Velazquez, L.A., Simal, S. and Rossello, C. (2014) Ultrasound-assisted extraction of pectins from grape pomace using citric acid: A response surface methodology approach. *Carbohydrate Polymers*, **106**: 179–189.

Nawirska, A. and Kwasniewska, M. (2005) Dietary fibre fractions from fruit and vegetable processing waste. *Food Chemistry*, **91**: 221–225.

Normand, F.L., Ory, R.L. and Mod, R.R. (1987) Binding of bile-acids and trace minerals by soluble hemicelluloses of rice. *Food Technology*, **41**: 86–99.

O'Shea, N., Arendt, E.K. and Gallagher, E. (2012) Dietary fibre and phytochemical characteristics of fruit and vegetable by-products and their recent applications as novel ingredients in food products. *Innovative Food Science and Emerging Technologies*, **16**: 1–10.

O'Shea, N., Rossle, C., Arendt, E. and Gallagher, E. (2015) Modelling the effects of orange pomace using response surface design for gluten-free bread baking. *Food Chemistry*, **166**: 223–230.

Ocen, D. and Xu, X. (2013) Effect of citrus orange (*Citrus sinensis*) by-product dietary fiber preparations on the quality characteristics of frozen dough bread. *American Journal of Food Technology*, **8**: 43–53.

Oreopoulou, V. and Tzia, C. (2007) Utilization of plant by-products for the recovery of proteins, dietary fibers, antioxidants, and colorants. In: *Utilization of By-Products and Treatment of Waste in the Food Industry*. Springer, London.

Pazmino-Duran, E.A., Giusti, M.M., Wrolstad, R.E. and Gloria, M.B.A. (2001) Anthocyanins from banana bracts (*Musa* X *paradisiaca*) as potential food colorants. *Food Chemistry*, **73**: 327–332.

Puravankara, D., Boghra, V. and Sharma, R.S. (2000) Effect of antioxidant principles isolated from mango (*Mangifera indica* L) seed kernels on oxidative stability of buffalo ghee (butter-fat). *Journal of the Science of Food and Agriculture*, **80**: 522–526.

Rabetafika, H.N., Bchir, B., Blecker, C., Paquot, M. and Wathelet, B. (2014) Comparative study of alkaline extraction process of hemicelluloses from pear pomace. *Biomass and Bioenergy*, **61**: 254–264.

Regand, A. and Goff, H.D. (2003) Structure and ice recrystallization in frozen stabilized ice cream model systems. *Food Hydrocolloids*, **17**: 95–102.

Rein, M. (2005) Copigmentation reactions and color stability of berry anthocyanins. University of Helsinki, PhD thesis.

Robbins, R.J. (2003) Phenolic acids in foods: An overview of analytical methodology. *Journal of agricultural and Food Chemistry*, **51**, 2866–2887.

Ruberto, G., Renda, A., Daquino, C., Amico, V., Spatafora, C. *et al.* (2007) Polyphenol constituents and antioxidant activity of grape pomace extracts from five Sicilian red grape cultivars. *Food Chemistry*, **100**: 203–210.

Saleem, M., Kim, H.J., Ali, M.S. and Lee, Y.S. (2005) An update on bioactive plant lignans. *Natural Product Reports*, **22**: 696–716.

Scalbert, A., Manach, C., Morand, C., Rémésy, C. and Jiménez, L. (2005) Dietary polyphenols and the prevention of diseases. *Critical reviews in Food Science and Nutrition*, **45**: 287–306.

Schieber, A., Stintzing, F.C. and Carle, R. (2001) By-products of plant food processing as a source of functional compounds – recent developments. *Trends in Food Science and Technology*, **12**: 401–413.

Schieber, A., Müller, D., Röhrig, G. and Carle, R. (2002) Effects of grape cultivar and processing on the quality of cold-pressed grape seed oils. *Mitteilungen Klosterneuburg*, **52**: 29–33.

Schieber, A., Hilt, P., Streker, P., Endreß, H.-U., Rentschler, C. and Carle, R. (2003) A new process for the combined recovery of pectin and phenolic compounds from apple pomace. *Innovative Food Science and Emerging Technologies*, **4**: 99–107.

Shalini, R. and Gupta, D.K. (2010) Utilization of pomace from apple processing industries: A review. *Journal of Food Science and Technology-Mysore*, **47**: 365–371.

Shi, J., Yu, J., Pohorly, J.E. and Kakuda, Y. (2003) Polyphenolics in grape seeds –Biochemistry and functionality. *Journal of Medicinal Food*, **6**: 291–299.

Singh, S., Gamlath, S. and Wakeling, L. (2007) Nutritional aspects of food extrusion: A review. *International Journal of Food Science and Technology*, **42**: 916–929.

Soukoulis, C., Lebesi, D. and Tzia, C. (2009) Enrichment of ice cream with dietary fibre: Effects on rheological properties, ice crystallisation and glass transition phenomena. *Food Chemistry*, **115**: 665–671.

Stojceska, V., Ainsworth, P., Plunkett, A., İbanoğlu, E. and İbanoğlu, Ş. (2008) Cauliflower by-products as a new source of dietary fibre, antioxidants and proteins in cereal-based ready-to-eat expanded snacks. *Journal of Food Engineering*, **87**: 554–563.

Sudha, M.L., Baskaran, V. and Leelavathi, K. (2007) Apple pomace as a source of dietary fiber and polyphenols and its effect on the rheological characteristics and cake making. *Food Chemistry*, **104**: 686–692.

Sun-Waterhouse, D. (2011) The development of fruit-based functional foods targeting the health and wellness market: A review. *International Journal of Food Science and Technology*, **46**: 899–920.

Treutter, D. (2006) Significance of flavonoids in plant resistance: A review. *Environmental Chemistry Letters*, **4**: 147–157.

Tsao, R. (2010) Chemistry and biochemistry of dietary polyphenols. *Nutrients*, **2**: 1231–1246.

Tsao, R., Yang, R. and Young, J.C. (2003) Antioxidant isoflavones in osage orange, *Maclura pomifera* (Raf.) Schneid. *Journal of Agricultural And Food Chemistry*, **51**: 6445–6451.

van der Sluis, A.A., Dekker, M., Skrede, G. and Jongen, W.M.F. (2002) Activity and concentration of polyphenolic antioxidants in apple juice. 1: Effect of existing production methods. *Journal of Agricultural and Food Chemistry*, **50**: 7211–7219.

Vasanthan, T., Jiang, G.S., Yeung, J. and Li, J.H. (2002) Dietary fiber profile of barley flour as affected by extrusion cooking. *Food Chemistry*, **77**: 35–40.

Vitrac, X., Monti, J.P., Vercauteren, J., Deffieux, G. and Merillon, J.M. (2002) Direct liquid chromatographic analysis of resveratrol derivatives and flavanonols in wines with absorbance and fluorescence detection. *Analytica Chimica Acta*, **458**: 103–110.

Voragen, A.G.J., Coenen, G.J., Verhoef, R.P. and Schols, H.A. (2009) Pectin, a versatile polysaccharide present in plant cell walls. *Structural Chemistry*, **20**: 263–275.

Wang, X. and Lu, X. (2014) Characterization of pectic polysaccharides extracted from apple pomace by hot-compressed water. *Carbohydrate Polymers*, **102**: 174–184.

Wang, X., Chen, Q.R. and Lu, X. (2014) Pectin extracted from apple pomace and citrus peel by subcritical water. *Food Hydrocolloids*, **38**: 129–137.

Watt, D.K., Brasch, D.J., Larsen, D.S. and Melton, L.D. (1999) Isolation, characterisation, and NMR study of xyloglucan from enzymatically depectinised and non-depectinised apple pomace. *Carbohydrate Polymers*, **39**: 165–180.

Wolfe, K., Wu, X. and Liu, R.H. (2003) Antioxidant activity of apple peels. *Journal of Agricultural and Food Chemistry*, **51**: 609–614.

Xie, D.Y. and Dixon, R.A. (2005) Proanthocyanidin biosynthesis – Still more questions than answers? *Phytochemistry*, **66**: 2127–2144.

Zhao, L.M., Jin, H.S., Sun, L.P., Piao, H.R. and Quan, Z.S. (2005) Synthesis and evaluation of antiplatelet activity of trihydroxychalcone derivatives. *Bioorganic and Medicinal Chemistry Letters*, **15**: 5027–5029.

3
Utilization of Waste from Tropical Fruits

H.K. Sharma[1] and Mandeep Kaur[2]

[1]*Sant Longowal Institute of Engineering and Technology, Punjab, India*
[2]*Amity University, Noida, India*

3.1 Introduction

Tropical fruits grow on trees native to the tropics, which are a rich source of many vitamins, phytochemicals and minerals, and eating a diet rich in these fruits may protect us from a number of serious diseases such as cancer, heart diseases, obesity, diabetes etc. (Kaur and Sharma, 2014). Tropical fruits are commonly used for juice extraction, preparation of jams, jellies and canned products, and the unused part is then discarded as waste. The waste in the fruit processing industry can be defined as the end-products of various processes, which is discarded as such and not used for any other purposes.

Fruit processing waste is polluting the environment and generation of this waste is unbalancing the ecological system (Anon, 1996). Although the type of waste depends upon the processing industry concerned, some part of the waste can be utilized for production of valuable products, whereas the remaining part will contribute to total waste production. The population is expanding rapidly, especially in Asia. India has Asia's largest fruit, vegetable and flower market, called the Koyambedu Market, Chennai, which is spread over an area of 60 acres and the waste generation rate of this market is 80 tons per day (Kameswari, 2007). Such waste generation from a single market explicitly reveals the importance of using the waste for value addition.

In the industrialized nations, more production of food products than demand or premature harvesting, leads to generation of waste (FAO, 2011). Poor transportation or infrastructure facilities also cause wastage (Stuart, 2009). Usually the cost of production of re-using the waste is higher than disposing of it, because of the lack of feasible and practicable technology in the field, therefore disposal of the waste is a cause of concern. However, waste can be reduced by the implementation of modern technologies and by improving the storage facilities and cold chains.

Food Processing By-Products and their Utilization, First Edition. Edited by Anil Kumar Anal.

3.1.1 Waste Utilization and Challenges

During processing of fruits, a number of waste materials are generated, for example peels, stones, meals, seeds etc. These wastes can be further utilized for preparation of other products or for extraction of valuable components, which will not only lead to development of new products but will also assist in reducing pollution problems. To utilize the waste and reduce pollution resulting from the disposal of by-products of fruit processing, joint efforts are required. There may be numerous ways by which waste from the fruit processing industry can be utilized. One of the approaches may be the adoption of fermentation processes, because of the high amount of sugar in waste materials and production of citric acid, acetic acid, alcohol etc.

Fruit processing waste contains high amounts of water, so is highly perishable in nature. The major problem in utilization of fruit processing waste is the collection, transportation and processing the waste within the same day. The economical conditions also influence the utilization of waste from fruits. The selling price of the final product should not exceed the cost of converting the waste into the final product. Such types of decisions can be supported by operational research to determine the economic value of by-products (Henningsson *et al.*, 2004). The other main factors affecting utilization of fruit processing waste are the type of waste, process technology and manpower involved. With improvements in technology, productivity with a higher yield can be achieved but attention is yet to be paid to the utilization of waste.

After extraction of juice, by-products may be utilized for extraction of fibre, seed oil, citrus compounds, pectin, vitamins, phenolic compounds etc. Dietary fibre obtained after extraction of juice plays an important role in the prevention of a number of diseases, such as cardiovascular diseases, colon cancer, diabetes etc (Gomez *et al.*, 2007). Also, fruit waste can be utilized for production of various enzymes. These enzymes play an important role in enhancing the yield of juice, by acting on the cell walls of fruits and vegetables (Kaur and Sharma, 2013)

Waste generated from the processing of tropical fruits can be minimized by adopting suitable and cost-effective technologies. The possible ways by which waste from the

Table 3.1 Utilization of by-products, from the selected tropical fruits into the value-added products

Fruit	Major products	By products	Value added products
Pineapple	Juice, juice concentrate, jams, puree, candies, canned/sliced pineapple, frozen and dried products etc.	Pomace, crown, peel etc.	Ethanol, vinegar, biogas, bromelain, citric acid, cellulose, jam, extruded products, yeast, lactic acid etc.
Guava	Jelly, juice, nectar, syrup, sorbet, sweet bar, canned guava halves etc.	Peel, pomace, seeds	Pectin, dietary fibre, alcohol, single cell protein, lycopene, supplementation in bakery products etc.
Papaya	Juice, cordial, nectar, papaya cherry, jams, jellies, puree, canned product etc.	Seeds, Peel, pomace, latex, leaves	Seeds as antioxidants, seed flour, papain, seed oil, alcohol, vinegar, carboxymethyl cellulose, amylase, single cell protein etc.

pineapple, guava and papaya processing industries can be minimized and could be used in an efficient way, are discussed further in this chapter. Table 3.1 indicates the possible ways to utilize the waste from these tropical fruits.

3.2 Pineapple

Pineapple (*Ananas comosus*) belongs to the family *Bromeliaceae* and is commonly used for extraction of juice, concentrates, preparation of jams, jellies and canned products, throughout the world (Tran, 2006). Pineapple slices can be preserved after freezing (Larrauri *et al.*, 1997). Being very popular due to its sweet, sour and juicy properties, juice from pineapple is among one of the highly preferred juices worldwide (Cabrera *et al.*, 2000). The height of the pineapple plant is usually 75–150 cm and it has spiny leaves. The estimated production of pineapple was about 21.8 million in 2011 (FAO, 2013). Approximately 40–80% of pineapple fruit is discarded as waste. It is estimated that the waste from tropical and sub-tropical fruit processing is higher than the part that is utilized (Schieber *et al.*, 2001). The major by-products from the pineapple juice industry are shown in Figure 3.1.

Figure 3.1 (a) Pineapple fruit; (b) pineapple cut into halves; (c) pineapple peel; and (d) pomace after juice extraction.

Table 3.2 The valuable constituents in pineapple waste

Parameter %Dry basis	[1]Fresh	[1]Dry	[1]Ensiled	[2]Skin	[2]Pulp	[2]Crown	[3]Peel
Pectin	6.7	7.1	5.1	–	–	–	–
Cellulose	11.2	12	9	14	14.3	29.6	19.8
Hemicellulose	7	6.5	4.7	20.2	22.1	23.2	11.7
Protein	3.13	3.3	0.91	4.1	4.6	4.2	–
Reducing sugar	25.8	27.8	5	–	–	–	–
Non-reducing sugar	5.7	4.9	1.7	–	–	–	–
Lignin	11.52	11	9	1.5	2.3	4.5	–

[1]Rani and Nand (2004)
[2]Ban-koffi and Hann (1990)
[3]Bardiya *et al.* (1996)

Pineapple waste consists of high amounts of reducing sugars, cellulose, protein and lignin content (Table 3.2), and therefore it could be utilized efficiently for the production of value-added products, which may further be used for fortification of other foods. The possible ways to utilize pineapple waste are given in Table 3.1 and valuable products are described below:

3.2.1 Bioethanol

The waste from pineapple consists of high amounts of sugar, therefore can be used for the production of bioethanol. Besides sugars, pineapple waste also produces lignocellulosics, that are composed of celluloses and hemicelluloses embedded in the plant cell walls surrounded by lignin (Himmel *et al.*, 2007; Sun and Cheng 2002). The lingo-cellulosics can be hydrolysed to release the sugars, which can further be fermented to produce ethanol (Wooley *et al.*, 1999; Yu and Zhang, 2004).

There are different methods for the production of ethanol. In the first method, enzymatic hydrolysis is carried out followed by fermentation, whereas in another method, enzymatic hydrolysis and fermentation is carried out simultaneously with the help of yeast. During the concurrence of hydrolysis and fermentation, care should be taken that the temperature chosen should suit both the processes. The fermentation is usually carried out below 40°C with the help of yeast, whereas enzymatic hydrolysis is carried out at 50°C (Tengborg *et al.*, 2001). Therefore, selection of temperature is very important. In this process, inhibition of enzymes due to some toxic compounds is also decreased because of fermentation (Tengborg *et al.*, 2001), hence it may result in enhanced yield of ethanol (Philippidis *et al.*, 1993; Soderstrom *et al.*, 2005).

The production of vinegar from pineapple waste is carried out in two successive fermentation processes – alcoholic fermentation and acetic fermentation. Yeast strain, *Saccharomyces cerevisiae* and acetic bacteria, *Acetobactor* sp., isolated from pineapple juice, can be used for the production of vinegar from pineapple waste. The production of vinegar to at least 4.5 acetic degrees, with 5.3°Brix, at pH 2.8, requires 23 to 25 days (Sossou *et al.*, 2009).

3.2.2 Biogas

Biogas is flammable in nature, which is produced when organic material is fermented under anaerobic conditions. Pineapple waste is mainly composed of organic substances, therefore it can be used for industrial applications, such as biogas production (Mbuligwe and Kassenga, 2004). The basic steps for conversion of waste into biogas are acidogenesis and methanogenesis by anaerobic conversion of organic waste. Major factors affecting production of biogas are the amount and nature of the organic material used, temperature, pH, carbon-nitrogen ratio, toxicity, rate of stirring, type of microorganism, presence of nutrients, total solids, volatile liquids etc. The biogas can be produced in 20 days with a methane content of 48% under the optimum conditions of pH 7, and a carbon to nitrogen ratio of 20:1 (Chulalaksananukul *et al.*, 2012). The lower rate of organic load imparts a higher production rate of biogas. Pineapple peels can yield biogas ranging from 0.41–0.67 mg/kg volatile solids, with methane content ranging from 41–65% (Upadhyay *et al.*, 2010).

3.2.3 Bromelain

A protein, called bromelain, can be extracted from pineapple waste. Bromelain is present in pineapple peel, core, leaves etc., but in relatively smaller quantities as compared to the stem (Sriwatanapongse *et al.*, 2000). Extract of crown (*Nang Lae* and *Phu Lae* cultivars) has higher proteolytic activity (Ketnawa *et al.*, 2012). Bromelain is a combination of proteinases having anti-inflammatory and antithrombotic properties and also acts as an anticancer compound (Bhui *et al.*, 2009; Chobotova *et al.*, 2009). Being inflammatory, it can be used for the treatment of arthritis (Brien *et al.*, 2004).

Bromelain also has a folk medicinal value, but it depends on the stability of the proteolytic compounds. Careless extraction and purification processes may lead to instability of bromelain, which affects its medicinal and culinary properties. Furthermore, its stability can be increased by immobilization processes (Gupta *et al.*, 2007). It can be directly sprinkled on uncooked meat as a culinary ingredient, acting as a meat tenderizer (Maurer, 2001). Bromelain is used in brewing, baking and for the production of protein hydrolysates (Ketnawa and Rawdkuen, 2011; Walsh, 2002). It is also used in tanning, the leather and textile industries, wool, skin softening, and for detergent formulations (Uhlig, 1998; Subhabrata and Mayura, 2006).

Extraction of bromelain from pineapple is an expensive process, since it is difficult to separate and purify the desired enzyme from the crude extract, which may contain different proteins with similar physical properties. Bromelain can be extracted in a crude form and then purified by ion exchange chromatography, aqueous two phase systems, gel filtration, metal affinity membranes etc. (Babu *et a*l., 2008; Nie *et al*, 2008). The techniques for extraction of bromelain are given below.

3.2.3.1 Reverse Micellar Extraction (RME) The Reverse Micellar Extraction (RME) technique employs thermodynamically stable reverse micellar structures, which can solubilize biomolecules in the micellar core. RME does not cause any loss to natural activity and has the potential for continuous operation (Harikrishna *et al.*, 2002). Optimized conditions for the extraction of bromelain from pineapple core using the RME system of cationic surfactant Cetyltrimethyl-ammonium bromide (CTAB) result in forward and back extraction efficiencies of 45% and 62%, respectively (Hebbar *et al.*, 2008). Using the CTAB system of RME from peel, crown and extended stem, can result in purification folds of 2.1, 1.7 and 3.5, respectively.

3.2.3.2 High-Speed Counter-Current Chromatography The High-Speed Counter-Current Chromatography technique is commonly used to obtain a higher purity of compound. Although a number of chromatographic techniques are available for extraction of bromelain, such as capillary electrochromatography, ion exchange chromatography, affinity chromatography etc., due to their lower separation capacity, High Speed Counter Current Chromatograph has been used. It is a form of liquid–liquid separation chromatography, which is based on hydrodynamic equilibrium of a two-phase solvent system in the separation column. This technique is also referred to as a hybrid technique of liquid chromatography and liquid–liquid counter-current distribution (Ito, 2005). The aqueous two-phase system used in this technique provides suitable conditions for protein separation, and high viscosity used in the process is helpful in low retention of the stationary phase, and poor resolutions and recoveries (Ito *et al.*, 2007). Crude protein from pineapple can be purified with the help of High-Speed Counter-Current Chromatography (Yin *et al.*, 2011). The mobile phase composed of a sodium phosphate buffer with an appropriate pH value and ionic strength and stationary phase with reverse micelles (0.10 g/mL CTAB/isooctane–hexylalcohol) can be used to separate bromelain from pineapple (Yin *et al.*, 2011).

3.2.3.3 Ethanol Precipitation Separation of a desired compound by the precipitation process is a traditional method and does not require complicated equipment. Extraction of bromelain is carried out by several different steps. Usually the purification step of enzymes is more expensive than isolation. Inexpensive precipitants can be used in the precipitation process, but care should be taken when selecting the organic solvents to avoid them causing adverse effects on the desired compounds. Ethanol precipitation can be used to purify bromelain from pineapple waste and bromelain can be recovered using 30% and 70% concentrations of ethanol; this process yields about 98% of the total enzymatic activity (Soare *et al.*, 2012).

3.2.4 Cellulase

Cellulase is the enzyme which converts cellulose into glucose. Cellulase can be produced by bacteria or fungus. A commonly-used fungus for the production of cellulase is *Trichoderma ressei*. Being a very important enzyme for the production of bioethanol and other important applications, the demand for cellulase is increasing. Pineapple waste can be a good source for the production of cellulase (Saravanan *et al.*, 2013).

Cellulase production from pineapple waste using *Trichoderma longibrachiatum*, *Aspergillus niger* and *Saccharomyces cerevisiae* suggest that the strain *T. longibrachiatum* produces the highest amount of glucose, 0.92 mg/0.5 ml under optimum conditions, 45°C, pH 4.5 on the 7th day of fermentation.

3.2.5 Citric Acid

Citric acid is a very useful compound in food industries (for acidification, preservation of products and flavour enhancement) and pharmaceutical industries. Pineapple waste is a good source of citric acid (2-hydroxy-1, 2, 3-propanetricarboxylic acid), which can be obtained with the help of the fungal strain *Aspergillus niger* (Alben and Erkmen, 2004). A higher concentration of citric acid is produced by *A. niger* as compared to *Saccharomyces cerevisiae* at an initial pH of 4.5 in the presence of methanol (Femi-Ola *et al.*, 2009). Addition of methanol was shown to stimulate biomass production, carbon IV oxide concentration and sugar utilization. However, *S. cerevisiae* is also found to be suitable for the production of citric acid from pineapple waste.

3.2.6 Extruded Product

Pineapple waste consists of about 76% fibre, with 99.2% as insoluble fibre and the remainder as soluble fibre (Martinez *et al.*, 2012). The dietary fibre prevents us from contracting various diseases, improves general health and promotes various metabolic processes, including absorbing cholesterol and glucose in our bodies (Rodriguez *et al.*, 2003); it possesses a high water binding capacity (Serena and Knudsen, 2007). Fibre if also useful in providing texture and consistency in food products, therefore, it can be used to develop food products rich in fibre content.

The development of extruded products from pineapple waste is a good option. Usually extruded products are not considered as beneficial to health due to their inferior nutritional value. But the use of dietary fibre in the development of the extruded product may lead to a healthy and nutritional product. Extrusion helps in depolymerisation of starch, which results in easily digestible carbohydrates (Brennan *et al.*, 2013). A number of researchers have worked on incorporation of fibre from tomato pomace, grape seed (Yağčı and Goğüs, 2008) and carrot pomace (Kumar *et al.*, 2010) etc. into the extrudates.

Pineapple pomace contains 0.61% fat and 45.22% fibre content, which indicates that it can be used to enhance the nutritional value of food products (Selani *et al.*, 2014). The extruded product, prepared by replacing corn flour with pineapple pomace, revealed lower microbial activity, lower water activity and lower risk of physicochemical deterioration, but the products do not expand much and are darker in colour than the control. The product with 10.5% pineapple pomace had same water absorption, hardness and bulk density as that of the control.

Broken rice flour, red gram powder and pineapple pomace powder can also be used in the preparation of the extruded product. Formulation is extruded at different combinations of temperature, screw speed and moisture content. The extrudates were best produced under optimum conditions at a screw speed of 315 rpm, a moisture content of 18.48% and a barrel temperature of 132.27°C (Kothakota *et al.*, 2013). An increased barrel speed results in higher expansion of extruded product, lower hardness and higher overall acceptability, whereas increased moisture content results in lower hardness, lower expansion and high overall acceptability.

3.2.7 Jam

Pineapple peels are usually discarded as waste (Table 3.1), but contain valuable nutrients and are a rich source of phytochemicals. Peels of fruits are also found to be rich in compounds having beneficial health effects and other aroma compounds (Paranthaman, 2009). The discarded portion of pineapple is very high, therefore the waste can be used either for extraction of pectin or in preparation of jam. Pectin can be extracted from single fruit peel or from a mixture with peels of other fruits. Jams, 68.5°Brix, prepared from peels of different fruits such as pineapple, pomegranate, banana and orange, reveal that the microbial growth can be prevented in the pH range of 4.4–5.9. Sensory evaluation indicated that pineapple peel jam was most acceptable by the judges (Chacko and Estherlydia, 2014). The peels of the pineapple can be a potential source of microorganisms, therefore the safety of the product is still an area of investigation.

3.2.8 Lactic Acid

Lactic acid is commonly used in the food industry as an acidity regulator, preservative and for other purposes. Pineapple waste is a good source of fructose (27.5 g/l), glucose (28 g/l) and sucrose (15.5 g/l), therefore it can be considered for fermentation to produce lactic acid (Abdullah and Hanafi, 2008; Abdullah, 2010). A number of researchers have worked on the production of lactic acid from pineapple waste, since it is a low-cost substrate. The lactic acid can be produced by using *Lactobacillus delbrueckii* under anaerobic conditions for 72 h (Idris and Suzana, 2006). Calcium alginate, when used as the immobilization matrix, gave a maximum yield of 0.7822–0.8248 g lactic acid/g glucose under varying conditions of pH and temperature.

Abdullah (2010) worked on production of lactic acid from pineapple cannery liquid waste, which is rich in glucose, sucrose fructose etc., by using *Lactobacillus delbrueckii subsp. delbrueckii ATCC 9649* under anaerobic conditions. To produce anaerobic conditions, flasks were flushed with nitrogen and then sealed with rubber stoppers. The effect of the nitrogen source on lactic acid production shows that the yeast extract gives the highest yield, followed by urea, corn steep liquor, malt sprout and ammonium sulphates, with yields of 78.52, 26.68, 19.14, 14.10 and 5.6%, respectively.

3.2.9 Animal Feed

Pineapple waste, including outer skin, core and peel, leaves and pomace after extraction of juice, can be used as feed for ruminants and non-ruminants. Due to high moisture content and the presence of soluble carbohydrates, rate of decay of pineapple waste is very high. Fresh pineapple waste is more acidic than fermented waste, therefore fermented pineapple waste is preferred for cattle. It is also being used as a dairy feed in China (Sruamisri, 2007). Drying and ensiling are the commonly-used preservation methods used by farmers. Dried waste increases the digestibility and weight of the animals (Costa *et al.*, 2007). Pineapple bran (solid residue of crown and skin of pineapple) can also be used as animal feed.

Figure 3.2 (a) Guava fruit on tree; (b) guava fruit; (c) guava seeds; and (d) guava peel.

3.3 Guava

The major parts of guava (*Psidium guajava* L.) are peel, pulp and seeds (Figure 3.2). A number of products are prepared from guava, such as guava-based jellies, pastes, sorbets, sweet bars etc. and a large quantity of waste is generated (Steinhaus *et al.*, 2008; Elizalde and Hernández, 2009). Guava juice is very popular in tropical regions, due to its flavour and viscosity. During extraction of guava juice, seeds of guava are usually discarded, but the seeds contain about 5–13% oil, which is rich in essential fatty acids (Sidhu, 2006). Guava seeds are also a rich source of essential amino acids (Table 3.3). If the amino acids from the guava seeds are compared with the amino acids of standard protein requirements for a pre-school-aged child, the essential amino acids are competitively matching in their presence and amounts.

After extraction of juice, guava cake consisting of peel, pulp and seeds accounts for nearly 30% of fruit weight, which is discarded as such or used as fertilizer or feed. However, it has been reported that guava cake is a good source of phenolic compounds and dietary fibre, which may provide protection against a number of diseases

Table 3.3 Amino acids composition of papaya and guava seeds

Type of amino acid	[1]Guava seeds (g/100g protein)	[1]Papaya seeds (g/100g protein)
Indispensable Amino Acid		
Isoleucine	3.21	3.21
Leucine	6.11	7.76
Methionine	4.09	1.34
Phenylalanine	2.83	3.44
Lysine	1.66	4.25
Threonine	3.92	2.88
Tyrosine	3.46	2.19
Valine	4.88	2.31
Cystine	2.18	1.21
Dispensable Amino Acid		
Glutamic	8.96	13.01
Aspartic	6.12	7.51
Serine	7.03	3.12
Proline	7.84	2.44
Glycine	4.16	2.42
Alanine	6.33	3.22
Histidine	1.87	2.23
Arginine	9.25	6.53

[1]El-Safy *et al.*, (2012)

(Jimenez-Escrig *et al.*, 2001). Waste from the guava processing industry may be utilized for the following purposes.

3.3.1 Pectin

Guava cake is commonly used as animal feed and fertilizer or discarded as waste, although it can be used for extraction of pectin. Furthermore, it can be used for fortification of other products. Insoluble fibre (cellulose, hemicelluloses) is helpful in bowel movement, whereas soluble pectin is helpful in lowering cholesterol. Pectin, being a polysaccharide, has very important properties like emulsion stabilization and ability to form a gel (Westerlund *et al.*, 1991). Therefore, it has potential for use in food processing, medicine and cosmetics industries.

Pectin can be extracted from guava fruit waste (Chakraborty and Ray, 2011). The procedure involved blanching of spent guava extract with boiling water and drying overnight at 60°C. Then distilled water was added, followed by heating at 90° for 45 min. The extraction was carried out with the help of nitric acid (pH 1.6, if required, maintained with the help of NaOH) and the desired concentration of the solution was attained at 90°C followed by filtration. Hereafter, cooling was done at room temperature and then centrifuged at 1200 rpm at 12°C. Again, centrifugation can be carried out with hot water at 60°C until dissolution of precipitate. Alcohol, 95% (v/v), was added under refrigerated conditions and kept for 4 h to form the precipitates. Again,

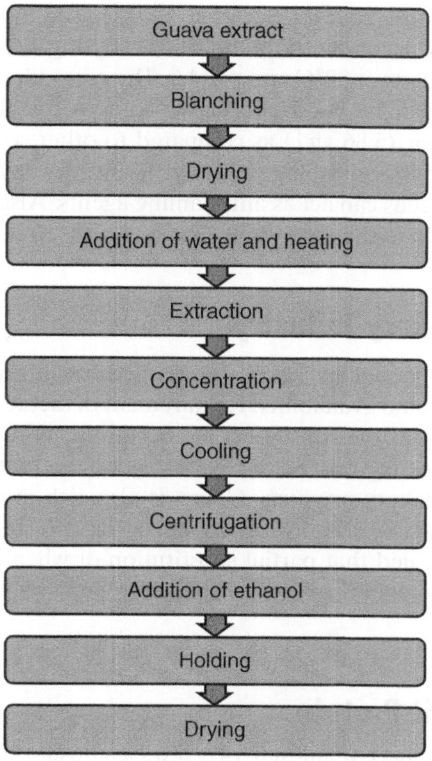

Figure 3.3 Flow chart for the extraction of pectin from guava.

after centrifugation, precipitates were dried at 40°C and ground. The different steps for extraction of pectin are shown in Figure 3.3.

The optimum conditions, time, temperature and pH for pectin extraction were 45 minutes, 80°C, and 5.0 respectively, with a yield of 0.06% in powdered form (Chakraborty and Ray, 2011).

3.3.2 Juice Fortified with Dietary Fibre

The nutritional value of food products can be increased with the fortification of dietary fibre. The guava juice can be fortified with fibre, extracted from guava cake (Thongsombat *et al.* 2007). Juice, fortified with pectin powder, 0.25%, obtained from guava only, was found to be the best sample in terms of sensory attributes. No significant difference was found between the control samples and samples fortified with guava pectin, except turbidity and odour of the fortified sample.

3.3.3 Alcoholic Fermentation

During processing of guava-based products, seeds are usually discarded, which constitute about 12% of the fresh fruit weight and are rich in protein content

(Bernardino-Nicanor *et al.*, 2006). Cock *et al.* (2013) used molasses and guava seed flour as a substrate for alcoholic fermentation, by using *Saccharomyces cerevisiae* under anaerobic conditions at 32°C and pH 4.6. The research indicates that molasses and guava seed flour can be good alternative sources for alcoholic fermentation. Ethanol yield was higher (5.86 gg^{-1}) as compared to other sources of nitrogen such as molasses, urea, diamine phosphate (MUDP), and dry mycelium of *Aspergillus niger*. Lipids in guava seeds can act as antifoaming agents. Also, the high vitamin and mineral content can help in the growth of yeast.

3.3.4 Use in Bakery Industry

Bakery products, mainly cookies, cakes, breads and biscuits, are becoming popular because of convenience and availability. Requirement of dietary fibre in food products is increasing day by day as the consumers are becoming more health conscious. Use of flour from fruit processing waste can be put to good use to increase the nutritional and sensory value of bakery products. Guava is also rich in omega-3 and omega-6 polyunsaturated fatty acids, along with vitamins A and C and dietary fibre. El-din and Yassen (1997) revealed that partial substitution of wheat flour with guava seed flour gives an improvement in volume, diameter and thickness of cookies after baking.

3.3.5 Single Cell Protein

Proteins are an essential part of our diet, therefore a deficiency of proteins may lead to a number of disorders. As the world population is increasing, requirement of protein sources is also increasing. Production of Single Cell Protein is an alternative way to meet these requirements.

Various combinations of fruit wastes can be used for production of Single Cell Protein (Yousufi, 2012). Out of the total 36 combinations formed for production of Single Cell Protein, pomegranate rind and guava peel are found to be the best combination by using *Rhizopus oligosporus*. Wastes were washed and a 1:1 combination of wastes was prepared. The combination of waste was autoclaved at 121°C for 15 min at 15 psi. Then the combination was transferred to sterilized Petri dishes and allowed to cool. In the next step, inoculation was carried out with *Aspergillus oryzae* and *R. oligosporus*. Then the samples were incubated at 28°C for 7 days. The mycelia obtained after 7 days was transferred to Whatmann filter paper no. 1 and to remove the adhering particles and impurities, washing was carried out with distilled water. The content was dried at 90°C to obtain a moisture-free fungal content (Yousufi, 2012). The different steps of the process are shown in Figure 3.4.

3.3.6 Lycopene

Pink guava is a good source of lycopene, which is an important antioxidant. Oxidative stress caused by different species in the body may lead to various inflammatory and degenerative diseases. The quenching ability of lycopene is higher than β-carotene and α-tocopherol. Therefore, lycopene may assist in protection from various diseases such as cardiovascular disease, cancer, atherosclerosis etc. Lycopene can be extracted by

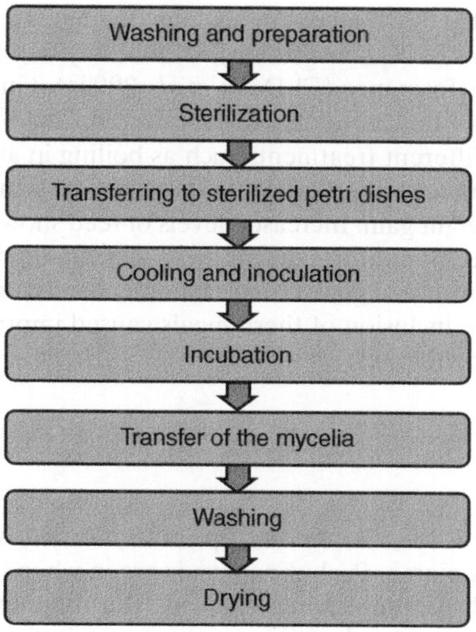

Figure 3.4 Flow chart for the production of Single Cell Protein.

different methods and the extraction of lycopene by the supercritical fluid extraction method is described below.

3.3.6.1 Supercritical Fluid Extraction The supercritical fluid extraction method is a popular and alternative method to conventional extraction process. Kong *et al.* (2010a) studied the effects of solvent and supercritical carbon dioxide extraction on antioxidant and cytotoxic activities of the lycopene-rich fractions obtained from pink guava waste. Supercritical fluid extraction was found to contain a higher lycopene-rich fraction, which can be further used for development of functional foods.

The lycopene content of pink guava is more than 80% of the total carotenoids present (Padula and Rodriguez-Amaya, 1986). During processing of pink guava into puree, the waste generated makes up about 25% of the total load weight. This waste can be used for extraction of lycopene, but care should be taken since thermal processes may affect the activity of lycopene. Kong *et al.* (2010b) optimized the oven drying conditions for lycopene content and lipophilic antioxidant capacity from the decanter-obtained pink guava waste. The optimum conditions found for oven drying were 43.8°C for 6.4 h, with lycopene content and antioxidant capacity of 14 mg/100 g and 21 μm LE/100 g, respectively.

3.3.7 Utilization as Feed

Guava waste has a high content of polyphenols (2.62–7.79%), dietary fibre (48.55–49.42%) and methoxylated pectin, which are sources of antioxidants (El-Deek *et al.*, 2009a). The anti-nutritional factors present in waste must be removed before using it as feed (Marquina *et al.*, 2008). Anti-nutritional factors may be eliminated by cooking in water, soaking in water, autoclaving etc.

It has been found that at 32–48 weeks of age, sun dried guava waste can be included in the diet of laying hens (15% substitution) without any harmful effect on the hens and their productive performance (El-Deek *et al.*, 2009a). Guava waste can also be incorporated into broiler finisher diets (El-Deek *et al.*, 2009b). In this study, guava waste had undergone different treatments such as boiling in alkaline solution, water and acid solution and autoclaving treatment. There was no significant effect of experimental diet on body weight gain. Increased levels of feed showed an effect on energy and protein intake as compared to control diets. Abdelhamin and Soliman, (2012) studied the effect of inclusion of guava tree leaves meal and camphor tree leaves meal in Nile tilapia diets. The inclusion of these meals caused improvements in weight of fish and also in their growth rates.

3.4 Papaya

Papaya (*Carica papaya L.*) is a common tropical fruit, which varies in colour, shape and taste with cultivation area. Peel of papaya is green when immature and changes to dark yellow to reddish orange when ripened. The different parts of papaya are shown in Figure 3.5. The natural antioxidants present in fruits and vegetables are

Figure 3.5 (a) Papaya fruit; (b) papaya cut into halves; (c) papaya seeds; and (d) papaya peel.

Table 3.4 Value-added products from papaya waste and their functions

Part of papaya	Important constituent	Function
Seed	Oil rich in fatty acids, crude fibre, benzylglucosinolate, myrosin, benzylthiourea, crude protein etc.	Useful for treatment of ringworm, enlarged liver and pectoral properties
Latex	Papain, chemopapain, proteolytic enzyme, glutamine cyclotransferase etc.	Useful for treatment of diarrhoea, stomachic and whooping cough, anathematic
Leaves	Choline, vitamin C, vitamin E, alkaloids, carpain, carposide	Treatment of jaundice, antifungal activity, anti-haemolytic activity
Unripe fruit	Malic acid, citric acid	Treatment of enlarged spleen and liver, to remove poison of snakebite, antibacterial activity

Boshra and Tajul (2013): Krishna *et al.* (2008)

helpful in the prevention of a number of diseases. Being a very good source of vitamins and minerals, it is well known for its health promoting effects. It possesses properties such as increase in thrombocyte count, antibacterial, antiviral, antifungal, antifertility, anti-inflammatory, wound healing, antitumor, neuroprotective and anti-sickling properties (Lim, 2012).

The health effects of papaya waste are shown in Table 3.4. Ripe papaya is consumed as fresh fruit and unripe papaya is used as a cooked vegetable. Unripe fruit contains malic and citric acids, which are helpful for treatment of enlarged spleen and liver, to remove poison from snakebite and possesses antibacterial activity. Unripe fruit is used as an ingredient in salad and cooked dishes in Thailand. After processing of papaya into puree, juice, canned products etc., the waste is discarded, which causes environmental concerns. Papaya seeds and peel are very good sources of functional constituents and nutrients, which can be processed for extraction of important nutrients. Phenolic content, ascorbic acid and β-carotene content of papaya peels are higher than papaya fruit (Maisarah *et al.*, 2014). Papaya leaves are used for treatment of diabetes, gonorrhea, syphilis and inflammation (Gill, 1992). Aqueous extract of papaya leaves has been found to possess anticancer activity (Morimoto *et al.*, 2008). It has been revealed that extract of papaya leaves is also beneficial for treatment of dengue fever, blood glucose level and wound-healing activity. A number of techniques for extraction of important nutrients and utilization of waste from papaya are discussed below.

3.4.1 Papaya Seeds as Antioxidants

Antioxidants present in different parts of papaya possess free radical scavenging properties. Free radicals produced in our bodies are very reactive in nature, therefore can damage lipids, proteins and DNA, leading to serious disorders. Papaya contains a number of phytochemicals, for example carotenoids present in fruits and seeds, phenolics present in fruits and leaves, glucosinolates present in seeds and fruits, and enzymes in latex (Maisarah, 2014). Papaya waste exhibits strong antioxidant activities. The ethyl acetate extract fraction from papaya seed is rich in antioxidants and phenolics and has a strong scavenging activity for super anion and hydrogen peroxide radicals, DPPH and hydroxyl free radicals (Zhou *et al.*, 2011).

3.4.2 Extraction of Papain

Enzymes are the proteins which catalyze a reaction and act by lowering the activation energy of that reaction, allowing it to proceed at a much faster rate than it otherwise would. Papain is the enzyme extracted from papaya, which consists of 212 amino acids stabilized by 3 disulphide bonds (Amri, 2012). Papain has many industrial applications such as tenderization of meat and meat products, manufacture of protein hydrolysate, chewing gum, cheese, cereals and pet foods. It can also be used in the aroma and perfume industries, textile and tanning industries, and the beer industry to remove cloudiness. It also finds usage in medicinal purposes for the treatment of inflammation, diarrhoea, parasites and other diseases.

Papain is extracted from papaya latex, which is obtained from unripe but mature papaya. An incision is made with a stainless steel knife and the latex is collected in glass vessels, as it flows from cuts. High humidity is required for the maximum collection of latex, therefore tapping of fruit should be done during high humidity periods, usually in the early morning. Incisions should be made in such a way that cuts should not be deeper than 2 mm to avoid dust and insects. Latex is mixed with $K_2S_2O_5$ at the rate of 0.05% for it to mature and to extend storage life. Latex should not come into direct contact with skin, since it may cause burning. Latex should be dried below the 5% moisture level at about 50°C (Gadekar and Rathi, 2007). Then the mixture is sieved, usually through a 10 mesh sieve, to remove the extraneous matter, then followed by centrifugation. It is stored in airtight containers at about 9°C for approximately 6 months. A flow chart for the extraction of papain is shown in Figure 3.6.

Papaya peel is also a very good source of papain. It has been reported that dried and ground green papaya peel mixed with spices acts as a low-cost meat tenderizer. Islam and Molinar Toribio (2013) found that the activity of enzymes found in papaya peel was 10 times lower than the enzymes found in latex. The spice batches were mixed with papaya peel to formulate the meat tenderizer. The meat, marinated with 30% green papaya peel, showed the highest level of acceptance for softness of texture and good flavour.

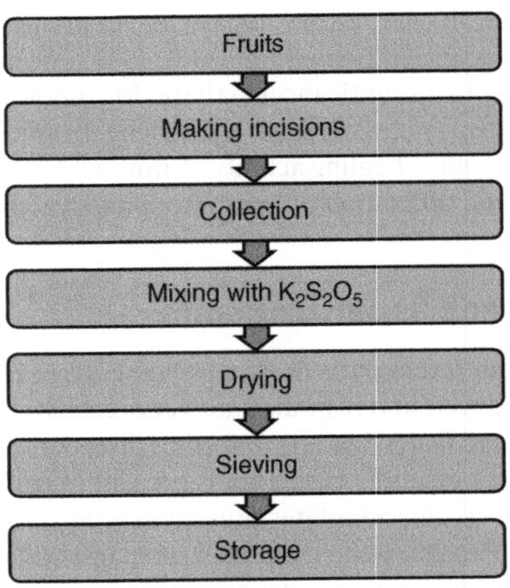

Figure 3.6 Flow chart for extraction of papain.

3.4.3 Extraction of Oil from Seeds

In order to use papaya waste, oil can be extracted from the seeds. Papaya seeds are rich in monounsaturated fatty acids and possess functional properties. The compound benzyl isothiocynate, found in the crushed seeds, has an effect on vascular contraction (Wilson *et al.*, 2002). Consumption of papaya seed oil can be effective in reducing the risk of coronary heart disease. It can also be used to increase the palatability of snacks, bakery products etc. (Corbett *et al.*, 2003).

Papaya seeds contain nearly 25.6% carbohydrates, 28.1% protein, 19.1% crude fibre and 8.2% ash content (Syed, 2011). The most commonly-adopted processes for extraction of seed oil are hydraulic press, solvent extraction, supercritical fluid extraction, enzyme-assisted extraction etc. In the hydraulic press method, further purification of oil is required, whereas in aqueous extraction, the yield of oil is very low. and in supercritical fluid extraction, a high investment is required (Syed, 2011; Wang and Weller, 2006). By using the solvent extraction method, 30.1% extraction of oil has been reported (Syed, 2011) and the majority of fatty acids are found as oleic (72.5%), palmitic (13.5%) and stearic acids (4.5%) (Syed, 2011).

Extraction of papaya seed oil by ultrasound-assisted extraction (UAE) and conventional techniques (Soxhlet method and solvent extraction, 12 h, 25°C) yielded 76.1% and 79.1% oil content, respectively (Samaram *et al.*, 2013). UAE significantly influenced the triacylglycerol profile of papaya seed oil. For the yield of papaya seed oil, 73% is reported under the optimum conditions of temperature 62.5°C, time 38.5 min at ultrasound power 700 W and a solvent to sample ratio of 7:1 w/v (Samaram *et al.*, 2014).

The oil extracted using supercritical fluid extraction (SCFE) has higher purity due to absence of any organic solvent. The process also gives an easy recovery of supercritical solvent by adjusting the temperature and pressure conditions. The extraction of oil from papaya seeds by using SCFE revealed good results at 40°C and 100 bar followed by 80°C and 200 bar. The efficiency increased with increase in pressure at the same temperature, but increase in temperature caused a decrease in the efficiency of process (Barroso *et al.*, 2012).

3.4.4 Alcohol and Vinegar

Papaya peel has crude protein content 1.24%, lipid 6.93%, crude fibre 1.72%, total acidity 5.62%, reducing sugar 7.49%, total sugar 9.13%, ash 3.23% and moisture content 75.13% (Vikas and Mridul, 2014). Alcohol can be produced from the papaya peel by using *S. Cerevisiae* at 28°C with a yield of 8.11% followed by acetic acid fermentation using *Acetobacter aceti* at 37°C (Vikas and Mridul, 2014).

3.4.5 Utilization of Seed Flour for Food Enrichment

The seeds discarded as waste from papaya puree plants constitute about 22% of waste (Marfo *et al.*, 1986). The flour of papaya seeds can be used as a nutrient source in food products. El-safy *et al.* (2012) determined the mineral composition and functional properties of different fruit seed flours including papaya.

Papaya seed flour contains considerable amounts of iron, calcium, phosphorus, copper, zinc magnesium and potassium and also possesses good functional properties

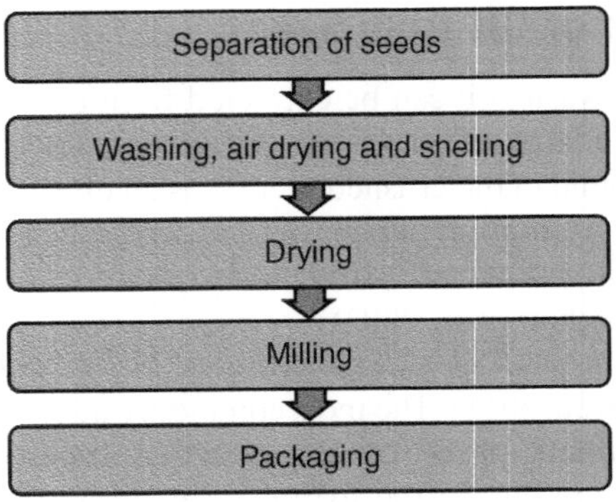

Figure 3.7 Flow chart for preparation of papaya seed flour.

The amino acid composition of papaya seed is shown in Table 3.3. Papaya seeds contain a good amount of indispensable amino acids, therefore papaya seed flour has the potential to be used as an effective food additive for preparation of different food products (El-Safy *et al.*, 2012). For preparation of seed flour, seeds were separated from pulp and washed thoroughly, followed by drying at about 50°C. The seed coat was removed and then milling was carried out to make the powder followed by sieving, packaging and storage. The different steps for preparation of papaya seed flour are shown in Figure 3.7.

3.4.6 Carboxymethyl Cellulose (CMC)

Carboxymethyl Cellulose (CMC) is a cellulose derivative, which has wider applications in the food industry as a viscosity modifier and emulsifier. CMC is produced from cellulose and chloroacetic acid in alkali catalysed reaction. After synthesis, purified CMC is a white or creamish-white coloured powder, which is tasteless and odourless (Hattori *et al.*, 2004).

Cellulose can be synthesized from papaya peel (Rachtanapun, 2007a,b). However, the mechanical properties of the film obtained from papaya CMC are reported as inferior when compared to commercial CMC. These properties can be improved by addition of corn starch/chitosan to papaya CMC (Xu *et al.*, 2005). The blended film of carboxymethyl cellulose obtained from papaya peel and corn starch showed an increase in tensile strength but decrease in elongation (Rachtanapun, 2009). Upon addition of glycerol (plasticizer), tensile strength decreased but elongation and water vapour permeability increased.

3.4.7 Single Cell Protein

Papaya waste can also be used for production of Single Cell Protein. Papaya extract prepared with 200 mL sterile water showed highest cell growth with *S. Cerevisiae* at 25°C for 5 days. Biomass was found to have 35.5% protein, 4.09% lipids, 40.7% saccharide, 9.63% moisture and 7.9% total ash (Ojokoh and Uzeh, 2005). Since the protein obtained from yeast is of high nutritional value, therefore the biomass can be used for poultry and cattle feed.

3.5 Summary and Future Trends

Tropical fruits are a rich source of functional compounds and play important roles against various diseases such cancer, cardiovascular disease, chronic diseases etc. These fruits are commonly processed for juice extraction, preparation of jams, jellies, canned products, nectar, syrups, puree etc., but the by-products generated during processing are discarded as waste. Numerous environmental and health problems may be created globally due to large volumes of wastes, which are rich in many vitamins, minerals, phytochemicals and dietary fibre. A major challenge in utilization of fruit processing wastes is their high moisture content, which makes them highly perishable in nature. Hence collection, transportation and processing of waste are the aspects to be dealt with. Since cost of utilizing the waste is usually higher than disposing of it, novel technologies need to be developed, and attention paid to improving existing technologies.

References

Abdelhamid, A.M. and Soliman, AA. (2012) Possibility of using dried leaves of guava and camphor trees in tilapia diets. *Journal of the Arabian Aquaculture Society*, **7**(1): 91–108.

Abdullah, A. and Hanafi M. (2008) Characterisation of solid and liquid pineapple waste. *Reaktor*, **12**(1): 48–52.

Abdullah, M.B. (2010) Effect of nitrogen source and initial sugar concentration on lactic acid fermentation of pineapple waste using *L. Delbrueckii. Teknik*, **3**(1): 31–34.

Alben, E. and Erkmen, O. (2004) Production of citric acid from a new substrate, undersized semolina, by *Aspergillus niger*. *Food Technology and Biotechnology*, **4**(1): 19–22.

Amri, M. (2012) Papain, a plant enzyme of biological importance: A review. *American Journal of Biochemistry and Biotechnology*, **8**(2): 99–104.

Anon (1996) Processing fruits. In *Science and Technology,* vol. 1, *Biology, Principles and Applications*. Technomic Publishing Co. Inc., Lanchester, PA, pp. 461–499.

Babu, B.R., Rastogi, N.K. and Raghavarao, K.S.M.S. (2008) Liquid–liquid extraction of bromelain and polyphenol oxidase using aqueous two-phase system. *Chemical Engineering and Processing: Process Intensification*, **47**: 83–89.

Ban-Koffi, L. and Han, Y.W. (1990) Alcohol production from pineapple waste. *World Journal of Microbiology and Biotechnology*, **6**: 281–284.

Bardiya, N., Somayaji, D. and Khanna, S. (1996) Biomethanation of banana peel and pineapple waste. *Bioresource Technology*, **58**: 73–76.

Barroso, P.T.W., Pessoa, F.L.P. and Mendes, M.F. (2012) Extraction of the oil present in *Carica papaya L.* seeds with supercritical CO_2, *10th International Symposium on Supercritical Fluids*, 13–16 May, San Francisco, CA.

Bernardino-Nicanor, A., Scilingo, A., Añón, M.C. and Dávila-Ortíz, G. (2006) Guava seed storage protein: Fractionation and characterization. *LWT – Food Science and Technology*, **39**: 902–910.

Bhui, K., Prasad, S., George, J. and Shukla, Y. (2009) Bromelain inhibits COX-2 expression by blocking the activation of MAPK regulated NF-kappa B against skin tumor initiating triggering mitochondrial death pathway. *Cancer Letters*, **28**: 167–176.

Boshra, V. and Tajul, A.Y. (2013) Papaya – an innovative raw material for food and pharmaceutical processing industry. *Environmental Health Journal*, **4**(1): 68–75.

Brennan, A.M., Derbyshire, E., Tiwari, B.K. and Brennan, C.S. (2013) Ready to eat snack products: The role of extrusion technology in developing consumer acceptable and nutritious snacks. *International Journal of Food Science and Technology*, **48**: 893–902.

Brien, S., Lewith, G. and Walker, A. (2004) Bromelain as a treatment for osteoarthritis: A review of clinical studies. *Evidence-based Complementary and Alternative Medicine: eCAM*, **1**(3): 251–257.

Cabrera, H.A.P., Menezes, H.C., Oliveira, J.V. and Batista, R.F.S. (2000) Evaluation of residual levels of benomyl, methyl parathion, diuron, and vamidothion in pineapple pulp and bagasse (Smooth cayenne). *Journal of Agricultural and Food Chemistry*, **48**: 5750–5753.

Chacko, C.M. and Estherlydia, D. (2014) Antimicrobial evaluation of jams made from indigenous fruit peels. *International Journal of Advanced Research*, **2**: 202–207.

Chakraborty, A. and Ray, S. (2011) Development of a process for the extraction of pectin from citrus fruit wastes viz. lime peel, spent guava extract, apple pomace etc. *Internet Journal of Food Safety*, **13**: 391–397.

Chobotava, K., Vernallis, A.B. and Majid, F.A.A. (2009) Bromelain's activity and potential as an anti-cancer agent: Current evidence and perspectives. *Cancer Letters*, **290**: 148–156.

Chulalaksananukul, S., Sinbuathong. N. and Chulalaksananukul, W. (2012) Bioconversion of pineapple solid waste under anaerobic condition through biogas production. *KKU Research Journal*, **17**(5): 734–742.

Cock, L.S., Ayala, J.D.M., López, J.E.A. and Schouben, A.L.G. (2013) Kinetics of alcoholic fermentation using guava (*Psidium guajava*) seed flour and dry mycelium of *Aspergillus niger* as nitrogen sources. *Dyna*, **80**(180): 113–121.

Corbett, P. (2003) It's time for an oil change! Opportunities for high oleic vegetable oils. *Inform*, **14**: 480–481.

Costa, R.G., Correia, M.X.C., Da Silva, J.H.V., De Medeiros, A.N. and De Carvalho, F F.R. (2007) Effect of different levels of dehydrated pineapple by-products on intake, digestibility and performance of growing goats. *Small Ruminant Research*, **71**: 138–143.

El-Deek, A.A., Asar, M.A, Hamdy, S.M, and Abdalla, A.A. (2009a) The utilization of guava by-products in broiler finisher diets. *Egyptian Poultry Science*, **29**(1): 53–79.

El-Deek, A.A., Hamdy, S.M., Attia, Y.A., El-Shahat, A.M. (2009b) Guava by-product meal processed in various ways and fed in differing amounts as a component in laying. *International Journal of Poultry Science*, **8**(9): 866–874.

El-Din, M.H.A.S. and Yassen, A.A.E. (1997) Evaluation and utilization of guava seed meal (*Psidium Guajava* L.) in cookies preparation as wheat flour substitute. *Food/Nahrung*, **41**(6): 344–348.

Elizalde, M.P. and Hernández, V. (2009) Guava seed as an adsorbent and as a precursor of carbon for the adsorption of acid dyes. *Bioresource Technology*, **100**: 2111–2117.

El-Safy, F.A., Salem, R.H. and El-Ghany, M.E.A. (2012) Chemical and nutritional evaluation of different seed flours as novel sources of protein. *World Journal of Dairy and Food Sciences*, **7**(1): 59–65.

FAO (2011) *Global Food Losses and Food Waste*. Food and Agriculture Organization of the United Nations, Rome.

FAO (2013) Food and Agriculture Organization of the United Nations. FAOSTAT Statistics Database – Agriculture, 2013. URL http://faostat.fao.org/; last accessed 10 May 2013.

Femi-Ola, T.O., Oluyege, J.O. and Gbadebo, A.O. (2009) Citric acid production from pineapple waste. *Continental Journal of Microbiology*, **3**: 1–5.

Gadekar, S.V. and Rathi, A. (2007) Manufacturing process of papain. *Chemical Product Finder*, **25**: 38–45.

Gill, L.S. (1992) *Ethnomedical Uses of Plants in Nigeria*. Uniben Press, Benin, Nigeria.

Gomez, M., Ronda, F., Coballera, A.P., Blanco, A.C. and Rosell, C.M. (2007) Functionality of different hydrocolloids on the quality and shelf life of yellow layer cakes. *Food Hydrocolloids*, **21**(2): 167–173.

Gupta, P., Maqbool, T. and Saleemuddin, M. (2007) Oriented immobilization of stem bromelain via the lone histidine on a metal affinity support. *Journal of Molecular Catalysis B: Enzymatic*, **45**: 78–83.

Harikrishna, S., Srinivas, N.D., Raghavarao, K.S.M.S and Karanth, N.G. (2002) Reverse micellar extraction for downstream processing of proteins/enzymes. *Advances in Biochemical Engineering/Biotechnology*, **75**: 119–183.

Hattori, K., Abe, E., Yoshida, T., Cuculo, J.A. (2004) New solvents for cellulose II ethylenediamine/thiocyanate salt system. *Polymer Journal*, **36**(2): 123–130.

Hebbar, H.U., Sumana, B. and Raghavarao, K.S.M.S. (2008) Use of reverse micellar systems for the extraction and purification of bromelain from pineapple wastes, *Bioresource Technology*, **99**: 4896–4902.

Henningsson, S., Hyde, K., Smith, A. and Campbell, M. (2004) The value of resource efficiency in the food industry: A waste minimisation project in East Anglia, UK. *Journal of Cleaner Production*, **12**(5): 505–512.

Himmel, M.E., Ding, S.Y., Johnson, D.K., Adney, W.S., Nimlos, M.R., *et al.* (2007) Biomass recalcitrance: Engineering plants and enzymes for biofuels production. *Science*, **315**(5813): 804–807.

Idris, A. and Suzana, W. (2006) Effect of sodium alginate concentration, bead diameter, initial pH and temperature on lactic acid production from pineapple waste using immobilized *Lactobacillus delbrueckii*. *Process Biochemistry*, **41**: 1117–1123.

Islam, M.N. and Toribio, E.M. (2013) Development of a meat tenderizer based on papaya peel. *RIDTEC*, **9**(2): 24-29.

Ito, Y. (2005) Golden rules and pitfalls in selecting optimum conditions for highspeed counter-current chromatography. *Journal of Chromatography A*, **1065**: 145–168.

Ito, Y., Clary, R., Sharpnak, F., Metger, H. and Powell, J. (2007) Mixer-settler countercurrent chromatography with multiple spiral disk assembly. *Journal of Chromatography A*, **172**: 151–159.

Jiménez-Escrig, A., Rincón, M., Pulido, R. and Saura-Calixto, F. (2001) Guava fruit (*Psidium guajava* L.) as a new source of antioxidant dietary fibre. *Journal of Agricultural and Food Chemistry*, **49**: 5489–5493.

Kameswari, S.A., Velmurugan, B., Thirumaran, K. and Ramanujam, R.A. (2007) Biomethanation of vegetable market waste-untapped carbon trading opportunities. *Proceedings of the International Conference on Sustainable Solid Waste Management*, 5–7 September, Chennai, India, pp. 415–420.

Kaur, M. and Sharma, H.K. (2013) Effect of enzymatic treatment on carrot cell wall for increased juice yield and effect on physicochemical parameters. *African Journal of Plant Science*, **7**(6): 234–243.

Kaur, M. and Sharma, H.K. (2014) Tropical fruits. In: *Functional Foods and Dietary Supplements: Processing Effect and Health benefits of Foods of Asian Origin* (eds A. Noornhorm, I. Ahmad and A.K. Anal). John Wiley & Sons Inc, Iowa, pp. 91–111.

Ketnawa, S. and Rawdkuen, S. (2011) Application of bromelain extract for muscle foods tenderization. *Food and Nutrition Sciences*, **2**: 393–401.

Ketnawa, S., Chaiwut, P. and Rawdkuen, S. (2012) Pineapple wastes: A potential source for bromelain extraction. *Food and Bioproducts Processing*, **90**: 385–391.

Kong, K.W., Rajab, N.F., Prasad, K.N., Ismail, A., Markom, M. and Tan, C.P. (2010a) Lycopene-rich fractions derived from pink guava by-product and their potential activity towards hydrogen peroxide-induced cellular and DNA damage. *Food Chemistry*, **123**: 1142–1148.

Kong, K.W., Ismail, A., Tan, C.P. and Rajab, N.F. (2010b) Optimization of oven drying conditions for lycopene content and lipophilic antioxidant capacity in a by-product of the pink guava puree industry using response surface methodology, *LWT – Food Science and Technology*, **43**: 729–735.

Kothakota, A., Jindal, N. and Thimmaiah, B. (2013) A study on evaluation and characterization of extruded product by using various by-products. *African Journal of Food Science.* **7**(12): 485–497.

Krishna, K.L., Paridhavi, M. and Patel, J.A. (2008) Review on nutritional and pharmacological properties of papaya (*Carica papaya* Linn.). *Natural Product Radiance*, **7**: 364–373.

Kumar, N., Sarkar, B.C. and Sharma, H.K. (2010) Development and characterization of extruded product using carrot pomace and rice flour. *International Journal of Food Engineering*, **6**: 1–24.

Larrauri, J.A., Ruperez, P. and Calixto, F.S. (1997) Pineapple shell as a source of dietary fibre with associated polyphenols. *Journal of Agricultural and Food Chemistry.* **45**: 4028–4031.

Lim, T. (2012) *Edible Medicinal and Non-Medicinal Plants,* vol. 1*: Fruits*. Springer Science & Business Media, New York.

Maisarah, A.M., Asmah, R. and Fauziah, O. (2014) Proximate analysis, antioxidant and antiproliferative activities of different parts of Carica Papaya. *Journal of Nutrition and Food Science*, **4**(2): 1–7.

Marfo, E.K., Oke, O.L. and Afolabi, O.A. (1986) Some studies on the proteins of Carica papaya seeds. *Food Chemistry*, **22**(4): 267–277.

Marquina, V., Araujo, L., Ruíz, J., Rodríguez-Malaver, A. and Vit, P. (2008) Composition and antioxidant capacity of the guava (*Psidium guajava L.*) fruit, pulp and jam. *Archive of Latinoamericana Nutrition*, **58**: 98–102.

Martinez, R., Torres, P., Meneses, M.A., Figueroa, J. G., Perez-Alvarez, J.A. and Viuda-Martos, M. (2012) Chemical, technological and *in vitro* antioxidant properties of mango, guava, pineapple and passion fruit dietary fibre concentrate. *Food Chemistry*, **135**: 1520–1526.

Maurer, H.R. (2001) Bromelain: Biochemistry, pharmacology and medical use. *Cell Molecular Life Science*, **58**: 1234–1245.

Mbuligwe, S.E. and Kassenga, G.R. (2004) Feasibility and strategies for anaerobic digestion of solid wastes for energy production in Dar es Salaam city, Tanzania. *Resources, Conservation and Recycling*, **42**: 183–203.

Morimoto, C., Dang, N.H and Dang, Y.S. (2008). Cancer prevention and treating composition for preventing, ameliorating, or treating solid cancers, e.g. lung, or blood cancers, e.g. lymphoma, comprises components extracted from brewing papaya. Patent number-WO2006004226-A1; EP1778262- A1; JP2008505887-W; US2008069907-A1.

Nie, H., Shubai, L., Zhou, Y., Chen, T., He, Z. *et al.* (2008) Purification of bromelain using immobilized metal affinity membranes. *Journal of Biotechnology*, **136**: 402–459.

Ojokoh, A.O. and Uzeh, R.E. (2005) Production of *Saccharomyces cerevisiae* biomass in papaya extract medium. *African Journal of Biotechnology*, **4**(11), 1281–1284.

Padula, M. and Rodriguez-Amaya, D.B. (1986) Characterisation of the carotenoids and assessment of the vitamin A value of Brazilian guavas (*Psidium guajava L.*). *Food Chemistry*, **20**(1), 11–19.

Paranthaman, R., Vidyalakshmi, R., Murugesh, S. and Singaravadivel, K. (2009) Effects of fungal co-culture for the biosynthesis of tannase and gallic acid from grape wastes under solid state fermentation. *Global Journal of Biotechnology and Biochemistry*, **4**(1): 29–36.

Philippidis, G.P., Smith, T.K., and Wyman, C.E. (1993) Study of the enzymatic hydrolysis of cellulose for production of fuel ethanol by the simultaneous saccharification and fermentation process. *Biotechnology and Bioengineering*, **41**(9): 846–853.

Rachtanapun, P. (2009) Blended films of carboxymethyl cellulose from papaya peel (CMCp) and corn starch. *Kasetsart Journal: Natural Sciences*, **43**: 259–266.

Rachtanapun, P., Kumthai, S., Yagi, N. and Uthaiyod, N. (2007a) Production of carboxymethylcellulose (CMC) films from papaya peel and their mechanical properties. *Proceedings of 45th Kasetsart University Annual Conference*, **4**: 790–799.

Rachtanapun, P., Tiwaratreewit, T. and Khumthai, S. (2007b) Effect of bleaching process on mechanical properties of carboxymethyl cellulose from papaya peel. *CMU Research Abstract*, 23–25 November, Chiang Mai, Thailand.

Rani, D.S. and Nand, K. (2004) Ensilage of pineapple processing waste for methane generation. *Waste Management*, **24**: 523–528.

Rodriguez, M.B.S., Megias, S.M. and Baena, B.M. (2003) Alimentos funcionales y nutricion optima. cerca o lejos? *Revista Española de Salud Pública*, **77**: 317–331.

Samaram, S., Mirhosseini, H., Tan, C.P. and Ghazali H.M. (2013) Ultrasound-Assisted Extraction (UAE) and solvent extraction of papaya seed oil: Yield, fatty acid composition and triacylglycerol profile. *Molecules*, **18**: 12474–12487.

Samaram, S., Mirhosseini, H., Tan, C.P., Ghazali, H.M., Bordbar, S. and Serjouie, A. (2014) Optimisation of ultrasound-assisted extraction of oil from papaya seed by response surface methodology: Oil recovery, radical scavenging antioxidant activity, and oxidation stability. *Food Chemistry*, **172**: 7–17.

Saravanan, P., Muthuvelayudham, R. and Viruthagiri, T. (2013) Enhanced production of cellulose from pineapple waste by response surface methodology. *Journal of Engineering*, **2013**: 1–8.

Schieber, A., Stintzing, F.C. and Carle, R. (2001) By-products of plant food processing as a source of functional compounds-recent developments. *Trends in Food Science and Technology*, **12**(11): 401–413.

Selani, M.M., Brazaca, S.G.C., Dias, C.T.S., Ratnayake, W.S., Flores, R.A. and Bianchini, A. (2014) Characterisation and potential application of pineapple pomace in an extruded product for fibre enhancement. *Food Chemistry*, **163**: 23–30.

Serena, A. and Knudsen, B. (2007) Chemical and physicochemical characterization of co-products from vegetable food and agro industries. *Animal Feed Science and Technology*, **139**: 109–124.

Sidhu, J.S. (2006) Tropical fruits: Guava, lychee, and papaya. In: *Handbook of Fruits and Fruit Processing* (ed. Y.H. Hui), Blackwell Publishing Professional, Ames, IO, pp. 597–634.

Soare, P.A.G., Vaz, A.F.M., Correia, M.T.S., Pessoa, A. and Carneiro-da-Cunha. M.G. (2012) Purification of bromelain from pineapple wastes by ethanol precipitation. *Separation and Purification Technology*, **98**: 389–395.

Soderstrom, J., Galbe, M. and Zacchi, G. (2005) Separate versus simultaneous saccharification and fermentation of two-step steam pretreated softwood for ethanol production. *Journal of Wood Chemistry and Technology*, **25**(3): 187–202.

Sossou, S.K., Ameyapoh, Y., Karou, S.D. and de Souza, C. (2009) Study of pineapple peelings processing into vinegar by biotechnology. *Pakistan Journal of Biological Sciences*, **12**(11): 859–865.

Sriwatanapongse, A., Balaban, M. and Teixeira, A. (2000) Thermal inactivation kinetics of bromelain in pineapple juice. *Transactions of the ASAE*, **43**: 1703–1708.

Sruamsiri, S. (2007) Agricultural wastes as dairy feed in Chiang Mai. *Animal Science Journal*, **78**: 335–341.

Steinhaus, M., Sinuco, D., Polster, J., Osorio, C. and Schieberle, P. (2008) Characterization of the aroma-active compounds in pink guava (*Psidium guajava*, L.) by application of the aroma extract dilution analysis. *Journal of Agriculture and Food Chemistry*, **56**(11): 4120–4127.

Stuart, T. (2009) *Waste – Uncovering the Global Food Scandal*. Penguin Books, London.

Subhabrata, S. and Mayura, D. (2006) Industrial and clinical applications excluding diagnostic clinical. *Enzymology*, **1**: 1–25.

Sun, Y. and Cheng, J. (2002) Hydrolysis of lignocellulosic materials for ethanol production: A review. *Bioresource Technology*, **83**(1): 1–11.

Syed, H.M., Kunte, S.P., Jadhav, B.A. and Salve R.V. (2011) Extraction and characterization of papaya seed oil. *Elixir Agriculture*, **37A**: 4102–4105.

Tengborg, C., Galbe, M. and Zacchi, G. (2001) Reduced inhibition of enzymatic hydrolysis of steam pretreated softwood. *Enzyme and Microbial Technology*, **28**(9–10): 835–844.

Thongsombat, W., Sirichote, A. and Chanthachum, S. (2007) The production of guava juice fortified with dietary fibre. *Songklanakarin Journal of Science and Technology*, **29**: 187–196.

Tran, A.V. (2006) Chemical analysis and pulping study of pineapple crown leaves. *Industrial Crops and Products*, **24**: 66–74.

Uhlig, H. (1998) *Industrial Enzymes and their Applications*, vol. **2**. John Wiley & Sons, New York, pp. 146–147.

Upadhyay, A., Lama, J.P. and Tawata, S. (2010) Utilization of pineapple waste: A review. *Journal of Food Science and Technology, Nepal*, **6**: 10–18.

Vikas, O.V. and Mridul, U. (2014) Bioconversion of papaya peel waste in to vinegar using *Acetobacter aceti*. *International Journal of Scientific Research*, **3**(11): 409–411.

Walsh, G. (2002) *Protein Biochemistry and Biotechnology*, vol. **1**. John Wiley & Sons, New York, pp. 51–210, 419–435.

Wang, L. and Weller, C.L. (2006) Recent advances in extraction of nutraceuticals from plants. *Trends in Food Science and Technology*, **17**: 300–312.

Westerlund, P.A., Anderson, J.W. and Rahman S.M.M. (1991) Chemical characterization of water-soluble pectin in papaya fruit. *Carbohydrate Research*, **15**: 67–78.

Wilson, R.K., Kwan, T.K., Kwan, C.Y. and Sorger, G.J. (2002) Effects of papaya seed extract and benzyl isothiocyanate on vascular contraction. *Life Sciences*, **71**: 497–507.

Wooley, R., Ruth, M., Glassner, D. and Sheehan, J. (1999) Process design and costing of bioethanol technology: A tool for determining the status and direction of research and development. *Biotechnology Progress*, **15**(5): 794–803.

Xu, Y.X., Kim, K.M., Hanna M.A. and Nag D. (2005) Chitosan-starch composite film: Preparation and characterization. *Industrial Crops Products*, **21**: 185–192.

Yagčı, S., and Gogůs, F. (2008) Response surface methodology for evaluation of physical and functional properties of extruded snack foods developed from food-by-products. *Journal of Food Engineering*, **86**: 122–132.

Yin, L., Sun, C.K., Han, X., Xu, L. *et al.* (2011) Preparative purification of bromelain (EC 3.4.22.33) from pineapple fruit by highspeed counter-current chromatography using a reverse-micelle solvent system. *Food Chemistry*, **129**: 925–932.

Yousufi, M.K. (2012) To determine protein content of single cell protein produced by using various combinations of fruit wastes and two standard food fungi. *International Journal of Advanced Biotechnology and Research.* **3**(1): 533–536.

Yu, Z. and Zhang, H. (2004) Ethanol fermentation of acid-hydrolyzed cellulosic pyrolysate with *Saccharomyces cerevisiae. Bioresource Technology*, **93**(2): 199–204.

Zhou, K., Wang, H., Mei, W., Li, X., Luo, Y. and Dai, H. (2011) Antioxidant activity of papaya seed extracts. *Molecules*, **16**: 6179–6192.

4

Valorization of Vegetable Wastes

Taslima Ayesha Aktar Nasrin[1] and Md. Abdul Matin[2]

[1]*Horticulture Research Centre, Bangladesh Agricultural Research Institute, Gazipur, Bangladesh*
[2]*Farm Machinery and Postharvest Process Engineering Division, Bangladesh Agricultural Research Institute, Gazipur, Bangladesh*

4.1 Introduction

Vegetable cultivation generates huge amounts of solid and liquid wastes during production, packaging, processing and final consumption. Usually, these wastes contain high amounts of water and organic matter such as proteins, carbohydrates and lipids. This water and organic matter decays and ferments easily, developing obnoxious odours. Thus, such wastes become a breeding ground for cockroaches, houseflies, mosquitoes, mice, and other vectors and vermin. In most cases, these wastes are disposed of improperly, occupying huge spaces, creating dumping problems, and producing toxic substances and poisonous gases upon decomposition, thus contaminating the environment.

Vegetable wastes may comprise of product off-cuts (leaf, stem and other portions that are not usually consumed), by-products such as products that cannot satisfy traders' specifications, products which are not harvested due to low value in the market, partially damaged products, peels, pomace (products after extracting juice), seeds etc. Vegetables yield 25–30% non-edible products (Ajila *et al.*, 2010). Different parts of vegetables, including peels, seeds, stems, leaves and stones, contain high levels of health protecting antioxidants, such as vitamins C and E, phenolic compounds including phenyl-propanoids and flavonoids, and carotenoids such as lycopene. The non-edible parts of vegetables can present similar or even higher contents of antioxidant and antimicrobial compounds than the final product (Ayala-Zavala *et al.*, 2010). Moreover, vegetable wastes represent an important source of sugars, minerals, organic acids, dietary fibre and phenolics, which have a wide range of benefits including antitumoral, antiviral, antibacterial, cardioprotective and antimutagenic activities in human health (Jasna *et al.*, 2009). Therefore, utilization of vegetable wastes is a requirement and a demand that can reduce overall nutritional and economic value, produce food additives or supplements with high nutritional value, animal feeds, biofertilizers, biofuels, and also generate less waste, resulting in less environmental pollution.

Food Processing By-Products and their Utilization, First Edition. Edited by Anil Kumar Anal.

4.2 Losses of Vegetables from Production to Consumption

During production at the farm, some vegetables are not harvested due to over- or under-maturity, unacceptable size and shape, and insect or pest infestation. Non-edible parts of the vegetables are also thrown away as waste.

During packaging, some harvested vegetables cannot satisfy consumers' or buyers' specifications and so are discarded as waste. Also, some vegetables are rejected during packaging due to over- or under-maturity, irregular size and shape, mechanical damage and partial injury by insects or other pests.

During postharvest handling and storage, vegetables are discarded due to spillage, mechanical damage and degradation during handling, storage and transportation between farm and market.

Wastes may also occur during sorting of vegetables, some considered not suitable to process, or during washing, peeling, slicing and boiling, or even due to process interruptions and accidental spoilage. Also, some wastes are produced during the processing. For example, carrot pomace is produced after extracting juice from the vegetable itself.

During distribution, some vegetables are rejected for over supply or quality grade (e.g. colour, size etc.) and spoilage in the market system (e.g. wholesale markets, supermarkets, retailers and wet markets).

Waste also occurs at the household level, due to rotten or uneaten vegetables.

4.3 Extent of Vegetable Losses

Recently, the Food and Agriculture Organisation (FAO) carried out a broad study on loss assessment of fresh food worldwide (Gustavsson *et al.*, 2011). Shockingly, the maximum losses were the most noticeable in developed countries like North America and Oceania, followed by Europe, even though they all have better food storage, transportation and well-organized supply chain facilities. People with higher earnings in the developed countries wasted more food than those with lower earnings in the developing countries. In this study, it was also reported that around one-third of produced food became waste between the farm and the retail store. A recent survey of seven European countries reported that consumers dispose of 20% of their purchased food (Cofresco, 2013). About 50% is fruit and vegetables, while this group symbolizes about 20% of purchased food (Cofresco, 2013). Production of fruits and vegetables in different regions are shown in Figure 4.1.

While losses happen across all type of foods, waste from fresh produce (fruits and vegetables) undoubtedly dominates the total. In developed countries, this waste occurs mainly due to strict quality standards which grade out some products as unfit, even though still suitable for consumption (Gustavsson *et al.*, 2011). More waste occurs at the farm, where the product is either left in the field or discarded soon after harvest, than at any other level in the supply chain. Table 4.1 shows losses of fruits and vegetables in different stages of the supply chain in each region.

On the other hand, as part of the USDA's dietary assessment of the US food supply, food losses were estimated based on the average reduction in weight from the farm to when it was consumed (Kantor, 1998). So it includes the non-edible parts that

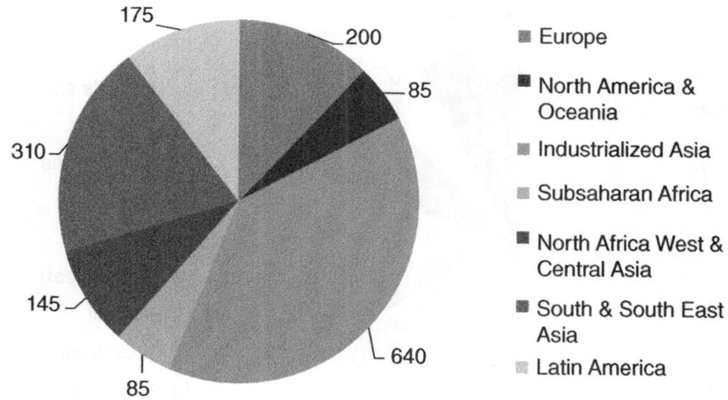

Figure 4.1 Production of fruits and vegetables in each region in million tonnes (adapted from Gustavsson *et al.*, 2011).

Table 4.1 Percentage of produced fruits and vegetables wasted at different levels throughout the supply chain (adopted from Gustavsson, *et. al.*, 2011)

Different regions	Fruits and vegetables wastage in different steps of supply chain (%)					
	Farm	Postharvest	Processing	Distribution	Consumption	Total
Europe	20	4	1.5	8	13	46.5
North America & Oceania	20	2.5	1.5	10	18.5	52.5
Industrialized Asia	10	8	1	6	12	37
Subsaharan Africa	10	8.5	20	10.5	2.5	51.5
South & South East Asia	15	8	19	5.5	4	51.5

are normally discarded (peel, core, stem etc.), as well as product waste. Figure 4.2 represents the percentage of non-edible parts of different commercial and important vegetables.

4.4 Reasons and Overall Prevention of Vegetable Wastes

Vegetables are wasted throughout the supply chain, from production down to final consumption. The highest amounts of vegetables are wasted in the medium and high income countries, meaning that it is thrown away, even if it is still fit for human consumption. In low income countries, vegetables are mostly lost during the production to processing stages.

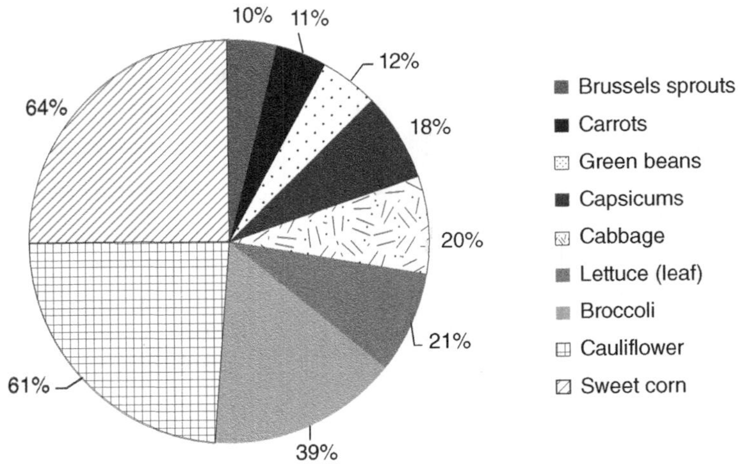

Figure 4.2 Percentage of non-edible portion of different vegetables (Kantor, 1998).

4.4.1 Production Exceeds Demand

In industrialized countries, to make sure of the delivery of the approved amounts of vegetables, while anticipating unpredictable weather or pest attacks, farmers sometimes produce more than is needed, even if conditions are normal. So these surplus crops are sold to processors at a low price or as animal feed, with some now being wasted because of its low value.

Communication and cooperation among farmers and processors may reduce this problem, by allowing surplus crops from one farm to solve a shortage on another (Stuart, 2009).

4.4.2 Premature Harvesting

Mostly in developing countries, poor farmers sometimes harvest vegetables too early due to shortage of food or the desperate need for cash. In this way, the crop does not contain its full nutritional and economic value, and may become waste if it is not suitable for consumption. Prevention of this problem could be if small farmers could be organized in groups to produce different types of cash crops. In this way they would receive credit from their organization or agricultural financial institutions, or advance payments from buyers of their produce.

4.4.3 Strict Quality Standards

Some vegetables are rejected by supermarkets, mostly in developed countries, due to precise quality standards regarding weight, size, shape, colour and appearance. Therefore, huge amounts of vegetables never leave the farms, although they are nutritionally acceptable with a small part of the outside making it considered unfit, such as heterogeneous size, shape or weight. Some rejected crops are used as animal feed.

The quality range of vegetables standardized by the supermarkets should be broader. Stuart (2009) reported in his survey report that consumers were willing to buy heterogeneous produce as long as the taste was not affected. Besides, if vegetables are sold at farmer's markets and farm shops without having to pass the strict quality standards fixed by supermarkets on weight, size and appearance, this may decrease the rate of rejected crops (Stuart, 2009).

4.4.4 Poor Storage Facilities

Due to lack of proper infrastructure facilities during handling, packaging, transportation and storage in developing countries, fresh vegetables can spoil in hot climates within a day (Rolle, 2006). Governments should improve the infrastructures during handling and storage of fresh vegetables. Consequently, private sector investments can improve storage and cool chain facilities, as well as transportation (Choudhury, 2006).

4.4.5 Unsafe Vegetables

Different factors make vegetables unsafe, such as naturally occurring toxins, over use of pesticides, fertilizers and hormones, and veterinary drug residues. Poor and unhygienic handling and storage conditions, washing with contaminated water, use of unsafe chemicals to increase shelf life of stored vegetables and lack of adequate temperature control, can also make vegetables unsafe.

Vegetable supply chain workers should know why a product has becomes unsafe and how to produce safer vegetables. So at every step of crop production and postharvest operations, good agricultural and strict hygienic practices should be followed to ensure safe vegetable production.

4.4.6 Throwing Rather than Using or Re-using

In industrialized countries, vegetables are usually trimmed to ensure the end product conforms to the exact shape and size required. Trimmings, in some cases, could be used for human consumption, but are usually disposed of. Errors during processing lead to final products with the wrong weight, shape or appearance, or damaged packaging, without affecting the safety, taste or nutritional value of the food. In a standardized production line, these products often end up being discarded (Stuart, 2009; SEPA, 2008).

Both commercial and charity organizations could collect and sell or make use of discarded 'sub-standard' vegetables that are still safe, with good taste and nutritional value (SEPA 2008).

4.4.7 Lack of Processing Facilities

In developing countries, sometimes the processing industry does not have the capacity to process and preserve huge amounts of fresh vegetables to meet demand. These problems usually happen due to seasonality of production. It also hinders the efficiency of the processing plant, as it is not possible to supply raw materials year round.

Association or linkage should be developed between food processors and growers, which may help to reduce fresh produce losses. Governments should create environments to encourage the private sector to invest in food industry and to act as a team with farmers to address supply issues.

4.4.8 Wide Range of Products/Brands

In industrialized countries, shops or stores order different types of processed products and brands from the manufacturer due to consumer's demand and to get the most beneficial prices. A wide range of products need more time to sell and leads to reaching their expiry date, resulting in them being wasted.

This problem may be reduced by giving discount on the products that are soon going to expire.

4.4.9 Inadequate Market Systems

If farm fresh vegetables reach the consumers in a very short time, under proper transportation and storage conditions, loss will be reduced. In developing countries, there are fewer shopping centres with proper storage facilities and sales conditions for food products. Most of the wholesale and retail markets in developing countries are often small, overcrowded and unsanitary and suffer a lack of cooling facilities (Kader, 2005).

Governments should develop marketing cooperatives where products from small farmers will be stored and prepared for suitable transportation to markets and other distribution channels. Moreover, a government body should encourage the private sectors to develop well-equipped wholesale and retail markets with improved facilities (Kader, 2005).

4.4.10 Abundance and Consumer Attitudes

In developed countries, vegetable waste at the consumption level is significantly higher than in developing countries. The most important reasons are that wealthier people simply can afford to waste food. In developing countries, most of the people buy foods including vegetables that are just enough or less for the day of purchase. On the other hand, in developed countries, many restaurants serve buffets at fixed prices that encourage people to fill their plates with more food than they need. Retail stores offer large packages with one free. Likewise, manufactures produce oversized ready-to-eat meals (Stuart, 2009).

Public awareness is the main factor that can reduce this huge amount of food, including vegetable wastes. Education on these matters in schools and political initiatives are possible starting points to change people's attitudes towards the current massive food wastage.

4.5 Loss Quantification of Some Important Vegetables after Harvest

An extensive survey of losses of vegetables in different parts in Australia was conducted by the Horticulture Australia Ltd (HAL) in 2013. Their findings are summarized here to give an overall idea regarding the losses of produced vegetables.

4.5.1 Cabbage

A percentage of cabbage may be sold at a high demand period, with 40–50% being left on the farm due to over-production at other times. Around 25% of cabbage is used for processing purposes. The main products are coleslaw, sauerkraut and other food service items and as additions in many cooked items, such as chiko rolls, spring rolls etc. The average annual loss of cabbages grown for processing is 20%, comprising discarded core and outer leaves (15% , and rejections because of damage due to insects, disease and other causes (5%). Cabbage wastes are generally used only as fresh animal feed or green manure and a large amount is disposed of in landfill.

4.5.2 Cauliflower

Around 37% and 24% of cauliflower from field to consumer goes to waste annually in Australia. Domestic oversupply and increasingly stricter market specifications result in high levels of waste during harvest and packaging. Besides deterioration or degradation in storage, insect infestation during storage, or damage during transport, increase the waste percentage. Cauliflower wastes are currently used only as green manure or as animal feed.

4.5.3 Broccoli

Annual loss of broccoli in the fresh market is around 28%. Wastes are caused by failing to satisfy the premium grade and the inadequate market place for second-grade products. Around 9% are not harvested due to low prices, which means 19% of crops do not meet the quality standards in terms of head size, shape and trimming, although they are still fit for human consumption. Some broccoli is produced mainly for processing purposes. Rarely, crops unfit for premium grade are also used for processing.

4.5.4 Sweet Corn

Sweet corn usually produces a high amount of waste, as only about 45% of the plant contains edible kernel (Williams, 2008). Waste is produced mainly from husks, tips, cores etc. Conversely, it is advantageous for its multipurpose utilization as fresh whole corn, boiled corn, trimmed pre-packs, canned corn, frozen corn and cobettes. Corn cores are high in fibre, so it can be extracted to use as a dietary enhancement. Tips and other trimmings are also important for their high protein content.

4.5.5 Carrots

Annually, 93% carrots are produced for fresh purposes and 7% for processing purposes. About 33% of carrots are discarded as waste in the fresh market. Carrot wastes are commonly used only as animal food. On the other hand, carrots are high in fibre, antioxidants (beta carotene) and other compounds of potential dietary or industrial use. So their waste can be used to extract beta carotene and other bioactive compounds.

4.5.6 Beetroot

Annually, 55% of beetroots are produced for fresh purposes and 45% for processing purposes. Beetroots for processing also have strict specifications, extra-large, small or uneven roots not being fit for canning. Moreover, roots are trimmed and peeled for canning, resulting in huge amounts of waste in the industry. These are generally used to produce compost, green manure and animal feed. Nevertheless, fresh wastes are a good source of dietary fibre and antioxidants (betalains). So this waste can be used to extract these compounds that could be incorporated as dietary supplements and natural colorants in many kinds of foods.

4.5.7 Lettuce

Annually, around 93% of lettuces are produced for fresh use and only 7% for processing purposes. Wastes from lettuce are comparatively very low, because when not fitting specifications or unfit for fresh markets, crops are mostly utilized in the processing industry. According to survey data, only 5% or less of lettuce is directly wasted. Some lettuces are grown specifically for processing; these are allowed to develop to a much larger size before harvest. The core and outer leaves representing around 10% waste are used as green manure or animal feed and cannot be used for other purposes due to their high water content and low nutritional value.

4.5.8 Capsicums

Capsicum is a high value and important crop in Australia. Around 46% of capsicums are fit for fresh market, 29% are used in the processing industry and up to 25% of the fruits are wasted. These symbolize a major financial loss for the industry due to its high value. Farm losses occur due to decay, cracking, sunburn and other damage. While, in processing, wastages occur due to strict industrial specifications, especially regarding size and shape.

4.5.9 Beans

Beans wasted in the fresh market represent 25%. In Australia, beans are now harvested by machine. It is obvious that machine harvesting helps to save labour, but amount of wastes increase by leaving unharvested crops in the field, causing damage during harvesting and by mixing the different maturities of crops. Bean waste contains high amounts of protein and dietary fibre, so can be used as important animal feed. It is commonly used as green manure.

4.6 Utilization of Vegetable Wastes

Vegetable losses from the farm to the consumer are above 40% and increasing (Hall *et al.*, 2009). This includes vegetables which are not harvested due to improper maturity, do not satisfy strict quality specifications by the supermarkets, and rejected crops or parts or fibre/pomace after extracting juice by processing industries etc. This waste represents a major cost to vegetable growers. The aim is searching for innovative techniques and markets, which can prospectively utilize vegetable waste, minimize qualitative and quantitative losses and reduce environmental pollution.

4.6.1 Utilization of Wastes by Priority Basis

Vegetable wastes such as crop residues, unfit for supermarket and trimmings, pomace and by-products from processing industries, can be utilized in many ways. These should be utilized on a priority basis from high-value fresh-cut vegetables used as ready to serve (salad mixed) or cooked food to low-value mass materials used to produce biofuel or for soil enhancement. Figure 4.3 shows the priority basis utilization of vegetable wastes in more detail. Utilization of wastes in different ways depends on some factors:

- quality and quantity of available waste materials;
- regularity of supply material (as wastes depend on production season, climate, market reliance etc.);
- cost of waste materials.

In fact, products not meeting strict specifications are often not a feasible option for processing. It usually requires a continual supply of known quality products and is not well adapted to utilizing occasional amounts of excess or variable product.

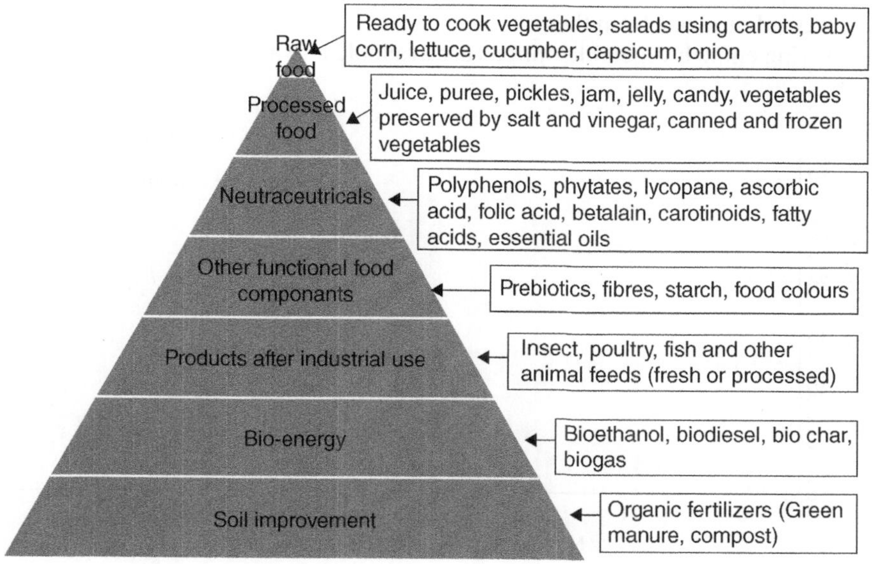

Figure 4.3 Vegetable waste pyramid.

4.6.2 Vegetable Demand should be Increased

In developing countries, main causes of vegetable wastes are the lack of well-equipped supply chain and storage facilities. Strict quality specifications force huge amounts of vegetables to become waste, as there is no market for imperfect products in the developed countries like Australia. Sometimes, if production of vegetables is high, market prices are so low that crops are simply abandoned, no matter how good their quality. Quality standards and specification requirements are applied strictly when the supply of vegetable crops is highest. For economically viable production and supply, vegetable quality needs to be good, supply must be adequate to meet demand and retail prices must be reasonable. If vegetable consumption increases, wastes will be minimized.

4.6.3 Vegetables for Better Health

The National Health and Medical Research Council, Australia, recommends that adults need to intake at least 2 servings of fruit and 5 servings of vegetables each day to ensure good quality nutrition and health. In a 2011/2012 Australian health survey, it was reported that only 8.3% people consumed vegetables up to the recommended intake (Figure 4.4). Enough vegetable consumption reduces the rate of cardiovascular disease, cancer, diabetes, obesity, stroke and many other diseases and disorders (Rogers *et al.*, 2013). Awareness of people regarding adequate vegetables consumption should be increased by campaigning or other governmental programs:

- Encouraging vegetable intake by advertising their immense health benefits;
- Eating enough vegetables makes people fresh and relaxed; otherwise people become tired and ill tempered;
- Providing information about storage of fresh vegetables;
- Demonstrating easy and tasty ways to prepare many vegetable dishes and different processed products from them;
- Ensuring enough vegetable options are available and affordable.

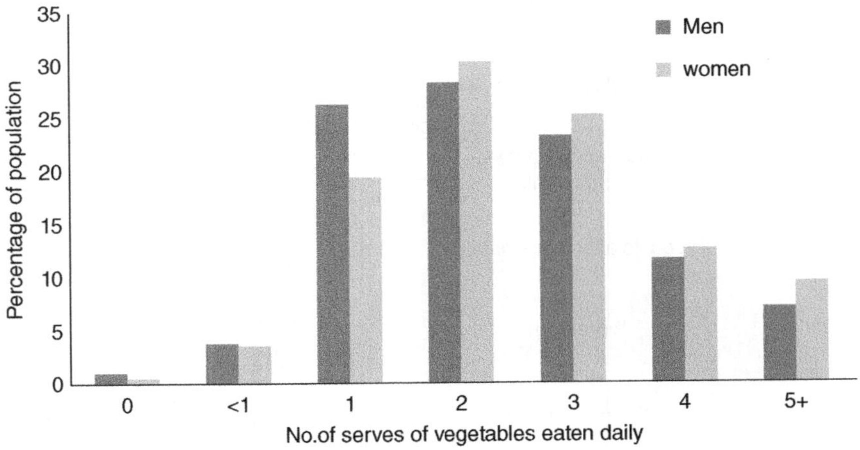

Figure 4.4 Vegetable consumptions by Australian adults (2011–2012 Australian health survey).

4.6.4 Bio Gas and Electricity Generation from Vegetable Wastes

In anaerobic digestion, vegetable wastes are broken down, producing solid wastes, liquid wastes and gas. Solid and liquid wastes can be used as fertilizers, whereas gas can be utilized in energy production. In a digester, vegetable wastes are crushed into small pieces, seeded with bacteria and certain temperatures are maintained under anaerobic conditions for biogas production. As an energy source, bacteria uses vegetable waste, a good resource of carbohydrates, oils, sugars and natural fibre. In the digestion process, alcohols, organic acids and carbon dioxide are produced. Then methanogen bacteria break down these compounds further into methane gas. Usually biogas has around 50–70% methane and the rest is CO_2. Then biogas can be converted into electrical power and heat using a generator. The heat produced can be used to heat the digester (Figure 4.5).

This process impedes the landfill of organic wastes and generates electricity as well. Its high (up to 90%) moisture content and its biodegradable ability, makes vegetable waste appropriate for anaerobic digestion (Bouallagui *et al.*, 2005). The biogas production ability of different vegetable wastes is seven times higher than that produced by pig or cow manure alone. Sridevi *et al.* (2012) reported that a mixture of carrots, beans and eggplant produced 0.58 m^3 biogas per kg of volatile solid (VS). Figure 4.6 shows the biogas content of different vegetable wastes, with their methane gas content being around 53% (Weiland, 2010). After biogas separation from vegetable wastes, solid and liquid residues can be used as animal feed, as they contains large amounts of nitrogen, potassium and calcium.

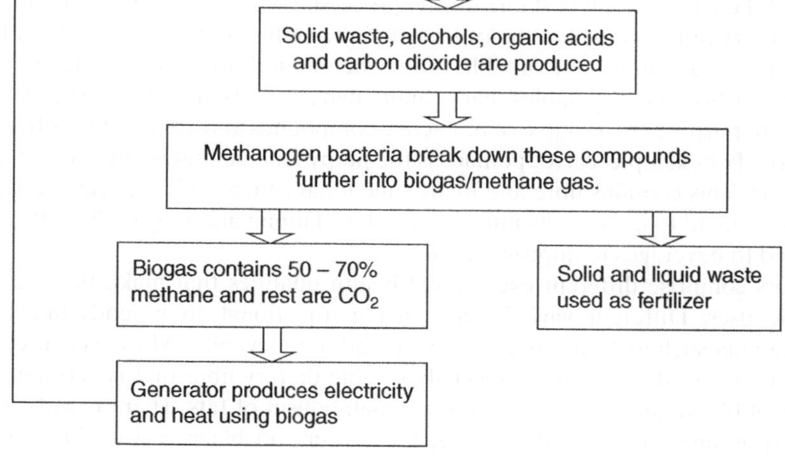

Figure 4.5 A conceptual method of biogas and electricity production.

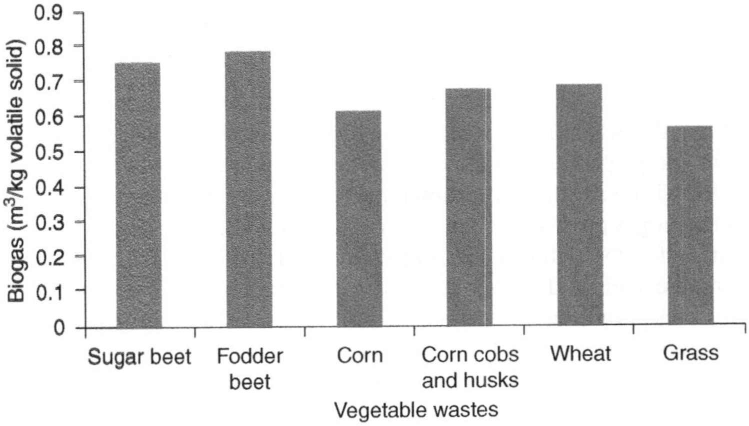

Figure 4.6 Biogas content of different vegetable wastes.

4.6.5 Bioactive Compounds Extraction from Vegetable Wastes

People who consume sufficient amounts of vegetables are less susceptible to serious diseases such as cardiovascular disease, stroke, cancer, respiratory diseases, diabetics, osteoporosis, cataracts, Alzheimers disease, macular degeneration and so on (Rogers *et al.*, 2013). Many researchers have revealed that different parts of vegetables contain some chemical compounds such as sulfur compounds, phenolics, betalains and carotenoids that have the ability to prevent these diseases. The 1996 Australian National Nutrition Survey reported that 26% of women and 15% of men regularly took a nutritional supplement. So, there are opportunities for either developing nutraceuticals based on bio-active compounds extracted from vegetables or replacing compounds in existing products.

Many phenolic compounds and carotenoids can be extracted from different vegetables and their waste, as it is stable during processing and consumption and maintain their activity in the body. On the other hand, the sulfur compounds found in brassicas such as broccoli are not suitable for extraction and use in supplements (Clarke *et al.*, 2011). So some vegetables that contain high amounts of certain compounds can serve both purposes as a source of bio-active compounds and are also fit for fresh consumption. For example, carrot pomace is the pulpy residue that is left over after juice extraction. This contains fibre and many functional compounds that can be incorporated into bread to increase its nutritional value (Filipini and Hogg, 1997). It can also be added to beverages to improve stability.

Carrots comprise different exceptional health qualities that make them ideal for versatile uses. Different varieties are rich in functional compounds like natural anti-oxidants such as β-carotene, lycopene and anthocyanins. Moreover, it contains high amounts of dietary fibre, especially soluble dietary fibre that is very important to control blood glucose levels and affect metabolism of fats. There is wide interest in incorporating natural anthocyanins, those found in black carrots, in a range of different food and drink products. However, there are a number of issues with these pigments. One is that they reversibly change their molecular structure as the pH of solutions changes from acidic to basic. These changes in structure result in a colour shift from red to purple to blue, as the pH changes from acidic to basic. Anthocyanins from black carrots are more stable over a wider pH range than anthocyanins from other sources. They can provide an excellent bright strawberry red shade at acidic pH values, making them an ideal choice for colouring fruit juices, soft drinks, conserves, jellies, confectionery and other low pH foods such as yoghurt and dairy products.

4.6.5.1 Polyphenols Polyphenols are one of the most important and largest groups of bioactive compounds produced in vegetables. Polyphenolic compounds have inhibitory effects on mutagenesis and carcinogenesis in humans by scavenging free radicals. Besides, they reduce the glycemic index, cholesterol, inflammation and microbial infestations (Friedman and Levin, 2009). Cieślik *et al.* (2006) reported that people should take 1–2 g polyphenols, including flavonoids, per day to get their beneficial effects.

Polyphenols include catechins (found in tea), flavonoids, flavanols and anthocyanins (Scalbert and Williamson, 2000). Plant polyphenols are responsible for the different colours of fruit and vegetables, help to germinate seeds and protect plants from pathogens and predators (Bravo, 1998). These compounds are usually intense in those parts (peels and stems) of the plants that are often discarded after harvest. For example, total phenolics are 10 times higher in eggplant peel than in its flesh (Jung *et al.*, 2011) and broccoli stems are high in polyphenols than its curd (Wijngaard *et al.*, 2009; Peschel *et al.*, 2006). Vegetables contain good amount of polyphenols, as shown in Figure 4.7.

In recent years, food technologists and nutritionists have become interested in incorporating polyphenols into the human diet, as there is proof of their function in decreasing susceptibility to cardiovascular diseases, cancers, arthritis, cataracts, Parkinson's disease, Alzheimers and other problems related to ageing (Cieślik *et al.*, 2006).

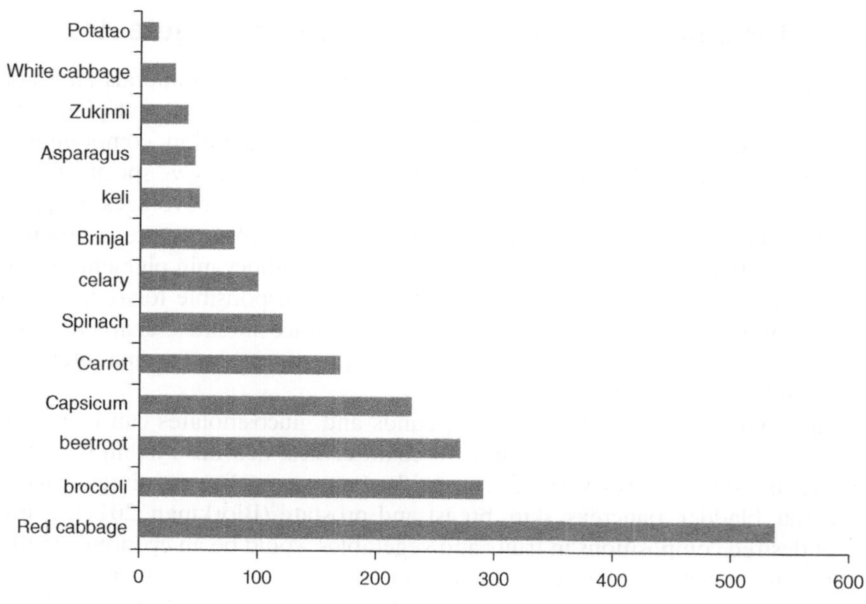

Figure 4.7 Average total polyphenols in some vegetables.

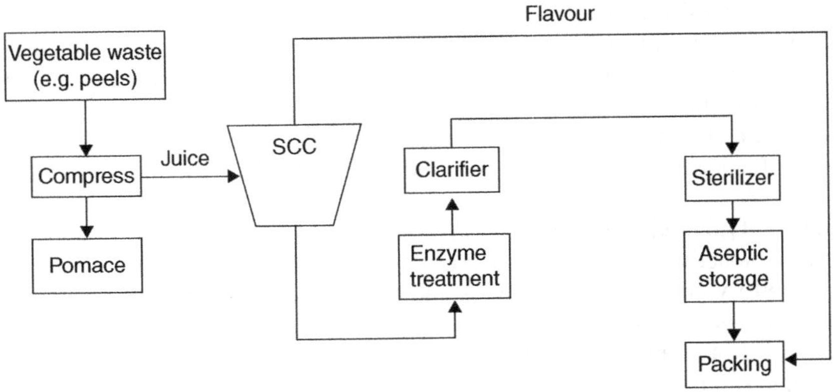

Figure 4.8 Aseptic extraction of flavour from vegetable waste by SCC.

4.6.5.2 Natural Flavours The cheapest and most abundant source of excellent natural flavours in the world is the discarded portion of fruits and vegetables. Vegetables such as celery, capsicums, chillies, herbs, onions, coffee and garlic are good sources of aromatic compound. But the fragile, delicate flavours of fruits and vegetables are lost or damaged during conventional processing.

Spinning Cone Column (SCC) is designed to execute pasteurisation, evaporation and flavour restoration of different fruit and vegetable juices and purees simultaneously. Aseptic process integration in SCC improves greatly to minimize unavoidable loss of aromatic compounds during heat treatment and storage. Figure 4.8 represents the procedure of flavour extraction and storage aseptically.

4.6.6 Increment of Bioactive Compounds in Vegetables

Vegetables containing proteins, lipids, carbohydrates, vitamins, minerals and bioactive compounds (e.g. phenolics and glucosinolates) are able to protect plants from pathogens and insects (Mazid, 2011) and absorb UV-radiation compounds for preventing serious leaf damage (Li, 1993). Phenolic compounds give specific flavours and colours to various plants widely utilized in foods and beverages. Capsaicin gives the pungent properties of the red peppers, alkylphenols are accountable for the distinctive taste and smell of clove oil and the anthocyanin pigments, such as the pelargonidins, the cyanidins and the delphinidins, responsible for red, blue and purple colours, respectively (Croteau, 2000). The glucosinolates, characteristic of cruciferous foods, also contribute the bitter taste (progoitrin) and aroma intensity to the vegetables (Pasini, 2011).

Human consumption of phenolic compounds and glucosinolates can control the oxidative processes in relation to cardiovascular, central nervous system health and neurodegenerative diseases, with a reduced risk of cancers of the gastrointestinal tract, lung, colon, bladder, pancreas, skin, breast and prostate (Bjorkman, 2011). Enrichment of desired compositions in fruits and vegetables would be an economic method for improving nutrition and disease prevention, as diet-induced health improvements do not include any extra expenses for the health sector. Additionally, it might help to reduce these costs (Kaur and Das, 2011).

Elicitors or stimulators are substances that encourage physiological changes in the plant. The use of stimulators are a tool to enhance the phytochemical content in plants, applied alone or in combinations at selected times of the vegetable growth. Biotic compounds, abiotic compounds, plant hormones (salicylic acid, jasmonates etc.) may be considered as stimulators (Angelova *et al.*, 2006, Radman *et al.*, 2003). Biotic stimulators originate from polysaccharides (pectin, cellulose, chitosan, chitin, alginate, Arabic gum etc.), oligosaccharides (galacturonides, guluronate), proteins (cellulase, cryptogein, glycoproteins, pectolyase, lactoferrin, fish protein etc.), fungal spores and pathogen toxins. On the other hand, modified gas composition, high or low temperatures, pressure, carbon dioxide, ozone, draughts, high or low osmolarity, UV irradiation, saline stress and wounding, are the abiotic stimulators. Plant hormones are the Jasmonic acid, methyl jasmonate (Rohwer and Erwin, 2008), methyl salicylate, salicylic acid, ethylene (Tomás-Barberán, 1997), cytokinin and gibberellin GA3 (Treutter, 2010) etc.

A stimulator can be applied both before, during and after harvest of vegetables. Preharvest stimulation could be performed by seed priming (Cho *et al.*, 2008), soaking seeds in stimulators solutions, or after seeding. It can be sprayed over the leaves (Baenas *et al.*, 2014) or applied in a hydroponic system (Wei *et al.*, 2011). Specific elicitor treatments have been used in postharvest practices to enhance the bioactive compounds and quality composition of many fruits and vegetables, such as the application of low or high temperature treatments (Lo Piero *et al.*, 2005), ultraviolet (UV) (Crupi *et al.*, 2014) or gas combinations before commercialization (Tomás-Barberán, 1997). The stimulator's nature, doses and time of treatment strongly affects the intensity of the plant's response.

4.6.7 Bioactive Compounds Affected by Stimulators

The most actively pursued strategies to increase the production of target natural products in plants are the applications of chemical elicitors and the study of the signal transduction pathways and transcription factors required for the expression of genes, involved in the biosynthesis of specific bioactive phytochemicals (Zhao *et al.*, 2005).

4.6.7.1 Phenolic Compounds Depending on the number and arrangement of carbon atoms, phenolic compounds can be divided into flavonoids and non-flavonoids. Flavonoids include flavonols, flavones, anthocyanidins, flavanones, isoflavones etc., while non-flavonoids comprise of phenolic acids, hydroxycinnamates, stilbenes etc., and they are usually conjugated to sugars and organic acids. Phenolic compounds are responsible for the flavour and colour characteristics of fruits and vegetables. An example of the production of phenolic compounds from vegetable waste using abiotic stress is shown in Figure 4.9.

Application of suitable elicitors on fruits and vegetables before or/and after harvest, shows that beneficial phytochemicals have been increased several times, as reported by many reviewers (Amarowicz *et al.*, 2009; Martínez-Ballesta *et al.*, 2008; Schreiner and Huyskens-Keil, 2006). For instance, 3-day-old sprouts of an alfalfa crop treated with high-light (700 μmol•m^{-2} s^{-1} for 1 day) and chilling (a growth chamber at 4°C with a light intensity of 120 μmol•m^{-2} s^{-1} for 1 day), gathered about 2.0 and 1.5 times higher concentrations of ferulic acid, respectively, than the control. Wounding (3 mm thick disks sliced) with ethylene (1000 ppm) and MeJA (250 ppm) in purple carrot

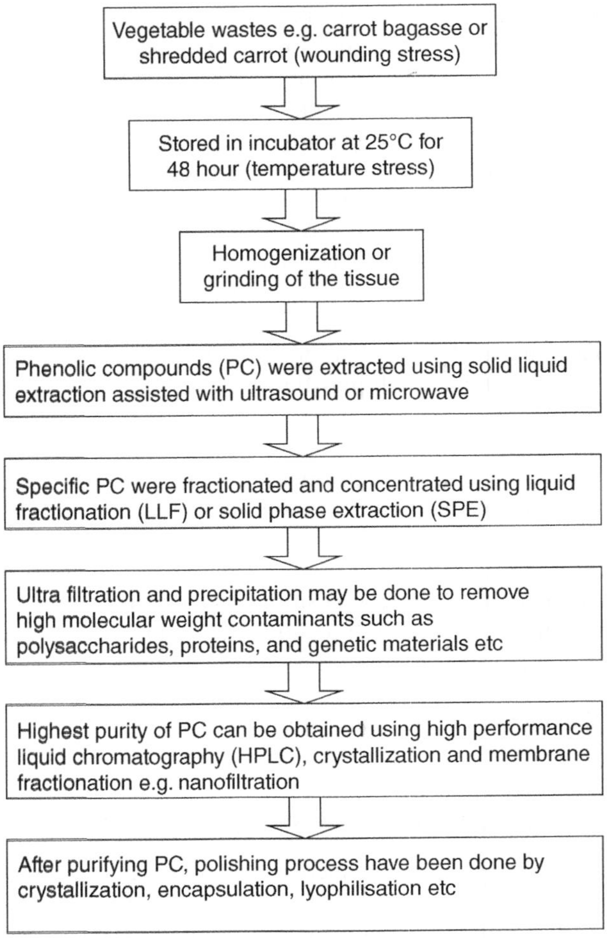

Figure 4.9 Abiotic stress-based bioprocesses for the production, separation, purification and polishing technologies of high-value phenolic compounds from vegetable wastes.

(*Daucus carota* L.) augmented the total phenolic content by about 176% and 210%, respectively, compared to the other treatments (Heredia and Cisneros-Zevallos, 2009).

Ethylene applied to butter leaf lettuce at 10 μL•L^{-1} in a flow of humid air for 3 days at 5°C, stimulated synthesis of phenolic compounds by 38%, although wounding increased by 87% (Tomás-Barberán, 1997). When butter lettuce, after 21 days of germination, were being sprayed by JA 1 μM, total phenolics contents were augmented by 280%, flavonoids by 133% and phenolic acids by 360% (Złotek *et al.*, 2014).

Total phenolics were enhanced by 60% when purple-flesh potatoes were wounded by a vegetable slicer after harvest (Reyes and Cisneros-Zevallos, 2003). When radish sprouts at 3, 5 and 7 days after sowing, were treated with NaCl (100 mM) in 0.5% agar media, the total phenolics contents were increased by 30% and 50% at 5- and 7-day-old sprouts, respectively (Yuan *et al.*, 2010). When sucrose and mannitol (176 mM) were added in a hydroponic system to 7-day-old broccoli sprouts, total anthocyanins and phenolics were increased by 40% and 50%, respectively (Guo *et al.*, 2011b). High intensity of light (700 μmol•m^{-2} s^{-1} for 1 day) applied to broccoli sprouts, induced synthesis of phenolic compounds by 83% more compared to the control (Oh and Rajashekar, 2009).

4.6.7.2 Glucosinolates Glucosinolates (GLS) consist of comparatively small but varied group of over 130 nitrogen- and sulfur-containing natural products found almost entirely in the cruciform group (Brown *et al.*, 2003). Glucosinolates protect plants against a range of pathogens and pests, and are usually accumulated in the parts that control the growth cycle of the plant (Brown *et al.*, 2003). Consumption of these compounds reduces the risk of different diseases like cancers, mainly in lungs and gastrointestinal tract, and also against cardiovascular diseases (Bjorkman *et al.*, 2011; Higdon *et al.*, 2007; Kristal and Lampe, 2009). Glucosinolates give some specialty in organoleptic properties of some vegetables, like cauliflower and mustards (Halkier and Gershenzon, 2006). Exogenous application of SA, JA and MeJA act as signalling molecules induced by pathogen infestation and mechanical wounding. This is very important, due to its ability in expression of genes involved in resistance responses, especially those that are related to biosynthesis of phytochemicals in plants (Awang *et al.*, 2013).

Baenas *et al.* (2014) reported that application of MeJA elicitor (25 μM) in 5 different 8-day-old Brassica and Raphanus sprouts, showed the glucoraphanin and glucoraphenin contents were increased by 50% than in the control. When 7-day-old sprouts of *Brassica napus* and *Brassica rapa*, JA were sprayed (5 n mol) topically, GLS was increased by 6-fold and 9-fold, respectively (Bodnaryk and Rymerson, 1994). In Turnip root, when MeJa (130 μM) was added in the hydroponic system for 10 days, Indole GLS was increased by 4-fold (Schreiner *et al.*, 2011).

When broccoli sprouts were treated with sucrose (146 mM) in 0.5% agar media for 5 days after sowing, total GLS was enhanced 2-fold (Guo *et al.*, 2011a). When glucose (5 g/100 mL) was added in the hydroponic system of 3-day-old sprouts of Chinese kale and pak choi after sowing, 150% and 60% gluconapin were increased, respectively (Wei *et al.*, 2011). When radish sprouts were treated with NaCl (100 mM) in 0.5% agar media for 5 and 7 days after sowing, 50% and 120% total GLS were increased in 5- and 7-day-old sprouts (Yuan *et al.*, 2010).

4.6.7.3 Carotenoids and Betalains Recently, carotenoids and betalains drew attention due to their beneficial effects on human health. Carotenoids have photo-protective properties that protect against cancers and cardiovascular diseases (Tapiero *et al.*, 2004). Beets are a unique source of phytonutrients called betalains. Betanin and vulgaxanthin are the two most studied betalains from beets, and both have been shown to provide antioxidant, anti-inflammatory and detoxification support.

Tomato fruits were treated with UV-B dosages of 0.075 and 0.15 Wh m^{-2} after different adaptation times of 22 and 44 h before harvest. In ripe tomatoes, the concentrations of carotenoids, lycopene and β-carotene were increased by a UV-B dosage of 0.075 Wh m^{-2} after 22 h of adaptation time (Pérez *et al.*, 2009). Betalains are used as additives in foods, drugs and cosmetics and determined by their antiradical activity. When hydrogen peroxide (H_2O_2) was sprayed and infiltrated with 0.1%, 0.33% and 1% in *Suaeda salsa* L. sprouts, the betacyanin content was increased significantly in its root initially and then was transferred to the leaves (Wang *et al.*, 2007). The increase in the micro-element Co^{2+} from 1–5 μM also resulted in a 60% increment in the production of betalains; however, Mo^{2+}, Fe^{2+} and Cu^{2+} presented a positive (10%

increment) but less marked effect on the production of betalains compared to the control medium (Trejo-Tapia *et al.*, 2001).

4.6.7.4 Nutrients with Biological Activity It is evident that elicitation or stimulation of vegetable plants helps to increase the nutraceutical potential of low-processed food and also the nutritional value (content of vitamins, bioactive peptides and carbohydrates). Vitamins do not provide energy, like carbohydrates, proteins and fats, but they contain vital compounds for growing and proper functioning of the body. Essential vitamins help to boost immunity, strengthen bones, heal wounds, support eyesight and assist the body in obtaining energy from food, among multiple other functions. Without adequate vitamin intake, people become exhausted, are vulnerable to infection and develop other serious complications that can endanger health and life. Hence, there is an increasing interest in fortifying many foods with vitamins. The content of vitamins in fruits and vegetables can be influenced by various factors such as genotypic differences, pre-harvest climatic conditions and cultural practices, maturity and harvesting methods, and postharvest handling procedures (Lee and Kader, 2000). Special treatments, including precursor feeding and elicitor applications, can be used to increase metabolite production.

There was a 2-fold increase of folate in coriander (*Coriandrum sativum*) foliage, when sprayed with MeJA and SA (250 μM). The treated plants presented a higher stability of folates than untreated foliage, during processing and storage (Puthusseri *et al.*, 2012). The application of 200, 300 μM of SA and 0.01% chitosan in 5–day-old broccoli sprouts were stimulated to increase vitamin C content by 26%, 18% and 54%, respectively (Pérez-Balibrea *et al.*, 2011). Broccoli sprouts cultivated in an environment with a 16 h light/8 h dark cycle were found to have 83% higher concentrations of vitamin C than those grown in the dark (Pérez-Balibrea *et al.*, 2008). Physical growth and phenolic content of pea (*Pisum sativum*) seedlings can be enhanced by the application of folic acid and vitamin C. Application of 50 μM folic acid and 500 μM vitamin C on days 8 and 10 of germinating seeds were the optimum for both agronomic and biochemical seed vigour parameters, and the highest level of phenolic content (Burguieres *et al.*, 2007). Lentil sprouts treated with 100 and 300 mM NaCl solution for germination was enhanced in starch content by 50%, as well as the *in vitro* digestibility and predicted glycemic index of the sprouts (Świeca *et al.*, 2013). Also, a decrease in total starch, high content of resistant starch and low starch bio-accessibility, a decrease in protein content and subsequent elevation of the non-protein nitrogen fraction were reported in lentil sprouts after an elicitation treatment with H_2O_2 (Świeca and Baraniak, 2014).

4.6.8 Extraction Techniques of Bioactive Compounds

The extraction conditions of different bioactive compounds are extremely important, due to their effects on the release of compounds from the matrix into the medium (Escarpa and Gonzalez, 2008). Hence, an overview of the main extraction systems is presented here.

4.6.8.1 Solid Liquid Extraction (SLE) In Solid Liquid Extraction (SLE), soluble constituents are extracted from a solid or semi-solid by means of suitable solvents in solid–liquid extraction. The most common example is the preparation of tea. Here, water (solvent) is used to remove tea flavours from the tea powder. The choice of the solvents is very important in order to reduce matrix interferences (Luthria, 2008). Moreover, the extraction conditions such as temperature, time, pH, solid–liquid ratio, particle size, stirring and solvent polarity should be optimum to obtain maximum extraction of the desired molecules. The main drawback of this method is that the solvent is costly, partly toxic and hazardous and it also takes a long time to extract the molecules (Proestos and Komaitis, 2008).

4.6.8.2 Soxhlet Extraction (SE) In Soxhlet Extraction (SE), repeated washing of the matrix by fresh solvent and application of hot solvent results in maximum extraction of desired materials compared to solid–liquid extraction. It takes less time, energy and input than other techniques. As with traditional solid–liquid extraction, the main disadvantage is the risk of degradation of heat sensitive materials (De Castro and Priego-Capote, 2010). It is generally used as an analytical technique in food testing and has only a limited application in food processing, such as extraction of essential oils and volatile compounds from botanical species.

4.6.8.3 Pressurized Fluid Extraction (PFE) Pressurized Fluid Extraction (PFE) is an advanced extraction technique using solvent at high pressures (e.g. 1500 psi) and temperatures (up to 200°C) near their supercritical region, to increase the rate and efficiency of the extraction process. PFE not only reduces the amount of time and solvent required, but also significantly improves extraction efficiency relative to traditional extraction techniques such as Soxhlet. This technique lessens the solvent volumes only from 15 to 40 mL and shortens the extraction times (15–20 min).

4.6.8.4 Supercritical Fluid Extraction (SFE) Supercritical Fluid Extraction (SFE) is the process of separating one component (the extractant) from another (the matrix) using supercritical fluids as the extracting solvent. Extraction is usually from a solid matrix, but can also be from liquids. Carbon dioxide is the most commonly-used supercritical fluid, sometimes modified by co-solvents such as ethanol that change its polarity. CO_2 is preferable as a supercritical fluid, as it is non-toxic, non-flammable, inexpensive, easy to remove from the product and its critical temperature and pressure are relatively low (critical temperature = 31.1°C, critical pressure = 72 bar) and it produces cleaner extracts than other conventional extraction techniques. The biggest hurdle in the commercial application of this technique is that it is too expensive for capital investment.

4.6.8.5 Ultrasound Assisted Extraction (UAE) Ultrasound Assisted Extraction (UAE) uses ultrasound waves that have frequencies of more than 20 kHz, creating mechanical vibrations in a solid, liquid and gas. In contrast to electromagnetic waves, sound waves must pass through a medium and it makes expansion and compression cycles during travelling. Expansion drags molecules apart and compression presses them together. The expansion can create bubbles in a liquid and produce negative pressure. The bubbles form, grow and finally collapse (Wang and Weller, 2006). Ultrasound not only provokes better diffusion of solvent into materials but also disrupts biological cell walls, resulting in higher mass transfer and release of contents.

UAE is an inexpensive, simple and efficient alternative to conventional extraction techniques. The main benefits involve the increase of extraction yield and faster kinetics. UAE can be used to extract heat-sensitive compounds, as it needs reduced operating temperatures. Compared with other novel extraction techniques, such as

Microwave Assisted Extraction (MAE), the ultrasound apparatus is cheaper and its operation is easier (Wang and Weller, 2006). Furthermore, UAE can be used for extracting a wide variety of natural compounds such as antioxidants, tocols, essential oils, steroids and lipids from plants.

4.6.8.6 Microwave Assisted Extraction (MAE) Microwave Assisted Extraction (MAE) uses microwaves, which are electromagnetic waves consisting of an electric and a magnetic field, which oscillate perpendicular to each other at frequencies ranging from 0.3 to 300 GHz. Microwave energy acts directly on molecules by ionic conduction and dipole rotation and thus only polar materials can be heated based on their dielectric constant (Eskilsson and Björklund, 2000). MAE has a number of advantages, such as shorter extraction time, less solvent, higher extraction rate and lower cost, over traditional extraction methods. The use of MAE in natural products extraction started in the late 1980s, and through technological development, it has now become one of the most popular and cost-effective extraction methods available today, and several advanced MAE instrumentations and methodologies have become available such as Pressurized Microwave Assisted Extraction (PMAE) and Solvent Free Microwave Assisted Extraction (SFMAE).

4.6.8.7 Pulsed Electric Field Extraction (PEF) Recently, the Pulsed Electric Field Extractions (PEF) treatment had become popular for enhancing pressing, drying, extraction and diffusion processes (Angersbach *et al.*, 2000; Vorobiev *et al.*, 2005; Vorobiev and Lebovka 2006). Less or non-destruction of plant tissues is an important characteristic of PEF extraction methods, helping to avoid many undesirable changes in pigments, vitamins and flavoring agents of the products, which are usual for other pre-treatment techniques, including thermal, chemical and enzymatic processes. Furthermore, the PEF treatment is also capable of inactivating microbes (Barbosa-Canovas and Vega-Mercado 1996 Toepfl *et al.*, 2007). PEF technology is based on pulsing power delivered to the product placed between two electrodes. Generated high-voltage electrical pulses are applied to the electrodes, which then conduct the high intensity electrical pulse to the product placed between the two electrodes. The food product experiences a force per unit charge (the electric field), which helps in the extraction of intracellular compounds.

4.6.8.8 Enzyme Assisted Extraction (EAE) Advanced extraction methods such as super critical extraction (SCE) or pressurized liquid extraction (PLE), Enzyme Assisted Extraction (EAE) has also fulfilled the demands such as small amount of organic solvents requirements, short time to extract and higher extraction yield. EAE is considered an environment-friendly extraction method based on the ability of enzymes to catalyze reactions with specificity and region selectivity, under mild processing conditions in aqueous solutions (Gardossi and Halling, 2009). Cellulases, hemicelullases and pectinases, as well as other enzymes, can be used to catalyze the hydrolysis of the cell wall polysaccharides, which results in better release and a more efficient extraction of bioactive compounds such as polysaccharides, oils, natural pigments, flavours, antioxidant and medicinal compounds (Pinelo and Meyer, 2008). The enzymes used can derive from bacteria, fungi, animal organs and vegetable extracts. In order to increase the extraction yields, operating conditions such as time, temperature, pH and enzyme-to-substrate ratio must be optimized. EAE of bioactive compounds from plants has some limitations:

1. the cost of enzymes;
2. inability of currently available enzyme preparations to hydrolyze some fractions of plant cell walls, limiting extraction yields of some compounds; and
3. scale-up to industrial scale may be troublesome, because local process conditions in the equipment may be difficult to maintain at the larger scale.

4.6.9 Dietary Fibres from Vegetable Waste

Dietary fibre (DF) comprises all parts of plants that the body cannot digest or absorb. Fats, proteins or carbohydrates can be broken down and absorbed by the body, but fibre is not digested, passes relatively intact through the stomach, small intestine and colon, and out of body. It is commonly classified as soluble or insoluble. Soluble fibre dissolves in water to form a gel-like material. It can help to lower blood cholesterol and glucose levels. Soluble fibre is found in oats, peas, beans, apples, citrus fruits, carrots and barley.

On the other hand, insoluble fibre helps the movement of material through the digestive system and increases stool bulk. Whole wheat flour, wheat bran, nuts, beans and vegetables, such as cauliflower, green beans and potatoes, are good sources of insoluble fibre. In addition to human health effects, DF is a useful ingredient in many processed foods. Among other purposes, it is used as a calorie-free bulking agent, a fat replacer and a gelling agent, with increased moisture content and improved stability of some processed products. Figure 4.10 represents the dietary fibre content of some important vegetables. Carrot pomace contains 63% DF, of which 13.5% is soluble fibre (Chau *et al.*, 2004). Beetroot, beans and bitter melon contain 2.4 g, 2.1 g and 3.1 g soluble fibre per 100 g portion (fresh weight), respectively. Beetroot contains 31% soluble fibre of its total DF content, whereas many grains and cereals contain much less soluble DF.

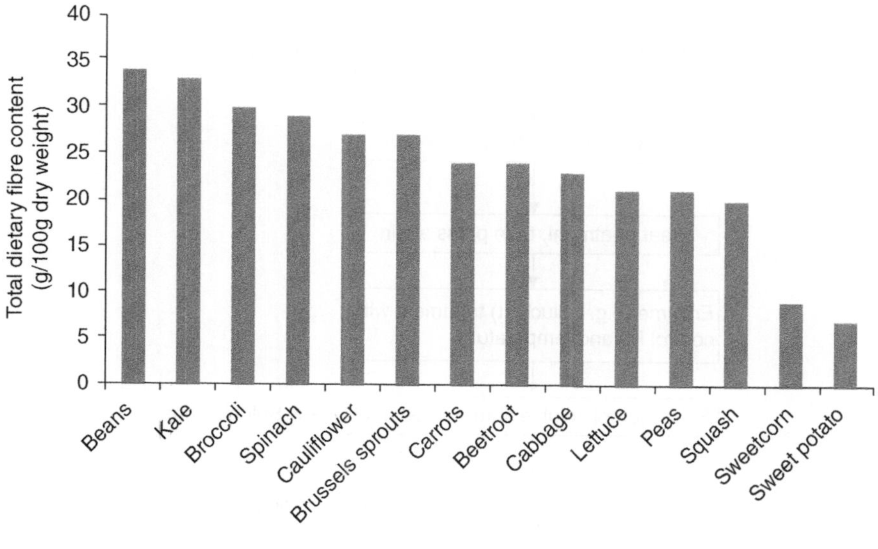

Figure 4.10 Dietary fibre contents of some important vegetables.

As people become increasingly more conscious of their health nowadays, many of them are aware of the health benefits of DF. So the demand for DF originated from natural sources with healthy properties. Many vegetables like cauliflower, bean, broccoli, carrot, kale etc. contain higher levels of DF than are found in pulses such as lentils and kidney beans and even that of some cereals (Anderson, and Bridges, 1988). There is good evidence that an Australian Company 'Kfibre' is producing insoluble DF from sugar cane. A key point of difference for this product is that it is totally natural, whereas wheat fibre is chemically modified to enable extraction.

Vegetable wastes, residues, trimmings and pomace (residues after juice extraction) are high in DF. Corn cores contain around 15–20% dietary fibre. Fibre extracted from some vegetable waste can have additional advantages, due to their other bio-active compounds content. The discarded outer leaves of cabbages normally left in the field are high in DF and also contain sulforafane that has many prospective health benefits. Sulforafane is not destroyed after blanching, drying and processing of leaves into DF powder (Tanongkankit, 2011). Cookies fortified with this DF powder are a good source of fibre content and overall other nutrition with suitable texture and flavour (Naik and Narayana, 2012).

Carrot pomace, usually thrown away after extracting juice, is a rich source of carotenoids, including β-carotene, lycopene (red carrots) and anthocyanins (purple carrots). Fibre extracted from carrot pomace can be incorporated into many food products for its suitable textural and structural qualities (O'Shea *et al.*, 2012). Cauliflower stems and leaf bases have more DF (3% by fresh weight) than in the florets. They also contain a range of flavanols, glucosinolates and other anti-oxidants. Fibre extracted from stem or leaf bases can be a model functional food component for fortifying other foods, due to its high nutritional content, light colour and mild flavour. Cauliflower flour can be added to snack foods to increase nutritional value

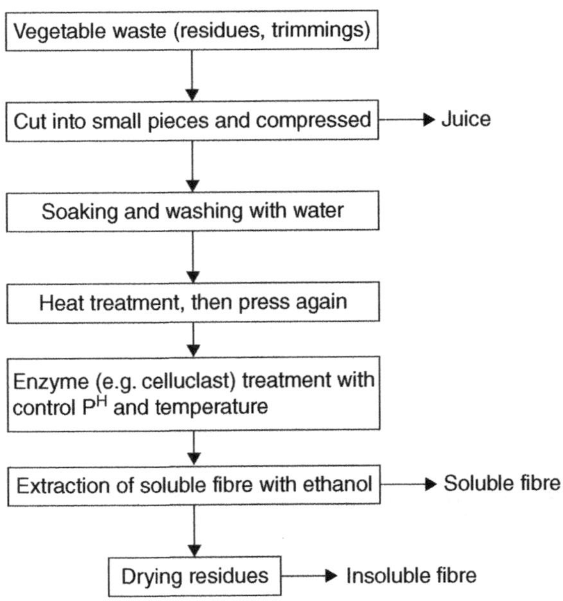

Figure 4.11 Example of procedure used to extract DF from vegetable waste.

without affecting the taste (Stojceska, 2008). In general, the process to extract dietary fibre from plant sources is simple and inexpensive that is shown in Figure 4.11.

4.6.10 Resistant Starch from Vegetable Waste

Resistant starch (RS) is not hydrolysed by amylase in the small intestine and passes to the large intestine where it is fermented by enteric bacteria in healthy persons. As RS is resistant to digestion or digested gradually, it helps to reduce the glycemic index, cholesterol and even body weight (Sajilata *et al.*, 2006). It produces short chain fatty acids (SCFA) in the large intestine that provide energy for cell functions and suppress the production of tumour or cancer cells. Physicochemical properties, particularly the low-water holding capacity, stability in high processing temperatures (type RS III), tastelessness, and white colour, provide more superior looking, texture and mouth feel than conventional fibre sources and improve expansion and crispness in certain food applications.

Resistant starch (RS) has also been found to encapsulate functional food ingredients such as probiotics, polyunsaturated fatty acids (PUFAs) and flavours for improving high processing temperature stability, controlled release ability and storage stability. Prospective physiological benefits and high quality of the food products make RS important to food formulators and nutritionists. Culled plantain and potato contain high amounts of starch, especially RS. Figure 4.12 illustrates an extraction technique of starch from culled potato or plantain. As culled plantain and potato starch contain

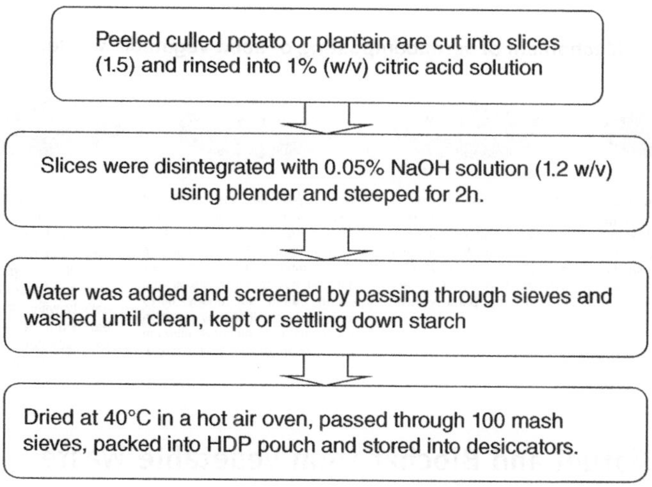

Figure 4.12 Flow diagram of extraction of starch from culled plantain or potato pulp.

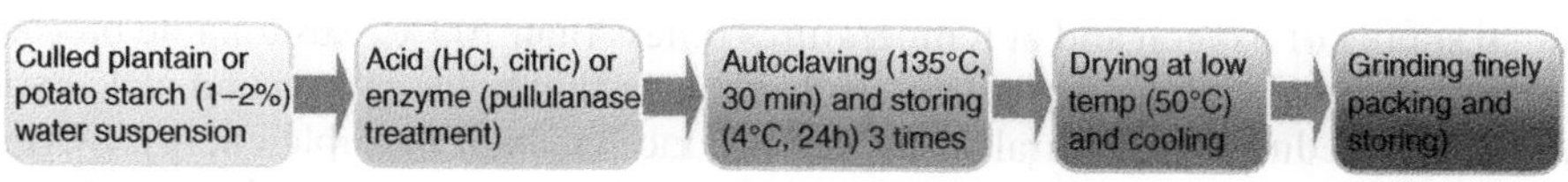

Figure 4.13 Overall procedures for the production of heat stable resistant starch from culled plantain or potato pulp starch.

good amounts of amylose and RSII, it can be used to produce RS III by enzyme treatment or lintnerization followed by auloclaving as shown in Figure 4.13.

4.6.11 Vegetable Waste as Vermicomposting Agent

Million of tons of solid waste produced from civilized society are ending up in landfills every day, creating unexpected economic and environmental problems for governments to handle and monitor for environmental safety. Large amounts of solid waste consist of vegetable wastes that are 'biodegradable' and can be turned into vermicompost, a highly nutritive biofertilizer. Vermicomposting technology, the bioconversion of organic waste into a biofertilizer through earthworm activity, is globally becoming a popular solid waste management technique (Manyuchi *et al.*, 2013; Thamaraj *et al.*, 2011). The earthworms feed on the vegetable waste and their gut acts as a bioreactor whereby the vermicasts are produced (Ansari and Sukhraj, 2011). It is rich in nitrogen (N), phosphorous (P) and potassium (K), as well as trace elements. depending on the feedstock type used (Gurav and Pathade, 2011; Thamaraj *et al.*, 2011). It is 4- to 5-fold more potent than conventional composts and even superior to chemical fertilizers for better crop growth and safe food production. Waste degradation and composting by earthworms is proving to be an economically and environmentally preferred technology over the conventional microbial degradation and composting technology, as it is a rapid and almost odorless process, reducing composting time by more than half and the end-product is both 'disinfected' and 'detoxified' (Ansari, and Sukhraj, 2011; Muthukumaravel *et al.*, 2008).

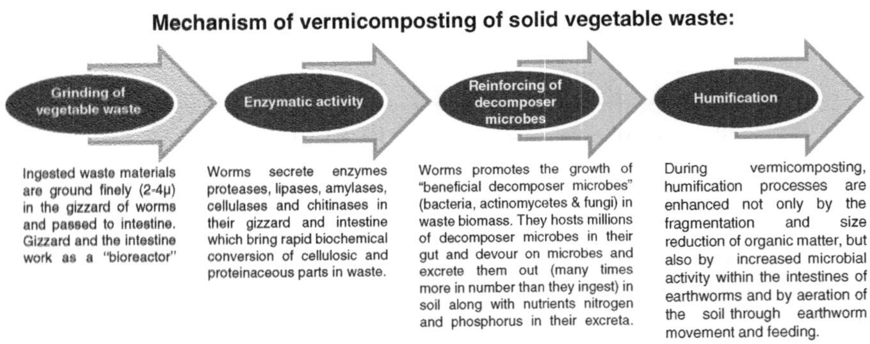

4.6.12 Biofuel and Biochar from Vegetable Waste

Biochar is the carbon-rich product when biomass, such as crop residues, vegetable wastes, manure or leaves, is heated with little or no available oxygen. In more technical terms, biochar is produced by thermal decomposition of organic material under limited supply of oxygen and at temperatures of less than 700°C. Sustainable biochar can control global warming, produce a soil enhancer that holds carbon and makes soil more fertile, reduce agricultural waste and produce clean, renewable energy. During biochar production, energy is generated which can be used directly as heat or turned into electricity. It can also produce biofuel, which can be used directly in suitable

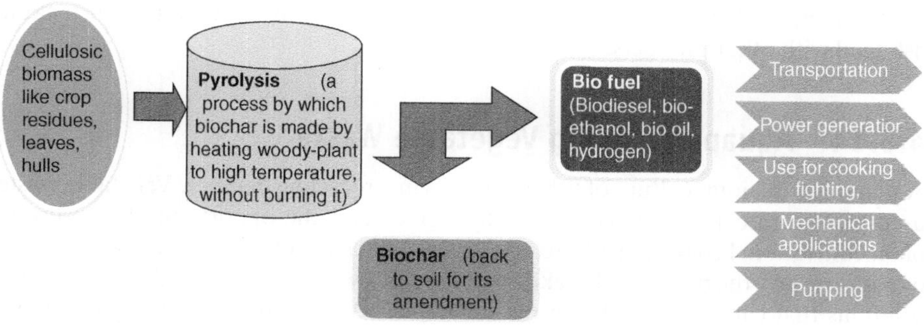

Figure 4.14 A theoretical representation of bio fuel and biochar production.

boilers or furnaces, or processed further to make biodiesel. Production of biofuel and biochar from vegetable waste is represented in Figure 4.14.

4.6.13 Fish Food from Vegetable Waste

People are very much more cautious about their healthy diet, preferring to consume fish or seafood protein rather than meat protein, as these contain high amounts of protein and are low in sodium and saturated fat. Moreover, seafood is also an excellent source of omega-3 fatty acids, a type of fat that is especially good for the heart. However, most wild fisheries are already maximized or in decline. The only way to meet the demand is through increased aquaculture. However, most aquaculture species are still fed, at least in part, on fishmeal made from wild caught fish. This is not only unsustainable, but a major factor limiting increased production. Plant-based fish food has problems with poor tastiness, presence of anti-nutrients (e.g. phytic acid), indigestion of complex carbohydrates and deficiencies in essential amino acids (Rogers *et al*., 2013). Animal-based foods are better but also have some problems, for instance poor consumer acceptance, high levels of saturated fats, absence of omega-3 fatty acids and issues with food safety from rendered animal meal (Rogers *et al.*, 2013).

Fish meal made from insect larvae has been proposed as an alternative. Insects are high in protein and fat, can be reared on vegetable wastes and are part of the natural diet of some farmed fish species. The main insect species that have been studied are yellow mealworm (*Tenebrio molitor*), black soldier fly (*Hermetia illucens*) and super worm (*Zophobas morio*). Each of these insects has its own characteristics, but all can be reared wholly or partly on vegetable wastes such as carrots, green leaves, pumpkin, eggplant, capsicum and processed vegetable sludge and plant stems (Ekman, 2014). The food conversion rate of these insects is also high; about 1 kg of mealworms can be produced from 2.2 kg of carrots, which is similar to chickens (1 kg from 2.3 kg feed) and much better than pigs or cattle (1 kg from 4–9kg feed) (Oonincx and de Boer, 2012). Mealworms retain 28% high-quality protein (59% after drying) and 12.5% fat from 190 mg of mature larvae (Ghaly and Alkoaik, 2009). Mature soldier fly larvae (around 110 g each) have approximately 44% dry matter, of which up to 42% is protein and is 35% fat, although this can be affected by the type and quantity of feed on which they are reared (Wuertz *et al.*, 2012). After harvest, insect larvae can be dried and milled, producing a high-protein 'flour' that is ready to use for making fish feed with other products.

4.6.14 Aquaponic using Vegetable Waste

Aquaponics is a joint culture of fish and plants in a recycling manner. Vegetable waste used to feed insects, finally consumed by fish, creates the opportunity for re-use in plant cultivation. From aquaculture, fish wastes, which are excreted directly by the fish or converted by the microbial breakdown of organic wastes, can be used as hydroponic solutions that can be utilized to grow vegetables. As the aquaculture effluent flows through the hydroponic component of the recirculating system, fish waste metabolites are removed by nitrification and direct uptake by the plants, thereby treating the water, which (clean water) flows back to the fish-rearing component as fresh for re-use. Aquaponics has several advantages over other recirculating aquaculture systems and hydroponic systems that mostly use inorganic nutrient solutions:

- The hydroponic component acts as biofilter, so a separate biofilter is not needed in recirculating systems.
- Aquaponic systems have the only biofilter that generates income, which is obtained from the sale of hydroponic produce such as vegetables, herbs and flowers.
- The nutrients provided by the fish would normally be discharged improperly, creating pollution. Removal of nutrients by plants prolongs water use and minimizes pollutions.
- Aquaponics increases profit potential due to free nutrients for plants, lower water requirements, elimination of a separate biofilter, less water quality monitoring and shared costs for operation and infrastructure.

4.6.15 Waste as Animal Feed

Currently, livestock is one of the fastest growing agricultural subsectors and demand is also rapidly increasing and feed deficits are faced in most developing countries. New unconventional alternate feed resources could play an important role in meeting this deficit. Vegetable processing, packing, distribution and consumption generate a huge quantity of waste that is usually disposed of either by composting or dumping into landfills/rivers, causing environmental pollution. Such resources can act as an excellent source of nutrients and help to bridge the gap between demand and supply of feedstuffs for livestock. Vegetable wastes are rich in many functional compounds, such as polyphenolics, carotenoids and dietary fibre, and have potential to be used in the production of animal feeds. Potential areas include processing vegetable waste into high protein stock feeds for cattle and dairy cows, and processing waste into silage. In addition, its use can also reduce the cost of animal feed, giving higher profits to farmers:

- Fresh banana foliage can be fed as such or after ensiling with broiler litter (40:60), or with wheat straw (75:25) to levels of up to 15% for lactating animals without affecting milk production.
- Banana peels can be added at 15% to 30% in the diet of lactating cows, with no adverse effect on palatability and milk production (Wadhwa and Bakshi, 2013).

- Fresh baby corn husk can be fed as fresh or preserved after wilting or after mixing with cereal straw. These are more suitable, delicious and nutritious as compared with conventional maize fodder.
- Bottle gourd is a widely produced vegetable and its juice has exclusive medicinal properties. Sun dried ground pulp can be added by up to 50% in ruminant feed with no adverse effects on animal health (Wadhwa and Bakshi, 2013).
- Fresh cauliflower and cabbage leaves with stems and cores are high in both macro- and micro-elements, having good digestibility and dry matter intake. These wastes, rich in proteins and soluble sugars, can be fed either raw, after drying or fortifying with cereal straws that have good palatability, nutrient utilization, and performance of livestock (Wadhwa and Bakshi, 2013).
- Fresh carrot has 88% water, protein, sugars, mostly sucrose and high levels of vitamin C and β-carotene. Culled carrots are appetizing and can be used as feed for up to 20–25 kg/day to dairy cows, helping to progress reproductive activities (Morel d'Arleux, 1990).
- Dehydrated carrots are a common commercial feed for horses. Carrot pomace is a rich source of soluble sugars (64.3%). Fortification of 4% to 8% of dried carrot meal in the diet of laying hens significantly improved yolk colour and did not affect egg production and their health (Sikder *et al.*, 1998).
- Pea vines can be fed as fresh or after ensiling. The pea straw contains high protein and less fibre, and has a higher nutritive value than cereal straws (Leclerc, 2003). The empty pea pods are rich both in macro- and micro-elements, especially crude protein (19.8%), soluble sugars and phenolics. Ruminants enjoy eating culled pea pods. It also has a high nutritive value and can be fed exclusively.
- Cull potatoes, a rich source of starch (60–70%), can be fed up to 15–20 kg/day in the raw form, without any adverse effect on the health of dairy and beef cows (De Boever *et al.*, 1983).
- Potatoes can also be chopped with forage and ensiled. Cooked potatoes are efficiently used by fattening and breeding sows, and can be fed up to 6 kg a day. Cooked potatoes can be used for poultry up to 40% of the total feed (Edwards *et al.*, 1986).
- Culled snow peas, an excellent source of protein, can be fed as fresh or after drying to ruminants. They are highly palatable and can be effectively utilized in the diet of livestock, with no effect on nutrient utilization or the health of the animals (Bakshi and Wadhwa, 2012).
- Sun-dried tomato pomace is a good source of lycopene. In dairy cows (26 kg milk/day), dried tomato pomace could be included up to 32.5% in the concentrate mixture without any adverse effect on health, milk yield and dry matter intake (Belibasakis, 1990). In the diet of rabbits, it can be incorporated up to 20–30% (Sayed and Abdel-Azeem, 2009), while in broiler and layer diets it should be added up to 5% and 10% (Sethi, 2012), respectively.
- Cauliflower and cabbage leaves can be used indirectly to increase protein levels in diets of cattle and poultry (Majid *et al.*, 1995). Cauliflower and cabbage leaves containing about 8.0% solid material (dry weight basis) were treated with local strains of *Aspergillus niger*, capable of hydrolysing cellulolytic materials and producing up to 2.7% monosaccharides (fresh weight basis). The sugars produced were then used to culture *Torulopsis utilis*, which when dried was used in the preparation of protein-rich feeds. The dried products contained 22.6% protein compared with 14.5% for the control culture.

4.6.16 Activated Carbon from Vegetable Waste

Activated carbon is a solid, porous, tasteless and black carbonaceous material. It is usually chars that have been subjected to reaction with gases during or after carbonization in order to increase porosity. Activated carbon is different from basic carbon by the taking away of all non-carbon impurities and the oxidation of the carbon surface. There is a diversified use of activated carbon as in gas purification, decaffeination, gold purification, metal extraction, water purification, medicine, sewage treatment, air filters in gas masks and respirators, filters in compressed air and so on. Activated carbon can be produced from corn residues (cobs). Suitable activated carbons were prepared by burning corn cobs at 500°C in the absence of oxygen, in addition with zinc chloride ($ZnCl_2$) (Tsai *et al.*, 1998). Corncobs can be converted to metal ion adsorbents for waste water treatment. Ground corncobs were treated with 0.6 M citric acid (CA) or 1.0 M phosphoric acid (PA) to improve adsorption capacity. Modified corncobs showed the same adsorption efficiency as Duolite GT-73 for cadmium, copper, nickel and zinc ions and had greater adsorption than CMC for nickel and zinc ions (Vaughan *et al.*, 2001). Tobacco stem was treated with a 10% $ZnCl_2$ solution for 24 h, air dried to remove excess solution and dried at 400°C for 1 h. After cooling, it was powdered and activated at 800°C for 1 h. This activated carbon is comparable with commercially-activated carbon (Jambulingam, 2007).

4.6.17 Biodegradable Plastic

Starch produced from culled potato or corn can be solubilized by α-amylase at high temperatures and glucoamylase is added to break down into glucose. Then Lactobacillus is used to ferment glucose to produce lactic acid. Equal amounts of hydroxyl and carboxyl groups are added to produce a condensed form of linear thermoplastic polyester, poly-lactic acid (PLA), a biodegradable plastic. It can be used as timed-release coatings for fertilizers, pesticides and agricultural mulch films, which degrade in the soil (Studt, 1990; Keeler, 1991).

4.6.18 Vegetable Wastes as Substrates in Citric Acid Production

Citric acid is a natural metabolite of plants and animals. It has versatile utilization in several industries like food, beverage, pharmaceutical, chemical, cosmetic etc., as acidulate, chelating agent, emulsifiers, antioxidant, flavour enhancer, preservatives, plasticizer and a synergistic agent (Kolicheski *et al.*, 1997, Soccol and Vandenberghe, 2003). Nowadays, citric acid is produced by the fermentation of starch or sucrose-based media using the filamentous fungus *Aspergillus niger* (Jianlong *et al.*, 2000; Kolicheski *et al.*, 1997). *A. niger* is the mostly used microorganism, but other microorganisms like *Aspergillus sp.*, yeasts, such as *Candida tropicalis, C. oleophila, C. guilliermondii, C. citroformans* and *Hansenula anamola* etc. (Vandenberghe *et al.*,1999; Yokoya, 1992), are also used for the production of citric acid through the surface (solid or liquid) and submerged fermentation of starch materials (Jianlong *et al.*, 2000).

A variety of vegetable wastes has been examined as substrates in solid state fermentation for citric acid production. Cassava begasse, coffee husk, corncobs, beet molasses, wheat bran, carob pod extract, citrus waste etc. have all been used as substrates in citric acid production (Vandenberghe *et al.*, 1999; Soccol and Vandenberghe, 2003). Citric acid production from cassava bagasse, beet molasses, corncob and carrot waste was carried out using a culture of *A. niger* NRRL 2001, *A. niger* ATTC 9142, *A. niger* NRRL 2270 and *A. niger* NRRL 2270 in solid state fermentation, resulting in 260, 109, 603.5 and 29 kg/m^3 of citric acid (Pandey *et al.*, 2011; Garg and Hang, 1995; Begum, 1988; Mahin *et al.*, 2007).

4.7 Conclusion

Vegetable wastes such as carrot pomace, culled potato and plantain, corn stem and husk, cauliflower and cabbage leaves, tomato pomace and seeds, pea pods and vines, culled snow peas, bottle gourd pulp, banana peels, lettuce leaves and potato peels etc. not only contain macro-nutrients such as carbohydrates, proteins, fat and dietary fibres, but also are good sources of health-protecting substances. These substances comprise vitamins C and E, phenolic compounds including phenyl-propanoids and flavonoids, and carotenoids such as lycopene, which have a wide range of effects, including antitumoral, antiviral, antibacterial, cardioprotective and antimutagenic activities. So these wastes can be utilized for the extraction of bioactive compounds such as phenolic compounds, glucosinolates, carotenoids, betalains, vitamins, flavours, dietary fibre and resistant starch etc. and can be incorporated into different food items such as breads, biscuits, fruit juices, soft drinks, conserves, jellies, confectionery and other low pH foods, such as yoghurt and dairy products etc. Vegetable waste can be used for the production of biogas, electricity, biochar, activated carbon, animal feeds, fish feeds and biodegradable plastics etc. The effective and efficient utilization of vegetable wastes will reduce the cost of vegetable production, nutritional loss, generate an array of value-added products and help in waste management and decrease of environmental pollution.

References

Ajila, C.M., Aalami, M., Leelavathi, K. and Rao, U.J.S.P. (2010) Mango peel powder: A potential source of antioxidant and dietary fibre in macaroni preparations. *Innovative Food Science and Emerging Technologies*, **11**: 219–224.

Amarowicz, R., Carle, R., Dongowski, G., Durazzo, A., Galensa, R. *et al.* (2009) Influence of postharvest processing and storage on the content of phenolic acids and flavonoids in foods. *Molecular Nutrition and Food Research*, **53**: S151–S183.

Anderson, J.W. and Bridges, S.R. (1988) Dietary fibre content of selected foods. *American Journal of Clinical Nutrition*, **47**: 440–447.

Angelova, Z., Georgiev, S. and Roos, W. (2006) Elicitation of plants. *Biotechnology and Biotechnological Equipment*, **20**: 72–83.

Angersbach, A., Heinz, V. and Knorr, D. (2000) Effects of pulsed electric fields on cell membranes in real food systems. *Innovative Food Science and Emerging Technologies*, **1(2)**: 135–149.

Ansari, A.A. and Sukhraj, K. (2011) Effect of vermiwash and vermicompost on soil parameters and productivity of okra (Abelmoschusesculentus) in Guyana. *African Journal of Agricultural Research*, **5(14)**: 1794–1798.

Awang, N.A., Islam, M.R., Ismail, M.R., Zulkarami, B. and Omar, D. (2013) Effectiveness of different elicitors in inducing resistance in chilli (*Capsicum annuum* L.) against pathogen infection. *Scientia Horticulturae*, **164**: 461–465.

Ayala-Zavala, J.F., Rosas-Domínguez, C., Vega-Vega, V. and González-Aguilar, G.A. (2010) Antioxidant enrichment and antimicrobial protection of fresh-cut fruits using their own by products: Looking for integral exploitation. *Journal of Food Science*, **75(8)**: 175–181.

Baenas, N., García-Viguera, C. and Moreno, D.A. (2014) Biotic elicitors effectively increase the glucosinolates content in brassicaceae sprouts. *Journal of Agricultural and Food Chemistry*, **62**: 1881–1889.

Bakshi, M.P.S. and Wadhwa, M. (2012) Utilization of residual biomass from fruit and vegetables. In: *Proceedings National Training on Fermentation Technology for Production of Value-Added Products from Agricultural Residues* (ed. H.S. Oberoi), Central Institute of Post-Harvest Engineering and Technology (CIPHET), Ludhiana 3–16 December.

Barbosa-Canovas, G. and Vega-Mercado, H. (1996) *Dehydration of Foods*. Chapman & Hall, New York.

Begum, A.A., Choudhury, N. and Islam, M.S. (1988) Effect of addition of methanol in molasses medium on the production of citric acid by *Aspergillus niger*. *Bangladesh Journal of Microbiology*, **5(1)**: 7–10.

Belibasakis, N.G. (1990) The effect of dried tomato pomace on milk yield and its composition, and on some blood plasma biochemical components in the cow. *World Review of Animal Production*, **25**: 39–42.

Bjorkman, M., Klingen, I., Birch, A.N., Bones, A.M., Bruce, T.J. *et al.* (2011) Phytochemicals of *Brassicaceae* in plant protection and human health – Influences of climate, environment and agronomic practice. *Phytochemistry*, **72**: 538–556.

Bodnaryk, R.P. and Rymerson, R.T. (1994) Effect of wounding and jasmonates on the physico-chemical properties and flea beetle defence responses of canola seedlings, *Brassica napus* L. *Canadian Journal of Plant Science*, **74**: 899–907.

Bouallagui, H., Touhami, Y., Ben Cheikh, R. and Hamdia, M. (2005) Bioreactor performance in anaerobic digestion of fruit and vegetable wastes: A review. *Process Biochemistry*, **40**: 989–995.

Bravo, L. (1998) Polyphenols: Chemistry, dietary sources, metabolism and nutritional significance. *Nutrition Reviews*, **56**: 317–333.

Brown, P.D., Tokuhisa, J.G., Reichelt, M. and Gershenzon, J. (2003) Variation of glucosinolate accumulation among different organs and developmental stages of *Arabidopsis thaliana*. *Phytochemistry*, **62**: 471–481.

Burguieres, E., McCue, P., Kwon, Y.-I. and Shetty, K. (2007) Effect of vitamin C and folic acid on seed vigour response and phenolic-linked antioxidant activity. *Bioresource Technology*, **98**: 1393–1404.

Chau, C.F., Chen, C.H. and Lee, M.H. (2004) Comparison of the characteristics, functional properties, and *in vitro* hypoglycemic effects of various carrot insoluble fibre-rich fractions. *LWT – Food Science and Technology*, **37**: 155–160.

Cho, M.H., No, H.K. and Prinyawiwatkul, W. (2008) Chitosan treatments affect growth and selected quality of sunflower sprouts. *Journal of Food Science*, **73**: S70–77.

Choudhury, M.L. (2006) Recent developments in reducing postharvest losses in the Asia-Pacific region. In: Postharvest Management of Fruit and Vegetables in the Asia-Pacific Region, *APO*, ISBN: 92-833-7051-1.

Cieślik, E., Gręda, A. and Adamus, W. (2006) Contents of polyphenols in fruit and vegetables. *Food Chemistry*, **94**: 135–142.

Clarke, J.D., Hsu, A., Riedl, K., Bella, D., Schwartz, S.J., *et al.* (2011) Bioavailability and interconversion of sulforaphane and erucin in human subjects consuming broccoli sprouts or broccoli supplement in a cross-over study design. *Pharmacological Research*, **64**: 456–463.

Cofresco (2013) Pan European study reveals disturbing numbers on European food waste. www.cofresco.de/en/unternehmen/save-food/ verschwendung.html; last accessed March 2013.

Croteau, R., Kutchan, T.M. and Lewis, N.G. (2000) Natural products (secondary metabolites). In: *Biochemistry and Molecular Biology of Plants* (eds B. Buchanan, W. Gruissem and R. Jones). ASPP: Rockville, MD, pp. 1250–1318.

Crupi, P., Pichierri, A., Milella, R.A., Perniola, R. and Antonacci, D. (2014) Role of the physical elicitors in enhancing postharvest antioxidant capacity of table grape cv redglobe (*Vitis vinifera* L.). *Journal of Food Research*, **3**: doi:10.5539/jfr.v3n2p61.

De Boever, J.L., Cottyn, B.G., Aerts, J.V., Vanacker, J.M. and Buysse, F.X. (1983) Chemical composition, digestibility and nutritive value of raw potatoes (*Solanum tuberosum*). *Revue de l'Agriculture*, **2**(36): 403–413.

De Castro, L.M.D. and Priego-Capote, F. (2010) Soxhlet extraction: Past and present panacea. *Journal Chromatography*, **1217**: 2383–2389.

Edwards, S.A., Fairbairn, C.B. and Capper, A.L. (1986) Liquid potato feed for finishing pigs: Feeding value, inclusion rate and storage properties. *Animal Feed Science and Technology*, **15**: 129–139.

Ekman, J. (2014) Production of fish food for aquaculture from vegetable waste feasibility study. *Horticulture Australia (HAL)*, Project Number: VG13050.

Escarpa, A. and Gonzalez, M.C. (2008) An overview of analytical chemistry of phenolic compounds in foods. *Critical* Reviews in *Analytical Chemistry*, **75**: 57–139.

Eskilsson, S.C. and Björklund, E. (2000) Review: Analytical-scale microwave-assisted extraction. *Journal of Chromatography*, **902**: 227–250.

Filipini, M. and Hogg, T. (1997) Upgrading of vegetable wastes and applications in the food industry. In: *Forum for Applied Biotechnology*. Gent (Belgium). 25–26 September. Mededelingen-Faculteit-Landbouwkundige-en-Toegepaste-Biologische-Wetenschappen-Universiteit-Gent (Belgium). **62**(4a): 1329–1331.

Friedman, M. and Levin, C. (2009) Analysis and biological activities of potato glycoalkaloids, calystegine alkaloids, phenolic compounds and anthocyanins. In: *Advances in Potato Chemistry and Technology* (eds J. Singh, and L. Kaur). Academic Press, Elsevier Inc., New York, pp. 127–161.

Gardossi, L. and Halling, P.J. (2009) Guidelines for reporting of biocatalytic reactions. *Trends in Biotechnology*, **28**: 171–180.

Garg, N. and Hang, Y.D. (1995) Microbial production of organic acids from carrot processing waste. *Journal of Food Science and Technology*, **32**: 119–121.

Ghaly, A.E. and Alkoaik, F.N. (2009) The yellow mealworm as a novel source of protein. *American Journal of Agricultural and Biological Sciences*, **4**: 319–331.

Guo, R., Yuan, G. and Wang, Q. (2011a) Effect of sucrose and mannitol on the accumulation of health-promoting compounds and the activity of metabolic enzymes in broccoli sprouts. *Scientia Horticulturae*, **128**: 159–165.

Guo, R., Yuan, G. and Wang, Q. (2011b) Sucrose enhances the accumulation of anthocyanins and glucosinolates in broccoli sprouts. *Food Chemistry*, **129**: 1080–1087.

Gurav, M.V. and Pathade, G.R. (2011) Production of vermicompost from temple waste (Nirmalya): A case study. *Universal Journal of Environmental Research and Technology*, **1(2)**: 182–192.

Gustavsson, J., Cederberg, C., Sonesson, U., van Otterdijk, R. and Maybeck, A. (2011) Global food losses and food waste: Extent, causes and prevention. *Food and Agriculture Organisation of the United Nations*, Rome.

Halkier, B.A. and Gershenzon, J. (2006) Biology and biochemistry of glucosinolates. *Annual Review of Plant Biology*, **57**: 303–333.

Hall, K.D., Guo, J., Dore, M. and Chow, C.C. (2009) The progressive increase of food waste in America and its environmental impact. *PLoS ONE*, **4**: e7940.doi10.1371.

Heredia, J.B. and Cisneros-Zevallos, L. (2009) The effects of exogenous ethylene and methyl jasmonate on the accumulation of phenolic antioxidants in selected whole and wounded fresh produce. *Food Chemistry*, **115**: 1500–1508.

Higdon, J.V., Delage, B., Williams, D.E. and Dashwood, R.H. (2007) Cruciferous vegetables and human cancer risk: Epidemiologic evidence and mechanistic basis. *Pharmacological Research*, **55**: 224–236.

Jambulingam, M., KarthIkeyan, S., Silvakumar, P., Kiruthika, S. and Maiyalagan, T. (2007) Characteristics of some activated carbon from agricultural waste. *Journal of Scientific and Industrial research*, **66**: 495–500.

Jasna, S.D., Brunet, C. and Cetkovic, G. (2009) By-products of fruits processing as a source of phytochemicals. *Chemical Industry and Chemical Engineering*, **15**: 191–202.

Jianlong, W., Xianghua, W. and Ding Z. (2000) Production of citric acid from molasses integrated with *in situ* product separation by ion-exchange resin adsorption. *Bioresource Technology*, **75**: 231–234.

Jung, E-J., Bae, M-S, Jo, E-K., Jo, Y-H. and Lee, S-C. (2011) Antioxidant activity of different parts of eggplant. *Journal of Medicinal Plants Research* **5**: 461–4615.

Kader, A.A. (2005) Increasing food availability by reducing postharvest losses of fresh produce, *Proceedings of the 5th International Postharvest Symposium,* Acta Horticulture. 682, ISHS.

Kantor, L.S. (1998) A dietary assessment of the US food supply (AER-772). *US Department of Agriculture,* Economic Research Service. Washington, DC, 52 pp.

Kaur, S. and Das, M. (2011) Functional foods: An overview. *Food Science and Biotechnology*, **20**: 861–875.

Keeler, R. (1991) Don't let food go to waste – Make plastic out of it. Research and Development Magazine, **February**: 52–57.

Kolicheski, M.B., Soccol, C.R., Martin, B., Medeiros, T. and Rainbault, M. (1997) Citric acid production on three cellulosic supports in solid state fermentation. In: *Advances in Solid State Fermentation* (eds S. Roussos, B.K. Lonsane, M. Raimbault and G. Viniegra-Gonzalez). Kluwer Academic Publishers, The Netherlands, pp. 449–462.

Kristal, A.R. and Lampe, J.W. (2009) Brassica vegetables and prostate cancer risk: A review of the epidemiological evidence. *Nutrition and Cancer*, **42**: 1–9.

Leclerc, M.C. (2003) *Protein Pea Straw*. Comite National des Coproduits, Fiche n°6 Coproduits riches en lignocellulose, Institut de l'Elevage.

Lee, S.K. and Kader, A.A. (2000) Preharvest and postharvest factors influencing vitamin C content of horticultural crops. *Postharvest Biology and Technology*, **20**: 207–220.

Li, J., Ou-Lee, T.M., Raba, R., Amundson, R.G. and Last, R.L. (1993) Arabidopsis flavonoid mutants are hypersensitive to UV-β irradiation. *Plant Cell*, **5**: 171–179.

Lo Piero, A.R., Puglisi, I., Rapisarda, P. and Petrone, G. (2005) Anthocyanins accumulation and related gene expression in red orange fruit induced by low temperature storage. *Journal of Agricultural and Food Chemistry*, **53**: 9083–9088.

Luthria, D.L. (2008) Influence of experimental conditions on the extraction of phenolic compounds from parsley (*Petroselinum crispum*) flakes using a pressurized liquid extractor. *Food Chemistry*, **107(2)**: 745–752.

Mahin, A.A., Hasan, S.M. and Khan, M.H.R. Begum. (2007) Citric acid production by *Aspergillus niger* through solid-state fermentation in sugarcane bagasse. *Bangladesh Journal of Microbiology*, **4(1)**: 9–13.

Majid, A., Haroon, S. and Joarder, G.K. (1995) High protein feed from vegetable waste. *Bangladesh Journal of Scientific and Industrial Research*, **30**: 1–11.

Manyuchi, M.M., Phiri, A., Muredzi, P. and Chirinda, N. (2013) Effect of drying on vermicompost macronutrient composition. *International Journal of Inventive Engineering and Sciences*, **1(**10**)**: 1–3.

Martínez-Ballesta, M., López-Pérez, L., Hernández, M., López-Berenguer, C., Fernández-García, N. and Carvajal, M. (2008) Agricultural practices for enhanced human health. *Phytochemistry Reviews*, **7**: 251–260.

Mazid, M., Khan, T.A. and Mohammad, F. (2011) Role of secondary metabolites in defense mechanisms of plants. *Biology and Medicine*, **3**: 232–249.

Morel d'Arleux, F. (1990) *Les Sous-Produits en Alimentation Animale*. Reseau national Experimentation Demonstration (RNED), ITEB.

Muthukumaravel, K., Amsath, A. and Sukumaran, M. (2008) Vermicomposting of vegetable wastes using cow dung. *E-Journal of Chemistry*, **5(4)**: 810–813.

Naik, P. and Narayana, C.K. (2012) Preparation of nutrient-rich fortified biscuit by incorporation of residual cabbage leaf powder. *Plant Archives*, **12**: 795–799.

Oh, M.M. and Rajashekar, C.B. (2009) Antioxidant content of edible sprouts: Effects of environmental shocks. *Journal of Science, Food and Agriculture*, **89**: 2221–2227.

Oonincx, D.G.A.B. and de Boer, I.J.M. (2012) Environmental impact of production of mealworms as a protein source for humans – A life cycle assessment. *PLoS ONE*, **7(**12**)**: e51145.

O'Shea, N., Arendt, E. K. and Gallagher, E. (2012) Dietary fibre and phytochemical characteristics of fruit and vegetable by-products and their recent applications as novel ingredients in food products. *Innovative Food Science and Emerging Technologies*, **16**: 1–10.

Pandey, A., Soccol, C.R., Rodriguez-Leon, J.A., and Nigam P. (2001) Production of organic acids by solid state fermentation. In: *Solid state fermentation in Biotechnology-Fundamentals and Applications*. Asiatech: Publishers Inc, New Delhi, pp. 113–126.

Pasini, F., Verardo, V., Cerretani, L., Caboni, M.F. and D'Antuono, L.F. (2011) Rocket salad (*Diplotaxis* and *Eruca* spp.) sensory analysis and relation with glucosinolate and phenolic content. *Journal of Science, Food and Agriculture*, **91**: 2858–2864.

Pérez, C.P., Ulrichs, C., Huyskens-Keil, S., Schreiner, M., Krumbein, A. *et al.* (2009) Composition of carotenoids in tomato fruits as affected by moderate UV-B radiation before harvest. *ISHS Acta Horticulture*, **821**: 217–222.

Pérez-Balibrea, S., Moreno, D.A. and García-Viguera, C. (2008) Influence of light on health-promoting phytochemicals of broccoli sprouts. *Journal of Science, Food and Agriculture*, **88**: 904–910.

Pérez-Balibrea, S., Moreno, D.A. and García-Viguera, C. (2011) Improving the phytochemical composition of broccoli sprouts by elicitation. *Food Chemistry*, **129**: 35–44.

Peschel, W., Sánchez-Rabanedab, F., Diekmann, W., Plescher, A. *et al.* (2006) An industrial approach in the search of natural antioxidants from vegetable and fruit wastes. *Food Chemistry*, **97**: 137–150.

Pinelo, M. and Meyer, A.S. (2008) Enzyme-assisted extraction of antioxidants, release of phenols from vegetal matrices. *Electronic Journal of Environmental, Agricultural and Food Chemistry*, **7**: 3217–3220.

Proestos, C. and Komaitis, M. (2008) Application of microwave-assisted extraction to the fast extraction of plant phenolic compounds. *LWT – Food Science and Technology*, **41**: 652–659.

Puthusseri, B., Divya, P., Lokesh, V. and Neelwarne, B. (2012) Enhancement of folate content and its stability using food grade elicitors in coriander (*Coriandrum sativum* L.). *Plant Food, Human and Nutrition*, **67**: 162–170.

Radman, R., Saez, T., Bucke, C. and Keshavarz, T. (2003) Elicitation of plants and microbial cell systems. *Biotechnology* and *Applied Biochemistry*, **37**: 91–102.

Reyes, L.F. and Cisneros-Zevallos, L. (2003) Wounding stress increases the phenolic content and antioxidant capacity of purple-flesh potatoes (*Solanum tuberosum* L.). *Journal of Agricultural and Food Chemistry*, **51**: 5296–5300.

Rogers, G., Ekman, J. and Titley, M. (2013) Identifying new products, uses and markets for Australian vegetables – A desktop study. *Horticulture Australia (HAL)*, Project Number: VG12046.

Rohwer, C.L., Erwin, J.E. (2008) Horticultural applications of jasmonates: a review, *J. Hortic. Sci. Biotechnol*, **83**: 283–304. 10.1016/j.biotechadv.2013.09.009.

Rolle, R.S. (2006) Improving postharvest management and marketing in the Asia-Pacific region: Issues and challenges. In: *Postharvest Management of Fruit and Vegetables in the Asia-Pacific Region, APO*, ISBN: 92-833-7051-1.

Sajilata, M.G., Singhal, R.S. and Kulkarni, P.K. (2006) Resistant starch: A review. *Comprehensive Reviews in Food Science and Food Safety*, **5**: 1–17.

Sayed, A.N. and Abdel-Azeem, A.M. (2009) Evaluation of dried tomato pomace as feedstuff in the diets of growing rabbit. *International Journal for Agro Veterinary and Medical Sciences*, **3**: 13–18.

Scalbert, A. and Williamson, G. (2000) Dietary intake and bioavailability of polyphenols. *Journal of Nutrition*, **130**: 20735–20855.

Schreiner, M. and Huyskens-Keil, S. (2006) Phytochemicals in fruit and vegetables: Health promotion and postharvest elicitors. *Critical Reviews in Plant Sciences*, **25**: 267–278.

Schreiner, M., Krumbein, A., Knorr, D. and Smetanska, I. (2011) Enhanced glucosinolates in root exudates of *Brassica rapa* ssp. *rapa* mediated by salicylic acid and methyl jasmonate. *Journal of Agriculture, Food and Chemistry*, **59**: 1400–1405.

SEPA (2008) Svinn I livsmedelskedjan – möjligheter till minskade mängder. *Swedish Environmental Protection Agency*, Bromma, Sweden, ISBN 978-91-620-5885-2.

Sethi, A.P.S. (2012) Effect of tomato pomace on the performance of broilers. In: *Proceedings National Seminar on New Frontiers and Future Challenges in Horticultural Crops* (eds W.S. Dhillon, A.S. Dhatt, P.P.S. Gill and N.P. Singh). Punjab Agricultural University, Ludhiana, 15–17 March, p. 219.

Sikder, A.C., Chowdhury, S.D., Rashid, M.H., Sarker, A.K. and Das, S.C. (1998) Use of dried carrot meal (DCM) in laying hen diet for egg yolk pigmentation. *Asian Australasian Journal of Animal Sciences*, **11**: 239–244.

Soccol, C.R. and Vandenberghe, L.P.S. (2003) Overview of applied solid-state fermentation in Brazil. *Biochemical Engineering Journal*, **13**: 205–218.

Sridevi, V.D., Srinivasan, S.V., Kayalvizhi, R. and Bhuvaneswari, R. (2012) Studies on conversion of carbohydrate content in the mixture of vegetable wastes into biogas in a single stage anaerobic reactor. *Research Journal Chemical Sciences*, **2**: 66–71.

Stojceska, V., Ainsworth, P., Plunkett, A., Ibanoglu, E. and Ibanoglu, S. (2008) Cauliflower by-products as a new source of dietary fibre, antioxidants and proteins in cereal-based ready-to-eat expanded snacks. *Journal of Food Engineering*, **87**: 554–563.

Stuart, T. (2009) *Waste – Uncovering the Global Food Scandal*. Penguin Books, London.

Studt, T. (1990) Degradable plastics. *Research and Development Magazine*, **March**: 50–56.

Świeca, M. and Baraniak, B. (2014) Influence of elicitation with H_2O_2 on phenolics content, antioxidant potential and nutritional quality of lens culinaris sprouts. *Journal of Science, Food and Agriculture*, **94**: 489–496.

Świeca, M., Baraniak, B. and Gawlik-Dziki, U. (2013) *In vitro* digestibility and starch content, predicted glycemic index and potential *in vitro* anti-diabetic effect of lentil sprouts obtained by different germination techniques. *Food Chemistry*, **138**: 1414–1420.

Tanongkankit, Y., Chiewchan, N. and Devahastin, S. (2011) Evolution of anticarcinogenic substance in dietary fibre powder from cabbage outer leaves during drying. *Food Chemistry*, **127**: 67–73.

Tapiero, H., Townsend, D.M. and Tew, K.D. (2004) The role of carotenoids in the prevention of human pathologies. *Biomedicine and Pharmacotherapy*, **58**: 100–110.

Thamaraj, K., Ganesh, P., Kolanjinathan, K., Suresh, K.R. and Anandan, A. (2011) Influence of vermicompost and vermiwash on physicochemical properties of rice cultivated soil. *Current Botany*, **2(3)**: 18–21.

Toepfl, S., Heinz, V. and Knorr, D. (2007) High intensity pulsed electric fields applied for food preservation. *Chemical Engineering and Processing*, **46(6)**: 537–546.

Tomás-Barberán, F.A., Loaiza-Velarde, J., Bonfanti, A. and Saltveit, M.E. (1997) Early wound- and ethylene-induced changes in phenylpropanoid metabolism in harvested lettuce. *Journal of the American Society for Horticultural Science*, **122**: 399–404.

Trejo-Tapia, G., Jimenez-Aparicio, A., Rodriguez-Monroy, M., de Jesus-Sanchez, A. and Gutierrez- Lopez, G. (2001) Influence of cobalt and other microelements on the production of betalains and the growth of suspension cultures of *Beta vulgaris*. *Plant Cell, Tissue and Organ Culture*, **67**: 19–23.

Treutter, D. (2010) Managing phenol contents in crop plants by phytochemical farming and breeding – Visions and constraints. *International Journal of Molecular Science*, **11**: 807–857.

Tsai, W.T., Chang, C.Y. and Lee, S.L. (1998) A low-cost adsorbent from agricultural waste corn cob by zinc chloride activation. *Bioresource Technology*, **64**: 211–217.

Vandenberghe, L.P.S., Soccol, C.R., Pandey, A. and Lebeault, J.M. (1999) Review: Microbial production of citric acid. *Brazilian Archives of Biology and Technology*, **42**: 263–276.

Vaughan, T., Seo, C.W. and Marshall, W.E. (2001) Removal of selected metal ions from aqueous solution using modified corncobs. *Bioresource Technology*, **78(2)**: 133–139.

Vorobiev, E. and Lebovka, N.I. (2006) Extraction of intercellular components by pulsed electric fields. In *Pulsed Electric Field Technology for the Food Industry. Fundamentals and applications* (eds J. Raso and V. Heinz). Springer, New York, pp. 153–194.

Vorobiev, E., Jemai, A.B., Bouzrara, H., Lebovka, N.I. and Bazhal, M.I. (2005) Pulsed electric field assisted extraction of juice from food plants. In: *Novel Food Processing Technologies* (eds G. Barbosa-Canovas, M.S. Tapia and M.P. Cano), CRC Press, New York, pp. 105–130.

Wadhwa, M. and Bakshi, M.P.S. (2013) *Utilization of Fruit and Vegetable Wastes as Livestock Feed and as Substrates for Generation of Other Value-Added Products* (ed. H.P.S. Makka). FAO: Rome, Available from: http://www.fao.org/docrep/018/i3273e/i3273e.pdf Last accessed on 30 August 2015.

Wang, C.Q., Chen, M. and Wang, B.S. (2007) Betacyanin accumulation in the leaves of C3 halophyte *Suaeda salsa* L. is induced by watering roots with H_2O_2. *Plant Science*, **172**: 1–7.

Wang, L. and Weller, C.L. (2006) Recent advances in extraction of nutraceuticals from plants. *Trends in Food Science and Technology*, **17**: 300–312, *Molecules* 2014, **19**: 14839.

Wei, J., Miao, H. and Wang, Q. (2011) Effect of glucose on glucosinolates, antioxidants and metabolic enzymes in brassica sprouts. *Science of Horticulture*, **129**: 535–540.

Weiland, P. (2010) Biogas production: Current state and perspectives. *Applied Microbiology and Technology*, **85**: 849–860.

Wijngaard, H.H., Rößle, C. and Brunton, N. (2009) A survey of Irish fruit and vegetable waste and by-products as a source of polyphenolic antioxidants. *Food Chemistry*, **116**: 202–207.

Williams, M.M. (2008) Sweet corn growth and yield responses to planting dates of the North Central United States. *Horticulture Science*, **43**: 1775–1779.

Wuertz, S., Pahl, C. and Kloas, W. (2012) Hermetia meal in fish meal substitution – Influences of substrates on Hermetia meal quality affecting fish nutrition of Tilapia. *Oreochromis hiloticus*. Aqua2012, 1–5 September, Prague, Czech Republic.

Yokoya, F. (1992) Citric acid production. In: *Industrial Fermentation Series* (ed. S.P Campinas), Brazil, pp. 1–82.

Yuan, G., Wang, X., Guo, R. and Wang, Q. (2010) Effect of salt stress on phenolic compounds, glucosinolates, myrosinase and antioxidant activity in radish sprouts. *Food Chemistry*, **121**: 1014–1019.

Zhao, J., Davis, L.C. and Verpoorte, R. (2005) Elicitor signal transduction leading to production of plant secondary metabolites. *Biotechnology Advances*, **23**: 283–333.

Złotek, U., Świeca, M. and Jakubczyk, A. (2014). Effect of abiotic elicitation on main health-promoting compounds, antioxidant activity and commercial quality of butter lettuce (*Lactuca sativa* L.), *Food Chemistry*, **148**: 253–260.

5 Application of Food By-Products in Medical and Pharmaceutical Industries

Muhammad Bilal Sadiq, Manisha Singh, and Anil Kumar Anal
Asian Institute of Technology, Klong Luang, Pathumthani, Thailand

5.1 Introduction

The food processing sector, like other natural resource-based processing industries, gives rise to organic waste, leading to environmental, economic and legal concerns regarding the disposal of food waste and by-products. This concern has led to the innovation of value-added bioactive compounds from food waste and by-products. These value-added compounds from food waste exhibit various physiological functions, such as antihypertensive, anti-oxidative, anti-thrombotic, mineral-binding and anti-amnesiac activities (Wilson, 2011).

Food by-products and waste are produced during the entire life cycle of food processing, starting from farm to fork. The distribution of food waste in developed countries is categorized as 42% by household, 39% in the food manufacturing industry, 14% by the food service sector and 5% in distribution and retail (Mirrabella *et al.*, 2014). In the current era, industrial ecology concepts such as cradle-to-cradle and circulatory economy are gaining considerable attention, aiming at a zero waste economy, in which waste materials are used as starting materials for development of new useful products and their applications.

In addition to loss of valuable bioactive materials, the food waste produced by the processing industries raises economic and environmental hazards. The agriculture and food processing sectors of food industry are considered as a sector for bio-refineries (Ghatak, 2011; Kamm and Kamm, 2004), with a special emphasis on potential use of food waste and by-products (Mahro and Timm, 2007). The first step in utilization of waste and by-products in the food processing sector is identification, characterization and quantification of residues (Rosentrater, 2005). Later, the food waste sources and recovery stages of biologically active and value-added ingredients should be characterized (Galanakis, 2012).

Food Processing By-Products and their Utilization, First Edition. Edited by Anil Kumar Anal.

The food industry produces huge amounts of fruits and vegetable waste that affects the environment and health sectors, because of high biodegradability and methane emissions (Misi and Forster, 2002). These residues contain large amounts of waste organic matter that includes approximately 75% sugars and hemicellulose, 9% cellulose, 5% lignin, small amounts of proteins and fats (Kosseva, 2011).

Nowadays, people are interested in healthy diets and lifestyles. Schieber *et al.* (2001) focused on functional and bioactive compounds such as polyphenols, tocopherols, carotenoids and others. They highlight the availability of untapped natural sources of bioactive compounds and micronutrients and their potential to be readily used as healthy ingredients. The biologically active ingredients from food processing by-products such as dietary fibers have received much attention, due to their potential as pharma-foods to reduce blood cholesterol, diabetes, coronary heart disease, ease of constipation and technological attributes such as water-binding, gelling, structure building and fat replacers (Telrandhe *et al.*, 2012).

The utilization of meat by-products is important for the profitability of the meat industry. In the past, meat by-products were a favorite food in Asia, but the health hazards associated with these by-products led to non-food uses, such as pharmaceuticals, cosmetics and animal feed (Rivera *et al.*, 2000). Due to potential components present in food by-products and wastes, the aim is to isolate and utilize high value-added components, such as proteins, peptides, polysaccharides, fibers, flavoring agents, phytochemicals and pharmacological ingredients.

5.2 Agroindustry By-Products and Potential Recovery of Bioactive Compounds

5.2.1 Fruits

The food industry produces huge amounts of fruit and vegetable wastes that present serious health and environmental hazards due to high biodegradability. Vegetable and fruit wastes are mainly comprised of hydrocarbons, comparatively small amounts of proteins and fat, with 80 to 90% moisture content. Therefore the concept of utilization of wastes to isolate and identify bioactive compounds is gaining interest worldwide. The examples of bioactive compounds isolated from fruit processing by-products are shown in Table 5.1.

5.2.1.1 Apples The important waste components of the apples processing industry are pomace and its extracts, which have a great potential for pharmaceutical and biotechnological applications. The waste comprises a heterogeneous mixture of pomace, seeds and peels with high water content and hemicellulose, cellulose and lignin. The apple juice industries utilize 75% of apple and the remaining 25% are by-products as pomace, peels and seeds (Shalini and Gupta, 2010). The apple pomace is composed of mainly simple carbohydrates (glucose, fructose and sucrose), pectin, fibers, proteins, vitamins and minerals (Kosseva, 2011). The flow diagram for

Table 5.1 Bioactive components isolated from byproducts of fruit processing

Components	Class	Source	Health benefits
Beta-carotene	Carotenoids	Various fruits pomace	Neutralizes free radicals and prevent cell damage, maintain cellular antioxidant defense
Lutein, Zeaxanthin		Citrus fruits peels and seeds	Contribute to maintain vision
Anthocyanidins		Berries, cherries, red grapes	Bolster cellular antioxidant defense, maintain brain function
Flavanols-catechins, epicatechins, procyanidins		Apples, grapes	Maintain normal cardiovascular functions
Flavanones		Citrus fruits	Bolster cellular antioxidant defense
Flavonols		Apples	Bolster cellular antioxidant defense
Hesperidine		Lemon and other citrus fruits	Treatment of rheumatoid arthritis
Proanthocyanidins	Flavonoids	Cranberries, apples, grapes	Maintain urinary tract and heart health
Pectin	Hetero-Polysaccharide	Apple, grape, citrus fruits	Decrease cholesterol and tryglyceride level, prevent colon and prostate cancer and diabetes

utilization of fruit and vegetable pomace in food and pharmaceutical industries is shown in Figure 5.1.

Several research studies have focused on biological properties of apple peel due to its high content of antioxidants and phenolic compounds. The high phenolic content in apple peels may prevent various chronic disorders such as cardiovascular diseases and cancers. Huber and Rupasinghe (2009) studied properties of apple peel antioxidants, and it was found that these antioxidants were able to prevent the lipid oxidation of polyunsaturated fatty acids that is a significant issue affecting food quality and health of customers. There is a high percentage of soluble fibers in apple by-products that results in the availability of pectin. Pectin is a polymer that can be used in food and pharmaceutical industries due to its unique characteristics of gelling, thickening and stabilizing. Pectin is also thought to lower cholesterol and delay gastric emptying (Hwang *et al.*, 1998; Rha *et al.*, 2011; Royer *et al.*, 2006).

These researchers also found significant levels of calcium, magnesium, zinc, iron and copper in apple peel. The iron, copper and magnesium work in synergy as an effective catalyst in the prevention of atherosclerosis and certain other diseases. Garcia *et al.* (2009) studied the presence of bioactive compounds in cider processing by-products of apple. Various important phenolic compounds were found, such as chlorogenic, protocatechui and caffeic acid, flavonols and flavanols. Schieber *et al.* (2003) reported similar findings and found a number of quercetin glycosides, with quercetin 3-galactoside

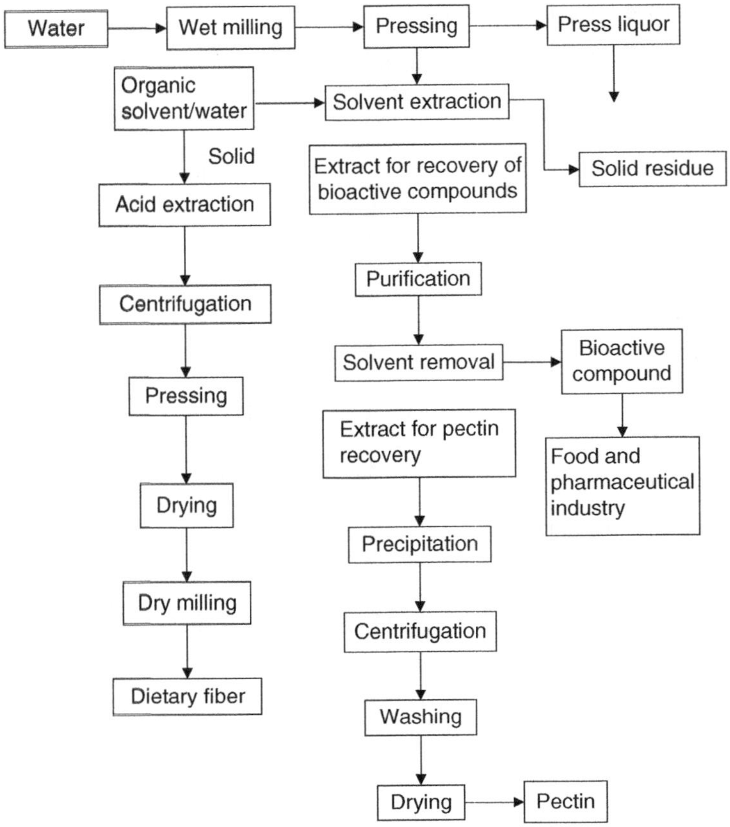

Figure 5.1 Flow chart for utilization of fruit pomace (Dilas *et al.*, 2009).

being the most abundant flavonol. Phloridzin was the most abundant compound in apple seeds.

Many health-enhancing effects are associated with the bioactive compounds found in apple by-products and waste. The predominant effects are inhibition of cancer cell proliferation, lipid oxidation and reduction in blood cholesterol levels. In turn, these health-enhancing effects present a preventive measure against chronic disorders, for example obesity, cancer and cardiovascular disorders. Procyanidins were found to have high antioxidant potential and inhibit oxidation of low-density lipoproteins. Quercetin is a major flavonoid compound present in apple peals, with a potential to reduce the incidence of breast cancer and leukemia (Boyer and Liu, 2004).

5.2.1.2 Berries Berries are used in the food industry for juices, jams and jellies preparations, especially raspberries and blackberries due to their high phenolic content, flavonoids, polyphenols and antioxidant effects. A large amount of waste and by-products are released from industrial manufacturing. These by-products can be successfully recovered for food and pharmaceutical industries. Bakowska-Barczak *et al.* (2009) found that Canadian blackcurrant seed residue proved to be a good source of antioxidants and value-added compounds.

5.2.1.3 Citrus Fruits In the case of citrus fruits, large amounts of waste materials such as peels and seeds are discarded during food and juice processing, because citrus fruits have a comparatively small edible portion. The by-products of citrus fruit juice processing comprises of various compounds, principally soluble sugars, organic acids, fibers, proteins, amino acids, minerals, oils, lipids, vitamins and also flavonoids (Fernandez-Lopez *et al.*, 2004). Viuda-Martos *et al.* (2011) determined physicochemical and antimicrobial characteristics of orange by-products after extraction of juice by determining pH, color, soluble solids, antioxidant capacity, phenolic compounds and sugars. Chau and Huang (2003) determined the dietary fiber content in orange peel and found that it contained mostly an insoluble dietary fiber fraction composed of pectic polysaccharides and cellulose. The dominant insoluble fraction provides health benefits, such as intestinal regulation and increased stool volume.

The by-products of lemon after juice extraction constitute almost 50% of the fruit and mainly consist of seeds, peels and fruit pulp. Normally this is left for animal feed or used for pectin extraction (Lario *et al.*, 2004). The lemon peel was found to be a good source of iron (Gorinstein *et al.*, 2001). The carotenoid and hesperidine were found in lemon by-products. The flavonoid hesperidine has a great potential in the treatment of rheumatoid arthritis (Gonzalez-Molina *et al.*, 2010).

5.2.1.4 Grape Grape pomace consists of stems, seeds and skin, and sometimes this by-product is used for extraction of grape seed oil. The grape pomace is also used for production of citric acid, ethanol, methanol and xanthan gum (Deng *et al.*, 2011). The components of dietary fiber in grape pomace are hemicellulose, cellulose and small portions of pectin (Kammerer *et al.*, 2005). The anthocyanins, flavonols and gallic acid are the major polyphenolic compounds found in grape pomace (Ruberto *et al.*, 2007).

The pharmacological benefits associated with polyphenols such as anthocyanins and flavonoids have been linked in the reduction of chronic disorders such as cardiovascular disease and some types of cancers. The mechanism behind this pharmacological activity is an increase in antioxidant activity of plasma that in turn results in inhibition of low-density lipoproteins oxidation. The bioactive compounds isolated from grape pomace have been found to reduce systolic blood pressure. The polyphenols such as gallic acid and flavonoids prevent lipid oxidation by reacting with superoxide anions, hydroxyl radicals and lipid peroxyl radicals. Ruberto *et al.* (2007) and Maier *et al.* (2009) demonstrated the availability of flavonoids and phenolic acids in grape pomace. These phenolic compounds have been linked with reducing heart diseases.

Spatafora and Tringali (2012) studied anti-proliferative compounds present in grape stem. The active constituents, betulinic acid, stilbenoid trans-resveratrol, trans-viniferin and a mixture of sitosterol 6'-O-acyl-glucosides, were obtained by Gas Chromatography-Mass Spectrometer (GC-MS) analysis. Some of these compounds showed a stronger anti-proliferative effect than 5-fluorouracil in human breast cancer cells.

5.2.1.5 Mango Mango peel has been reported as a rich source of dietary fiber (Ajila *et al.*, 2008). The peel (ripe and unripe) is a good source of polyphenols. Beta carotene was found to be the prominent carotene available in mango peel (Pott *et al.*, 2003). The ripened mango peel is also a good source of antioxidants such as vitamin C and vitamin E (Ajila *et al.*, 2007). The polyphenolic compounds present in mango peel present a wide range of applications in the food and pharmaceutical industries. These phytochemicals prevent DNA oxidative damage, inhibition of cell communication and free radical scavenging. Free radicals have been seen to drain immune system antioxidants, and catalyse unusual proteins, which are known to start progressive disease and ageing and change gene expression (Masibo and He, 2008).

Abdalla *et al.* (2007) studied the composition of mango seed kernels and found amino acids, phenolic compounds, the characteristics of the extracted oil including unsaponifable matter constituents, lipid classes and fatty acids. Results showed that mango seed kernels contained a substantial content of total phenolic compounds, total lipid, unsaponifiable matter, and a low amount of crude protein, but the quality of protein was good because it was rich in all the essential amino acids. The combination of both mango seed kernel extract and oil had optimum antioxidant potency higher than each one alone, and could be used as a natural antibacterial and antioxidant in food and pharmaceutical products.

5.2.1.6 Miscellaneous Fruits Canteri *et al.* (2010) found passion fruit waste (peel) rich in pectin that is used as a stabilizer and gelling agent, generated in high amounts in the Brazilian food industry. In pineapple processing, the fruit stem is a major waste material and found to be the main source of the enzyme bromelain. Upadhyyay *et al.* (2012) studied biological activities of pineapple stem waste and found antioxidant, antimicrobial, inhibitions against 15-lipoxygenase and advanced glycation end-product formation by pineapple stem waste.

5.2.2 Vegetables

5.2.2.1 Carrots Carrots (*Daucus carota L.*) are normally used, after peeling, in the industrial and household sector, and the peels are generally considered as a waste. Carrot peels contain high amounts of dietary fibers and phenolic acids, especially anthocyanin and carotenoids (Goncalves *et al.*, 2010). Chau *et al.* (2004) reported that by-products (carrot pomace) of the carrot juice industry contained a high content of dietary fibers. Carrots pomace is rich in anthocyanins and carotenoids that are responsible for the aroma, bitterness and color of carrots (Goncalves *et al.*, 2010). As carrots are a good source of carotenoids, they are recommended for healthy eyesight and due to a high phenolic content, they have strong antioxidant potential. Carrot anthocyanins reduce/inhibit inflammation and lipid peroxidation, thus leading to prevention of heart diseases (Arscott and Tanumihardjo, 2010).

5.2.2.2 Cauliflower Many food dishes, such as soups and stews, contain cauliflower (*Brassica oleracea* L. var. *botrytis*) as an important component, but only 40% of the vegetable is used, while the rest is discarded as waste (Femenia *et al.*, 1998). The by-products of cauliflower, such as the floret and stem, were analyzed for non-starch polysaccharides. The stem contained non-starch polysaccharides as 3.11% of fresh weight compared to the floret that was 2.31% of fresh weight (Femenia *et al.*, 1997). The insoluble fraction was dominant, and the soluble fraction in both by-products and major components identified was pectin in a non-starch polysaccharide fraction. Llorach *et al.* (2003) studied the antioxidant activities of cauliflower by-products and found that hydroxycinnamic acid and flavonoids were the main antioxidant compounds present in by-products. They further evaluated that the edible part of the cauliflower was much lower in these phenolic compounds compared to the by-products.

5.2.2.3 Onions Onions (*Allium cepa L.*) are commonly used in fast foods, restaurant food and salads, due to their strong flavor (Griffiths *et al.*, 2002). Currently in Europe, 500 000 tonnes of onion by-products are produced annually (Benitez *et al.*, 2011). Onion by-products mainly consist of skins, roots, fleshy scales, undersized, diseased or damaged onions. Different layers of onions contain dietary fibers in different proportions. The highest levels of total dietary fibers were found in the skin of onions, which is normally considered as waste (Jaime *et al.*, 2002). The phytochemical analysis of onions revealed that phenolic and flavonoid were the main components (Ng *et al.*, 2000). Protocatechuric acid and ferulic acid were in abundance in the papery scales of onions compared to the fleshy scale. The onions by-products were found to be rich in flavonoids that indicated their potential in reducing inflammation, heart disease and cancers. Onions by-products are also rich in organo-sulphur compounds that are associated with various pharmacological activities, which indicates under-utilization of onion by-products (Lanzotti, 2006).

5.2.2.4 Potato Potatoes (*Solanum tuberosum L.*) contain nutrients such as fibers, carbohydrates, minerals and phenolic compounds (Abu-Ghannam and Crowley, 2006). The processing and preparation of potatoes generate tonnes of potato peel and pulp as by-products. Liu *et al.* (2007) studied the composition of three species of potatoes and their peels and found no significant difference in fiber content and other nutrients. Potatoes are rich in chlorogenic acid (soluble fraction) and caffeic acid (insoluble fraction). Mattila and Hellstrom (2007) found that the chlorogenic acid and its derivative compounds were the main soluble phenolic compounds present in potato peels. The phenolic compounds present in potatoes contribute to various health benefits, for example chlorogenic acid exhibits antioxidant activities, tumor suppression and lowering of hyperglycemia. Due to the nutraceutical benefits of potatoes by-products, the concept between phyto-constituents present in potato peel and health benefits is gaining more attention.

5.2.2.5 Tomato Tomato (*Solanum lycopersicum L.*) is used in the preparation of many food products, such as ketchup, sauces, pasta, various food dishes and juices. The processing of tomatoes generates many by-products that still contain numerous phytochemicals, for example lycopene that belongs to the family of carotenoids and helps to reduce the risk of cancers (Chang *et al.*, 2006). Tomato pomace contains fiber as a main ingredient that contributes 50% of the by-product on a dry weight basis (Del Valle *et al.*, 2006). Lycopene is present in abundance in tomato by-products and lowers the risk of prostate, pancreas and stomach cancers (Chang *et al.*, 2006). Lycopene is a cancer-reducing phytochemical and is present in abundance in tomato peel, and it can be included as a nutraceutical in food and pharmaceutical preparations.

5.2.2.6 Olives Olives (*Olea europaea*) are used for the production of olive oil that generates large quantities of waste (Lozano-Sanchez *et al.*, 2011). The by-products of the olive oil industry contain various bioactive and pharmaceutical ingredients with strong antioxidant potential. The waste products generated during the preparation of extra virgin olive oil were analyzed for the presence of phenolic compounds and flavones, and lignans and secoiridoids were found in abundance

(Lozano-Sanchez *et al.*, 2011). The quantitative and qualitative analysis of these phenolic compounds suggested that olive by-products could be considered as a vital source of phenolic compounds and after purification could be used as antioxidants in food and nutraceutical products due to their biological and technological properties. Ramos *et al.* (2013) studied the antioxidant and anti-cancer activities of olive pomace extracts from Portuguese industries. The extracts were found to be potent for antioxidant and breast cancer anti-proliferative activities, proving olive by-products as a potential source of antioxidants in food supplements and nutraceuticals.

5.3 By-Products from Animal Origin

Regulation (EC) No. 1096/2009 has defined animal by-products as the whole body of an animal or parts from the animal body or other products derived from animals which can be but are not meant for consumption by humans. The culture, tradition and religion of country influence the type and generation of by-products. Under-utilization of by-products presents a problem to the environment and economic losses to the producers (Lafarga and Hayes, 2014). Here, the by-products from meat processing, poultry, fish and seafood and their possible applications in the pharmaceutical industry are included.

5.3.1 By-Products from Meat Processing

Both slaughterhouses and meat-processing industries, along with the wholesalers, retailers and renderers, produce tons of solid and liquid by-products annually. The disposal of these by-products has created a huge problem for the producers. They mainly comprise of blood, trimming, skin, bones etc., which contain many bioactive components that can be extracted by using suitable techniques and utilized in food, feed or pharmaceutical industries. There have been several studies on the conversion of by-products into the more value-added components or products (Mora *et al.*, 2014; Toldra *et al.*, 2012). The potential pharmaceutical and medical application of meat by-products are discussed further below (Figure 5.2).

5.3.1.1 Animal Blood Animal blood from slaughterhouses is one of the major by-products of the meat production food chain and constitutes about 4% of the weight of the live animal or around 6–7% of the lean meat of the carcass (Wismer-Pedersen, 1988). Mainly of these blood products are either processed as blood meal, which is a low-value product, or discarded as effluent. However, they can be converted into other products which have high biological value (Anderson and Yu, 2003). From animal blood, different valuable components have been recovered and used in medical and pharmaceutical applications. Some of these are listed in Table 5.2

Several studies have been done on the isolation of bioactive components from animal blood. It has been revealed that these bioactive compounds, mainly from the bovine and porcine blood fraction, possess antioxidant and antimicrobial properties along with angiotensin I-converting enzyme (ACE)-inhibitory activity, ability to bind mineral and opioid activity. The bioactive peptides that have been generated from bovine and porcine blood have shown ACE inhibitory activity (Bah *et al.*, 2013). Yu *et al.* (2006) mentioned that the peptides obtained from food sources like meat, sea

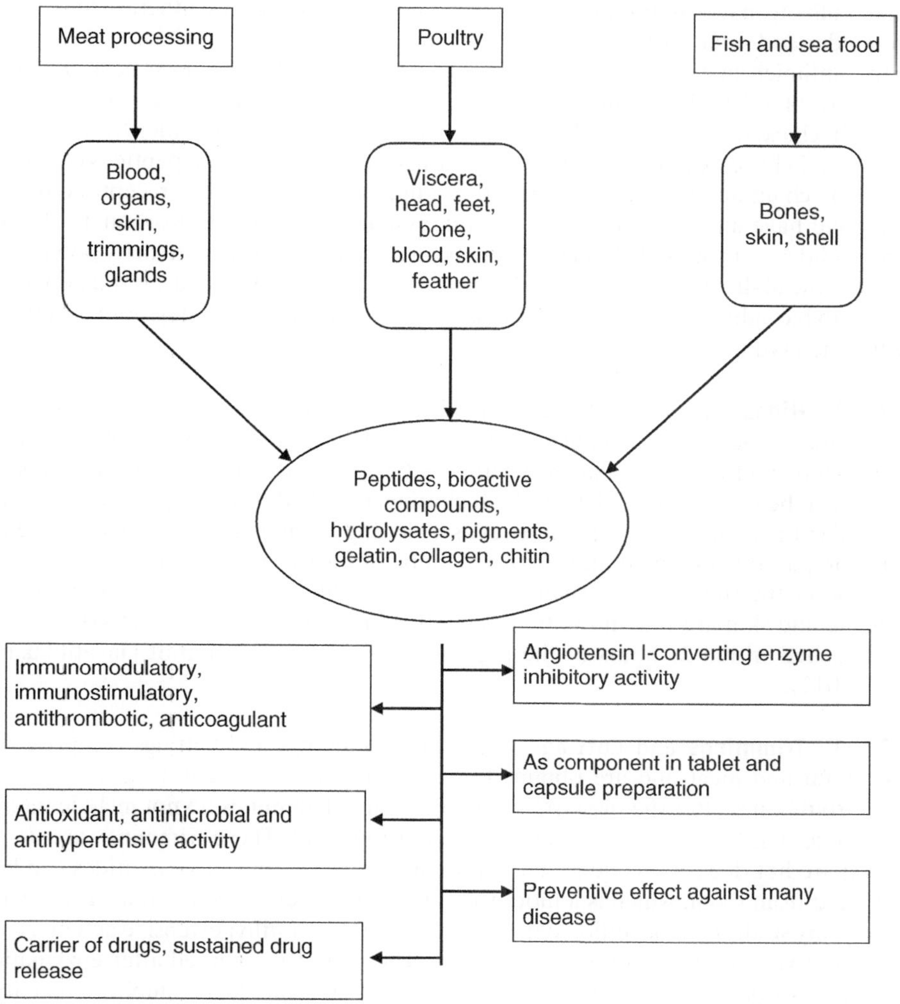

Figure 5.2 By-products from food of animal origin and their possible pharmaceutical and medical applications.

Table 5.2 Application of blood components in medical and pharmaceutical field (adapted from Bah *et al.*, 2013)

S.N	Components from blood	Applications
1	Bovine hemoglobin powder	Production of porphyrin derivative
2	Bovine fibrinogen	As blood clotting reagent in serology laboratories
3	Bovine prothrombin	For production of thrombin as a precursor and its purification
4	Bovine thrombin	Wound treatment, helps in coagulating blood, for holding skin grafts
5	Porcine plasmin enzyme	In heart attack patients, for digesting the fibrin in blood clots

food, milk etc. have ACE inhibitory activity and as compared to synthetic drugs, they are milder and safer for use.

The antimicrobial peptides such as cathelicidins and defensins have been derived from the blood of both animals (pigs, goat, sheep, deer and cattle) and poultry (ostrich, chicken and turkey) (Yu *et al.*, 2010). The hydrolyzed peptides from bovine hemoglobin have exhibited opioid activity (Zhao *et al.*, 1997). These peptides containing short-chain amino acids are capable of binding with the opioid receptors in the brain and have an influence on the nervous system (Pihlanto and Korhonen, 2003). The peptides from animal blood have also been linked with antigenotoxicity potential, which is the ability to prevent DNA damage. The Comet assay of hydrolyzed peptides from bovine albumin, globulin and plasma demonstrated the antigenotoxic effect (Park and Hyun, 2002).

5.3.1.2 Hides and Skins Hides and skins of animals are one of the important by-products generated by animal meat processing. They contain considerable amounts of collagen and from this collagen, gelatin can be extracted. The collagen-based compound can be used for stimulating blood clotting, mainly during surgery. Gelatin can be used in the pharmaceutical industries as a binding agent and a compounding agent in the preparation of tablets and also for manufacturing the outer coatings of capsules. Because of the similarity between the skins of pigs and humans, it can be used for dressing and skin grafting in the treatment of skin burns and ulcers. The skin has to be detached within 24 hours of the pig's death to use as a skin graft (Jayathilakan *et al.*, 2012).

5.3.1.3 Trimmings and Cuttings Generally, trimmings and cuttings comprise of gristle, fat and meat and are obtained as the leftovers while removing the muscle meat from bones after the preparation of prime cuts. Trimmings do not include internal organs, ligaments, tendons and the parts of the head. These trimmings are used to prepare hot dogs, a secondary meat product. However, bioactive peptides can be extracted from them, which is a more valuable and cost-effective method. There has been many studies on the antioxidant, antimicrobial and antihypertensive activities of the bioactive peptides from muscle proteins. The peptides obtained after enzymatic hydrolysis of porcine skeletal muscle proteins indicated the ACE-inhibitory activity and this effect was proved *in vivo* by experimenting on hypertensive rats (Mora *et al.*, 2014).

5.3.1.4 Glands and Organs Different glands and organs from the healthy animal have been traditionally used for medicinal purposes in many countries like India, Japan and China etc. A rapid freezing technique is used to preserve the glands, as it can inhibit the breakdown of tissue by microbial contamination. Prior to freezing, the glands should be properly cleaned, and fat and connective tissues should be removed and wrapped in waxed paper. Finally, they will be stored at less than or equal to −18 °C. For pharmaceutical applications, these glands are either vacuum dried or extracted with suitable solvents. The excess fat content of the gland is removed by using solvents like ethylene, acetone, gasoline or light petroleum. The dried and defatted extract is then powdered and used for capsule preparation or in the liquid form (Jayathilakan *et al.*, 2012).

Liver extract is considered as a good source of vitamin B_{12}. Therefore, it can be used as a nutrient supplement for the treatment of different forms of anemia. Liver extract is also used as a raw material for various pharmaceutical applications. Bile juice obtained from the gall bladder of the animal can be used to treat indigestion, constipation, problems related to the bile tract and to improve liver functioning. Insulin from the pancreas can be used for diabetic patients to regulate blood glucose levels. Similarly, glucagon hormone from the pancreas can maintain blood glucose levels and overdose of insulin. Heparine, an anticoagulant agent, can be extracted from the liver, lungs and small intestine linings. It has been utilized in the thinning of blood and delaying of blood coagulation during organ transplant surgery. The threads for surgical sutures can be prepared from the sheep and calves intestine (Rahman *et al.*, 2014).

5.3.1.5 Poultry By-Products Only one-third portion of the total weight of poultry birds is converted into meat during slaughtering and processing, whereas the remainder is by-products and waste. Therefore, these wastes and by-products should be efficiently processed and utilized to produce high-value products. Many by-products can be obtained from the poultry industry, including egg shell, shell membranes, feathers, heads, blood, feet, intestine, glands, gizzards and proventriculus. They can be a good source of calcium, gelatin, collagen, protein hydrolysates, enzymes etc., which can be utilized in many applications as in the food, feed, fertilizer, cosmetics, biochemical and pharmaceutical industries (Lasekan *et al.*, 2013; Mishra *et al.*, 2015).

Egg shell is a natural and rich source of calcium and almost 90% of it is absorbable. Calcium supplementation using egg shell powder along with vitamin D and magnesium have shown positive results on bone mineral density in osteoporosis conditions. Egg shell membrane consists of collagen which has higher biosafety and lower risk of allergic reactions. Collagen from egg shell membranes can be used as a treatment aid in orthopedics, dentistry and for cancer patients for improving the skin and thickening of hair (Kingóri, 2011).

Chicken by-products vary in their compositions, such as the viscera, which is rich in lipids, whereas the head, skin and feet have high protein content (collagenous and keratinous). Thus hydrolysates with high bioactive properties can be obtained from these by-products. These hydrolysates have shown antioxidant and antihypertensive potential. The keratinous hydrolysates have shown better antioxidant properties, but the collagenous fraction exhibited strong antihypertensive potential. The hydrolysates from chicken leg bone, viscera, acid hydrolyzed feather and blood have indicated good ACE inhibitory activity. The ACE inhibitory activity of hydrolysates is mostly affected by the type of enzyme, degree of hydrolysis (DH) and the peptide's molecular weight (Lasekan *et al.*, 2013). Collagen extracted from chicken neck, feet and other by-products can have various medical applications in the fields of surgery, cardiology, dermatology, orthopaedic, ophthalmology, urology and vascular departments (Silvipriya *et al.*, 2015).

5.3.2 Fish and Seafood Processing

In recent years, aquaculture has been considered the fastest growing food-production sector in the world (Kandra *et al.*, 2012). The by-products from fish and shellfish processing may contribute to as much as 70% of the initial weight of the catch (Olsen *et al.*, 2014). Most of the by-products from fish processing are used for producing fish meal, fish oil, fish silage, animal food and fertilizer, which holds low economic profitability. Many researches have revealed that fish and shellfish by-products, such as muscle protein, collagen, gelatin, oil, bone, shells of shellfish and crustaceans, comprise of numerous valuable bioactive compounds. They can be extracted and purified as desired using suitable technologies and finally employed in nutraceutical, biotechnological and pharmaceutical applications (Kim and Mendis, 2006).

The important and valuable constituents from fish by-products belong to proteins, lipids and other compounds. The protein by-product includes collagen, gelatin, fish protein isolates, fish protein hydrolysates, enzymes, bioactive peptides etc., whereas lipid components consist of omega-3 fatty acids, marine oils, cholesterol, vitamins etc. Calcium, astaxanthin, taurine and bioactive principles are classified under other compounds (Rustad *et al.*, 2011). Similarly, Kandra *et al.* (2012) have mentioned that shrimp waste is also a rich source of chitin, minerals, protein and astaxanthin.

5.3.2.1 Bioactive Peptides from Fish Protein Hydrolysates The bioactive peptides extracted from fish protein have been proved to be associated with several bioactive properties such as antioxidative, antithrombotic, antihypertensive and immunomodulatory activities. They have shown strong antioxidant activities in many oxidative systems (Kim and Mendis, 2006). Rajapakse *et al.* (2005) have reported that enzymatic hydrolysis of marine fish can yield proteins with anticoagulant and antiplatelet properties. With the formation of the inactive complex, these proteins were capable of inhibiting coagulation factor XII (FXIIa). A few fish peptides have also been associated with antihypertensive activity that inhibited the ACE (Kim *et al.*, 2000). It has been reported that some fish protein have growth factors and hormone-like peptides, which are capable of accelerating calcium absorption and can be used for treating calcium deficiency disorders such as osteoporosis and Paget's disease (Fouchereau-Peron *et al.*, 1999). The acidic peptides from Atlantic cod hydrolysates have shown strong immunostimulatory effects, which is related with the bactericidal power of phagocytes (Gildberg *et al.*, 1996).

5.3.2.2 Fish Oil Omega-3 fatty acids, eicosapentaenoic acid (EPA) and docosahexaenoic acid (DHA) are the important bioactive components that can be isolated from the fish oil by the process called transesterification (Olsen *et al.*, 2014). Intake of omega-3 fatty acids has been associated with numerous health benefits for combating many diseases. Consumption of fish oil raises EPA and DHA levels in the blood. As a result, there is a reduction in the rate of coronary heart diseases through various modes of action. It also helps in the prevention of atherosclerosis by hindering the formation of atherosclerotic plaques (Schacky, 2000). Omega-3 fatty acids have shown the ability to lower blood pressure (Appel *et al.*, 1993), benefits for diabetic patients (Sheehan *et al.*, 1997), effective against asthma (Broughton *et al.*, 1997), preventive against arrhythmias (Christensen *et al.*, 1997) and also protect against chronic obstructive pulmonary diseases (Shahar *et al.*, 1994). Gogos *et al.* (1998) concluded that intake of omega-3 fatty acids along with supplementation of antioxidant may prolong the life of cancer patients with general malignancy through the immunomodulating effect. EPA and DHA can be transformed into prostaglandins and leukotrienes that have natural anti-inflammatory action (Belch and Muir, 1998). Therefore, EPA and DHA can be used for treating kidney diseases (Donadio *et al.*, 1994) and Crohn's disease (Belluzzi *et al.*, 1996).

5.3.2.3 Gelatin and Collagen from Fish Skin and Bones Almost 30% of the waste from the fish processing industry includes fish bones, skin and fins. Collagen and gelatin having diverse applications can be isolated from the skin and bones. The contribution of collagen from these sources is in the range of 36% to 54%. Thus, the collagen obtained has potential for use in the food, pharmaceutical and cosmetic industries (Blanco *et al.*, 2007). There is an increase in demand for collagen and gelatin for industrial uses (Karim and Bath, 2009). Structurally, collagen has a triple helix arrangement made from three extended protein chains wrapped around one another. When collagen is partially hydrolyzed, gelatin can be obtained. So collagen and gelatin are different forms of the same macromolecule (Gomez-Gillen *et al.*, 2002; Kim *et al.*, 1994). In general, acid treatment is used to extract collagen followed by solubilization, so that the triple helix structure is not lost. The heat treatment of collagen breaks the triple helix structure and converts collagen into gelatin. Hence, a simple hot water treatment can be applied for solubilizing collagen from the source to obtain gelatin (Djabourov *et al.*, 1993). Both collagen and gelatin have good biodegradability and biocompatibility in the physiological environments, so are used widely in the pharmaceutical and medical industries (Young *et al.*, 2005).

The gelatin extracted from fish skin by enzymatic hydrolysis has shown antioxidant and antihypertensive properties, which was assumed to be present due to the unique amino acid sequence Gly–Pro–Ala in gelatin peptides. Along with this, further research on animals has suggested that these gelatin peptides can increase the bioavailability of calcium by accelerating the calcium absorption (Kim and Mendis, 2006). Gelatin has been used in the pharmaceutical industries as a stabilizer in live attenuated viral vaccines (mumps, rabies, measles, diphtheria etc.) (Burke *et al.*, 1999), as intravenous infusions (Saddler and Horsey, 1987) and injectable microspheres for drug delivery (Rao, 1995).

Pharmaceutical application of collagen includes its use as carrier molecules for drugs, proteins and genes (Lee *et al.*, 2001). For cancer treatment, microfibrous collagen sheets are utilized as drug carriers (Sato *et al.*, 1996). In addition, collagen is also used in the manufacture of vitreous implants and wound dressings (Takeshi and Suzuki, 2000). Moskowitz (2000) has mentioned that collagen can help in the functional expression of cells and tissue and organ formation. Further research has shown that intake of gelatin or collagen hydrolysates are effective in reducing the pain of osteoarthritis patients and they also take part in cartilage matrix synthesis.

5.3.2.4 Chitin, Chitosan and their Oligomers from the Shell of Crustacean Chitin is the ubiquitous abundant polysaccharide, which can be obtained from many organisms, including cell wall of fungi and algae, exoskeleton of insects, endoskeleton of cephalopods and shells of crustacean. But the major commercial source is contributed by marine organisms. Around 10^{10}–10^{11} tonnes of chitin are produced annually by living organisms (Revathi *et al.*, 2012; Merzendorfer, 2011). Chitin is a polymer of β-(1→4)-linked 2- acetamido-2-deoxy-D-glucopyranose, whose structure is similar to that of cellulose, where the hydroxyl group of cellulose in C-2 is replaced by the acetamido group in chitin (Kurita, 2006). It is white to cream in color, odorless, with varying degrees of acetylation (Meyers *et al.*, 2008). Chitin has three different polymorphic forms: α, β and γ forms. The most abundant and stable form is α-chitin. β-chitin, found in squid pen. is more reactive than the α- form.

Table 5.3 Application of chitin and chitosan biomedical and pharmaceutical field (adapted from Hamed *et al.*, 2016; Kim *et al.*, 2008)

Field	Uses
Biomedical	Manufacturing of surgical stitches Fabrication of scaffolds for tissue regeneration Barrier against microbial infections
Pharmaceuticals	Encapsulation of gene and active constituents Carrier of ingredients and controlled drug release For improvement of arrival of nucleic acids to the therapeutic targets Wound healing properties Non-viral vectors for delivery of genes

γ-chitin is a combination of the *α* and *β* form, which is comparatively rare and only found in the squid's stomach and cocoons of two genera of beetles (Hamed *et al.*, 2016).

The exoskeletons of crustacean are demineralized and deproteinized in order to obtain chitin. One of the main drawbacks of chitin is its strong hydrophobicity. Therefore chitosan, a deacetylated derivative of chitin, is produced as it is water soluble and the only known natural cationic polysaccharide (Du *et al.*, 2014). Both chitin and chitosan are biodegradable, renewable, biocompatible and non-toxic compounds that are linked with several biological activities that include anti-cancer (Salah *et al.*, 2013), antimicrobial (Goy *et al.*, 2009) and antioxidant potentials. The possible pharmaceutical and biomedical application of chitosan and its derivatives are listed in Table 5.3.

5.3.2.5 Astaxanthin from Crustacean Shell The ketocarotenoid astaxanthin, 3,3′-dihydroxy-*β*, β-carotene-4, 4′-dione, is the oxidized derivative of β-carotenoid. It is the natural substance with properties like highly versatile, non-toxicity and solubility in both water and lipid, which allows it to be used for many applications. It is also considered in food and medical applications due to its attractive pink color, high antioxidant activity and functionality as a precursor of vitamin A. Astaxanthin can be synthesized or extracted from many sources such as plants, bacteria, algal cells, crustacean shells and egg yolk. One of the richest sources is shrimp shell waste. Astaxanthin is one of the major pigments in crustacean shells, constituting 74–98% of the total pigments. Therefore, crustacean shell can be used for extracting astaxanthin along with the recovery of chitin (Ferraro *et al.*, 2010).

The antioxidative effect of astaxanthin is stronger than β-carotene, vitamins E and vitamin C. In addition, it was found to enhance the immune system, and protect against cancers and ultraviolet radiation damage. It has been demonstrated that astaxanthin inhibits prostate cancer and modulating immune responses against tumor cells. It also inhibits bladder carcinogenesis. Astaxanthin can be extracted from the crustacean shell through various methods such as by fermentation, enzymatic process, using hydrochloric acid, edible oils and organic solvents. Many vegetable oils such as sunflower, coconut, palm, ground nut, soybean, ginger and rice bran oils have been used for extracting astaxanthin from fish by-products and crustaceans. The solution of proteolytic enzymes such as pepsin, trypsin and papain dissolved in citrate phosphate has been used for precipitating astaxanthin present in the crustacean shell. Currently, the lactic acid fermentation process is one of the feasible methods to obtain partially concentrated astaxanthin from crustacean shells. This method also preserves the biomass from bacterial decomposition (Ferraro *et al.*, 2010; Kandra *et al.*, 2012).

5.4 Conclusion

Food processing operations generate a huge amount of by-products, which have a large variety of applications but are generally handled with low economic value. Mostly, they are used in the field of food, feed, cosmetics, fertilizers etc. However, valuable components can be recovered from these by-products and exploited as high value-added compounds. Several studies have proven that the bioactive constituents from food by-products, including the agroindustry, meat, poultry and fish processing, have numerous beneficial properties. Therefore, they can be utilized in medical and pharmaceutical applications. Various physiological functions such as antioxidant activity, antihypertensive, anti-thrombotic, mineral-binding and anti-amnesiac activities have been found to be associated with the bioactive compounds from food by-products. Hence, this shows that there is a great potential of food by-products for effective utilization in the field of medical and pharmaceutical applications that will be more cost-effective and efficient.

References

Abdalla, A.E., Darwish, S.M., Ayad, E.H. and El-Hamahmy, R.M. (2007) Egyptian mango by product. 1: Compositional quality of mango seed kernel. *Food Chemistry*, **103**(4): 1134–1140.

Abu-Ghannam, N. and Crowley, H. (2006) The effect of low temperature blanching on the texture of whole processed new potatoes. *Journal of Food Engineering*, **74**(3): 335–344.

Ajila, C.M., Bhat, S.G. and Rao, U.P. (2007) Valuable components of raw and ripe peels from two Indian mango varieties. *Food Chemistry*, **102**(4): 1006–1011.

Ajila, C.M., Leelavathi, K. and Rao, U.P. (2008) Improvement of dietary fiber content and antioxidant properties in soft dough biscuits with the incorporation of mango peel powder. *Journal of Cereal Science*, **48**: 319–326.

Anderson, R.C. and Yu, P.L. (2003) Isolation and characterisation of proline/arginine-rich cathelicidin peptides from ovine neutrophils. *Biochemical and Biophysical Research Communications*, **312**: 1139–1146.

Appel, L.J., Miller, E.R., Seidler, A.J. and Whelton, P.K. (1993) Does supplementation of diet with fish oil reduce blood pressure? A meta-analysis of controlled clinical trials. *Archives of Internal Medicine*, **153**: 1429–1438.

Arscott, S.A. and Tanumihardjo, S.A. (2010) Carrots of many colors provide basic nutrition and bioavailable phytochemicals acting as a functional food. *Comprehensive Reviews in Food Science and Food Safety*, **9**(2): 223–239.

Bah, C.S.F., Bekhit, A.E.A., Carne, A. and McConnell, M.A. (2013) Slaughterhouse blood: An emerging source of bioactive compounds. *Comprehensive Reviews in Food Science and Food Safety*, **12**: 314–331.

Bakowska-Barczak, A.M., Schieber, A. and Kolodziejczyk, P. (2009) Characterization of Canadian black currant (*Ribes nigrum L.*) seed oils and residues. *Journal of Agricultural and Food Chemistry*, **57**(24): 11528–11536.

Belch, J.J.F. and Muir, A. (1998) n-6 and n-3 essential fatty acids in rheumatoid arthritis and other rheumatic conditions. *Proceedings of the Nutrition Society*, **57**: 563–569.

Belluzzi, A., Brignola, C., Campieri, M., Pera, A., Boschi, S. and Miglioli, M. (1996) Effect of an enteric-coated fish-oil preparation on relapses in Crohn's disease. *The New England Journal of Medicine*, **334**: 1557–1560.

Benítez, V., Mollá, E., Martín-Cabrejas, M.A., Aguilera, Y., López-Andréu, F.J. *et al.* (2011) Characterization of industrial onion wastes (*Allium cepa L.*): Dietary fiber and bioactive compounds. *Plant Foods for Human Nutrition*, **66**(1): 48–57.

Blanco, M., Sotelo, C G, Chapela, M.J. and Perez-Martin, R.I. (2007) Towards sustainable and efficient use of fishery resources: Present and future trends. *Trends in Food Science Techology*, **18**: 29–36.

Boyer, J. and Liu, R.H. (2004) Apple phytochemicals and their health benefits. *Nutrition Journal*, **3**(5): 12.

Broughton, K.S., Johnson, C.S., Pace, B.K., Liebman, M. and Kleppinger, K.M. (1997) Reduced asthma symptoms with n-3 fatty acid ingestion are related to 5-series leukotriene production. *American Journal of Clinical Nutrition*, **65**: 1011–1017.

Burke, C.J., Hsu, T.A. and Volkin, D.B. (1999) Formulation, stability and delivery of live attenuated vaccines for human use. *Critical Reviews in Therapeutic Drug Carrier Systems*, **16**(1): 1–83.

Canteri, M. H., Scheer, A.P., Wosiacki, G., Ginies, C., Reich, M. and Renard, C.M. (2010) A comparative study of pectin extracted from passion fruit rind flours. *Journal of Polymers and the Environment*, **18**(4): 593–599.

Chang, C.H., Lin, H.Y., Chang, C.Y. and Liu, Y.C. (2006) Comparisons on the antioxidant properties of fresh, freeze-dried and hot-air-dried tomatoes. *Journal of Food Engineering*, **77**(3): 478–485.

Chau, C.F. and Huang, Y.L. (2003) Comparison of the chemical composition and physicochemical properties of different fibers prepared from the peel of *Citrus sinensis L. Cv. Liucheng. Journal of Agricultural and Food Chemistry*, **51**(9): 2615–2618.

Chau, C.F., Chen, C.H. and Lee, M.H. (2004) Comparison of the characteristics, functional properties, and *in vitro* hypoglycemic effects of various carrot insoluble fiber-rich fractions. *Lebensmittel-Wissenschaft and Technologie (LWT)*, **37**: 155–160.

Christensen, J. H., Korup, E., Aaroe, J., Toft, E., Moller, J., *et al.* (1997). Fish consumption, n-3 fatty acids in cell membranes, and heart rate variability in survivors of myocardial infraction with left ventricular dysfunction. *American Journal of Cardiology*, **79**: 1670–1673.

Del Valle, M., Cámara, M. and Torija, M.E. (2006) Chemical characterization of tomato pomace. *Journal of the Science of Food and Agriculture*, **86**(8): 1232–1236.

Deng, Q., Penner, M.H. and Zhao, Y. (2011) Chemical composition of dietary fiber and polyphenols of five different varieties of wine grape pomace skins. *Food Research International*, **44**(9): 2712–2720.

Dilas, S., Canadanovic-Brunet, J. and Cetkovic, G. (2009) By-products of fruits processing as a source of phytochemicals. *Chemical Industry and Chemical Engineering Quarterly*, **15**(4): 191–202.

Djabourov, M., Lechaire, J.P. and Gaill, F (1993) Structure and rheology of gelatin and collagen gels. *Biorheology*, **30**: 191–205.

Donadio, J. V. Jr., Bergstrahl, E.J. and Offord, K.P. (1994) A controlled trial of fish oil in IgA nephropathy. *New England Journal of Medicine*, **331**: 1194–1199.

Du, J., Tan, E., Kim, H.J., Zhang, A., Bhattacharya, R. and Yarema, K.J. (2014) Carbohydrate polymers. Comparative evaluation of chitosan, cellulose acetate, and polyethersulfone nanofiber scaffolds for neural differentiation. *Carbohydrate Polymers*, **99**: 483–490.

Femenia, A., Lefebvre, A.C., Thebaudin, J.Y., Robertson, J.A. and Bourgeois, C.M. (1997) Physical and sensory properties of model foods supplemented with cauliflower fiber. *Journal of Food Science*, **62**(4): 635–639.

Femenia, A., Robertson, J.A., Waldron, K.W. and Selvendran, R.R. (1998) Cauliflower (*Brassica oleracea L*), globe artichoke (*Cynara scolymus*) and chicory witloof (*Cichorium intybus*) processing by-products as sources of dietary fiber. *Journal of the Science of Food and Agriculture*, **77**(4): 511–518.

Fernández-López, J., Fernández-Ginés, J.M., Aleson-Carbonell, L., Sendra, E., Sayas-Barberá, E. and Pérez-Alvarez, J.A. (2004) Application of functional citrus by-products to meat products. *Trends in Food Science and Technology*, **15**(3): 176–185.

Ferraro, V., Cruz, I.B., Jorge, R.F., Malcata, F.X., Pintado, M.E. and Castro, P.M.L. (2010) Valorisation of natural extracts from marine source focused on marine by-products: A review. *Food Research International*, **43**: 2221–2233.

Fouchereau-Peron, M., Duvail, L., Michel, C., Gildberg, A., Batista, I. and Gal, Y.I. (1999) Isolation of an acid fraction from a fish protein hydrolysate with a calcitonin–gene-related-peptide-like biological activity. *Biotechnology and Applied Biochemistry*, **29**: 87–92.

Galanakis, C.M. (2012) Recovery of high added-value components from food wastes: Conventional, emerging technologies and commercialized applications. *Trends in Food Science and Technology*, **26**(2): 68–87.

García, Y.D., Valles, B S. and Lobo, A.F. (2009) Phenolic and antioxidant composition of by-products from the cider industry: Apple pomace. *Food Chemistry*, **117**(4): 731–738.

Ghatak, H.R. (2011) Biorefineries from the perspective of sustainability: Feedstocks, products, and processes. *Renewable and Sustainable Energy Reviews*, **15**(8): 4042–4052.

Gildberg, A., Bogwald, J., Johansen, A. and Stenberg, E. (1996) Isolation of acid peptide fractions from a fish protein hydrolysate with strong stimulatory effect on Atlantic salmon (*Salmo salar*) head kidney leucocytes. *Comparative Biochemistry and Physiology*, **11**: 97–101

Gogos, C.A., Ginopoulos, P., Salsa, B., Apostolidou, E., Zoumbos, N.C. and Kalfarentzos, F. (1998) Dietary omega-3 polyunsaturated fatty acids plus vitamin E restore immunodeficiency and prolong survival for severely ill patients with generalized malignancy. *Cancer*, **82**: 395–402.

Gomez-Gillen, M.C., Turnay, J., Fernandez-Diaz, M.D., Ulmo, N., Lizarbe, M.A. and Montero, P. (2002) Structural and physical properties of gelatin extracted from different marine species: A comparative study. *Food Hydrocolloids*, **16**: 25–34.

Gonçalves, E.M., Pinheiro, J., Abreu, M., Brandão, T.R.S. and Silva, C.L. (2010) Carrot (*Daucus carota L.*) peroxidase inactivation, phenolic content and physical changes kinetics due to blanching. *Journal of Food Engineering*, **97**(4): 574–581.

González-Molina, E., Domínguez-Perles, R., Moreno, D.A. and García-Viguera, C. (2010) Natural bioactive compounds of Citrus limon for food and health. *Journal of Pharmaceutical and Biomedical Analysis*, **51**(2): 327–345.

Gorinstein, S., Martín-Belloso, O., Park, Y.S., Haruenkit, R., Lojek, A., Ĉíž, M. and Trakhtenberg, S. (2001) Comparison of some biochemical characteristics of different citrus fruits. *Food Chemistry*, **74**(3): 309–315.

Goy, R.C., de Britto, D. and Assis, O.B.G. (2009) A review of the antimicrobial activity of chitosan. *Polimeros*, **19**: 241–247

Griffiths, G., Trueman, L., Crowther, T., Thomas, B. and Smith, B. (2002) Onions – A global benefit to health. *Phytotherapy Research*, **16**(7): 603–615.

Hamed, I., Ozogul, F. and Regenstein, J.M. (2016) Industrial applications of crustacean by-products (chitin, chitosan and chitooligosaccharides): A review. *Trends in Food Science and Technology*, **48**: 40–50.

Huber, G.M., and Rupasinghe, H.P.V. (2009) Phenolic profiles and antioxidant properties of apple skin extracts. *Journal of Food Science*, **74**(9): C693–C700.

Hwang, J.K., Kim, C.J. and Kim, C.T. (1998) Extrusion of apple pomace facilitates pectin extraction. *Journal of Food Science*, **63**(5): 841–844.

Jaime, L., Mollá, E., Fernández, A., Martín-Cabrejas, M.A., López-Andréu, F.J. and Esteban, R.M. (2002) Structural carbohydrate differences and potential source of dietary fiber of onion (*Allium cepa L.*) tissues. *Journal of Agricultural and Food Chemistry*, **50**(1): 122–128.

Jayathilakan, K., Sultana, K., Radhakrishna, K. and Bawa, A.S. (2012) Utilization of by-products and waste materials from meat, poultry and fish processing industries: A review. *Journal of Food Science Technology*, **49**(3): 278–293.

Kamm, B. and Kamm, M. (2004) Principles of biorefineries. *Applied Microbiology and Biotechnology*, **64**(2): 137–145.

Kammerer, D. R., Schieber, A. and Carle, R. (2005) Characterization and recovery of phenolic compounds from grape pomace: A review. *Journal of Applied Botany and Food Quality*, **79**(3): 189–196.

Kandra, P., Challa, M.M. and Jyothi, H.K.P. (2012) Efficient use of shrimp waste: Present and future trends. *Applied Microbiology Biotechnology*, **93**: 17–29.

Karim, A.A. and Bath, R. (2009) Fish gelatin: Properties, challenges and prospects as an alternative to mammalian gelatin. *Food Hydrocolloids*, **23**: 563–576.

Kim, I.Y., Seo, S.J., Moon, H.S., Yoo, M.K., Park, I.K. *et al.* (2008) Chitosan and its derivatives for tissue engineering applications. *Biotechnology Advances*, **26**: 1–21.

Kim, S. and Mendis, E. (2006) Bioactive compounds from marine processing byproducts: A review. *Food Research International*, **39**: 383–393

Kim, S.K., Byun, H.G. and Lee, E.H. (1994) Optimum extraction conditions of gelatin from fish skins and its physical properties. *Journal of Korean Industrial and Engineering Chemistry*, **5**: 547–559.

Kim, S.K., Choi, Y.R., Park, P.J., Choi, J.H. and Moon, S.H. (2000) Screening of biofunctional peptides from cod processing wastes. *Journal of the Korean Society of Agricultural Chemistry and Biotechnology*, **43**: 225–227.

Kingóri, A. M. (2011) A review of the uses of poultry eggshells and shell membranes. *International Journal of Poultry Science*, **10**(11): 908–912.

Kosseva, M.R. (2011) Wastes from agriculture, forestry and food processing: Management and processing of food wastes. In: *Comprehensive Biotechnology*, 2nd edition (ed. M. Moo-Young), vol. **6**, Elsevier, Spain, pp. 557–593.

Kurita, K. (2006) Chitin and chitosan: Functional biopolymers from marine crustaceans. *Marine Biotechnology*, **8**: 203–226.

Lafarga, T. and Hayes, M. (2014) Bioactive peptides from meat muscle and by-products: Generation, functionality and application as functional ingredients. *Meat Science*, **9**: 227–239.

Lanzotti, V. (2006) The analysis of onion and garlic. *Journal of Chromatography A*, **1112**(1–2): 3–22.

Lario, Y., Sendra, E., Garcí, J., Fuentes, C., Sayas-Barberá, E. *et al.* (2004) Preparation of high dietary fiber powder from lemon juice by-products. *Innovative Food Science and Emerging Technologies*, **5**(1): 113–117.

Lasekan, A., Bakar, F.A. and Hashim, D. (2013) Potential of chicken by-products as sources of useful biological resources. *Waste Management*, **33**: 552–565.

Lee, H.C., Singla, A. and Lee, Y. (2001) Biomedical applications of collagen. *International Journal of Pharmaceutics*, **221**: 1–22.

Liu, Q., Tarn, R., Lynch, D. and Skjodt, N.M. (2007) Physicochemical properties of dry matter and starch from potatoes grown in Canada. *Food Chemistry*, **105**(3): 897–907.

Llorach, R., Espín, J.C., Tomás-Barberán, F.A. and Ferreres, F. (2003) Valorization of cauliflower (*Brassica oleracea L. var. botrytis*) by-products as a source of antioxidant phenolics. *Journal of Agricultural and Food Chemistry*, **51**(8): 2181–2187.

Lozano-Sánchez, J., Giambanelli, E., Quirantes-Piné, R., Cerretani, L., Bendini, A. *et al.* (2011) Wastes generated during the storage of extra virgin olive oil as a natural source of phenolic compounds. *Journal of Agricultural and Food Chemistry*, **59**(21): 11491–11500.

Mahro, B. and Timm, M. (2007) Potential of biowaste from the food industry as a biomass resource. *Engineering in Life Sciences*, **7**(5): 457–468.

Maier, T., Schieber, A., Kammerer, D.R. and Carle, R. (2009) Residues of grape (*Vitis vinifera L.*) seed oil production as a valuable source of phenolic antioxidants. *Food Chemistry*, **112**(3): 551–559.

Masibo, M. and He, Q. (2008) Major mango polyphenols and their potential significance to human health. *Comprehensive Reviews in Food Science and Food Safety*, **7**(4): 309–319.

Mattila, P. and Hellström, J. (2007) Phenolic acids in potatoes, vegetables, and some of their products. *Journal of Food Composition and Analysis*, **20**(3): 152–160.

Merzendorfer, H. (2011) Chitin. In: *The Sugar Code: Fundamentals of Glycosciences* (ed. H.J. Gabius). John Wiley & Sons, Weinheim, Germany, p. 597.

Meyers, M.A., Chen, Po.-Yu., Yu-Min Lin, A. and Seki, Y. (2008) Biological materials: Structure and mechanical properties. *Progress in Materials Science*, **53**: 1–206.

Mirabella, N., Castellani, V. and Sala, S. (2014) Current options for the valorization of food manufacturing waste: a review. *Journal of Cleaner Production*, **65**: 28–41.

Mishra, J., Biswas, S., Sarangi, N.R., Mishra, R.P., Kumar, N. and Mishra, C. (2015) Efficient utilization of poultry by-products for economic sustainability – The need of the hour. *International Journal of Livestock Research*, **5**(10): 1–9.

Misi, S.N. and Forster, C.F. (2002) Semi-continuous anaerobic co-digestion of agro-wastes. *Environmental Technology*, **23**(4): 445–451.

Mora, L., Reig, M. and Toldra, F. (2014) Bioactive peptides generated from meat industry by-products. *Food Research International*, **65**: 344–349.

Moskowitz, R.W. (2000) Role of collagen hydrolysate in bone and joint disease. *Seminars in Arthritis Rheumatism*, **30**: 87–99.

Ng, A., Parker, M.L., Parr, A.J., Saunders, P.K., Smith, A.C. and Waldron, K.W. (2000) Physicochemical characteristics of onion (*Allium cepa L.*) tissues. *Journal of Agricultural and Food Chemistry*, **48**(11): 5612–5617.

Olsen, R.L., Toppe, J. and Karunasagar, I. (2014) Challenges and realistic opportunities in the use of by-products from processing of fish and shellfish. *Trends in Food Science and Technology* **36**: 144–151.

Park, K.J. and Hyun, C.K. (2002) Antigenotoxic effects of the peptides derived from bovine blood plasma proteins. *Enzymes and Microbial Technology*, **30**: 633–638.

Pihlanto, A. and Korhonen, H. (2003) Bioactive peptides and proteins. In: *Advanced Food Nutrition Research* (ed. S.L. Taylor). Elsevier Inc, San Diego, US, pp. 175–276.

Pott, I., Breithaupt, D.E. and Carle, R. (2003) Detection of unusual carotenoid esters in fresh mango (*Mangifera indica* L. cv. 'Kent'). *Phytochemistry*, **64**(4): 825–829.

Rahman, U., Sahar, A. and Khan, M.A. (2014) Recovery and utilization of effluents from meat processing industries. *Food Research International*, **65**: 322–328.

Rajapakse, N., Jung, W.K., Mendis, E., Moon, S.H. and Kim, S.K. (2005) A novel anticoagulant purified from fish protein hydrolysates inhibits factor XIIa and platelet aggregation. *Life Sciences*, **76** : 2607–2619.

Ramos, P., Santos, S.A., Guerra, Â.R., Guerreiro, O., Felício, L. *et al.* (2013) Valorization of olive mill residues: Antioxidant and breast cancer antiproliferative activities of hydroxytyrosol-rich extracts derived from olive oil by-products. *Industrial Crops and Products*, **46**: 359–368.

Rao, K.P. (1995) Recent developments of collagen-based materials for medical applications and drug delivery systems. *Journal of Biomaterials Science Polymer Edition*, **7**(7): 623–645.

Regulation (EC) No. 1069/2009 of the European Parliament and of the Council of 21 October 2009. Laying down health rules as regards animal by-products and derived products not intended for human consumption and repealing Regulation (EC) No 1774/2002 (Animal by-product Regulation) [2009] OJEU L300/1.

Revathi, M., Saravanan, R. and Shanmugam, A. (2012) Production and characterization of chitinase from Vibrio species, a head waste of shrimp *Metapenaeus dobsonii* (Miers, 1878) and chitin of *Sepiella inermis* (Orbigny, 1848). *Advances in Bioscience and Biotechnology*, **3**: 392–397.

Rha, H.J., Bae, I.Y., Lee, S., Yoo, S.H., Chang, P.S. and Lee, H.G. (2011) Enhancement of anti-radical activity of pectin from apple pomace by hydroxamation. *Food Hydrocolloids*, **25**(3): 545–548.

Rivera, J.A., Sebranek, J.G., Rust, R.E. and Tabatabai, L.B. (2000) Composition and protein fractions of different meat by-products used for pet food compared with mechanically separated chicken (MSC). *Meat Science*, **55**(1): 53–59.

Rosentrater, K.A. (2005) Strategic methodology for advancing food manufacturing waste management paradigms. *International Journal of Environmentally Conscious Design and Manufacturing*, **13**(1): 1–13.

Royer, G., Madieta, E., Symoneaux, R. and Jourjon, F. (2006) Preliminary study of the production of apple pomace and quince jelly. *LWT – Food Science and Technology*, **39**(9): 1022–1025.

Ruberto, G., Renda, A., Daquino, C., Amico, V., Spatafora, C. *et al.* (2007) Polyphenol constituents and antioxidant activity of grape pomace extracts from five Sicilian red grape cultivars. *Food Chemistry*, **100**(1): 203–210.

Rustad, T., Storro, I. and Slizyte, R. (2011) Possibilities for the utilisation of marine by-products. *International Journal of Food Science and Technology*, **46**: 2001–2014.

Saddler, J.M. and Horsey, P.J. (1987) The new generation gelatins. A review of their history, manufacture and properties. *Anesthesia*, **42**: 998–1004.

Salah, R., Michaud, P., Mati, F., Harrat, Z., Lounici, H., *et al.* (2013) Anticancer activity of chemically prepared shrimp low molecular weight chitin evaluation with the human monocyte leukaemia cell line, THP-1. *International Journal of Biological Macromolecules*, **52**: 333–339.

Sato, H., Kitazawa, H., Adachi, I. and Horikoshi, I. (1996) Microdialysis assessment of microfibrous collagen containing a p-glycoprotein mediated transport inhibitor, cyclosporine A, for local delivery of etopocide. *Pharmacological Research*, **13**: 1565–1569.

Schacky, C.V. (2000) n-3 fatty acids and the prevention of coronary atherosclerosis. *American Journal of Clinical Nutrition*, **71**: 224–227.

Schieber, A., Stintzing, F.C. and Carle, R. (2001) By-products of plant food processing as a source of functional compounds: Recent developments. *Trends in Food Science and Technology*, **12**(11): 401–413.

Schieber, A., Hilt, P., Streker, P., Endress, H. U., Rentschler, C. and Carle, R. (2003) A new process for the combined recovery of pectin and phenolic compounds from apple pomace. *Innovative Food Science and Emerging Technologies*, **4**(1): 99–107.

Shahar, E., Folsom, A.R., Melnick, S.L., Tockman, M.S., Comstock, G.W., *et al.* (1994) Dietary n-3 polyunsaturated fatty acids and smoking-related chronic obstructive pulmonary disease. *New England Journal of Medicine*, **331**: 228–233.

Shalini, R. and Gupta, D. K. (2010) Utilization of pomace from apple processing industries: A review. *Journal of Food Science and Technology*, **47**(4): 365–371.

Sheehan, J. P., Wei, I. W., Ulchaker, M. and Tserng, K.Y. (1997) Effect of high fiber intake in fish oil-treated patients with non-insulin-dependent diabetes mellitus. *American Journal of Clinical Nutrition*, **66**: 1183–1187.

Silvipriya, K.S., Kumar, K.K., Bhat, A.R., Kumar, B.D., John, A. and Lakshmanan, P. (2015) Collagen: Animal sources and biomedical application. *Journal of Applied Pharmaceutical Science*, **5**(3): 123–127.

Spatafora, C. and Tringali, C. (2012) Valorization of vegetable waste: Identification of bioactive compounds and their chemo-enzymatic optimization. *The Open Agriculture Journal*, **6**(1): 9–16.

Takeshi, N. and Suzuki, N. (2000) Isolation of collagen from fish waste material – Skin, bone and fins. *Food Chemistry*, **68**: 277–281.

Telrandhe, U.B., Kurmi, R., Uplanchiwar, V., Mansoori, M.H., Jain, V., *et al.* (2012) Nutraceutical – A phenomenal resource in modern medicine. *International Journal of Universal Pharmacy and Life Sciences*, **2**(1): 179–195.

Toldra, F., Aristoy, M.C., Mora, L. and Reig, M. (2012) Innovations in value-addition of edible meat by-products. *Meat Science*, **92**: 290–296.

Upadhyay, A., Chompoo, J., Araki, N. and Tawata, S. (2012) Antioxidant, antimicrobial, 15-LOX, and AGEs inhibitions by pineapple stem waste. *Journal of Food Science*, **77**(1): H9–H15.

Viuda-Martos, M., Fernandez-Lopez, J., Sayas-Barbera, E., Sendra, E. and Perez-Alvarez, J.A. (2011) Physicochemical characterization of the orange juice waste water of a citrus by-product. *Journal of Food Processing and Preservation*, **35**: 264–271.

Wilson, J., Hayes, M. and Carney, B. (2011) Angiotensin-I-converting enzyme and prolyl endopeptidase inhibitory peptides from natural sources with a focus on marine processing by-products. *Food Chemistry*, **129**(2): 235–244.

Wismer-Pedersen, J. (1988) Use of haemoglobin in foods: A review. *Meat Science*, **24**: 31–45.

Young, S., Wong, M., Tabata, Y. and Mikos, A.G. (2005) Gelatin as a delivery vehicle for the controlled release of bioactive molecule. *Journal of Controlled Release*, **109**: 256–274.

Yu, P. L., van der Linden, D.S., Sugiarto, H. and Anderson, R.C. (2010) Antimicrobial peptides isolated from the blood of farm animals. *Animal Production Science*, **50**: 660–669.

Yu, Y., Hu, J., Miyaguchi, Y., Bai, X., Du, Y. and Lin, B. (2006) Isolation and characterization of angiotensin I-converting enzyme inhibitory peptides derived from porcine hemoglobin. *Peptides*, **27**: 2950–2956.

Zhao, Q., Garreau, I., Sannier, F. and Piot, J.M. (1997) Opioid peptides derived from hemoglobin: Hemorphins. *Biopolymers*, **43**: 75–98.

6
Dietary Fibers, Dietary Peptides and Dietary Essential Fatty Acids from Food Processing By-Products

Seema Medhe, Manisha Anand, and Anil Kumar Anal
Asian Institute of Technology, Klong Luang, Pathumthani, Thailand

6.1 Introduction

Food processing industries are a major sector of the economy worldwide. Agricultural crops and animal products are the main raw materials used in these industries. All food groups such as oil seeds, milk and milk products, meat and meat products, cereals and pulses, fruits and vegetables serve as raw materials for food processing. After production of the desired product, the processing industries are giving out huge amounts of by-products or waste.

After processing, essential compounds are left in the waste or by-products. These can be recovered by further reprocessing. However, by-products are product specific. Plant-based foods produce by-products like pomace, pills, bran, hulls, skin of fruit, cores, seeds, leaves, cake etc. Animal food processing results in by-products comprised of blood, gut waste, skin, milk, hops, hair etc. Some by-products, such as fruit skin, grain and bran are rich in dietary fiber. Oil seed processing by-products such as de-oiled cake are rich sources of protein, as the oil seeds are protein rich. This chapter deals with fiber recovery from food processing by-products. Oil seed, meat and seafood processing industries produce by-products such as de-oiled cakes, meals, fish bones, fish skin, blood, whey and skimmed milk, which are rich sources of dietary proteins and peptides. These by-products can be reprocessed to obtain dietary proteins, peptides and essential amino acids.

Food Processing By-Products and their Utilization, First Edition. Edited by Anil Kumar Anal.

6.2 Dietary Fiber from Food Processing By-Products

Fruits and vegetables are important sources of dietary fibers. One-third of fruits and vegetables are in the form of peel, pips and skin. During preparation and processing, they can be discarded, therefore creating a 'waste'. Wastes are full of dietary fiber, carotenoids, polyphenols, tocopherol etc. One of the main nutrients in waste is dietary fiber. Researchers are developing new techniques to use these functional compounds in value-added products from such wastes. In recent years, dietary fiber has received much positive attention with regards to its potential as a pharma food, due to its ability to reduce cholesterol, diabetes and coronary heart disease and to ease constipation (Telrandhe *et al.*, 2012).

In 1972, Trowell stated that: 'Dietary fiber derived from cellular walls of plant are poorly digested by humans but improves the human health.' The classification of dietary fiber is based on its solubility in a buffer at a defined pH, and/or its fermentation ability in an *in-vitro* system using an aqueous enzyme solution representative of the human alimentary canal. Thus, most appropriately dietary fiber is classified into two categories, soluble and insoluble dietary fibers. Cellulose, hemicellulose and lignin are insoluble in water, whereas gums, pectins and mucilages become softened in water. Table 6.1 represents dietary fiber classified on the basis of solubility and fermentation ability.

6.2.1 Structural Features of Dietary Fiber

The structure of fibers is in a complex form. Mostly, plant originated fibers are non-digestable carbohydrate, lignin and other substances, whereas fibers from animal origin are modified or synthetic non-digestable carbohydrate polymers.

Table 6.1 Classification of dietary fiber components based on water solubility

Dietary fibers	Sources	Features
Soluble fibers		
Gum	Oatmeal, haricot bean, legumes	Generally composed of monomers of hexose and pentose
Pectin	Whole grains, apple, legumes, cabbage, root vegetables	Mainly composed of galacturonic acid, rhamnose, arabinose, high content of galactose, intermediate laminate on the primary wall
Mucilages	Food additives	Compounds which are synthesized in plants, containing glycoprotein
Insoluble fibers		
Lignin	Is a component of cell walls and mainly consists of aromatic alcohols	Vegetables and flour
Cellulose	Is the main component of cell walls, consisting of glucose monomers	Whole grains, root vegetables, bran, peas, beans family, apples
Hemicellulose	Primary and secondary in cell walls	Bran, whole grains

In non-digestable carbohydrate, there are the following polysaccharides: cellulose, hemicellulose gums, mucilage, ß-glucan, pectin, inulin, resistant starch; oligosaccharides: fructo-oligosaccharides, oligofructose, polydextrose, galacto-oligosaccharides; and soybean oligosaccharides: stachyose and raffinose.

6.2.2 Technological Functionality of Dietary Fiber

6.2.2.1 Solubility Solubility is related to the structure of polysaccharides; they can be set regularly (insoluble) or irregularly (soluble) on the backbone or as side chains. The presence of a substitution group such as COOH or SO_{4}^{2-} increases solubility. Solubility is also affected by temperature and ionic strength. The soluble and insoluble nature of dietary fiber involves differences in their technological functionality and physiological effects. Soluble fibers are characterised by their capacity to increase viscosity, and to reduce the glycemic response and plasma cholesterol. Insoluble fibers are characterized by their porosity, their low density and by their ability to increase faecal bulk and decrease intestinal transit.

6.2.2.2 Hydration Properties and Oil-Binding Capacity Water-holding capacity (WHC) is defined as the amount of water that is retained by 1 g of dry fiber under specified conditions of temperature, soaking time, and duration and speed of centrifugation (Table 6.2).

6.2.2.3 Viscosity Most polysaccharide solutions exhibit non-Newtonian flow and an increased shear rate can increase or decrease viscosity (Sanderson, 1981). Water soluble fibers are the major component that will increase the viscosity of a

Table 6.2 Water-holding capacity (WHC) and oil holding capacity (OHC) of some fruit and cereal processing by-products

Fruits and other by-products	WHC (g water/g)	OHC (g oil/g)	References
Date dietary fiber concentrate	15.6	9.75	Elleuch *et al.* (2008)
Orange dietary fiber concentrate	7.3	1.27	Grigelmo-Miguel and Martina-Belloso (1999a)
Peach dietary fiber concentrate	12.1	1.09	Grigelmo-Miguel and Martina-Belloso (1999a)
Lime peel	6.96–12.84		Abdul-Hamid and Luan (2000)
Defatted rice bran	4.89	4.54	Sangnark and Noomhorm (2003)
Sugarcane bagasse (>0.3 mm)	7.5	11.3	Vergara-Valencia *et al.* (2007)
Mango dietary fibers concentrate	11	1	Eim *et al.* (2008)
Carrot dietary fiber	18.6	5.5	Fuentes-Alventosa *et al.* (2009)
Asparagus by-product	11.4–20.3	5.28–8.53	

solution, showing that defatted rice bran has low viscosity (~1.25 cps at 7% fiber in water), because it contains only 9% soluble fiber. Viscosity increases with an increased fiber concentration but decreases with the temperature of a solution of dietary fiber. For example, date dietary fiber suspensions (between 20 and 50 g/l) exhibit pseudoplastic behaviour ($n < 1$). This index passes from 0.326 to 0.13 when the concentration of dietary fiber suspension increases from 20 to 50 g/l.

6.2.2.4 Antioxidant properties Antioxidant properties suggest possibilities for the use of fiber with high antioxidant activities as ingredients that allow the stabilization of fatty foodstuffs, thereby improving their oxidative stability and prolonging their shelf life. These high-fiber products have numerous technological properties (water holding capacity, water swelling capacity, water-solubility, fat binding capacity, viscosity and antioxidant properties), that warrant their utilization as ingredients by the food industry.

6.2.3 Health Benefits of Dietary Fibers

Since the mid-1970s, interest in the role of dietary fibers in health and nutrition has prompted a wide range of research and received considerable public attention. Fiber reduces cholesterol, diabetes, coronary heart disease and also prevents obesity. Food full of fiber reduces the chances of cancer, because it absorbs carcinogenic agents. So it is recommended to consume plant fibers in diet, which have the ability to link hydrophobic carcinogenic agents. Food rich in fiber improves the digestion and slows the absorption of glucose in the intestinal tract, which reduces diabetes. In addition, using too much fiber can prevent adsorption of the minerals iron, zinc, calcium and magnesium by the body. Using high fiber in the short term may cause gas, bloating and abdominal cramps. Therefore, gradually increasing the use of fiber in nutrition will protect from these side effects. The World Health Organization has recommended the total fiber intake as 25 g/day (WHO/FAO, 2003; Food and Nutrition Board, Institute of Medicine, 2001). However, actual fiber is 14 to 29 g/day. Recommended intake for children and different age groups are given below in Table 6.3.

Table 6.3 Dietary fiber reference intake values for total fiber by age group

Age group	Adequate intake (g/day)	
	Male	Female
1–3 years	19	19
4–8 years	25	25
9–13 years	31	26
14–18 years	38	26
19–30 years	38	25
31–50 years	38	25
51–70 years	30	21
Over 70 years	30	21
Pregnancy		29
Lactation		29

6.2.4 Dietary Fiber from Fruits and Vegetables

Fruit processing industry wastes are rich sources of dietary fiber and also bioactive compounds are present in these by-products. The concept of antioxidant dietary fiber (ADF) was first proposed by Saura-Calixto in 1988 with some criteria, such that 1 g of ADF should have a DPPH free radical scavenging capacity equivalent to at least 50 mg vitamin E and dietary fiber content higher than 50% dry matter from the natural constituents of the material. In wine industry waste, seeds and skin are rich in dietary fiber and phenolic compounds, and the definition of ADF was proved when these compounds were stored under vacuum for 16 weeks at 15°C (Tseng and Zhao, 2013).

Dietary fibers from grapes show higher reducing efficacy in lipid profile and blood pressure than oat fiber or psyllium, due to the combined effects of dietary fiber and antioxidants. ADF and wine grape pomace (WGP) have pharmaceuticals value, reducing not only the low-density lipoprotein oxidation *in vitro* but also helping to enhance the gastrointestinal health of the host by promoting a beneficial microbiota profile. ADF and WGP are also used as antioxidant and antimicrobial agents to extent the shelf life of food products. For example, ADF has strong possibilities to be incorporated with flour, making high dietary fiber bakery goods, while the polyphenols in ADF could act as antioxidants for improving color, taste and aroma of the product. There are many examples of ADF applications outlined by Tseng and Zhao (2013). Mango peel powders used to enhance the antioxidant properties in macaroni (Ajila *et al.*, 2007); grape pomace used with sour dough for rye bread (Mildner-Skudlar *et al.*, 2011, as cited by Tseng and Zhao, 2013); grape seed flour used for cereal bars, noodles and pancakes (Rosales Solo *et al.*, 2012, as cited by Tseng and Zhao, 2013); and apple pomace incorporated into wheat flour to improve the rheological characteristics of cake due to dietary fiber (Sudha *et al.*, 2007, as cited by Tseng and Zhao, 2013).

WGP was mixed with minced fish and chicken breast to delay lipid oxidation. WGP was incorporated into beef patties and exhibited an antimicrobial effect against foodborne pathogens (Sagdic, 2011, as cited by Tseng and Zhao, 2013). Moreover, research has indicated that WGP seed extracts show better antioxidant activities than those of synthetic antioxidants of butylated hydroxyanisole (BHA) and butylated hydroxytoluene (BHT) (Baydar *et al.*, 2007, as cited by Tseng and Zhao, 2013).

Orange, apple, peach and olive are mainly use for the extraction of their juices. They all contain a by-product from which can be recovered different high-added value compounds; among those, it is remarkable that the fiber fraction has a great potential in the preparation of functional foods. Also there are several vegetables, such as pepper, artichoke, onion and asparagus, which during their processing produce waste, for example, in the case of asparagus spears, this waste can represent up to 40–50% of their fresh weight. They also contain both types of fiber compounds, soluble and insoluble fibers, that can be used to design new 'functional foods' (Rodríguez *et al.*,1999).

Cauliflower has a high waste index and is a high source of protein (16.1%), cellulose (16%) and hemicellulose (8%). It is considered a rich source of dietary fibre among other vegetables and also has both antioxidant and anti-carcinogenic properties. Phenolic compounds and vitamin C are the major antioxidants of this family (brassica) vegetable (Podsedek, 2007). The purpose of this study is to incorporate cauliflower by-products as a source of dietary fiber into ready-to-eat snacks. Dried and milled cauliflower at levels of 5–20% was added to the formulation mix. The results came from the analysis of the extrudates and are discussed in terms of the increase in nutritional and textural characteristics, and the effects of the processing conditions.

Citrus fibers are considered to be of higher quality than those from cereals, due to a better balance of soluble and insoluble dietary fibers and they also have a higher value of water-holding and oil-holding capacities. One more advantage of citrus dietary fiber is their content of associated bioactive compounds (flavonoids and vitamin C) with antioxidant properties, which may exert higher health promoting effects than the dietary fiber itself. Among all citrus fruits, lemons possess the highest antioxidant potential. Dietary fibers are not good from a nutritional aspect, but are useful for their functional and technological properties.

Carrot residue contains a large amount of fiber, with cellulose making up the largest portion (51.6%), followed by lignin (32.2%), hemicellulose (12.3%) and pectin (3.88%) (Nawirska and Kwasniewska, 2005), and hence most studies are focused on fiber recovery and application. The yield of alcohol-insoluble dietary fiber was raised from 3.3% to 41.7% after subjecting dried carrot pomace to 1% NaOH and 2% acetic acid treatment, followed by the addition of β-galactosidase, endoglucanase, exoglucanase and β-glucosidase (Yoon *et al.*, 2005). The insoluble fiber-rich fraction prepared from carrot pomace was found to be effective in lowering hamster serum triglyceride, serum total cholesterol and liver cholesterol, and meanwhile it resulted in a significantly higher HDL, and a total cholesterol ratio as well as higher levels of fecal lipids, cholesterol and bile acids (Hsu *et al.*, 2006). This outcome demonstrated the great potential of carrot insoluble fiber as being protective against coronary heart disease. Carrot fiber increases the water-binding capacity and enhances the porosity in rye and white bread.

Sweet potato peel is reported to contain around 79% carbohydrate, 12% fat, 7% fiber 6% ash and 3.06% protein. After starch isolation, DF was extracted from sweet potato residues by a sieving method (Mei *et al.*, 2010). The approximate composition of DF in sweet potato residues was investigated, such as chemical composition, monosaccharide composition and physicochemical properties. The average yield and DF content were 9.97% and 75.19%, respectively. Average contents of cellulose, lignin, pectin and hemicellulose were 31.19, 16.85, 15.65 and 11.38 g/100 g of dry matter in DF products, respectively. Swelling, water-holding, oil-holding and glucose absorption capacities were in the ranges of 8.11–12.56 mL/g, 3.54–6.15 g/g, 1.43–2.48 g/g and 0.54–1.27 mmol/g, respectively.

6.2.5 Dietary Fiber from Legumes

The functional compounds in three legume by-products, okara from soybean (*Glycine max* L.), pea pod (*Pisum sativum* L.) and broad bean pod (*Vicia faba* L.), were investigated. These by-products are rich in dietary fiber, because it is their major constituent, i.e. more than 50% for pea pod and okara and 40% for broad bean pod (Mateos-Aparico *et al.*, 2010). Okara is the main by-product from soymilk and tofu preparation. Moreover, it contains a high quantity of vegetable protein, more than 30% present in okara and is also rich in fats such as linoleic and oleic acids. Due to these characteristics, these by-products can be used for food fortification. Both pod by-products are rich in vegetable protein and also contain a remarkable concentration of potassium, while in pea pod, iron is higher than potassium. In addition, broad bean pod and okara contain α-glucosidase. The presence of nutritional compounds, such as inulin in okara, provide an important added value. Soluble dietary fiber of these by-products plays an important role in the reduction of cholesterol levels in some hyperlipidemic individuals, and it can also be used to improve glucose tolerance in diabetics. On the other hand, insoluble dietary fiber has properties to increase faecal bulk and reduce the gastrointestinal transit time. Moreover, it has a positive effect on diarrhoea and constipation and is used as a treatment for irritable bowel syndrome (Bosaeus, 2004).

6.2.6 Dietary Fiber from Cereals

Cereals have high amounts of dietary fiber and comprise of various layers. Bran is composed of various tissues, such as aleurone and pericarp, and accounts for most of the micronutrient, phytochemical and fiber content of the grain. Bran is particularly rich in dietary fiber and, for rice bran, omega fatty acids, and contains significant quantities of starch, protein, vitamins and dietary minerals. At the industrial level, bran, and in particular wheat bran, is currently used for the production of low-value products, like composts and livestock feed, and significant amounts are disposed of in landfills. Dry fractionation technologies allow bran to be separated and recovered to efficiently separate high- and low-value components to develop nutritionally-enhanced ingredients and products. Wheat bran has started to be used to enrich the fiber content of breads and breakfast cereals, especially for the benefit of those wishing to increase their intake of dietary fiber, such as for alleviating constipation. The dietary fiber content of cereal bran is usually based on its pentosan content, mainly as arabinoxylans and xylans (Table 6.4). The presence of phenolic acids, in particular ferulic acid, esterified to arabinoxylans gives this fraction the potential to be exploited as a source of polymers and oligosaccharides with anti-oxidant properties for food use.

Wheat by-products have unique functional and nutritional characteristics related to color and cooking performance and to their content of DF (Dexter *et al.*, 1994a b). DF contains cellulose and lignin, hemicellulose, pectin, gums and other polysaccharides and oligosaccharide associated with plants. It is normally defined as: 'edible parts of plants or analogous carbohydrates that are resistant to digestion and absorption in the human small intestine with complete or partial fermentation in the large intestine' (Mongeau, 2003). Until now, the use of wheat milling by-products are in 'all-bran' breakfast extruded products (Eastman *et al.*, 2001). These products mainly contain IDF (the average amount of SDF is 1.5%), while many recent nutritional studies state that the average dietary intake of SDF is far below the optimum (Cui *et al.*, 1999).

Table 6.4 Technological and physiological properties of dietary fiber products

Technological property	Physiological functionality
Water holding capacity	Laxative
Water swelling capacity	Reduction of blood cholesterol
Water retention capacity	Reduction of blood glucose
Water solubility Oil holding capacity	Reduction the risk of chronic disorder, e.g. coronary heart disease, diabetes, obesity and some forms of cancer
Viscosity	
Texturizing	
Stabilizing	
Gel-forming capacity	
Antioxidant capacity	

The milling process can be modulated to obtain cereal by-products enriched in fiber and also decrease the ratio between insoluble and soluble fibers (Dexter and Wood, 1996). Recently developed technological advances in the processing of durum wheat allows separating different fractions indicated as durum wheat by-products. Some of these products mainly containing IDF, which can be used to hold free water, while others containing a large amount of SDF have a great nutritional value (Dexter and Wood, 1996). Approaches other than milling modulation were attempted to obtain high SDF-enriched fractions from cereals.

Oat processing by-products contain bran or oat fiber, which appears to be a suitable fat replacement in ground beef and pork sausage products, due to its ability to retain water and emulate particle definition in ground meat in terms of both color and texture (Verma and Banerjee, 2010). Researchers investigated how to develop low salt, low fat and high fiber functional chicken nuggets. Verma *et al.* (2009) incorporated various fiber sources like pea hull flour, gram hull flour, apple pulp and bottle gourd, in different combinations at the 10% level.

6.2.7 Coffee, Tea and Cocoa

Food companies are responding to the demand for foods with enhanced safety and lower energy density, by allocating more resources to the development of dietary fiber-enriched products, since with special fiber blends, physiological as well as functional properties can be met. In this context, the chocolate industry's major waste product is cocoa husks, which have the potential to serve as a valuable source of dietary fiber for the low-calorie food segment Cocoa husks appear to be a good source of DF that might be used as a supplement for other sources of fiber or food products. Both cocoa husks and cocoa bean shell contain a ready source of dietary fiber, which is inexpensive and contains approximately 40% (DF) of the total dietary fiber. Mostly these fibers are insoluble as comprising glucose and uronic acid with smaller amounts of galactose, arabinose, xylose and mannose. Cocoa husk contains all the essential amino acids, which represent 44.57% (w/w) of the total identified amino acids. The total protein content and the amino acid score value would allow cocoa husk to be used as a source of dietary fiber (Collar *et al.*, 2009). This cocoa fiber have antioxidant properties and physicochemical properties, which make it a suitable product for use in the preparation of low-calorie, high-fiber foods such as chocolate cookies, chocolate cakes, diebetic chocolate supplements, and so on, where the color and flavor of this cocoa fiber might be advantageous for these food products.

It was also reported that the extract of cocoa contained mostly insoluble dietary fiber, which accounted for 80% of the total dietary fiber and half of the total dry weight. Ten percent of the total dietary fiber was made up of soluble fiber. It was observed that hydration properties (swelling and water-holding capacity) and the glucose retardation index of cocoa fiber were similar to other natural commercial insoluble fibers (Lecumberri *et al.*, 2006). Moreover, DF also retains important amounts of polyphenolic compounds with antioxidant activity. Furthermore, the dietary fiber-rich cocoa shell has already been considered as an additive for non-baking human food applications. The dietary fibers derived from cocoa shell at different pH, adding up to 6%, can be technologically feasible, acceptable to the senses, and increase the shelf life, which give storable innovative fiber-enriched wheat bread formulations (Collar *et al.*, 2009).

6.2.8 Spices

Cumin contains total dietary fiber (TDF) of 59.0%, mostly insoluble dietary fiber (IDF) of 48.5%, and soluble dietary fiber (SDF) of 10.5%, while the spent residue from cumin processing (after oil and oleoresin extraction) was found to contain 62.1% TDF, 51.7% IDF and 10.4% SDF. The spent residue also has 7.7% starch and 5% bound fat. Thus, the residue from cumin does not have much commercial value at present, but can be a rich source of useful dietary fiber and can be used in food applications. The way of utilizing industrial waste can be effective from the point of view of environmental pollution from the residues of spice processing industries.

6.2.9 Utilization of Dietary Fiber in Different Food Industries

1. There are many reports about dietary fibers used in food products, such as baked goods, beverages, confectionery, dairy, frozen dairy, meat, pasta and soups. Most dietary fibers used in baking products increase shelf life, freshness, capacity to retain water, and softness of the dough. It increases the nutritional level. In addition, introduction of dietary fiber into meat products improves cooking yield, fat binding, texture and water binding (Cofrades *et al.*, 2000).
2. Citrus dietary fiber added to a food product performs in a satisfactory manner as a food ingredient (Figuerola *et al.*, 2005). From a functionality perspective, citrus fiber has many functions:
   i. it can be used to improve texture;
   ii. as a bulking agent in reduced-sugar applications;
   iii. it reduces the use of fat by managing moisture;
   iv. it gives color; and
   v. it works as a natural antioxidant. It is also used in marmalade and jam. The most common fibers contain pectins with different degrees of esterification.
3. Dietary fibers can be added in fish products. These fibers are from algae such as carrageenan, guar or xanthan from seeds. It improves the water-binding activity, thickening, and emulsion and gelling properties. Mostly, these fibers are incorporated into minced fish muscle or raw muscle, and it also causes rigidity and elasticity in muscle protein. Insoluble fiber, such as wheat fiber, is mixed into fish muscle gels to protect surimi from loss of gel strength and gives hardness during freezing, but the adverse effect is that it may reduce the gel property and cohesiveness throughout thermal gelling. Grape pomace, which has more antioxidant properties, is added to minced fished, which increases the shelf life, improves the flavor and protects from lipid oxidation (Sánchez-Alonso *et al.*, 2007).
4. Addition of fibers into meat products like sausage and burgers does not affect their organoleptic properties. Citrus fibers with bioactive compounds added to meat products inhibit lipid oxidation and also reduce nitrite levels. Grape pomace added to chicken hamburgers improves oxidative stability and increases shelf life.
5. Dietary fibers are commonly added into dairy products, mostly inulin incorporated into dairy products, which improves body and mouth feel in cheese products and ice cream. Fiber improves the smoothness and texture, gives resistance to melting, improves the handling properties as temperatures fluctuate during storage, and maintains crystallization or re-crystallization controllers in frozen

dairy products. Fibers reduce synaeresis in yoghurt and other fermented milk products. Nowadays, the use of orange fiber as a fat replacer in ice cream led to a 70% reduction of fat, without causing any significant changes in the product's attributes such as color, odor and texture. Addition of orange fiber proved to be a promising alternative as a fat replacer in ice cream production.

6.3 Dietary Proteins and Peptides from Food Processing By-Products

Chains of amino acids linked by peptide bonds are known as proteins. Normally up to 100 amino acids are linked together to form protein structure. As proteins are serving as a fuel source for bodily functions and are the main constituents of bodily tissue, they can be termed as building blocks of body tissue. Similar to carbohydrates, 1 g of protein consumption provides 4 kcal energy.

Body tissues are mainly composed of proteins. Enzymes, antibodies and hormones are important components of human biological processes, which are also proteins. Proteins are composed of typically much longer chains of (>100) amino acids similarly linked by peptide bonds. They are responsible for various biochemical reactions within cells. There are many differences between proteins and peptides, but they are simply defined by long chains or short chains. During the digestion process in the human stomach, hydrochloric acid and protease play key roles in degradation of protein into polypeptides. Essential amino acids and non-essential amino acids are two categories of amino acids. Essential amino acids must be obtained from food or diet, but non-essential amino acids can be produced by the body with no need to consume food. Deficiency of essential amino acids may lead to protein energy malnutrition. Tryptophan, leucine, lysine, histidine, valine, threonine, phenylalanine and methionine are the essential amino acids, while alanine, glutamic acid, asparagine, serine and aspartic acids are non-essential amino acids. To treat cancer, Huntington's, Parkinson's and Alzheimer's diseases, around 800 drugs are being developed and more than 40 approved peptide-based drugs are in use today. The high specificity and potency nature with a combination of low toxic incidence of antibody drugs under peptide and protein therapeutics are extremely attractive. Furthermore, the scope for protein-based drugs is also market expandable. Therefore, with mainstreaming of peptide and protein drugs, the need for developments in manufacturing inclusive of synthetic, transgenic and recombinant methods are essential.

6.3.1 Oil Seed Processing By-Products Valorization to Produce Proteins

Among basic food groups, nuts and oil seeds are one important group consisting of oil seeds providing edible oil like cottonseed, sesame, peanut, soybean, rapeseed, safflower and sunflower, as well as non-edible oil seeds such as castor beans. Oil seeds are a good source of protein carbohydrates and fat.

6.3.1.1 Peanut Cake The peanut is a very popular edible oil seed containing about 48% oil and rich in protein (25–30%), vitamins (E, K, and B_1) and minerals (Ca, Mg, P, and K) (Grosso and Guzmán, 1995). Peanut cake and peanut haulm are by-products of peanut processing. Peanut cake is obtained after extracting the oil and it has been used for human food and also incorporated into animal feeds (Savage and Keenan, 1994). As groundnut cake has a good amount of protein, it is being utilized in the production of edible protein-enriched flour containing low fat. This is further used in baby foods, nutritious flour and so on. Groundnut meal is found to have 47% crude protein, followed by 12% crude fiber (max.), 10% moisture (max.), 2.5% acid insoluble ash (max.) and 1.5% crude fat (max.).

The protein extraction process from residual oil-cake is affected by time and number of extraction stages, the extraction and precipitation pH, concentration of NaCl and extraction temperature (Yu *et al.*, 2007; Barbosa *et al.*, 2006). The minimum water/flour ratio (10:1) is recommended to obtain protein isolate from peanut flour. At pH 10, optimum protein recovery from peanut meal is possible (Yu *et al.*, 2007). Wu *et al.* (2009) found that 4.5 was the isoelectric pH for precipitation of peanut protein for preparing protein concentrates and 40°C was the best extraction temperature for preparing peanut protein isolates from peanut oil-cake flour.

6.3.1.2 Sunflower Cake After oil extraction, the meal left behind from sunflower seeds has high value as a nutritional by-product in the livestock feed industry. Sunflower meal is reported as having less calories and high fiber. The sunflower oil cake (SuOC) is a good source of highly digestible proteins, as well as containing a number of essential amino acids (González-Pérez and Vereijken, 2008). Limited use of SuOC is caused by presence of higher amounts of phenolic compounds like chlorogenic acid and caffeic acid, which are responsible for lowering the solubility of protein and undesired sensory qualities (Pringent *et al.*, 2003; González-Pérez and Vereijken, 2008). SuOC is giving out protein isolates and soluble protein concentrates with different concentrations of phenolic compounds. Antioxidant activity and disease preventing actions of SuOC with phenolic trace were observed by Raskin *et al.* (2002). Further research supported the value of SuOC as a source of highly water-soluble protein, with good antioxidant properties as well as essential physiochemical mechanisms (Salgado *et al.*, 2011).

6.3.1.3 Safflower Cake De-oiled safflower cake is a good source of protein. It is documented that solvent-extracted cake has 64.2% protein followed by de-oiled cake gives 56.3%. To separate protein from de-oiled safflower cake solvent is extracted and powdered to make flour (size 75 μm). From this, wet protein isolate can be prepared by dissolving safflower protein in water at pH 9.0. Then the slurry should be acidified to obtain pH 4.0 to precipitate proteins, which are later filtered. High fiber content and strong bitter flavor limits the edible use of safflower cake. Ethanol extraction of de-oiled cake or meal is helping to minimize the bitterness of cake and protein isolate recovered.

6.3.1.4 Soyabean Processing Waste Soybean (*Glycine max* (L.) Merr) is protein-rich oilseed with on average 40% protein, 35% carbohydrate, 20% oil and 5% ash present on a dry weight basis. It has been reported that soybean is an inexpensive source of protein, and has been effective in improving protein-energy malnutrition in many developing countries. Soy protein contains some natural phytochemicals like isoflavones, which have antioxidant properties (Conforti and Davis, 2006). Intake of products high in soy protein can lower total serum cholesterol, decrease risk of several cancers and coronary heart disease and inhibit bone resorption (Zind, 1998; Messina, 2001; Gallagher *et al.*, 2000). Soy flour or flakes is a major by-product remaining after oil extraction from soya beans.

As soya bean is a rich source of protein, the de-oiled soy flour or meal is being utilized to obtain protein. Soy protein concentrate and soy protein isolate are the major valued products recovered from de-oiled soya flakes. Soy-protein isolate is prepared by extracting the proteins with an aqueous medium, which may vary in pH from near neutrality to an alkaline pH. Centrifugation is the next step to separate the aqueous extract and fibrous residue. The pH of the clarified extract is then adjusted to about 4.5 with food grade acid to precipitate the proteins. Then the volume of precipitate should be reduced and washed with water and the concentrated protein can then be dried.

6.3.1.5 Industrial Enzymes Production by using Different Oil Cakes as Substrate

Oil cakes are documented as a potential substrate for solid state fermentation processing in the production of different industrial enzymes such as inulinase, proteases, amylases, glucoamylase, phytase, lipases and tannase (Sumitra *et al.*, 2007).

6.3.1.5.1 Lipase A wide range of fungal strains was documented as using different oil cakes to produce enzyme lipase. Ramakrishnan and Banerjee (1952) worked on the isolation of *Penicillium chrysogenum* S1 strain from sesame oil cake, *Penicillium simplicissimum* using SBC as a substrate in SSF (Di Luccio *et al.*, 2004), *Candida rugose* using coconut oil cake (Benjamin and Pandey, 1996), *Aspergillus niger* using a gingelly oil cake (Kamini *et al.*, 1998), *Rhizomucor pusillus*, and *Rhizopus rhizopodiformis* using OOC (Cordova *et al.*, 1998). Gingelly oil cake is recorded as a good carbon source for production of lipase enzyme by *Rhizopus* sp. BTS-24 under conditions such as agitation speed of 100 rpm, incubation at 28°C and an initial pH of 5.0 (Bapiraju *et al.*, 2004). A cotton seed cake as substrate produces the highest amount of extracellular lipase (400 U/mL) in 50 h by *Pseudomonas* strain X.

6.3.1.5.2 Phytase Solid state fermentation of different oil cakes, such as COC, SOC, PKC, groundnut oil cake, CSC and OOC as substrate by *Rhizopus oligosporus NRRL 5905*, *Rhizopus oryzae* NRRL 1891 and *R. oryzae* NRRL 3562 to produce phytase enzyme, is well-documented. COC and SOC are proved to be effective in increasing the phytase production by two times when used together in mixed substrate fermentation (Ramachandran *et al.*, 2005). When COC is used as a substrate by *R. oligosporus*, it produces a maximum amount of phytase (14.29 IU/g of dry substrate) (Sabu *et al.*, 2002). *Mucor racemosus* was found to produce phytase by using COC and CaOC as substrate. *M. racemosus* NRRL 1994 was also reported to produce phytase enzyme using SOC and GOC (Roopesh *et al.*, 2006). *Aspergillus ficuum* produced phytase using CaOC in an SSF process (Ebune *et al.*, 1995).

6.3.1.5.3 Protease Proteases are extracellular enzyme, produced naturally by microorganisms. Protease plays an important role in fragmenting complex polypeptides into peptides and amino acids, making them easily absorbable. It has great commercial importance as an industrial enzyme. It has great demand in food industries for cheese-making, brewing, baking, meat processing and some nonfood industries like pharmaceutical, protein hydrolysis, detergent and leather industries. It has also been included in human and animal feeds as digestive aids (Synowiecki, 2010;

Seiffzadeh *et al.*, 2008; Dias *et al.*, 2008). Soybean oil cake was documented as a fair source of carbon and nitrogen and solid matrix for SSF by *Penicillium* sp. to produce protease (Germano *et.al.*, 2003). Previously, Joo *et al.* (2002) documented the production of extracellular alkaline protease enzyme by *Bacillus horikoshii* by using soybean oil cake as a substrate.

6.3.1.5.4 Other Enzymes Ramachandran *et al.* (2004a) studied the production of amylase by *Aspergillus oryzae* using two different oil cakes as substrate and recorded that COC supports the maximum amylase production. COC was also used for the production of glucoamylase enzyme by *A. niger* NCIM 1245 and under optimized conditions maximal enzyme production (194 IU/g dry fermented substrate) was recorded after 96 h with initial substrate pH 4.5–4.7 and 65% moisture content. CaOC was used as a substrate in solid state fermentation by *Trichoderma reesi* to produce xylanase. Selvakumar and Pandey (1999) documented the use COC in SSF by *Kluyveromyces marxianus* and *Staphylococcus* sp. RRL-1 in production of inulinase enzyme. Increased concentration of SuCO in the medium increases the enzyme production (Federici *et al.*, 1988) Wheat bran and SOC have been used as a substrate by the salinity-tolerant yeast *Zygosaccharomyces rouxii* NRRL-Y 2547 to produce glutaminase (Kashyap *et al.*, 2002). It has been reported that enzyme tannase can be produced by *A. niger* ATCC 16520 using PKC in SSF (Sabu *et al.*, 2005, 2006). Defatted copra is a good source of nitrogen and has been used to produce extracellular mannanase by shaken flask cultures of *A. niger* NCH-189. Presence of oil in copra reduces mannanase production by fungus (Lin and Chen, 2004).

6.3.2 Proteins from Dairy Waste

6.3.2.1 Skim Milk Skim milk is a by-product formed during the manufacture of cream and has a high nutritional value. but it is fat-free solid material. This non-fat solid material is being used to obtain value products like milk protein hydrolysate and casein. Skim milk obtained from cows immunized against a polyvalent human gut bacterial vaccine has been proved as having a triglyceride-lowering effect and is hypocholestrolmic as well. This effect is due to increased immunoglobin G (IgG), which has an effect on bacterial antigens (Stolle and Beck, 1987). Skim milk also serves as a key ingredient in infant formula milk. This formula milk fortified with skim milk has shown growth promoting effects. This is because of the high content of saturated fatty acid, which interferes with fat absorption (Samsudin and Williams, 1967).

6.3.2.2 Whey Protein The processing of milk products, such as in shrikhand, cheese, caseins, channa and paneer manufacturing, produce huge amounts of a by-product called whey. Whey protein has a great market demand as a potential dietary protein supplement. It is documented as helpful in preventing heart disease and osteoporosis. It has some beneficial properties that make it even more valuable, such as antimicrobial, immune modulation, antioxidant, antihypertensive, antitumoral, hypolipidemic, antiviral, antibacterial and as a chelating agent (Marshall, 2004). Different industrial processing technologies can be adopted to convert whey into varied whey products. These processes are microfiltration, reverse osmosis, ultra-filtration and ion-exchange. Every finished whey product varies in the amount of nutrients present, such as protein, carbohydrates, immunoglobulins, lactose, mineral and fat. Whey protein concentrates (WPC) contain between 50% and 85% protein on a dry weight basis, whey protein isolate (WPI) contain between 90% and 98% protein and very small amounts of lactose and fat, reduced lactose whey, demineralized whey and hydrolyzed whey (Huffman, 1996).

High solubility, water absorption, gelatinization and emulsifying capacities are some functional properties that make whey protein applicable in food industries (Gunasekaran *et al.*, 2007). The high nutritional value of whey protein is due to the presence of a large amount of essential amino acids. Whey protein mainly consists of ß-lactoglobulin (ß-LG) 50%, a-lactalbumin (aLA) 20%, bovine serum albumin (BSA) 10%, and immunoglobulin (IG) 10%. Besides these, whey also contains numerous minor proteins, called low-abundance proteins, such as lactoferrin (LF), lactoperoxidase (LP), proteose peptone (PP) and osteopontin (OPN), lisozyme (LZ), among others; LF and LP are the most abundant minor proteins (Santos *et al.*, 2012; Jovanovic *et al.*, 2007). The whey produced during cheese and paneer manufacturing is categorized as acid whey and rennet whey. Whey protein is a mixture of proteins that have a biological value of 100, which is higher than the value for casein, soy protein, beef or wheat gluten. Mold, yeast, actinomycetes, fungi, algae and bacteria are grown in large-scale culture for use as a protein source in human food or animal feed. Dried cells of these microorganisms are termed as single cell proteins. Whey is served as the main component of substrate for the production of single cell protein with a suitable microorganism. Whey lactose is the main source of carbon for biomass production (Ghaly and Kamal, 2004). Whey proteins, such as ß-lactoglobulin, serum albumin and lactoferrin, contain high levels of cysteine and glutamylcysteine, which are precursors for glutathione synthesis.

6.3.3 Proteins from Sugar Industry Waste

The complex unicellular organism called yeast is a rich source of protein (65–70%). Food processing by-products rich in carbohydrates and hydrocarbons can be used as a substrate for the production of single cell protein. Sugarcane bagasse has been documented to be of good commercial value in yeast production. Molasses, a sugar industry waste, contains a good range of minerals, organic compounds, vitamins and, most importantly, a good amount of sugar. It can act as a fair substrate for baker's yeast production. Approximately 1 kg of active baker's yeast can be obtained by using 4 kg of molasses. In the fermentation process of yeast production, fine aeration is required. *Saccharomyces cerevisiae* (baker's yeast) and *Torula utilis* (feed yeast) are mainly utilized for single cell protein production. Baker's yeast was the first microorganism to be produced in aerobic stirred fermentation using molasses and even today it is still produced (White, 1954; Chen and Chinger, 1985).

6.3.4 Proteins from Marine Waste

Onshore and aquaculture fish processing generates significant quantities of waste. It has been reported that up to 25% of fish comprise waste. This waste can be categorized as trimmings, viscera, fins, bones, head, skin, under-sized fish and shellfish, damaged shellfish and shells. Good quantity and quality proteins are present in this waste. Waste obtained from fish processing is a rich source of protein, which can be hydrolyzed by mild processing techniques to obtain biologically active peptides (Kim and Mendis, 2006). Studies have shown that 10–20% (w/w) of total fish protein can be found in waste components (Kristinsson, 2008). In addition to trimmings, frames and heads contain residual meat and thus are a good source of muscle proteins. Frames, fins and skin are an excellent source of collagen and gelatin (Venugopal, 2009). Fish waste-derived peptides are having positive health benefits. Angiotensin-I-converting enzyme (ACE) peptides are found to be effective in the treatment of hypertension (Kim and Wijesekara. 2010). It has been reported that protein hydrolysates obtained from marine waste produce peptides that show antioxidative, antimicrobial, antiproliferative and antithrombotic functions (Himaya *et al.*, 2012; Ngo *et al.*, 2011). Such bioactive peptides can be utilized as neutraceuticals and further use in the formulation of functional foods, and could hence reduce the population's need for drugs and other medical therapies. Enzymatic hydrolysis converts the microalgae, fish and shellfish processing waste proteins into bioactive protein hydrolysates and peptides. An inherent proteolytic enzyme, such as calcium-activated calpains and lysosomal cathepsins, acts upon muscle proteins of fish and shellfish during regular postmortem storage to produce peptides (Bauchart *et al.*, 2007).

For the generation of bioactive peptides from fish muscle (marine waste), the enzymatic hydrolysis method is suitable. Proteolytic enzymes should be utilized or sometimes a combination of exopeptidases and endopeptidases is chosen. It is very important to select a proper enzyme and provide suitable conditions for the chosen enzyme (Pihlanto and Korhonen, 2003). Gastrointestinal enzyme acts upon skeletal muscle protein during digestion and proteolytic enzymes during the postmortem period. Bioactive peptides are liberated by enzymatic hydrolysis or by fermentation of sarcoplasmic and myofibrillar proteins. Bioactive peptides can be obtained from fish waste by treatment of fish muscle protein by enzymatic treatment, aging and gastrointestinal proteolysis (Escudero *et al.*, 2010). Diagrammatic presentation of the process of obtaining bioactive peptides from marine waste is given in Figure 6.1.

6.3.5 Antimicrobial Peptides from Marine By-Products

Fish and fish processing waste are being considered as novel sources of antimicrobial peptides. These peptides have a range of vital biotechnological applications in operations such as drug delivery and antimicrobial treatment. Fish processing by-products, such as fish skin and mucus, have been newly determined as potential sources of antimicrobial peptides (Alvarez-Pellitero, 2008). Penaeidins are antimicrobial peptides originally isolated from shrimps and prawns (Destoumieux *et al.*, 2000). A penaeidin homologue was isolated from spider crab and named hyastatin (Sperstad *et al.*, 2009). Pacific white shrimp (*Litopenaeus vannamei*), Scallop (*Chlamys farreri*) and Abalone (*Halitotis discus discus*) are found to be a source of Histone H2A antimicrobial peptides (Patat *et al.*, 2004; Li and Zhao, 2007; De Zoysa *et al.*, 2009). The skin mucus of yellow catfish has been discovered to be a source of a novel linear antimicrobial peptide called pelteobagrin (Su, 2011).

6.3.6 Peptides from Meat and Meat Processing Waste

In the world's food industry era, meat and meat processing industries contribute widely. Meat has a great importance as a human food. Modernized eating and varied consumer demands have given rise to the meat processing industry. The growing meat

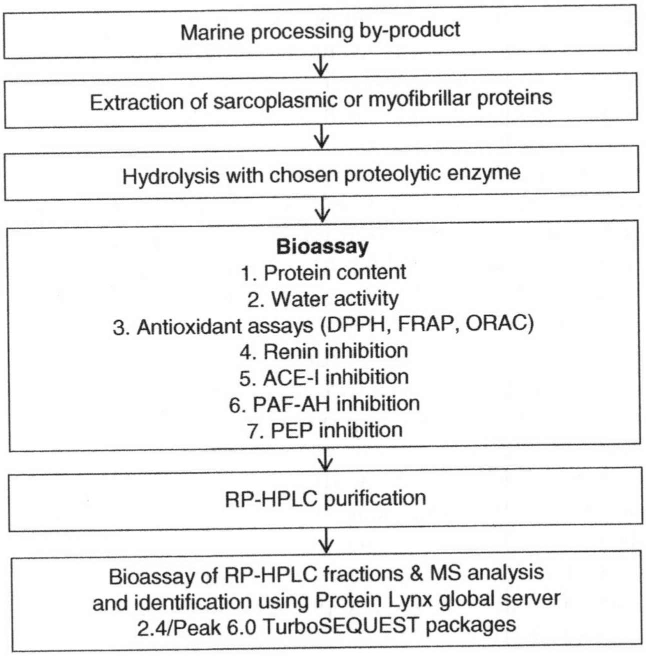

Figure 6.1 Flow diagram of the main steps in the isolation of bioactive peptides from marine by-products/co-product processing streams.

processing plants output high amounts of residual meat as well as meat processing by-products (Liang and Hultin, 2003). This huge range of waste and by-products can be valorized to obtain biomolecules with great economic importance. Kidney, liver, heart and blood are edible meat by-products, while horns, paunch, hoofs and bones are non-edible by-products. Keratin and collagen are the structural protein, which implies different industrial applications. The non-edible meat waste or by-products like horns, hoofs and bones are documented as a good source of these structural proteins (Morimura *et al.*, 2002). Meat by-products are rich in protein, as they serve as a potential source of bioactive peptides. Bioactive peptides are short sequences of approximately 2–20 amino acids in length, which exert physiological benefits when consumed (Di Bernardini *et al.*, 2011). A functional property of peptides depends upon the numbers and sequence of amino acids present. Bioactive peptides cover the antioxidant antithrombotic, antimicrobial and antihypertensive roles.

6.4 Dietary Essential Fatty Acids

Fatty acids are the building blocks of lipids, which can be subdivided into essential and non-essential fatty acids. The essential fatty acid is a concept referring to the group of fatty acids that cannot be synthesized in the human body or are under-synthesized but important for bodily functions, and need to be taken into the body through food. Non-essential fatty acids can be synthesized in the body. Essential fatty acids comprise polyunsaturated fatty acids (PUFA). Fatty acids, in which one are more carbons are connected by a double bond, is termed an unsaturated fatty acid. There are two families of EFA, omega-3 (ω-3) and omega-6 (ω-6). Marine food like shrimp, mussel and fish, and plant material like sesame, walnut, flax seed and soybean, are well known as potential sources of omega-3 polyunsaturated fatty acids (PUFAs) (Namiki, 2007).

6.4.1 Health Benefits of Omega Fatty Acids

The omega fatty acids or polyunsaturated fatty acids have been a subject of attraction and attainment because of their health promoting effects. Among these essential fatty acids, EPA, DHA and ALA are considered the most valuable. It is well documented that the intake of omega-6:omega-3 should be 2:1 for adults. Much scientific evidence proved that consumption of fish and fish oil was found effective in treatment and recovery of nerve disorders, cardiac arrhythmias, Alzheimer's, diabetes, dermatitis, hypertension, photoreception (vision), skin disorders and preventive against cancers (Shahidi and Kamil, 2001). Brain cell and some other vital organs of the human body have EPA and DHA as cell membrane elements. It provides as medium for the passage of important components to freely communicate with the neurons. Eicosanoids are considered as vital compounds in regulatory physiology. Anti-cachexia and anti-inflammatory properties of EPA make it applicable to treat rheumatoid arthritis (Rennie *et al.*, 2003). It has also been used in kidney diseases and Crohn's disease, as it is anti-inflammatory. The incidences of cardiac arrhythmias, heart attack and atherosclerosis can be reduced by limiting the levels of low density lipoproteins in blood and EPA is proved to be effective in lowering the LDL levels (Williams, 2000). Its properties to deplete the fibrinogen levels help to control thrombus formations in blood vessels. Dietary omega fatty acids help to maintain normal blood sugar utilization, hence they are beneficial in diabetes mellitus. Moreover, they can activate PPAR-a and PPAR-γ, which controls lipid metabolism and insulin sensitivity (Shahidi and Zhong, 2005).

6.4.2 Essential Fatty Acids from Marine Waste

A huge amount of fish processing leftovers, such as trimmings, backbones, skin, heads, stomach, liver, roe and viscera are thrown away as waste. Fish is a well-known source of protein as well as essential fatty acids. Fish oil, especially of marine fish, is an important source of PUFAs with enormous health benefits (Lordan *et al.*, 2011). Docosahexaenoic acid (DHA) and Eicosapentanoic acid (EPA) are among the polyunsaturated fatty acids that have potential health benefits. The human body has the ability to convert linoleic acids to form EPA (eicosapentaenoic acid) and DHA (docosahexaenoic acid) and later to polyunsaturated fatty acids. The valuable components from fin-fish processing by-products include enzymes, bioactive peptides, minerals such as calcium, collagen, lipid- and long-chain omega-3 polyunsaturated fatty acids (n-3 or ω-3 PUFAs).

6.4.3 Methods of Extraction of Omega Fatty Acid

Traditional methods for fish oil extraction include pressing, cooking and centrifugal separation. These methods are very harsh on lipids, causing deterioration and affecting proteins that are the by-product of this processing (Dauksas *et al.*, 2005). Most commonly, chemical methods or enzymatic procedures are used to obtain fish oil from fish waste or fish. In the Goldfisch method, extraction by using chloroform/methanol/water as a solvent and digestion with acid are some chemical methods used for oil extraction. Chemical methods are not considered wise, as they produce some undesirable by-products.

Neutrase, flourzyme, alcalase and protamex are some enzymes used for fish oil extraction (Yang *et al.*, 2012). To avoid decomposition, the molecular separation method is suggested as a good alternative method. Apart of this, some other innovative techniques have been developed to obtain high yields and low production costs, such as chromatographic methods, urea complexation, microwave assisted extraction, distillation methods, low-temperature crystallization, enzymatic methods and supercritical fluid extraction.

6.4.3.1 Chromatographic Method Chromatographic separation can be done on the basis of the degree of unsaturation of fatty acids and number of carbon atoms; a proper adsorbent and solvent need to be chosen. PUFA concentrated has been produced by silver resin chromatography and the HPLC method (Beebe *et al.*, 1988). Adolf *et al.* (1985) investigated the isocratic segregation of DHA and EPA by using solvent acetonitrile-acetone. Silica gel chromatography technique was performed by Guerrero and Belarbi (2001) to separate PUFA from edible oils. Belarbi *et al.* (2000) carried out argenated silica gel chromatography to obtain EPA concentrates.

6.4.3.2 Distillation Method As we know, DHA, EPA and omega fatty acids are long-chain structures and because of this structural feature, they are less volatile than short-chain fatty acid. Distillation is a suitable method for extraction of these less volatile unsaturated fatty acids. Another property of these PUFAs is they have higher boiling points and can be degraded in quality when treated at high temperatures. Low temperature, minimum residence time and low pressure are preferable conditions and these more suitable conditions can be achieved by molecular distillation or short path distillation (Wijesundera *et al.*, 1989). Due to varied boiling points, vacuum distillation can be used for separation of fatty acids. PUFA, especially n-3, is highly susceptible to oxidation, polymerization or isomerization at higher temperatures and end up with undesirable compounds (Fournier *et al.*, 2006).

6.4.3.3 Enzymatic Hydrolysis Method The enzymatic hydrolysis method has been used for obtaining fish oil. This method involves the use of different lipases for the esterification process. Bending of fatty acids due to the presence of cis-olefinic bonds is the basis for this separation process. This method is carried out under low temperatures and does not result in any by-product. Alcalase enzyme is reported as an efficient enzyme to use in the enzymatic hydrolysis method. In the enzymatic hydrolysis method, lipases are unable to reach the ester linkage between these FAs and glycerol, because structural features of EPA and DHA create steric hindrance. But in the case of mono-unsaturated fatty acids and saturated fatty acids, such barriers cannot be formed and could be hydrolyzed easily by lipase (Halldorsson *et al.*, 2003).

6.4.3.4 Low-Temperature Crystallization It is suggested that the low temperature crystallization technique can be adopted for segregation of PUFA. The principle of this technique is solubility of fatty acids or esters of fatty acids in organic solvent and crystallization below a temperature of 0°C (Brown and Kolb, 1955). It is not compulsory to use solvent in the low temperature crystallization method. If using solvent, the selection of solvent should be appropriate, as the concentration of PUFA will be affected by it. Crystallization in the absence of solvent, such as the tritiaux process, produces slurry of solid and liquid components, the latter being enriched in omega-3 fatty acids. Acetone and methanol are the commonly-used solvents or separation of PUFA. At lower temperatures, long-chain saturated FA has compact packing of its straight chain rather than unsaturated FA; the former becomes crystallized and concentrates of mono-unsaturated fatty acids (MUFA) or PUFA remain in the solution (Patil and Nag, 2011).

6.4.3.5 Supercritical Fluid Extraction In this method of separation of PUFA, the molecular size of fatty acid is the main point to be considered, instead of the degree of unsaturation of PUFA (Mishra *et al.*, 1993).

6.4.3.6 Urea Complexation PUFA can be separated from a mixture of saturated and MUFA by using Urea Complexation. This is reported as a more suitable and simpler method for gaining PUFA concentrates (Hamam and Shahidi, 2008).

References

Abdul-Hamid, A. and Luan, Y.S. (2000) Functional properties of dietary fiber prepared from defatted rice bran. *Food Chemistry*, **68**: 15–19.

Adolf, R.O. and Emiken, E.A. (1985) The isolation of omega-3 fatty acids and methyl esters of fish oil by silver resin chromatography. *Journal of the American Oil Chemists' Society*, **62**: 1592–1595.

Ajila, C.M., Naidu, K.A., Bhat, S.G. and Prasada Rao, U.J.S. (2007) Bioactive compounds and antioxidant potential of mango peel extract. *Food Chemistry*, **105**: 982–988

Ajila, C.M., Aalami, M., Leelavathi, K. and Rao, U.P. (2010) Mango peel powder: A potential source of antioxidant and dietary fiber in macaroni preparations. *Innovative Food Science & Emerging Technologies*, **11(1)**: 219–224.

Alvarez-Pellitero, P. (2008) Fish immunity and parasite infections: From innate immunity to immunoprophy-lactic prospects. *Veterinary Immunology and Immunopathology*, **126**: 171–198.

Bapiraju, K.V.V.S.N., Sujatha, P., Ellaih, P. and Ramana, T. (2004) Screening of oil seed cakes for isolation of lipolytic fungi and parametric optimization of lipase production by the selected isolate *Rhizopus* sp. BTS-24. *Asian Journal of Microbiology, Biotechnology and Environmental Sciences*, **6**: 735–740.

Bauchart, C., Chambon, C., Mirand, P.P., Savary-Auzeloux, I., Rémond, D. and Morzel, M. (2007) Peptides in rainbow trout (*Oncorhynchus mykiss*) muscle subjected to ice storage and cooking. *Food Chemistry*, **100**(4): 1566–1572.

Beebe, L.M., Brown, P.R. and Turcotte (1988) Preparative scale high performance liquid chromatography of omega-3 fatty acid esters derived from fish oils. *Journal of Chromatography*, **495**: 369–378.

Belarbi, H., Molina, E. and Chisty, Y. (2000) A process for high yield and scalable recovery of high purity eicosapentaenoic acid esters from microalgae and fish oils. *Enzyme and Microbial Technology*, **16**: 516–529.

Benjamin, S. and Pandey, A. (1996) Lipase production by *C. rugosa* on copra waste extract. *Indian Journal of Microbiology*, **45**: 452–456.

Bosaeus, I. (2004) Fibre effects on intestinal functions (diarrhoea, constipation and irritable bowel syndrome). *Clinical Nutrition Supplements*, **1(2)**: 33–38.

Brown, J.B. and Kolb, D.K. (1955) Applications of low temperature crystallization in the separation of the fatty acids and their compounds. *Progress in the Chemistry of Fats and other Lipids*, **3**: 57–94.

Chen, S.L. and Chinger, M. (1985) Production of baker's yeast. In: *Comprehensive Biotechnology* (eds H W. Blanch, S. Drew and D I.C. Wang), vol. **3**. Pergamon, Oxford, p. 429.

Cofrades, S., Guerra, M.A., Carballo, J., Fernández-Martín, F. and Jiménez-Colmenero, F. (2000) Plasma protein and soy fiber content effect on Bologna sausage properties as influenced by fat level. *Journal of Food Science*, **65(2)**: 281–287.

Collar, C., Rosell, C. M., Muguerza, B. and Moulay, L. (2009) Bread-making performance and keeping behavior of cocoa-soluble fiber-enriched wheat breads. *Food Science and Technology International*, **15(**1**)**: 79–87.

Conforti, F.D. and Davis, S.F. (2006) The effect of soya flour and flaxseed as a partial replacement for bread in yeast bread. *International Journal of Food Science Technology*, **41**, 95–101.

Cordova, J., Nemmaoui, M., Ismai li-Alaoui, M., Morin, A., Roussos, S., *et al.* (1998) Lipase production by solid-state fermentation of olive cake and sugar cane bagasse. *Journal of Molecular Catalysis B: Enzymatic*, **5**: 75–78.

Cui, W., Wood, P.J., Weisz, J. and Beer, M.U. (1999) Nonstarch polysaccharides from preprocessed wheat bran: Carbohydrate analysis and novel rheological properties. *Cereal Chemistry*, **76(**1**)**: 129–133.

Dauksas, E., Falch, E., Slizyte, R. and Rustad, T. (2005) Composition of fatty acids and lipid classes in bulk products generated during enzymatic hydrolysis of cod (*Gadus morhua*) by-products. *Process Biochemistry*, **41**: 2659–2670.

De Zoysa, M., Nikapitiya, C., Whang, I., Lee, L. S. and Lee, J. (2009) Abhisin: A potential antimicrobial peptide derived from histone H2A of disk abalone (*Haliotis discus discus*). *Fish and Shellfish Immunology*, **27**: 639–646.

Destoumieux, D., Munoz, M., Cosseau, C., Rodriguez, J., Bulet, P. *et al.* (2000) Penaeidins, antimicrobial peptides with chitin-binding activity are produced and stored in shrimp granulocytes and released after microbial challenge. *Journal of Cell Science*, **113**: 461–469.

Dexter, J.E. and Wood, P.J. (1996) Recent applications of debranning of wheat before milling. *Trends in Food Science & Technology*, **7(2)**, 35–41.

Dexter, J.E., Martin, D.G., Sadaranganey, G.T., Michaelides, J., Mathieson, N., *et al.* (1994a) Preprocessing: effects on durum wheat milling and spaghetti-making quality. *Cereal Chemistry*, **71(**1**)**: 10–15.

Dexter, J.E., Symons, S.J. and Martin, D.G. (1994b) Enhancement of durum wheat quality by preprocessing and an evaluation of fluorescence imaging as a rapid technique for monitoring preprocessing efficiency. *Association of Operative Millers-Bulletin*, 6415–6420.

Dias, D.R., Vilela, D.M., Silvestre, M.P.C. and Schwan, R.F. (2008) Alkaline protease from Bacillus sp. isolated from coffee bean grown on cheese whey. *World Journal of Microbiology and Biotechnology*, **24(**10**)**, 2027–2034.

Di Bernardini, R., Harnedy, P., Bolton, D., Kerry, J., O'Neill, E., *et al.* (2011) Antioxidant and antimicrobial peptidic hydrolysates from muscle protein sources and by-products. *Food Chemistry*, **124(4)**, 1296–1307.

Di Luccio, M., Capra, F., Ribeiro, N.P., Vargas, G.D.L.P., Freire, D.M.G. and De Oliveira, D. (2004) Effect of temperature, moisture, and carbon supplementation on lipase production by solid-state fermentation of soy cake by *Penicillium simplicissimum. Applied Biochemistry and Biotechnology: Part A: Enzyme Engineering and Biotechnology*, 113–116, 173–180.

Eastman, J., Orthoefer, F. and Solorio, S. (2001) Using extrusion to create breakfast cereal products. *Cereal Foods World*, **46(10)**: 468–471.

Ebune, A., Al-Asheh, S. and Duvnjak, Z. (1995) Effects of phosphate, surfactants and glucose on phytase production and hydrolysis of phytic acid in canola meal by *Aspergillus ficuum* during solid-state fermentation. *Bioresource Technology*, **54**: 241–247.

Eim, V.S., Simal, S., Rosselló, C. and Femenia, A. (2008) Effects of addition of carrot dietary fibre on the ripening process of a dry fermented sausage (sobrassada). *Meat Science*, **80(2):** 173–182.

Elleuch, M., Besbes, S., Roiseux, O., Blecker, C., Deroanne, C., *et al.* (2008) Date flesh: Chemical composition and characteristics of the dietary fiber. *Food Chemistry*, **111**: 676–682.

Escudero, E., Sentandreu, M.A., Arihara, K. and Toldra, F. (2010) Angiotensin I-converting enzyme inhibitory peptides generated from *in vitro* gastrointestinal digestion of pork meat. *Journal of Agricultural and Food Chemistry*, **58**: 2895–2901.

Federici, F., Montedoro, G.F., Sevili, M. and Petruccioli, M. (1988) Pectic enzyme production by *Cryptococcus albidus* var. *albidus* on olive oil vegetation waters enriched with sunflower calathide meal. *Biological Wastes*, **25**, 291–301.

Figuerola, F., Hurtado, M.L., Estévez, A.M. Chiffelle, I.. and Asenjo, F. (2005) Fiber concentrates from apple pomace and citrus peel as potential fiber sources for food enrichment. *Food Chemistry*, **91**: 395–401.

Fournier, V., Destaillats, F., Juanéda, P., Dionisi, F., Lambelet, P., *et al.* (2006) Thermal degradation of long-chain polyunsaturated fatty acids during deodorization of fish oil. *European Journal of Lipid Science and Technology*, **108**: 3342.

Fuentes-Alventosa, J.M., Rodríguez-Gutiérrez, G., Jaramillo-Carmona, S., Espejo-Calvo, J.A., Rodríguez-Arcos, R., *et al.* (2009) Effect of extraction method on chemical composition and functional characteristics of high dietary fibre powders obtained from asparagus by-products. *Food Chemistry*, **113(2)**, 665–671.

Gallagher, J.C., Rafferty, K., Haynatzka, V. and Wilson, W. (2000) The effect of soy protein on bone metabolism. *Journal of Nutrition*, **130**: 6675–6735.

Germano, S., Pandey, A., Osaku, C.A., Rocha, S.N. and Soccol, C.R. (2003) Characterization and stability of proteases from *Penicillium* sp. produced by solid-state fermentation. *Enzyme and Microbial Technology*, **32**: 246–251.

Ghaly A.E. and Kamal M.A. (2004) Submerged yeast fermentation of acid cheese whey for protein production and pollution potential reduction. *Water Research*, **38**: 631–644.

González-Pérez, S. and Vereijken, J.M. (2008) Sunflower proteins: Overview of their physicochemical, structural and functional properties. *Journal of the Science of Food Agriculture*, **87**: 2172–2191.

Grigelmo-Miguel, M., Gorinstein, S. and Martin-Belloso, O. (1999a) Characterizations of peach dietary fiber concentrate as a food ingredient. *Food Chemistry*, **65**: 175–181.

Grigelmo-Miguel, N. and Martina-Belloso, O. (1999b) Influence of fruit dietary fiber addition on physical and sensorial properties of strawberry jams. *Journal of Food Engineering*, **41**: 13–21.

Grosso, N.R. and Guzmán, C.A. (1995) Chemical Composition of aboriginal peanut (*A. hypogaea*) seeds from Perú. *Journal of Agricultural Food Chemistry*, **43**: 102–105.

Guerero, J.L. and Belarbi. H. (2001) Purification process for cod liver oil polyunsaturated fatty acids *Journal of the American Oil Chemists' Society*, **78**: 477–484.

Gunasekaran, S., Ko, S. and Xiao L. (2007) Use of whey proteins for encapsulation and controlled delivery applications. *Journal of Food Engineering*, **83**: 31–40.

Halldorsson, A., Kristinsson, B., Glynn, C. and Haraldsson, G. (2003) Separation of EPA and DHA in fish oil by lipase-catalyzed esterification with glycerol. *Journal of the American Oil Chemists' Society*, **80**: 915–921.

Hamam, F. and Shahidi, F. (2008) Incorporation of selected long-chain fatty acids into trilinolein and trilinolenin. *Food Chemistry*, **106(1)**, 33.

Himaya, S.W.A., Ngo, D.H., Ryu, B. and Kim, S.K. (2012) An active peptide purified from gastrointestinal enzyme hydrolysate of Pacific cod skin gelatin attenuates angiotensin-1 converting enzyme (ACE) activity and cellular oxidative stress. *Food Chemistry*, **132**: 1872–1882.

Hsu, P. K., Chien, P.J., Chen, C.H. and Chau, C.F. (2006) Carrot insoluble fiber-rich fraction lowerslipid and cholesterol absorption in hamsters. *LWT – Food Science Technology*, **39**: 337–342.

Huffman, L.M. (1996) Processing whey protein for use as a food ingredient. *Food Technology*, **50**: 5049–5052.

Joo, H.S., Kumar, C.G., Park, G.C., Kim, K.T., Paik, S.R. and Chang, C.S. (2002) Optimization of the production of an extracellular alkaline protease from *Bacillus horikoshii. Process Biochemistry*, **38**: 155–159.

Jovanovic, S., Barac, M., Macej, O., Vucic, T. and Lacnjevac, C. (2007) SDS–PAGE analysis of soluble proteins in reconstituted milk exposed to different heat treatments. *Sensors*, **7**: 371–383.

Kamini, N.R., Mala, J.G.S. and Puvanakrishnan, R. (1998) Lipase production from *Aspergillus niger* by solid-state fermentation using gingelly oil cake. *Process Biochemistry*, **33**: 505–511.

Kashyap, P., Sabu, A., Pandey, A., Szakacs, G. and Soccol, C.R. (2002). Extra-cellular L-glutaminase production by *Zygosaccharomyces rouxii* under solid-state fermentation. *Process Biochemistry*, **38(3)**, 307–312.

Kim, S.K. and Mendis, E. (2006) Bioactive compounds from marine processing byproducts: A review. *Food Research International*, **39**: 383–393.

Kim, S.K. and Wijesekara, I. (2010) Development and biological activities of marine-derived bioactive peptides: A review. *Journal of Functional Foods*, **2(1)**: 1–9.

Kristinsson, H.G. (2008) Functional and bioactive peptides from hydrolyzed aquatic food proteins. In: *Marine Nutraceuticals and Functional Foods* (eds C. Barrow and F. Shahidi). CRC Press, Boca Raton, FL, pp. 229–246.

Lecumberri, E., Mateos, R., Izquierdo-Pulido, M., Rupérez, P., Goya, L. and Bravo, L. (2007) Dietary fibre composition, antioxidant capacity and physico-chemical properties of a fibre-rich product from cocoa (*Theobroma cacao* L.). *Food Chemistry*, **104(3)**, 948–954.

Li, C. and Zhao, S.J. (2007) Preliminary study on a potential antibacterial peptide derived from histone H2A in hemocytes of scallop *Chlamys farreri*. *Fish and Shellfish Immunology*, **22**: 663–672.

Liang, Y. and Hultin, H.O. (2003) Functional protein isolates from mechanically deboned turkey by alkaline solubilization with isoelectric precipitation. *Journal of Muscle Foods*, **14(3)**, 195–205.

Lin, T.C. and Chen, C. (2004) Enhanced mannanase production by submerged culture of *Aspergillus niger* NCH-189 using defatted copra based media. *Process Biochemistry*, **39(9)**: 1103–1109.

Liu, K. (1997) Chemistry and nutritional value of soybean components. In: *Soybeans*. Springer. US, pp. 25–113.

Lordan, S., Ross, R.P. and Stanton, C. (2011) Marine bio-actives as functional food ingredients: Potential to reduce the incidence of chronic diseases. *Marine Drugs*, **9**: 1056–1100.

Marshall. K. (2004) Therapeutic applications of whey protein. *Alternative Medical Review*, **9(2)**: 136–156.

Mateos-Aparicio, I., Redondo-Cuenca, A., Villanueva-Suárez, M. J., Zapata-Revilla, M. A., & Tenorio-Sanz, M. D. (2010). Pea pod, broad bean pod and okara, potential sources of functional compounds. *LWT – Food Science and Technology*, **43(9)**: 1467–1470.

Mei, X., Tai-Hua Mu and Jun-Juan, H. (2010) Composition and physicochemical properties of dietary fiber extracted from residues of 10 varieties of sweet potato by a sieving method. *Journal of Agricultural and Food Chemistry*, **58**: 7305–7310.

Messina, M. (2001) Noteworthy evidence mounts on soy and human health. *Journal of the American Dietetic Association*, **9**: 1–3.

Mishra, V.K., Temelli, F. and Ooraikul, B. (1993) Extraction and purification of [omega]-3 fatty acids with an emphasis on supercritical fluid extraction – A review. *Food Research International*, **26**: 217–226.

Mongeau, R. (2003) Dietary fiber: Determination. In: *Encyclopedia of Food Sciences and Nutrition*. pp. 1823–1833.

Morimura, S., Nagata, H., Uemura, Y., Fahmi, A., Shigematsu, T. and Kida, K. (2002). Development of an effective process for utilization of collagen from livestock and fish waste. *Process Biochemistry*, **37**(12): 1403–1412.

Namiki, M. (2007) Nutraceutical functions of sesame: A review. *Critical Reviews in Food Science and Nutrition*, **47**: 651–673.

Nawirska, A. and Kwasniewska, M. (2005) Dietary fiber fractions from fruit and vegetable processing waste. *Food Chemistry*, **91**: 221–225.

Ngo, D.H., Ryu, B., Vo, T.S., Himaya, S.W.A., Wijesekara, I. and Kim, S.K. (2011) Free radical scavenging and angiotensin-I converting enzyme inhibitory peptides from Pacific cod (*Gadus macrocephalus*) skin gelatin. *International Journal of Biological Macromolecules*, **49**: 1110–1116.

Patat, S.A., Carnegie, R.B., Kingsbury, C., Gross, P.S., Chapman, R. and Schey, K.L. (2004) Antimicrobial activity of histones from hemocytes of the Pacific white shrimp. *European Journal of Biochemistry*, **271**: 4825–4833.

Patil, D. and Nag, A. (2011) Production of PUFA concentrates from poultry and fish processing waste. *Journal of the American Oil Chemists' Society*, **88(4)**, 589–593.

Pihlanto, A. and Korhonen, H. (2003) Bioactive peptides and proteins. *Advances in Food Nutrition Research*, **47**: 175–276.

Podsedek, A. (2007) Natural antioxidant capacity of brassica vegetables: A review *LWT – Food Science and Technology*, **40**: 1–11

Pringent, S.V.E., Gruppen, H., Visser, A.J.W.G., van Koningsveld, G.A., DeJong, G.A.H., and Voragen, A.G.J. (2003) Effects of non-covalent interactions with 5-O-caffeoyl quinic acid (Chlorogenic Acid) on the heat denaturation and solubility of globular proteins. *Journal of Agricultural and Food Chemistry*, **51**: 5088–5095.

Ramakrishnan, C.V. and Banerjee, B.N. (1952) Studies on mold lipase-comparative study of lipases obtained from molds grown on *Sesamum indicum. Archives of Biochemistry and Biophysics*, **37**: 131–135.

Ramachandran, S., Patel, A.K., Nampoothiri, K.M., Francis, F., Nagy, V., *et al.* (2004) Coconut oil cake – A potential raw material for the production of α-amylase. *Bioresource Technology*, **93**: 169–174.

Ramachandran, S., Roopesh, K., Nampoothiri, K.M., Szakacs, G. and Pandey, A. (2005) Mixed substrate fermentation for the production of phytase by *Rhizopus* spp. using oilcakes as substrates. *Process Biochemistry*, **40**: 1749–1754.

Raskin, I., Ribnicky, D.M., Komarnytsky, S., et al. (2002) Plants and human health in the twenty-first century. *Trends in Biotechnology*, **20**: 522–531.

Rennie, K.L., Hughes, J., Lang, R. and Jebb, S.A. (2003) Nutritional management of rheumatoid arthritis: A review of the evidence. *Journal of Human Nutrition and Dietetics*, **16**: 97–109.

Rodríguez, R., Jiménez, A., Guillén, R., Heredia, A. and Fernández-Bolaños, J. (1999) Postharvest changes in white asparagus during refrigerated storage. *Journal of Agricultural and Food Chemistry*, **47**: 3551–3557.

Roopesh, K., Ramachandran, S., Nampoothiri, K.M., Szakacs, G. and Pandey, A. (2006) Comparison of phytase production on wheat bran and oilcakes in solid-state fermentation by *Mucor racemosus. Bioresource Technology*, **97**: 506–511.

Sabu, A., Sarita, S., Pandey, A., Bogar, B., Szakacs, G. and Soccol, C.R. (2002) Solid-state fermentation for production of phytase by *Rhizopus oligosporus. Applied Biochemistry and Biotechnology: Part A: Enzyme Engineering and Biotechnology*, **102–103**: 251–260.

Sabu, A., Pandey, A., Daud, M.J. and Szakacs, G. (2005) Tamarind seed powder and palm kernel cake: two novel agro residues for the production of tannase under solid state fermentation by *Aspergillus niger* ATCC 16620. *Bioresource Technology*, **96(11)**, 1223–1228.

Sabu, A., Augur, C., Swati, C. and Pandey, A. (2006) Tannase production by Lactobacillus sp. ASR-S1 under solid-state fermentation. *Process Biochemistry*, **41(3)**, 575–580.

Salgado, P.R., Molina-Ortiz, S.E., Petruccelli, S. and Mauri, A.N. (2011) Sunflower protein concentrates and isolates prepared from oilcakes have high water solubility and antioxidant capacity. *Journal of the American Oil Chemist's Society*, **88**: 351–360.

Samsudin, M.D. and Williams, M.I. (1967) Rational use of skim milk in a complete infant formula. *American Journal of Clinical Nutrition*, **20(12)**: 1308–1311.

Sánchez-Alonso, I., Jiménez-Escrig, A., Saura-Calixto, F. and Borderias, A.J. (2007) Effect of grape antioxidant dietary fiber on the prevention of lipid oxidation in minced fish: Evaluation by different methodologies. *Food Chemistry*, **101**: 372–378.

Sanderson, G.R. (1981) Polysaccharides in foods. *Food Technology*, **35(1)**: 50–57.

Sangnark, A. and Noomhorm, A. (2003) Effect of particle sizes on functional properties of 730 dietary fiber prepared from sugarcane bagasse. *Food Chemistry*, **80**: 221–229.

Santos M.J., Teixeira J.A. and Rodrigues L.R. (2012) Fractionation of the major whey proteins and isolation of beta-Lactoglobulin variants by anion exchange chromatography. *Separation and Purification Technology*, **90**: 133–139.

Savage, G.P. and Keenan, J.L. (1994) The composition and nutritive value of groundnut kernels. In: *The Groundnut Crop: A scientific basis of improvement* (ed. J. Smart), Chapman & Hall, London, pp. 173–213.

Seifzadeh, S., Hassan Sajedi, R. and Sariri, R. (2008) Isolation and characterization of thermophilic alkaline proteases resistant to sodium dodecyl sulfate and ethylene diamine tetraacetic acid from Bacillus sp. GUS1. *Iranian Journal of Biotechnology*, **6(4)**, 214–221.

Selvakumar, P. and Pandey, A. (1999). Solid state fermentation for the synthesis of inulinase from *Staphylococcus* sp. and *Kluyveromyces marxianus*. *Process Biochemistry*, **34**(8): 851–855.

Shahidi, F. and Kamil, Y.V.A.J. (2001) Enzymes from fish and aquatic invertebrates and their application in the food industry. *Trends in Food Science and Technology*, **12**: 435–464.

Shahidi, F. and Zhong, Y. (2005) *Marine Mammal Oils. Bailey's Industrial Oil and Fat Products*, 6th edition. John Wiley & Sons, Chichester, UK.

Sperstad, S.V., Haug, T., Vasskog, T. and Stensvog, K. (2009) Hyastatin, a glycine-rich multi domain antimicrobial peptide isolated from the spider crab (*Hyas araneus*) hemocytes. *Molecular Immunology*, **46**: 2604–2612.

Stolle, R.J. and Beck, L.R. (1987) Inventors: Stolle Research and Development Co, assignee. Method for treating disorders of the vascular and pulmonary systems. US patent 4,636,384. 13 January 1987, 17 p. Int. Cl A6 1K 39/ 40; A61K 35/20.

Su, Y. (2011) Isolation and identification of pelteobagrin, a novel antimicrobial peptide from the skin mucus of yellow catfish (*Pelteobagrus fulvidraco*). *Comparative Biochemistry and Physiology B*, **158**: 149–154.

Sudha, M.L., Baskaran, V. and Leelavathi, K. (2007) Apple pomace as a source of dietary fiber and polyphenols and its effect on the rheological characteristics and cake making. *Food Chemistry*, **104**: 686–692.

Sumitra, R., Sudheer, K.S., Christian, L., Carlos, R.S. and Ashok, P. (2007) Oil cakes and their biotechnological applications: A review. *Bioresource Technology*, **98**: 2000–2009.

Synowiecki, J. (2010) Some applications of thermophiles and their enzymes for protein processing. *African Journal of Biotechnology*, **9(42)**: 7020–7025.

Telrandhe, U.B., Kurmi, R., Uplanchiwar, V., Mansoori, M.H., Raj, V.J., et al. (2012) Nutraceuticals-phenomenal resource in modern medicine. *International Journal of Universal Pharmacy and Life Sciences*, **2(1)**: 179–195.

Trowell, H. (1972) Fschaemic heart disease and dietary fiber. *American Journal of Clinical Nutrition*, **25**: 926–932.

Tseng, A. and Zhao, Y. (2013) Wine grape pomace as antioxidant dietary fiber for enhancing nutritional value and improving storability of yoghurt and salad dressing. *Food Chemistry*, **138**: 356–365.

Venugopal, V. (2009) Seafood proteins: Functional properties and protein supplements. In: *Marine Products for Healthcare: Functional and Bioactive Nutraceutical Compounds from the Ocean*. CRC Press, Boca Raton, pp. 51–102.

Vergara-Valencia, N., Granados-Pérez, E., Agama-Acevedo, E., Tovar, J., Ruales, J. and Bello-Pérez, L.A. (2007) Fibre concentrate from mango fruit: Characterization, associated antioxidant capacity and application as a bakery product ingredient. *LWT –Food Science and Technology*, **40(4)**: 722–729.

Verma, A.K., Sharma, B.D. and Banerjee, R. (2009) Quality characteristics and storage stability of low fat functional chicken nuggets with salt substitute blend and high fiber ingredients. *Fleischwirtsch International*, **24(6)**: 54–57.

Verma, A.K., Banerjee, R. and Sharma, B.D. (2011) Quality of low fat chicken nuggets: effect of sodium chloride replacement and added chickpea (*Cicer arietinum* L.) hull flour. *Asian-Australasian Journal of Animal Sciences*, **25**(2): 291–298.

White, J. (1954) *Yeast Technology*. Chapman & Hall Ltd, London.

WHO/FAO (2003) Food energy – Methods of analysis and conversion factors. Report of a technical workshop, *FAO Food and Nutrition Paper – 77*, Rome.

Wijesundera, R., Ratnayake, W. and Ackman, R. (1989) Eicosapentaenoic acid geometrical isomer artifacts in heated fish oil esters. *Journal of the American Oil Chemists' Society*, **66**: 1822–1830.

Williams, C.M. (2000) Dietary fatty acids and human health. *Ann De Zootechnie*, **49**: 165–180.

Wu, H., Wang, Q., Ma, T. and Ren, J. (2009) Comparative studies on the functional properties of various proteins concentrate preparations of peanut protein. *Journal of Food Research International*, **42**: 343–348.

Yang, Z.M., Galsius, M. and Xu, X. (2012) Enzymatic transesterification of ethyl ferulate with fish oil and reaction optimization by response surface methodology. *Food Technology and Biotechnology Journal*, **50**: 88–97.

Yoon, K.Y., Cha, M., Shin, S.R., and Kim, K.S. (2005) Enzymatic production of a soluble-fiber hydrolysate from carrot pomace and its sugar composition. *Food Chemistry*, **92**: 151–157.

Yu, J., Aahmedna, M. and Goktepe, I. (2007) Peanut protein concentrates: Production and functional properties as affected by processing. *Food Chemistry*, **103**: 121–129.

Zind, T. (1998) Making the case for soy. *Food Processing*, **25**, 31–34.

7

Prebiotics and Dietary Fibers from Food Processing By-Products

Santad Wichienchot[1] and Wan Rosli Bin Wan Ishak[2]
[1]*Prince of Songkla University, Hat Yai, Songkhla, Thailand*
[2]*Universiti Sains Malaysia Health Campus, Kubang Kerian, Kota Bharu, Kelantan, Malaysia*

7.1 Introduction

The abundance of agricultural wastes or by-products from industrial and domesticated food processing is the main cause of environment problems. These by-products are generally managed by disposal or even sold at a cheaper price. Disposal of these underutilized by-products are commonly done in inappropriate ways, i.e. discharge effluent into rivers or by burning in the open, which may cause air and water pollutions. Presently, scientific investigation on the benefits or functional properties of waste and by-products from industrial food processing, which produces a large amount of by-products, is necessary in the search for possible ways for their utilization (Vanesa *et al.*, 2011). Three main groups of by-product from food processing, classified according to their main chemical compositions, are carbohydrate and dietary fibers, protein and lipids. The most common by-products are generated by the food industry, in particular the beverage, starch and flour industries. These items are classified under carbohydrate and dietary fiber groups. They are further divided into four sub-groups: monosaccharides, disaccharides, oligosaccharides and polysaccharides.

Dietary fibers are a class of non-starch polysaccharides (i.e. cellulose, dextrins, chitins, pectins, β-glucans and waxes) and lignin, which are able to modulate the transit time through the gut. Thus, it provides similar beneficial effects to those of inulin-type fructans. These compounds are commonly found in many foods such as cereal, nuts, fruits and vegetables etc. They are also partially susceptible to bacterial gut fermentation and may induce changes in bacterial microbiota, particularly in the numerous bifidobacteria and lactobacilli. These soluble dietary fibers have been shown to exert additional beneficial effects, for instance by improving gut barrier function *in vitro* and *in vivo*, which could be partially a consequence of their effect on the gut microbiota composition (Laparra and Sanz, 2010).

Food Processing By-Products and their Utilization, First Edition. Edited by Anil Kumar Anal.

Prebiotics are non-digestible food ingredients that beneficially affect the host by selectively stimulating the growth and/or activity of one or a limited number of bacteria in the colon (Gibson and Roberfroid, 1995). However, the definition of prebiotics has changed several times. In 2007, WHO redefined prebiotics as a non-viable food component that confers a health benefit on the host associated with modulation of the microflora. Thus, the prebiotic approach advocates the administration of non-viable entities, and therefore overcomes any viability problems of probiotics in the upper gastro intestinal tract (GIT). In addition, a prebiotic can be considered as a growth substrate that fortifies the multiplication of beneficial intestinal microflora. However, it differs from the classical dietary fibers in which is selectively stimulates the growth and/or activity of beneficial bacteria, i.e. bifidobacteria and lactobacilli species. To be classified as a prebiotic, a food ingredient must:

1. be neither hydrolyzed nor absorbed in the upper part of the GIT;
2. be able to alter the colonic microflora in favor of a healthier composition through selective fermentation; and
3. induce luminal or systemic effects that are beneficial to the host's health (Cummings *et al.*, 2001).

The non-digestible and fermentable oligosaccharides are the most popular compounds, which gained specific interest because they had been reported to possess significant prebiotic properties, such as galacto-oligosaccharides (GOS), fructo-oligosaccharides (FOS), gentio-oligosaccharides, gluco-oligosaccharides, isomalto-oligosaccharides (IMO), mannan-oligosaccharides, chito-oligosaccharides (COS), melibiose, pectic oligosaccharides (POS) and xylo-oligosaccharides (XOS) (Roberfroid, 2007; Gibson *et al.*, 2004; Gibson and Roberfroid, 2008; Fastinger *et al.*, 2008; Calame *et al.*, 2008). Consumption of these prebiotics provides a beneficial physiological effect on the host by selectively stimulating the favorable growth or activity of a limited number of indigenous bacteria, namely bifidobacteria and lactobacilli (Gibson and Roberfroid, 1995). Metabolites from fermentation of prebiotics by microbiota are short-chain fatty acids (SCFA), principally acetate, propionate and butyrate and vitamins B and folic acid (Cummings, 1981).

Resistant starch is the other potential prebiotic candidate that resists small intestinal digestion and enters the large bowel in humans in normal health. The amount of resistant starch is dependent of the source. It is thought that approximately 10% of the total dietary starch may escape digestion in the human small intestine. More importantly, some types of resistant starch display prebiotic properties naturally. This resistant starch will contribute to the SCFA production, increase stool weight, and decrease faecal pH. Several studies have shown that resistant starch favors butyrate production (Cummings and Macfarlane, 1997). Butyrate is the preferred substrate for the colonocytes, even when competing substrates such as glucose are present, and it is probably protective against colorectal cancer (Topping and Clifton, 2001).

Potential sources of by-products from food processing for production of prebiotics are derived from various food industries and some from agricultural wastes or by-products. For example, pectic oligosaccharide and fructo-oligosaccharide are generated from fruit and vegetable processing and the sugar cane industry, xylo-oligosaccharide from wood and corncob, chito-oligosaccharide from seafood processing, ß-glucan from cereal and mushroom by-products, while soybean oligosaccharide is a by-product from the tofu and soy protein industry. Production of these prebiotics

Table 7.1 Potential sources of by-products for prebiotics production

Source	Prebiotic	Production approach	Reference
Fruits and vegetables processing industry	Inulin, fructo-oligosaccharide Pectic oligosaccharide	Hot water extraction Acidified water extraction, enzymatic depolymerization of pectin	Bornet *et al.* (2002) Wichienchot *et al.* (2010)
Seafood industry	Chito-oligosaccharide	Hot acid extraction and enzymatic depolymerization of chitosan	Knaul *et al.* (1999)
Agriculture	Xylo-oligosaccharide	Thermal extraction, enzymatic depolymerization of xylans from corncobs, sawdust	Imaizumi *et al.* (1991); Lamtumluk (2014)
Dairy industry	Galacto-oligosaccharide	Enzymatic transgalactosylation of whey lactose	Teuri and Korpel (1998)
	Lactulose	Isomerization of whey lactose	Salminen and Salminen (1997)
Sugar cane industry	Fructo-oligosaccharide	Enzymatic fructosyltransferase from molasses	Sangeetha *et al.* (2005)
Flour/starch industry	Isomalto-oligosaccharide	Enzymatic glycosyltransferase from flour/starch	Morgan *et al.* (1992)

occurs through three main approaches; direct extraction and/or purification, enzymatic depolymerization/synthesis and chemical synthesis (Wichienchot and Chinachoti, 2011), as summarized in Table 7.1.

Currently, the top production of prebiotics at high volumes is inulin, followed by galacto-oligosaccharide. Prominent raw materials or substrates for production of these prebiotics are roots of chicory and Jerusalem artichoke and lactose from the agriculture and dairy industries, respectively. However, due to high demand, the cost of these raw materials is exponentially increasing every year. Thus, many manufacturers of prebiotics are shifting the utilization of their raw materials from other alternative sources with lower cost. Due to this situation, various by-products derived from food processing, including agricultural-based items, are of much interest in this solution.

Functional food was first defined in Japan in 1983, when the Japanese government launched the first functional beverage in 1988, a dietary fiber (polydextrose)-containing soft drink. In Japan, according to the food regulation of Food for Specified Health Uses (FOSHU), oligosaccharides are classified as dietary fiber and can be health claimed (Weststrate *et al.*, 2002). Today, prebiotic ingredients are mostly added to dairy food products such as yoghurts, yoghurt drinks, spreads, fresh cheeses and milk. Other emerging food products are kefir, sport products, functional waters and beverages, nutrition bars, weight loss products, soymilk, green foods, mineral supplements, nutraceuticals and also pet foods (Kaur and Gupta, 2002; Niness, 1999).

7.2 Oligosaccharides from Food Processing By-Products

7.2.1 Pectic Oligosaccharide (POS)

Pectic oligosaccharides (POS) are one of potential prebiotic groups of carbohydrate. Their common feature is D-galacturonic acid, the backbone majority of POS. The major components of pectic oligosaccharide are uronic acid, neutral sugars such as L-rhamnose, D-galactose, D-xylose and L-arabinose. The chemical structures of these natural sugars are shown in Figure 7.1.

Generally, pectic oligosaccharide is not in commercial production as a prebiotics ingredient. However, previous reports show that POS are one type of prebiotic groups. It has been reported that it could be an excellent candidate for second-generation prebiotics (Hotchkiss *et al.*, 2003). This is because it can increase the growth of probiotic bacteria in the large intestine (Manderson *et al.*, 2005). Pectic oligosaccharides can be produced from pectin by enzyme depolymerization of natural carbohydrates. Pectin is defined as a mixture of heteropolysaccharides, mainly found in higher plant middle lamellae and primary cell walls (Albersheim *et al.*, 1996). Potential sources of pectin are waste biomass or low-cost by-products such as apple pomace, orange peels, sugar beet pulp and olive by-products. In a very recent study by Wichienchot *et al.* (2015), the percent yields of by-products from different cultivars of orange are varied (Table 7.2). The orange juice industry generated by-products containing pectin varying between

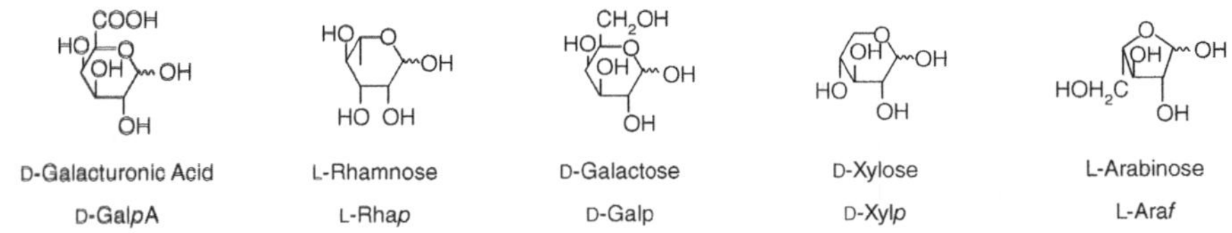

Figure 7.1 Some structures of neutral sugars.

Table 7.2 Percent yields of by-products of orange with different cultivars

Cultivar of orange	Yield (%)		
	Peel	Pulp residue	Total
Ora Mandarin	21.09	32.92	54.01
Navel	25.41	44.07	69.48
Brazil mandarin	21.97	37.97	59.94
Small Chinese orange dip	21.94	42.96	64.90
Honey queen	22.34	47.39	69.73
Chinese orange dip	17.92	24.62	42.54
Sweet orange	28.46	37.91	66.37
Mini *honey queen*	14.42	33.92	48.34

Table 7.3 Example of enzymes for pectic oligosaccharide production

Enzyme	Source	Working conditions	
		Temperature (°C)	pH
EPG-M2	*A. aculeatus*	40	5.5
Pectinase	*A. niger*	20–60 (40)	3–5.0 (3.8)
Viscozyme L	*A. aculeatus*	25–55 (35)	3.3–5.5 (4.0)
Pectinex Ultra SP-L	*A. aculeatus*	Below 45 (35)	4.5–6 (4.5)
Pectinase 62L	*A.* sp.	10–55 (37)	3.0–5.0 (5.0)
Macer8 FJ	*A.* sp.	40–60 (40)	3.0–5.0 (5.0)

Value in bracket is an optimal value

40% and 70% of the raw material. These by-products have potential for production of prebiotic pectic oligosaccharide.

Pectic oligosaccharides are normally obtained according to three main strategies:

1. Extracting pectin from plants, such as peel or pulp residue of orange. The depolymerization is carried out by either chemical or biological methods (e.g. by pre-hydrolysis or enzymatic hydrolysis). This method generates a variety of undesired, non-volatile impurities (i.e. mineral salts, monosaccharides or lignin-derived products) in the liquid phase of the reaction media, which must be processed to obtain a refined product suitable as a food ingredient (Gómez *et al.*, 2014).
2. Synthesis by some enzymes such as glycosyltransferases, acetyltransferases and methyltransferases. These enzymes were responsible for synthesizing the polymers in plants, and esterases, lyases and various glycosylases degrade and modify the oligosaccharide. If the assumption is made that one discrete glycosyltransferase is responsible for each individual glycosidic linkage found in the pectic-polysaccharides (Clausen and Madsen, 2003).
3. Depolymerization of the polysaccharides. Spagnuolo *et al.* (1997) studied the synergistic effects of cellulolytic and pectinolytic enzymes in the hydrolysis of the polysaccharides. They also suggest that pectinase is the most important enzyme, because it favors the degradation of cellulose and hemicellulose of pectic-polysaccharides and makes them become pectic oligosaccharides. Enzymes and their working and optimal conditions used for production of pectic oligosaccharide are summarized in Table 7.3 (Agnan *et al.*, 2012).

Applications and benefits of pectic oligosaccharides also protect colonocytes against *Escherchia coli* verocytotoxins and stimulate apoptosis in human colonic adenocarcinoma cells (Olano *et al.*, 2003). Manderson *et al.* (2005) studied pectic oligosaccharide production from orange peel and reported that extraction of POS from Hamlin orange peel produced 158.5 g of the polar supernatant fraction from 4.542 kg wet peel (976.5 g, dry weight). The yield was 16%, and the monosaccharide composition of POS consisted mainly of glucose and arabinose, with lesser amounts of galactose, galacturonic acid, xylose, rhamnose and fucose (Table 7.4).

The dragon fruit pulp and peel consists of highly viscous carbohydrate fibers (cellulose, hemicellulose and simple sugars), vitamin C, minerals and non-starch polysaccharides and pectin (Nuráliaa *et al.*, 2010). Dragon fruit juices are turbid, very viscous

Table 7.4 POS monosaccharide composition

Monosaccharide	% of POS
Glucose	48.12
Arabinose	31.19
Galactose	9.59
Xylose	2.44
Rhamnose	2.13
Fucose	0.24
Galacturonic acid	6.29

Table 7.5 Chemical composition of white flesh dragon fruit peel

Proximate content (%)	Percent (dry basis)
Protein	0.83 ± 0.02
Crude Fat	0.03 ± 0.02
Moisture	92.65 ± 0.03
Ash	2.14 ± 0.01
Crude Fiber	1.84 ± 0.01
Carbohydrate	4.35 ± 0.02

and contain colloidal suspension. Thus, enzymatic treatment by using pectinase is an effective way to reduce the pectin in the fruit juices or pectin contained by-products, because pectinase has the ability to hydrolyze pectin (Sin *et al.*, 2006) and also obtain oligosaccharides.

Dragon fruit pulp had also been reported as a potential source for pectic oligosaccharide production (Kewyok, 2015). The chemical compositions of dragon fruit peel (white flesh cultivar) were moisture as the major component and 4.35% carbohydrate, 2.14% ash, 1.84% crude fiber, 0.83% protein and 0.03% crude fat (Table 7.5).

The optimal conditions for extraction of extracts (main component was pectin) were peel paste and acidified water (pH 3.5) at a ratio of 1 and 2 at 85°C for 5 h. Under these optimal conditions, the highest yield was obtained (26.82% dry basis). The chemical composition of pectin extract powder consisted of carbohydrate as a major component (72.42%); and moisture, ash, crude fiber, protein and crude fat were 13.30, 11.71, 0.26, 2.52 and 0.05, respectively (Table 7.6) (Kewyok, 2015).

The purity of pectin in the dragon fruit peel was 60.07%, based on detection by galacturonic acid content. Production of pectic oligosaccharide from dragon fruit pectin was succeeded by enzymatic approach using pectinase depolymerization in an enzymatic membrane reactor (EMR). The optimal conditions for production of pectic oligosaccharide from dragon fruit peel pectin were 5 kDa MWCO membrane, 0.5 bar, 0.1 U/ml commercial *A. oryzae* pectinase, 9 h in continuous mode. The highest pectic oligosaccharide (25.71%) was obtained. In addition, pectic oligosaccharide from dragon fruit peel showed prebiotic properties, since it was resistant to upper gut digestion and promoting growth of lactobacilli in *in vitro* batch feacal fermentation (Kewyok, 2015).

Table 7.6 Chemical composition of pectin extract from white flesh dragon fruit peel

Proximate content (%)	Percent (dry basis)
Protein	2.52 ± 0.02
Crude Fat	0.05 ± 0.02
Moisture	13.30 ± 0.03
Ash	11.71 ± 0.01
Crude Fiber	0.26 ± 0.01
Carbohydrate	72.42 ± 0.02

Pectic oligosaccharide is also produced from dragon fruit pulp. Dasaesamoh *et al.* (2016a; 2016b) reported that dragon fruit pulp is a potential source of natural prebiotics since it contains oligosaccharides. The oligosaccharides from dragon fruit pulp are non-digestible oligosaccharides, since they were resistant to hydrolysis by human α-amylase in the mouth, artificial human gastric juice in the stomach and human α-amylase with sucrase giving maximum hydrolysis of 6.7%, 0.6% and 4.81%, respectively. Faecal fermentation of the oligosaccharides showed increases in the populations of *Bifidobacteria* and *Lactobacillus* and decreases in the populations of *Bacteroides* and *Clostridium*. It has a positive prebiotic effect with a prebiotic index (PI) of 0.41. Acetic, lactic, propionic and butyric acids were produced in concentrations of 860, 265, 15.95 and 29.63 mM, respectively. The mixture of these short-chain fatty acids could inhibit Caco-2 cells (Dasaesamoh *et al.*, 2016a).

The optimum conditions for oligosaccharide extraction from dragon fruit pulp were 85°C for 5 h. The highest yield of oligosaccharide was 43.98% (dry basis). The molecular weight of oligosaccharides derived from the extraction with water was 790 Daltons. The optimum conditions for extraction with pectinase were a pulp to water ratio of 1:2, pectinase of 124 units/g solid of extract, at 40°C and extraction time of 45 min. The yield of oligosaccharides was 41.92% (dry basis). The molecular weights of oligosaccharides derived from the extraction with pectinase were in the range of 1609 Daltons. The yields of oligosaccharides extraction by water and pectinase were not significantly different ($P > 0.05$). However, the extraction of oligosaccharide by pectinase could reduce the reaction time very significantly ($P < 0.05$) (Dasaesamoh *et al.*, 2016b). Therefore, by-products from dragon fruit processing have potential for production of prebiotic and dietary fiber.

7.2.2 Xylo-Oligosaccharide (XOS)

Xylo-oligosaccharide (XOS) is the great prebiotic and is one of emerging prebiotic sources. The chemical structure of XOS is β-1,4-(D-xylose); n = 2~6, that are sugar oligomers made up of xylose units. They consist of a main backbone of xylose linked by β 1→4 bonds. Xylo-oligosaccharides are mostly composed of non-digestible xylose-based disaccharide (xylobiose, 35%) and 60% total fiber (non-digestible oligomers; n = 3 – 6) with xylose backbones (Aachary and Prapulla, 2009). For food applications, xylobiose (DP = 2) is considered to be a xylo-oligosaccharide (Vazquez *et al.*, 2000), as depicted in Figure 7.2.

Xylo-oligosaccharides can be used as ingredients in foods and beverages, such as dairy products, gelatin desserts or salads, cereal-based products, energy bars,

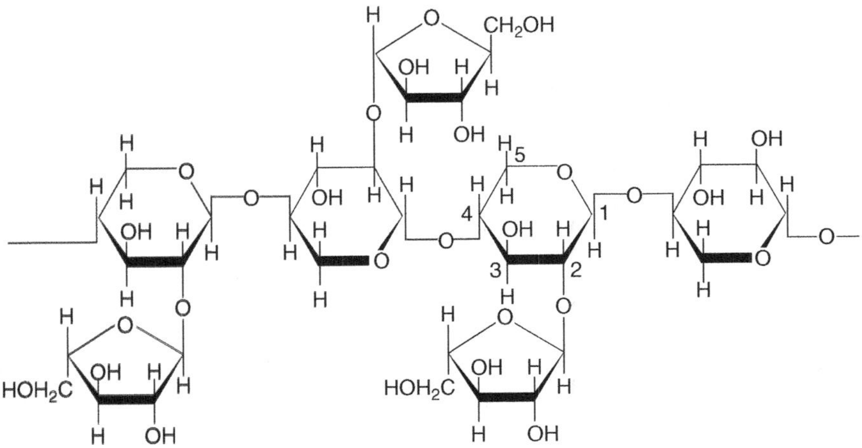

Figure 7.2 Chemical structure of xylo-oligosaccharide.

processed fruits, functional drinks, chewing gums and medical foods (Vazquez *et al.*, 2000). Xylo-oligosaccharides are produced by chemical methods, autohydrolysis, direct enzymatic hydrolysis of susceptible substrates, or a combination of chemical and enzymatic treatments (Moure *et al.*, 2006; Vazquez *et al.*, 2000). The substrates for production of XOS could come from by-products of industrial food and agriculture processing, for example corncobs, hard woods, hazelnut skins, wood sawdust and barley solid by-products. In commercial production of xylo-oligosaccharides, major sources of raw materials are corncobs and hardwoods, particular birchwood and larchwood. The chemical structure of hard and soft woods consisted of 30–35% and 25–30% hemicelluloses, respectively, which was composed of xylans. Hemicellulose is non-soluble dietary fiber, which is of benefit to gut health by forming bulk in stools and help the regularity of bowel transit (Perez-Hidalgo *et al.*, 1997).

Xylan is a major component in hemicellulose and is the starting compound that can be converted to xylo-oligosaccharide by action of xylanase. Xylan is a carbohydrate polymer of xylose connected by β-1,4-linkage in backbone and branched by substitution with various sugars such as arabinose and glucuronic acid (Sunna and Antranikian, 1997).

Process for commercial production of XOS from hardwoods is shown in Figure 7.3 (Van Laere *et al.*, 2000). It has several steps by chemical and/or enzymatic approaches. The first step is degradation of cellulose (lignocelluloses) in wood, which is called the 'lignocelluloses method (LCM)'. This method is used for removal of cellulose, hemicellulose and lignin. The second step is the hemicellulose degrading step by various treatments using (i) high pressure steam, and (ii) as a pretreatment for the enzymatic hydrolysis of cellulose (Dekker and Wallis, 1983; Grethlein and Converse, 1991). The final step is separation and purification of xylo-oligosaccharide by the chromatographic method.

Combination of thermal treatment coupled with solvent extraction for production of XOS is proposed, as shown in Figure 7.4 (Vazquez *et al.*, 2005; Swennen *et al.*, 2005; Vegas *et al.*, 2005). Solvent extraction can remove non-saccharide compounds to yield both a refined aqueous phase and a solvent-soluble fraction that mainly contains most of the phenolics and extractive-derived compounds. The recovery yields and the

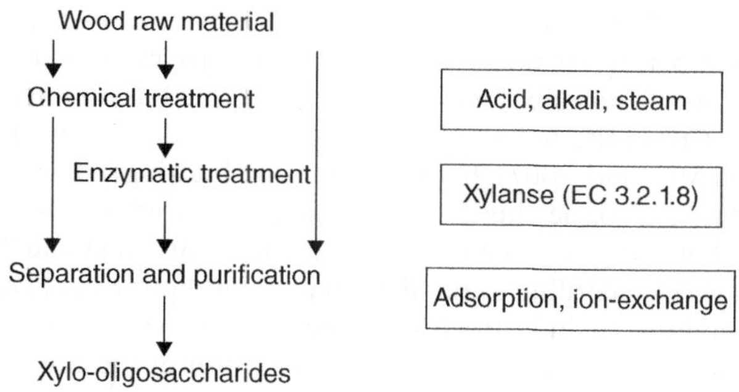

Figure 7.3 Xylo-oligosaccharides production from wood raw material.

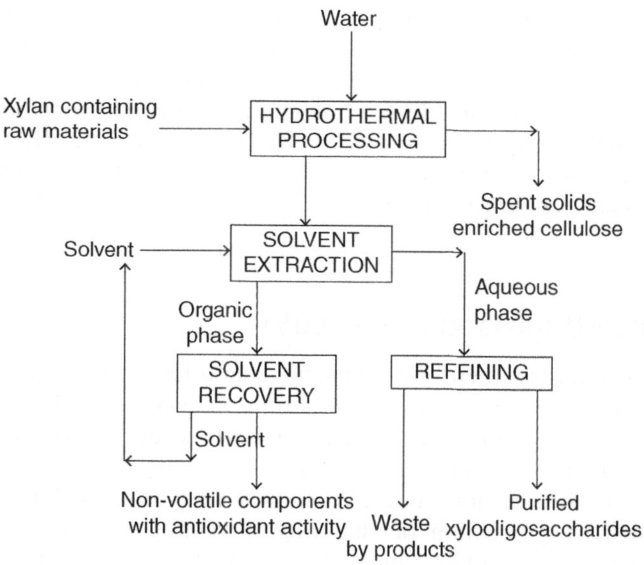

Figure 7.4 Hydrothermal treatment coupled with solvent extraction for production of purified XOS from xylan-containing lignocellulosic materials.

degree of purification depend on the solvent employed for extraction, with ethanol, acetone and 2-propanol being the most common choices to refine crude XOS solution.

In the food market, maize is the third-most widely grown cereal crop in the world. The total production of maize is around 822 million tons from approximately 160 million hectare areas of land (FAO, 2008). It is estimated that corncobs account for 15–20% of stover yield (John and Antonio, 2007). The availability of corncobs as a by-product is about 123–164 million tons per year. Corncobs are a major raw material used for commercial production of XOS. The method used is the enzymatic approach using endo-1,4,-β-xylanase to degrade xylans in corncobs to xylobiose, xylotriose and xylotetraose (Chen *et al.*, 1997).

Barley (*Hordeum vulgare* L.) was one of the first cultivated grains and is now grown worldwide. Barley grain is a staple starch-based food item, which has been widely consumed in both America and Europe. There are two major non-starch polysaccharides in barley, i.e. mixed linkage of β-glucans and arabinoxylans (Henry, 1987). β-glucans are dietary fiber and have been accepted as an immune modulator (Du *et al.*, 2015). Currently, β-glucans are being confirmed on their prebiotic properties (Laroche and Michaud, 2007). In the malting and brewing industry, an abundance of solid by-products, i.e. barley husks, fragments of barley grains, spent barley grains, malt bagasse and brewery residual yeast are obtained (Brennan and Tudorica, 2007; Mathias *et al.*, 2014). According to the three main world producers (America, China and Brazil) in the brewing industry, it generated about 16.9 million tonnes per year of malt bagasse and 21.1 million tonnes per year of brewery residual yeast (Mathias *et al.*, 2014). These barley by-products are rich in arabinoxylans and β-glucans, respectively, which can be used as functional ingredients and dietary supplements. Barley xylans consist of D-xylopyranose units connected in a series as backbone and it is substituted with arabinose, uronic acid moieties (or their 4-O-methyl ethers), and acetyl groups as branch chains (Garrote *et al.*, 2004). Production of xylo-oligosaccharide from barley husks has several steps. First, barley husks are treated in an aqueous media under suitable operating conditions (autohydrolysis reaction). The catalytic action of hydronium ions resulting from water ionization and from *in situ* generated acids (such as acetic acid coming from acetyl groups) leads to the cleavage of the xylan heterocyclic ether bonds. Second, unbound xylans are further hydrolysed, resulting in the formation of xylo-oligosaccharides (Rodolfo *et al.*, 2005).

7.2.3 Chito-Oligosaccharide (COS)

In the seafood industry, there are abundant quantities of by-products generated from food processing of crustacean such as shrimp, crab and shell. These by-products are shrimp shells and crab shells. In the last few decades, these have been considered to be physically solid waste and useless, but later have become used as biofertilizer. However, in more recent years, they have been used as raw materials for the commercial production of chito-oligosaccharide (COS). Applications of chito-oligosaccharide are suitable for use in both the food and non-food industry. In the food industry, it is being produced in the form of a dietary supplement; however, it is not allowed for consumption in some countries due to its stringency limit on solubility properties (Lodhi *et al.*, 2014). Chitosan is a linear biopolymer composed of randomly distributed β-(1,4)-linked D-glucosamine (deacetylated unit) and N-acetyl-D-glucosamine (acetylated unit). Chitosan is commonly found in the shells of animals, especially in the crustacean groups, for example crabs, shrimp, prawns and krill. Chitosan is produced commercially by deacetylation of chitin, with molecular weights between 3800 and 20000 Daltons. A common method for production of chitosan is the deacetylation of chitin using sodium hydroxide in excess (Shahidi and Synowiecki, 1991). The chito-oligosaccharide is commercially manufactured from chitosan by chemical and enzymatic hydrolysis with glycosyl hydrolases, such as chitinases or chitosanases (Knaul *et al.*, 1999). Chito-oligosaccharide is a low molecular weight and partially water soluble version of chitosan. Its chemical name is 2-amino-β-1,4-glucose polymer and its molecular formula is $(C_6H_{11}O_4N)_n$. Chemical structures of chitin, chitosan and chito-oligosaccharide are compared in Figure 7.5 (Lodhi *et al.*, 2014).

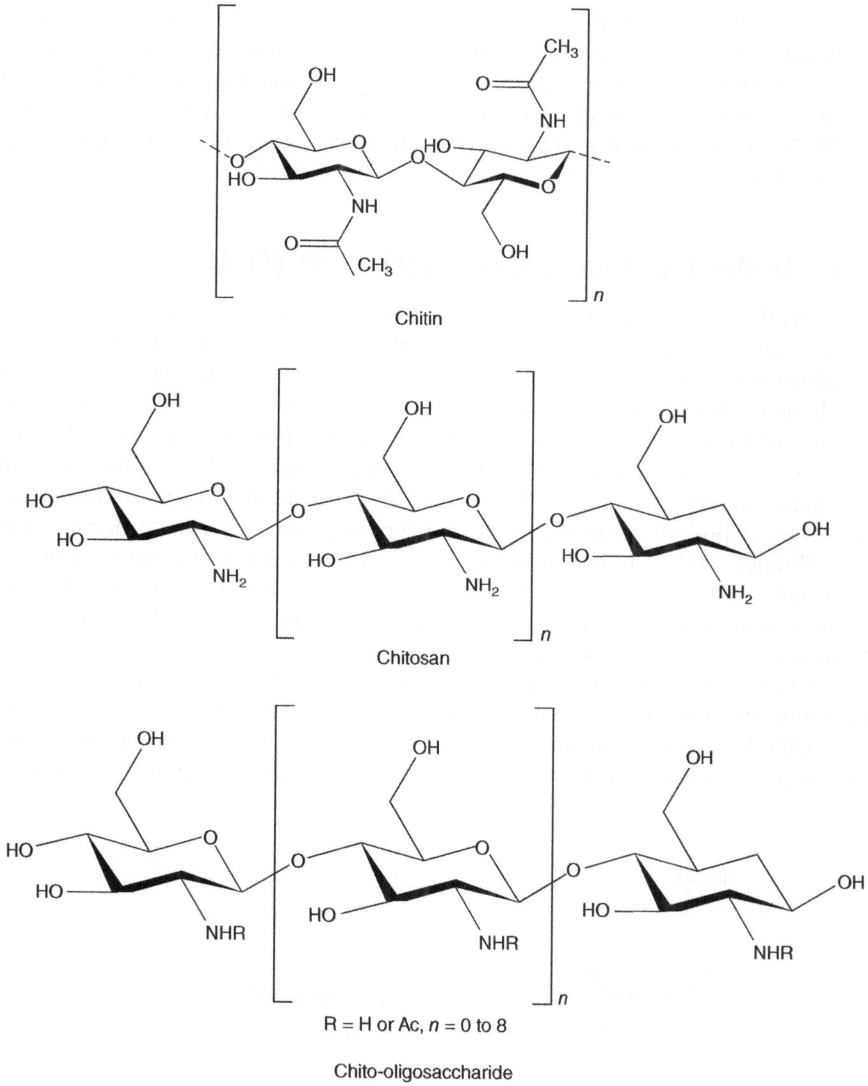

Figure 7.5 Chemical structure of chitin, chitosan and chito-oligosaccharide.

The low viscosity and greater solubility of COS at neutral pH has attracted the interest of many researchers to utilize chitosan in its oligosaccharide form. Chito-oligosaccharide and its derivatives have shown various functional properties that have made it possible to use them in many fields, including foods and biological applications. Applications in functional food, the disaccharides and trisaccharides, have various functions such as regulating blood pressure, reducing cholesterol, increasing immunity and improving water retention of food (Shahidi *et al.*, 1999), in cosmetics (Kumar, 2000) and biological applications (Lodhi *et al.*, 2014). In medicine, COS has physiological functions in various respects. Especially hexasaccharide can be used as a medicine for treating cancer in its early stages, by controlling the producing of the inner wall of tumor blood capillaries (Felt *et al.*, 1998). In the agricultural sector, it can be used as an adjusting agent for plant growth and new biological pesticides. Tetrasaccharide, pentasaccharide and hexasaccharide of COS have outstanding functions in anti-plant diseases and promoting growth (Yamada *et al.*, 1993), environmental protection (Peniche-Covas *et al.* 1992), and wastewater treatment (Jeuniaux, 1986).

7.2.4 Inulin and Fructo-Oligosaccharide (FOS)

Nutraceuticals and functional foods are an emerging promising market for health-conscious consumers. These heathfood products and dietary supplements contain various potent bioactive compounds that ameliorate a wide range of significant health benefits. Prebiotic ingredients are essential items of bioactive compounds, which are added to functional food or consumed in supplementary form. Inulin is one type of polysaccharide that has shown some potential prebiotic properties and is beneficial for enhancement of human gut health. Inulin molecules consist of a linear series of β-(2→1) fructose units, and typically have a terminal non-reducing glucose (Franck, 2002), as shown in Figure 7.6. In native inulin, the total number of fructose units can go up to over 60. Inulin is affordable and unsweetened, because the degree of polymerization (DP) is higher, whereas lower DP (oligofructose and FOS) increases sweetness (Crittenden and Playne, 1999).

Researchers have been investigating the improvement of quality of inulin by hydrolyzing the structure of inulin to a smaller form that is fructo-oligosaccharide (FOS), called oligofructose or oligofructan, used as an alternative sweetener. Fructo-oligosaccharide exhibits sweetness levels between 30 and 50% of sugar

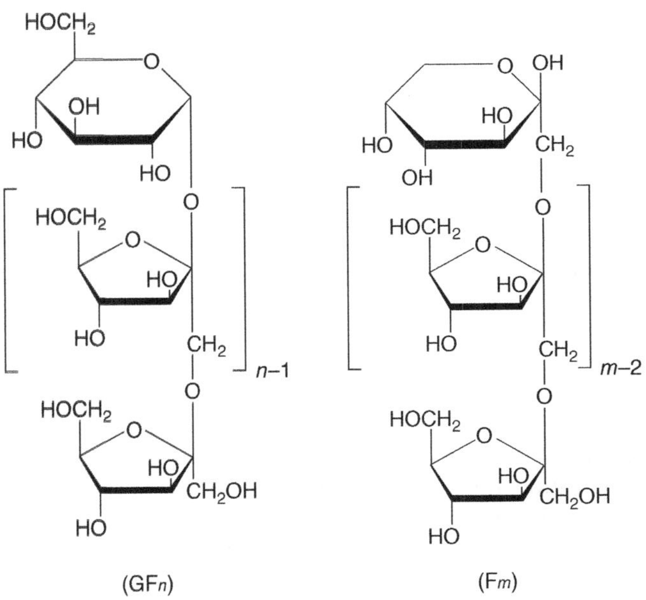

Figure 7.6 Chemical structure of inulin.

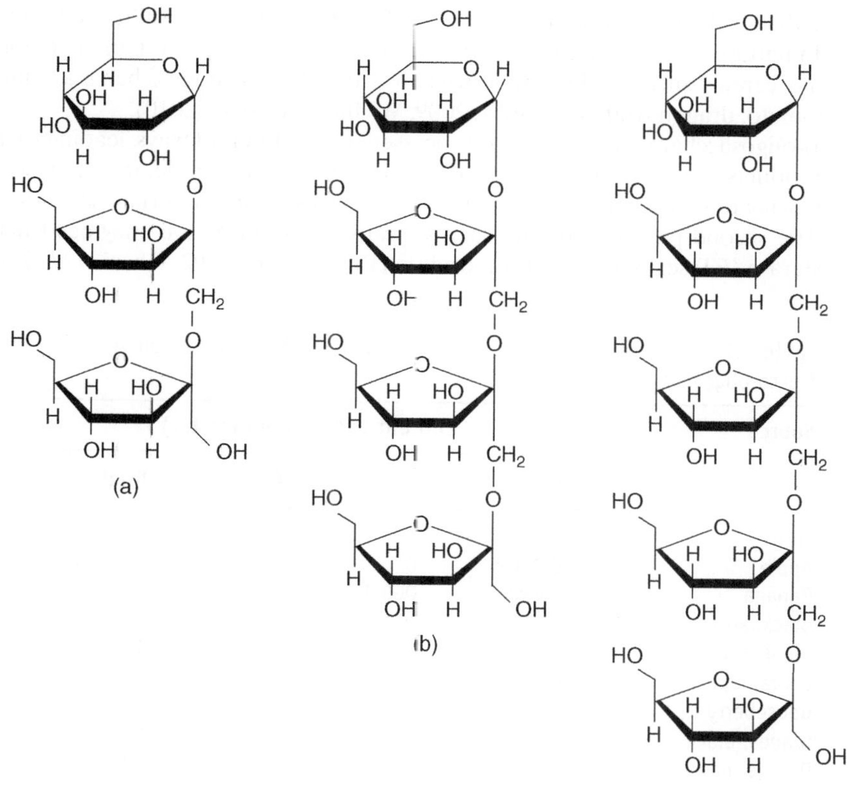

Figure 7.7 Chemical structure of fructo-oligosaccharide: a) 1-kestose (GF_2); b) nystose (GF_3); and c) 1-β-fructofuranosylnystose (GF_4).

in commercially prepared syrups (Joseph, 2008). Fructo-oligosaccharide has low molecular weight with degrees of polymerization of up to 10 linked by β (2-1) fructose oligomers of D-fructose and it may contains one molecule of D-glucose at the reducing-end (Lima *et al.*, 2011). In the FOS structure, fructosyl-glucose linkage is always β (2↔1), whereas fructosyl-fructose linkages are β (2→1) (Waterhouse and Chatterton, 1993). The main components of FOS are 1-kestose (GF_2), nystose (GF_3) and 1-β-fructofuranosylnystose (GF_4), as shown in Figure 7.7 (Lima *et al.*, 2011). Both functional and technological properties of FOS depend on the degree of polymerization of the molecular structure of the compounds (the number of glucose and/or fructose units).

Fructo-oligosaccharide can be applied not only for human consumption but also for animal feed formulation to enhance the growth and production efficiency of broilers (Wu and Wu, 1999). They are classified as non-digestible oligosaccharides because of the β-linkages between fructose monomers, which cannot be hydrolyzed by human endogenous enzymes of endogenous origin (Xu *et al.*, 2002). They are shown to enhance the growth of Bifidobacterium and Lactobacillus, but inhibit *Escherichia coli* and Salmonella in the large intestine (Fukata *et al.*, 1999; Xu *et al.*, 2002). They also promote calcium and magnesium absorption in both animals and the human gut, and increase the levels of phospholipids, triglycerides and cholesterol (Gülizar and Sibel, 2011). In nature, FOS is especially found in plants and a large portion of vegetables, fruits and cereals such as chicory, Jerusalem artichoke, asparagus, banana, onion, garlic, tomato, dragon fruit etc. (Yildiz, 2009; Wichienchot *et al.*, 2010).

Fructo-oligosaccharide production is processed in two different ways, leading to different products. The first production process is based on inulin degradation, when the products are longer fructo-oligomer chains. It is also call 'oligofructose' (Franck, 2002). The second process employs sucrose transformation by the enzymes fructosyltransferase (FTase) or β-fructofuranosidase (Bekers *et al.*, 2002; Ghazi *et al.*, 2007;

Table 7.7 Potential by-products from fruits and vegetables for production of fructo-oligosaccharide

Source	Content (g/kg of dry matter)			
	GF_2[a]	GF_3[b]	GF_4[c]	Total[d]
Fruit				
Apple	0.2–0.6	0	0	0.2–0.6
Banana	1.8–8.6	0–0.1	0–2.3	1.8–10.9
Blackberry	0	0	1.2	1.2
Blueberry	0.2	0.1	0	0.3
Cantaloupe	0.3	0	0.4	0.7
Gooseberry	0.6	0	0.6	0.8
Grapes, black	0.5	0	0.6	1.1
Orange, navel	1.7	0	1.1	2.8
Peach	3.5	0	0	3.5
Pear	0.3–0.8	0	0–1.1	0.8–1.4
Plantain	1.1	0	0	1.1
Plum	1.8	0.2	0	2.0
Raspberry	1.4	0.1	0	1.5
Watermelon	2.8	0	0.1	2.9
Vegetables				
Acorn squash	1.4	0	1.9	3.3
Artichoke, globe	13.4	5.5	2.8	21.8
Carrot	1.4–2.2	0	0	1.4–2.2
Chicory root	9.1	6.1	5.9	21.0
Garlic	8.7	1.2	0.4	10.3
Jerusalem artichoke	93.9	94.3	98.1	286.2
Leek	3.4	0.6	0.7	4.8
Lettuce	4.9	1.9	1.1	7.9
Onions	5.8–17.1	2.1–8.8	0.9–6.1	13.4–32.0
Pea	0.2–1.0	0–0.2	5.4–8.0	6.5–8.4
Shallot	28.2	14.2	10.6	52.9

Note :
[a]1-Kestose.
[b]Nystose.
[c]1^F-β-Fructofuranosylnystose.
[d]Total fructo-oligosaccharide.

L'Hocine *et al.*, 2000; Sangeetha *et al.*, 2004). Currently, FOS is mainly produced on an industrial scale by enzyme FTase, an enzyme responsible for the production of fructo-oligosaccharides from sucrose, using banana peel, corncobs, coffee husk, spent coffee, spent tea, sugarcane bagasse, cassava bagasse and cereal brans (from wheat, rice and oat), and cork oak as solid substrates (Sangeetha *et al.*, 2005). Current commercial production of FOS is made by depolymerization of inulin with inulinase, using chicory and Jerusalem artichoke inulin as starting materials. Alternatively, the commercial production of FOS by fructosyltransferase was done by using sucrose as a substrate. The main incursion of production cost comes from raw materials or substrates. Technically, the possible alternative way to reduce production costs can be accomplished by using any type of agricultural by-products. Agricultural by-products that can be used for the production of FOS by fermentation are from fruits and vegetables processing and agricultural by-products. Potential sources of fruits and vegetables that contain a high content of FOS are summarized in Table 7.7 (Franck and Leenheer, 2005).

Onion by-products also have potential commercial production of FOS. Fructo-oligosaccharides present in onion are about 40% (in dry weight), with a degree of polymerization (DP) ranging from DP 3 to DP 11 (Darbyshire and Henry, 1978). The commercial production of FOS by FTase immobilization is depicted as a flow chart in Figure 7.8 (Katarína *et al.*, 2008).

7.2.5 Soybean Oligosaccharide (SOS)

Many types of beans have been progressively grown around the world, but only soybean production reached 264.25 million metric tonnes in 2010/2011. The USA, Brazil and Argentina contribute about 92.75% of all soybean production (USDA, 2011). Soybeans (*Glycine max* L. Merrill) were first grown as a dominant crop in China about 5000 years ago (Liu *et al.*, 2008). They have been widely-consumed as traditional folk medicines in China, India, Japan and Korea for hundreds of years (Jeng *et al.*, 2010). There are many kinds of soybean cultivars possessing different biological compositions and profitably impacting the economy. According to the consensus recommendations of the Organization for Economic Cooperation and Development, soybean nutrients (i.e. amino acids, fatty acids, isoflavones) and anti-nutrients (i.e. phytic acid, raffinose and stachyose) are important markers in assessment of the nutritional quality of soybean varieties (Jiao *et al.*, 2012). Presently, soybean production has been expanded to most of the continents and 90% of the world's soybean production is concentrated in tropical and semi-arid tropical regions, which are characterized by high temperatures and low or erratic rainfall. In the tropics, most of the crops are near their maximum temperature tolerance (Thuzar *et al.*, 2010).

Soybean also contains a wide range of micronutrients and phytochemicals, including minerals, vitamins, phytic acids (1.0–2.2%), sterols (0.23–0.46%) and saponins (0.17–6.16%) (Kang, 2010). The primary inorganic compounds of soybeans are minerals. Potassium is found in the highest concentration, followed by phosphorus, magnesium, sulfur, calcium, chloride and sodium, which vary in concentration depending on the variety, growing location and season (Erickson, 1995). On average, oil and protein together constitute about 60% of dry soybeans. The remaining dry matter is composed of mainly carbohydrates (~35%) and ash (5%) (Liu, 1997).

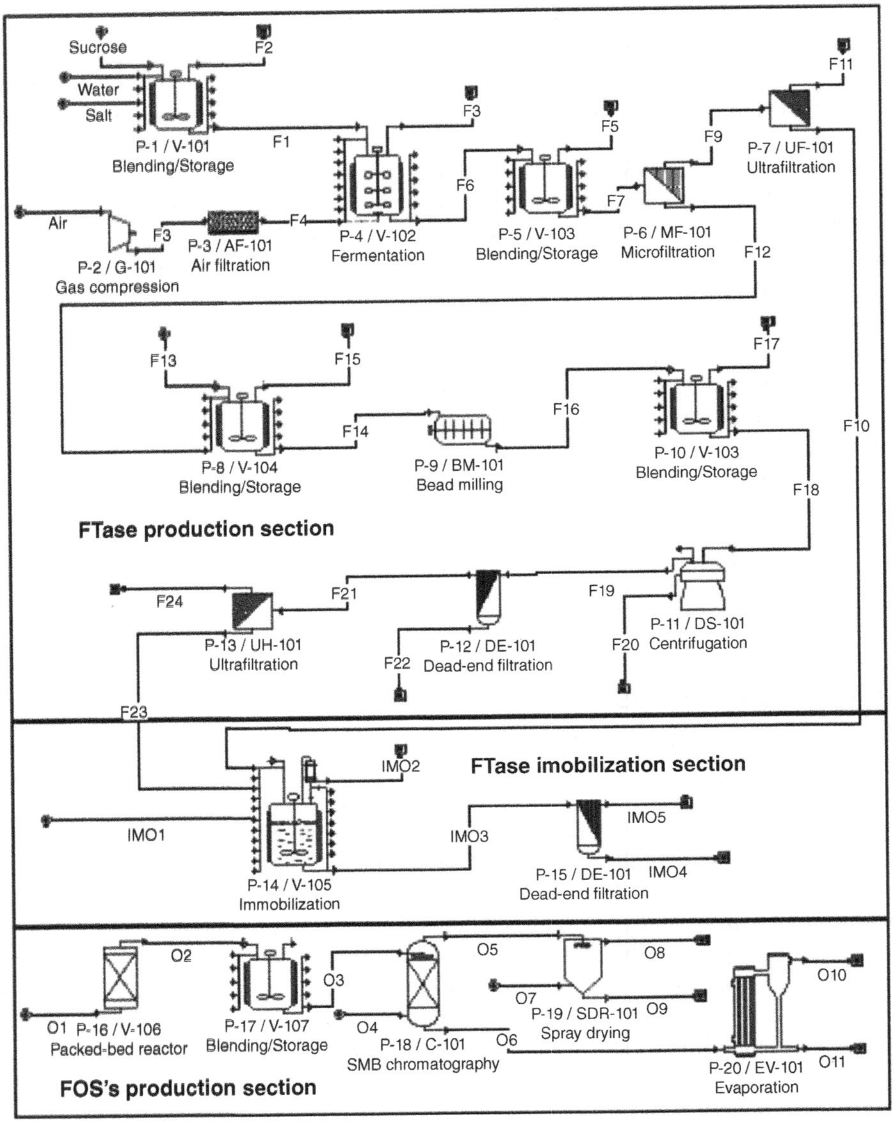

Figure 7.8 Process flow sheet of the industrial production of FOS consisted of FTase production section, FTase immobilization section, and FOS production section.

Soybean processing (Figure 7.9) begins with solvent extraction, which involves washing the soybeans to remove foreign material and drying them to approximately 9.5% moisture content (Toda *et al.*, 2000). The beans are tempered where the moisture is allowed to equilibrate and loosen the hull, which enables the dehulling process. The dehulled soybeans are known as 'meats'. There are several advantages for removing the hull; less total volume to process, which means greater extraction output per day, and improved protein content of the meal. The cracked soybean meats are heated to about 165° F to soften them prior to flaking. The flakes are

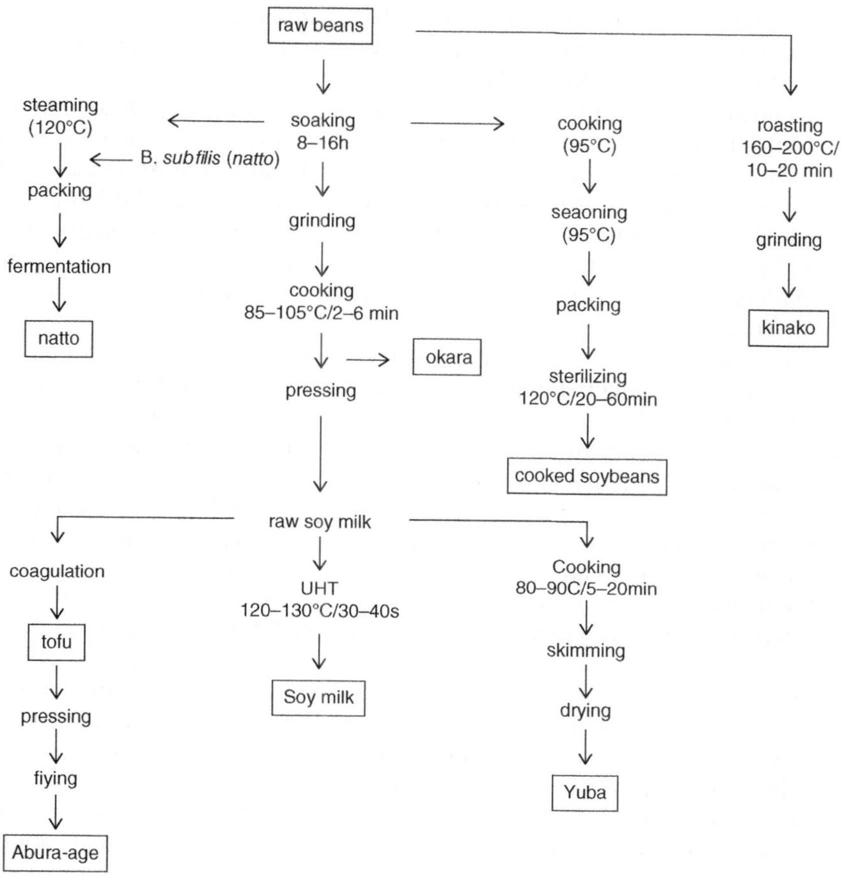

Figure 7.9 Flow diagram illustrating the processing of soybeans for commercial products.

washed in a countercurrent manner with hexane, a petroleum distillate. This solvent solubilizes the soybean lipid material, and through a few series of subsequent steps, the hexane-oil mixture is then separated. The oil-rich extract called 'miscella' is evaporated and the solvent is recycled to the extractor. The soybean flakes, with the oil removed, are called 'spent flakes'. From the extractor, the spent flakes are conveyed to a desolventizer-toaster for removing any undrained hexane. The process involves heating the spent flakes to evaporate the hexane and sparging steam to carry away hexane vapors. This process also provides for toasting of the meal to inactivate enzymes like urease and trypsin inhibitors that may reduce the digestibility and nutritional value of the meal. From the desolventizer-toaster, the meal will go to a dryer-cooler system where it is dried to about 13–14% moisture and cooled for safe storage. The finished meal from dehulled soybeans will contain less than 1.5% crude fat and about 48% protein, and is referred to as high protein meal.

The United Soybean Board has estimated that about 46% of the soybean meal produced in the USA is used to feed broilers, layers and turkeys. Swine use another 25% of soy meal production. Ruminants, beef cattle, dairy and sheep account for about 21% and the remaining 8% finds its way into a large variety of uses (pet foods, aquaculture, minor feed uses, food and industrial uses etc.).

In the nutrition market, processed soy bean known as soymilk has a good nutrient content for human consumption. It can be used in the preparation of a few cooked foods such as beverages and tofu. Tofu is a soymilk curd produce made through a non-fermented process from soymilk (soybean processing) (Fukushima, 1981). Although the word 'tofu' is derived from Japanese, this food item seems to have originated in ancient China, where the original Mandarin term is doufu (Obatolu, 2008). With respect to nutrition, tofu is cholesterol free, low in saturated fat and high in protein (Li *et al.*, 2013).

Tofu, also known as soybean curd, is a soft, cheese-like food made by curdling fresh hot soymilk with a coagulant. It is produced traditionally by curdling fresh hot soymilk with either $CaCl_2$ or $CaSO_4$, or an acid (glucono-δ-lactone). The coagulant produces a soy protein gel, which traps water, soy lipids and other constituents in the matrix forming curds. The curds are then pressed into solids (Obatolu, 2008). It is inexpensive, nutritious and versatile meat or cheese substitute, with bland taste and porous texture. On the moisture-free basis, tofu contains about 50% protein and 27% oil, and the remaining constituents are carbohydrate and minerals. Several processing factors affect the quality and texture of tofu, such as cultivar of soybean, processing methods and type of coagulants (Rekha and Vijayalakshmi, 2011). During the processing of soybean, such as *miso* (soybean paste), *shoyu* (soy sauce), *natto* (fermented soybeans) and *tofu,* there is some waste that becomes the main cause of water pollution. The waste water reported by Kawashima (1983) has protein, sugar, fat and ash content. The sugar derived from soybean prepared by steaming and/or boiling techniques is generally known as soybean oligosaccharides (SOS).

Soybean oligosaccharides (SOS) are a group of soluble low molecular weight oligosaccharides in soybean seeds, which include sucrose, stachyose and raffinose. SOS are defined as non-digestible oligosaccharides (NDOS) or non-digestible sugars (NDS), except sucrose (Trugo *et al.*, 1995; Voragen, 1998), since the human gastrointestinal tract does not possess α-galactosidase enzymes essential for hydrolysis of the α-1,6 galactosyl linkages (Reddy and Salunkhe, 1980).

Soybean oligosaccharides are oligosaccharides of the raffinose family or simply α-galactosides. They are water-soluble, low-molecular weight, containing ≥50% stachyose, ≥14% raffinose and ≥16% verbascose. In matured seeds, oligosaccharides are formed by successive addition of galactosyl moieties to a sucrose primer. α galactosides are characterized by the presence of $\alpha(1{\rightarrow}6)$ linkages between galactose moieties, which are bonded via $\alpha(1{\rightarrow}3)$ to terminal sucrose (Figure 7.10) (Choct *et al.*, 2010). Unlike other oligosaccharides, soybean α-galactosides can be extracted directly from the raw material and do not require enzymatic manufacturing processes. Soybean oligosaccharides comprise approximately 4% of the soybean dry matter (DM) and during processing in the preparation of soybean meal (SBM), they are not removed or destroyed. Therefore, in SBM, α-galactosides represent approximately 5–6%, but could be as high as 8% DM. Other processed soybean products, however, may contain significantly lesser oligosaccharide content than SBM. The oligosaccharide content of soy protein concentrates (SPC) is as low as 3% DM, while soy protein isolates (SPI) contain only trace amounts of oligosaccharides. Recently, Sriniworn *et al.* (2015) reported that the production of soybean oligosaccharide by the hybrid process from tofu whey had higher recovery yield than commercial processing at 78.6%.

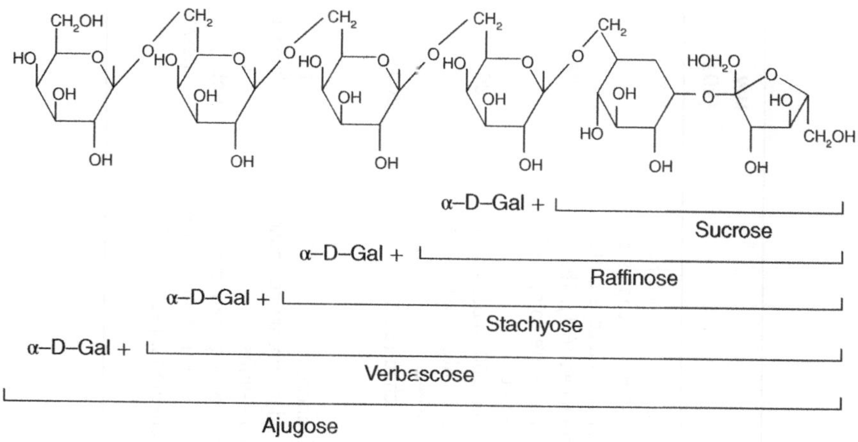

Figure 7.10 Chemical structure of soybean oligosaccharides.

7.3 Polysaccharides from Food Processing and Agricultural By-Products

7.3.1 β-Glucans

β-Glucans are conserved glucose polymers, which can be found in some macro-fungi (i.e. mushrooms), and unicellular fungi such as yeast (baker's yeast and *Candida albicans*) (Samuelsen *et al.*, 2014). All β-glucans are glucose polymers linked together through a 1 → 3 linear β-glycosidic chain, and they differ from each other by their lengths and branching structures (Stone and Clarke, 1992). The branches derived from the glycosidic chain are highly variable, and the two major groups of branching are 1 → 4 or 1 → 6 glycosidic chains. These branching assignments appear to be species-specific, fungal β-glucans having 1 → 6 side branches (Chan *et al.*, 2009). β-Glucans from various fungi have diverse immuno-modulatory activities both *in vitro* and *in vivo*. They have the capacity to activate innate immunity, thereby enhancing defense barriers. β-Glucans can vary in solubility, molecular mass, tertiary structure, polymer charge and solution conformation. Various β-glucan glycosidic linkages are shown in Figure 7.11 (Mantovani, 2008). The natural β-glucans occur in the form

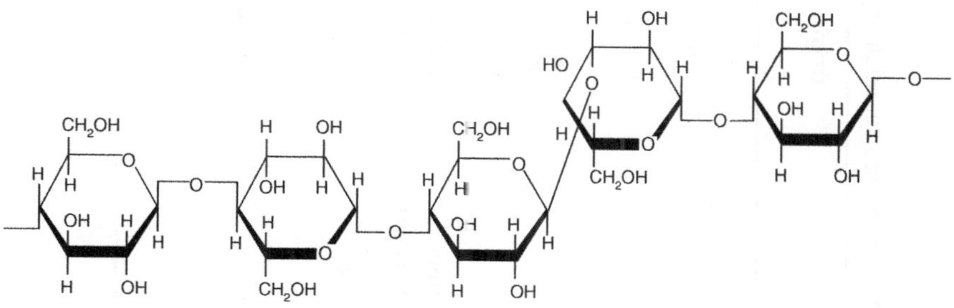

Figure 7.11 Chemical structure of (1,3) β-glucans.

Table 7.8 Some source, solubility and structure of the β-glucans

Name	Origin	Source	Solubility	Structure	MW (kDa)
Laminarin	*Laminaria digitate*	Algae, e.g. brown seaweeds	Soluble	β-1,3 with some β-1,6 branching (30:1). The β-1,6-side chains are composed of two glucose-units	7.7
Scleroglucan	*Sclerotium rolfsii*	Fungus	Soluble	β-1,3 β-1,6 branched (6:1). The β-1,6-side chains are composed of two glucose-units	1020
Curdlan	*Alcaligenes faecalis*	Gram negative bacteria	Particulate	β-1,3 unbranched 100. Glucan from *Euglena gracilis*. Algae particulate β-1,3 unbranched	500
Glucan from *Saccharomyces cerevisiae*	*Saccharomyces cerevisiae*	Yeast	Particulate	β-1,3 β-1,6 branched (30:1)	200
Macrogard	*Saccharomyces cerevisiae*	Yeast	Particulate	β-1,3 β-1,6 branched (10:1 or 20:1). The β-1,6-sidechainsare composed of 2 or 3 glucose-units	Not known
Zymosan	*Saccharomyces cerevisiae*	Yeast	Particulate	Crude extract with β-glucan, mannan and proteins, non-uniform branches and backbone units. β-1,3 β-1,6 branched	Not known

of Lentinans, Schizophyllan and PSK (Krestin). Lentinan is mushroom-extracted and has a triple helix structure with five (1,3)-β-glucose linear residues and two (1,6)-β-glucopyranoside side branches. Schizophyllan, from the mushroom of *Schizophyllum commune*, has β-glucopyranosyl 1-6 linkage every 3rd or 4th interval between the 1,3 units. It also had a triple-helix structure. PSK (Krestin) is composed of 25–38% protein residues and is a 1,4-β-glucan with 1,6-β-glucopyranosidic branch chains (Zimmerman, 1998).

β-Glucans are from various sources, as shown in Table 7.8 (Eva *et al.*, 2010). A potential prebiotic source is spent mushroom waste (SMW), which is a by-product derived from mushroom cultivation. Total mushroom production worldwide has increased more than 6-fold in the last three decades, from about 1.2 million metric tonnes in 1980 to about 7.3 million metric tonnes in 2010 (Sendi *et al.*, 2013). Worldwide, mushrooms are produced on natural materials obtained from agricultural sources, woods and the food industry (Danny, 2002). Some of these wastes include sawdust, banana leaves, peanut hulls, corn leaves and husks, sugarcane leaves, rice and wheat straw, cotton wastes, paper wastes, cocoa shells, wheat, and various other wastes (Cangy and Peerally, 1995; Jonathan, 2002). The widely used substrate for cultivation of mushroom in Asia is rice straw base, while in the Southeast Asian countries, sawdust is more commonly used nowadays (Thomas *et al.*, 1998).

The process for manufacturing β-glucans is shown in Figure 7.12 (Behall *et al.*, 2006). An initial process is monitoring mushroom production and harvesting by certified technicians to ensure optimal maturity as well as minimal contamination. After β-glucan is extracted, the subsequence unique process is initiated by isolating the β-glucans from the fungi cell walls in different stages. This procedure maximizes the production of β-glucans and preserves the effective polysaccharide linkage. The final process is purification of β-glucan to render the quality of the final product. Due to the traceable components adhering to the glycosidic chains of the β-glucan, the purification process is required to 'cleanse' the final product to decrease the risk of hypersensitivity associated with naturally-occurring impurities.

The health benefits of β-glucan are known to slow both gastric emptying and digestive transit rate in the small intestine, impede the actions of starch and other hydrolases and, consequently, improve the rate of nutrient absorption in the upper gut (Behall

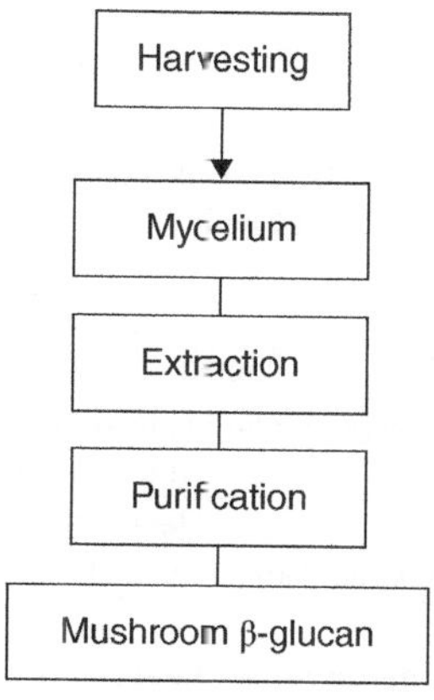

Figure 7.12 Flow chart of the β-glucans manufacturing process.

et al., 2006). Liatis *et al.* (2009) showed that consumption of β-glucan in the long term reduced fasting glucose and insulin levels and improved insulin sensitivity. Long-term studies in rats also showed an improvement in insulin sensitivity after consumption of a diet containing β-glucan for 12 weeks (Brockman *et al.*, 2012). However, concomitant reductions in body weight and adiposity may have been responsible for the observed improvements in insulin sensitivity (Choi *et al.*, 2010).

7.3.2 Non-Starch Dietary Fibers

Dietary fiber consists of non-digestible carbohydrate in polysaccharide group. They are called by many names such as either dietary fiber, or roughage, or bulk or non-starch polysaccharides. Dietary fiber is derived from plants and includes all parts of plant foods. This fiber is not digested in the body, but instead passes relatively intact through the stomach, small intestine, colon and out from the body. Soluble dietary fiber is able to dissolve in water to form a gel-like material and is readily or partially fermented in the colon, where it produces gases and physiologically active by-products. It can be a source of prebiotic and viscous or insoluble fiber that promotes the movement of material through the digestive system and increases stool bulk, so it can benefit individuals who struggle with constipation. In addition, these materials ameliorate irregular stools, as they are metabolically inert and provide bulking, or metabolically ferment in the large intestine. Bulking fibers absorb water as they move through the digestive system, and then ease defecation (Eastwood and Kritchevsky, 2005). Dietary fibers are also classified according to the methods used for determination of their dietary fiber content (Figure 7.13). Example of dietary fibers are arabinoxylans, cellulose and many other plant components, such as resistant starch, resistant dextrins, inulin, lignin, waxes, chitins, pectins, β-glucans and oligosaccharides (Anderson *et al.*, 2009; Farvid *et al.*, 2014).

Dietary fibers has attracted attention because of the significant health benefits they offer to consumers (Cheung, 2013; Tungland and Meyer, 2002; Lunn and Buttriss, 2007). Prospective studies were inconclusive regarding the protective effect of high dietary fiber intake on the risk of type 2 diabetes mellitus (Hopping *et al.*, 2010; Barclay *et al.*, 2007). Inconsistency in the results may be explained partly by differences in the analytical method used to estimate the dietary fiber intake and the errors arising from

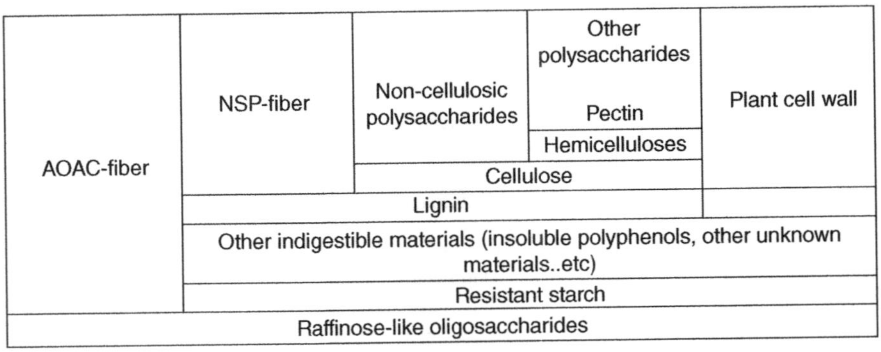

Figure 7.13 Constituents of total dietary fiber measured by the Association of Official Agricultural Chemists (AOAC) method and non-starch polysaccharides (NSP) measured by the Englyst method.

Table 7.9 Soluble (SDF), insoluble (IDF) and total (TDF) dietary fiber content of some fruit and vegetable wastes (values in g/100 g dry product) were obtained by the Association of Official Agricultural Chemists (AOAC) method)

Origin of the waste	SDF	IDF	TDF	Reference
Orange peel after pectin extraction	21.3	62.9	97.6	Aravantinos-Zafiris *et al.* (1994)
Apple pomace	18.6	69.9	88.5	Renard and Thibault (1991)
Apple pomace after pectin extraction	10.3	78.3	88.6	Renard and Thibault (1991)
Grape pomace	9.5	68.4	77.9	Valiente *et al.* (1995)
Beet pulp	12.5	60.3	72.8	Thibault *et al.* (1994)
Pea hulls	4.1	87.4	91.5	Ralet *et al.* (1993)

the dietary assessment tool that is commonly used in prospective studies. Nowadays, most dietary ingredients (e.g. cereals-based, fruits-based and legumes-based dietary fiber) are derived from their processing by-products (e.g. milling, juice extraction or dehulling), followed by different refining steps (e.g. grinding, sieving, bleaching or defatting) to accommodate a wide range of customer requirements (McKee and Latner, 2000; O'Shea *et al.*, 2012; Robin *et al.*, 2012). Due to the highly competitive market of fiber-enriched food products, exploration of alternative sources of dietary fiber are urgently required (Wong and Cheung, 2009). Our dietary fiber intake derives primarily from cereal products, fruits and vegetables, which are traditionally consumed after being prepared via various conventional preparation methods. However, foods can also be supplemented with dietary fiber. Fibers from cereals (bran essentially), or from fruits or vegetables, are produced and may be added to various foodstuffs in order to increase their dietary fiber content. Indeed, some industries have developed programs to obtain fiber from agricultural or agro-industrial by-products (Gelroth and Ranhotra, 2001). Table 7.9 shows the soluble, insoluble and total dietary fiber content of some typical wastes and by-products from fruit and vegetable processing.

Dietary fiber from corn is a high export value product, because there are many commercial products of corn such as dent corn, flint corn, flour corn, popcorn, sweet corn, waxy corn and pod corn. As early as 1880, the USA grew over 62 million acres of corn. By 1900, this reached approximately 95 million acres; by 1910, it was over 100 million acres. The highest acreage ever recorded in the USA was 111 million acres in 1917. In 2000, US farmers planted over 79 million acres of corn. More than 40% of the world's corn is produced in the USA (Lance and Garren, 2002).

There are four parts of corn kernel; the pericarp, endosperm, germ and tip cap. The pericarp (also known as the seed coat) consists of the bran component of the kernel (hull and fiber). The endosperm consists mostly of starch (70% of the kernel) and the gluten fraction, which is high in protein. Of the other components, the germ fraction is high in oil. There are two main industrial processes that are responsible for producing corn by-products for the dairy industry, which involve dry and wet milling. Approximately 40% of fuel ethanol is produced by wet-mills after the starch has been separated from the corn kernel, and these plants produce wet or dried corn gluten feed, corn gluten meal, and corn germ meal as the primary by-products. Corn is the primary grain used in wet mills and dry-grind ethanol plants, because of its high fermentable starch content compared to other feed stocks. Schematic diagrams comparing the corn wet-milling and the corn dry-grind processes, and the corn by-products produced from these types of ethanol plants, are shown in Figure 7.14 (Shurson *et al.*, 2005).

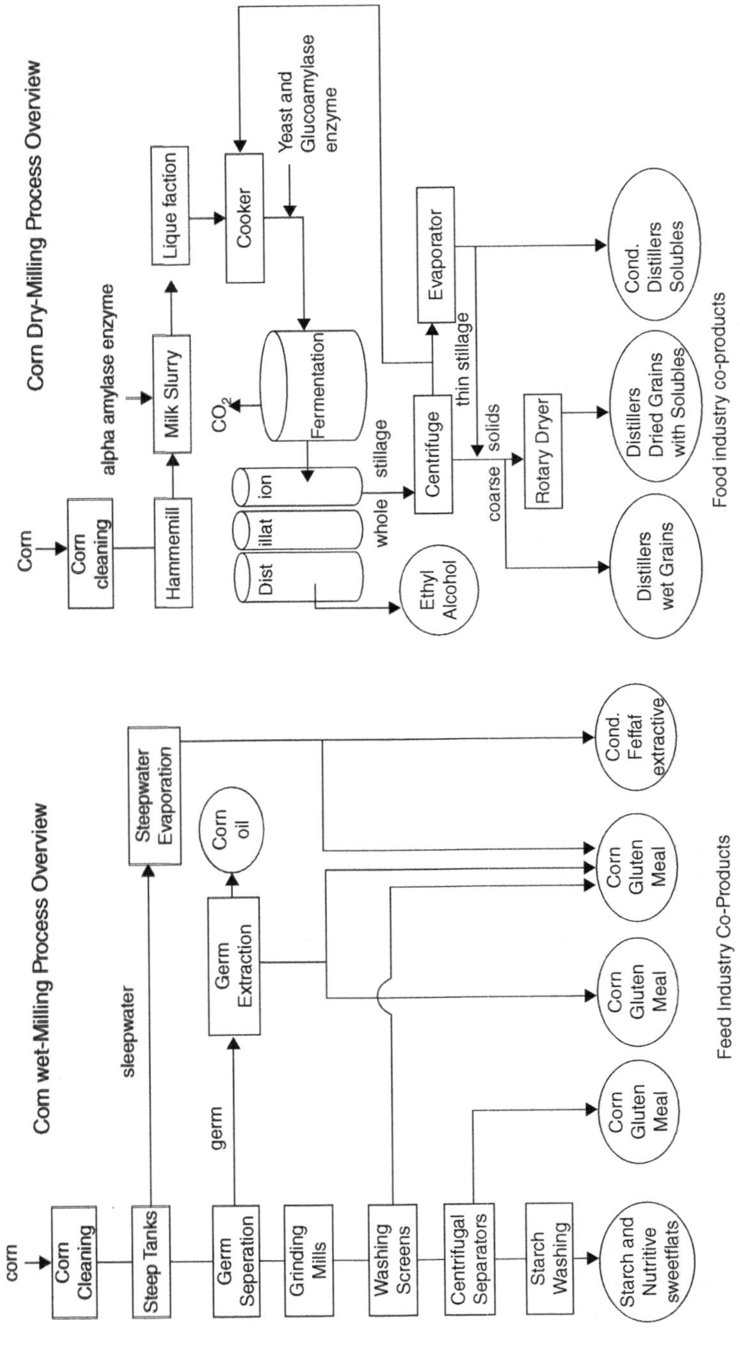

Figure 7.14 Comparison of corn wet-milling and dry-grind processes and the by-products produced.

The by-products from corn can be used in many applications. One of the agricultural by-products of corn is corn hairs, commonly known as cornsilk, which refers to the yellowish threads found inside the husks of young corn. These are the collection of the stigmas from the female flowers of the maize plant (Wan Rosli *et al.*, 2007). Dried cornsilk (also known as *Maydis stigma*) is rich in a variety of chemical composition such as protein (±13.0%), ash (±5.3%), fat (±1.3%), and particularly total dietary fiber (TDF) (±38.4%) (Wan Rosli *et al.*, 2011). Traditionally, cornsilk is well known for its use in folk medicine for diuretic treatment. Cornsilk have been reported to be useful in treating urinary infection and cystitis (Steenkamp, 2003). Besides, it is also thought to act against gout by inhibiting the xanthine oxidase (XOD) (Jiang *et al.*, 2011) and hyperglycemic effects (Li *et al.*, 2004). However, kernels of young corn are the only components for consuming purposes, while the other parts such as cornsilk are usually discarded during processing of young corn, due to the lack of knowledge of their benefits.

The use of cornsilk, which is rich in antioxidants and dietary fibers, has grown in the current decade and led to the development of a large market for antioxidant and fiber-rich ingredients and products, and some local species are consumed as tea, as food additives and as flavorings agents, in several regions of the world (Yesilada and Ezer, 1989). In the baking industry, the main ingredient for making biscuits, cookies or the like is flour. Currently, many bakery products in the market are formulated with oatmeal and wheat bran, which are expensive, to form high dietary fiber bakery products. Nevertheless, it increases the production cost of the bakery products, because if the flour has been substituted with a desired amount of cornsilk, the production cost can be reduced.

The other interesting product in corn production is baby corn or young corn, or cornlettes (young *Zea mays*), a type of vegetable commonly found in Asian cuisines, which has a sweet and succulent taste. Cornlettes are referred to as candle corn and appear yellow in colour with a size that is finger-length. Cornlettes are a unique product of corn, since the harvested part is the immature female inflorescence rather than mature plant (Hallauer, 2001). It is usually hand-picked prior to fertilization and before or just after the emergence of the cornsilk. Nutrients in cornlettes are comparable with other vegetables such as cauliflower, cabbage, tomato, eggplant and cucumber (Yodpet, 1979). Hooda and Kawatra (2013) reported that cornlettes are rich in dietary fiber (27.12 g/100 g). Natural dietary fiber in cornlettes may potentially be incorporated into food products as a source of functional ingredient for health promotional purposes. Other than being nutritious, it is also free from pesticides, because the cob is tightly protected in its husk (Chutkaew and Paroda, 1994). Cornlettes are a valuable and attractive product in both national and international markets.

Baby corn may be consumed raw or used as an ingredient in various preparations such as soups, deep fried with meat or rice, sautéed with other vegetables, pickles etc. The nutritive value of baby corn is comparable to several high-priced vegetables such as cauliflowers, cabbage, French beans, spinach, lady finger, brinjal, tomato, radish etc. (Hooda and Kawatra, 2013). The calorific and nutritional values in 100 g of baby corn are 26 kCal, 2.5 g protein, 3.1 g carbohydrate, 0.4 g fat and 1.7 g fiber. Total soluble sugars and reducing sugars content were found to be 23.43 and 1.96 g/100 g, respectively. Non-reducing sugars content is 21.47 g/100 g. Baby corn contained comparatively lesser amounts of total soluble and reducing sugars and higher non-reducing sugar than tomato (Vashisht, 1998) and cucumber (Verma, 2000). In addition, the starch content of baby corn was 15.60 g/100 g (Santosh and Asha, 2013) (Table 7.10).

Table 7.10 Carbohydrate content of baby corn (dry matter basis)

Parameters	Content (g/100 g)
Total soluble sugars	23.43 ± 0.23
Reducing sugars	1.96 ± 0.03
Non-reducing sugars	21.47 ± 0.13
Starch	15.60 ± 0.02

Note: Values are mean ± SE of three independent determinations

7.3.3 Resistant Starch

In the dietary fiber market, apart from non-starch dietary fibers from various fruits and vegetables, resistant starch can also be found in a wide variety of foods rich in starch, such as legumes, grains, roots and also derived from food ingredient added (Panlasigui *et al.*, 1991). The typical starches are digested rapidly and likely to raise blood sugar levels. However, high-amylose rice is less likely to increase blood glucose levels, due to it retarding amylase digestion that is useful for diabetes prevention (Bonithon-Kopp *et al.*, 2000; Soral-Oemietana and Wronkowska, 2004; Asp, 1992). Resistant starch is one type of dietary fiber that can increase faecal bulk (Cairns *et al.*, 1995). Resistant starch has four types; RS type I, II, III and IV, and is believed to account for 30% of the total fiber fraction in the diet (Englyst, 1989). It enters the large intestine where it is partially or wholly fermented by the resident bacteria, producing a variety of end-products including short-chain fatty acids (SCFA) that in turn provide a range of physiological benefits (Topping and Clifton, 2001). These short-chain fatty acids are useful as an energy source of cellular tissue particular, and butyrate is a major energy source of colonocyte (Schwiertz *et al.*, 2002). Butyrate has also been shown to facilitate other important physiological changes and modulate the apoptosis cycle to prevent normal cell change to cancer cell (Ferguson *et al.*, 2000; Mentschel and Claus, 2003). Importantly, resistant starch is one of the best substrates for the production of butyrate, producing twice that of wheat fiber and four times that of pectin (Champ, 2004). Interestingly, recent investigation showed that commercial RS type III was higher in the butyrogenic (butyrate production) effect than commercial RS type II and IMO, respectively. However, the bifidogenic (promote bifidobacterial growth) effect of IMO is higher than RS. Phrukwiwattanakul (2013) have reported on the production of resistant starch from different types of starch. It was found that resistant starch from jackfruit seed was 26.99% higher than rice starch and mung bean starch, which were 0.31% and 4.04%, respectively. The retrograded starch from jackfruit seed showed high resistance to simulated upper gut conditions of up to 96.34%.

Presently, resistant starch has been categorized into four groups. Resistant starch type I (RS1) is starch bound by indigestible plant cell walls occurring for example in not completely ground cereal grain. It is unavailable to amylolytic enzymes, since the gastrointestinal tract lacks enzymes capable of degrading cellulose, hemicelluloses, lignins, and other constituents of plant cell walls. Therefore, such starch together with fragments of plant tissue, passes through the small intestine in its intact form as found in beans, grains and seeds.

RS2 is starch that is intrinsically indigestible in the native (uncooked) state due to its high amylose content. It is found in potatoes, bananas and plantains, and the resistance

Table 7.11 Examples of naturally origins and its by-products of resistant starch

Food	Serving size	Resistant starch (g)
Banana flour, from green bananas	1/4 cup, uncooked	10.5–13.2
Banana, raw, slightly green	1 medium, peeled	4.7
High amylose RS2 corn resistant starch	1 tablespoon (9.5 g)	4.5
Oats, rolled	1/4 cup, uncooked	4.4
Green peas, frozen	1 cup, cooked	4.0
White beans	1/2 cup, cooked	3.7
Lentils	1/2 cup cooked	2.5
Cold pasta	1 cup	1.9
Pearl barley	1/2 cup cooked	1.6
Cold potato	1/2" diameter	0.6–0.8
Oatmeal	1 cup cooked	0.5

of raw potato starch has been attributed to large sizes of its granules, hence the limited area of their availability to enzymes (Ring *et al.*, 1988). Still, fine-grain high-amylose maize starch demonstrates the same resistance to enzymatic activity as coarse-grain potato starch (Planchet *et al.*, 1995). Starch hydrolysis requires enzyme adsorption on the surface of the starch granules (Leloup *et al.*, 1992).

RS3 is retrograded starch, when some starches have been cooked then cooled (fridge or freezer), which changes the structure precipitated from paste or starch gel in the retrogradation process. During starch gelatinization, the starchy substance is partly depolymerized as a result of the temperature–water interaction, which passes from swollen granules into solution.

RS4 does not occur naturally and has to been chemically or physically treated (mainly by thermal treatment), or with both those treatments. The resistance of chemically-modified starches to the activity of amylolytic enzymes results from changes in the composition and structure of starch particles proceeding modification. The resistance of products of starch thermal depolymerization, dextrins, to enzymatic activity results from changes in their structure compared to starch. Upon heating of starch, depolymerization, transglucosidation and repolymerization proceed in the interior of the particles. With elongation of the dextrinization process, an increase is observed in the number of 1,3 and 1,2 linkages between glucoside residues of resultant dextrins (Ohkuma *et al.*, 1990). Table 7.11 shows some sources of resistant starch that can be found naturally (Murphy *et al.*, 2008). By-products from the food processing industry, particular the flour industry, of these raw materials are potentially a source for production of resistant starch.

The structure of resistant starch (Figure 7.15) can be formed through the retrogradation phenomenon. Generation of RS after hydrothermal treatment is primarily due to increased interactions between starch polymers. The hydrogen bonds within hydrated starch interact and result in physical-chemical changes; however, no permanent chemical bond is created (Berry, 1986). The linear fragments formed after starch de-branching can contribute to high RS content, and the length of the linear chains significantly influences retrogradation (Eerlingen and Delcour, 1995). α-Polyglucan chains with a degree of polymerization (DP) of about 20 are optimal for high RS outputs (Schmiedl *et al.*, 2000), and if the α-1,6 linked side-chains are long enough RS can be produced from amylopectin chains (Lehmann *et al.*, 2002). Then

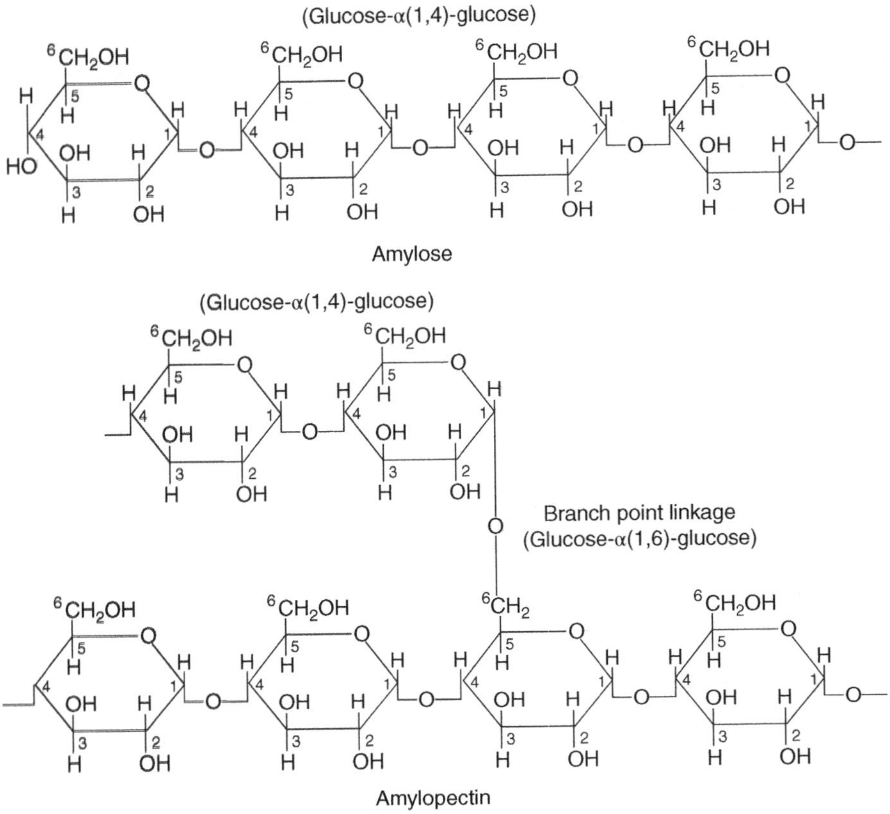

Figure 7.15 Chemical structure of amylose and amylopectin in resistant starch.

amylopectin retrogrades very slowly. The higher the amylose content, the greater the retrogradation occurrence. It is also found that high amylose starch is more resistant to digestion than amylopectin, due to its compact linear structure (Rashmi and Urooj, 2003).

7.4 Conclusion

A large amount of by-products are abundantly produced annually worldwide from food processing and agricultural by-products. The disposal of these by-products confers negative impacts on the ecosystem, while giving less opportunity for wealth creation. The agricultural and food processing by-products contain high nutritive values that can be used as main source of functional or nutraceutical ingredients. Potential resources of by-products from food processing for production of prebiotics and dietary fibers are generally derived from various food industries and some are from agricultural wastes or by-products. Dietary fibers are complex carbohydrates, non-starch polysaccharides (i.e. cellulose, dextrins, chitins, pectins, β-glucans and waxes) and lignin, which can modulate the transit time through the gut and form bulk in stools. These compounds are found naturally in many food items such as cereals, nuts, fruits and vegetables. Prebiotics are non-digestible but fermentable (mainly oligosaccharides) food ingredients that beneficially affect the host by selectively stimulating the growth and/or activity of one or a limited number of bacteria in the colon.

In addition, a prebiotic differs from the classical dietary fibers, specifically soluble dietary fibers that selectively stimulates the growth and/or activity of beneficial bacteria including probiotics, i.e. bifidobacteria and lactobacilli species. These soluble dietary fibers have been shown to exert additional beneficial effects. They will improve gut barrier function *in vitro* and *in vivo*, which could be partially a consequence of their effect on the microflora composition. For example, pectic oligosaccharide (POS) and fructo-oligosaccharide (FOS) formed from fruits and vegetables processing and the sugar cane industry, xylo-oligosaccharide (XOS) from woods and corncobs, chito-oligosaccharide (COS) from seafood processing, β-glucan from mushroom by-products, and soybean oligosaccharide (SOS) from tofu and soy protein industry. These by-products have high potential as resources of prebiotic ingredients and dietary fibers production, which can be achieved by three main approaches: direct extraction and/or purification, enzymatic depolymerization and enzymatic syntheses.

References

Aachary, A.A. and Prapulla, S.G. (2009) Value addition to corncob: Production and characterization of xylooligosaccharides from alkali pretreated lignin-saccharide complex using *Aspergillus oryzae* MTCC 5154. *Bioresource Technology*, **100**(2): 991–995.

Agnan, M.M.C., Mario, A., Dorothée, G., Bernard, W. and Michel, P. (2012) Enzymatic production of pectic oligosaccharides from polygalacturonic acid with commercial pectinase preparations. *Food and Bioproducts Processing*, **90**: 588–596.

Albersheim, P., Darvill, A.G., O'Neill, M.A., Schols, H.A. and Voragen, A.G.J. (1996) Pectins and pectinases. In: *Progress in Biotechnology* (eds J. Visser and A.G.J. Voragen), Elsevier, Amsterdam, The Netherlands, pp. 47–55.

Anderson, J.W., Baird, P. and Davis, R.H. (2009) Health benefits of dietary fiber. *Nutrition Reviews*, **67**(4): 188–205.

Aravantinos-Zafiris, G., Oreopoulou, V., Tzia, C. and Thomopoulos, C.D. (1994) Fibre fraction from orange peel residues after pectin extraction. *Lebens Wiss Technology*, **27**: 468–471.

Asp, N.G. (1992) Resistant starch. Proceedings from the second plenary meeting of EURESTA: European FLAIR Concerted Action, 11 on physiological implications of the consumption of resistant starch in man. Preface', *European Journal of Clinical Nutrition*, **46**: 1.

Barclay, A.W., Flood, V.M., Rochtchina, E., Mitchell, P. and Brand-Miller, J.C. (2007) Glycemic index, dietary fiber, and risk of type 2 diabetes in a cohort of older Australians. *Diabetes Care*, **30**(11): 2811–2813.

Behall, K.M., Scholfield, D.J., Hallfrisch, J.G. and Liljeberg-Elmstahl, H.G. (2006) Consumption of both resistant starch and beta-glucan improves postprandial plasma glucose and insulin in women. *Diabetes Care*, **29**: 976–081.

Bekers, M., Laukevics, J., Upite, D., Kaminska, E., Vigants, A., *et al.* (2002) Fructooligosaccharide and levan producing activity of *Zymomonas mobilis* extracellular levansucrase. *Process Biochemistry*, **38**: 701–706.

Berry, C.S. (1986) Resistant starch formation and measurement of starch that survives exhaustive digestion with amylolytic enzymes during the determination of dietary fiber. *Journal of Cereal Science*, **4**: 301–314.

Bonithon-Kopp, C., Kronborg, O., Giacosa, A., Rath, U. and Faivre, J. (2000) Calcium and fibre supplementation in prevention of colorectal adenoma recurrence: A randomized intervention trial. *Lancet*, **356**: 1300–1306.

Bornet, F.R.J., Brouns, F., Tashiro, Y. and Duviller, V. (2002) Nutritional aspect of short-chain fructooligosaccharides: Natural occurrence, chemistry, physiology and health implications. *Digestive and Liver Disease*, **34**: 111–120.

Brennan, C.S. and Tudorica, C.M. (2007) Fresh pasta quality as affected by enrichment of nonstarch polysaccharides. *Journal of Food Science*, **72**: 659–665.

Brockman, D.A., Chen, X. and Gallaher, D.D. (2012) Consumption of a high beta-glucan barley flour improves glucose control and fatty liver and increases muscle acylcarnitines in the Zucker diabetic fatty rat. *European Journal of Nutrition*, **52**: 1743–1753.

Cairns, P., Sun, L., Morris, V.J. and Ring, S.G. (1995) Physicochemical studies using amylose as an *in vitro* model for resistant starch. *Journal of Cereal Science*, **21**: 37–47.

Calame, W., Weseler, A.R. and Viebke, C. (2008) Gum arabic establishes prebiotic functionality in healthy human volunteers in a dose-dependent manner. *British Journal of Nutrition*, **9**: 1–7.

Cangy, C. and Peerally, A. (1995) Studies of Pleurotus production on sugarcane bagasse. *African Journal of Mycology and Biotechnology*, **3**: 67–79.

Champ, M.J. (2004) Adapted from physiological effects of resistant starch *and in vivo measurements. Journal of the Association of Official Analytical Chemists International*, **87**(3): 749–755.

Chan, G.C., Chan, W.K. and Sze, D.M. (2009) The effects of beta-glucan on human immune and cancer cells. *Journal of Hematology and Oncology*, **2**: 25.

Chen, C., Chen, J.L. and Lin, T.Y. (1997) Purification and characterization of a xylanase from Trichoderma longibrachiatum for xylooligosaccharide production. *Enzyme and Microbial Technology*, **21**: 91–96.

Cheung, P.C.K. (2013) Mini-review on edible mushrooms as source of dietary fiber: Preparation and health benefits. *Food Science and Human Wellness*, **2**(3–4): 162–166.

Choct, M., Dersjant-Li, Y., McLeish, J. and Peisker, M. (2010) Soy oligosaccharides and soluble non-starch polysaccharides: A review of digestion, nutritive and anti-nutritive effects in pigs and poultry Asian-Australasia. *Journal of Animal Science*, **23**(10): 1386–1398.

Choi, J.S., Kim, H., Jung, M.H., Hong, S. and Song, J. (2010) Consumption of barley beta-glucan ameliorates fatty liver and insulin resistance in mice fed a high-fat diet. *Molecular Nutrition and Food Research*, **54**: 1004–1013.

Chutkaew, C. and Paroda, R.S. (1994) Baby corn production in Thailand. In: *A Success Story, Asia-Pacific Association of Agricultural Research Innovation*, Bangkok.

Clausen, M.H. and Madsen, R. (2003) Synthesis of hexasaccharide fragments of pectin. *Chemistry*, **9**(16): 3821–3832.

Crittenden, R.G. and Playne, M.J. (1999) Continuous preparation of fructose syrups from Jerusalem artichoke tuber using immobilized intracellular inulinase from *Kluyveromyces* sp. Y-85. *Process Biochemistry*, **34**: 643–646.

Cummings, J.H. (1981) Cellulose and the human gut. *Gut*, **25**: 805–810.

Cummings, J.H. and Macfarlane, G.T. (1997) Role of intestinal bacteria in nutrient metabolism. *Clinical Nutrition*, **16**: 3–11.

Cummings, J.H., Macfarlane, G.T. and Englyst, H.N. (2001) Prebiotic digestion and Fermentation. *American Journal of Clinical Nutrition*, **73**: 415S–420S.

Danny, L.R. (2002) Handling and using 'spent' mushroom substrate around the world. In: *Mushroom Biology and Mushroom Product* (ed. J.E. Sanchez), UAEM, pp. 43–60.

Darbyshire, B. and Henry, R.J. (1978) The distribution of fructans in onions. *New Phytology*, **81**: 29–34.

Dasaesamoh, R., Youravong, W. and Wichienchot, S. (2016a) Digestibility, fecal fermentation and anti-cancer of dragon fruit oligosaccharides. *International Food Research Journal*, **23**(6): 2581–2587.

Dasaesamoh, R., Youravong, W. and Wichienchot, S. (2016b) Optimization on pectinase extraction and purification by yeast fermentation of oligosaccharides from dragon fruit (*Hyloceus undatus*). *International Food Research Journal*, **23**(6): 2601–2607.

Dekker, R.F.H. and Wallis, A.F.A. (1983) Autohydrolysis-explosion as pretreatment for the enzymic saccharication of sunflower seed hulls. *Biotechnology Letters*, **5**: 311–316.

Du, B., Lin, C., Bian, Z. and Xu, B. (2015) An insight into anti-inflammatory effects of fungal beta-glucans. *Trends in Food Science and Technology*, **41**(1): 49–59.

Eastwood, M. and Kritchevsky, D. (2005) Dietary fiber: How did we get where we are? *Annual Review of Nutrition*, **25**: 1–8.

Eerlingen, R.C. and Delcour, J.A. (1995) Formation analysis, structure and properties of type III enzyme resistant starch. *Journal of Cereal Science*, **22**: 129–137.

Englyst, H. (1989) Classification and measurement of plant polysaccharides. *Animal Feed Science and Technology*, **23**: 27–42.

Erickson, D.R. (1995) *Practical Handbook of Soybean Processing and Utilization*. AOCS, USA.

Eva, S., Edith, S., Bruno G. and Eric, C. (2010) The effect of ß-glucans on porcine leukocytes. *Veterinary Immunology and Immunopathology*, **135**: 199–207.

FAO (2008) Food and Agricultural Organization fao blog, web log post, 24 November 2013, Last accessed 10 January 2015, http://www.fao.org

Farvid, M.S., Homayouni, F., Shokoohi, M., Fallah, A. and Farvid, M.S. (2014) Glycemic index, glycemic load and their association with glycemic control among patients with type 2 diabetes. *European Journal of Clinical Nutrition*, **68**: 459–463.

Fastinger, N.D., Karr-Lilienthal, L.K. and Spears, J.K. (2008) A novel resistant maltodextrin alters gastrointestinal tolerance factors, fecal characteristics and fecal microbiota in healthy adult humans. *Journal of the American College of Nutrition*, **27**: 356–366.

Felt, O., Buri, P. and Gurny, R. (1998) Chitosan: A unique polysaccharide for drug delivery. *Drug Development and Industrial Pharmacy*, **24**(1): 979–993.

Ferguson, L.R., Tasman-Jones, C. and Englyst, H. (2000) Comparative effects of three resistant starch preparations on transit time and short-chain fatty acid production in rats. *Nutrition and Cancer*, **36**(2): 230–237.

Franck, A. (2002) Technological functionality of inulin and oligofructose. *British Journal of Nutrition*, **87**: 287–291.

Franck, A. and De Leenheer, L. (2005) Inulin. In: *Polysaccharides and Polyamides in the Food Industry: Properties, production, and patents* (eds A. Steinbuchel and S.K. Rhee). Wiley-Blackwell, Oxford, pp. 281–322.

Fukata, T., Sasai, K., Miyamoto, T. and Baba, E. (1999) Inhibitory effects of competitive exclusion and fructooligosaccharide, singly and in combination, on Salmonella colonization of chicks. *Journal of Food Protection*, **62**(3): 229–233.

Fukushima, D. (1981) Soy protein for foods centering around soy sauce and tofu. *Journal of the American Oil Chemists' Society*, **58**(3): 346–354.

Garrote, G., Cruz, J.M., Domínguez, H. and Parajó, J.C. (2004) Production of xylooligosaccharides by autohydrolysis of lignocellulosic materials. *Trends in Food Science and Technology*, **15**: 115–120.

Gelroth, J. and Ranhotra, G.S. (2001) Food uses of fiber. In: *Handbook of Dietary Fiber and Functional Foods* (eds S. Cho and M.S. Dreher). Taylor & Francis, New York, pp. 345–452.

Ghazi, I., Fernandez-Arrojo, L., Garcia-Arellano, H., Ferrer, M., Ballesteros, A. and Plou, F.J. (2007) Purification and kinetic characterization of a fructosyltransferase from *Aspergillus aculeatus*. *Journal of Biotechnology*, **128**: 204–211.

Gibson, G.R. and Roberfroid, M.B. (1995) Dietary modulation of the human colonic microbiota: Introducing the concept of prebiotics. *Journal of Nutrition*, **125**: 1401–1412.

Gibson, G.R. and Roberfroid, M.B. (2008) *Handbook of Prebiotics*, vol. **504**. Taylor & Francis Group (eds), CRC Press, Boca Raton, FL.

Gibson, G.R., Probert, H.M., Van Loo, J.A. and Roberfroid, M.B. (2004) Dietary modulation of the human colonic microbiota: Updating the concept of prebiotics. *Nutrition Research Reviews*, **17**: 257–259.

Gómez, B., Gullón, B., Remoroza, C., Schols, H.A., Parajo, J.C. and Alonso, J.L. (2014) Purification, characterization and prebiotic properties of pectic oligosaccharides from orange peel wastes. *Journal of Agricultural and Food Chemistry*, **62**(40): 9769–9782.

Grethlein, H.E. and Converse, A.O. (1991) Common aspects of acid prehydrolysis and steam explosion for pretreating wood. *Bioresource Technology*, **36**: 77–82.

Gülizar, K. and Sibel, Y. (2011) Extraction of fructo-oligosaccaride components from banana peels. *Gazi University Journal of Science*, **24**(4): 877–882.

Hallauer, A.R. (2001) *Specialty Corns*. 2nd edition. CRC Press, Washington. DC.

Henry, R.J. (1987) Pentosan and (1 → 3), (1 → 4)-β-glucan concentrations in endosperm and wholegrain of wheat, barley, oats and rye. *Journal of Cereal Science*, **6**: 253–258.

Hooda, S. and Kawatra, A. (2013) Nutritional evaluation of baby corn (*Zea mays*). *Nutrition and Food Science*, **43**: 68–73.

Hopping, B.N., Erber, E., Mead, E., Sheehy, T., Roache, C. and Sharma, S. (2010) Socio-economic indicators and frequency of traditional food, junk food, and fruit and vegetable consumption amongst Inuit adults in the Canadian Arctic. *Journal of Human Nutrition and Diet*, **23**(Suppl. 1): 51–58.

Hotchkiss, A.T., Olano-Martin, E., Grace, W.E., Gibson, G.R. and Rastall, R.A. (2003) Pectic oligosaccharides as prebiotics. In: *Oligosaccharides in Food and Agriculture*, (eds G. Eggleston and G.L. Cote). ACS Publications, Washington, DC, pp. 54–62.

Imaizumi, K., Nakatsu, Y., Sato, M., Sedarnawati, Y. and Sugano, M. (1991) Effects of xylooligosaccharides on blood glucose, serum and liver lipids and caecum short-chain fatty acids in diabetic rats. *Agriculture Biology and Biochemistry*, **55**: 199–205.

Jeng, T.L., Shih, Y.J., Wua, M.T. and Sung, J.M. (2010) Comparisons of flavonoids and anti-oxidative activities in seed coat, embryonic axis and cotyledon of black soybeans. *Food Chemistry*, **123**: 1112–1116.

Jeuniaux, C. (1986) Chitosan as a tool for the purification of waters. In: *Chitin in Nature and Technology*. Plenum Press, New York. pp. 551–570.

Jiang, T.W., Xu, J., Wang, X.M., Chen, X.S. and Qiu, Z.D. (2011) Screening of active ingredient of *Stigmata maydis* inhibiting xanthine oxidase and its effect. *Journal of Jilin University Medicine Edition*, **37**(3): 433–436.

Jiao, Z., Si, X.X., Zhang, Z.M., Li, G.K. and Cai, Z.W. (2012) Compositional study of different soybean (*Glycine max* L.) varieties by 1H NMR spectroscopy, chromatographic and spectrometric techniques. *Food Chemistry*, **135**: 285–291.

John, S. and Antonio, M. (2007) Nutrient removal when harvesting corn stover. Integrated Crops news, web log post 24 November 2013. Last accessed 20 January 2015: http://www.ipm.iastate.edu/ipm/icm/2007/8-6/nutrients.html

Jonathan, S.G. (2002) Vegetative growth requirements and antimicrobial activities of some higher fungi in Nigeria. PhD dissertation, University of Ibadan.

Joseph, O. (2008) Using inulin and oligofructose with high-intensity sweeteners. web log post 14 July 2012. Last accessed 15 January 2015, http://newhope360.com/sweeteners/using-inulin-and-oligofructose-high-intensity-sweeteners

Kang, J., Badger, T.M., Ronis, M.J. and Wu, X. (2010) Non-isoflavone phytochemicals in soy and their health effects. *Journal of Agricultural and Food Chemistry*, **58**(14): 8119–8133.

Katarína, V., Zdenka, O., Monika, A. and Milan, P. (2008) Design and economics of industrial production of fructooligosaccharides. *Chemical Papers*, **62**(4): 375–381.

Kaur, N. and Gupta, A.K. (2002) Applications of inulin and oligofructose in health and nutrition. *Journal of Biosciences*, **27**(7): 703–714.

Kawashima, M. (1983) *Handbook of Membrane Application Technology* (eds D. Nomura and H.S. Ohya), Tokyo, pp. 176–187.

Kewyok, K. (2015) Production of pectic oligosaccharide from dragon fruit peel. MSc Thesis, Prince of Songkla University, Thailand.

Knaul, J.Z., Hudson, S.M. and Creber, K.A.M. (1999) Improved mechanical properties of chitosan fibers. *Journal of Applied Polymer Science*, **72**: 1721–1732.

Kumar, M.N.V.R. (2000) A review of chitin and chitosan applications. *Reactive and Functional Polymers*, **46**(1): 1–27.

Lance, G. and Garren, B. (2002) Origin, History, and Uses of Corn (*Zea mays*). web log post January 2002. Last accessed 18 January 2015: http://agron-www.agron.iastate.edu/Courses/agron212/readings/corn_history.htm

Lamtumluk, D. (2014) Production of xylo-oligosaccharide from rubber wood sawdust. MSc Thesis, Prince of Songkla University, Thailand.

Laparra, J.M. and Sanz, Y. (2010) Interactions of gut microbiota with functional food components and nutraceuticals. *Pharmacological Research*, **61**(3): 219–225.

Laroche, C. and Michaud, P.H. (2007) New developments and properties for β-(1,3)-glucans. *Recent Patents on Biotechnology*, **1**: 59–73.

Lehmann, U., Jacobasch, G. and Schmiedl, D. (2002) Characterization of resistant starch type III from banana (*Musa acuminata*). *Journal of Agriculture Food Chemistry*, **50**: 5236–5240.

Leloup, V.M., Colonna, P. and Ring, S.G. (1992) Physico-chemical aspects of resistant starch. *Journal of Cereal Science*, **16**: 253–266.

L'Hocine, L., Wang, Z., Jiang, B. and Xu, S. (2000) Purification and partial characterization of fructosyltransferase and invertase from *Aspergillus niger* AS0023. *Journal of Biotechnology*, **81**: 73–84.

Li, J., Qiao, Z., Tatsumi, E., Saito, M. and Cheng, Y. (2013) A novel approach to improving the quality of Bittern solidified tofu by w/o controlled released coagulant in tofu processing and product evaluation. *Food and Bioprocess Technology*, **6**(7): 1801–1808.

Li, W.L., Zheng, H.C., Bukuru, J. and De Kimpe, N. (2004) Natural medicines used in the traditional Chinese medical system for therapy of diabetes mellitus. *Journal of Ethnopharmacology*, **92**(1): 1–21.

Liatis, S., Tsapogas, P., Chala, E., Dimosthenopoulos, C., Kyriakopoulos, K. and Kapantais, E. (2009) The consumption of bread enriched with betaglucan reduces LDL-cholesterol and improves insulin resistance in patients with type 2 diabetes. *Diabetes Metab*, **35**: 115–120.

Lima, D.M., Fernandes, P., Nascimento, D.S., Rita de Cássia, L., Ribeiro, F. and de Assis, S.A. (2011) Fructose syrup: A biotechnology asset. *Food Technology and Biotechnology Journal*, **49**(4): 424–434.

Liu, K. (1997) *Soybeans: Chemistry, Technology, and Utilization*. Chapman & Hall, New York.

Liu, X., Jin, J., Wang, G. and Herbert, S.J. (2008) Soybean yield physiology and development of high-yielding practices in Northeast China. *Field Crops Research*, **105**: 157–171.

Lodhi, G., Kim, Y.S., Hwang, J.W., Kim, S.K., Jeon, Y.J. and Je, J.Y. (2014) Chitooligosaccharide and its derivatives: preparation and biological applications, *BioMed Research International*, **10**: 1155.

Lunn, J. and Buttriss, J.L. (2007) Carbohydrates and dietary fibre. *Nutrition Bulletin*, **32**(1): 21–64.

Manderson, K., Pinart, M., Tuohy, K.M., Grace, W.E., Hotchkiss, A.T., *et al.* (2005) *In vitro* determination of prebiotic properties of oligosaccharides derived from an orange juice manufacturing by-product stream. *Applied and Environmental Microbiology*, **71**: 8383–8398.

Mantovani, M.S. (2008) ß-Glucans in promoting health: Prevention against mutation and cancer. *Mutation Research*, **658**(3): 154–161.

Mathias, T.R.S., Mello, P.P.M. and Servulo, E.F.C. (2014) Solid wastes in brewing process: A review. *Journal of Brewing and Distilling*, **5**(1): 1–9.

McKee, L.H. and Latner, T.A. (2000) Under-utilized sources of dietary fiber: A review. *Plant Foods and Human Nutrition*, **55**(4): 285–304.

Mentschel, J. and Claus, R. (2003) Increased butyrate formation in the pig colon by feeding raw potato starch leads to a reduction of colonocyte apoptosis and a shift to stem cell compartment. *Metabolism*, **52**(11): 1400–1405.

Morgan, A.J., Mul, A.J., Beldman, G. and Voragen, A.G.J. (1992) Dietary oligosaccharides: new insights. *AGRO Food Industry High Technology*, **3**: 35–38.

Moure, A., Gullón, P., Domínguez, H. and Parajó, J.C. (2006) Advances in the manufacture, purification and applications of xylo-oligosaccharides as food additives and nutraceuticals. *Process Biochemistry*, **41**: 1913–1923.

Murphy, M., Douglass, J.S. and Birkett, A. (2008) Resistant starch intake in the United States. *Journal of the American Dietetic Association*, **108**: 67–78.

Niness, K.R. (1999) Nutritional and health benefits of inulin and oligofructose. *Journal of Nutrition*, **129**: 1402S–1406S.

Nuráliaa, A., Siti, M.M., Taip, F. and Liew, A.A. (2010) Response surface optimization for clarification of white pitaya juice using a commercial enzyme. *Journal of Food Process Engineering*, **33**(2): 333–347.

Obatolu, V.A. (2008) Effect of different coagulants on yield and quantity of tofu from soymilk. *European Food Research and Technology*, **226**(3): 467–472.

Ohkuma, K., Matsuda, I., Katta, Y. and Hanno, Y. (1990) Pyrolysis of starch and its digestibility by enzymes-characterization of indigestible dextrin. *Denpun Kagaku*, **37**: 107–114.

Olano, M.E., Rimbach, G.H., Gibson, G.R. and Rastall, R.A. (2003) Pectins and pectic oligosaccharides induce apoptosis in *in vitro* human colonic adenocarcinoma cells. *Anticancer and Research*, **23**: 334–346.

O'Shea, N., Arendt, E.K. and Gallagher, E. (2012) Dietary fiber and phytochemical characteristics of fruit and vegetable by-products and their recent applications as novel ingredients in food products. *Innovation of Food Science and Emerging Technology*, **16**: 1–10.

Panlasigui, L.N., Thompson, L.U., Juliano, B.O., Perez, C.M., Yiu, S.H. and Greenberg, G.R. (1991) Rice varieties with similar amylose content differ in starch digestibility and glycemic response in humans. *The American Journal of Clinical Nutrition*, **54**: 871–877.

Hooda, S. and Kawatra, A. (2013) Nutritional evaluation of baby corn (*zea mays*). *Nutrition & Food Science*, **43**(1): 68–73.

Peniche-Covas, C., Alvarez, L.W. and Arguelles-Monal, W. (1992) Adsorption of mercuric ions by chitosan. *Journal of Applied Polymer Science*, **46**(7): 1147–1150.

Perez-Hidalgo, M., Guerra-Hernandez, E. and Garcia-Villanova, B. (1997) Determination of insoluble dietary fiber compounds: cellulose, hemicellulose and lignin in legumes. *Ars Pharmaceutica*, **38**(4): 357–364.

Phrukwiwattanakul, P. (2013) Production of resistant starch from different types of starch and determination of their prebiotic properties. MSc Thesis, Prince of Songkla University, Thailand.

Planchet, V., Colonna, P., Galant, D.J. and Bouchet, B. (1995) Extensive degradation of native starch granules by alpha-amylase from *Aspergillus fumigatus*. *Journal of Cereal Science*, **21**: 163–171.

Ralet, M.C., Dellavalle, G. and Thibault, J.F. (1993) Raw and extruded fibre from pea hulls. Part I: Composition and physico-chemical properties. *Carbohydrate Polymers*, **20**: 17–23.

Rashmi, S. and Urooj, A. (2003) Effect of processing on nutritionally important starch fractions in rice varieties. *International Journal of Food Science and Nutrition*, **54**: 27–36.

Reddy, N.R. and Salunkhe, D.K. (1980) Changes in oligosaccharide during germination and cooking of black gram and fermentation of black gram/rice blend. *Cereal Chemistry Journal*, **57**: 356–360.

Rekha, C.R. and Vijayalakshmi, G. (2011) Influence of processing parameters on the quality of soycurd (tofu). *Journal of Food Science and Technology*, **50**(1): 176–180.

Renard, C.M.G.C. and Thibault, J.F. (1991) Composition and physicochemical properties of apple fibres from fresh fruits and industrial products. *Lebensmittel-Wissenschaft and Technologie*, **24**: 523–527.

Ring, S.G., Gee, J.M., Whittam, M., Orford, P. and Johnson, I.T. (1988) Resistant starch: Its chemical form in foodstuffs and effect on digestibility *in vitro*. *Food Chemistry*, **28**: 97–109.

Roberfroid, M. (2007) Prebiotics: The concept revisited. In: *The Journal of Nutrition Effects of Probiotics and Prebiotics*, **137**: 831S–837S.

Robin, F., Schuchmann, H.P. and Palzer, S. (2012) Dietary fiber in extruded cereals: Limitations and opportunities. *Trends in Food Science and Technology*, **28**: 23–32.

Rodolfo, V., José Luis, A., Herminia, D. and Juan, C.P. (2005) Manufacture and refining of oligosaccharides from industrial solid wastes. *Industrial and Engineering Chemistry Research*, **44**: 614-620.

Salminen, S. and Salminen, E. (1997) Lactulose, lactic acid bacteria, intestinal microecology and mucosal protection. *Scandinavian Journal of Gastroenterology*, **32**: 45–48.

Samuelsen, A.B., Schrezenmeir, J. and Knutsen, S.H. (2014) Effects of orally administered yeast-derived beta-glucans: A review. *Molecular Nutrition and Food Research*, **58**: 183–193.

Sangeetha, P.T., Ramesh, M.N. and Prapulla, S.G. (2004) Production of fructo-oligosaccharides by fructosyltransferase from *Aspergillus oryzea* CFR 202 and *Aureobasidium pullulans* CFR 77. *Process Biochemistry*, **39**: 755–760.

Sangeetha, P.T., Ramesh, M.N. and Prapulla, S.G. (2005) Recent trends in the microbial production, analysis and application of fructo-oligosaccharides, *Trends in Food Science and Technology*, **16**: 442–457.

Santosh, H. and Asha, K. (2013) Nutritional evaluation of baby corn (*Zea mays*). *Nutrition and Food Science*, **43**(1): 68–73.

Schmiedl, D., Bäuerlein, M., Bengs, H. and Jacobasch, G. (2000) Production of heat stable, butyrogenic resistant starch. *Carbohydrate Polymer*, **43**: 183–193.

Schwiertz, A., Lehmann, U., Jacobasch, G. and Blaut, M. (2002) Influence of resistant starch on the SCFA production and cell counts of butyrate producing *Eubacterium* spp. in the human intestine. *Journal of Applied Microbiology*, **93**(1): 157–162.

Sendi, H., Mohamed, M.T.M., Anwar, M.P. and Saud, H.M. (2013) Spent mushroom waste as a media replacement for peat moss in Kai-Lan (*Brassica oleracea* var. Alboglabra) production. *The Scientific World Journal*, **258562**: 1–8.

Shahidi, F. and Synowiecki, J. (1991) Isolation and characterization of nutrients and value-added products from snow crab (*Chionoecetes opilio*) and shrimp (*Pandalus borealis*) processing discards. *Journal of Agricultural and Food Chemistry*, **39**(8): 1527–1532.

Shahidi, F., Arachchi, J.K.V. and Jeon, Y.J. (1999) Food applications of chitin and chitosans. *Trends in Food Science and Technology*, **10**(2): 37–51.

Shurson, J., Noll, S. and Goihl, J. (2005) Corn by-product diversity and feeding value to non-ruminants. *66th Minnesota Nutrition Conference*, pp. 20–21.

Sin, H.N..S, Yusof, N., Sheikh, A.H. and Rahman, R.A. (2006) Optimization of enzymatic clarification of sapodila juice using response surface methodology. *Journal of Food Engineering*, **73**: 313–319.

Soral-Oemietana, M. and Wronkowska, M.G. (2004) Resistant starch-nutritional and biological activity. *Polish Journal of Food and Nutrition Sciences*, **13**(54): 51–64.

Spagnuolo, M., Crecchio, C., Pizzigallo, M. and Ruggiero, P. (1997) Synergistic effects of cellulolytic and pectinolytic enzymes in degrading sugar beet. *Bioresource Technology*, **60**: 215–222.

Steenkamp, V. (2003) Phytomedicines for the prostate. *Fitoterapia*, **74**(6): 545–552.

Stone, B.A. and Clarke, A.E. (1992) *Chemistry and Biology of (1,3)-D-glucans*. La Trobe University Press, Victoria, Australia.

Sunna, A. and Antranikian, G. (1997) Xylanolytic enzymes from fungi and bacteria, *Critical Review in Biotechnology*, **17**(1): 39–67.

Sriniworn, P., Youravong, W. and Wichienchot, S. (2015) Permeate flux enhancement in ultrafiltration of tofu whey using pH-shifting and gas-liquid two-phase flow. *Separation Science and Technology*, **50**: 2367–2374.

Swennen, K., Courtin, C.M. and Van der Bruggen, B. (2005) Ultrafiltration and ethanol precipitation for isolation of arabinoxylo-oligosaccharides with different structures. *Carbohydrate Polymers*, **62**(3): 283–292.

Teuri, U. and Korpel, R. (1998) Galacto-oligosaccharides relieve constipation in elderly people. *Annals of Nutrition and Metabolism*, **42**: 319–327.

Thibault, J.F., Renard, C.M.G.C. and Guillon, F. (1994) Physical and chemical analysis of dietary fibres in sugar beet and vegetables. In: *Modern Methods of Plant Analysis*, **16**: 23–55.

Thomas, G.V., Prabhu, S.R., Reeny, M.Z. and Bopaiah, B.M. (1998) Evaluation of lignocellulosic biomass from coconut palm as substrate for cultivation of *Pleurotus sajor-caju* (Fr.) Singer. *World Journal of Microbiology and Biotechnology*, **14**(6): 879–882.

Thuzar, M., Puteh, A.B., Abdullah, N.A.P., Lassim, M.B.M. and Jusoff, K. (2010) The effects of temperature stress on the quality and yield of soya bean [(*Glycine max* L.) Merrill. *Journal of Agricultural Science*, **2**: 172–179.

Toda, T., Sakamoto, A., Takayanagi, T. and Yokotsuka, K. (2000) Changes in isoflavone compositions of soybean during soaking process. *Food Science and Technology Research*, **6**: 314–319.

Topping, D. and Clifton, P. (2001) Short-chain fatty acids and human colonic function: roles of resistant starch and non-starch polysaccharides. *Physiological Reviews*, **81**: 1031–1064.

Trugo, L.C., Farah, A. and Cabral, L. (1995) Oligosaccharides distribution in Brazilian soya bean cultivars. *Food Chemistry*, **52**, 385–387.

Tungland, B.C. and Mayer, D. (2002) Nondigestible oligo- and polysaccharides (dietary fiber): their physiology and role in human health and food. *Comprehensive Reviews in Food Science and Food Safety*, **1**(3): 90–109.

Valiente, C., Arrigoni, E., Esteban, R.M. and Amado, R. (1995) Grape pomace as a potential food fiber. *Journal of Food Science*, **60**: 818–820.

Van Laere, K.M.J., Hartemink, R., Bosveld, M., Schols, H.A. and Voragen, A.G.J. (2000) Fermentation of plant cell wall derived polysaccharides and their corresponding oligosaccharides by intestinal bacteria. *Journal of Agricultural and Food Chemistry*, **48**: 1644–1652.

Vanesa, B., Esperanza, M., María, A.M.C., Yolanda, A., Francisco, J.L.A., *et al.* (2011) Characterization of industrial onion wastes (*Allium cepa* L.). *Plant Foods for Human Nutrition*, **66**: 48–57.

Vashisht, A. (1998) Development of ready-to-drink tomato juice and instant soup from newly evolved tomato cultivars. MSc Thesis, CCS Haryana Agricultural University, Hisar.

Vazquez, M.J., Alonso, J.L, Domınguez, H. and Parajo, J.C. (2000) Xylooligosaccharides: manufacture and applications. *Trends in Food Science and Technology*, **11**: 387–393.

Vazquez, M.J., Garrote, G. and Alonso, J.L. (2005) Refining of autohydrolysis liquors for manufacturing xylooligosaccharides: evaluation of operational strategies. *Bioresource Technology*, **96**(8): 889–896.

Vegas, R., Alonso, J.L., Dominguez, H. and Parajo, J.C. (2005) Manufacture and refining of oligosaccharides from industrial solid wastes. *Industrial and Engineering Chemistry Research*, **44**(3): 614–620.

Verma, S. (2000) Nutritional evaluation and utilization of pumpkin. MSc Thesis, CCS Haryana Agricultural University, Hisar.

Voragen, A.G.J. (1998) Technological aspects of functional food-related carbohydrates. *Trends in Food Science and Technology*, **9**: 328–335.

Wan Rosli, W.I., Jesmine, K., Mohsin, S.S.J. and Farid, C.G. (2007) High-resolution morphological characterization and micro-minerals composition of corn silk (*Zea mays* – Maydis stigma). *Proceedings of the International Conference on Chemical Sciences*, UGM, Yogyakarta, Indonesia: 24–26 May.

Wan Rosli, W.I., Nurhanan, A.R., Solihah, M.A. and Mohsin, S.S.J. (2011) Cornsilk improves nutrient, physical traits and unaffected sensory properties of chicken patties. *Sains Malaysiana*, **40**(10): 1165–1172.

Waterhouse, A.L. and Chatterton, N.J. (1993) Science and technology of fructans. In: *Glossary of Fructan Terms* (eds. M. Suzuki and N.J. Chatterton), CRC Press, Boca Raton, FL, pp. 1–8.

Weststrate, J., Van Poppel, G. and Vershuren, P. (2002) Functional foods, trends and future. *British Journal of Nutrition*, **88**: S233–S235.

Wichienchot, S. and Chinachoti, P. (2011) Prebiotic oligosaccharides: origins and production, health benefits and commercial applications. In: *Oligosaccharides: Sources, Properties and Applications* (eds N.S. Gordon). Nova Science Publishers, Inc., New York, pp. 59–83.

Wichienchot, S., Jatupornpipat, M. and Rastall, R.A. (2010) Oligosaccharides of Pitaya (dragon fruit) Flesh and their prebiotic properties. *Food Chemistry*, **120**(3): 850–857.

Wichienchot, S., Hemmaratchirakul, J., Lamtumluk, D. and Inpun, R. (2015) The by-products from different cultivars of orange. *Nutraceutical and Functional Food Research and Development Center (NFF)*, Prince of Songkla University, Thailand.

Wong, K.H. and Cheung, P.C.K. (2009) Enzymatic preparation of mushroom dietary fiber: A comparison between analytical and industrial enzymes. *Food Chemistry*, **115**: 795–800.

Wu, D. and Wu, L.Y. 1999) Effects of fructooligosaccharide on the broiler production. *Acta Agriculture Zhejiangensis*, **11**(1): 85–87.

Xu, Z.R., Hu, C.H. and Wang, M.Q. (2002) Effects of fructooligosaccharide on conversion of L-tryptophan to skatole and indole by mixed populations of pig fecal bacteria. *The Journal of General and Applied Microbiology*, **48**(2): 83–89.

Yamada, A., Shibbuya, N. and Komada, O. (1993) Induction of phytoalexin formation in suspension-cultured rice cells by N-acetylchitooligosaccharides. *Bioscience, Biotechnology and Biochemistry*, **57**(3): 405–409.

Yesilada, E. and Ezer, N. (1989) The anti-inflammatory activity of some Sideritis species growing in Turkey. *International Journal of Crude Drug Research*, **27**(1): 38–40.

Yildiz, S. (2009) The metabolism of fructooligosaccharides and fructo-oligosacchariderelated compounds in plants. *Food Reviews International*, **27**(1): 16-50.

Yodpet, C. (1979) Studies on sweet corn as potential young cob corn (*Zea mays* L.). PhD Thesis, University of Philippines, Quezon City.

Zimmerman, J.W. (1998) A novel carbohydrate-glycosphingolipid interaction between a beta-(1-3)-glucan immunomodulator, PGG-glucan, and lactosylceramide of human leukocytes. *The Journal of Biological Chemistry*, **273**(34): 22014–22020.

8

Utilization of By-Products from Food Processing as Biofertilizers and Biopesticides

Avishek Datta, Hayat Ullah, and Zannatul Ferdous

Asian Institute of Technology, Pathumthani, Thailand

8.1 Introduction

The agricultural sector plays a major role in improving economic growth in terms of raising people's income and reducing poverty in the developing countries, along with fulfilling food and nutritional needs for a growing world population. About 1 billion people are facing severe problems in accessing nutritionally adequate and safe food worldwide. By 2050, the world population is projected to be 9 billion and food production needs to be raised by 70% between 2005/07 and 2050 to feed such a huge population (FAO, 2009). Globally, more than 870 million people are hungry and the majority of them are small-scale farmers living in rural areas and often managing marginal lands (Godfray *et al.*, 2010; FAO, 2012). At present, agricultural production is facing increasing challenges such as water shortage, climate change and volatility, raising the risk of production shortfalls. The future vision of the world would be to adopt new methods to increase sustainability of agricultural production.

Most of agricultural food production was dependent on chemical inputs such as fertilizers and pesticides before the 1990s, and these synthetic compounds have played a key role in increasing food production. Since the 1990s, food demand has increased sharply for which farmers need increasing use of these chemicals. As a result, a major risk is created to the environment and human health (Gottlieb and Joshi, 2010). In order to reduce damage to the environment and health, some major classes of pesticides were banned, thus creating a market for biopesticides and microbial inoculants.

The rapidly increasing rate of the world population and simultaneously growing trend of industrialization, urbanization, agricultural activities and excessive use of natural resources for better and suitable life styles in recent years, have generated a large amount of waste (Semple *et al.*, 2001). A major portion of this waste is released

Food Processing By-Products and their Utilization, First Edition. Edited by Anil Kumar Anal.

to the soil, water and air, posing a great threat to the environment and human life (Semple *et al.*, 2001; Ahmad *et al.*, 2007). Consequently, there is a serious concern in communities to effectively manage these wastes. One of the best ways of management is the re-use and transformation of these wastes into beneficial end-products (Larney and Hao, 2007; Mohee *et al.*, 2008). For this purpose, composting is one of the economical and environmentally-friendly methods in managing these organic wastes and using them for the production of utilizable materials (Mohee *et al.*, 2008; An *et al.*, 2012).

By the second half of the 20th century, crop yields started to increase sharply, which was associated with the 'green revolution', for which one of the reasons was the development and utilization of inorganic fertilizers and synthetic pesticides. However, excessive use of pesticides and chemicals has unrecoverable environmental and health implications. Therefore, the current paradigm of almost exclusively relying on chemicals for crop production and protection may need to be reconsidered. New pesticides, including natural product-based pesticides, are being discovered and developed to replace and/or supplement the chemical one.

In the past 20 years, increased population pressure, rapid urbanization, industrialization and mechanized agricultural activities have contributed to an increased accumulation of organic waste materials in the environment as 'solid organic waste'. On the one hand, tropical soils are largely deficient in essential plant nutrients, and on the other hand, large quantities of such nutrients are available in food by-products. Utilization of food by-products has, therefore, become an essential component worldwide. The alarming rate of increase in their quantity and the mixing of bio-degradable and non-biodegradable wastes at the generation points make it complicated to handle them with limited resources. Therefore, the overall objective of this chapter is to discuss the use of food processing by-products as biofertilizers and biopesticides, in order to minimize the adverse environmental impacts and to reduce the use of chemicals (fertilizer and pesticides) for crop production.

8.2 Concept of Food Processing By-Products

The clear borderline between a useful food product and a by-product is rather thin, because the by-product is generated in one step of the food processing, may be used as a raw material for other steps or an important compound for industry. Any product produced during the processing for the generation of the main product is called a by-product; such products may not be of direct interest to the manufacturer, but can be used elsewhere. Therefore, at the source they may be regarded as waste, but could be considered as a product at some other process. In general, food processing by-products include the remains or residues that are produced after removing the desirable portion for further processing or direct consumption as food. The biochemical constituents of these residues are the same as food and they are suitable to consider for further utilization. Not only the residues but the other portions such as husk, skin or hulls, which are usually removed during the processing, are also suitable for consideration due to their chemical constituents.

The food residues, husk of cereals and pulses, fruit and vegetable peels, unused portion of animals such as bones, eggs, shells and many others usually considered as waste, also contain a considerable amount of reusable materials and energy, although they pose serious environmental and economic challenges. Most of the food processing materials contain components that could be utilized in a variety of biotechnological and chemical processes, as well as biofertilizers for increasing soil productivity.

Russ and Meyer-Pittroff (2004) classified food wastes into some major groups which are grain products (bran, broken grains, seed, seed husk, fine dust, chaff, straw), noodles (dough wastes, egg shells), potato (*Solanum tuberosum*) processing industry (potato peels), coffee (*Coffea arabica*) roasters (silver skin), sugar industry (molasses, beet pulp, beet leaves and stem, carbonation sludge), dairy processing industry (whey, cheese residues, waste from milk production), meat industry (slaughterhouse wastes), egg production (egg shells), beer production (malt dust, spent grains, break materials), and wine cellars (pomace, clarification sediments, yeast sediments).

8.2.1 Existing Methods of By-Product/Wastes Management Practiced by Food Industries

Different methods are used for the management of food industry wastes, including physical, chemical and biological methods. Whatever the method, the concept is that of three 'R's, i.e. reduce, reuse and recycle. This is well-recognized and appreciated by most of the food industries and other industries that generate voluminous solid and liquid wastes in disposing of their by-products and waste.

8.2.1.1 Composting and Land Spreading The conversion of by-products of the food industry into useful soil additives is known as compost, which can be successfully used as a soil ameliorating agent or biofertilizer. Composting is an aerobic biological process in which organic wastes are converted by the activities of microorganisms into the simpler, stable and soil-like compounds and reduce the volume of the wastes (Semple *et al.*, 2001; Khalil *et al.* 2008). It is evident that mature compost has a positive effect on the soil and environment and is used as an organic fertilizer (Szanto *et al.*, 2007). Compost has a high water-holding capacity, high amount of nutrients and improves soil structure and aeration, which lead to an increase in agricultural production (Huang *et al.*, 2004).

Composting is one of the most efficient ways to reduce bulky food waste and release the valuable components to be reused. Food waste is the largest component of municipal waste streams after the recyclables are separated and require high disposal costs. In most cases, the first option is to use the food by-product as livestock feed, but when it is not possible then composting is the best alternative.

8.2.1.2 Vermicompost Vermicompost is another good alternative for recycling of beneficial materials. This is the process by which earthworms convert organic materials into a humus-like material known as vermicompost. This is an efficient way of solid waste management and has emerged as an innovative and low-cost biotechnology for the conversion of agro-industrial wastes into valuable products that can be utilized for improving soil structure and enhancing soil fertility. During the processing of food wastes through vermicomposting, plant nutrients present in them, such as N, P, K and Ca, are released. These nutrients are then converted into more soluble forms available for the plant absorption. The final product is called vermicompost, which can be used as organic fertilizer to add nutrients to the soil. Hence, vermicomposting technology has enormous potential in the management of food wastes and by-products in a sustainable manner.

8.3 Plant-Based Food By-Products and their Importance as Biofertilizers

Plant-based food industries are important in producing a large amount of solid waste along with food products. Many industries which produce plant-based food materials also generate valuable by-products and these can be used as biofertilizers. Being rich in lignocelluloses, food industry wastes can act as substrates for mushroom cultivation, cellulosic-rich tea (*Camellia sinensis*) production unit wastes and citrus processing wastes have been found to support the biomass production of the edible fungi Basidiomycetes. An excellent production of edible mushroom, such as *Morchella crassipes* and *Agaricus bisporus*, has been reported on lime waste (Das and Singh, 2004). Some of the important industries and their products are discussed below.

8.3.1 Sugarcane By-Products

The primary purpose of the sugarcane (*Saccharum* spp.) industry is to produce sugar for which sugarcane is the major source in tropical countries. However, during the process of sugar formation, many by-products are also produced, for example trash (dried leaf bases), bagasse, molasses and pressmud. All of these by-products constitute about 40% of the weight of crushed cane, besides furnace ash.

8.3.1.1 Importance of Pressmud as a Plant Growth Promoter Pressmud is a soft, amorphous blackish brown, spongy and bulky by-product of the sugar industry. The production of pressmud amounts to about 3% of cane crushed in the sugar factory (Shankaraiah and Kalyana Murthy, 2005). Pressmud is primarily used both as a soil reclamation agent and as a soil conditioner. This material has the potential to supply nutrients to the soil, besides its favorable effects on physico-chemical and biological properties of the soil (Shankaraiah and Kalyana Murthy, 2005). Pressmud has a pH of 5.0 and can be used for amendment of alkaline soil. Its various inorganic constituents, such as N, P and K, are anticipated to supplement fertility of the soil and thereby reduce the dependence on chemical fertilizers. Furthermore, its micronutrients, notably Zn and Cu, can supplement micronutrient fertility without additional cost. The addition of pressmud to the soil has been reported to improve microbial activity and as such regarded as an effective source of carrier material for microbial inoculants (Jauhri, 1990). Others also report a significant increase in cane and sugar yields (21%), besides economy in fertilizer NPK up to 50% due to an integrated use of enriched pressmud cake at 15 t ha^{-1} with the recommended fertilization (Shankaraiah and Kalyana Murthy, 2005). The synergistic effect due to addition of Azatobacter in combination with N fertilizer and pressmud was also observed (Ramalinga Swamy *et al.*, 1996). Pressmud is processed into compost, which is required for the formation of useful fertilizers.

Recently, some sugar factories in India with distilleries are producing pressmud biocompost. This is an excellent source of organic manure for sugarcane crops and is used for higher cane yield and sugar recovery. Some major advantages of using this manure for soil application include:

(i) very low power requirement;
(ii) zero discharge to inland water resources;

(iii) free from river or groundwater pollution;
(iv) rich in micronutrients and can reduce the requirement of chemical fertilizers;
(v) provides bacteria for N fixing and solubilization of phosphates;
(vi) keeps the soil healthy and develops the self-reclamation cycle;
(vii) free from all pathogens, harmful bacteria, weeds and seeds; and
(viii) free from flowing and easy to handle, pack and transport.

8.3.2 Utilization of Oilseed Processing By-Products as Biofertilizer

Biofertilizer is one of the important components of an integrated nutrient management system. It plays a key role in maintaining productivity and sustainability of soil and in protecting the environment as eco-friendly and cost-effective inputs for the farmers. It can be regarded as a renewable source of plant nutrients to supplement chemical fertilizers in sustainable agricultural production systems. Non-edible cakes such as castor (*Ricinus communis*) cake, karanja (*Pongamia pinnata*) cake, neem (*Azadirachta indica*) cake and mustard (*Brassica juncea*) cake are produced during the extraction of oils from these oilseed plants. These cakes can be used as biofertilizers due to their high N, P and K contents.

Some of these oil cakes are helpful in increasing N uptake ability of the plants. The value of these cakes lies in their N content, which varies from 3% to more than 9% in some cases. These cakes help in retarding soil nitrification. Nitrogen content is dependent upon the type of oil cake, its oil content and hull/husk present in it. Seeds dehulled/decorticated before the extraction provide a product that is richer in N than those which are not dehulled or decorticated. The C:N ratio of oil cake is usually narrow, being 3:15 for most of the oil cakes.

8.3.2.1 Uses of Cuphea Oil Processing By-Products and Processed Oils Seeds of *Cuphea* spp. have high levels of medium-chain fatty acids (MCFA) and have a good potential to be commercially cultivated. During the processing of cuphea oil, a number of by-products are generated. Some of these by-products are solid residue fractions, including presscake, bin trash, stem trash and seed trash. If these by-products are commercially used, it will help in improving the economics of cuphea cultivation. Presently, two tropical plants such as coconut (*Cocos nucifera*) and oil palm (*Elaeis guineensis*) are the major industrial sources of the world's plant derived MCFA (~5.5 million tonnes of lauric acid) (Cermak and Isbell, 2004). These MCFA, such as capric, lauric and myristic acids, are widely used in manufacturing various chemical products including cosmetics, confections, lubricants and detergents. As an alternative to these tropical plants, tropical and temperate Cuphea spp. also produce MCFA (Cermak and Isbell, 2004; Cermak *et al.*, 2007). Currently, cuphea is grown on an experimental basis in the USA. Different types of by-products such as bran, husk, bagasse and seeds as well as processed oil cakes are employed as substratum for microorganisms (Ramachandran *et al.*, 2007) and/or a soil amendment to improve plant growth (Lee and Lee, 2010; Vaughn *et al.*, 2010). Tisserat *et al.* (2012) investigated the effect of cuphea oil by-products as growth regulators and solid residues as organic soil amendments on Calabrese broccoli (*Brassica oleracea*) seedling growth. It has been observed that ground presscake and bin trash could be utilized as an organic soil amendment up to 1% without detrimental effects on broccoli. Moreover, certain oil by-product fractions from cuphea oil processing can be used as 'environmentally-friendly' herbicidal sprays.

8.3.2.2 Spent Bleaching Earth as a Bioorganic Fertilizer Spent bleaching earth (SBE) is produced in considerable amounts during the processing and refining of crude palm oil (CPO), for which the process involves degumming and bleaching (Loh *et al.*, 2013). Bleaching earth is a fine powder with silicon dioxide as its main constituent (~57% and more, depending on the type). Bleaching earth is primarily used to absorb dark color matter and odor-causing substances in crude and vegetable oils. It has been estimated that worldwide oil production was about 60 million tonnes from where about 600 000 metric tonnes or more of this bleaching earth was utilized in the refining process (Park *et al.*, 2004).

It was found that SBE, when co-composted with some agricultural and palm oil milling by-products, has a positive impact on soil physical attributes for plant growth and microbial rejuvenation (Loh *et al.*, 2013). This is due to the presence of adequate amounts of beneficial mineral elements (N, P and K), improved organic carbon (7.1–16%), greater cation exchange capacity (8.0–33.8 cmol/kg), water-holding capacity (6.1–16.5 ml/100 g) and higher C:N ratio (290). Although SBE contains high amounts of C and less N, the bio-availability of C could be low due to slow decomposition, thus the direct application of SBE can have a detrimental effect on plant growth by decreasing N availability. Loh *et al.* (2013) reported a significant increase in the productivity of okra (*Abelmoschus esculentus*), swamp morning glory (*Ipomoea aquatica*) and groundnut (*Arachis hypogaea*) var. magenta with 2-fold increases (35–60%) on average in fresh and dry matter production when treated with composted SBE.

8.3.2.3 Jatropha By-Products and its Uses in Agriculture *Jatropha curcas* is a small tree or large perennial shrub, with a life expectancy of about 50 years. Morphologically, this plant can reach a height of 3 m and under favorable conditions can even reach 5–6 m. *J. curcas* seeds are used for oil extraction. After oil extraction, the remaining is the *J. curcas* seed or presscake, which has a high content of nutrients and proteins with high calorific value and is used in a variety of ways, such as organic fertilizer or fuel, for biogas production or, after conversion, as high protein animal feed (Ye *et al.*, 2009). The cake is nutrient rich, therefore is suitable to be used as fertilizer. When it is applied as fertilizer (0.75–3 t ha^{-1}), it significantly increases the seed yield of a *J. curcas* plantation by 13–120% (Ghosh *et al.*, 2007). It also help increase yields in other crops such as *Pennisetum glaucum* (5 t ha^{-1}), *Brassica oleracea* (2.5 t ha^{-1}), *Oryza sativa* (10 t ha^{-1}) by 46, 40–113, and 11%, respectively, when used as a fertilizer (Achten *et al.*, 2008). Seed cake contains phorbol esters, which are completely biodegradable in soil and their degraded products appear to be innocuous. The seed cake application as fertilizers has no adverse impact on the beneficial microbial communities, insects, invertebrates and plant or animal communities (Devappa *et al.*, 2010).

8.3.2.4 Olive Oil Mill By-Products as a Source of Biofertilizer Olive oil is a major product in many countries of the world, particularly Mediterranean countries such as Spain, Syria, Italy, Turkey, Greece and Tunisia, where more than 95% of the world's olive oil is produced (Aktas *et al.*, 2001). Olive oil production causes serious environmental problems due to the production of high amounts of by-products. The major by-products are olive pomace (OP) and olive mill wastewater (OMW) (for three-phase systems) and two-phase olive mill waste (TPOMW) (for two-phase systems) within the short production season (Roig *et al.*, 2006). Alfano *et al.* (2008) reported that 2-P systems produce a lingo-cellulosic olive humid husk (OHH), which is a watery solid by-product with a high water content (57–75%), phenols (0.6–2.4%) and lipids (Ranalli *et al.*, 2002; Vlyssides *et al.*, 2004). In order to use these hazardous wastes as a beneficial compound for supplying nutrients to the soil, composting is the most feasible way. Compost produced from these materials is free of pathogens and provides maximum benefits to the crops. The chemical composition of OMW varies and depends on the variety of olive, the harvesting period and the extraction techniques (Magdich *et al.*, 2012). It includes various simple and complex phenolic compounds, which result in the generation of antimicrobial and phytotoxic effects (Quaratino *et al.*, 2007; Hachicha *et al.*, 2009). It has been reported that some of the OMW components could potentially be used for agricultural purpose, since they contain 83–96% of water, 3.5–15% of organic matter and 0.5–2% of minerals (Ammar *et al.*, 2005). Furthermore, OMW contains a high amount of K and appreciable levels of N, P, Ca, Mg and Fe (Sellami *et al.*, 2008), suggesting that OMW spreading would offer an attractive beneficial disposal way fertilizing agricultural lands (Sierra *et al.*, 2001, 2007). This would induce microbial community change (Mechri *et al.*, 2008) and soil-crop systems, increasing its nutrient contents and microbial community growth.

8.3.2.5 Soybean Meal as Fertilizer Soybean (*Glycine max*) is a very important crop cultivated worldwide. The major by-product produced during soybean oil production process is soybean meal, which is produced during oil extraction from seed (Rosenthal *et al.*, 2001). It has been reported that the application of rapeseed cake, soybean meal, or other meals can increase the amount and activity of microbes in soil, boost the growth rate of plants and improve the quality of produce when used as fertilizers (Lima *et al.*, 2011). Composting of fermentation meal fertilizers is still in use in many parts of China, but there are certain problems associated with this process, including a long period of fermentation (1–6 months) and a high N loss (nearly 30–50% during the process of composting) (Xu *et al.*, 2000). Therefore, N cannot be used efficiently and sufficiently for plant growth, which makes the use of meal for this purpose wasteful. In order to overcome this shortcoming, the use of a strain that produces a high yield of proteases in solid state fermentation (SSF) can be used as an alternative (Pandey, 2003). SSF through microorganisms has been widely used to hydrolyze soybean meal protein (Kuo *et al.*, 2006; Terlabie *et al.*, 2006). However, the required conditions during the SSF process are not easy to control, which restricts its large-scale industrialization (Mitchell *et al.*, 2003; Hölker *et al.*, 2004; Xiao *et al.*, 2007). The optimization process of SSF is critical for the synthesis of a high-quality fertilizer, especially water-soluble fertilizer with free amino acids, which has been widely used for promoting the growth and quality of plant.

8.3.2.6 Castor Meal and Castor Husk as Biofertilizers Castor is an oilseed crop commonly found in tropical regions. The oil content of castor seeds (40–60%) is higher compared with sunflower (*Helianthus annuus*) (38–48%), soybean (18–19%) and cotton (*Gossypium hirsutum*) (15–19%) (Nass *et al.*, 2007; Baldwin and Cossar, 2009). Castor oil is primarily used in the chemical industry and in the production of biodiesel (Baldwin and Cossar, 2009). Castor oil is different from other crop oils, due to the fact that it is composed of almost 90% of ricinoleic fatty acid (O'Brien, 2000). In the food industry, castor oil is utilized as a food preservative, a food additive and as a medicine.

Two major by-products of castor oil processing are castor meal and castor husk (Lima *et al.*, 2011).

It has been reported that about 1.13 tonnes of castor husk and 1.31 tonnes of castor meal as by-products are generated from the production of 1 tonne of castor oil (Lima *et al.*, 2011). Castor meal and castor husk are utilized as a biofertilizer for increasing crop yields and reducing environmental pollution. These by-products when used as organic fertilizer can improve soil microbial activity, soil physical properties and immobilize toxic elements such as Al and supply various micro- and macro-nutrients (Costa *et al.*, 2009). Castor meal contains about 7.54% N, which is greater than the N content of other traditional by-products used as fertilizers, such as sugarcane bagasse (0.24% N), animal manure (0.77% N) and wood ash (0.55% N) (Gupta *et al.*, 2004). Castor meal has a high amount of N (7.54%), with a very low amount of K (0.66%), whereas castor husk has a low amount of N (1.86%) and a high amount of K (4.50%) (Lima *et al.*, 2011).

8.3.3 Food Processing Industrial Sludge as Sources of Biofertilizers

Waste water from the food industry is generally disposed of and associated with certain environmental issues. Sludge is a by-product of wastewater treatment processes and its use as a source of biofertilizer is a good alternative (Mahapatra *et al.*, 2013). Sludge contains essential plant nutrients and organic matter that can help increase crop production. Many researches revealed a positive effect of sludge on crop yields and the physico-chemical properties of soil. It has been reported that sludge application positively affects crop productivity and significantly improves soil quality (Mahapatra *et al.*, 2013). The authors also mentioned that the application of sludge to soil increased its pH, EC, CEC and other nutrients such as organic matter. Sigua *et al.* (2005) observed no negative effect on forage or soil quality with the consistent application of sludge. The authors recommended that successive use of sludge for three years followed by no use of sludge for at least two years is a good practice to enhance forage productivity and to minimize the accumulation of trace elements. de Silva *et al.* (2011) studied the effect of wet and dry sewage sludge on the growth and nutrient cycling of *Eucalyptus grandis* plantations in Brazil and reported that sewage sludge application has a positive effect on leaf litter production, with an increased amount of nutrient transfer among the components of the ecosystem. Barriquelo *et al.* (2003) reported an enhancement in crop yield followed by the application of sludge to agricultural land as sludge contains various macro- and micronutrients. Due to the presence of a high concentration of nutrients, sludge is recommended for its agricultural application (Dolgen *et al.*, 2007; Kidd *et al.*, 2007). The application of sludge is not recommended for soil with a pH of less than 5.0. This is due to the fact that at lower pH, micronutrients can accumulate, causing toxicity for plant growth. In acidic soil, phytotoxic elements are more likely available, which can damage the plant cells of the crop.

8.3.4 Rice Straw and Rice Bran

Rice straw is used for various purposes such as animal feed, energy source, construction materials, paper production, incorporation in soil and compost. Rice bran is one of the major by-products of milling in the production of refined grains and accounts for 5–8% of the weight of rough rice. It is rich in minerals and helps enhance composting process by increasing the compost pile temperature. A proportion of the Japanese organic farmers use rice bran (200 gm^{-2}) for weed control and as a fertilizer for transplanted rice (Khan *et al.*, 2007). The straw residue of rice is rich in organic matter (80%) with an oxidizable organic C of 34% and has a high C/N ratio (on an average of 50). These properties make it a feasible source of C for microorganisms, especially under composting conditions. The physical properties of rice straw impart it with a good-quality structure for mixing with sewage sludge. The resulting materials generally have a suitable C:N ratio, pH and moisture values, which are required for microorganisms involved in the composting process. The physical properties of rice straw are suitable as biofertilizer and when combined with sewage sludge it can be used for composting (Iranzo *et al.*, 2004).

8.3.5 Coffee Processing By-Products

Coffee is the most popular beverages worldwide (Daglia *et al.*, 2000). Coffee ranks second in commercial trading worldwide with the production of a huge amount of by-products and residues (solid and liquid wastes) during processing (Nabais *et al.*, 2008; Mussatto *et al.*, 2011). These by-products are a source of severe contamination associated with negative environmental impacts in coffee-producing countries. The major types of by-products are coffee pulp, husk and effluent and their improper disposal can cause pollution of water and land around the processing units. Coffee pulp is rich in moisture and the disposal of coffee pulp is becoming an emerging environmental problem due to its putrefaction (Corro *et al.*, 2014). Traditionally, coffee pulp and husk have found only limited applications as fertilizer, livestock feed and compost. There has been a renewed interest in utilizing these by-products as substrates in bioprocesses and vermicomposting.

8.3.6 Tea Processing Wastes

Globally, tea ranks second in consumption after water and is one of the most popular beverages due to its certain properties such as attractive aroma, pleasant taste and health-promoting effects (Hilal and Engelhardt, 2007). The important part of the tea plant is the leaf and when it enters processing, it is converted into one product or other. For this reason, the wastes or by-products of tea processing are considered insignificant. During the past several years, there has been an increasing trend of ready-made tea drinks in bottles, packs and cans. These ready-made beverage companies producing different tea drinks generate a considerable amount of tea leaf wastes annually. According to FAO statistics, 352 million tonnes of tea was produced in 2005 and finding a use for these wastes is important due to economic and environmental concerns (Morikawa and Saigusa, 2008). Some possible solutions for this waste management are provided below.

8.3.6.1 Use as a Raw Material for Composting When green tea is produced from processed leaves, tea leaf waste is generally burned or dumped into landfills, which are not good waste management practices and have serious environmental consequences (Kondo *et al.*, 2004). Only a small amount of tea leaf waste is converted into raw compost material. Green tea waste contains more than 90% organic matter and 5–7% N, which might be used as a good source of composting materials. Upon mixing this green tea waste with rice bran in a 30:70 ratio (v/v, on a dry basis), it was found to produce a good-quality compost with an excellent weed control potential and growth enhancing ability for spinach (*Spinacia oleracea*) (Khan *et al.*, 2007). Composting of tea by-products is one of the most cost-effective options for utilizing the tea waste, because dumping creates environmental problems and burning is expensive (Khan *et al.*, 2007).

8.3.6.2 Uses as a Material for Improving Soil Nutrient Status The major causes for nutrient loss from soil are soil erosion, rapid decomposition and mineralization of organic matter, particularly due to intensive agricultural activities. Restoration and even increase of these nutrients is required for the continuity of agricultural activities; therefore, the tea waste containing plant nutrients and organic matter could be used to supplement commercial fertilizers for crop production.

8.3.7 Turmeric Solid Waste

Turmeric (*Curcuma longa*) powder is used for different purposes, and during its processing, various types of waste are produced, which are generally thrown away causing environmental problems. Therefore, efforts are being made to use these wastes as biofertilizers after passing through a compost process using suitable bacteria. The solid waste is yellowish in color with a moisture content of about 10%. Its main product is its curcumin content, but in addition it also contains plant nutrients and when mixed with bacterial consortium, it is converted into biofertilizer. The moisture content of the solid waste is retained by up to 35–40% with the application of bacterial inoculums. The plant oil of turmeric has a potential to be used in eco-friendly pest management.

8.3.8 Cassava Processing By-Product as Biofertilizers

Cassava processing by-products from industry are considered to be one of the major environmental pollutants, as well as an aesthetic nuisance (Ubalua, 2007). Cassava sievates, which is obtained from the processing of 'garri', and cassava offal, which is obtained during the production of 'fufu', are the two major wastes produced during cassava processing. Cassava peels are considered a by-product of cassava-fermented products and are thrown away to rot. Peel waste constitutes about 20–30% of tuber weight and poses a serious problem as an environmental pollutant (Ubalua, 2007). Cassava by-products are utilized as biofertilizer by converting cassava peel waste into phosphate biofertilizer through the use of phosphate solubilizing fungi with locally available inoculum. Use of cassava by-products as biofertilizers makes the farming economical, especially for developing countries (Ogbo, 2010). Cassava processing by-products can be grouped into four different categories:

1. peeling by-products from cassava;
2. fibrous by-products from cassava;
3. residues of starch; and
4. waste water or effluent.

Liquid (biol) and solid (biosol) fractions obtained through anaerobic decomposition of cassava by-products (cassava peels) are considered as a good source of biofertilizer for various crops. Biosol supplies many essential nutrients for crops, which are required for their growth and development, such as Ca, K, P and N. In addition to essential nutrients, it also provides auxin and gibberellin (growth regulators), which have beneficial effects on crops. Biosol also contains similar nutrients.

8.4 Importance of Plant-Based Food Processing By-Products as Biopesticides

Biopesticides are certain types of compounds (pesticides) derived from natural materials such as animals, plants bacteria and certain minerals. Among these compounds, pyrethrins I and II, cinerins, jasmolins, azadirachtin, spinosad and rotenone are some of the key molecules. In recent years, alternatives to synthetic pesticides for effective pest control in different agro-systems have been investigated. Amongst them, plant-derived products proved to be eco-friendly, residue-free, biodegradable and cost-effective.

8.4.1 Maize Gluten Meal

Gluten meal is a by-product of maize (*Zea mays*) obtained during the milling process. Commercially, it can be used both as biofertilizer and as pre-emergence herbicide on various high-value crops and lawns (McDade and Christians, 2000). The commercial products available on the market are sold under different trade names and contain between 50 and 100% of maize gluten. In order to use it as herbicides for controlling grasses and other weed species, application at a high rate (2 t ha^{-1}) is required and thus becomes cost prohibitive. The mode of action of maize gluten is different from other herbicides. Maize gluten has no effect on existing weeds, but prevents germination and growth of young broadleaf and grassy weeds (Gough and Carlstrom, 1999). Upon hydrolysis by soil microbes, maize gluten releases several phytotoxic dipeptides and a phytotoxic pentapeptide. The exact mode of action of these oligopeptides is still not fully understood; however, they have an effect on cell wall formation, membrane integrity and nuclear development (Unruh *et al.*, 1997). Maize gluten meal is considered as a slow-release pro-herbicide. This is due to the fact that it must be hydrolyzed for release of active ingredients (Dayan *et al.*, 2009). Some of the major types of phytotoxic oligopeptides in the hydrolysate of maize gluten meal are glutaminyl-glutamine, glycinyl-alanine, alanyl-glutamine, alanyl-asparagine, alaninyl-alanine and leucine-serine-proline-alanine-glutamine (Christians *et al.*, 2010).

8.4.2 Cuphea Oil

As non-selective contact herbicides, petroleum oils (petroleum distillates, gasoline, kerosene and diesel fuel) have been used in the past to control weeds; however, their use could pose environmental concern because they are moderately toxic and slow to degrade in the environment (Vaughn and Holser, 2007). Presently, as an alternative to petroleum-based oils, vegetable oils and the esters synthesized from them are employed for crop protection with high biodegradability and low environmental toxicity (Vaughn and Holser, 2007). Vegetable oils containing highly saturated fatty acid mixtures may behave as non-selective contact herbicides (Vaughn and Holser, 2007). Some species of Cuphea are used to produce cuphea oil, which contains high levels of MCFA. Cuphea oil fractions possess high concentrations of capric acid, a C10 (50–82% v/v) saturated fatty acid. This distilled fatty acid mixture (81.9% capric acid) obtained from refined cuphea oil was found to be highly effective as a biodegradable herbicide (Tisserat *et al.*, 2012).

8.4.3 Jatropha Oil

The seeds of *Jatropha curcas* contain 30–40% oil that can be easily converted into biodiesel for use in diesel engines. Due to the presence of fungicidal, molluscicidal, insecticidal and nematicidal properties, jatropha oil can be used as a biopesticide (Kumar and Sharma, 2008). For example, jatropha oil has been used in controlling insect pests of cotton (e.g. cotton bollworm) and on pests of pulses, potato and maize (Kumar and Sharma, 2008). Jatropha oil, extracts and phorbol esters, used as natural pesticides in controlling insect pests, could be an excellent alternative to hazardous chemicals (Gübitz *et al.*, 1999). Jatropha oil and extracts have no harmful effect on the beneficial arthropods compared with other chemical sprays. YuChuan *et al.* (2009) pointed out that water crude extract from *J. curcas* shell, branch or leaf has a herbicidal activity.

8.4.4 Olive Compounds

Washing waters of green olives and the acidified preservation solutions of black olives contain compounds with antimicrobial properties, which can be used in agriculture for pest management (Medina *et al.*, 2008; Brenes *et al.*, 2011). It has been reported that glutaraldehyde-like compounds obtained from olive processing have a strong anti-bacterial and antifungal activity against plant phytopathogens (Brenes *et al.*, 2011). The bactericidal properties of storage solutions of ripe black olives were observed against different species of *Agrobacterium, Erwinia, Pseudomonas* and *Clavibacter*, while the washing water from Spanish style green olives were less effective (Brenes *et al.*, 2011). A good antimicrobial activity has also been found in olive leaf extracts, which has the potential to provide remedies for microbial diseases (Lee and Lee, 2010). The strong bactericidal activity of olive oil is due to the presence of dialdehydic form of decarboxymethyl elenolic acid linked to tyrosol (TyEDA) and hydroxytyrosol (HyEDA) (Medina *et al.*, 2006). These extracts are also used against various phytopathogenic microorganisms because of their bactericidal and fungicidal properties when salt-free table olive solutions were used (Brenes *et al.*, 2011). Several researchers have indicated the phytotoxic effects of olive mill residue, which are particularly evident during the germination and seedling development of higher plants (Casa *et al.*, 2003; Sampedro *et al.*, 2004). The potential use of olive mill waste and compost as biopesticides against weeds, fungi and nematodes have been evaluated by some investigators for the recovery of biological activity in soil after its application. For example, Piotrowska *et al.* (2006) observed a complete recovery of seed germination at 42 days after olive mill wastewater application at the rate of

40 m^3 ha^{-1}, suggesting that olive mill wastes could be used as pre-plant biopesticide or in crops on which they exert no phytotoxic effect. Cayuela *et al.* (2008) reported a substantial inhibition in the germination of two weed species (*Amaranthus retroflexus* and *Solanum nigrum*) with the sterile water extracts of two-phase olive mill waste (TPOMW) and TPOMW composts. The authors also mentioned a strong inhibitory effect on egg hatch and second-stage juvenile motility of the root-knot nematode *Meloidogyne incognita* with TPOMW extracts.

8.4.5 Plant Extracts Classified as Minimal Risk Pesticides

Some of the common plants and their extracts that can be used as minimal risk pesticides are castor oil, cinnamon and cinnamon oil, cedar oil, citric acid, citronella oil and citronella, cloves and clove oil, maize oil, maize gluten meal, cotton seed oil, garlic and garlic oil, eugenol, geraniol, germanium oil, linseed oil, lauryl sulfate, lemongrass oil, malic acid, mint and mint oil, peppermint and peppermint oil, rosemary and rosemary oil, sesame (includes ground sesame plant) and sesame oil, soybean oil, sodium lauryl sulfate, thyme and thyme oil and white pepper (Cavoski *et al.*, 2011).

8.4.6 Rotenone as Biopesticide

Rotenone is a group of natural products (alkaloid) present in the Leguminosae family, especially extracted from the root of some tropical plants such as *Derris elliptica*, *Tephrosia vogelli* and *Lonchocharpus nicou* (El-Wakeil, 2013). Rotenone is used as a toxin against arthropods, insects, mites and fish (El-Wakeil, 2013). Its mode of action is similar to pyrethrin, but with a stronger action and a more persistence (Crombie, 1999). Rotenone has the efficiency of exerting rapid neurotoxic action against insects, called the 'knock-down effect'; and due to this property it is used to control aphids, suckers, thrips and other insects mainly on organic fruit and vegetables (Isman, 2006; El-Wakeil, 2013).

8.5 Concluding Remarks

Different types of plant- and animal-based food by-products can be used as a source of plant nutrients for satisfactory crop growth and production and also as a natural agent for protecting a crop against insects, diseases and weeds. Usually these processing products are disposed of in the soil or water, causing severe environmental consequences. In the changing scenarios of global climate, we need sufficient and quality food for a rapidly increasing population without compromising environmental sustainability, which is equally important for prosperous life and for our future generations. In order to achieve both of these goals, we need to change the traditional ways of systems of cropping by minimizing the use of synthetic fertilizers and pesticides and replacing or substituting them with biofertilizers and biopesticides. For this purpose, food processing by-products have increasingly been viewed as a great option in the production of biofertilizers and biopesticides, which can be effectively used in increasing agricultural production in a sustainable manner.

References

Achten, W.M.J., Verchot, L., Franken, Y. J., Mathis, E., Singh, V.P. *et al.*, (2008) Jatropha bio-diesel production and use. *Biomass and Bioenergy*, **32**(12): 1063–1084.

Ahmad, R., Jilani, G., Arshad, M., Zahir, Z.A. and Khalid, A. (2007) Bio-conversion of organic wastes for their recycling in agriculture: an overview of perspectives and prospects. *Annals of Microbiology*, **57**(4), 471–479.

Aktas, E.S., Imre, S. and Ersoy, L. (2001) Characterization and lime treatment of olive mill wastewater. *Water Research*, **35**(9): 2336–2340.

Alfano, G., Belli, C., Lustrato, G. and Ranalli, G. (2008) Pile composting of two-phase centrifuged olive husk residues: Technical solutions and quality of cured compost. *Bioresource Technology*, **99**(11): 4694–4701.

Ammar, E., Nasri, M. and Medhioub, K. (2005) Isolation of Enterobacteria able to degrade simple aromatic compounds from the wastewater from olive oil extraction. *World Journal of Microbiology and Biotechnology*, **21**(3): 253–259.

An, C.J., Huang, G.H., Yao, Y., Sun, W. and An, K. (2012) Performance of in-vessel composting of food waste in the presence of coal ash and uric acid. *Journal of Hazardous Materials*, **203–204**: 38–45.

Baldwin, B.S. and Cossar, R.D. (2009) Castor yield in response to planting date at four locations in the south-central United States. *Industrial Crops and Products*, **29**(2): 316–319.

Barriquelo, M.F., Juliatti, M.A., Silva, M.S.D. and Lenzi, E. (2003) Lead behavior in soil treated with contaminated sewage sludge and cultivated with maize. *Brazilian Archives of Biology and Technology*, **46**(4): 499–505.

Brenes, M., García, A., De los Santos, B., Medina, E., Romero, C., *et al.* (2011). Olive glutaraldehyde-like compounds against plant pathogenic bacteria and fungi. *Food Chemistry*, **125**(4): 1262–1266.

Casa, R., D'Annibale, A., Pieruccetti, F., Stazi, S.R., Sermanni, G.G. and Cascio, B.L. (2003) Reduction of the phenolic components in olive-mill wastewater by an enzymatic treatment and its impact on durum wheat (*Triticum durum* Desf.) germinability. *Chemosphere*, **50**(8): 959–966.

Cavoski, I., Caboni, P. and Miano, T. (2011) *Natural Pesticides and Future Perspectives*. INTECH Open Access Publisher.

Cayuela, M.L., Millner, P.D., Meyer, S.L.F. and Roig, A. (2008) Potential of olive mill waste and compost as biobased pesticides against weeds, fungi, and nematodes. *Science of the Total Environment*, **399**(1): 11–18.

Cermak, S.C. and Isbell, T.A. (2004) Synthesis and physical properties of cuphea-oleic estolides and esters. *Journal of the American Oil Chemists' Society*, **81**(3): 297–303.

Cermak, S.C., John, A.L. and Evangelista, R.L. (2007) Enrichment of decanoic acid in cuphea fatty acids by molecular distillation. *Industrial Crops and Products*, **26**(1): 93–99.

Christians, N., Liu, D. and Unruh, J.B. (2010) The use of protein hydrolysates for weed control. In: Protein Hydrolysates in Biotechnology). Springer, The Netherlands, pp. 127–133.

Corro, G., Pal, U. and Cebada, S. (2014) Enhanced biogas production from coffee pulp through deligninocellulosic photocatalytic pretreatment. *Energy Science and Engineering*, **2**(4): 177–187.

Costa, F.X., Beltrão, N.D.M., Júnior, E.N., Guimarães, M.M. and Damace, F.V. (2009) Efeito do lixo orgânico e torta de mamona nas características de crescimento da mamoneira (*Ricinus communis* L.). *Engenharia Ambiental: Pesquisa e Tecnologia*, **6**(1): 259–269.

Crombie, L. (1999) Natural product chemistry and its part in the defence against insects and fungi in agriculture. *Pesticide Science*, **55**(8): 761–774.

Daglia, M., Papetti, A., Gregotti, C., Bertè, F. and Gazzani, G. (2000) *In vitro* antioxidant and *ex vivo* protective activities of green and roasted coffee. *Journal of Agricultural and Food Chemistry*, **48**(5): 1449–1454.

Das, H. and Singh, S.K. (2004) Useful byproducts from cellulosic wastes of agriculture and food industry: A critical appraisal. *Critical Reviews in Food Science and Nutrition*, **44**(2): 77–89.

da Silva, P.H.M., Poggiani, F. and Laclau, J.P. (2011) Applying sewage sludge to *Eucalyptus grandis* plantations: effects on biomass production and nutrient cycling through litter fall. *Applied and Environmental Soil Science,* Article ID 710614, 11 p.

Dayan, F.E., Cantrell, C.L. and Duke, S.O. (2009) Natural products in crop protection. *Bioorganic and Medicinal Chemistry*, **17**(12): 4022–4034.

Devappa, R.K., Makkar, H.P. and Becker, K. (2010) Biodegradation of *Jatropha curcas* phorbol esters in soil. *Journal of the Science of Food and Agriculture*, **90**(12): 2090–2097.

Dolgen, D., Alpaslan, M.N. and Delen, N. (2007) Agricultural recycling of treatment-plant sludge: A case study for a vegetable-processing factory. *Journal of Environmental Management*, **84**(3): 274–281.

El-Wakeil, N.E. (2013) Botanical pesticides and their mode of action. *Gesunde Pflanzen*, **65**(4): 125–149.

FAO (2009) *High Level Expert Forum – How to feed the world in 2050*. FAO, Rome.

FAO, WFP, IFAD (2012) *The State of Food Insecurity in the World 2012: Economic growth is necessary but not sufficient to accelerate reduction of hunger and malnutrition*. FAO, Rome.

Ghosh, A., Patolia, J.S., Chaudhary, D.R., Chikara, J., Rao, S.N., Kumar, D. and Zala, A. (2007) Response of *Jatropha curcas* under different spacing to Jatropha de-oiled cake. In: Expert Seminar on *Jatropha curcas* L. Agronomy and Genetics, March, pp. 26–28.

Godfray, H.C.J., Beddington, J.R., Crute, I.R., Haddad, L., Lawrence, D., *et al.* (2010) Food security: The challenge of feeding 9 billion people. *Science*, **327**(5967): 812–818.

Gottlieb, R. and Joshi, A. (2010) *Food Justice*. MIT Press.

Gough, R.E. and Carlstrom, R. (1999) Wheat gluten meal inhibits germination and growth of broadleaf and grassy weeds. *HortScience*, **34**(2): 269–270.

Gübitz, G.M., Mittelbach, M. and Trabi, M. (1999) Exploitation of the tropical oil seed plant *Jatropha curcas* L. *Bioresource Technology*, **67**(1): 73–82.

Gupta, A.P., Antil, R.S. and Narwal, R.P. (2004) Utilization of deoiled castor cake for crop production. *Archives of Agronomy and Soil Science*, **50**(4–5): 389–395.

Hachicha, R., Hachicha, S., Trabelsi, I., Woodward, S. and Mechichi, T. (2009) Evolution of the fatty fraction during co-composting of olive oil industry wastes with animal manure: Maturity assessment of the end product. *Chemosphere*, **75**(10): 1382–1386.

Hilal, Y. and Engelhardt, U. (2007) Characterisation of white tea – Comparison to green and black tea. *Journal für Verbraucherschutz und Lebensmittelsicherheit*, **2**(4): 414–421.

Hölker, U., Höfer, M. and Lenz, J. (2004) Biotechnological advantages of laboratory-scale solid-state fermentation with fungi. *Applied Microbiology and Biotechnology*, **64**(2): 175–186.

Huang, G.F., Wong, J.W.C., Wu, Q.T. and Nagar, B.B. (2004) Effect of C/N on composting of pig manure with sawdust. *Waste Management*, **24**(8): 805–813.

Iranzo, M., Canizares, J.V., Roca-Perez, L., Sainz-Pardo, I., Mormeneo, S. and Boluda, R. (2004) Characteristics of rice straw and sewage sludge as composting materials in Valencia (Spain). *Bioresource Technology*, **95**(1): 107–112.

Isman, M.B. (2006) Botanical insecticides, deterrents, and repellents in modern agriculture and an increasingly regulated world. *Annual Review of Entomology*, **51**: 45–66.

Jauhri, K.S. (1990) Modified sugarcane pressmud: A potential carrier for commercial production of bacterial inoculants. *Indian Journal of Agricultural Research*, **24**(4): 189–197.

Khalil, A., Domeizel, M. and Prudent, P. (2008) Monitoring of green waste composting process based on redox potential. *Bioresource Technology*, **99**(14): 6037–6045.

Khan, M.A.I., Ueno, K., Horimoto, S., Komai, F., Tanaka, K. and Ono, Y. (2007) Evaluation of the upland weed control potentiality of green tea waste-rice bran compost and its effect on spinach growth. *American Journal of Agricultural and Biological Sciences*, **2**(3): 142.

Kidd, P.S., Dominguez-Rodriguez, M.J., Diez, J. and Monterroso, C. (2007) Bioavailability and plant accumulation of heavy metals and phosphorus in agricultural soils amended by long-term application of sewage sludge. *Chemosphere*, **66**(8): 1458–1467.

Kondo, M., Kita, K. and Yokota, H. O. (2004) Effects of tea leaf waste of green tea, oolong tea, and black tea addition on Sudan grass silage quality and *in vitro* gas production. *Journal of the Science of Food and Agriculture*, **84**(7): 721–727.

Kumar, A. and Sharma, S. (2008) An evaluation of multipurpose oil seed crop for industrial uses (*Jatropha curcas* L.): A review. *Industrial Crops and Products*, **28**(1): 1–10.

Kuo, L.C., Cheng, W.Y., Wu, R.Y., Huang, C.J. and Lee, K.T. (2006) Hydrolysis of black soybean isoflavone glycosides by *Bacillus subtilis* natto. *Applied Microbiology and Biotechnology*, **73**(2): 314–320.

Larney, F.J. and Hao, X. (2007) A review of composting as a management alternative for beef cattle feedlot manure in southern Alberta, Canada. *Bioresource Technology*, **98**(17): 3221–3227.

Lee, O.H. and Lee, B.Y. (2010) Antioxidant and antimicrobial activities of individual and combined phenolics in *Olea europaea* leaf extract. *Bioresource Technology*, **101**(10): 3751–3754.

Lima, R.L., Severino, L.S., Sampaio, L.R., Sofiatti, V., Gomes, J.A. and Beltrão, N.E. (2011) Blends of castor meal and castor husks for optimized use as organic fertilizer. *Industrial Crops and Products*, **33**(2): 364–368.

Loh, S.K., James, S., Ngatiman, M., Cheong, K.Y., Choo, Y.M. and Lim, W.S. (2013) Enhancement of palm oil refinery waste – Spent bleaching earth (SBE) into bioorganic fertilizer and their effects on crop biomass growth. *Industrial Crops and Products*, **49**: 775–781.

Magdich, S., Jarboui, R., Rouina, B.B., Boukhris, M. and Ammar, E. (2012) A yearly spraying of olive mill wastewater on agricultural soil over six successive years: Impact of different application rates on olive production, phenolic compounds, phytotoxicity and microbial counts. *Science of the Total Environment*, **430**: 209–216.

Mahapatra, K., Ramteke, D S., Paliwal, L.J. and Naik, N.K. (2013) Agronomic application of food processing industrial sludge to improve soil quality and crop productivity. *Geoderma*, **207**: 205–211.

McDade, M.C. and Christians, N.E. (2000) Corn gluten meal – A natural pre-emergence herbicide: effect on vegetable seedling survival and weed cover. *American Journal of Alternative Agriculture*, **15**(04): 189–191.

Mechri, B., Mariem, F.B., Baham, M., Elhadj, S.B. and Hammami, M. (2008) Change in soil properties and the soil microbial community following land spreading of olive mill wastewater affects olive trees key physiological parameters and the abundance of arbuscular mycorrhizal fungi. *Soil Biology and Biochemistry*, **40**(1): 152–161.

Medina, E., de Castro, A., Romero, C. and Brenes, M. (2006) Comparison of the concentrations of phenolic compounds in olive oils and other plant oils: correlation with antimicrobial activity. *Journal of Agricultural and Food Chemistry*, **54**(14): 4954–4961.

Medina, E., Romero, C., de Castro, A., Brenes, M. and García, A. (2008) Inhibitors of lactic acid fermentation in Spanish-style green olive brines of the Manzanilla variety. *Food Chemistry*, **110**(4): 932–937.

Mitchell, D.A., von Meien, O.F. and Krieger, N. (2003) Recent developments in modeling of solid-state fermentation: Heat and mass transfer in bioreactors. *Biochemical Engineering Journal*, **13**(2): 137–147.

Mohee, R., Mudhoo, A. and Unmar, G.D. (2008) Windrow co-composting of shredded office paper and broiler litter. *International Journal of Environment and Waste Management*, **2**(1): 3–23.

Morikawa, C.K. and Saigusa, M. (2008) Recycling coffee and tea wastes to increase plant available Fe in alkaline soils. *Plant and Soil*, **304**(1–2): 249–255.

Mussatto, S.I., Machado, E.M., Martins, S. and Teixeira, J.A. (2011) Production, composition, and application of coffee and its industrial residues. *Food and Bioprocess Technology*, **4**(5): 661–672.

Nabais, J.M.V., Nunes, P., Carrott, P.J., Carrott, M.M.L.R., García, A.M. and Díaz-Díez, M.A. (2008). Production of activated carbons from coffee endocarp by CO_2 and steam activation. *Fuel Processing Technology*, **89**(3): 262–268.

Nass, L.L., Pereira, P.A.A. and Ellis, D. (2007) Biofuels in Brazil: An overview. *Crop Science*, **47**(6): 2228–2237.

O'Brien, R.D. (2000) Fats and oils: An overview. *Introduction to Fats and Oils Technology*, **2**: 1–19.

Ogbo, F.C. (2010) Conversion of cassava wastes for biofertilizer production using phosphate solubilizing fungi. *Bioresource Technology*, **101**(11): 4120–4124.

Pandey, A. (2003) Solid-state fermentation. *Biochemical Engineering Journal*, **13**(2): 81–84.

Park, E.Y., Kato, A. and Ming, H. (2004) Utilization of waste activated bleaching earth containing palm oil in riboflavin production by *Ashbya gossypii*. *Journal of the American Oil Chemists' Society*, **81**(1): 57–62.

Piotrowska, A., Iamarino, G., Rao, M.A. and Gianfreda, L. (2006) Short-term effects of olive mill waste water (OMW) on chemical and biochemical properties of a semiarid Mediterranean soil. *Soil Biology and Biochemistry*, **38**(3): 600–610.

Quaratino, D., D'Annibale, A., Federici, F., Cereti, C.F., Rossini, F. and Fenice, M. (2007) Enzyme and fungal treatments and a combination thereof reduce olive mill wastewater phytotoxicity on *Zea mays* L. seeds. *Chemosphere*, **66**(9): 1627–1633.

Ramachandran, S., Singh, S.K., Larroche, C., Soccol, C.R. and Pandey, A. (2007) Oil cakes and their biotechnological applications – A review. *Bioresource Technology*, **98**(10): 2000–2009.

Ramalinga Swamy, K., Naidu, M.R. and Mallikarjuna Rao, T.K.V.V. (1996) Studies on the effect of fertilizer nitrogen, pressmud cake and Azotobacter on the uptake, yield and quality of sugarcane ratoon. *Cooperative Sugar*, **27**: 351–354.

Ranalli, G., Principi, P., Zucchi, M., Da Borso, F., Catalano, L. and Sorlini, C. (2002) *Pile Composting of Two-Phase Centrifuged Olive Husks: Bioindicators of the process.* Springer, Berlin Heidelberg, pp. 165–175.

Roig, A., Cayuela, M.L. and Sánchez-Monedero, M.A. (2006) An overview on olive mill wastes and their valorisation methods. *Waste Management*, **26**(9): 960–969.

Rosenthal, A., Pyle, D.L., Niranjan, K., Gilmour, S. and Trinca, L. (2001) Combined effect of operational variables and enzyme activity on aqueous enzymatic extraction of oil and protein from soybean. *Enzyme and Microbial Technology*, **28**(6): 499–509.

Russ, W. and Meyer-Pittroff, R. (2004) Utilizing waste products from the food production and processing industries. *Critical Reviews in Food Science and Nutrition*, **44**(1): 57–62.

Sampedro, I., Aranda, E., Martın, J., Garcıa-Garrido, J.M., Garcıa-Romera, I. and Ocampo, J.A. (2004) Saprobic fungi decrease plant toxicity caused by olive mill residues. *Applied Soil Ecology*, **26**(2): 149–156.

Sellami, F., Hachicha, S., Chtourou, M., Medhioub, K. and Ammar, E. (2008) Maturity assessment of composted olive mill wastes using UV spectra and humification parameters. *Bioresource Technology*, **99**(15): 6900–6907.

Semple, K.T., Reid, B.J. and Fermor, T.R. (2001) Impact of composting strategies on the treatment of soils contaminated with organic pollutants. *Environmental Pollution*, **112**(2): 269–283.

Shankaraiah, C. and Murthy, K.K. (2005) Effect of enriched pressmud cake on growth, yield and quality of sugarcane. *Sugar Technology*, **7**(2–3): 1–4.

Sierra, J., Martı, E., Montserrat, G., Cruanas, R. and Garau, M.A. (2001) Characterization and evolution of a soil affected by olive oil mill wastewater disposal. *Science of the Total Environment*, **279**(1): 207–214.

Sierra, J., Martí, E., Garau, M.A. and Cruañas, R. (2007) Effects of the agronomic use of olive oil mill wastewater: Field experiment. *Science of the Total Environment*, **378**(1): 90–94.

Sigua, G., Adjei, M. and Rechcigl, J. (2005) Cumulative and residual effects of repeated sewage sludge applications: Forage productivity and soil quality implications in South Florida, USA. *Environmental Science and Pollution Research*, **12**(2): 80–88.

Szanto, G.L., Hamelers, H.V.M., Rulkens, W.H. and Veeken, A.H.M. (2007) NH3, N2O and CH_4 emissions during passively aerated composting of straw-rich pig manure. *Bioresource Technology*, **98**(14): 2659–2670.

Terlabie, N.N., Sakyi-Dawson, E. and Amoa-Awua, W.K. (2006) The comparative ability of four isolates of *Bacillus subtilis* to ferment soybeans into dawadawa. *International Journal of Food Microbiology*, **106**(2): 145–152.

Tisserat, B., O'Kuru, R.H., Cermak, S.C., Evangelista, R.L. and Doll, K.M. (2012) Potential uses for cuphea oil processing byproducts and processed oils. *Industrial Crops and Products*, **35**(1): 111–120.

Ubalua, A.O. (2007) Cassava wastes: Treatment options and value addition alternatives. *African Journal of Biotechnology*, **6**(18): 2065–2073.

Unruh, J.B., Christians, N.E. and Horner, H.T. (1997) Herbicidal effects of the dipeptide aianinyl-alanine on perennial ryegrass (*Lolium perenne* L.) seedlings. *Crop Science*, **37**(1): 208–212.

Vaughn, S.F. and Holser, R.A. (2007) Evaluation of biodiesels from several oilseed sources as environmental friendly contact herbicides. *Industrial Crops and Products*, **26**(1): 63–68.

Vaughn, S.F., Deppe, N.A., Berhow, M.A. and Evangelista, R.L. (2010) Lesquerella press cake as an organic fertilizer for greenhouse tomatoes. *Industrial Crops and Products*, **32**(2): 164–168.

Vlyssides, A.G., Loizides, M. and Karlis, P.K. (2004) Integrated strategic approach for reusing olive oil extraction by-products. *Journal of Cleaner Production*, **12**(6): 603–611.

Xiao, Z.J., Liu, P.H., Qin, J.Y. and Xu, P. (2007) Statistical optimization of medium components for enhanced acetoin production from molasses and soybean meal hydrolysate. *Applied Microbiology and Biotechnology*, **74**(1): 61–68.

Xu, X., Zhou, L., Van Cleemput, O. and Wang, Z. (2000) Fate of urea-15N in a soil-wheat system as influenced by urease inhibitor hydroquinone and nitrification inhibitor dicyandiamide. *Plant and Soil*, **220**(1–2): 261–270.

Ye, M., Li, C., Francis, G. and Makkar, H. P. (2009) Current situation and prospects of *Jatropha curcas* as a multipurpose tree in China. *Agroforestry Systems*, **76**(2): 487–497.

YuChuan, L., QiaoSheng, G., QingSong, S., Pei, Z. and XiaoLei, D. (2009) Bioassay on herbicidal activity of extracts from *Jatropha curcas*. *Journal of Plant Resources and Environment*, **18**(4): 72–78.

9

Banana Peels and their Prospects for Industrial Utilization

Prerna Khawas, Arup Jyoti Das, and Sankar Chandra Deka
Tezpur University, Napaam, Sonitpur, Assam, India

9.1 Introduction

The banana, one of the most favorite fruits, widely grown in many countries, being the fourth-most important food crop in the world (Ganapathi *et al.*, 1999) is a staple food and export commodity. It contributes to the food security of millions of people in the developing world. Globally, banana (*Musa* species) is grown over a 4.8 million ha area producing 99.99 million tonnes of bananas and plantains, with 20.8 MT/ha productivity. India is the largest producer of plantains and bananas, with an annual production of 29.78 MT from an area of 0.83 million ha with 35.9 MT/ha productivity accounting for 29% of the world's production (Kumar *et al.*, 2011). In India, the *Musa* species is well adapted to the regions varying from the tropics to the humid sub-tropics and semi-arid subtropics. Both plantains and bananas are the staple foods of rural and urban consumers in India and an important source of income (Khawas *et al.*, 2014a; Seenappa *et al.*, 1986).

The peels are often the wasted part of various fruits and these wastes have not generally gained much attention with a view to being used or recycled. This might be due to their unknown benefits of commercial application. The peel of banana represents 40% of the total weight of fresh banana (Tchobanoglous *et al.*, 1993) and has been underutilized and discarded as waste. Like its pulp flour counterpart, banana peel flour can potentially be used in new products with standardized composition for various industrial and domestic uses (Emaga *et al.*, 2007). Peels are the major by-products of all fruits and vegetables obtained during processing; some studies show that these are good sources of polyphenols, carotenoids and other bioactive compounds, which possess various beneficial effects on human health (Zhang *et al.*, 2005). But these wastes are either uneconomically utilized or disposed of as they are, thereby causing

Food Processing By-Products and their Utilization, First Edition. Edited by Anil Kumar Anal.

serious pollution problems. Of particular interest is the finding that banana peel extract contains higher antioxidant compounds than that of the pulp (Someya *et al.*, 2002), thus promising a more intense utilization of the peels in food and nutraceuticals. Potential applications of banana peel however depend on its chemical composition as well as physicochemical and functional properties (Emaga *et al.*, 2007). Banana peel waste is a by-product of processing during the production of food such as banana chips and baby foods. The edible part of banana constitutes only 12 wt.% of the plant; the remaining parts become agricultural waste and cause environmental problems (Elanthikkal *et al.*, 2010). However, the problem can be recovered by utilizing its high-added value compounds, including the dietary fibre fraction that has a great potential in the preparation of functional foods (Khawas *et al.*, 2014b; Wachirasiri *et al.*, 2009).

9.2 Chemical Properties and Bioactive Compounds Present in Banana Peel

The peels of a variety of fruits and vegetables have gained much attention as a natural source of antioxidants and phytochemical contents, which are rich in compounds with free radical scavenging activity. Banana and plantain peels are the major agricultural wastes that have been used as medicine, animal feeds, blacking of leathers, soap-making, and fillers in rubber. Banana peel, being a key source of many functionally important bioactive compounds, is still underutilized and very little scientific effort has been put into identifying its functionality in terms of application to food and nutraceuticals. Banana peel can potentially offer new products with standardized composition for various industrial and domestic uses (Essien *et al.*, 2005; Annadurai *et al.*, 2002).

9.2.1 Nutrients

As reported by Emaga *et al.* (2007, 2008), banana peel is a rich source of dietary fibre (50% on a dry matter basis), protein (8–11%), crude fat (3.8–11%), lipid (2.2–10.9%) pectin, essentials amino acids (leucine, valine, phenylalanine and threonine), polyunsaturated fatty acids (mainly linoleic acid and α-linolenic acid) and micronutrients (potassium, phosphorous, calcium, magnesium etc.) The authors have also reported that all essential amino acids contents are higher than the FAO standard, except for lysine. Pectin extracted from banana peel also contains glucose, galactose, arabinose, rhamnose and xylose (Table 9.1) Maturation of fruits involves increase in soluble sugar, decrease in starch and hemicelluloses, and slight increase in protein and lipid content in fruit peel. Degradation of starch and hemicelluloses by endogenous enzymes may explain increase in soluble sugar content.

Archibald (1949) has reported that banana peel can also be utilized for extraction of banana oil (amyl acetate), which can be potentially used for food flavoring. Banana peels are also a good source of lignin (6–12%), pectin (10–21%), cellulose (7.6–9.6%), hemicelluloses (6.4–9.4%) and galactouroninc acid.

Table 9.1 Peel composition (fresh weight basis) of dessert bananas

Parameters	Amount
Moisture, %	83.5
Proteins, %	1.8
Fat*, %	1.7*
Glucose, %	2.4
Fructose, %	6.2
Sucrose, %	2.6
Maltose, %	0
Starch, %	1.2
Cellulose, %	8.4
Total Sugar, %	29
Potassium (K), mg/100 g	78.1*
Calcium (Ca), mg/100 g	19.2
Sodium (Na), mg/100 g	24.3
Iron (Fe), mg/100 g	0.61
Manganese (Mn), mg/100 g	76.20
Bromine (Br), mg/100 g	0.04
Rubidium (Rb), mg/100 g	0.21
Strontium (Sr), mg/100 g	0.03
Zirconium (Zr), mg/100 g	0.02
Niobium (Nb), mg/100 g	0.02

*dry weight basis (Archibald, 1949; Lustre, 1976; Adisa and Okey, 1987; Wall, 2006: Anhwange, 2008)

9.2.2 Phytochemicals and Antioxidants

Banana peel has been considered as a potential source of phytochemicals and antioxidants compared to its pulp (Someya *et al.*, 2002; Kondo *et al.*, 2005; Sulaiman *et al.*, 2011). Someya *et al.* (2002) investigated the total phenolic contents to be more abundant in peel (907 mg/100 g dry wt.) than in pulp (232 mg/100 g dry wt.), which was consistent with the antioxidant activity (Table 9.2). The peel extract showed 2.2 times stronger antioxidant activity than the pulp extract when the incubation times were compared. Gallocatechin content found in fruit peel (158 mg/100 g dry wt.) was also in higher amounts than in pulp (29.6 mg/100 g dry wt.). The higher gallocatechin content of the banana peel may account for the better antioxidant effects.

9.2.3 Flavonoids and Polyphenols

Singhal and Ratra's (2013) research on banana (*Musa acuminata*) peel extract indicated that banana peel is a potential source of bioactive compounds like flavonoids and polyphenols, with wide range of medicinal properties, in particular high free radical scavenging activity. The study also reported that banana peel extract helps to increase the total leukocytes and the percentage of lymphocytes, which show good biological activities and can be effective in treating various diseases.

Table 9.2 Antioxidant and antinutrient components present in the different varieties of banana peels (per kg) (adapted from Nagarajaiah and Prakash, 2011)

Components	*Pachabale*	*Yelakkibale*	*Nendranbale*
Insoluble dietary fibre (g)	418	499.3	356
Soluble dietary fibre (g)	41.3	68.3	22.3
Tannins (g)	5.2	10.7	11.1
Phytic acid (mg)	693.1	901.3	497.9
Total carotenes (mg)	31.2	23.5	12.7
β-carotene (mg)	18.6	15.2	4.9
Vitamin C (mg)	17.9	178.3	18.0
Total oxalate (mg)	0.9	28.3	10.5
Water soluble oxalate (mg)	0.5	18.6	1.6

Table 9.3 Total antioxidant activity and phytochemical content in peel of different banana varieties (adapted from Baskar *et al.*, 2011)

Banana varieties	Total antioxidant activity (mM AAE/g)	Total polyphenols (mg CE/g)	Flavonoids (mg rutin/g)
Kadali	4.9	0.2	13.8
Karpooravalli	3.5	0.2	16.9
Monthan	4.7	0.3	11.9
Nendran	2.6	0.5	21.7
Poovan	3.6	0.4	22.8
Pachainadan	5.9	0.3	18.8
Rasthali	3.4	0.6	21.3
Robusta	4.8	0.2	17.9
Sevvazhai	4.0	0.5	13.0

Fatemeh *et al.* (2012) found the content of total polyphenols and total flavonoids of peel to be on the higher side compared to the fruit pulp in all stages of fruit ripening. Similarly, the ability of banana peel extracts to scavenge DPPH radicals was reported with the higher value, which is associated with stronger antioxidant activity (Table 9.3).

Studies on inhibition of lipid peroxidation by ethanolic extract of a few varieties of banana peel by Baskar *et al.* (2011) reported the moderate percentage inhibition with Poovan variety of peel extract exhibiting the highest inhibition. Lipid peroxidase are unstable and decomposes to form reactive carboxyl compounds, which are responsible for cancer, causing age-related diseases and also damage to the DNA, which can be significantly controlled by using peel of the banana. Furthermore, the authors have also concluded that flavonoids, the strong antioxidant with phenolic hydroxyl group, are also related to anti-lipid peroxidation.

9.2.4 Micronutrient

The micronutrient contents (iron and zinc) found in banana peels was comparatively higher than that of the fruit (Davey *et al.*, 2009). The study on banana peel done by

Anhwange (2008) found that peel contains reasonable amounts of minerals including potassium, calcium, sodium, iron and manganese, among which the content of potassium was highest. The author also suggested that consumption of banana peel may help in the regulation of body fluids and maintenance of normal blood pressure. It may also help in controlling kidney failure, heart oddities and respiratory flow. Feming (1998) have stated that the percentage of iron content in banana peel is an ideal source for carrying oxygen to the cells and production of energy, synthesis of collagen and for proper functioning of the immune system, cell growth and heart.

9.2.5 Bioactive Components

The study by Waghmare and Kurhade (2014), to determine the possible bioactive components from ethanolic extract of banana peel using gas chromatography-mass spectrometry (GC-MS), reported the presence of eight compounds, which included estragole (11.18%), hexadecanoic acid ethyl ester (9.76%), epicatechin (9.97%), gallocatechin (8.58%), p-coumaric acid ethyl ester(4.2%), 1,2 benzene dicarboxylic acid mono (2-ethylhexyl) ester (13.47%), beta-tocopherol (11.37%) and vitamin E (31.35%). These compounds have many biological effects, including antioxidant and antimicrobial activities.

9.3 Utilization of Banana Peel

9.3.1 Yellow Noodles

Ramli *et al.* (2009) developed yellow noodles by partial substitution of wheat flour with green banana peel flour and their study reported that partial substitution of banana peel into noodles may be useful for controlling starch hydrolysis of yellow noodles. Banana peel noodles had lower estimated glycemic index values as compared to noodles prepared with wheat flour. The modified noodle product described in their study may broaden the range of low glycemic index food products and increase utilization of waste products from banana agro-industries.

9.3.2 Dietary Fibre Concentrate

Wachirasiri *et al.* (2009), in their research on developing banana peel dietary fibre concentrate, reported that banana peel is a good source of dietary fibre, exhibiting 50 g/100 g dry matter. Their results indicated that dietary fibre concentrate obtained from banana peel provides an opportunity to enhance the functionality (Table 9.4). The use of banana peel dietary fibre concentrate as a low-caloric functional ingredient for fibre enrichment and incorporation of them into the food system may give high value-added food products.

9.3.3 α-amylase

Amylases are well known for applications ranging from starch and food processing industries to medical applications. Krishna *et al.* (2012) reported the potential of

Table 9.4 Concentration of NDF (neutral detergent fibre), ADF (acid detergent fibre), cellulose, hemicellulose and lignin (% dry matter) in banana and plantain peels (adapted from Emaga *et al.*, 2008)

Variety	Stage	NDF	ADF	Cellulose	Hemi-celluloses	Lignin
Grande Naine	I	26	20	8	6	12
	II	23	16	10	7	6
	III	23	15	8	8	7
French Clair	I	23	21	7	2	14
	II	26	25	8	1	17
	III	22	22	6	0.6	15

Note:
Stage I = Green
Stage II = More yellow than green
Stage III = Yellow with brown spots

banana peel evaluated for α-amylase production using the fungal culture of *Aspergillus niger* NCIM 616 in solid submerged (SmF) and solid state (SSF) fermentation. The effect of different parameters, such as substrate concentration, water content, layer thickness and external salt addition was studied in terms of amylase activity. This study suggests that banana peel could be used as a potential raw material for α-amylase production. Paul and Sumathy (2013) suggested that banana peel could be employed as a promising substrate for the production of amylase by *Bacillus subtilis*.

9.3.4 Xylose

Rehman *et al.* (2013) have confirmed that banana peel is proved to be a good source for producing xylose. Xylitol is the first rare sugar that has a global market for having beneficial health properties and being an alternative to current conventional sweeteners. Banana peel was used as a substrate for xylitol production by acid hydrolysis. Detoxification of peel hydrolysate by neutralization, charcoal treatment and vacuum evaporation increased the xylitol yield. Xylitol extracted from banana peel can be used to replace sugars in different products such as bakery and confectionary products, without affecting their physico-chemical characteristics and shelf stability.

9.3.5 Lipase

Jadhav *et al.* (2013) utilized banana peel for producing lipase, an important enzyme which is extensively used in the food and dairy industry for the hydrolysis of milk fat, cheese ripening, flavour enhancement and lipolysis of butter fat and cream. The authors concluded that peel of banana can potentially be used for the production of bacterial enzymes such as lipase and amylase, which hold an important place in the food industry.

9.3.6 Wine Vinegar

The attempt made by Byarugaba-Bazirake *et al.* (2014) for producing wine vinegar from banana peels took 28 days and the final product obtained (wine vinegar from banana) had physiochemical characteristics of 6.0% (v/v) acetic acid, 5.0°Brix, and pH of 2.9, which complied with the standard ranges of brewed vinegar after complete fermentation. The aroma of the vinegar produced was appreciated by the consumers who were acquainted with vinegar. The study therefore showed that banana (*matooke*) peels can be used as an ideal substrate for production good-quality vinegar. This will not only increase the economical and food value of banana peel, but also provide a way of utilizing banana waste in Uganda. Simmonds (1966) reported that vinegar has been prepared by fermenting a mash of banana pulp and peel. Vinegar production from banana may minimize cost of production and be eco-friendly.

9.3.7 Wine

Isitua and Ibeh (2010) produced wine using banana and pineapple peel. Treatment of substrates for wine production with pumpkin leaves infusion increased alcoholic content of wine produced after fermentation from 0.035 to 0.57 % and 0.21 to 0.57 % for banana and pineapple peel wine, respectively. These fruit wastes can be considered as a potential source for wine production, which are currently viewed as a waste that is normally discarded. These materials may provide good adjuncts for wine and alcohol production.

9.3.8 Feed

Several studies have shown that artificial feeding of starchy banana peels can improve chicken production, seen from weight gain, feed intake, feed conversion, serum cholesterol levels in the blood, meat, liver, feces, and alleviating severe digestive organ problems. Artificial feeding starchy banana peel can also produce broiler meat with low cholesterol levels. To avoid health issues caused by high cholesterol content food, many people avoid consuming meat and meat products. Therefore, it is required that poultry products, healthy and low in cholesterol, can be achieved by using banana peel as a poultry feed. Therefore, peels could be a good feed material for cattle and poultry (Emaga *et al.*, 2008; Dormond *et al.*, 1998a,b; Adeniji *et al.*, 2008).

9.3.9 Sustainability

Memon *et al.* (2008) have described the importance of banana peel for the removal of cadmium from the environmental and industrial wastewater. The authors have explored new, cheaper, economical and selective adsorbent as an alternative to costly adsorbents for the removal of cadmium (II) ions. FT-IR analysis of banana peel showed the presence of various functional groups, indicating the complex nature of banana peel. Scanning electron microscopic analysis demonstrated the microporous structure of the material, while energy dispersive X-ray analysis confirmed the sorption of cadmium on banana peel. Therefore, the use of banana peel in removal of cadmium ions from environmental and industrial wastewater samples will have the advantages of low cost, rapid attainment of phase equilibration and high sorption capacity values.

The study by Achak *et al.* (2009) reported that banana peel has proved to be a promising material for the removal of contaminants from olive mill wastewaters. The authors also found banana peel a highly efficient abundant cheap adsorbent for removing phenolic compounds from olive mill wastewaters. The banana peel showed a high adsorption capacity of phenolic compounds (689 mg/g), revealing that banana peel could be employed as a promising adsorbent for phenolic compounds adsorption.

Bardiya *et al.* (1996) studied biomethanation of banana and pineapple peels at various hydraulic retention times (HRT) ranging from 10 to 40 days. The authors reported that the results obtained on biomethanation of both banana peel and pineapple-processing wastes suggest their potential and suitability for economically viable waste treatment technology through anaerobic digestion. The energy generated in the form of methane when utilized efficiently not only improves the overall economy of these fruit-processing industries but also provides on-site solutions to waste-management problems.

9.3.10 Bioethanol

Singh *et al.* (2014), in their study on bioethanol production from banana peel, reported that the maximum ethanol yield was 6.540 % by using a fungal strains *Aspergillus niger* and *Saccharomyces cerevisiae*. The amount of ethanol content increased with the increase in fermentation time. The findings of their work suggest that banana peels could be a good substrate for ethanol production.

Another study on production of bio-ethanol from banana peel by Kumar *et al.* (2011) indicated that bio-ethanol obtained from banana peel waste, using industrial yeast with good fermentation attributes, could enhance ethanol yield and minimize cost of production. Banana peels are always available in abundance in India and thus serve as a readily available raw substrate for the production of ethanol. At present, ethanol is produced from molasses, a by-product of the sugar industry. Therefore, sources like banana wastes and other agriculture wastes could be used for bio-ethanol production.

9.3.11 Alkali

Olabanji *et al.* (2012) reported on the use of banana peel as a cheap source of alkali in the soap industry. Neat soap, having physical properties as pure potassium hydroxide soap, is derivable from the water extract of ashes of plantain and banana peels.

Fruit peels have been a valuable source for maintaining human health. The use of fruit peels extracts for antimicrobial properties can be of great significance in therapeutic treatments. The study by Chabuck *et al.* (2013), to evaluate the antibacterial activity of banana peel, found that aqueous extracts of fresh yellow banana peels showed effective antibacterial effects against *S. aureus, S. pyogenes, E. aerogenes, K. pneumoniae, E. coli, M. catarrhalis* and *Candida albicans*. As a result, extracts of fresh yellow banana peels could be considered as a good antibacterial agent against both Gram-positive and Gram-negative bacteria, to replace the synthetic medicines in treatment of diseases caused by these bacteria. Similar studies on antimicrobial properties of banana peel by Ehiowemwenguan *et al.* (2014) have also reported that the secondary metabolites such as glycosides, alkaloids, saponins, volatile oil, flavonoids and tannins present in *Musa sapientum* peels play an important role in therapeutic activity.

9.3.12 Biogas

Apart from the importance of banana peel in food and pharmaceutical applications, banana peel can also be potentially used for the production of biogas. The anaerobic digestion of banana waste could generate important amounts of energy, which could be used to cover essential needs of either households or to meet the requirements of the processing industry in developing countries (Zafar and Jamil, 2014; Zhengyun *et al.*, 2013).

9.4 Conclusion

Peels of plantains and banana, which are often considered as waste and discarded, is a richer source of many functionally important nutrients and bioactive compounds. They are rich in antioxidants, polyphenols, minerals, vitamins, sugars, amino acids, dietary fibre and pectin. Peels has a potential application in food, pharmaceutical industries and cattle feed formulation because of their functionality. They can be exploited as an excellent source for production of bio-ethanol and biogas. Therefore, looking at the backdrop of promising utilization of plantains and banana peel, they cannot be considered as a 'waste', rather they can be prospectively utilized for developing economic and eco-friendly value-added products and substrate for many industrial production.

References

Achak, M., Hafidi, A., Ouazzani, N., Sayadic, S. and Mandia, L. (2009) Low cost biosorbent 'banana peel' for the removal of phenolic compounds from olive mill wastewater: Kinetic and equilibrium studies. *Journal of Hazardous Materials*, **166**(1): 117–125.

Adeniji, T.A., Barimalaa, I.S., Tenkouano, A., Sanni, L.O. and Hart, A.D. (2008) Antinutrients and heavy metals in new Nigerian *Musa* hybrid peels with emphasis on utilization in livestock production. *Fruits*, **63**(2): 65–73.

Adisa, V.A. and Okey, E.N. (1987) Carbohydrate and protein composition of banana pulp and peel as influenced by ripening and mold contamination. *Food Chemistry*, **25**(2): 85–91.

Anhwange, B.A. (2008) Chemical composition of *Musa Sapientum* (banana) peels. *Journal of Food Technology*, **6**(6): 263–266.

Annadurai, G., Juang, R.S. and Lee, D.J. (2002) Use of cellulose-based wastes for adsorption of dyes from aqueous solution. *Journal of Hazardous Material*, **92**(3): 263–274.

Archibald, J.G. (1949) Nutrient composition of banana skins. *Journal Dairy Science*, **32**(11): 969–971.

Bardiya, N., Somayaji, D. and Khanna, S. (1996) Biomethanation of banana peel and pineapple waste. *Bioresource Technology*, **58**(1): 73–76.

Baskar, R., Shrisakthi, S., Sathyapriya, B., Shyampriya, R., Nithya, R. and Poongodi, P. (2011) Antioxidant potential of peel extracts of banana varieties (*Musa sapientum*). *Food and Nutrition Sciences*, **2**(10): 1128–1133.

Byarugaba-Bazirake, G.W., Byarugaba, W., Tumusiime, M. and Kimono, D.A. (2014) The technology of producing banana wine vinegar from starch of banana peels. *African Journal of Food Science and Technology*, **5**(1): 1–5.

Chabuck, Z.A.G., Al-Charrakh, A.H., Hindi, N.K.K. and Hindi, S.K.K. (2013) Antimicrobial effect of aqueous banana peel extract, Iraq. *Research Gate: Pharmaceutical Sciences*, **1**: 73–75.

Davey, M.W., Van-den Bergh, I., Markham, R., Swennen, R. and Keulemans, J. (2009) Genetic variability in *Musa* fruit provitamin A carotenoids, lutein and mineral micronutrient contents. *Food Chemistry*, **115**(3): 806–813.

Dormond, H., Boschini, C. and Rojas, A. (1998a) Effect of two levels of ripe banana peel on milk production by dairy cattle. *Agronomía Costarricense*, **22**(1): 43–49.

Dormond, H., Boschini, C. and Zuniga, A.M. (1998b) Effect of four levels of ripe banana peel on rumen degradability of dry matters from Kikuyo grass (*Pennnisetum clandestinum*) and African Star grass (*Cynodon nlemfluensis*) in Jersey cows. *Agronomía Costarricense*, **22**(2): 163–172.

Ehiowemwenguan, G., Emoghene, A.O. and Inetianbor, J.E. (2014) Antibacterial and phytochemical analysis of banana fruit peel. *IOSR Journal of Pharmacy*, **4**(8): 18–25.

Elanthikkal, S., Gopalakrishnapanicker, U., Varghese, S. and Guthrie, J.T. (2010) Cellulose microfibres produced from banana plant wastes: Isolation and characterization. *Carbohydrate Polymers*, **80**(3): 852–859.

Emaga, T.H., Andrianaivo, R.H., Wathelet, B., Tchango J.T. and Paquot, M. (2007) Effects of the stage of maturation and varieties on the chemical composition of banana and plantain peels. *Food Chemistry*, **103**(2): 590–600.

Emaga, T.H., Robert, C., Ronkart, S.N., Wathelet, B. and Paquot, M. (2008) Dietary fibre components and pectin chemical features of peels during ripening in banana and plantain varieties. *Bioresource Technology*, **99**(10): 4346–4354.

Essien, J.P., Akpan, E.J. and Essien, E.P. (2005) Studies on mould growth and biomass production using waste banana peel. *Bioresource Technology*, **96**(13): 1451–1455.

Fatemeh, S.R., Saifullah, R., Abbas, F.M.A. and Azhar, M.E. (2012) Total phenolics, flavonoids and antioxidant activity of banana pulp and peel flours: Influence of variety and stage of ripeness. *International Food Research Journal*, **19**(3): 1041–1046.

Feming, D.J. (1998) Dietary determination of iron stones in a free-living elderly population. The Framingham Heart Study. *American Journal of Clinical Nutrition*, **67**: 722–733.

Ganapathi, T.R., Suprasanna, P., Bapat, V.A, Kulkarni, V.M. and Rao, P.S. (1999) Somatic embryogenesis and plant regeneration from male flower buds in banana. *Current Science*, **79**(10): 1228–1231.

Isitua, C.C. and Ibeh, I.N. (2010) Novel method of wine production from banana (*Musa acuminata*) and pineapple (*Ananas comosus*) wastes. *African Journal of Biotechnology*, **9**(44): 7521–7524.

Jadhav, S., Chougule, D. and Rampure, S. (2013) Lipase production from banana peel extract and potato peel extract. *International Journal of Research in Pure and Applied Microbiology*, **3**(1): 11–13.

Khawas, P., Das, A.J., Sit, N., Badwaik, L.S. and Deka, S.C. (2014a) Nutritional composition of culinary *Musa* ABB at different stages of development. *American Journal of Food Science and Technology*, **2**(3): 80–87.

Khawas, P., Das, A.J., Dash, K.K. and Deka, S.C. (2014b) Thin-layer drying characteristics of *kachkal* banana peel (*Musa* ABB) of Assam, India. *International Food Research Journal*, **2**(3): 975–982.

Kondo, S., Kittikorn, M. and Kanlayanarat, S. (2005) Preharvest antioxidant activities of tropical fruit and the effect of low temperature storage on antioxidants and jasmonates. *Postharvest Biology and Technology*, **36**(3): 309–318.

Krishna, R.P., Srivastava, A.K., Ramaswamy, N.K., Suprasanna, P. and D'Souza, S.F. (2012) Banana peel as substrate for α-amylase production using *Aspergillus niger* NCIM 616 and process optimization. *Indian Journal of Biotechnology*, **11**(3): 314–319.

Kumar, B., Mistry, N.C., Singh, B. and Gandhi, P.C. (2011) Indian Horticulture Database-Aristo Printing Press, India, 4–5. (http://nhb.gov.in/area-pro/database-2011.pdf).

Kumar, R.S., Moorthy, G., Rajeswari, R. and Harikrishnan, H. (2011) Utilization of waste ripe banana and peels for bioethanol production using *Saccharomyces cerevisiae*. *Journal of Bioscience Research*, **2**(2): 67–71.

Lustre, A.O., Sorano, M.S., Morga, N.S., Balagot, A.H. and Tunac, M.M. (1976) Physico-chemical changes in 'Saba' bananas during normal and acetylene-induced ripening. *Food Chemistry*, **1**(2): 125–137.

Memon, R.J., Memon, Q.S., Bhanger, M.I., Memon, Z.G., El-Turki, A. and Allen, C.G. (2008) Characterization of banana peel by scanning electron microscopy and FT-IR and its use for cadmium removal. *Colloids Surfaces B: Biointerfaces*, **66**(2): 260–265.

Nagarajaiah, S.B. and Prakash, J. (2011) Chemical composition and antioxidant potential of peels from three varieties of banana. *Asian Journal of Food and Agro-Industry*, **4**(1): 31–46.

Olabanji, O., Oluyemi, A.E. and Ajayi, S.O. (2012) Metal analyses of ash derived alkalis from banana and plantain peels (*Musa* spp.) in soap-making. *African Journal of Biotechnology*, **11**(99): 16512–16518.

Paul, S.M. and Sumathy, H.J.V. (2013) Production of amylase from banana peels with *Bacillus subtilis* using solid state fermentation. *International Journal of Current Microbiology and Applied Sciences*, **2**(10): 195–206.

Ramli, S., Alkarkhi, A.F.M., Yong, Y.S. and Easa, A.M. (2009) Utilization of banana peel as a functional ingredient in yellow noodle. *Asian Journal of Food and Agro-Industry*, **2**(3): 321–329.

Rehman, S., Nadeem, M., Ahmad, F. and Mushtaq, Z. (2013) Biotechnological production of xylitol from banana peel and its impact on physicochemical properties of rusks. *Journal of Agricultural Science and Technology*, **15**(4): 747–756.

Seenappa, M., Laswai, H.S.M. and Ferando, S.P.F. (1986) Availability of L-ascorbic acid in Tanzanian banana. *Journal of Food Science and Technology*, **23**(5): 293–295.

Simmonds, N.W. (1966) *Bananas*, 2nd edition. Longmans, London.

Singh, A.K., Rath, S., Kumar, Y., Masih, H., Peter, J.K., Benjamin, J.C., Singh, P.K. and Singh, P.D. (2014) Bio-ethanol production from banana peel by simultaneous saccharification and fermentation process using cocultures *Aspergillus niger* and *Saccharomyces cerevisiae*. *International Journal of Current Microbiology and Applied Sciences*, **3**(5): 84–96.

Singhal, M. and Ratra, P. (2013) Antioxidant activity, total flavonoid and total phenolic content of *Musa acuminata* peel extracts. *Global Journal of Pharmacology*, **7**(2): 118–122.

Someya, S., Yoshiki, Y. and Okubo, K. (2002) Antioxidant compounds from bananas (*Musa* Cavendish). *Food Chemistry*, **79**(3): 351–354.

Sulaiman, F.S., Yusoff, A.N.M., Eldeen, M.I., Seow M.E., Sajak A.A.B., *et al.* (2011) Correlation between total phenolic and mineral contents with antioxidant activity of eight Malaysian bananas (*Musa* sp.) *Journal of Food Composition and Analysis*, **24**(1): 1–10.

Tchobanoglous, G., Theisen, H. and Vigil, S. (1993) *Integrated Solid Waste Management Engineering Principals and Management Issues*. McGraw-Hill, Inc., Boston, MA.

Wachirasiri, P., Julakarangka, S. and Wanlapa, S. (2009) The effects of banana peel preparations on the properties of banana peel dietary fibre concentrate. *Songklanakarin Journal of Science and Technology*, **31**(6): 605–611.

Waghmare, S.J. and Kurhade, H.A. (2014) GC-MS analysis of bioactive components from banana peel (*Musa sapientum* peel). *European Journal of Experimental Biology*, **4**(5): 10–15.

Wall, M.M. (2006) Ascorbic acid, vitamin A, and mineral composition of banana (*Musa* sp.) and papaya (*Carica papaya*) cultivars grown in Hawaii. *Journal of Food Composition and Analysis*, **19**(5): 434–445.

Zafar, Z.A. and Jamil, M. (2014) Evaluation of mixing cow dung with apple and banana peels on biogas yield. *World Applied Sciences Journal*, **29**(7): 903–907.

Zhang, P., Whistler, L.R., BeMiller, N.J. and Hamake, R.B. (2005) Banana starch: Production, physicochemical properties, and digestibility: A review. *Carbohydrate Polymer*, **59**(4): 443–458.

Zhengyun, Z., Rui, X., Huanyun, D., Qiuxia, W., Bin, Y., *et al.* (2013) Biogas yield potential research of the wastes from banana manufacturing process under mesophilic anaerobic fermentation. *Research Journal of Applied Sciences, Engineering and Technology*, **5**(19): 4740–4744.

10
Utilization of Carrot Pomace

H.K. Sharma[1] and Navneet Kumar[2]

[1]*Sant Longowal Institute of Engineering and Technology, Punjab, India*
[2]*Anand Agricultural University, Gujarat, India*

10.1 Introduction

The total production of food is estimated at 4 billion tonnes and about 30–50% (or 1.2–2 billion tonnes) of the total food produced on the planet is lost before consumption (Aggidis *et al.*, 2013). The large quantity of waste is produced by the food industries. Proper disposal of such waste biomass is crucial for the reduction of environmental pollution. On the other hand, such residual biomass still contains significant amounts of value-added constituents, including polysaccharides, polyphenols, carotenoids and other important functional components that could be recovered and then re-used in other production chains (Donato *et al.*, 2014).

Fruit and vegetable wastes are cheap, available in large amounts, and characterized by high dietary fiber content, leading to high water-binding capacity and relatively low enzyme digestible organic matter (Serena and Knudsen, 2007). Dietary fiber concentrates from vegetables showed better insoluble/soluble dietary fiber ratios than cereal bran (Grigelmo-Miguel and Martin-Belloso, 1999). Due to the presence of high dietary fiber content, this by-product can be used to change physicochemical properties of diets. In the production of carrot juice, pomace is available in good quantities, which can be exploited into carrot insoluble fiber-rich fractions (IFRF) as a promising hypocholesterolemic ingredient to fulfill the increasing demand of functional ingredients in developing fiber-rich food products (Hsu *et al.*, 2006). The incorporation of cellulose and IFRF into the fiber-free diet significantly reduces the serum total cholesterol level by 17.3% and 33.5%, respectively (Hsu *et al.*, 2006).

The food industry is now increasingly developing newer products, such as bread, snacks and other highly consumed products by all groups of consumers, enriched with ingredients with scientifically proven beneficial effects on human health. The incorporation of different by-products from the food industry may provide an economical viable solution for their utilization. Along with justifiable incorporation they may prove to be helpful in substantially increasing the nutritional and functional value of food products. The by-products from the food industry, such as apple pomace

Food Processing By-Products and their Utilization, First Edition. Edited by Anil Kumar Anal.

and spent grains, are a rich source of polyphenols, fiber and β-glucan, whereas carrot pomace is a rich source of β-carotene and crude fiber.

10.1.1 Carrot

The carrot (*Daucus carota*), a root vegetable, is usually orange, red, purple, white or yellow in colour. It is a rich source of bioactive compounds such as dietary fibers and carotenoids, along with good amounts of other functional components, possessing significant health-beneficial properties (Table 10.1). It is considered a rich source of β-carotene and vitamins (thiamine, riboflavin, folic acid and vitamin B-complex) and minerals (calcium, copper, magnesium, potassium, phosphorus and iron).

The composition of carrot and its processed products (baby carrot, cooked, frozen, canned, canned and drained, juice and dehydrated forms) are given in Table 10.1. The consumption pattern of carrot/carrot products is increasing, which may be due to the presence of natural antioxidants having significant effects in the prevention of chronic diseases. The total production of carrot and turnips is estimated as nearly 38.8 million tonnes worldwide during the year 2014 (FAO, 2017).

According to the shape and size, the major varieties of carrots (Figure 10.1) are Imperator (long, tapered roots), Nantes, (sweet, crisp, 6–7″ cylindrical carrots, with blunt tips), Chantenay (short and stout, with broad, 4–8 cm crowns) and Mini Carrots (Short, 3–4″ (7–10 cm) length).

10.1.2 Processing of Carrot

The carrot can be consumed raw, converted to juice drink, cooked as a vegetable dish and used to make sweet dishes. In recent years, fruit and vegetable juices have become important over the traditional caffeine-containing beverages. Carrot juice has particularly good amounts of β-carotene, vitamins and minerals. The juice can find applications as natural colorants in a number of food products. The conventional carrot juice extractor results in poor juice yield due to the hard texture of the root, which can be increased by enzymes or heat processing to soften the tissue (Tingting *et al.*, 2013). In conventional extraction systems, the presence of insoluble matters makes the juice cloudy and thus poses solid–liquid separation with storage problems and hence affects the overall acceptability (Giacomo and Taglieri, 2009). The hydraulic pressed juice is almost clear with minimum insoluble solids that prevent sedimentation problems (Bazhal *et al.*, 2001) and thus increases the stability of the clarity (Sinchaipanit and Kerr, 2007). Extraction of the juice using a hydraulic press has a potent advantage of controlled extraction and thus can be employed for the development of value-added products with designer components. A juice yield up to 62% can be obtained using pressure, 14.061 × 105 kg/m^2 (Haq *et al.*, 2013), whereas the yield can be enhanced up to 71.26% using enzymatic treatment (Kaur and Sharma, 2013). Carrot juice is one of the most popular juices, which has a unique demand in the market but the utilization of the by-product, pomace of the carrot juice industry, has still not found the commercial importance to make this industry more economically competitive. Degradation of waste, especially that of vegetable waste, has become a mounting challenge owing to its generation of huge amounts of waste (Patil *et al.*, 2012).

Table 10.1 Nutritional composition of raw and processed carrot

Nutrient	Unit per 100g	Raw	Baby	Cooked	Frozen	Canned	Canned and drained	Juice	Dehydrated
Proximates									
Water	g	88.29	90.35	90.17	90.04	92.99	92.95	88.87	4.00
Energy	kcal	41	35	35	36	23	25	40	341
Protein	g	0.93	0.64	0.76	0.78	0.58	0.64	0.95	8.10
Total lipid (fat)	g	0.24	0.13	0.18	0.46	0.14	0.19	0.15	1.49
Carbohydrate, by difference	g	9.58	8.24	8.22	7.9	5.37	5.54	9.28	79.57
Fiber, total dietary	g	2.8	2.9	3	3.3	1.8	1.5	0.8	23.6
Sugars, total	g	4.74	4.76	3.45	4.76	2.46	2.48	3.91	38.82
Minerals									
Calcium, Ca	mg	33	32	30	36	31	25	24	212
Iron, Fe	mg	0.3	0.89	0.34	0.44	0.52	0.64	0.46	3.93
Magnesium, Mg	mg	12	10	10	12	9	8	14	118
Phosphorus, P	mg	35	28	30	33	20	24	42	346
Potassium, K	mg	320	237	235	235	173	179	292	2540
Sodium, Na	mg	69	78	58	68	240	242	66	275
Zinc, Zn	mg	0.24	0.17	0.2	0.33	0.29	0.26	0.18	1.57
Vitamins									
Vitamin C, total ascorbic acid	mg	5.9	2.6	3.6	2.5	2	2.7	8.5	14.6
Thiamin	mg	0.066	0.03	0.066	0.044	0.019	0.018	0.092	0.534

(*continued overleaf*)

Table 10.1 *(continued)*

Nutrient	Unit per 100g	Raw	Baby	Cooked	Frozen	Canned	Canned and drained	Juice	Dehydrated
Riboflavin	mg	0.058	0.036	0.044	0.037	0.027	0.03	0.055	0.417
Niacin	mg	0.983	0.556	0.645	0.464	0.421	0.552	0.386	6.567
Vitamin B_6	mg	0.138	0.105	0.153	0.095	0.112	0.112	0.217	1.04
Folate, DFE	μg	19	27	14	10	8	9	4	55
Vitamin B_{12}	μg	0	0	0	0	0	0	0	0
Vitamin A, RAE	μg	835	690	852	710	613	558	956	3423
Vitamin A, IU	IU	16706	13790	17033	14210	12264	11170	19124	68466
Vitamin E (alpha-tocopherol)	mg	0.66		1.03	0.57	0.73	0.74	1.16	5.45
Vitamin D ($D_2 + D_3$)	μg	0	0	0	0	0	0	0	0
Vitamin D	IU	0	0	0	0	0	0	0	0
Vitamin K (phylloquinone)	μg	13.2	9.4	13.7	17.6	9.8	9.8	15.5	108
Lipids									
Fatty acids, total saturated	g	0.037	0.023	0.03	0.047	0.025	0.036	0.027	0.256
Fatty acids, total monounsaturated	g	0.014	0.007	0.006	0.014	0.006	0.009	0.007	0.076
Fatty acids, total polyunsaturated	g	0.117	0.065	0.089	0.263	0.065	0.092	0.071	0.73
Cholesterol	mg	0	0	0	0	0	0	0	0

Source: USDA Nutritional database (2011).

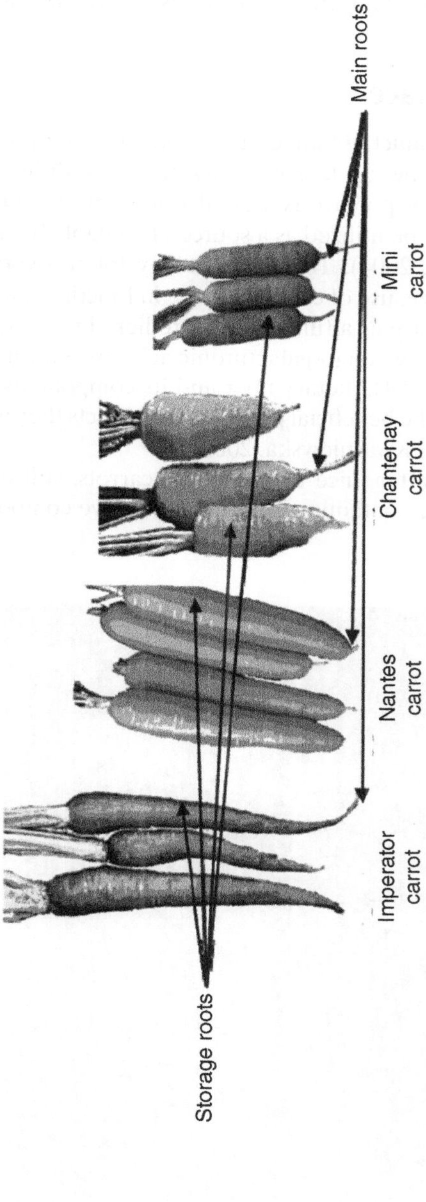

Figure 10.1 Major carrot varieties.

10.1.3 Carrot By-Products

The main by-product of carrot is carrot pomace, generated during juice extraction. The peel removed prior to processing (epidermis and lateral roots mainly) and discarded carrots due to inferior quality can be utilized as cattle feed and for the extraction of other important functional components, i.e.. pectin, organic acids, biogas etc. The utilization of carrot pomace into different products is described below.

10.1.4 Carrot Pomace

Carrot pomace is a by-product obtained during carrot juice processing (Figure 10.2). A large amount of carotene, up to 80%, may be lost with leftover carrot pomace (Bohm *et al.*, 1999). Carrot pomace is a good source of vitamins, dietary fiber and carotenoids, therefore can be utilized as a source of valuable bioactive and functional compounds (Schweiggert, 2004). The methodology for recovery of bioactive compounds from carrot pomace, use of carrot pomace in functional foods and application of hydrolyzed carrot pomace as a functional ingredient in beverages, are the aspects being explored. The bioactive oligo-galacturonic acids by enzymic breakdown of carrot pectin (Schweiggert, 2004), dietary fiber and its components are also recognized priority areas, because of the beneficial physiological effects they may exert on humans and animal (Nawirska and Kwasniewska, 2005).

The left-over pomace, after juice extraction of carrots, still does not find proper utilization. Carrot pomace containing different bio-active compounds can be used in

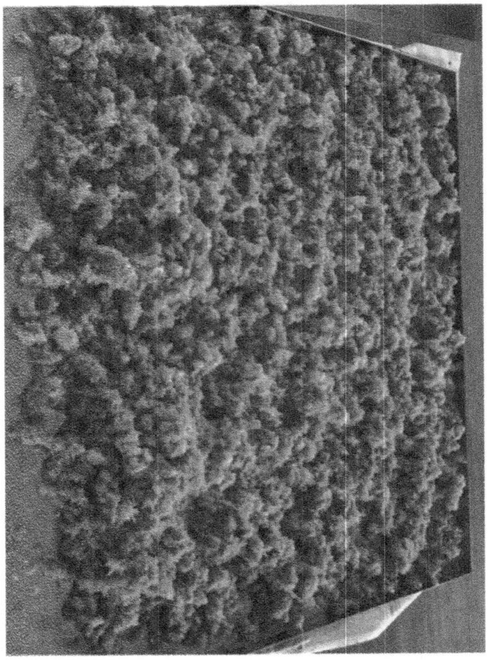

Figure 10.2 Fresh carrot pomace.

different products such as cake, bread, biscuits etc., to enhance their nutritional and functional values.

10.1.4.1 Gazrella (Carrot-Based Condensed Milk Product) A carrot-based condensed milk product (gazrella, an Indian sweetmeat) can be prepared from the incorporation of carrot pomace. The by-product (carrot pomace) was treated osmotically (Singh *et al.*, 2006) in two different ways: First, by dipping in 65°Brix sucrose syrup and second, by adding 35% sucrose (dry powder) to the pomace. The pomace was then dehydrated convectively at 60°C to 4–5% moisture content (wet basis) and packaged under vacuum in aluminum laminated packages (100 gauge) (Singh *et al.*, 2006). A carrot-based condensed milk product was prepared from the dehydrated product, which was stored at an ambient temperature (28–42°C) for 6 months. The product, so obtained from osmo-convectively dehydrated pomace, was observed to have moderate to excellent overall acceptability (Singh *et al.*, 2006).

10.1.4.2 Carrot Pomace Powder Carrot pomace powder (CPP) can be prepared from the dehydration of carrot pomace by different dryers. An important and promising way to store the carrot pomace is in its dried form, which can be utilized for the development of bakery products and extrudates, which are now becoming very popular. Drying of materials having high moisture content is a complicated process, involving simultaneous heat and mass transfer (Yilbas *et al.*, 2003). The effects of different micronization methods, such as ball milling, jet milling and high-pressure, on the characteristics and various functional properties of carrot insoluble fiber-rich fraction to improve the functionality of carrot insoluble fiber, suggests micronization treatments would provide an opportunity to improve the functionality of the carrot insoluble fiber-rich fraction (IFRF) in food applications (Chau *et al.*, 2007). Several researchers have reported various drying equipments, i.e. hot air oven and tray dryer to obtain dried pomace (Upadhyay *et al.*, 2008; Kumar *et al.*, 2012a).

Upadhyay *et al.* (2008) extracted juice using a domestic juicer-mixer-grinder. The post juice extraction remains/shreds of carrots were collected. The drying of carrot pomace was carried out in a tray drier. For drying of pomace, the bed thickness was 0.5 cm and the temperature varied from 60 to 80°C at an interval of 5°C. The drying temperature of 65°C provides dried pomace with optimal retention of both ascorbic acid and β-carotene for the removal of moisture from carrot pomace.

A modified electric forced convection dryer was used for drying of carrot pomace, which could regulate the desired drying air temperature from room temperature to 100°C with ±1°C accuracy (Kumar *et al.*, 2011). The major parts of the dryer were preheating (fitted with 2 KW electric heater) and main heating (fitted with 4 KW electric heater) chamber with thermostat-based control unit, an electrical centrifugal fan, measurement sensors, an air-duct and a plenum chamber (Figure 10.3). The inner dimensions of the drying chamber were 0.42 m × 0.32 m × 0.40 m, which was made of mild steel sheet with the provision of a tray with a 0.113 m^2 perforated area and a 0.10 m sidewall to accommodate the sample. The heated air mixed well in the chamber before passing through the sample. The air velocity was controlled by a gate valve provided at the blower outlet to obtain the desired air velocity above the tray surface, which was measured using a hot wire digital anemometer (precision 0.1 m/s). Thin-layer drying characteristics and modeling for carrot pomace with an air velocity of 0.7 ± 0.1 m/s was performed over a temperature range of 60 to 75°C (Kumar *et al.*, 2012a). Drying experiments were performed at 60, 65, 70 and 75°C in

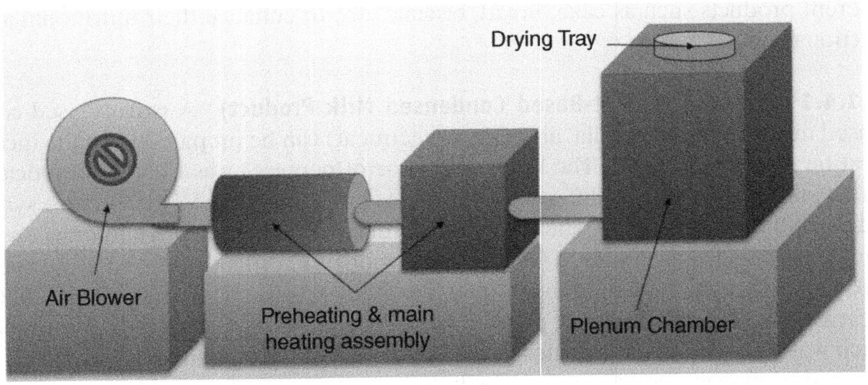

Figure 10.3 Drying set-up used for drying studies of carrot pomace.

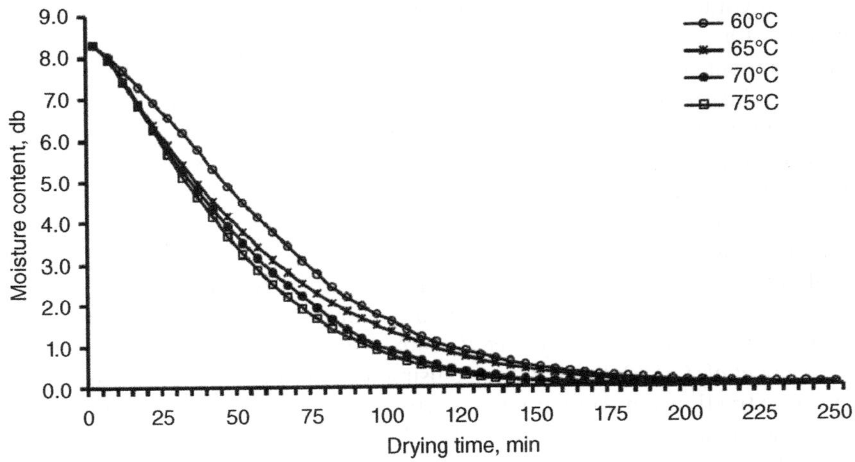

Figure 10.4 Thin layer drying curves of carrot pomace at selected temperatures and 0.7 m/s air velocity.

the forced convection dryer. The thickness of the sample was 1 cm, with a load of 7.2 ± 0.1 kg/m^2.

The weight of the sample was 205 ± 1 g per trial. The weight of the tray with the sample was measured with a digital balance and recorded at 5 min intervals at different temperatures. The moisture content versus drying time for carrot pomace at the selected temperatures is shown in Figure 10.4. The increase in drying temperature speeds up the drying process and shortens the drying time. The drying time to reach the final moisture content for the fresh carrot pomace varied from 235 to 325, 200 to 250 and 165 to 205 min at air velocities of 0.5, 0.7 and 1.0 m/s, respectively, for selected temperature range, 60–75°C (Kumar *et al.*, 2012a). Several mathematical models were fitted. The experimental drying data of carrot pomace at different temperatures were fitted into commonly used thin-layer drying models. Midilli *et al.* (2002) model fitted better for drying of carrot pomace at air velocity and temperature, ranging from

0.5–1.0 m/s and 60–75°C, respectively. The model was successfully applied for carrot pomace drying:

$$MR = a.\ \exp(-k\ .\ t^{n}) + b.\ t \qquad (10.1)$$

where MR and t are moisture ratio and drying time (min) respectively; and k, n, a and b are constants at different temperatures. The coefficients of the Midilli *et al.* (2002) model were further analysed for temperature dependency. A third-order polynomial represented the variation of coefficient with temperature. The regression equations of these parameters against drying temperature, θ (°C) are:

$$k,\ n,\ a,\ b = f\ (\theta) \qquad (10.2)$$

$$k = 2.40 \times 10^{-6}\theta^{3} - 5.02 \times 10^{-4}\theta^{2} + 3.49 \times 10^{-2}\theta - 8.05 \times 10^{-1} \qquad (10.3)$$

$$n = -2.24 \times 10^{-4}\theta^{3} + 4.63 \times 10^{-2}\theta^{2} - 3.17\ \theta + 73.5 \qquad (10.4)$$

$$a = 3.45 \times 10^{-5}\theta^{3} - 6.97 \times 10^{-3}\theta^{2} + 4.68 \times 10^{-1}\theta - 9.45 \qquad (10.5)$$

$$b = 9.15 \times 10^{-8}\theta^{3} - 1.86 \times 10^{-5}\theta^{2} + 1.25 \times 10^{-3}\theta - 2.82 \times 10^{-2} \qquad (10.6)$$

The effective diffusivity increases with the increase in drying temperature. The effective diffusivities were higher in the second falling rate period (Kumar *et al.*, 2011), whereas no significant variation was observed in the overall activation energy (Table 10.2).

The pretreatment (citric acid blanching) followed by convective drying at 65°C has been considered the best drying combination for retaining the quality attributes of carrot pomace (Alam *et al.*, 2013). Hernández-Ortega *et al.* (2013) dried carrot pomace to obtain carrot pomace powders by microwave and hot air drying. Carrot pomace dried with microwave had the highest amount of the bioactive compounds. The infrared drying of carrot pomace was completed in 310, 160, 100 and 60 min at the infrared power levels of 83,125, 167 and 209 W, respectively, indicating that the power level affected the drying rate and time (Doymaz, 2013).

The drying of carrot pomace was performed by convective, microwave-convective, infrared-convective and freeze-drying. The freeze-dried material had the lowest apparent density (422 kg/m^3) and adsorbed vapour more quickly than the other dried

Table 10.2 Final effective diffusivity and activation energy of carrot pomace using hot air dryer at air velocity of 0.7 m/s

S. No.	Temperature, °C	Periods					
		First Falling		Second Falling		Overall	
		$D_{eff,}$ m^2/s	Ea, kJ/mol	$D_{eff,}$ m^2/s	Ea, kJ/mol	$D_{eff,}$ m^2/s	Ea, kJ/mol
1	60	2.74E-09	18.61	3.12E-09	24.45	2.98E-09	22.70
2	65	3.01E-09		3.75E-09		3.47E-09	
3	70	3.41E-09		4.06E-09		3.82E-09	
4	75	3.62E-09		4.64E-09		4.27E-09	

variants, probably due to high porosity and relatively low degree of structural damage in freeze drying (Janiszewska *et al.*, 2013).

10.2 Value-Added Products from Carrot Pomace Powder

Dried carrot pomace powder, which can be stored for a long time, can change the textural attributes of the food products along with nutrient potential. Carrot pomace can be incorporated into extrudates, biscuits, cookies, wheat rolls, wheat breads, fish sausages and several other products.

10.2.1 Biscuits

The biscuit is a twice baked bread having a hard and crunchy texture. Carrot pomace powder (CPP) at 10, 20 and 30% levels was incorporated into wheat flour to prepare high fiber sweet and sweet 'n' salty biscuits (Kumari and Grewal, 2007). Incorporation of CPP increased the spread ratio of both types of biscuits, whereas no significant difference in mean scores for all sensory attributes was observed by the addition of CPP up to 20%. Furthermore, 30% CPP significantly decreased the mean scores for both types of biscuits.

The biscuits prepared from composite flours by incorporating 5, 8 and 10 parts of germinated chickpea flour (GCF) and similar corresponding parts of carrot pomace powder (CPP) into wheat flour, revealed that the spread ratio of the biscuits increased from 6.1 to 8.4 with the increase of CPP and GCF in the blends. With increase in proportion of CPP and GCF, protein, ash and crude fiber content were increased. The crude fiber content of the biscuits supplemented with 10% CPP and GCF was the highest (3.2%). The biscuits, supplemented with CPP and GCF up to the 8% level were rated as acceptable in terms of sensory attributes (Baljeet *et al.*, 2014).

10.2.2 Cookies

The cookie is a soft and squashy bread, sometimes known as a slightly moist biscuit. Bioactive compounds are considered to exert protective effects against many diseases due to inhibition of oxidative damage. Dietary fiber is the main focus area for human nutrition, and increased fiber intake is a dietary recommendation for reducing the risk of the number of chronic diseases. The cookies, prepared from the incorporation of black carrot pomace at different levels of 0–15% into flour, showed that increased level of fiber may result in enhancement of polyphenol content and antioxidant activity of the cookies (Turksoy *et al.*, 2011). The polyphenol content and antioxidant activity increased 5 and 5.5 times, respectively, from the addition of 15% pomace. The general acceptability was good at the level of 10%. The cookies prepared from the incorporation of carrot pomace contained β-carotene, epicatechin, gallic and ferulic acids (Hernández-Ortega *et al.*, 2013).

Kumar and Kumar (2011) incorporated dried carrot pomace in different proportions (0–9%) into fine wheat flour, shortening, sugar and water. The formulation was baked at 175°C for 12 min in a gas oven with air circulation. The hardness of cookies ranged from 41.05 to 116.1 N, with an average value of 81.50 N. It was further observed that moisture content, hardness and colour, L*, a* values increased with the increase in proportion of carrot pomace in cookies. Sensory scores for all the cookies ranged from fair to very good, whereas the cookies with the 6% dried carrot pomace proportion gained the highest score.

10.2.3 Wheat Rolls

The incorporation of carrot pomace powder in the preparation of wheat rolls showed higher values of hydration properties. Addition of carrot pomace influences farinographic characteristics (increase in water absorption, dough development time and dough stability and decrease in mixing tolerance index) of dough and qualitative parameters of final products (decrease in loaf volume and cambering). On the basis of sensory attributes, loaves incorporated with carrot pomace powder of up to 3% were the most acceptable (Kohajdova *et al.*, 2012).

10.2.4 Wheat Bread

Whole-wheat bread can be made by using flour that is partly or entirely milled from whole/ almost-whole wheat grains. This may also be referred to as brown bread. The addition of 5, 7.5 and 10% of dried carrot pomace showed the incorporation supplemented carotenoids, fiber and mineral components in bread (Tanska *et al.*, 2007). At the same time, properties of the bread were not adversely changed. The addition of carrot pomace at 5% increased dough water absorption by 8% and bread volume by 10%. Crumb of this bread was elastic and attained the highest sensory scores (Kumar and Kumar, 2012).

Kumar and Kumar (2012) prepared small-sized wheat bread buns by incorporating carrot pomace (Figure 10.5). Product development was done by conventional methods utilizing carrot pomace in different proportions of (0, 2.5, 5, 7.5 and 10%). All the ingredients were mixed to obtain good a consistency dough. Later, these were baked in a hot air-oven at 177 ± 2°C for 40 min to a golden brown colour. With increase in

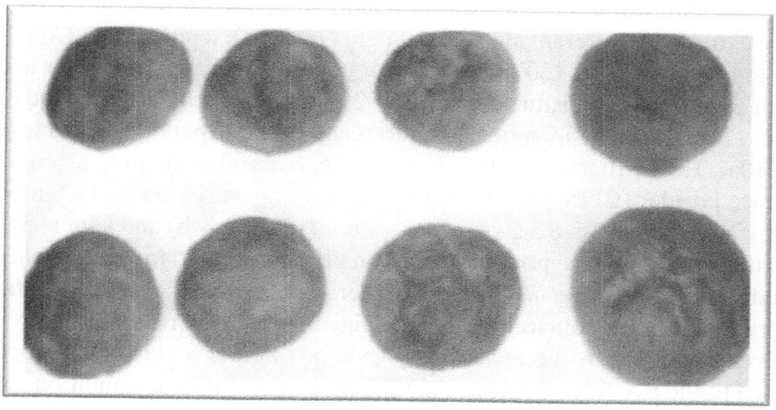

Figure 10.5 Carrot pomace-based wheat bread (buns).

pomace proportion, the expansion, water solubility and absorption index decreased, whereas bulk density and moisture content increased. The expansion, water solubility, absorption index and bulk density showed significant correlation to the pomace proportion. The buns, prepared with carrot pomace at 2.5% levels, showed very good sensory attributes.

10.2.5 Fish Sausage

Sausage refers to a food usually made from ground meat with a skin around it. Fouda *et al.* (1999) used a fine powdered mixture of apple and carrot fibers (at a ratio of 1:1 w/w) prepared from their pomace to increase the level of fiber in bolti (*Tilapia ylilotica*) fish sausage. The different levels of fiber mixture (0, 6, 9, 12 and 18%) were added in a selected formula based on organoleptic evaluation. Defatted soya flour (DSF) was also added at the levels of 4, 8, 12 and 16%. The results revealed that organoleptic properties were narrowly changed upon the addition of fiber powder. Incorporation of DSF and fiber powder, 8–9%, improved the texture of fibre-enriched fish sausages. DSF raised the percent satisfaction value of essential amino acids when added to fiber mixtures (FM) samples. With addition of FM, the cooking yield and water holding capacity (WHC) were increased, whereas plasticity decreased. The cooking yield, WHC and plasticity of cooked sausages with 9% FM+8% DSF were higher than in the control sample (9% FM).

10.2.6 Extrudates

Extrusion is a process which combines several unit operations including mixing, cooking, kneading, shearing, shaping and forming. It is a continuous and efficient system that ensures uniformity of the final product. Generally, starch is the major component of extrudates, which comparatively lacks in nutritional value. Therefore, attempts were being made to incorporate nutritional rich components to achieve the desired quality of the extrudates. Dehydrated carrot pomace powder can be used as a potential ingredient. Extrudates prepared from the addition of carrot pomace are described below.

The formulations were prepared from the incorporation of carrot pomace powder and were extruded at different moisture content (17–21%), screw speed (270–310 rpm) and die temperature (110–130°C). The carrot pomace, 11.75% in rice flour under the optimum conditions of moisture content, 19.92%, screw speed, 249.1 rpm and die temperature, 114.3°C, can be incorporated into ready-to-eat expanded products (Kumar *et al.*, 2010a). The extrudates can be obtained from rice and gram flour mixed with carrot pomace powder (CPP). The data obtained from the variation of die temperature (65–125°C), feed rate (2.5–8.5 g/s), feed moisture (10–30%, wb) and CPP (1.5–15.5%) showed that carrot pomace powder at 5% can be incorporated for acceptable quality of extrudates (Upadhyay *et al.*, 2010). Another process to make extrudates from the incorporation of carrot pomace powder and pigeon pea powder and rice flour has also been developed. The carrot pomace and pigeon pea were mixed in equal proportions and added to the rice flour at a rate of 16.5% under the optimum conditions of moisture content of 19.23%, screw speed of 310 rpm and temperature of 110°C (Kumar *et al.*, 2010b).

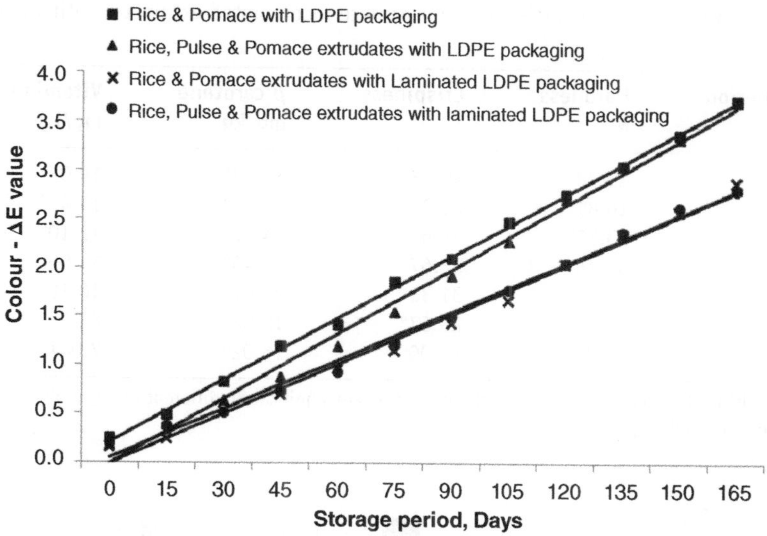

Figure 10.6 Change in colour during storage of carrot pomace-based extrudates.

Extrudates prepared under the optimum conditions of rice flour, pulse powder and carrot pomace-CPPP, moisture content, screw speed and die temperature, were packed in LDPE bags and aluminum-laminated LDPE bags. The moisture content increased with the increase in storage period in all the cases; however, minimum gain in moisture was observed in case of aluminum-laminated LDPE. The average rates of increase in moisture content (wb, per cent) were observed as 0.032 and 0.034 per day for carrot pomace and rice flour-based extrudates and rice flour, carrot pomace and pulse powder-based extrudates in LDPE bags respectively, whereas the rates were 0.021 and 0.028 per day for carrot pomace and rice flour-based extrudates and rice flour, carrot pomace and pulse powder-based extrudates in aluminum-laminated LDPE bags, respectively (Kumar *et al.*, 2012b). Minimum overall change in colour was observed in aluminum-laminated LDPE bags as compared to LDPE bags (Figure 10.6).

Textural functional and micro-structural changes in extrudates prepared from rice flour, pulse powder and carrot pomace during storage revealed decrease in colour, L values, b values, crispiness, β-carotene and vitamin C; and increase in a-value and hardness during storage at a temperature of 30°C and relative humidity of 50%. The harder and compact texture in the micro-structural study of extrudates was observed with increase in temperature. The hardness of the product was also increased with a decrease in crispness, which may be due to increase in moisture with storage time (Table 10.3). The significant decrease in β-carotene and vitamin C with storage period was observed. The decrease of vitamin C during storage may be due to oxygen, heat, light and storage temperature (Dar *et al.*, 2014a).

The deep-fat frying process removes moisture from the extrudates and imparts a crisp and softer texture. The temperature and time for frying operations was recommended as 180°C for 15 s on the basis of overall acceptability (Figure 10.7) for carrot pomace, pulse powder and rice flour-based extrudates (Dar *et al.*, 2014b). The minimum changes in crispiness and L-values were observed during storage for fried extrudates as compared to unfried extrudates (Dar *et al.*, 2013). The retention of the

Table 10.3 Variation in properties of carrot pomace incorporated extrudates with storage period

Storage period months	Hardness N	Crispiness	β carotene mg/kg	Vitamin C mg/kg
0	13.78[f]	40.33[a]	34.23[a]	14.07[a]
1	16.81[e,f]	36.33[a]	26.23[b]	13.43[b]
2	23.25[e]	33.67[a]	24.47[c]	12.10[c]
3	33.76[d]	32.67[a]	22.43[d]	11.05[d]
4	40.18[c]	31.33[ab]	20.07[e]	10.19[e]
5	41.55[b]	28.67[b]	18.02[f]	8.17[f]
6	45.80[a]	22.00[c]	16.02[g]	7.04[g]

a-f means within the same column with different letters are significantly different at ($p \leq 0.05$)
Source: Dar *et al.* (2014a).

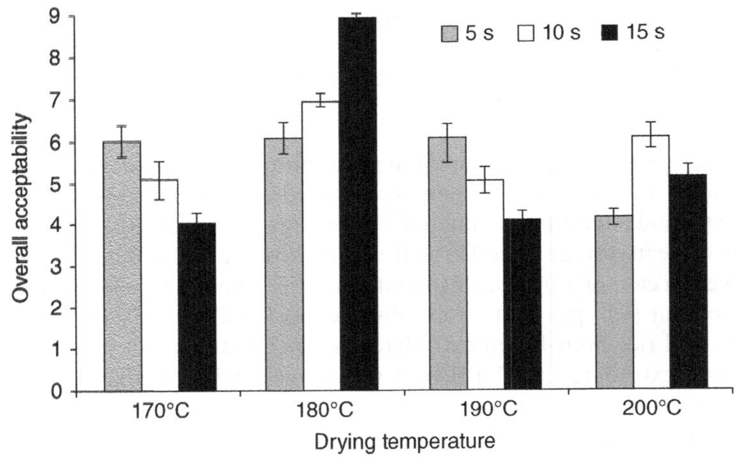

Figure 10.7 Overall acceptability of carrot pomace incorporated extrudates during frying at different times and temperatures.

crispiness in the fried extrudates may be due to the removal of moisture from the surface of the extrudate sample. The oil absorption increased from 5.37 to 14.0%, with increase in temperature and time. The high temperature causes partial evaporation of the water, which moves away from the food and through the surrounding oil. The oil is absorbed by the food, replacing the lost water during evaporation. This increase may be attributed to the high permeability of the thin outer layer surface that provides weak resistance to the internal vapour pressure as frying time increases, coupled with additional thermal degradation of structural components such as proteins and starch.

Dar *et al.* (2013) applied seasoning to the extrudates prepared from carrot pomace, pulse powder and rice flour to enhance the flavor and taste. Seasoning is the process of adding slurry of oils with salt, herbs, or spices to food to enhance the flavor followed by proper mixing with the extrudates. Seasoned extrudates retained crispness better than unfried and fried extrudates (Table 10.4). Fried extrudates with seasoning showed the better crispiness, even after six months of storage as compared to fried and unfried

Table 10.4 Effect of frying and seasoning on various properties of extrudate properties with storage period

Colour	Unfried extrudates		Fried extrudates		Seasoned extrudates	
	Fresh	After 6 month	Fresh	After 6 month	Fresh	After 6 month
L-value	66.22 ± 0.13	62.53 ± 0.49	38.58 ± 0.03	36.38 ± 0.51	34.95 ± 0.19	33.78 ± 0.09
a-value	6.31 ± 0.28	6.59 ± 0.13	18.45 ± 0.78	22.15 ± 0.06	18.48 ± 0.46	20.21 ± 0.03
b-value	23.66 ± 0.69	22.28 ± 0.35	20.06 ± 1.63	17.59 ± 0.26	20.99 ± 0.13	19.08 ± 0.12
Hardness, N	13.78 ± 0.03	45.80 ± 0.08	8.24 ± 0.06	19.04 ± 0.07	8.62 ± 0.04	21.85 ± 0.02
Crispiness	40.33 ± 1.53	22.00 ± 2.00	57.00 ± 1.00	33.00 ± 1.00	59.00 ± 1.00	41.33 ± 0.58

samples. This may be due to the frying of extrudates and absorption of moisture by seasoning during storage, thereby retaining the crispiness.

The fried and seasoned extrudates had lower L-value and higher a-value, indicating darkness and redness due to the destruction of heat sensitive pigments by thermal treatment in the form of frying and some non-enzymatic browning reactions, Maillard reaction and retrogradation. The darkness and redness of the extrudates were also increased with the increase in storage period (Table 10.4). The increase in hardness of the extrudates prepared from incorporation of carrot pomace may be due to recombination of structure and strength enhancement as a result of moisture inclusion in the structure of extrudates. In the fried and seasoned extrudates, hardness of the extrudates increased to a lesser extent as compared to the unfried extrudates, because of the removal of moisture from the surface as well as from the inner cellular components of the extrudates during the frying process (Table 10.4).

Crispiness of the extrudates was better observed in fried and seasoned extrudates. The results are in line with the fact that frying of the extrudates had considerably improved the crispiness by removing the moisture from the surface and intercellular components of the extrudates sample as compared to unfried extrudates. The crispness decreased with the storage period. Unfried structures showed the presence of protruding structures in the control sample of the unfried extruded, probably indicating the abundance of starch (Figure 10.8). The photomicrograph for the six months of storage

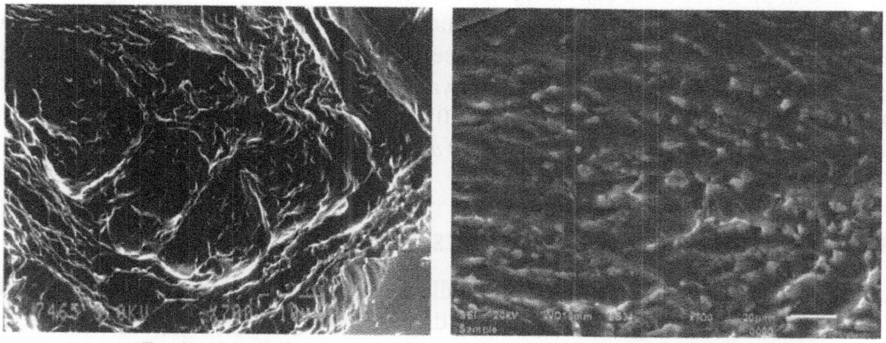

Fresh extrudates　　Stored extrudates (Six months)

Figure 10.8 Micrograph of carrot pomace incorporated extrudates.

revealed the presence of relatively more compact and dense structural orientation with numerous globules, which may have led to the increase in the hardness of the extrudates and may be the cause for the loss of crispiness.

10.2.7 Fiber

Soluble fiber, which dissolves in water, is readily fermented in the colon into gases and physiologically active by-products, and can be prebiotic and viscous (Lieberman, 2010). Soluble fibers tend to slow down the movement of food through the system. In a process to separate alcohol soluble and insoluble dietary fiber, carrot pomace was treated with a mixture of 1% NaOH and 2% acetic acid, and the pretreated sample was hydrolyzed by edible crude enzyme. The supernatant portion was then treated with 85% ethanol and separated into alcohol-soluble dietary fiber (ASDF) and alcohol-insoluble dietary fiber (AIDF) (Yoon *et al.*, 2005). AIDF increased from 3.3 to 41.7 g/100 g carrot pomace after 96 h of reaction. AIDF contained rhamnose, mannose, arabinose, glucose galactose, and a small amount of xylose. Monosaccharides (glucose, fructose, galactose, arabinose), cellooligosaccharides (cellopentaose, cellotetraose, cellobiose, cellotriose) and galactooligosaccharides (galactotetraose, galactotriose) were detected in the ASDF (Yoon *et al.*, 2005).

10.2.8 Bio-ethanol

Bio-ethanol is mainly produced by the sugar fermentation process. Major sources of sugar, which are mainly used to produce ethanol, come from energy crops. Kasavi *et al.* (2012) studied the feasibility for utilizing used industrial *Saccharomyces cerevisiae* strains for ethanol production using biological residues (including carrot, tomato and potato peel) and sugar beet pulp, starch and sugar beet molasses. The choice of strain not only depends on the ethanol tolerance but also on the preferential utilization of the carbon resources. Aimaretti *et al.* (2012) used carrot discards for the generation of ethanol and reported the production of 77.5 L of ethanol for each tonne of discarded carrots. Yu *et al.* (2013) used carrot pomace as a feedstock for bio-ethanol production by fermentation with the thermo-tolerant yeast, *Kluyveromyces marxianus*. The carrot pomace, treated with AccelleraseTM 1000 and pectinase at 50°C for 84 h, converted 42% of its mass into fermentable sugars, mainly glucose, fructose and sucrose. The simultaneous saccharification and fermentation (SSF) at 42°C on 10% (w/v) carrot pomace yielded ethanol of 18 g/L. However, the highest ethanol concentration of 37 g/L was observed with an additional charge of 10% (w/v) supplemented to the original 10% (w/v) of carrot pomace after 12 h with the corresponding yield of 0.185 g/g (Yu *et al.*, 2013).

10.2.9 Functional Components

Functional food components are potentially beneficial components found naturally in foods, which may include carotenoids, dietary fiber, fatty acids, flavonoids, isothiocyanates, phenolic acids, sterols, polyols, prebiotics and probiotics, phytoestrogens, soy protein, vitamins and minerals (Raquel *et al.*, 2009). Carle and Schieber (2006) explored the possibilities of the extraction of carotenoid-rich hydrolysate and oligogalacturonic acids from carrot pomace. Enzymatic treatment of carrot pomace led to the release of oligogalacturonic acids, which may be promising for the treatment of bacterial diarrhea. A novel process for the recovery of carrot pomace hydrolyzate has also been developed (Stoll *et al.*, 2003). The process includes techniques of fine grinding, enzymatic hydrolysis, finishing, homogenization and concentration. An enzyme mixture (pectolytic and cellulolytic, 1:1 ratio), with a dosage of 1500 ppm at 50°C and pH 4 achieved the best degradation, with a reduction in hydrolysis time to 1 h at the pilot plant scale. Total carotene content of the concentrated hydrolyzate (3.8°Bx) was 64 mg per kg (Stoll *et al.*, 2003). Plant tissues such as the bagasse can respond to abiotic stresses (i.e. UV-light, temperature, wounding) in synthesizing phenolic compounds (PC), which are chemicals of high pharmaceutical and nutraceutical value. PC can be beneficial in the prevention of different chronic diseases such as cancer, diabetes and cardiovascular. The application of abiotic stresses in carrot bagasse can be used to produce nutraceutical (Sanchez-Rangel *et al.*, 2013). The bagasse was exposed to UV-C light (30, 60 and 120 min), and then incubated at 25°C for 48 h. The samples stored for 48 h showed a higher accumulation of PC. Likewise, the highest ORAC value was observed in the bagasse treated with UV-light for 120 min and 48 h of incubation. The value increased to 183% compared to the 0 h samples.

10.2.10 Citric Acid Production

The microbial production of organic (citric and lactic) acids from carrot processing solid waste under solid state fermentation was investigated. The strain, *A. niger*, produced 29 g citric acid per kg of carrot waste with a 36% yield (Garg and Hang, 1995). Citric acid exists in large amounts in different fruits and vegetables, mostly in citrus fruits. Lemons and limes have particularly high concentrations of citric acid; it can constitute as much as 8% of the dry weight of these fruits (Penniston *et al.*, 2008).

10.2.11 Animal Feed

The livestock sector is one of the fastest growing agricultural subsectors in the developing world. The demand for livestock products is rapidly increasing and there are times when the ratio of demand and supply is not balanced in some countries. New unconventional alternate feed resources could play an important role in balancing the ratio of demand and supply. Fruit and vegetable processing generate a huge quantity of wastes. These wastes may be utilized as feed. Pozdnyakova *et al.* (1986) incorporated carrot waste at the levels of 20–25% in feeding pigs. Carrot by-product has 11.3 MJ/kg dry matter metabolizable energy and is almost equivalent to those of high-quality spring pasture, which shows the feasibility to use it as a feed for dairy cows (Bramwell *et al.*, 1993). Onodera *et al.* (1997) prepared silages consisting of barley shochu by-product and pulps (juice residues) of Mandarin oranges and carrots. The silages had 40% crude protein and 30% crude fiber (on dry matter). Silages containing 60 and 70% moisture had nitrogen-free extract values of 10.0 and 18.5%, respectively. The dried carrot waste of up to 50% can be used in growing rabbit diets in place of the more conventional ingredients, without any adverse effects on the productive performance, nutrient digestibility, blood components and economic efficiency (El-Medany *et al.*, 2008).

10.2.12 Composting and Biogas

The conversion of carrot residue into compost is one of the simplest ways. Vogtmann *et al.* (1993) prepared compost using wastes of carrot and other vegetables. The mixture of vegetable wastes (carrot along with other vegetable wastes) can be used for biogas production in a single-stage anaerobic reactor (Sridevi *et al.*, 2012). The production of biogas from carrot waste with cow dung slurry has been reported as 888.7 cc per alternate day (Parvathi, 2004).

10.3 Nutritional, Functional and Medicinal Value of Carrot and Carrot By-Products

The increased intake of antioxidant-rich fruits and vegetables assist in reducing the risks of cancer and cardiovascular diseases. Carrot is a rich source of calcium pectate; an extraordinary pectin fiber that has the cholesterol lowering properties. The properties can also be used to reduce the risk of high blood pressure, stroke, heart disease and cancer (Bakhru, 1993). The dietary carotenoids have shown to have anti-cancer effects, due to their antioxidant power in reducing free radicals in the body. The use of carrot juice extract is effective to kill leukemia cells and inhibit their propagation (Zaini *et al.*, 2011). The smokers have three times more risk for developing lung cancer in comparison to smokers having carrot in their diets once a week (Pisani *et al.*, 1986). The beta-carotene plays a protective role against prostate cancer (Wu *et al.*, 2004) and the colon cancer (Okuyama *et al.*, 2014). The role of vitamin A in the visual cycle is well established. The vitamin A (Retinol), 835 microgram may easily fulfill the DRI (daily requirement intake) of 900 and 700 microgram per day for males and females respectively.

In situ ruminal digestive characteristics indicate that total digestible nutrients, digestible energy and metabolizable energy in carrot residue were 93.0%, 15.66 MJ/kg and 14.11 MJ/kg (on dry matter basis), respectively (Enishi *et al.*, 2004). Carrot juice residue can be used as a promising new energy feed resources. The dietary fibre and pectin reduce cholesterol levels and liver enzyme activity (El-Bastawesy and Hareedy, 2004). Insoluble fiber fractions in carrot might exert favorable effects on improving intestinal functions and health and could be exploited as functional ingredients in fiber-rich food products (Chau *et al.*, 2005). Hypocholesterolemic effects of insoluble fiber-rich fractions may partly be attributed to the reduced absorption of lipid and cholesterol (Hsu *et al.*, 2006). Moreover, dietary carrot pomace powder can be used to reduce body weight and reduce hypercholesterolemic complications (Afify *et al.*, 2013).

Carrot pomace contains low amounts of pectin but higher amounts of lignin (Nawirska and Kwasniewska, 2005). The carrot pulp, which also contains 68.9% N-free extract and 17.9% crude fibre, shows the greatest fermentability (Swanson *et al.*, 2001). Carrot pomace is rich in insoluble FRFs (50.1–67.4 g/100 g), which are mainly composed of pectic polysaccharides, hemicellulose and cellulose (Chau *et al.*, 2004). The ability of these FRFs to adsorb glucose and reduce amylase activity implies that they might help to control post-prandial serum glucose level. The insoluble fibers from carrot pomace might exert favorable effects on improving intestinal functions and health, therefore can be exploited as functional ingredients in fibre-rich food products (Chau and Chen, 2006). The peonidin and gliadin glycosides, anthocyanins and non-anthocyanin phenolicscan are recovered from black carrots (Kammerer, 2005). The carrot anthocyaninscan serves as natural food colorants, therefore the application of pomace in this respect can further be explored.

References

Afify, A.E.M.R., Romeilah, R.R.M., Osfor, M.M.H. and Elbahnasawy, A.S.M. (2013) Evaluation of carrot pomace (*Daucus carota* L.) as hypocholesterolemic and hypolipidemic agent on albino rats. *Notulae Scientia Biologicae*, **5**(1): 7–14.

Aggidis, G., Arbon, I., Brown, C., Clarke, C., Earp, J., *et al.* (2013) *Global Food Waste Not, Wants Not.* Institute of Mechanical Engineers, London, pp. 36.

Aimaretti, N.R., Ybalo, C.V., Rojas, M.L., Plou, F.J. and Yori, J.C. (2012) Production of bioethanol from carrot discards. *Bioresource Technology*, **123**: 727–732.

Alam, M.S., Gupta, K., Khaira, H. and Javed, M. (2013) Quality of dried carrot pomace powder as affected by pretreatments and methods of drying. *Agricultural Engineering International: CIGR Journal*, **15**(4): 236–243.

Bakhru, H.K. (1993) *Foods that Heal: The natural way to good health.* Orient Paper Backs, Delhi.

Baljeet, S.Y., Ritika, B.Y. and Reena, K. (2014) Effect of incorporation of carrot pomace powder and germinated chickpea flour on the quality characteristics of biscuits. *International Food Research Journal*, **21**(1): 217–222.

Bazhal, M.I., Lebovka, N.I. and Vorobiev, E.I. (2001) Pulsed electric field treatment of apple tissue during compression for juice extraction. *Journal of Food Engineering*, **50**: 129–139.

Bohm, V., Otto, K. and Weissleder, F. (1999) *Yield of juice and carotenoids of the carrot juice production.* In: Symposium Jena-Thuringen, Germany, pp. 115–119.

Bramwell, N.F., Parker, W.J., Gray, D.I. and McCutcheon, S.N. (1993) The potential of industry by-products as supplementary feeds for dairy cows in the Hawkes Bay region. *Proceedings of the New Zealand Society of Animal Production*, **53**: 55–59.

Carle, R. and Schieber, A. (2006) Functional food components obtained from carrot and apple juice processing waste. In: *Ernährungs Umschau: Nutrition Survey.* Umschau Zeitschriftenverlag Breidenstein GmbH, pp. 18–24.

Chau, C.F. and Chen, C.H. (2006) Effects of two pomace insoluble fibres on the activities of faecal bacterial enzymes and intestinal health. *European Food Research and Technology*, **222**(5/6): 681–685.

Chau, C.F., Chen, C.H. and Lee, M.H. (2004) Comparison of the characteristics, functional properties, and *in vitro* hypoglycemic effects of various carrot insoluble fiber-rich fractions. *LWT – Food Science and Technology*, **37**(2): 155–160.

Chau, C.F., Chien, P.J. and Chen, C.H. (2005) Influence of insoluble fiber fractions from carambola and carrot on intestinal enzymes and fecal bacterial enzymes in hamsters. *Nutrition Research*, **25**(10): 947–957.

Chau, C.F., Wang, Y.T and Wen, Y.L. (2007) Different micronization methods significantly improve the functionality of carrot insoluble fibre. *Food Chemistry*, **100**(4): 1402–1408.

Dar, A.H., Kumar, N. and Sharma, H.K. (2013). Physical and micro-structural changes in carrot pomace-based extrudates. *Italian Journal of Food Science*, **25**(3): 313–321.

Dar, A.H., Sharma, H.K. and Kumar, N. (2014a) Effect of extrusion temperature on the microstructure, textural and functional attributes of carrot pomace-based extrudates. *Journal of Food Processing and Preservation*, **38**(1): 212–222.

Dar, A.H., Sharma, H.K. and Kumar, N. (2014b) Effect of frying time and temperature on the functional properties of carrot pomace, pulse powder and rice flour-based extrudates. *International Journal of Food Engineering*, **10**(1): 139–147.

Donato, P.D., Finore, I., Anzelmo, G., Lama, L., Nicolaus, B. and Poli, A. (2014) Biomass and biopolymer production using vegetable wastes as cheap substrates for extremophiles. *Chemical Engineering Transactions*, **38**: 163–168.

Doymaz, I. (2013) Determination of infrared drying characteristics and modelling of drying behaviour of carrot pomace. *Tarim Bilimleri Dergisi*, **19**(1): 44–53.

El-Bastawesy, A.M. and Hareedy, L.A. (2004) Utilization of food processing wastes as sources of dietary fibers and its effect on lipoproteins in rats. *Egyptian Journal of Agricultural Research*, **82**(3): 1311–1332.

El-Medany, N.M., Hashem, N.A. and Abdel-Azeem, F. (2008) Effect of incorporating dried carrot processing waste in growing rabbit diets. *Egyptian Journal of Nutrition and Feeds*, **11**(1): 25–37.

Enishi, O., Yoshioka, T. Nakashima, K., Saeki, M. and Kawashima, T. (2004) Analysis of in situ ruminal digestive characteristics and nutritive value of pumpkin and carrot juice residue for ruminant feeds. *Grassland Science*, **50**(4): 360–365.

FAO (2017) Production Statistics, Food and Agriculture Organization of the United Nations: http://www.fao.org/faostat/en/#data Last accessed on 26 June 2017.

Fouda, Z.M.A. (1999) Quality attributes of fish sausage as affected by adding dietary fibers. *Annals of Agricultural Science (Moshtohor)*, **37**(2): 1287–1298.

Garg, N. and Hang, Y.D. (1995) Microbial production of organic acids from carrot processing waste. *Journal of Food Science and Technology (Mysore)*, **32**(2): 119–121.

Giacomo, G. and Taglieri, L. (2009) A new high yield process for the industrial production of carrot juice. *Food Bioprocess Technology*, **2**: 441–446.

Grigelmo-Miguel N. and Martin-Belloso O. (1999) Comparison of dietary fibre from by-products of processing fruits and greens and from cereals. *LWT – Food Science and Technology*, **38**: 503–508.

Haq, R., Singh, Y., Kumar, P. and Prasad, K. (2013) Quality of dehydrated carrot shreds as affected by partial juice extraction through hydraulic press. *International Journal of Agriculture and Food Science Technology*, **4**(4): 331–336.

Hernández-Ortega, M., Kissangou, G., Necoechea-Mondragón, H., Sánchez-Pardo, M.E. and Ortiz-Moreno, A. (2013) Microwave dried carrot pomace as a source of fiber and carotenoids. *Food and Nutrition Sciences*, **4**(10): 1037–1046.

Hsu, P.K., Chien, P.J., Chen, C.H. and Chau, C.F. (2006) Carrot insoluble fiber-rich fraction lowers lipid and cholesterol absorption in hamsters. *LWT – Food Science and Technology*, **39**(4): 338–343.

Janiszewska, E., Witrowa-Rajchert, D., Kidon, M. and Czapski, J. (2013) Effect of the applied drying method on the physical properties of purple carrot pomace. *International Agrophysics*, **27**(2): 143–149.

Kammerer, D.R. (2005) *Characterization of Phenolic Compounds from Grape Pomace and Black Carrots in High-Performance Liquid Chromatography – Electrospray ionization multistage mass spectrometry*. Shaker Verlag GmbH, p. 176.

Kasavi, C., Finore, I., Lama, L., Nicolaus, B., Oliver, S.G., *et al.* (2012) Evaluation of industrial *Saccharomyces cerevisiae* strains for ethanol production from biomass. *Biomass and Bioenergy*, **45**: 230–238.

Kaur, M. and Sharma, H.K. (2013) Effect of enzymatic treatment on carrot cell wall for increased juice yield and effect on physicochemical parameters. *African Journal of Plant Science*, **7**(6): 234–243.

Kohajdova, Z., Karovicova, J., Jurasova, M., Wydawnictwo, A., Rolniczej, P and Poznan, P. (2012) Influence of carrot pomace powder on the rheological characteristics of wheat flour dough and on wheat rolls quality. *Acta Scientiarum Polonorum – Technologia Alimentaria*, **11**(4): 381–387.

Kumar, K. and Kumar, N. (2012) Development of vitamin and dietary fibre enriched carrot pomace and wheat flour based buns. *Journal of Pure and Applied Science and Technology*, **2**(1): 108–116.

Kumar, N. and Kumar, K. (2011) Development of carrot pomace and wheat flour based cookies. *Journal of Pure and Applied Science and Technology*, **1**(1): 4–10.

Kumar, N., Sarkar, B.C. and Sharma, H.K. (2010a) Development and characterization of extruded product using carrot pomace and rice flour. *International Journal of Food Engineering*, **6**(3): Article 7.

Kumar, N., Sarkar, B.C. and Sharma, H.K. (2010b) Development and characterization of extruded product of carrot pomace, rice flour and pulse powder. *African Journal of Food Science*, **4**(11): 703–717.

Kumar, N., Sarkar, B.C. and Sharma, H.K. (2011) Effect of air velocity on kinetics of thin layer carrot pomace drying. *Food Science and Technology International*, **17**(5): 459–469.

Kumar, N., Sarkar, B.C. and Sharma, H.K. (2012a) Mathematical modelling of thin layer hot air drying of carrot pomace. *Journal of Food Science and Technology (Mysore)*, **49**(1): 33–41.

Kumar, N., Sarkar, B.C., Sharma, H.K. and Jha, S.K. (2012b) Colour kinetics and storage characteristics of carrot, pulse and rice by-product based extrudates. *British Food Journal*, **114**(9): 1279–1296.

Kumari, S. and Grewal, R.B. (2007) Nutritional evaluation and utilization of carrot pomace powder for preparation of high fiber biscuits. *Journal of Food Science and Technology (Mysore)*, **44**(1): 56–58.

Lieberman, M.A. (2010) Study guide for *Marks Basic Medical Biochemistry*, 4th edition. Cram101, Pallit Publications, p. 510.

Midilli, A., Kucuk, H. and Yapar, Z.A. (2002) New model for single-layer drying. *Drying Technology*, **20**(7): 1503–1513.

Nawirska, A. and Kwasniewska, M. (2005) Dietary fibre fractions from fruit and vegetable processing waste. *Food Chemistry*, **91**(2): 221–225.

Okuyama, Y., Ozasa, K., Oki, K., Nishino, H., Fujimoto, S. and Watanabe, Y. (2014) Inverse associations between serum concentrations of zeaxanthin and other carotenoids and colorectal neoplasm in Japanese. *International Journal of Clinical Oncology*, **19**(1): 87–97.

Onodera, R., Kawamura, O., Inazawa, A., Izumi, T., Okuda, M., *et al.* (1997) Preparation of silages consisting of barley shochu distillery byproduct and pulps (juice residues) of Mandarin oranges and carrots. *Bulletin of the Faculty of Agriculture (Miyazaki University)*, **43**(2): 145–150.

Parvathi, S. (2004) Micro-biological degeneration of organic kitchen wastes into energy. *Journal of Ecotoxicology and Environmental Monitoring*, **14**(4): 277–283.

Patil, R.C., Tushar, P.M. and Shamim, A.S. (2012) Extraction of pectinase from pectinolytic bacteria isolated from carrot waste. *International Journal of Pharma and Bio Sciences*, **3**(1): 261–266.

Penniston, K.L., Nakada, S.Y., Holmes, R.P. and Assimos, D.G. (2008) Quantitative assessment of citric acid in lemon juice, lime juice, and commercially-available fruit juice products. *Journal of Endourology*, **22**(3): 567–570.

Pisani, P., Berrino, F., Macaluso, M., Pastorino, U., Crosignani, P. and Baldasseroni, A. (1986) Carrots, green vegetables and lung cancer: A case-control study. *International Journal of Epidemiology*, **15**(4): 463–468.

Pozdnyakova, A. and Evtushenko, V. (1986) Food wastes – an important source of feed for pigs. *Svinovodstvo*, **1**: 24–27.

Raquel, G., Maria, J. and Lima, M.J.B. (2009) Role and health benefits of different functional food components. *Millenium*, **37**: 1–17

Sánchez-Rangel, J.C., Benavides, J., Jacobo-Velázquez, D.A., Abdullah, H. and Latifah, M.N. (2013) Production of nutraceuticals in carrot bagasse using abiotic stresses. *Acta Horticulturae*, **1012**: 1475–1479.

Schweiggert, U. (2004) Carrot pomace as a source of functional ingredients. *Flussiges Obst.*, **71**(3): 136–140.

Serena, A. and Kundsen, B. (2007) Chemical and physicochemical characterisation of co-products from vegetable food and agro industries. *Animal Feed Science and Technology*, **139**: 109–124.

Sinchaipanit, P. and Kerr, W.L. (2007) Effect of reducing pulp particles on the physical properties of carrot juice. *Asian Food Journal*, **3**: 205–214.

Singh, B., Panesar, P.S. and Nanda, V. (2006) Utilization of carrot pomace for the preparation of a value added product. *World Journal of Dairy and Food Sciences*, **1**(1): 22–27.

Sridevi, V.D., Srinivasan, S.V., Kayalvizhi, R. and Bhuvaneswari, R. (2012) Studies on conversion of carbohydrate content in the mixture of vegetable wastes into biogas in a single stage anaerobic reactor. *Research Journal of Chemical Sciences*, **2**(6): 66–71.

Stoll, T., Schweiggert, U., Schieber, A. and Carle, R. (2003) Application of hydrolyzed carrot pomace as a functional food ingredient to beverages. *Journal of Food, Agriculture and Environment*, **1**(2): 88–92.

Swanson, K.S., Grieshop, C.M., Clapper, G.M., Shields, R.G.J., Belay, T., *et al.* (2001) Fruit and vegetable fiber fermentation by gut microflora from canines. *Journal of Animal Science*, **79**(4): 919–926.

Tanska, M, Zadernowski, R. and Konopka, I. (2007) The quality of wheat bread supplemented with dried carrot pomace. *Polish Journal of Natural Science*, **22**: 126–136.

Tingting, M., Chengrui, T., Jiyang, L., Rui, Z., Xiangyu, S. and Jinjin, M. (2013) Influence of technical processing units on polyphenols and antioxidant capacity of carrot (*Daucus carota* L.) juice. *Food Chemistry*, **141**(3): 1637–1644.

Turksoy, S., Keskin, S., Ozkaya, B. and Ozkaya, H. (2011) Effect of black carrot (*Daucus carota* L. ssp. sativus var. atrorubens Alef.) fiber addition on the composition and quality characteristics of cookies. *Journal of Food, Agriculture and Environment*, **9**(3): 57–60.

Upadhyay, A., Sharma, H.K. and Sarkar, B.C. (2010) Optimization of carrot pomace powder incorporation on extruded product quality by response surface methodology. *Journal of Food Quality*, **33**(3): 350–369.

Upadhyay, A., Sharma, H.K. and Sarkar, B.C. (2008) Characterization and dehydration kinetics of carrot pomace. *Agricultural Engineering International: CIGR Journal*, **10**: 1–9.

USDA Nutritional Database (2011) United States Department of Agriculture, Agricultural Research Service, National Nutrient Database for Standard Reference Release 27, Nutrient Data Laboratory USDA-ARS, Beltsville, http://ndb.nal.usda.gov/ndb/

Vogtmann, H., Matthies, K., Kehres, B. and Meier-Ploeger, A. (1993) Enhanced food quality: Effects of composts on the quality of plant foods. *Compost Science and Utilization*, **1**(1): 82–100.

Wu, K., Erdman, J.W., Schwartz, S.J., Platz, E.A., Leitzmann, M., *et al.* (2004) Plasma and dietary carotenoids, and the risk of prostate cancer: A nested case-control study. *Cancer Epidemiology Biomarkers and Prevention*, **13**: 260–269.

Yilbas, B.S., Hussain, M.M. and Dincer, I. (2003) Heat and moisture diffusion in slab products to convective boundary condition. *Heat and Mass Transfer*, **39**: 471–476.

Yoon, K.Y., Cha, M.H., Shin, S.R. and Kim, K.S. (2005) Enzymatic production of a soluble-fibre hydrolyzate from carrot pomace and its sugar composition. *Food Chemistry*, **92**(1): 151–157.

Yu, C.Y., Jiang, B.H. and Duan, K. (2013) Production of bioethanol from carrot pomace using the thermo-tolerant yeast *Kluyveromyces marxianus*. *Energies*, **6**(3): 1794–1801.

Zaini, R., Clench, M.R. and Le Maitre, C.L. (2011) Bioactive chemicals from carrot (*Daucus carota*) juice extracts for the treatment of leukemia. *Journal of Medicinal Food*, **14**(11): 1303–1312.

11

Processing and Utilization of Soy Food By-Products

M.K. Tripathi[1] and Rahul Shrivastava[2]

[1]*Agro Produce Processing Division, ICAR-CIAE, Bhopal, India*

[2]*Maulana Azad National Institute of Technology, Bhopal, India*

11.1 Introduction

Soybean and soy products are relatively inexpensive sources of proteins that are widely recognized for their high nutritional and excellent functional properties. Soybean is a grain legume and major source of human nutrition and livestock feed. It originated in Manchuria, China and from there spread to many countries of the world, especially in Asia, Central and South America, Europe and Africa. The present world production of soybean is about 230 million tones, accounting for nearly 57% of the global oilseed production. Soybean provides approximately 60% of vegetable protein and 30% of oil to the world. The major soybean producing countries of the world are the USA (41%), Brazil (26%), Argentina (14%), China (7%), India (4%), Paraguay (2%) and others (6%). The production and utilization of soybean is growing globally. Soybean has been and continues to be a major source of well-being for people in different regions of the world. Its use has increased in human nutrition and health, edible oil, livestock feed, biofuel, industrial and pharmaceutical applications.

Soybean is an excellent source of nutrition and health-promoting phytochemicals and has tremendous potential to be transformed into a number of healthy foods, suiting to individual requirements, across the globe (Danji, 2000). Soy protein is the least expensive and best plant protein. In 1993, the US Food & Drug Administration (FDA) adopted the Protein Digestibility Corrected Amino Acid Score (PDCAAS), a new and more accurate method for evaluating protein quality. The highest PDCAAS value that any protein can receive is 1.00. The PDCAAS value of soy protein is 1.00 and that of egg white, casein and meat are 1.00, 1.00 and 0.92, respectively. Soybean, being rich in protein and calories, has a great potential to tackle the problem of protein-calorie malnutrition (PCM) prevalent in India and other developing countries of the world.

Soybean contains twice as much protein as pulses, groundnut, meat and fish; three times as much as egg and more than 10 times that of milk. Soybean does not contain lactose, hence soymilk and other dairy analogs are better suited by lactose-intolerant

Food Processing By-Products and their Utilization, First Edition. Edited by Anil Kumar Anal.

persons. Soybean is also a very good source of food for diabetics. Incorporation of soy foods into the daily diet is a superior way to improve human nutrition and health. The FDA has also approved a health claim stating that 25 g of soy protein in the daily diet low in saturated fat and cholesterol can help reduce total and LDL cholesterol (FDA, 1999, FAO/WHO, 1990). Soybean has now become a preferred vegetable protein for food applications due to its multiple functional properties (FAO, 2009). It is a cost-effective and high-quality ingredient that can replace dairy, egg and meat proteins as consumers search for ever-increasing variations to diet staples. Global population is increasing, and with it the demand for plant and animal origin food commodities and its ingredients.

Soybean is a very rich and economical source of food and nutrition for human beings, as well as livestock. Hence, more soybeans will be needed in the future. The average income of people in Asia and Africa is increasing, and with it the demand for livestock products. Soymeal is the major source of protein for livestock feed around the world, therefore the demand for soymeal will rise. New industrial applications will further increase the demand for soybean and its products. Soybean also enriches the soil and protects the environment. New findings on the role of soybean in human health will boost its direct food uses globally and the majority of people in the world will eat soybean by 2025. The backward linkage of industry with farmers will ensure a regular supply of good-quality soybean that suits specific end-products and needs (Gandhi *et al.*, 2008). Soybean helps people feel better and live longer, with an enhanced quality of life. Soybean, therefore, has an excellent future and it will have a significant impact on human nutrition and health globally, and thereby on its economy. Production and utilization of soybean need to be encouraged worldwide, as it is in the interest of a greener globe and better human health resulting in well-being and longevity.

Among various soy foods, soymilk and tofu are becoming more popular as low-cost substitutes for traditional dairy products for the common masses and an ideal nutritional supplement for lactose intolerants (Biranjan *et al.*, 1982). The by-product obtained during processing of soybean is either under-utilized or un-utilized. Okara (insoluble residue from tofu or soymilk) and whey are two major by-products of the tofu-making process. Both can be put to a number of creative uses, which can serve as potential sources of income and utilization of their nutrients. Curds can be sold at low prices for institutional use or as a ready-to-serve snack. Hulls can be ground to make soy bran. Soybean seeds fiber is supposed to have good nutritional and functional properties. The neutral taste and the absence of colour of such fibers makes them suitable for incorporation into food products without any change in their quality, unlike those from wheat fibers. Ma *et al.* (1997) reported high recovery of protein from okara in both highly acidic and alkaline pH ranges. Soybean molasses is a by-product in protein-concentrate production and vinases is the waste product of ethanol recovery by distillation. Ethanol is produced by the alcoholic fermentation of soybean molasses, an industrial residue, produced concomitantly with the soybean protein-concentrate that results from the extraction of sugars from de-oiled soybean meal.

11.1.1 Soybean: Global Scenario and its Future

Soybean is one of the most important crops in the world today and it is extraordinarily rich in protein (40%), oil (20%) and phytochemicals. It originated in China and has been cultivated for more than 5000 years. It is believed that with the development of sea and land trades, soybean moved out of China to nearby countries such as Myanmar, Japan, India, Indonesia, Malaysia, Nepal, Philippines, Thailand and Vietnam during the first century AD and 1100 AD. However, it remained a minor crop everywhere, except in China. With its introduction into the USA in the 18th century, and its systematic breeding in that country in the 1940s and 1950s, soybean was transformed from an inefficient fodder-type crop to highly productive erect plant type and the USA has been the largest producer of soybean in the world ever since. As of now, there are more than 85 countries that produce soybeans. The five major world soybean-producing countries of the world are the USA, Brazil, Argentina, China and India, and all the five put together produce more than 90% of the world's soybeans.

China was the world's largest soybean producer and exporter during the first half of the 20th century. In the 1950s, soybean production developed rapidly in the USA, which is now the largest soybean-producing country of the world. In the 1970s, soybean production developed in Brazil, which is now the second-largest soybean-producing country. Since then, soybean production developed in Argentina and now it is the third-largest soybean-producing country of the world. Soybean is processed into a very wide range of products to realize its astonishing potential as food, feed, pharmaceutical and industrial products. Traditionally, soybean has been utilized mainly as fermented (sauce, miso, natto etc.) and non-fermented (oil, milk, tofu, flour etc.) foods. During the 20th century, however, with the increased demand for meat and eggs, the use of soy-products/derivatives as feed has been extensively developed, mainly in the Western countries and to a lesser extent in Asia. In addition, during the 1980s, 1900s and 2000s, there have been tremendous improvements in soybean processing and utilization technologies and significant developments in marketing. Technologies are constantly being updated to produce better-quality dairy analogs and soy-protein isolates. New findings also include the physiological functions of soybean for human nutrition and health. The global goal of soybean processing and utilization is to strengthen the development of new foods, feeds, pharmaceuticals, cosmetics and industrial products, including co-products and ingredients for specialty applications.

On account of increased health awareness among people, there is a tremendous demand for functional foods that contain enhanced levels of phytochemicals that are beneficial for human health. In soybean, such phytochemicals are isoflavones, fatty acids, amino acids, phytic acids, phytoestrogens, glucosides and saponin glycosides. Isoflavones have several health benefits, such as in breast cancer, prostate cancer and cardiovascular diseases. In the near future, improvement in functional foods is likely to become one of the top priorities of soybean molecular breeding programs. Soybean is a grain legume and provides high-quality nutrition (Table 11.1), nutraceuticals and therapeutic ingredients that help people feel better and live longer, with an enhanced quality of life (Table 11.2). Soybean has tremendous potential to be transformed into a number of healthy foods (Table 11.3) and pharmaceuticals suited to each individual, not only in India but across the globe. Soy protein is one of the best proteins (Table 11.4) and least expensive.

The likely utilization pattern of soybean in the 21st century is in direct food uses, mechanically expressed and physically refined soy oil, livestock and aqua feed and pharmaceutical and other industrial products. The strategy would be for the complete utilization of soybean constituents for food, feed and pharmaceutical products. This would require needs-based high-quality research and development in soybean

Table 11.1 The approximate food value of 100g of edible soybean

Constituents	Value, g
• Protein	40
• Carbohydrates	23
• Fat	20
• Moisture	8
• Minerals	5
• Fiber	4
• Energy	430 kcal

Table 11.2 Major health benefits from the regular use of soybean in the daily diet

Soy-based nutrients	General health benefits on regular eating
Protein	Lowers blood cholesterol
Carbohydrates	Relieves constipation, good for diabetics
Fat	Prevents cardiovascular diseases
Mineral	Overall health promotion
Vitamin	Overall health promotion
Phytochemicals	Prevents cancer, helpful in menopause and osteoporosis

Table 11.3 Some of the food uses of soybean and its derivatives

Soybean and its derivatives	Food/Feed/Fuel uses
• *Whole soybean* (containing all nutrients and phytochemicals)	Soybean dairy analogs, full fat soy flour, snack foods, sprouted bean, Tempeh, Miso, Natto and sauce
• *Soybean Oil* (20% of wholegrain)	Cooking/shortening oil, salad oil, margarine, bread spread, and soy lecithin
• *Soyproteins* (about 50% in edible soy meal)	Texturised soy protein, soy protein concentrate, hydrolysates and isolates, defatted soy flour, and dietary fiber
• *Soybean byproducts* (hull and okara)	Single Cell Protein (SCP), dietary fiber, livestock feed, snacks
• *Soybean crop residue* (leaf, branches and stems)	Animal feed, fuel and manure

production to consumption value chain. Soybean helps people feel better and live longer with an enhanced quality of life. Soybean, therefore, has an excellent future and will have a significant impact on human nutrition and health globally, and thereby, on its economy. Production and utilization of soybean need to be encouraged worldwide, as it is in the interest of greener globe and better human health resulting into happiness and longevity (Acasio, 1997).

Table 11.4 Protein digestibility corrected amino acids score (PDCAAS) of some selected proteins

Protein source	PDCAAS Value
• Egg white	1.00
• Casein	1.00
• Soy protein	1.00
• Meat	0.92
• Pea flour	0.69
• Lentil	0.52
• Wheat	0.40

11.1.2 Post-Production Management of Soyabean

Soybean is a grain legume and a source of high-quality nutrition for human diet and livestock feed, at a low cost. In 2010/11, the global production of soybean was approximately 230 million tonnes with average world productivity of about 2.45 t/ha. The USA, Brazil, Argentina, China and India are the major soybean producing countries of the world. Soybean is grown in various sequential and inter/mixed cropping systems, as it has moisture stress tolerance, contributes to soil fertility and is leguminous in nature (Ghosh and Jayas, 2010). India produced about 10.0 Mt of soybean in 2010/11 from a 9 Mha of crop area with an average productivity of 1100 kg/ha. The extent of post-harvest loss in soybean from field to factory, in India, is about 16% in the conventional system of soybean harvesting, threshing storage and marketing. The break-up is as follows. At Farm level, 10.8%; at grain Market, 1.0%; and at Factory level, 4.0%. However, in mechanized and modernized systems of harvesting, drying, storage, marketing and processing, the loss could be brought down to 3–5%.

11.1.2.1 Harvesting Soybean is harvested when the grains are mature and the foliage is dry. Its normal production cycle is 90–120 days from planting to harvesting. Typically, soybeans are harvested at 13–15% grain moisture, wet basis, to avoid field losses due to shattering of over-dried seeds and to reduce drying time to achieve safe storage moisture content. Harvesting at 1% or low moisture contents creates seed cracks and shattering losses. Mechanized harvesting is carried out by most of the farmers. Soybean seed size depends on the genotype and production environment. Its spherecity is in the range of 0.8–0.9.

11.1.2.2 Drying Drying soybean grains to a storage moisture level prevents microbial growth, slows enzymatic changes and enhances grain storability. Drying also reduces the grain mass and facilitates transportation and handling. Soybean drying is accomplished by forcing air through the bulk of grains at different temperatures, as under:

- Natural air drying (ambient air)
- Near – ambient drying (1–5°C above ambient)
- Low temperature drying (5–15°C above ambient)
- High temperature drying (50–200°C)

The maximum permissible moisture content in the harvested soybean is 18% wb. The drying temperature for soybean to be used as food and oil should not exceed 49°C and that for seed should be limited to 43°C. The maximum safe drying temperature for in-storage drying of soybean is 38°C. The relative humidity (RH) of the drying air is another major factor and determines the moisture content to which the crop will dry. This moisture is referred to as the equilibrium moisture content (EMC) of the grain. Soybeans are hygroscopic materials that either lose or gain moisture from the surrounding air.

11.1.2.3 Storage Soybean grains after harvest are stored until processed or consumed. Storage can vary from a few months to more than a year. Soybean at about 14% moisture, wb and maintained at 5–8°C, can be stored for 2 years or more without any mould damage. Whereas, soybean stored at 30°C are susceptible to mould growth within a few weeks and become severely damaged in 6 months. Fungal growth in the stored soybean can be dangerous, because it may result in loss of germination, spot heating and production of mycotoxins, which are poisonous when eaten or inhaled. It is better to remove foreign materials and fines from soybean and inspect them for moulds and insect-damaged grains before drying and storage, for a longer period of storage.

Soybean are usually stored in cylindrical steel or concrete silos, having aeration and mechanized handling facilities, while in most of the developing countries, it is done in bags but now shifting towards bulk handling and storage system (Kong *et al.*, 2008 and Hou *et al.*, 1998). Periodical aeration is required to prevent moisture migration and heating of grains in storage and to maintain a uniform temperature to retain grain quality. Even at safe storage moisture content, a non-uniform temperature in the bulk storage creates inter-grain air currents, resulting in moisture migration and deterioration of the grain. Typically, insects develop and reproduce between 27 and 35°C, but they become inactive or die at temperatures of 16°C or less. In cold weather, soybean can be stored at as high as 14% moisture content, but for safe storage during summer months, soybean should not have more than 11–12% moisture, wb.

11.1.2.4 Quality Preservation of soybean quality is very important for marketing. The major quality-degrading factors in soybeans are moisture, splits, foreign materials and damaged grains. A reduced moisture level can sometimes be responsible for a greater number of splits. Splitting soybeans accelerates the growth of moulds and changes in composition. Hot-air drying also causes the seed coat and cotyledon to crack, which increases breakage during subsequent handling or conveying. Soybeans are susceptible to cracks and breakage when grain moisture falls below 11%, wb. Impact during handling of soybean also results in physical damage and therefore soybean needs careful handling at a desired moisture level.

During storage, seeds undergo physical, physiological and chemical changes such as increased free fatty acid (FFA) content, decreased seed viability, increased moisture and decomposition of phospholipids or denaturation of protein. FFA increases with storage moisture and temperature, as fats in grains are readily broken down by lipases (Narayan *et al.*, 1988a). Bag storage helps in self-ventilation and cooling, but uneven airflow during aeration causes non-uniform seed temperature, which deteriorates seed quality. The bulk storage is easy to aerate and cool and, therefore, the desired quality of the seed can be maintained. Image processing or machine vision systems are gaining importance in grain handling and storage. NIR spectroscopy is used to obtain chemical constituents from solid or liquid samples. It is fast and non-destructive and requires little or no sample preparation. It can provide simultaneous determination of multiple components per measurement with a remote sampling capacity. The combination of high grain moisture and temperature during soybean storage accelerates the deterioration process of the sub-products of soybean.

Soybeans are mainly composed of about 40% protein and 20% oil/lipids and are susceptible to qualitative deterioration processes via degradation of these compounds when stored improperly and can result in serious damage to the food industry (Liu, 1997a). Physical, chemical and biochemical alterations may occur in soybean, depending on conditions and storage time. The qualitative changes of soybean during storage contribute to the loss of oil and meal quality, as well other derivatives such as soymilk and tofu (Narayan *et al.*, 1998b). The sensory qualities of products made from stored soybean are directly related to the various physico-chemical and biological changes occurring in grains during storage. Soy-products such as soymilk, tofu and soy-nuts made from soybeans, were stored for different periods of time and then evaluated for their sensory properties along with some chemical qualities and it was found that total solids and protein in soy-milk, as in tofu, decreased with time. The overall organoleptic score of all the products was found to decrease with an increase in storage time of soybeans (Saio *et al.*, 1980).

11.1.3 Soybeans Product History

Historically, soybeans have played an important part in Asian culture, both as a food and as a medicine (Messina, 1995). The soybean is widely believed to have originated 4000–5000 years ago in the north and central regions of China. In Asian countries, peoples utilize soybeans mainly for producing traditional soy food. The soy foods, those prepared from whole soybeans, are typically categorized into two groups: non-fermented and fermented. Traditional non-fermented soy foods include fresh green soybeans, whole dry soybeans, whole-fat soy flour, soymilk, tofu, okara, soy nuts, soy sprouts and yuba. Whole-fat soy flour, prepared from whole dried soybeans, is used in bakery applications in the place of milk powder or as a substitute for whole-wheat flour. Preparations of many soya products, i.e. soymilk, tofu and yuba, are initiated with the soaking of whole soybeans, and followed by rinsing, grinding and filtering. The fermented soy foods include tempeh, miso, soy sauces, natto and fermented tofu (sufu) and soymilk products. Tempeh is a soy product produced from the fermentation by *Rhizopus oligosporus*. Tempeh is mainly served as a meat substitute that, when sliced and deep-fried, has a nutty flavor, pleasant aroma and crunchy texture (Liu, 1997b).

Soy sauce is a salty, sharp tasting, dark-brown liquid extracted from fermented mixture of soybeans and wheat. Natto is a sweet, aromatic product made from the fermentation of whole soybeans with *Bacillus natto*. A number of traditional fermented soy food products also exist that are made from the fermentation of tofu and soymilk. Sufu, or Chinese cheese, is produced by the fermentation of fresh tofu by fungi, such as *Mucor hiemalis* or *Actinomucor elegans*. Soymilk is used to produce soybean yogurt, which is similar to Western dairy yogurt, and is basically a mixture of soymilk, whey and sucrose that has been cooked and cooled and inoculated with *Lactcoccus acidophilus* (Liu, 1997c). Although soybeans were introduced to Europe in the 1700s, little interest developed until the early 1900s, primarily because of the plant's inferior flavor quality compared to the native oil and meal products. Soybeans contain about 40% fat, and oil from soybeans is the world's leading vegetable oil, and accounts for well over half the fats and oils going into food products in the USA. Industrial applications for soybean oil also exist and include soap manufacture, paints, resins and drying oil products.

The soybean plant (*Glycine max*) originated in China, and Samuel Bowen introduced the beans to the American colonies in 1765 (Hymowitz and Harlen, 1983). He used soybeans to prepare soy sauce and soy noodles (vermicelli) for export from Georgia to England. Large-scale production of soybeans in the USA appears to have started during the 1850s (Smith and Circle, 1972). In 1999, farmers planted soybeans on 72 million acres, amounting to 27% of the total crop area in the USA. Soybeans are an economically important crop, which serves as a source of good-quality protein for animals and humans. Nitrogen fixation by root nodule bacteria in soybean plants is the ultimate source of soy proteins. The seeds contain up to 48% protein and up to 22% oil, which is widely consumed as a cooking oil. The storage soy proteins consist of a mixture of proteins ranging in molecular weight from about 140 000 to 300 000 Da and differing in physicochemical and other properties. The seeds also contain bioactive proteins including α-amylase, cytochrome c, lectin, lipoxygenase, urease, the Kunitz inhibitor of trypsin (KTI), and the Bowman-Birk inhibitor of chymotrypsin and trypsin, as well as secondary metabolites including isoflavones, saponins, phytic acid, flatus-producing oligosaccharides, and goitrogens. Processing of soy meal improves its digestibility and destroys much of the inhibitors, but may also result in the formation of unnatural amino acids, including heat-induced fructosyllysine derived from protein-carbohydrate browning reactions, as well as lysinoalanine and D-amino acids formed at high pH.

Soybean proteins are used in human foods in a variety of forms, including infant formulas, flours, protein isolates and concentrates, and textured fibers. Soy foods include cheese, drinks, miso, tempeh, tofu, salami and vegetarian meat substitutes. New soy foods are continually being developed. Consumption of soy foods is increasing because of reported beneficial effects on nutrition and health. These effects include lowering of plasma cholesterol, prevention of cancer, diabetes, and obesity, and protection against bowel and kidney disease.

11.1.3.1 Soybeans and Legumes Legumes have been cultivated for thousands of years, although many of the varieties of beans and peas that are commonplace today were unknown until relatively recent times. Legumes have played an important role in the traditional diets of many regions throughout the world. In fact, bean intake has actually declined during the past century in many European countries (Hellendoorn, 1976). In the USA, the availability of dry beans, peas, nuts and soybeans combined has remained fairly constant at 7.3 kg (16 lb), 7.3 kg (16 lb) and 8.2 kg (18 lb) per person per year during the time periods 1909–1913, 1967–1969 and 1985, respectively (CDH, 1989). The most popular dried bean in the USA is the pinto bean, followed by the navy, kidney, great Northern and lima bean (annual kg per person for 1995: 1.5 (3.3 lb), 0.8 (1.7 lb), 0.3 (0.6 lb), 0.2 (0.4 lb) and 0.1 (0.2 lb), respectively) (CAD, 1997).

Beans tend to have a poor image and one that stands in stark contrast to the nutritional value they offer. Beans have been called the 'poor man's meat', a metaphor which is consistent with the inverse relation between bean intake and income. For US males aged 20 years, the frequency of bean intake during a 3-day period was 36.3%, 32.3% and 25.7% among men with incomes <131%, 131–350%, and >350%

of the poverty level, respectively. Given the important role of beans in populations that consume plant-based diets, it is not surprising that legume intake is higher in vegetarians than in non-vegetarians, although the data are limited (Ridout *et al.*, 1988). Certainly, one would expect the consumption of beans to increase with the elimination of meat and eggs from the diet by lacto vegetarians and vegans. Appropriately, the vegetarian food guide pyramid, recently developed by Loma Linda University, places legumes in their own group at the bottom of the pyramid (Whitten *et al.*, 1997). Beans have long been recognized for their protein content and more recently have been noted for their soluble-fiber content, but in general there has been relatively little research and discussion about the nutritional attributes of legumes. The glaring exception to this is the soybean, which has been investigated intensively during the past 5–10 years. This is largely because soybeans are a unique dietary source of a group of phytochemicals called isoflavones. Isoflavones are thought to exert a myriad of biological effects and it has been hypothesized that they reduce the risk of a number of chronic diseases.

11.1.4 Nutrient Composition Soyabean

11.1.4.1 Protein The protein content of beans is generally between 20% and 30% of energy. A serving of beans (<90 g or 1/2 cup cooked beans) provides <7–8 g protein or <15% of the recommended dietary allowance (RDA) of protein for a 70-kg adult. Although legumes are recognized as being high in protein, the quality of bean protein is often underestimated. This is because the protein-efficiency ratio, which is based on the growth of laboratory animals (most commonly rats), was the standard method of evaluating protein quality until recently. Rats have a methionine requirement that is <50% higher than that of humans (Sarwar *et al.*, 1989). Consequently, because bean proteins are relatively low in sulphur amino acids (SAAs), the protein-efficiency ratios of beans are quite low (Sarwar *et al.*, 1985). However, the World Health Organization (WHO) and the US Food and Drug Administration have adopted an alternative method for evaluating protein quality, called the protein digestibility corrected amino acid score (PDCAAS) (FDA, 1991). This method uses the amino acid score (based on the Food and Agriculture Organization estimated amino acid requirement for 2–5-year-old children) and a correction factor for digestibility to arrive at a value for protein quality. The PDCAASs of most beans are reasonably good, although their overall value is reduced somewhat by their lower digestibility. Some types of soy protein products have PDCAASs of close to 1, the highest value possible. Some concerns have been raised about the use of the PDCAAS, but it certainly represents an improvement over the protein efficiency ratio. Ironically, the relatively low SAA content of beans may actually provide an advantage in terms of calcium retention. The reported hypercalciuric effect of protein is likely to be at least partially due to the metabolism of SAAs.

The skeletal system serves as one of the main buffering systems in the body; as a result, the hydrogen ions produced from the metabolism of SAAs cause demineralization of bone and excretion of calcium in the urine (Remer and Manz, 1994). Thus, bean protein may improve calcium retention relative to animal and grain proteins. In general, it has been estimated that every gram of protein consumed causes the loss of 1 mg Ca (Kerstetter and Allen, 1989; Anderson *et al.*, 1999). Although this may appear to be a trivial amount, every additional milligram of calcium excreted may markedly increase dietary calcium requirements, because net calcium absorption is substantially less than the average calcium absorption (<30%) from foods. Human studies showed that the consumption of soy protein is associated with a markedly lower urinary calcium excretion compared with the consumption of similar amounts of whey protein or a mixture of animal proteins. With regard to bone health, the nutritional significance of substituting bean protein for animal protein depends on the relative amounts consumed. In general, this process would appear to play a minor role because legume protein, even among populations eating plant-based diets, comprises only a small percentage of total protein intakes. However, the hypocalciuric effect of bean proteins may be important for some individuals, such as those substituting soy protein for animal protein because of its reported hypocholesterolemic effect (Anderson *et al.*, 1999) and athletes using soy protein supplements. It should be noted, however, that not all studies are in agreement about the effects of protein on calcium balance.

11.1.4.2 Fat Most beans are very low in fat, generally containing <5% of energy as fat. The primary exceptions are chickpeas and soybeans, which contain <15% and 47% fat, respectively. The predominant fatty acid in beans is linoleic acid, although beans also contain the n23 fatty acid, a-linolenic acid (USDA, 1998). However, because the overall fat content of most beans is so low, the dietary contribution of beans to a-linolenic acid intake is generally minor. As noted, soybeans are high in fat, and the consumption of full-fat soy foods contributes significantly to a-linolenic acid intake. The ratio of linoleic to a-linolenic acid in soybeans is <7.5:1 (a-linolenic acid makes up <7–8% of the total fat) (USDA, 1998). Polyunsaturated fatty acids, especially eicosapentaenoic acid (EPA) and docosahexaenoic acid (DHA), are being studied for their health benefits. Adequate DHA status is particularly important for infants.

11.1.4.3 Micronutrients Beans are an excellent source of folate, which in addition to being an essential nutrient, is thought to reduce the risk of neural tube defects (Daly *et al*, 1995). One serving of beans provides more than half of the current RDA for folate. Beans are also high in iron; 1 serving provides <2 mg. This compares favorably with the iron RDAs of 10 and 15 mg for adult men and premenopausal women, respectively. However, iron bioavailability from legumes is poor and thus their value as a source of iron is diminished. In acute studies, the addition of vitamin C to foods markedly increased non-heme iron absorption, but in longer-term studies, the effects of vitamin C intake on iron absorption and status were much less pronounced. In general, single-meal studies overestimate the effects of both inhibitors and enhancers of non-heme iron absorption. Also, many beans are good sources of calcium, providing on average <50 mg Ca/serving, although there is quite a bit of variation among the legumes. Calcium bioavailability from soybeans and soy foods is good, essentially equivalent to calcium bioavailability from milk, despite the fact that soybean is high in phytate and oxalate (Weaver and Plawecki, 1994).

11.1.4.4 Fiber Beans are an important source of dietary fiber; 1 serving provides 2–4 g of a mix of soluble and insoluble fiber. High fiber, high-bean diets were shown to lower serum cholesterol in hypercholesterolemic individuals (Anderson *et al.*, 1984). In addition, beans have very low glycemic indexes. This has been attributed to many factors including their fiber, tannin and phytic acid contents. Although neither the American Diabetes Association nor the American Dietetic Association endorse the glycemic index as a tool for constructing diets for individuals with diabetes, research published during the past decade makes a persuasive argument that the glycemic index of foods is one factor affecting the overall quality of the diet. In support of this statement are findings from a prospective study showing that women who consumed diets with a high glycemic index were <40% more likely to develop diabetes than those consuming low-glycemic-index diets, even after controlling for several diabetes risk factors. Thus, beans may be a particularly important food for individuals with diabetes and those with an elevated risk of developing diabetes.

11.1.4.5 Nonnutritive Components Beans contain several components that are considered to be antinutrients, such as trypsin inhibitors, phytate (inositol hexaphosphate), oligosaccharides and saponins. More recent information suggests, however, that the antinutrient label may be an over-simplification, especially in the case of oligosaccharides and saponins. Trypsin inhibitors from beans can certainly interfere with protein digestion and in some species of animals do cause pancreatic enlargement and enhance chemically induced pancreatic tumors. However, boiling dry beans generally reduces the trypsin inhibitor content by 80–90% (Rayas-Duarte *et al.*, 1992) and there is little reason to think that the amount of trypsin inhibitors obtained by eating commonly-consumed beans would exert any adverse effects in humans. In contrast to the trypsin inhibitor, the trypsin and chymotrypsin inhibitor (Bowman-Birk inhibitor) found in beans, especially soybeans, has been studied as an anticancer agent. As noted above, phytate is thought to contribute to the poor mineral bioavailability of beans. On average, the phytate concentration in beans is between 1% and 2%. Although the effect of phytate in reducing mineral bioavailability in plant foods is an important consideration, it has also been postulated that phytic acid may play a role in reducing cancer risk, possibly because of its antioxidant effects. Specifically, it has been suggested that phytic acid may lower the risk of colon cancer and perhaps breast cancer.

More than 40 years ago, diets containing beans were first shown to markedly increase flatulence (Grant, 1989). In 1970, it was reported that the oligosaccharides in beans were responsible for gas production. The oligosaccharide content of dry beans is <25–50 mg/g. Because there is no a-galactosidase in the human intestinal mucosa to cleave the a-(1–6) galactose linkage present in galactoside-containing oligosaccharides, such as raffinose and stachyose, these oligosaccharides pass into the large intestine where bacteria metabolize them and form large amounts of carbon dioxide, hydrogen and sometimes methane. Because of the discomfort and social embarrassment associated with flatulence, some people opt to avoid beans entirely. Commercial products, such as Beano (AK Pharma Inc, Pleasantville, NJ), a digestive aid that contains a-galactosidase, are available so that individuals can eat beans without discomfort.

Additionally, it is possible to remove substantial amounts of oligosaccharides and to markedly reduce flatulence by changing the water in which beans are boiled one or more times. However, there may be some beneficial effects associated with oligosaccharide consumption. The oligosaccharides, because of their growth-promoting effect on bifidobacteria, have been hypothesized to promote the health of the colon, increase longevity, and decrease colon cancer risk. In fact, for these reasons, researchers in Japan have actually suggested that soybean oligosaccharides be used as a substitute for common table sugar. Saponins are glycosides composed of a lipid-soluble aglycone that consists of either a sterol or, more commonly, a triterpenoid structure attached to water-soluble sugar residues that differ in their type and amount. The major sources of dietary saponins are legumes, and many types of saponins can be present in the same bean.

Saponins are very poorly absorbed. Most saponins form insoluble complexes with 3-b-hydroxysteroids and are known to interact with and form large, mixed micelles with bile acids and cholesterol. Although saponins were shown to lower cholesterol in some animal species, the hypocholesterolemic effects of saponins in humans are more speculative. Saponins may have anticancer properties, as suggested by a recent rodent study that found that a saponin-containing diet (3% by wt) inhibited by about two-thirds the development of azoxymethane-induced preneoplastic lesions in the colon. Isoflavones make up another group of phytochemicals found in beans, but for practical purposes the soybean is the only nutritionally relevant source of these compounds. Soybeans and soy products contain <1–3 mg isoflavones/g protein; 1 serving of traditional soyfoods provides <25–40 mg isoflavones.

Isoflavones have received considerable attention in recent years. They are being studied for their potential role in the prevention and treatment of a number of chronic diseases including certain forms of cancer, osteoporosis and heart disease, and also for their ability to relieve menopausal symptoms. The primary isoflavones in soybeans are genistein (49,5,7-trihydroxyisoflavone) and daidzein (49,7-dihydroxyisoflavone) and their respective b-glycosides, genistin and daidzin (sugars are attached at the 7 position of the A ring). Much lower amounts of glycitein (7,49-dihydroxy-6-methoxyisoflavone) and its glycoside, glycitin, are present in soybeans (Wang and Murphy, 1994). In non-fermented soyfoods, the isoflavones appear mostly as the conjugate, whereas in fermented soy products such as miso, the aglycones dominate. In addition to the isoflavones found in soybeans, the intestinal microflora can convert daidzein into several different products, including the isoflavonoids equol (7-hydroxyisoflavan), dihydrodaidzein and O-desmethylangolensin.

11.2 Soy Products and Human Diet

11.2.1 Nutritionally Balanced Diets

Clearly, a major aim of nutritional and medical sciences is to encourage and promote the consumption of nutritionally balanced diets: balanced diets in terms of total protein, amino acid pattern, fat, carbohydrate, fiber, vitamins, minerals and total calories. Health professionals have at their disposal the Recommended Dietary Allowances (RDA) and requirements published by the US National Academy of Science (1984) and the World Health Organization (FAO/WHO/UNO, 1985) (Table 11.5).

One desirable way to alter typical American diet patterns to meet the above dietary recommendations involves partial replacement of foods of animal origin with cereals and legumes. Moderate substitutions of cereals and legumes in the diet can improve nutritional status with regard to desirable weight, total fat, saturated fatty acids, cholesterol, starch and fiber, i.e. most of the USDA dietary recommendations for Americans. Soy-protein foods fit well into a modified diet. Soy foods contain a good amino acid profile, an array of minerals and vitamins and in full-fat products, a high polyunsaturated fatty acid content. The Food and Nutrition Board (CDA, 1980) or the Food and Agriculture Organization (FAO/WHO/UNO, 1985) demonstrate adequate quantities

Table 11.5 Composition of typical soy protein products (adapted from USDA, 1986)

Nutrient	Soy flour full fat roasted	Soy flour defatted	Soy protein concentrate	Soy protein isolate
Proximate (g)				
Protein (N × 6.25)	38.1	51.5	63.6	88.3
Total Carbohydrate	30.4	33.9	25.4	–
Lipid	21.9	1.0	0.5	3.4
Crude fiber	2.2	4.3	3.8	0.3
Ash	5.9	6.2	4.7	3.6
Minerals (mg)				
Ca	188	241	363	178
Fe	5.8	9.2	10.8	14.5
Mg	369	290	315	39
P	476	674	839	776
Zn	3.6	2.5	4.4	4.0
Vitamins (mg)				
Thiamine	0.41	0.70	0.32	0.18
Riboflavin	0.94	0.25	0.14	0.10
Niacin	3.29	2.61	0.72	1.44
Vit B_6	0.35	0.57	0.13	–
Folacin	0.23	0.31	0.34	0.18

of the essential amino acids histidine, isoleucine, leucine, lysine, phenylalanine plus tyrosine, threonine, tryptophan and valine. Only the combined amount of the essential sulfur amino acids methionine plus cysteine fall below recommended patterns in most soy proteins. Lysine concentration is particularly high in soy protein. Its abundance in this inexpensive legume thrusts soy protein into the forefront of protein nutriture worldwide.

Lysine is limited in most staple sources of plant protein consumed in areas where protein malnutrition is found. Complementation of cereals, such as rice, wheat or corn, with soy protein improves protein status. The use of soy protein as the sole source of protein has been investigated in many studies. Rat bioassays, such as the protein efficiency ratio (PER), generally underestimate the protein quality of soy protein for humans, because the rat has higher relative amino acid requirements for the sulfur amino acids methionine and cysteine, the limiting amino acids of soy protein. A number of human studies have been published, which investigated protein quality of soy products for adults, children and infants. The appropriateness of soy protein for children and infants was evaluated by Torun (1981), who concluded from his literature review that well-processed soy-protein products were comparable with milk in protein quality for pre-school and older children. For full-term infants, it is generally accepted that modern soy-protein-based infant formulas promote growth to the same extent as do cow-milk-based formulas (Fomon *et al.*, 1979). In addition, soy formulas may be less allergenic than cow's milk. Soy-protein-based formulas offer the benefit of being lactose free and provide digestible fat. However, soy-based formula may not be the ideal protein source for these infants.

It can be concluded that, except for premature infants, soy protein can serve as a sole protein source in the human diet. This conclusion is based on the assumption that the soy protein is adequately processed. Heat processing of soy protein is necessary for optimizing digestibility of the protein and reducing heat-labile protein inhibitors. Digestibility of protein refers to the ability of man or animal to hydrolyze proteins with stomach acid and proteolytic enzymes to yield small peptides and free amino acids. Essentially, all food proteins, especially plant proteins, require some heat treatment to denature the protein sufficiently to allow an adequate enzyme cleavage and digestion. Digestibility can be measured in humans by feeding known quantities of protein and measuring faecal losses of N. After adjustment for normal turnover of protein in the intestine, a percent digestibility can be estimated.

Another positive affect of heat treatment of soy protein is the reduction of protease inhibitors. Raw soy proteins, in common with other unheated plant and animal protein (Fomon *et al.*, 1979), contain a family of proteins that bind to proteolytic digestive enzymes and reduce the intestinal digestive process. Although these protease inhibitors inhibit the action of a wide variety of enzymes besides trypsin, they are generically referred to as trypsin inhibitors. Raw potato contains twice the trypsin inhibitor activity of raw soy flour, whereas raw egg contains comparable amounts. Cooked potatoes and hard-boiled eggs contain relatively higher levels than toasted soy flours. Nutritionally complete diets based on raw, unheated soy protein, will produce little to no growth when fed to rats. Proper heating increases growth markedly. It has been estimated that 60% of the increased growth in rats fed heat-processed soy protein is due to the heat-induced improved digestibility of the protein and 40% is due to decreased trypsin-inhibitor activity (Kakade *et al.*, 1973). For monogastric animals, a reduction of 80% of the trypsin-inhibitor activity by heat processing is sufficient to produce optimal protein nutriture and growth.

Additional heat treatment to completely remove trypsin-inhibitor activity may result in overheating, giving rise to adverse chemical changes in the protein and reduced growth. Commercially available soy-protein products are known to contain from 5–20% residual trypsin inhibitor activity. Although this quantity is not considered to be detrimental to growth or protein status, researchers have extensively investigated the effects of residual trypsin-inhibitor activity on the pancreas. In order for there to be pancreas enlargement in humans, there should be a feedback-control system involving intestinal release of the hormone cholecystokinin, which stimulates the pancreas. Whether or not such a control system is in place in humans is controversial (Toskes, 1986). A recent abstract reported that in three of five normal subjects, such a feedback control mechanism could be demonstrated and the preponderance of studies supports its presence. Pancreas enlargement per se is not in itself a major concern for human health. Of ultimate concern is whether prolonged exposure of the human pancreas to trypsin inhibitors from soy and other sources may eventually result in pancreatic cancer. Heat treatment, however, essentially abolishes the potentiating effect of the raw soy protein on chemically-induced pancreatic cancers. It is unrealistic to expect that humans would consume diets high in raw soy flour but humans, especially infants fed soy-based infant formula, vegetarians or those consuming soy protein for its hypolipidemic affects, could be chronically exposed to low levels of trypsin inhibitors (Morgan *et al.*, 1986). The long-term effect of low intake of trypsin inhibitors has not been delineated and is now being investigated in humans.

11.2.2 Lipid Metabolism

Carroll *et al.* (1978) demonstrated that systematic substitution of various vegetable protein for animal proteins in low-fat, cholesterol-free, semi-purified diets generally produced a hypocholesterolemic response. Although soy protein is only one of the hypocholesterolemic proteins of plant origin, this legume protein has received by far the greatest research attention. Carroll *et al.* (1978) found significant reduction in serum cholesterol levels with soy-protein feeding to female volunteers with normal blood lipid levels. In addition, he and his co-workers (Wolfe *et al.*, 1981) fed hypercholesterolemia men either a conventional low-cholesterol diet containing mixed protein from meat, dairy products and plant sources or an all-plant-protein test diet in which the meat and the dairy protein of the former diet was replaced by soy-protein meat analogs and soy milk. Both total serum cholesterol (13%) and serum triglycerides (23%) were significantly depressed in the plant-protein diet. van Raaij *et al.* (1981) evaluated the effects of substituting soy isolate or soy concentrate for casein in diets of healthy volunteers with normal blood lipid levels. They found no significant effect of soy on total serum cholesterol, but reported that soy protein isolate caused a small but significant improvement (increase) in the serum HDL-LDL cholesterol ratio.

11.2.3 Glucose Tolerance

A number of factors affect serum glucose response to a food or meal. They include composition of the food, digestibility of the components of the food, presence of some forms of fiber in the food, physical form of the food and interaction of components of the food (e.g. starch-protein interactions). Legumes were shown to be particularly useful in decreasing blood glucose responses compared with other high-fiber foods. The beneficial effects of legumes are thought to be derived from various components of the beans, including their slow starch digestion and their fiber content. Unlike many legumes, soybeans do not contain significant quantities of starch. However, recent results suggest that soy holds promise for diabetics. Madar (1983), with streptozotocin induced diabetic rats, reported beneficial blood glucose effects of soybean fiber. Eizirik and Migliorini (1985) showed protective and restorative effects of high-protein diets containing soy protein isolate versus casein against the diabetogenic action of streptozotocin-treated rats.

11.2.4 Caloric Reduction

Defatting full-fat soy flour reduced the caloric content of 100 g flour from 439 to 327 kcal. Soy-protein concentrates and isolates also contain 330-kcal/100 g. These defatted products are ideal for extenders or replacements of meat products, which are considerably higher in caloric content when compared on an equivalent protein basis. For example, a 100 g cooked all-beef hamburger contains 300 kcal, 23% fat and 22% protein. A small 20% replacement of beef with rehydrated soy concentrate results in a beef soy product with similar protein content but with a 9% reduction of calories (Miles *et al.*, 1987). Additional calorie savings can be gained by using soy fiber supplementation in appropriate food systems. This can be achieved with a soybean hull fiber or cotyledon fiber source. Mineral bioavailability with an increased utilization of a specific food in the diet, means consideration should be given to the effects that the food may have on the overall nutrition. This becomes a very important concern if that food provides a significant quantity of the calories of a diet or if that food replaces another that contributed important amounts of nutrients to the diet. As discussed earlier, soy protein products provide a good balance of nutrients, including minerals. Of concern, however, is the relative bioavailability of those minerals for humans. It is well recognized that minerals from cereals, legumes and other plant foods, in contrast to minerals from animal sources, are generally poorly-utilized by humans and other monogastric animals (Forbes and Erdman, 1984). In this regard, soy and wheat are probably the most studied of the plant foods. The results from numerous publications demonstrate low lower mineral bioavailability from these two food sources compared with animal protein sources.

11.2.5 Zinc Bioavailability

Zinc bioavailability is an important issue, because consumption of foods containing highly available Zn is the key to adequate Zn nutritive. There are many published reports with experimental animals that point to a general poor bioavailability of Zn from soy-based foods. A study by Ketelsen *et al.* (1984) demonstrated that Zn bioavailability is not low for all soy-protein products. The type of food processing utilized in the preparation of soy products is not benign, but affects mineral bioavailability from the final product. Thus, fortification of soy protein foods should overcome the lower bioavailability of the zinc endogenous to the soy. The Committee on Nutrition, American Academy of Pediatrics (1983), in their examination of soy-protein formulas for children, suggests that Zn absorption appears to be lower from soy formulas than from cow's milk or human milk.

11.2.6 Iron Bioavailability

Fe deficiency is less common in areas where meat is commonly eaten. Fe, like Zn, is poorly absorbed from legumes and cereals. Not only does meat (meat, fish and poultry) contain a portion of its Fe as heme Fe, which is much better absorbed than non-heme Fe, meat also enhances the absorption of non-heme Fe (Taylor *et al.*, 1986). Soy protein per se has received considerable attention since the initial reports of Cook *et al.* (1981) from the University of Kansas Medical Center, showing an inhibitory effect of a variety of different types of soy-protein products on Fe absorption. As pointed out by Lynch *et al.* (1985), the influence of soy protein on Fe balance is not entirely negative, because the absolute quantity of non-heme Fe contributed to a meal by soy is usually high, because most soy products contain considerable amounts of Fe. Lynch *et al.* (1985) found another positive effect of soy: a higher proportion of the highly bioavailable heme Fe from a mixed meat and soy meal was absorbed when soy was included in the meal. Recently, Lykken *et al.* (1987) reported that Fe from soy hulls was highly bioavailable for humans and suggested that soy hulls would be an excellent source of Fe for bread enrichment. From the above, it is easy to see that the prediction of the net Fe absorption from meals containing soy protein is extremely difficult.

The soy-extended meat example points out the complexity of the factors that come into play to affect total Fe absorption from a meal.

11.3 Functionality of Soyabean in Various Food Products

11.3.1 Fermented Products

The main objections to soybean products by some consumers are the associated intrinsic flavor, which has been described as beany, and the phenomenon of flatulence. Indeed, the first benefit of soy fermentation is the reduction of its beany flavor and chalkiness. Furthermore, soybeans are high in phytate and oxalate substances that can block the uptake of essential minerals such as calcium, magnesium, iron, copper and zinc in the intestinal tract. Fermentation can reduce antinutritional factors (whole beans, soy meals or soy flakes) and addition of objectionable flavor and flatulent sugars, namely stachyose and raffinose. This can be achieved by using microbial cultures in the form of mono- and multi-cultures; particularly using lactic acid bacteria. However, it has been reported that acceptability of soy bean products may be enhanced by modification of processing methods. For instance, in soymilk preparation, application of heat, soaking of soybean in ethanol or alkali, and acid grinding are some of modifications in cold-water soy extraction soybeans.

Soy is considered a good substrate for functional foods. Fermentation by probiotics has the potential to reduce the levels of some carbohydrates possibly responsible for gas production in the intestinal system, and increase free isoflavone levels and flavour desirable changes in bacterial populations in the gastrointestinal tract. During fermentation and microbial growth, the functional properties of soy products like other foods are formed as follows: protein is hydrolyzed to amino acids and peptides by proteolytic enzymes, oligosaccharides are hydrolyzed to monosaccharides, and phytic acid degraded to inorganic phosphates. These properties are fundamental physicochemical characteristics, which affect the behaviour of proteins in food systems during processing, manufacturing, storage and preparation (Riaz, 1999). The most well-known soy fermented products are miso, tempeh, soy sauces, Natto and fermented tofu.

Miso, meaning fermented bean paste, is used as a base for soups or as a flavoring. Varieties of miso are rice miso, barley miso and soybean miso. The production of miso starts with rice, barley or soybeans that have been soaked, cooked, cooled and inoculated with a mixture of strains of *Aspergillus oryzae* and *Aspergillus soyae*. The product is fermented and ripened prior to blending and mashing to form the final product (Liu, 1997c). Tempeh is a product produced from the fermentation of dehulled, boiled soybeans by *Rhizopus oligosporus*. This fermentation yields a cake-like product with a clean yeasty odor, covered completely by mycelium. Tempeh is usually a main dish or a meat substitute in a vegetarian or Asian diet. When it is sliced and deep fried, it has a pleasant aroma, crunchy texture and nutty flavor.

Soy sauce is one of the most widely-consumed products in Asia. In Japan, per capita consumption is more than 10 liters per person per year, or more than 30 g per day. There are different types of soy sauces, each having their own particular characteristics and specific manufacturing procedures. The manufacture of traditional soy sauce or shoyu starts with preparation of raw material (whole beans, soy meals or soy flakes) and addition of koji (source of enzymes, containing bacteria, mold and yeast). After addition of salt brine, mashing is done. Fermentation, pressing and refining, pasteurization and packaging are the next steps.

There are two types of fermented tofu: pickled and stinky tofu. Pickled tofu consists of cubes of dried tofu soaked in salt water, Chinese wine, vinegar and minced chilies, or a unique mixture of whole rice, bean paste and soybeans. Stinky tofu is a soft tofu that does not have a fixed formula for starter bacteria. The traditional method for producing stinky tofu is to prepare brine made from fermented milk, vegetables and meat; the brine can also include dried shrimp, amaranth greens, mustard greens bamboo shoots and Chinese herbs. This kind of tofu has strong pungent odor, described by some as rotten and faecal.

11.3.2 Dairy Type Products

The demand for alternatives to dairy products is growing due to problems with intolerance and allergies, desire for vegetarian alternatives and so on, and hence the interest in soy-based foods has developed (Granato *et al.*, 2010). Soybeans provide an alternative source of protein for people who are intolerant to milk protein. Soy protein is highly digestible (92–100%) and contains all essential amino acids. Although relatively low in methionine, it is a good source of lysine. Soy dairy-free products are lactose free and a good source of essential fatty acids. They contain no cholesterol and little or no saturated fat. These alternatives can be good sources of high-quality protein, B vitamins, potassium, iron, dietary fiber and bio-active components, including isoflavones. Nowadays, dairy soy products free of lactose and milk protein, including soy milk, cheese alternatives, yogurt (sogurt), soy sour cream, soy cream cheese and frozen soymilk desserts are available on the market. Many soymilks are fortified with calcium, vitamins A and D, riboflavin, zinc and vitamin B_{12}.

Soy cheese alternatives may be made from tofu or soymilk or a combination of soymilk with tofu and soy protein isolates. Soy cheese alternatives are now available in a number of traditional cheese varieties, such as mozzarella, Cheddar, pepper jack and jalapeño, which can be found pre-sliced, shredded or in blocks. Soy Parmesan cream cheese and sour cream alternatives are also on the market. Tofu, also called soybean curd, is manufactured by adding a coagulant to soymilk extracted from ground soaked soybean. Two types of coagulants (salts and acids) are used commercially. The third type of coagulant, enzymes, is not yet used commercially but shows potential for producing both firm and soft tofu. Depending on the amount of water that is extracted from the tofu curds, fresh tofu can be divided into three main varieties: soft/silken, firm and dried tofu. Extra firm tofu is best used for marinating and cutting into cubes for stir-frying. The softer version is used for desserts or other foods that require wetter consistency. Among the few calcium-fortified foods, tofu provides calcium in concentrations (128 mg/ 100 g) comparable to milk (117 mg/ 100 g) in an acceptable serving-size portion (Weaver, 2002).

11.3.3 Cereal-Based Products

In bakery products, soy protein ingredients are being used for a variety of functional and nutritional reasons. Soy protein can help reduce fat, increase protein content and improve overall baking characteristics of baked goods, such as breads, crackers, doughnuts, cakes, pies, muffins and other bakery items (Riaz, 1999). Soy protein is a rich source of lysine, the first limiting essential amino acid in most cereal proteins.

11.3.4 Meat and Seafood Products

Soy products also are used as a low-cost substitute in meat, fish and poultry products. Soy protein binds, emulsifies and hydrates meat products. It makes meat products juicier, which improves flavor, colour, texture, shelf life, slice-ability and yield. It can be used to lower the percentage of fat in ham, sausage, luncheon meats and hot dogs. Textured soy protein ingredients have been advocated as seafood extenders. Many processors of seafood items, including surimi products, have become interested in soy concentrates because of their water-binding and holding capacity. The fiber-like structure, for example, adds texture and mouthfeel to poultry rolls. The largest area of current domestic food utilization is in emulsified meats (frankfurters) and coarse ground meats.

11.3.5 Beverages

The beverage industry has currently realized the application and marketing benefits of soy proteins. Beverages seem to be an easy way for consumers to add soy protein into their diets. Soy protein concentrates and isolates are an ideal source of highly digestible protein in beverages (Riaz, 1999). Solubility is one of the most important properties of a soy protein when used in a beverage application. Other requirements include that the protein should form a clear and translucent solution that is bland, possesses low viscosity and demonstrates stability over a wide range of pH, ionic strength and temperature conditions. The interactions of soy proteins with water are important in relation to dispersibility, water absorption and swelling, viscosity, gelation and surfactant and swelling, viscosity, gelation and surfactant important functions of soy proteins in beverages (Kinsella, 1979). The viscosity contributed by soy isolate makes it ideal for other nutritious liquid products, such as infant formulas, creamers, milk replacers and spray-dried products.

11.3.6 Daily Intake

According to the Food and Drug Administration (FDA), foods containing soy protein may have heart-healthy benefits of soy protein. The FDA recommends that consumers incorporate 4 servings of at least 6.25 g of soy protein into their daily diet, with a total of at least 25 g of soy protein each day. Numerous clinical investigations have found that daily intake of up to 50 g of soy protein is not only safe, but may also be effective in improving risk factors for chronic disease such as some types of cancer, diabetes and cardiovascular disease. Even if nutraceutical and functional properties of soy are in demand, flavor is still the ultimately drive to consumer liking, and the majority, approximately 33% of US consumers, seek out products containing soy and approximately 31% of consumers are aware of specific health benefits of soy in their diet. On an aided basis, consumers are most aware of the health benefits of soy in relation to weight management (31%), reduced risk of heart disease (27%) and some cancers (20%). Over 34% of consumers said they were aware of the FDA claim that consuming 25 g of soy protein per day reduced the risk of coronary heart disease, and 84% of consumers perceive soy products as healthy (USB, 2009).

11.3.7 Soybean in Meals

Soybeans have been a part of traditional food for the human population in Asian and Eastern parts of the world for several decades. Recently, it has become popular in other parts of the world. They are high in protein, fiber and unsaturated fat, and rich in vitamins and minerals. They also show many anticarcinogenic properties related to the unique benefits of soy isoflavones, phytochemicals which exert biological effects in humans and other animals (de Mejia *et al.*, 2003). In recent years, breakthroughs in food science and processing have made it possible to use soybean ingredients in new ways, creating foods that are familiar to consumers, incorporating parts of the soybean for functional or nutritional purposes. This has greatly expanded the food processing industries to use soybeans and soy-based ingredients.

11.4 Processing and Soyabean Composition

Processing of soybeans, whether by traditional or Western processes, is necessary to produce animal feed ingredients, food for human consumption or functional protein ingredients intended for incorporation into a wide variety of human foods. During the course of soybean processing to change properties, the chemical composition may change as well. In this section, we summarize the compositional changes that occur in trypsin inhibitors, phytate, saponins and isoflavones as a result of processing.

11.4.1 Proteins

The content and properties of oil and protein are the primary determinants of whole soybean value (Table 11.6). Oil and protein are separated from each other early in most Western processing, In some traditional Oriental soy foods, oil and protein remain together, whereas in others they are separated The proteins of soy contribute to the nutritional value of foods and feeds and are responsible for a number of functional characteristics in a variety of foods. Heating is necessary to attain maximum nutritional value and to modify functional properties of soybean protein. Kakade *et al.* (1973) reported that approximately 40% of the growth-depressing effect of the unheated extract was due to the trypsin inhibitors. The remainder of the growth depression was attributed to poor digestibility of the under-natured protein.

11.4.2 Soybean Processing and Trypsin Inhibitors

Although soybean protein products require heat processing to achieve maximum nutritional value, partly through trypsin inhibitor denaturation, trypsin inhibitors also display anticarcinogenic properties (Messina and Barnes, 1991). The predominant trypsin inhibitors in soybeans and derived materials are proteins and they are located, for the most part, within the main storage proteins in the protein bodies of the cotyledon. This being the case, trypsin inhibitors tend to fractionate with the milieu

Table 11.6 Uses of soy flour

Type	Uses
Full-fat flours (40% protein)	Production primarily in Europe and Asia for the baking industry and the production of soy milk
High enzyme flours (52–54% protein)	Increasing mixing tolerance and bleaching in bread; preparation of functional concentrates and isolates
Defatted flours (52–54% protein)	Varied uses requiring a wide range of protein solubilities
Lecithinated flours	Improving water dispersibility and emulsifying capability in baking applications

of storage proteins as soybeans are processed into ingredients and foods. There are exceptions to this generality that will be discussed as they arise. Rackis *et al.* (1986) reviewed the literature throughout 1985 on the protease inhibitor content of some plant foods. Based on 19 varieties and strains, whole soybeans were reported to contain 17–27 mg of trypsin inhibitor/g (Hafez, 1983) according to activity assays, with the results expressed as though all activity resided in the main species, the Kunitz trypsin inhibitor.

Soybeans contain approximately 40% protein. After analyzing 108 strains and cultivars, it was reported that soybean contain 35–123 mg trypsin inhibitor/g of protein. When soybeans are processed into raw defatted flour, a process in which the hulls and oil are removed, none of the trypsin inhibitors are removed but they become enriched in the flour to levels of 28–32 mg/g of flour, which equates to approximately 58 mg trypsin inhibitor/g of protein. Trypsin inhibitors, being proteins, are subject to denaturation and inactivation by heat. Heated or toasted flours are produced having a range of trypsin inhibitor activities, depending upon their intended use. Fully toasted soy flour will have a trypsin inhibitor level of only 8–9 mg/g of flour or approximately 16 mg/g of protein. Soy protein concentrates are produced by washing defatted flours with either aqueous ethanol or acidified water. In these processes, oligosaccharides and other constituents are removed, but most of the protein is not; the protein content is therefore increased to more than 65%.

Depending upon the amount of washing of the precipitated curd, as well as the trypsin inhibitor content of the starting soybeans, isolates contain trypsin inhibitor levels in the range of 1–30 mg/g or, because nearly all of this product is protein, the same range per gram of protein. Oriental and other soybean foods are generally low in the trypsin inhibitor. Soy sauce, produced by enzymic or acidic hydrolysis of a mixture of soybeans and wheat, is low in trypsin inhibitor content at 0.3 mg/g of sample or 3.3 mg/g of protein equivalent. It is unlikely that these values represent activity from the protein inhibitors, but the origin of the measured activity is uncertain. Miso is a product made by fermentation of soaked and steamed soybeans. The value found in miso is of 4 mg/g of sample or 23 mg/g of protein. However, when lipids are removed from miso by extraction, the trypsin inhibitor level falls to 4 mg/g of protein and it will decrease further if the miso is heated. Tofu, which is made from heated soybean milk by coagulation of the proteins with a salt such as calcium sulfate followed by cooling, has a trypsin inhibitor content (9 mg/g of protein) near to that associated with a fully toasted soy flour. Soy infant formula is made from soy protein isolate with the addition of other nutrients.

The trypsin inhibitor level in these products falls in the range of 0.3–3 mg/g of sample or 2–16 mg/g of protein. The Kunitz soybean trypsin inhibitor present was quantified by rocket immuno-electrophoresis and the Bowman-Birk inhibitor was measured by chymotrypsin inhibition. Soy flours were found to contain from 1.1 to 19.6 mg/g and from <0.2 to 4.9 mg/g of Kunitz and Bowman-Birk inhibitors, respectively. Concentrate samples contained a range of <0.5–6.1 mg/g of Kunitz and <0.2–1.0 mg/g of Bowman-Birk inhibitors, whereas soy protein isolate inhibitor contents ranged from <0.5 to 3.6 mg/g and from 0.3 to 2.0 mg/g, respectively, for Kunitz and Bowman-Birk inhibitors. A dehydrated soymilk was found to contain 11.3 mg Kunitz inhibitor/g and 1.9 mg/g Bowman-Birk inhibitor, whereas a wheat-soy pancake mix had 1.0 mg/g of Kunitz and 0.4 mg/g of Bowman-Birk inhibitors. A variety of other soy containing foods were found to have no more than approximately 1% of the quantity of these inhibitors present in raw soy flours. The ratio of Kunitz inhibitor to Bowman-Birk inhibitor ranged between 0.5 and 8.1.

11.4.3 Soybean Processing and Phytic Acid Composition

Phytic acid has long been recognized to interfere with the absorption of minerals, especially zinc. Recent work suggests that phytic acid may have anticarcinogenic properties as well (Messina and Barnes, 1991); consequently, there is renewed interest in this compound. Phytic acid content of soybeans can vary considerably. Field type cultivars, which are the usual items of commerce, fall into the range of 1.0–2.3%. The hulls contain only 0.1–0.5% phytic acids, whereas the hypocotyls contain 0.9%. However, the hulls (8%) plus the hypocotyl (2%) represent only approximately 10% of the seed, thus the bulk of the phytic acid occurs in the cotyledons. The cotyledons contain 1.6% phytic acid, which is localized in the protein bodies and appears to be distributed between soluble protein-phytate salts and insoluble globoid inclusions. Full-fat flour, which is essentially the cotyledon fraction, contains 1.5–1.8% phytic acid. Values for defatted flours, concentrates and isolates likewise are in the range of 1–2% phytic acid, despite considerable additional processing as compared with full-fat flours. Texturized flours and concentrates are also in this range; phytic acid is stable in cooking and probably is not degraded during texturization by extrusion. The phytic acid values fall in the range of 1–3%, except for okara and tempeh, which contain only 0.5–1.2% phytic acid.

11.4.4 Soybean Processing and Saponins Composition

Most listings of soybean antinutritional factors in the past included saponins, although with little or no justification. More recently, hypocholesterolemic and anticarcinogenic effects have been attributed to soybean saponins (Messina and Barnes, 1991). Soybean saponins are complex compounds consisting of nonpolar triterpenoid alcohol aglycones linked to one or more polar oligosaccharides, resulting in molecules with amphiphilic properties similar to detergents. Early studies resulted in confusion about the number and structures of the aglycones, but it is now believed that only three aglycones, soyasapogenol A, B and E, occur in intact saponins.

Other soyasapogenols reported earlier evidently were artifacts of hydrolysis conditions. Five major saponins have been isolated and structurally characterized (Kitagawa *et al.*, 1988). This diversity of structure and composition has also made it difficult to analyze soybean saponins. Soybeans contain 0.1–0.5% saponins; the high end of the range depends on a value determined by analysis of the sapogenols. The cotyledons contain 0.2–0.3% saponins, whereas the level in the hypocotyl approaches 2%. However, the hypocotyl represents only approximately 2% of the seed; hence, it has little effect on the composition of flours and related products that may contain this fraction. The seed coat (hull) contains no saponins. Unheated and heated full-fat flours contain approximately 0.5% saponins; the results indicate that the saponins are heat stable. No saponins were detected in a protein concentrate prepared by aqueous alcohol extraction, as would be expected based on the solubility properties of the saponins. Isolates have been reported to contain 0.8% saponins. Although water soluble, the saponins apparently complex with the proteins and hence are retained with the proteins during the isolation process. Saponin content of several Oriental foods, such as soymilk, yuba, tofu and natto, is in the range of 0.3–0.4%, about the same as in soybeans. Okara (the insoluble residue remaining after preparation of soymilk), miso (fermented soybean paste) and natto (fermented, cooked soybeans) tend to be somewhat lower in saponin content at 0.1–0.3%. With the exception of alcohol-extracted concentrates, the saponins tend to remain with the protein products derived from soybeans. Fermentation, as in the preparation of miso, results in some degradation of the saponins and thus lower saponin content as compared with soybeans.

11.4.5 Soybean Processing and Isoflavones

Although long considered to be antinutritional factors, because of their estrogenic properties, soybean isoflavones are now of great interest because of their anticancer activities (Coward *et al.*, 1993). Soybeans contain two major isoflavones, genistein and daidzein, and a minor one, glycetein. In the seed, the isoflavones are present primarily as 3-glucosides and a portion of the glucosides also is substituted on the C-6 hydroxyl of the glucose by a malonyl group, especially in the hypocotyl, which represents only approximately 2% of the seed. Acetyl derivatives may also be degradation products resulting from decarboxylation of themalonylisoflavones during extraction and working of the extracts. The glucoside forms of the isoflavone predominate. In contrast, the aglycones (genistein, daidzein and glycetein) make up only 2% of the total isoflavones. Soy flour, made by defatting dehulled flakes and grinding, has an isoflavone profile approximating that of soybeans. Granules and textured vegetable protein products contain appreciable quantities of the 6″-Oacetyl derivatives of genistin and daidzin, presumably because of heat treatment during processing, which decarboxylates the malonyl glucosides. Protein isolates contain reduced levels of isoflavones (600–1000 mg/g) as compared with soybeans and flours, as a result of the aqueous processing used during manufacture. A concentrate made by aqueous alcohol extraction is very low in isoflavones (73 mg/g), because the isoflavones are soluble in aqueous alcohol and are thus largely removed during processing.

Additional data on flours, concentrates and isolates were reported by Coward *et al.* (1993), although these authors did not report values for the acetylated or malonylated derivatives. Roasted soybeans are low in the malonyl derivatives but higher in the acetyl forms as a result of the heat treatment. Beverage products approximate the composition for soybeans, except for reduced levels of the malonyl derivatives. Tofu is greatly reduced in isoflavones because of the aqueous processing during manufacture. Tempeh is somewhat lower in isoflavone content than tofu and contains elevated levels of the aglycones formed by enzymatic hydrolysis during fermentation. Similar trends are noted for bean paste (miso) and the other fermented soy foods. Content of isoflavones in Western-style soy foods falls into the range of 43 ug/g for a Cheddar cheese-like product to 386 ug/g for a tempeh burger. The values for these foods are considerably lower than those of soybeans, because soy is only one of several ingredients used, including fat and water (Coward *et al.*, 1993). It is apparent that processing generally does not remove the isoflavones during fermentation but the isoflavones remain. Lowering of isoflavone from soybean protein products and foods occurs, except for concentrates prepared by alcohol extraction. Their form may be altered by heat treatment and by enzyme reactions contents of some soybean foods, which may also result from dilution because of the addition of other ingredients, for example salt in miso.

11.5 Raw Soy and Soybean Inhibitors in Digestive Enzymes of the Pancreas

In lifetime feeding studies of soybean-derived inhibitors, rats developed pancreatic lesions leading to neoplasia or tumor formation. This raises the question of whether it is safe to consume such inhibitors, which are present in all legumes and to lesser extents in cereals and vegetables such as potatoes and tomatoes and in processed soy products. An examination of the reported effects of inhibitors on the pancreases of different animal species outlined below could help provide an answer to this question. McGuiness *et al.* (1987) showed that long-term feeding of raw soy flour to rats results in the development of pancreatic adenoma. To help elucidate the significance of these findings, a related long-term (95-week) feeding study was carried out in this laboratory. Soy flour or diets supplemented with soy or potato trypsin inhibitors caused dose-dependent pancreatic pathologies consisting of nodular hyperplasia and acinar lesions. However, the lesions described in rats were not observed with mice following long-term feeding. Liener and Hasdai (1986) also did not observe tumor formation in the mouse pancreas following the long-term feeding of raw soy flour. In related studies, it was shown that rats that consumed a soybean trypsin inhibitor concentrate for 14 weeks developed pancreatic hyperplasia and/or hypertrophy. Myers *et al.* (1991) found that:

(a) soybean trypsin inhibitor concentrate promoted the growth of azaserine-induced lesions of the rat pancreas to a greater extent than did corn oil;
(b) heat-treated soy protein isolate neither increased azaserine-induced carcinogenesis nor raised plasma cholecystokinin (CCK) levels;
(c) the concentrate alone without azaserine did not initiate pancreatic lesions;
(d) formation of inhibitor-induced rat pancreatic lesions appears to be mediated by CCK; and
(e) the high protein and fat content in some of the diets used in long-term rat feeding studies could be partly or fully responsible for the observed pancreatic lesions.

Studies with hamsters revealed that feeding raw soy-based diets induced short-term trophic effects on the pancreas as evidenced by increases in pancreatic weights, DNA, RNA and protein content. However, the same diets did not potentiate pancreatic cancer in hamsters treated with the cancer initiator N-nitroso-bis (2-oxopropyl) amine (BOP).

11.6 Soybean Inhibitors and Inactivation of Digestive Enzymes

Inhibitors of enzymes such as trypsin, chymotrypsin, carboxypeptidase, elastase and R-amylase appear in many agricultural products including legumes, cereals, potatoes and tomatoes (Friedman *et al.*, 1986). BBI and KTI appear to act as transition-state analogue inhibitors of chymotrypsin and trypsin (Turner *et al.*, 1975). Soy-based diets containing active protease inhibitors depress growth in rats compared to analogous diets utilizing inhibitor-free soybeans. Growth inhibition and the accompanying pancreatic hypersecretion of trypsin and chymotrypsin, pancreatic hypertrophy and pancreatic adenoma are presumably partly due to the antitryptic and antichymotryptic activities of the inhibitors. Toasting soy flour largely prevents pancreatic enlargement and related changes. Although the possible significance of trypsin inhibitors for human health is yet to be resolved, any concern could be avoided by eliminating the inhibitors from the diet. Heat alone does not completely inactivate all inhibitory activity. Whether there are adverse or beneficial effects of residual inhibitors in food is not known. In addition, heat used to inactivate inhibitors may also destroy certain essential amino acids, such as cystine, methionine and lysine. These considerations prompted us to examine a possible synergistic effect of heat and disulfide bond modification on the activity of soybean inhibitors, both in pure form and in soy flour.

11.7 Beneficial Effects of Soy-Containing Diets

11.7.1 Cholesterol-Lowering

Elevated plasma levels of low-density lipoproteins (LDL) and triglycerides present a risk of cardiovascular disease. By contrast, high-density lipoproteins (HDL) are beneficial. A low ratio of LDL to HDL and low plasma triglyceride levels decrease the risk of cardiovascular disease. It has long been recognized that consumption of plant proteins often results in significant lowering of LDL and total cholesterol levels. Thus, Mokady and Liener (1982) found that rats fed soy- and wheat-based diets had lower serum cholesterol concentrations than those on a control casein diet. Total cholesterol was significantly decreased in rats fed five plant proteins (alfalfa, fava, gluten, pea and soy) compared to those fed three animal proteins (casein, lactalbumin and ovalbumin). It has also been recognized that vegetarians generally suffer less from cardiovascular diseases than non-vegetarians. A possible mechanism of the cholesterol-lowering effect of soy protein is its ability to modulate LDL receptor levels in the liver. The animal data on hypocholesterolemic effects of soy proteins have been confirmed by numerous human studies.

11.7.2 Soybean Bowman Birk Inhibitor as an Anticarcinogen

Factors in soy foods that may inhibit tumors in rats include phytoestrogens (diadzein and genistein), deficiency of methionine, and enzyme (protein kinase) inhibitors. Troll *et al.* (1986) proposed possible pathways by which dietary protease inhibitors might contribute to the prevention of human cancer. They may block the formation of active oxygen species by stimulated neutrophils, inhibit tumor promotion, and prevent the digestion of proteins to amino acids, thus depriving rapidly growing cancer cells of essential amino acids. The pioneering studies by these investigators on the reduction of breast and skin tumors in rodents by dietary protease inhibitors stimulated extensive studies on possible beneficial anticarcinogenic effects of BBI in animals and humans. Some of these are briefly outlined below. Evidence from case-control studies suggests that consumption of soy-containing products may protect against breast cancer in women. Whether the protection is due to the interaction of soy isoflavones such as genistein with estrogen receptors or to other factors such as BBI and phospholipids (Foehr *et al.*, 1999) is not known.

The suppression of colon carcinogenesis in mice by BBI may be due to the observed uptake of the inhibitor by intestinal epithelial cells. Such internalization of BBI by colonic cells could facilitate the inhibition of intracellular proteases associated with the transformation of normal to malignant cells. Studies in progress have begun to address whether the reported activity of BBI in cell culture and in animals is relevant to human therapy. The protective effect of BBI in human oral cancer (leukoplakia) may be due to the inhibition of serine proteases, which cleave the neu-oncogene protein, a cell surface biomarker for human cancer. The selective inhibition of serine proteases important for the growth of tumor cells may be a key step in cancer prevention by BBI. The effect of BBI may also be related to its ability to reach intracellular target molecules. Although the biochemical and molecular bases for the beneficial effect of BBI need to be further elucidated, it raises the question of how a protein can prevent cancer. One possibility is that the inhibitor-protease complexes formed *in vivo* can act as free radical traps, whereby the free electrons of hydroxyl or other radicals that can damage DNA are dissipated to sulfur atoms of the cysteine-rich inhibitors or complexes (Ware *et al.*, 1999). The high sulfur amino acid content of the BBI is similar to that of human hair and wool, with one of every five residues being half-cysteine. Another possibility is that the insoluble inhibitor complex acts as an insoluble dietary fiber, physically adsorbing carcinogens during passage through the digestive tract, thus preventing induction of colon cancer.

11.7.3 Soybean Lectins

Liener (1991) discovered a class of bioactive glycoproteins in soybeans, which agglutinate red blood cells, the so-called hemagglutinins or lectins. Although the lectins inhibited the growth of rats, removal of the lectins from a soy extract by passage through a column of concanavalin A did not result in significant improvement in the PER. The lectins are largely inactivated by heat treatment of the soybean meal. However, lectins in other legumes such as black beans, kidney beans and lima beans are not as heat-labile as the soybean lectins (Pusztai *et al.*, 1997). They are also more growth-inhibiting in rats than the soybean lectins. As mentioned earlier, unlike lima bean lectins, soybean lectins do not contain disulfide bonds and their hemagglutinating activity was not affected by SH-containing amino acids. Because soybean lectins have a strong affinity for carbohydrates on cell surfaces, they are used in biochemistry and medicine in a variety of applications. These include fractionation of bone marrow cells, treatment of ulcerative colitis, as a diagnostic reagent for stomach cancer, and to quantify *Bacillus anthracis* microorganisms (Kalinin *et al.*, 1993).

11.8 Traditional Soy-Foods

11.8.1 Tofu

Tofu is a regular part of the diet in many Asian countries and is available across the USA and in most Western nations (Liu, 2004). It is a soft, white and almost cheese-like food favored for its versatility, mild flavor and high nutritional value. Since it is naturally processed, it retains the soybean's important nutrients such as the isoflavones. It is made from soymilk by adding a coagulant (calcium sulfate) to the milk to form curds that are shaped and pressed into cakes. Depending on the coagulant used, tofu is rich in minerals and is an excellent source of high-quality protein (10–15%), polyunsaturated fats viz. linoleic and linolenic acids (5–9%) and B vitamins. Tofu is versatile and nutritious as it can be used in soups, salads, pastries, sandwiches and spreads. It is relatively low in carbohydrates and fiber (as the pulp is removed), making it easy to digest.

11.8.2 Soy Milk

Liquid extract of soybean is used in the preparation of tofu, or as a nutritious beverage. It can be consumed by people who are dairy sensitive (lactose intolerant) or by strict vegetarians who eat no animal proteins. Soymilk is an excellent source of protein, B-vitamins and iron. There are also a number of soymilks available that are fortified with vitamins and minerals, such as β-carotene, calcium, docosahexaenoic acid and omega-3 fatty acid. It has low levels of saturated fat and no cholesterol. The popularity of soymilk has grown in the USA and Europe since the 1980s, when refrigerated soymilks came onto the market.

11.8.3 Green Vegetable Soybeans

Green vegetable soybeans are a simple and nutritious soy food, which is really just the whole soybean picked at its peak of green maturity, at a time when it is high in sucrose and chlorophyll, and has a firm texture. It is harvested in the pod, and sold either in the pod or shucked, after being blanched and frozen. Because they are picked when their sugar levels are high, green vegetable soybeans are very sweet. Green vegetable soybeans contain approximately 13% protein, the same as in tofu, and are naturally high in calcium.

11.8.4 Tempeh

Tempeh is a traditional fermented soy food from Indonesia and is unique in its texture, flavor and versatility. It is made from the whole soybean, which has been de-hulled, cracked and cooked in water with added vinegar to reduce the pH. Once cooked, the soybeans are mixed with the spores of the *Rhizopus oligosporus* fungus and left to incubate for 24 h at around 88°F. Various grains or seeds may be mixed in during processing to vary the taste and texture of the final product. It contains approximately 19% protein, is higher in fiber than tofu, and is a significant source of vitamins and minerals.

11.8.5 Miso

Miso is a rich and flavorsome paste made from fermented and aged whole soybeans, or from soybeans in combination with wheat, barley or rice using fungal starter *Aspergillus oryzae* under specific conditions. Its salty paste is a treasured soup-base and flavoring ingredient used throughout Japan, Korea, Taiwan, Indonesia and China. It has unique medicinal properties and is believed to help reduce the effects of environmental poisons on the body, and contains enzymes and bacteria that can aid in digestion. It is high in protein but contains large amount of sodium, therefore should be consumed sparingly.

11.8.6 Soy Sauce

Soy sauce is the most well-known and popular traditional soyfood, used extensively as a flavoring ingredient in most Asian cooking. There are a variety of soy sauces on the market: tamari (naturally processed) and shoyu (processed with a fermented wheat starter) etc. Much of the soy sauce sold today is made with hydrolyzed vegetable protein (HVP) with added sugar, color and preservatives. All soy sauces are high in sodium due to which it has been precluded to use as nutritional food. From a nutritional prospective, tamari has the highest protein level followed by shoyu and then the HVP-based soy sauce.

11.8.7 Natto

Natto is a whole soybean food (very popular in Japan) that is produced by fermenting small, cooked soybeans with *Bacillus natto* until they develop a sticky, viscous coating. It has a strong taste and aroma and is definitely for the more adventurous eater. It can be found frozen, or fresh, and will last for about a week if refrigerated.

11.8.8 Okara

Okara is the fibrous remains of the soybean after it has been processed to make soymilk. It is very high in moisture content and contains the insoluble carbohydrates and dietary fiber of the soybean, as well some remaining protein and fat. If fully cooked, it is bland in flavour and is an excellent ingredient to add to breads and other baked goods. It is not usually sold in stores, as it is very wet, heavy and highly perishable.

11.8.9 Soy Sprouts

Fresh, crisp soy sprouts of germinated soybeans are sold after having been grown for 5 to 7 days. They are a traditional food of Korea and eaten either raw or in prepared food dishes. They are high in protein and fiber, and contain vitamin C.

11.8.10 Soybean Oil

Soybean oil is produced from soybeans by mechanical or solvent extraction. Crude soybean oil is further filtered to produce salad and cooking oils. In the USA, most of what is sold as 'vegetable oil' in the stores is really pure soybean oil. In other countries, it is often labelled as soybean oil, or soya oil. It is the primary oil used in the food processing industry and is found in vegetable oil shortenings and margarine. Soybean oil is high in polyunsaturated fats, low in saturated fats, and a good source of linoleic and linolenic acids, the omega-3 and omega-6 fatty acids also found in fish oils.

11.8.11 Second-Generation Soy-Foods

Soy-foods manufacturers have been very responsive to consumer demands for convenient, healthful foods. As a result, a wide array of 'second-generation' soy-foods is available in the marketplace today, catering to different tastes and use preferences. Here are some second-generation soy-foods that are currently available.

11.8.12 Soy Nuts

Crunchy nuts have been prepared by dry roasting or oil roasting of whole or split soybeans that have first been soaked in water. They can be sold with salt or other flavoring ingredients added as a coating. Soy-nut butter is a paste of ground soy-nuts that have been prepared in a similar manner to peanut butter, and may have salt, sweeteners and oil added.

11.8.13 Meat Alternatives

Meat alternatives include the category made from tofu, tempeh, textured soy flour, textured soy concentrate, isolated soy protein and wheat gluten. Products may take the shape of burgers, hot dogs, sausages, luncheon meats, ground meat and meatballs. Most products are made with a combination of vegetable protein ingredients to achieve the best texture and are flavored for a particular use.

11.8.14 Cheese Alternatives

Block, sliced, spreadable and grated cheese alternatives may be made from soymilk, tofu or other vegetable protein ingredients. They can be flavored with cheese, for example American cheeses, mozzarella, Cheddar, Monterey jack, Parmesan etc. Most of these products are made with some amount of casein (from cow's milk), which is responsible for the melting action in cheese when heated.

11.8.15 Soymilk Yogurt

Soymilk yogurt is made in the same manner as cow's milk yogurt. by inoculating pasteurized soymilk with cultures like *Acidophilus, Bifidobacterium* etc. It tastes very similar to cow's milk yogurt and is available in a variety of styles and flavors. It is very high in protein, a great source of isoflavones and can be used as it is or in recipes calling for yogurt.

11.8.16 Non-Dairy Frozen Desserts

Non-dairy frozen desserts are produced in the same manner as their dairy counterparts. They may be prepared from a base of soymilk, soymilk yogurt, tofu or isolated soy protein. Some brands, such as the pioneering Tofutti brand, have created a loyal niche for themselves, but the category as a whole still struggles to reach a broad market.

11.9 Source of Various Enzymes having Industrial Significance

Soybean has enormous economic significance due to its high food value, role in oil industries, as well in bio-diesel production. Leguminous plants are a source of various industrial enzymes, particularly found during seed imhibition and germination, as well as those involved in nitrogen metabolism (Dey and Campillo, 1984). A recent approach emphasizes the use of soybean, which is a legume, for extraction of enzymes at the industrial scale, thus helpful for its economic significance. Most importantly, extract prepared from soybean, used for enzyme isolation while leaving behind the residues, is useful for various purposes, i.e. animal fodder, oil extraction, bio-diesel production etc. For various industrial applications, enzymes so extracted from leguminous seeds are immobilized onto suitable matrices, leading to improvement in physico-chemical properties (Kumar *et al.*, 2009).

11.9.1 Cellulases

Cellulases are used commercially for food processing in the coffee and textile industries, and also in laundry detergents, paper and pulp industries (Kemmerer and Tucker, 1994). They are even used for pharmaceutical applications, fermentation of biomass into biofuels, and used as treatment for phytobezoars, a form of cellulose bezoar found in the human stomach.

11.9.2 α- and β-Amylases

Amylases find application in bread-making and in clothing and dishwasher detergents to dissolve starches from fabrics and dishes, and production of sugars from starch, such as in making high-fructose corn syrup (Tripathi *et al.*, 2004).

11.9.3 Proteases

Proteases are used to dissolve gelatin off scrap film, allowing recovery of its silver content, and to remove cloudiness produced during storage of beers. They are used in the manufacture of biscuits, lowering the protein levels of flour (Oh *et al.*, 2004).

11.9.4 Phytases

Phytases have been used as animal feed additives largely for swine and poultry, and to some extent for fish (Hegeman *et al.*, 2001). They help to improve the utilization of phosphate from phytate. Therefore, including adequate amounts of phytase in the diets of simple-stomached animals reduces the need for orthophosphate supplementation of the feed. As a result, the environment is protected from pollution by phosphorus-containing manure, because of the reduced (~50%) faecal phosphate excretion by the animals. There is also a great potential for the use of phytases in processing and manufacturing of food for human consumption, but up to now, no phytase product for relevant food application has found its way onto the market. Technical improvements by adding phytases during food processing have been reported for bread-making, production of plant protein isolates, corn wet milling and the fractionation of cereal bran.

11.9.5 Transglutaminases

Transglutaminases have a significant role in the food industry, used in cold bond meat pieces, attaching bacon to the surface of meat, and improving texture of cheese and yogurt by reducing water loss etc. (Lilley *et al.*, 1998).

11.9.6 Ureases

Ureases are used in the field of diagnosis to determine urea in the blood serum, and to decompose urea in the artificial kidneys (Kumar *et al.*, 2009). In the food industry, urea has been marked as an undesirable substance, particularly in biologically fermented food products such as sake, beer, wine, soy sauce etc. Presence of urea in food products gives a bitter taste, causes discoloring or deterioration of the flavor and lowering of food safety. Urease is being used in treatment of waste water plants, particularly from agriculture, rich in urea, and in secondary and tertiary oil recovery etc.

11.9.7 Peroxidases

Peroxidases are used for treatment of industrial waste waters; i.e. phenols (important pollutants) are oxidized to phenoxy radicals, which participate in reactions leading to production of polymers and oligomers that are less toxic than phenols by the enzyme (Chatfield and Dalton, 1993). Furthermore, peroxidases can be an alternative option to a number of harsh chemicals, eliminating harsh reaction conditions. There are many investigations on the use of peroxidase in many manufacturing processes such as adhesives, computer chips, car parts, and linings of drums and cans.

11.9.8 α-Galactosidases

The α-galactosidase enzyme works in the digestive tract to break down the complex or branching sugars (polysaccharides and oligosaccharides) in foods such as legumes (beans and peanuts) and cruciferous vegetables (cauliflower, broccoli, cabbage, Brussels sprouts, among others). The enzyme breaks down these complex sugars to simple sugars, making these foods more digestible, and reducing intestinal gas. The polysaccharides and oligosaccharides found in these foods otherwise pass through the small intestine unaffected. Once in the large intestine, those sugars may be metabolized by intestinal flora, fermenting to produce the gases that cause discomfort. The approach is moving to using oral α-galactosidase for solving gastric problems (Herman and Shannon, 1985).

11.10 Major Soybean By-Products

11.10.1 Okara and its Uses

Every pound of dry soybeans made into tofu or soy milk generates about 1.1 pounds or 2.5 cups (0:15 gallons) of well pressed, firmly packed okara, containing 76–80% moisture, 20–24% solids, ~23.6–24% protein, 8.1–15.2% fats, and 12.0–14.5% crude fiber (Figure 11.1). Okara is not only a rich source of dietary fiber but also contains higher-quality protein (as measured by PER, protein efficiency ratio) than any other fraction in the tofu-making process (Hackler *et al.*, 1967). Lysine is the first-limiting amino acid in okara. In the majority of poorer countries, okara is used as food. With the growing awareness of the importance of crude fiber in the human diet, okara can be utilized in foods specially formulated with high fiber content. With the addition of suitable flavorings, okara can easily be dehydrated in a drum dryer to make a dry staple. Dried okara can be milled to produce high-fiber soy flour, which could be used in breads. The major uses of okara are described as follows.

11.10.2 Livestock Fodder

Most of the okara from Japanese and American tofu shops and soy dairies is fed to dairy cattle. In China, many makers of tofu, soymilk or yuba run a hog farm adjacent to their shop and use the okara as a primary fodder source. Ruminants (cows, sheep and buffalo) can digest crude fiber (okara) and use it as a nutrient source. Monogastrics (chickens, hogs) cannot utilize nearly as much. While this way of using okara may be most convenient, it is relatively wasteful of nutrients and brings only a low financial return.

11.10.3 Organic Compost

Added to a compost pile or turned directly into the soil, okara adds valuable nitrogen and organic nutrients, while helping to aerate and lighten heavy soils.

11.10.4 Pet Food

Okara makes a good addition to household or commercial dog or cat foods. In Japan, some large factories sell much of their output to pet food manufacturers.

11.10.5 Soysage

Soysage is a tasty sausage-like food, developed by The Farm in Tennessee, is now produced commercially at a number of US tofu shops and soy dairies, and may be the simplest and best way to utilize okara. Packed with nutrition, it can easily be produced in large quantities. To serve, they are sliced into thin rounds and fried in oil on both sides until crisp and golden brown, then served like breakfast sausages (with scrambled tofu), used in sandwiches or pizza, or mashed and used like a vegetable pate. Several tofu shops makes several hundred pounds of soysage each week and it contains okara, whole wheat flour, wheat germ, nutritional yeast, oil, soy sauce, honey, herbs and spices.

11.10.6 Baked Goods

Many tofu shops connected to bakeries use their okara like bran to add natural fiber and protein to their products, and to give a crumbly texture to breads, muffins, brownies (usually made with carob), fudge, cookies (okara coconut macaroons or okara peanut butter cookies), okara and tofu gingerbread, or other baked goods.

11.10.7 Okara Tempeh

Okara tempeh is one of the most delicious and creative ways to use large amounts of okara, and is a popular Indonesian fermented food that consists of okara (alone or mixed with soybeans) bound together into cakes or patties by a fragrant mycelium of *Rhizopus* mold. The concept of a tofu and tempeh shop operating together is most interesting, and several tofu shops in North America are now preparing both okara and soy tempeh.

11.10.8 Okara Party Mix

Develop your own okara party mix using toasted okara, nuts, sunflower seeds, shredded coconut, raisins etc.

11.10.9 Soysage Paté

Soysage paté is delicious served in sandwiches and the ingredients include soysage, tofu, soymilk mayonnaise, mustard, celery, pickles and sunflower seeds. It can be served on onion-herb bread with alfalfa sprouts, tomato and pickle.

11.10.10 Okara and Vegetable Saute

A number of tofu shops in Hawaii and Japan use their okara to prepare okara and vegetable saute, which is sold chilled through delicatessens or natural food stores.

11.10.11 Okara Burgers

Okara burgers are also called okara and grain burgers and are developed by Swan Foods in Miami, Florida. Ingredients include organic brown rice, okara, carrots, onions, garlic, whole-wheat flour, corn oil, sesame seeds, rolled oats and sea salt.

11.10.12 Okara Onchom

Okara onchom is a popular Indonesian fermented food (also spelled ontjom) and is a close relative of tempeh, made and served in the same way but using a different starter culture, *Neurospora*. The finished product has overtones of flavor reminiscent of walnuts or almonds.

11.10.13 Other Food Uses

Okara can be used in salads (with mayonnaise, sliced vegetables and seasonings), in miso as a substitute for part of the whole soybeans, or as a primary ingredient in dry breakfast cereals.

11.11 Tofu Whey and its Uses

Each pound of soybeans used to make tofu generates about 8 pounds or slightly less than 1 gallon of whey. Tofu whey consists of only 1–2 % solids, 59% of which are uncoagulated soy proteins; 9% of the protein in the original soybeans is lost in the whey, together with a large percentage of the soybean's water-soluble B vitamins, lecithin, 60–80% of the oligosaccharide sugars (believed to cause flatulence), 70% of the added coagulant, and 76% of the trypsin inhibitors (Figure 11.1). Rackis *et al.* (1986) reported that if the whey protein is isolated and heated, it obtains a protein quality higher than soybean meal. The high nutritive value of toasted whey proteins are attributed to a good balance of essential amino acids, particularly tryptophan, lysine, methionine and cystine. Thus, in very large-scale tofu production, it might be profitable to extract the whey proteins (as by ultra-filtration) and toast them to produce foods.

Some health and environmental protection agencies may object to large tofu plants discharging extremely large amounts of whey into public sewers or watercourses, since it raises the Biochemical Oxygen Demand (BOD) and therefore causes

Figure 11.1 By-products of soybean.

eutrophication. Large Japanese factories have developed ingenious recycling ponds in which the whey solids are consumed by microorganisms and the clear water filtered for reuse.

11.11.1 Natural Organic Soap

Hot whey cuts oils, fats and grease, and cleans like a mild soap. Tofu whey contains lecithin from the lecithin-rich soybean, the same lecithin used by doctors to cut through another type of fatty substance (cholesterol) in the bloodstream. Most Japanese tofu shops save their whey and use it to wash all tools and utensils before giving them a final rinsing in cold water.

11.11.2 Livestock Fodder

Livestock can digest the oligosaccharides in whey; they love it and it helps them to gain weight.

11.11.3 Organic Fertilizer

When applied to houseplants or gardens, whey adds soil nutrients and stimulates growth.

11.11.4 Fuel Alcohol

A growing number of organic wastes are now being used to charge simple fermenters, where they are fermented to produce fuel alcohol (ethanol). A mixture of okara and whey might make a good fermentation base. The residue could still be used for fodder, or composted.

11.11.5 Soymilk Curds

The process of making fresh soymilk curds into tofu takes time and money. They could be sold as they are, packed in sausage casings, or vacuum packed in bulk at reduced rates to restaurants, schools, hospitals or food processors, to be made into any or a number of other foods such as tofu burgers, salads, dairy-like products, and other items. At tofu shops, curds can be mixed with other ingredients (grated or minced green peppers, carrots, onions, seeds, herbs, spices etc.) to make various types of Veggie Tofu or Veggie tofu cutlets. In Japan, several varieties of five-colored tofu are sold commercially. In America, Swan Foods developed a Caraway Tofu, a relative of dairy cheeses containing caraway seeds. Pontecorvo and Bourne (1978) developed simple methods of extending the shelf life of tofu by adding salt, herbs, chilies and spices to the curds, then making an aged cheese similar to Mexican cotija and it kept for three months or more without decomposition. In China, soymilk curds (called toufu-fa or 'flowers of tofu') are a popular part of the traditional breakfast or a snack food sold from small carts by street vendors. They can be topped with a brown sugar sauce, a sweetened peanut sauce, or soy sauce.

11.11.6 Soybean Hulls or Seed Coats

Soybean hulls or seed coats are often removed in the process of making soymilk. They have long been used as livestock fodder, but recently they have been sold commercially by Daniels Midland as Soy Bran. The hulls contain about 3% of the total protein of the bean and have a surprisingly high PER. To make soy bran, the dried hulls are simply ground to a fine powder, which contains 10% protein, 38% crude fiber and 38% other carbohydrates. It can be used in bread, crackers, cookies, muffins, cereals, dry mixes and snacks.

11.12 Applications of important soybean products

11.12.1 Okara as Source of Dietary Fiber in Functional Food Development

Soybean seeds fiber is supposed to have good nutritional and functional properties. The neutral taste and the absence of colour of such fibers makes them suitable for incorporation into food products without any change in their quality, unlike those from wheat fibers (Surel and Couplet, 2005). The most notable attributes of soybean seeds are their oil and protein content; however, they are also a good source of dietary fiber and isoflavones. The production of soymilk and tofu results in a by-product called okara. In any country, discarding okara as waste is potentially an environmental problem, because okara is highly susceptible to putrefaction. Okara also has a high moisture content (nearly 80%), making it difficult to handle and too expensive to dry by conventional means. In light of its content in dietary fiber, okara is a suitable candidate for nutritional enrichment of cereal-based products. Finding convenient ways to incorporate okara into food could eliminate a possible source of pollution and add economic value to this currently valueless product.

Soybean seeds fiber is supposed to have good nutritional and functional properties. The neutral taste and the absence of colour of such fibers makes them suitable for incorporation into food products without any change in their quality, unlike those from wheat fibers (Surel and Couplet, 2005). Dietary fiber has all the characteristics required to be considered as an important ingredient in the formulation of functional foods. The importance of food fiber has led to the development of a large potential market for fiber-rich products and ingredients and, in recent years, there has been a trend to find new sources of dietary fiber that can be used as ingredients in the food industry. The most common fiber sources are bran from wheat, barley, corn, rice and oats; citrus fruits, grape, apple and sugar-beet fiber; and soybean, peanut, pea and sunflower hull. Many of these sources of dietary fiber have been used in bread-making and other cereal-based products (Anil, 2007).

11.12.2 Okara as Source of Protein in Functional Food Development

Soybean and soy products are a relatively inexpensive source of proteins that are widely recognized for their high nutritional and excellent functional properties (Karki *et al.*, 2009). Alkaline extraction is commonly employed in soy protein isolate production and extractions are generally carried out under pH conditions of 7.5–9 and up to 80°C (Berk, 1992). Acidifying the protein extract to pH 4.5 causes selective recovery of proteins due to their precipitation and subsequent washing of the curd (precipitated proteins) for the removal of non-protein soluble, neutralizing (pH 7) and spray drying produces protein concentrate or isolate. Okara is a by-product obtained during processing of soybean for soymilk, which is either underutilized or unutilized. Wang and Cavins (1989) reported that the ratio of essential amino acid to total amino acids in okara was similar to soymilk and tofu. Although the high quality of protein fraction of okara suggests that Okara Protein Isolate (OPI) could be produced from okara. Ma *et al.* (1997) reported high recovery of protein from okara in both highly acidic and alkaline pH ranges, while at pH values between 3 and 5, the protein recovery was low. At pH 9.0, the amount of protein extracted increased with increase in temperature (25–90°C), but the extract viscosity was found to increase sharply above 80°C.

11.12.3 Production of Natural Cellulose Fibers from Soybean Straw

About 220 million tonnes of soybeans and an equivalent amount of by-products (straw) are produced every year (FAO, 2009). Although soybean straw is an inexpensive, abundant and annually renewable source suitable for producing technical fibers, no previous reports are available on the study of the suitability of using soybean straw for fibers or other industrial applications. Also, there is limited literature available on the composition, structure and properties of the soybean straw and especially on the properties of cellulose in the straw (Galletti, 1991). Using soybean straw as a source of fibers will not only add value to the soybean crops but also provide a sustainable source for fibers. However, it will be necessary to develop appropriate fiber production conditions and understand the structure and properties of the soybean straw fibers in order to utilize the soybean straw for various industrial applications.

The major cereal crops, corn, wheat, rice, soybeans and sorghum, have annual world production of 700, 630, 619, 214 and 58 million tonnes, respectively (FAO, 2009). This means availability of about 2200 million tonnes of by-products that are suitable for fiber production. Even using 10% of the available by-products for fiber production and assuming that about 20% of the by-product by weight can be obtained as high-quality technical fibers, more than 40 million tonnes of fibers can be obtained from these by-products every year. The 40 million tonnes of fibers from the currently limited value by-products is substantial since, it is about 57% of the current total annual world consumption of fibers (70 million tonnes).

11.12.4 Recovery of Phytosterols from Waste Residue of Soybean Oil Deodorizer Distillate

Soybean oil deodorizer distillate (SODD) is a by-product of the deodorization step during the refining of soybean oil. In the refining of soybean oil, most bioactive compounds, such as free phytosterols, fatty acid steryl esters (FASEs), tocopherols and squalenes, are concentrated (Verleyen *et al.*, 2001). Furthermore, SODD also contains free fatty acids (FFAs), polycyclic hydrocarbons and the bioactive compounds (Bockisch, 1998). Thus, SODD is a natural source of phytosterols (or fatty acid sterol esters, FASEs), free fatty acids (or fatty acid methyl esters, FAMEs, namely biodiesel), tocopherols and squalenes. Several strategies have been developed to recover and purify these compounds from the deodorizer distillate of vegetable oils. Phytosterols play major roles in several areas, namely in pharmaceuticals (production of therapeutic steroids), nutrition (anti-cholesterol additives in functional foods, anti-cancer properties) and cosmetics (creams, lipstick) (Fernandes and Cabral, 2007). Phytosterols are known to have cholesterol-lowering effects. Especially, many methods have been proposed to recover phytosterols or FASEs from the deodorizer distillate. Torres *et al.* (2007) reported that a two-step enzymatic procedure could be used to obtain FASEs and FAMEs from soybean oil deodorizer distillate. Sun *et al.* (1998) developed a process including esterification, water bathing, alcoholysis, cooling precipitation and distillation to extract tocopherols and sterols from by-products of refined vegetable oil. Sterols, fatty acid methyl ester and tocopherols can be separated efficiently from DD. Although the process has no effect on extracting fatty acid methyl ester from SODD, it is still commonly used in industry.

11.12.5 Production of α-Galactosidase from Soybean Vinasse

Soybean molasses is a by-product in protein-concentrate production. Vinasse is the waste product of ethanol recovery by distillation. Ethanol is produced by the alcoholic fermentation of soybean molasses, an industrial residue, produced concomitantly with the soybean protein-concentrate that results from the extraction of sugars from de-oiled soybean meal. The soybean vinasse contains a considerable amount of residual sugars (oligomers linked by an α-1,6 bond). There are several reports which demonstrate the possibility to produce α-galactosidase from the soybean vinasse with satisfactory yields (7.60 U/mL in shake-flasks and 11.07 U/mL in the fermenter). The enzyme a-D-galactoside galactohydrolase (EC 3.2.1.22), also named a-galactosidase, which catalyzes the hydrolysis of α-1,6-galactosidic bonds, releasing a-D-galactose. These bonds are found in oligosaccharides such as melibiose, raffinose and stachyose, which are associated to flatulence in humans (Suarez *et al.*, 1999). The consumption of soy-derived products has been limited by the presence of these non-digestible oligosaccharides (NDOs), since most mammals, including humans, lack pancreatic α-galactosidase. However, such NDOs can be fermented by gas-producing microorganisms present in the cecum and large intestine, thus inducing flatulence and other gastrointestinal disorders in sensitive individuals (LeBlanc *et al.*, 2004). α-Galactosidases are applied in biocatalytic processes to improve products destined for human nutrition and also as a component in animal diets to increase digestibility and reduce the fermentation of NDOs.

11.12.6 Production of Bio-Ethanol from Soybean Molasses

Soybean molasses is a co-product generated in the production of protein-concentrate soybean meal. The protein concentrate is soy bran with around 70% proteins (in dry basis), obtained by the extraction of sugars from de-oiled soybean meal using a mixture of water/ethanol as the solvent. An important Brazilian soybean-processing company produces 600 tonnes per day of protein-concentrate, generating 220 tonnes per day of molasses (Siqueira *et al.*, 2006). Siqueira *et al* (2006) reported that soybean molasses is an attractive raw material for the production of bio-ethanol by using *Saccharomyces cerevisiae* at laboratory, pilot and industrial scales. Since it provided the necessary nitrogen and magnesium and the appropriate hydrogen balance for fermentation, there was no need for supplementing any molecule additionally or making any pH adjustment. The residue of bio-ethanol production, called soybean vinasse, did not present any environmental problems, since it was employed as a raw material for energy co-generation. Natural energy resources such as petroleum and coal have been consumed at high rates over the last decades. The heavy reliance of the modern economy on these fuels is bound to end, due to their environmental impact and due to the fact that they might eventually run out. Therefore, alternative resources such as ethanol are becoming more important. Bio-ethanol is one of the most important renewable fuels contributing to the reduction of negative environmental impacts generated by the worldwide utilization of the fossil fuels (Cardona and Sánchez, 2007).

11.12.7 Production of Citric Acid from Okara

Large amounts of okara or soy-residue are generated as a by-product during soymilk and tofu production (Matsuo, 1990). This poses disposal problems and adds to existing pollution. Utilization of okara by microbial fermentation has recently been shown to be promising in the production of riboflavin (Kinoshita *et al.*, 1985), lipase (H-Kittikun and Tani, 1986), fructofuranosidase (Hayashi *et al.*, 1992) and single-cell protein (Khare *et al.*, 1994). Khare *et al.* (1995) showed that okara, a cellulosic byproduct of the soymilk and tofu (soypaneer) industry can be utilized for the production of citric acid by solid-state fermentation using a cellulolytic *Aspergillus terreus* and citric-acid producing *Aspergillus niger*. They found that okara, supplemented with ammonium sulphate and fermented by *A. niger* with simultaneous saccharification using *A. terreus* at pH 8.3 and incubation temperature of 30°C, resulted in production of 5.10 g citric acid/ 100 g drysolids (gds) by the 11th day.

11.12.8 Antioxidant Extraction from Soybean By-Products

This content has been retracted at the request of the Editor.

References

Acasio, U.A. (1997) Handling and storage of soybean and soymeal. In: *Technical Bulletin of ASA American Soybean Association*. St Louis, MO.

American Diabeties Association (1994) Nutrition recommendations and principles for people with diabetes mellitus. *Diabetes Care*, **17**: 519–522.

Amin, I. and Mukhrizah, O. (2006) Antioxidant capacity of methanolic and water extracts prepared from food-processing by-products. *Journal of the Science of Food and Agriculture*, **86**: 778–784.

Anderson, J.W., Story, L., Sieling, B. and Chen, W-J.L. (1984) Hypocholesterolemic effects of high-fibre diets rich in water-soluble plant fibres. *Journal of the Canadian Dietetic Association*, **32**: 140–145.

Anderson. J.W., Smith, B.S. and Washnock, C.S. (1999) Cardiovascular and renal benefits of dry beans and soybean intake. *American Journal of Clinical Nutrition*, **70**: 464S–474S.

Anil, M. (2007) Using of hazelnut testa as a source of dietary DF in bread-making. *Journal of Food Engineering*, **80**: 61–67.

Berk, Z. (1992) Isolated soybean protein. In: *Technology of Production of Edible Flours and Protein Products from Soybeans* (ed. Z. Berk). Technicon, Israel Institute of Technology, Haifa, pp. 83–96.

Biranjan, J.R., Ingole, D.R. and Sanyal, S.R. (1982) Utility of soymilk in dietotherapy of under-nourished children. *Indian Medical Gazette*, **9**: 313–315.

Bockisch, M. (1998) *Fats and Oils Handbook*. American Oil Chemists Society, Champaign, IL. pp. 613–718.

CAD (Commercial Agricultural Division) (1997) Economic Research Service, US Department of Agriculture. *Dry Edible Beans: US per capita use for selected classes, 1970–1997*. US Government Printing Office, Washington, DC.

Cardona, C.A. and Sánchez, O.J. (2007) Fuel ethanol production: Process design trends and integration opportunities. *Bioresource Technology*, **98**: 2415–2457.

Carroll, M.C., Giovannetti, P.M., Moase, O., Roberts, D.C.K. and Wolfe, B.M. (1978) Hypocholesterolemic effect by substituting soybean protein for animal protein in the diet of healthy young women. *American Journal of Clinical Nutrition*, **31**: 1312–1321.

CDA (Committee on Dietary Allowances) (1980) Food and Nutrition Board, National Research Council. *Recommended Dietary Allowances*, 9th edition. National Academy Press, Washington DC.

CDH (Committee on Diet and Health) (1989) National Research Council. *Diet and Health: Implications for reducing chronic disease risk*. National Academy Press, Washington DC.

Chatfield, M. and Dalton, D.A. (1993) Plant Gene Register, ascorbate peroxidase from soybean root nodules. *Plant Physiology*, **103**: 661–662.

Cook, J.D., Morck, T.A. and Lynch, S.R. (1981) The inhibitory effect of soy products on nonheme iron absorption in man. *American Journal of Clinical Nutrition*, **34**(12): 2622–2629.

Coward, L., Barnes, N.C., Setchell, K.D.R. and Barnes, S. (1993) Genistein, Daidzein and their beta-Glycoside conjugates: Antitumor isoflavones in soybean foods from American and Asian diets. *Journal of Agricultural Food Chemistry*, **41**(11): 1961–1967.

Daly, L.E., Kirke, P.N., Molloy, A., Weir, D.G. and Scott, J.M. (1995) Folate levels and neural tube defects: Implications for prevention. *JAMA*, **274**(21): 1698–702.

Danji, F. (2000) Recent progress in research and technology for processing and utilization of soybean In: *Processing of Third International Soybean Processing and Utilization Conference (ISPUC-111)*. Tsukuba, Tokyo, pp. 11–16.

de Mejia, E.G., Bradford, T. and Hasler, C. (2003) The anticarcinogenic potential of soybean lectin and lunasin. *Nutrition Reviews*, **61**(7): 239–246

Dey, P.M. and Campillo, E.D. (1984) Biochemistry of the multiple forms of glycosidases in plants. In: *Advances in Enzymology and Related Areas of Molecular Biology*, vol. **56** (ed. A. Meister). John Wiley & Sons, Inc., Hoboken, NJ.

Eizirik, D.L. and Migliorini, R.H. (1985) Protection against streptozotocin diabetes by high protein diets: Effect of a vegetable protein source (Soya). *Nutrition Reports International*, **32**: 41–47.

FAO (2009) *Food Outlook – Global market analysis*. Available at http/: www.fao.org

FAO/WHO (1990) *Protein Quality Evaluation FAO/WHO Nutrition Meetings Report Series 51*. FAO/WHO, Rome.

FAO/WHO/UNU (1985) *Expert Consultation: Energy and protein requirements*. World Health Organization, Geneva (WHO technical report, series 724).

FDA (Food and Drug Administration) (1991) Department of Health and Human Services. Food labeling: General requirements for health claims for food. *Federal Register*, **56**: 60537–60566.

FDA (1999) Food labeling, health claim, soyprotein and coronary heart diseases. *Federal Register*, **57**: 699–733.

Fernandes, P. and Cabral, J.M.S. (2007) Phytosterols: Applications and recovery methods. *Bioresource Technology*, **98**: 2335–2350.

Foehr, M.W., Tomei, L D., Goddard, J.G., Pemberton, P.A. and Bothurst, I.C. (1999) Anti-apoptotic activity of Bowman-Birk inhibitor can be attributed to co-purified phospholipids. *Nutrition and Cancer*, **34**(2): 199–205.

Fomon, S.J., Ziegler, E.E., Filer Jr, L.J., Nelson, S E. and Edwards, B.B. (1979) Methionine fortification of a soy protein formula fed to infants. *American Journal of Clinical Nutrition*, **32**(12): 2460–2471.

Forbes, R.M., Parker, H.M., Erdman, J.W. (1984) Effects of dietary phytate, calcium and magnesium levels on zinc bioavailability to rats. *Journal of Nutrition*, **114**(8): 1421–1425.

Friedman, M.E.D. (1986) *Nutritional and Toxicological Significance of Enzyme Inhibitors in Foods: Advances in Experimental Medicine and Biology*. Plenum, New York, pp. 1–572

Galletti, G.C. (1991) *Production and Utilization of Lignocellulosics*, 1st edition. Elsevier Science Publishers, Essex, UK, pp. 339–374.

Gandhi, A.P., Kulkarni, S.D. and Jha, K. (eds) (2008) *Highlights of Achievements of Soybean Processing and Utilization Centre*. Technical Bulletin No: CIAE/2008/142. CIAE, Bhopal-462038, MP, India.

Ghosh, P.K. and Jayas, D.S. (2010) Storage of soybean. In: *The Soybean Botany, Production and Uses* (ed. G Singh). Cabi, London, pp. 247–275

Granato, D., Branco, G.F., Nazzaro, F., Cruz, A.G. and Faria, J.A.F. (2010) Functional foods and nondairy probiotic food development: Trends, concepts and products. *Comprehensive Reviews in Food Science and Food Safety*, **9**(3): 292–302.

Grant, G. (1989) Anti-nutritional factors of soybean: A review. *Progress in Food Nutrition Science*, **13**(3–4): 317–348.

Hackler, L.R., Stillings, B.R. and Polimeni, R.J. (1967) Correlation of amino acid indices with nutritional quality of several soybean fractions. *Cereal Chemistry*, **44**: 638.

Hafez, Y.S. (1983) Nutrient composition of different varieties and strains of soybean. *Nutrition Reports International*, **28**: 1197–1206.

Hayashi, S., Matsuzaki, K., Kawahara, T., Takasaki, Y. and Imada, K. (1992) Utilization of soybean residue for the production of fructofuranosidase. *Bioresources Technology*, **41**: 231–233.

Hegeman, C.E., Elizabeth, A. and Grabau, E.A. (2001) A novel phytase with sequence similarity to purple acid phosphatases is expressed in cotyledons of germinating soybean seedlings. *Plant Physiology*, **126**: 1598–1608

Hellendorn, E.W. (1976) Beneficial physiologic action of beans. *Journal of the American Diet Association*, **69**: 248–253.

Herman, E.M. and Shannon, L.M. (1985) Accumulation and subcellular localization of α-galactosidase-hemagglutinin in developing soybean cotyledons. *Plant Physiology*, **77**: 886–890.

H-Kittikun, A. and Tani, Y. (1986) Preliminary study on microbial production of lipase using soybean residue from tofu-making process. *Annual Report ICBiotech*, **9**: 275–276.

Hou, H.J. and Chang, K.C. (1998) Yield and quality of soft tofu as affected by soybean physical damage and storage. *Journal of Agricultural Food Chemistry*, **4**(12): 4798–4805.

Hymowitz, T. and Harlan, J.R. (1983) Introduction of soybeans to North America by Samuel Bowen in 1765. *Economic Botany*, **37**: 371–379.

Kakade, M.L., Hoffa, D.E. and Liener, I.E. (1973) Contribution of trypsin inhibitors to the deleterious effects of unheated soy beans fed to rats. *Journal of Nutrition*, **103**: 1772–1778.

Kalinin, N.L., Shakhalina, K.L. and Kuljakina, M.N. (1993) Quantitation of *Bacillus anthracis* by using of soybean agglutinin conjugates. *Molekuliarnaia genetika, mikrobiologiia i virusologiia*, **6**: 27–29.

Karki, B., Lamsal, B.P., Jung, S., Leeuwen, J., Pometto, A.L., *et al.* (2009) Enhancing protein and sugar release from defatted soy flakes using ultrasound technology. *Journal of Food Engineering*, **96**: 270–278.

Kemmerer, E.C. and Tucker, M.L. (1994) Comparative study of cellulases associated with adventitious root initiation, apical buds, and leaf, flower, and pod abscission zones in soybean. *Plant Physiology*, **104**: 557–562.

Kerstetter, J.E and Allen, L.H. (1989) Dietary protein increases urinary calcium. *Journal of Nutrition*, **120**: 134–136.

Ketelsen, S.M., Stuart, M.A., Weaver, C.M., Forbes, R.M. and Erdman, J.W. (1984) Bioavailability of zinc to rats from defatted soy flour, acid-precipitated soy concentrate and neutralized soy concentrate as determined by intrinsic and extrinsic labeling techniques. *Journal of Nutrition*, **14**: 536–542

Khare, S.K., Jha, K. and Gandhi, A.P. (1994) Single-cell protein production by solid-state fermentation of okara (soy-residue). *Pollution Research*, **13**: 417–420.

Khare, S.K., Jha, K. and Gandhi, A.P. (1995) Citric acid production from okara (soy-residue) by solid-state fermentation. *Bioresource Technology*, **54**: 323–325.

Kim, J.A., Jung, W.S., Chun, S.C., Yu, C.Y., Ma, K.H. and Gwag, J.G. (2006) A correlation between the level of phenolic compounds and the antioxidant capacity in cooked-with-rice and vegetable soybean (*Glycine max* L.) varieties. *European Food Research and Technology*, **224**: 259–270.

Kinoshita, S., H-Kittikun, A. and Pinthong, R. (1985) Production of riboflavin from waste of tofu (soybean curd). *Annual Report IC Biotechology*, **8**: 322–324.

Kinsella, J.E. (1979) Functional properties of soy proteins. *Journal of the American Oil and Chemists' Society*, **56**: 242–249.

Kitagawa, I., Wang, H.K., Taniyama, T. and Yoshikawa, M. (1988) Saponin and Sapogenol. XLI. Re-investigation of the structures of soyasapogenols A, B and E, Oleanene-Sapogenols from soybean, structures of soyasponins I, II and III. *Chemical and Pharmaceutical Bulletin*, (Tokyo), **36**: 153–161.

Kong, F., Chang, S.K., Liu, Z. and Wilson, L.A. (2008) Changes of soybean quality during storage as a related to soymilk and tofu-making. *Journal of Food Science*, **73**(3): S134–S144.

Kong, K.W., Amin, I., Tan, C.P. and Rajab, N.F. (2010) Optimization of oven drying conditions for lycopene content and lipophilic antioxidant capacity in a by-product of the pink guava puree industry using response surface methodology. *LWT – Food Science and Technology*, **43**(5): 729–735.

Kumar, S., Dwevedi, A. and Kayastha, A.M. (2009) Immobilization of soybean (*Glycine Max*) urease on alginate and chitosan beads showing improved stability: Analytical applications. *Journal of Molecular Catalysis B: Enzymatic*, **58**: 138–145.

LeBlanc, J.G., Garro, M.S. and Giori, G.S. (2004) Effect of pH on *Lactobacillus fermentum* growth, raffinose removal, α-galactosidase activity and fermentation products. *Applied Microbiology Biotechnology*, **65**: 119–123.

Liener, I.E. (1991) From soybeans to lectins: A trail of research revisited. *Carbohydrate Research*, **213**: 1–5.

Liener, I.E. and Hasdai, A. (1986) The effect of the long-term feeding of raw soy flour on the pancreas of the mouse and hamster. In: *Nutritional and Toxicological Significance of Enzyme Inhibitors in Foods*. Springer, US, pp. 189–197.

Lilley, G.R., Skill, J., Griffin, M. and Bonner, P.L.R. (1998) Detection of Ca^{2+}-dependent transglutaminase activity in root and leaf tissue of monocotyledonous and dicotyledonous plants. *Plant Physiology*, **117**: 1115–1123.

Liu, K. (1997a) Chemistry and nutritional value of soybean complements. In: *Soybeans: Chemistry, Technology, and Utilization* (ed. K. Liu). Chapman & Hall: New York, pp. 25–113.

Liu, K. (1997b) Non-fermented oriental soyfoods. In: *Soybeans: Chemistry, Technology, and Utilization* (ed. K. Liu). Chapman & Hall: New York, pp. 137–217.

Liu, K. (1997c) Fermented oriental soyfoods. In: *Soybeans: Chemistry, Technology, and Utilization* (ed. K. Liu). Chapman & Hall: New York, pp. 218–296.

Liu, K. (2004) *Soybeans as Functional Foods and Ingredients*. AOCS Publishing, Champaign, IL.

Lykken, G., Hunt, J.R., Nielsen, E.J. and Dintzi, F.R. (1987) Availability of soybean hull iron fed to humans in a mixed, Western meal. *Journal of Food Science*, **52**(6): 1545–1548

Lynch, S.R., Dassenko, S.A., Morck, T.A., Beard, J.L. and Cook, J.D. (1985) Soy a protein products and heme iron absorption in humans. *American Journal of Clinical Nutrition*, **41**: 13–20

Ma, C.Y., Liu, W.S., Kwok, K.C. and Kwok, F.K. (1997) Isolation and characterization of proteins from soymilk residue (okara). *Food Research International*, **29**(8): 799–805.

Madar, Z. (1983) Effects of brown rice and soybean dietary fiber on the control of glucose and lipid metabolism in diabetic rats. *American Journal of Clinical Nutrition*, **38**: 3883–3893.

Mateos-Aparicio, I., Redondo Cuenca, A., Villanueva-Suarez, M. J. and Zapata-Revilla, M. A. (2008) Soybean, a promising health source. *Nutricion Hospitalaria*, **23**: 305–312.

Matsuo, M. (1990). Effect of rice bran on properties of okara tempeh. *Journal of the Agricultural Chemical Society of Japan*, **64**: 1237–1239.

McGuiness, E.E., Morgan, R.G. and Wormsley, K.G. (1987) Fate of pancreatic nodules induced by raw soya flour in rats. *Gut*, **28**: 207–212.

Messina, M. (1995) Modern applications for an ancient bean: Soybeans and the prevention and treatment of chronic disease. *Journal of Nutrition*, **125**: 567S–569S.

Messina, M.A. and Barnes, S. (1991) The role of soy products in reducing risk of cancer. *Journal of the National Cancer Institute*, **83**: 541–546.

Miles, C.W., Bodwell, C.E., Morris, E., Ziyad, J.A., Prather, E.S., *et al.* (1987) Long-term consumption of beef extended with soy protein by men, women and children. I: Study design, nutrient intakes, and serum zinc levels. *Plant Foods for Human Nutrition*, **37**(4): 341–359.

Mokady, S. and Liener, I.E. (1982) Effect of plant proteins on cholesterol metabolism in growing rats fed atherogenic diets. *Annals of Nutrition and Metabolism*, **26**: 138–144.

Morgan, R.G.H., Crass, R.A. and Oates, P.S. (1986) Dose effects of raw soybean flour on pancreatic growth. In: *Nutritional and Toxicological Significance of Enzyme Inhibitors in Foods* (ed. M. Friedman). Plenum Press, New York, pp. 81–91.

Myers, B.A., Hathcock, J., Shiekh, N. and Roebuck, B.D. (1991) Effects of dietary soybean trypsin inhibitor concentrate on initiation and growth of putative preneoplastic lesions in the pancreas of the rat. *Food and Chemical Toxicology*, **29**: 437–443.

Namiki, M. (1990) Antioxidants/antimutagens in foods. *CRC Critical Reviews in Food Science and Nutrition*, **29**(4): 273–300.

Narayan, R., Chauhan, G.S. and Verma, N.S. (1988a) Changes in the quality of soybean during storage. IL Effect of storage on some physico-chemical properties of soybean. *Food Chemistry*, **27**(i): 12–23.

Narayan, R., Chauhan, G.S. and Verma, N.S. (1988b) Changes in the quality of soybean during storage. 2: Effect of soybean storage on the sensory qualities of the products made there–from. *Food Chemistry*, **30**(3): pp. 181–190.

Oh, C.J., Lee, H., Kim, H.B. and An, C.S. (2004) Isolation and characterization of a root nodule-specific cysteine proteinase cDNA from soybean. *Journal of Plant Biology*, **47**: 216–220.

Pontecorvo, A.J. and Bourne, M.C. (1978) Simple methods for extending the shelf life of soy curd (tofu) in tropical areas. *Journal of Food Science*, **43**(3): 969–972.

Pusztai, A., Grant, G., Bardocz, S., Gelencser, E. and Hajos, G. (1997) Novel dietary strategy for overcoming the antinutritional effects of soyabean whey of high agglutinin content. *British Journal of Nutrition*, **77**: 933–945.

Rackis, J.J., Wolf, W.J. and Abaker, E.G. (1986) Protease inhibitors in plant foods: Content and inactivation. In: *Nutritional and Toxicological Significance of Enzyme Inhibitors in Foods*. Plenum Press, New York, pp. 299–347.

Rayas-Duarte, P., Bergeron, D. and Nielsen, S.S. (1992) Screening of heat-stable trypsin inhibitors in dry beans and their partial purification from great Northern beans (*Phaseolus vulgaris*) using a hydrotrypsin-sepharose affinity chromatography. *Journal of Agriculture and Food Chemistry*, **40**: 32–42.

Remer, T. and Manz, F. (1994) Estimation of the renal net acid excretion by adults consuming diets containing variable amounts of protein. *American Journal Clinical Nutrition*, **59**: 1356–1361.

Riaz, M.N. (1999) Soybeans as functional foods. *Cereal Foods World*, **44**(2): 88–92.

Ridout, C.L., Wharf, G., Price, K.R., Johnson, L.T. and Fenwick, G.R. (1988) UK mean daily intakes of saponins – Intestine-permeabilizing factors in legumes. *Food Science and Nutrition*, **142F**: 111–116.

Saio, K., Nikkuni, I., Ando, Y., Osturu, M., Terauchi Y. and Kito, M. (1980) Soybean quality changes during model storage studies. *Cereal Chemistry*, **57**(2): 77–82.

Sarwar, G., Peace, R.W. and Botting, H.G. (1985) Corrected relative net protein ratio (CRNPR) method based on differences in rat and human requirements for sulfur amino acids *Journal of the American Oil and Chemists Society*, **68**: 689–693.

Sarwar, G., Peace, R.W., Botting, H.G. and Brule, D. (1989) Relationship between amino acid scores and protein quality indices based on rat growth. *Plant Foods for Human Nutrition*, **39**(1): 33–44.

Siqueira, P. F., Karp, S.G., Carvalho, J.C., Sturm, W., José, A., *et al.* (2006) Production of bio-ethanol from soybean molasses by *Saccharomyces cerevisiae*. *Bioresource Technology*, **99**: 8156–8163.

Smith, A.K. and Circle, S.J. (1972) *Soybean Chemistry and Technology*. Avi. Westport, CT.

Suarez, F.L., Springfield, J., Furne, J.K., Lohrmann, T., Kerr, P.S. and Levitt, M.D. (1999) Gas production in humans ingesting soybean flour derived from beans naturally low in oligosaccharides. *American Journal of Clinical Nutrition*, **69**(1): 135–139.

Sun, D., Qu, D., Liu, F. and Lu, Z. (1998) A new process to extract tocopherols and sterols from by-product of refined vegetable oil. CN Patent ZL 93100259.

Surel, O. and Couplet, B. (2005) Influence of the dehydration process on active compounds of okara during its fractionation. *Journal of the Science of Food and Agriculture*, **85**: 1343–1349.

Taylor, P.G., Martinez-Torres, C. and Romano, E.L.M. (1986) The effect of cysteine-containing peptides released during meat digestion on iron absorption in humans. *American Journal of Clinical Nutrition*, **43**: 68–71.

Torres, C.F., Torrelo, G., Seňorans, F.J. and Reglero, G. (2007) A two steps enzymatic procedure to obtain sterol esters, tocopherols and fatty acid ethyl esters from soybean oil deodorizer distillate. *Process Biochemistry*, **42**: 1335–1341.

Torun, B. (1981) Soybeans and soy products in the feeding of children. *Journal of the American Oil and Chemists Society*, **58**: 460–464.

Toskes, P.P. (1986) Negative feedback inhibition of pancreatic exocrine secretion in humans. In: *Nutritional and Toxicological Significance of Enzyme Inhibitors in Foods* (ed. M. Friedman). Plenum Press, New York, pp. 143–152.

Tripathi, P., Dwevedi, A. and Kayastha, A.M. (2004) Purification and partial characterization of α-amylase from soybean (*Glycine max*). *Oriental Pharmacy and Experimental Medicine*, **4**; 227–234.

Troll, W., Frankel, K. and Wiesner, R. (1986) Protease inhibitors: Their role as modifiers of carcinogenic processes. In: *Nutritional and toxicological significance of enzyme inhibitors in foods*. Springer US, pp. 153–165.

Turner, R., Liener, I.E. and Lovrien, R.E. (1975) Equilibrium of Bowman-Birk inhibitor association with trypsin and R-chymotrypsin. *Biochemistry.* **14**: 275–282.

USB (United Soy Board) (2009) Consumer attitudes toward soy foods. Available from: http://www.soyfoods. org/wp/wpcontent/uploads/2009/ConsumerAttitud es2009.pdf. Last accessed 12 July 2009.

USDA (United States Department of Agriculture) (1986a) *Composition of Foods*. USDA Handbook, Washington, DC, pp. 8–16.

USDA (United States Department of Agriculture) (1988) *Nutrient Data Research Branch, Nutrition Monitoring Division.* Provisional table on the content of omega-3 fatty acids and other fat components in selected foods. Human Nutrition Information Service. Hyattsville, MD: (Publication HNIS/PT-103).

van Raaij, J.M., Katan, M.B., Hautvast, J.G. and Hermus, R.J. (1981) Effects of casein versus soy protein diets on serum cholesterol and lipoproteins in young healthy volunteers. *American Journal of Nutrition*, **34**(7): 1261–1271.

Verleyen, T., Verhe, R., Garcia, L., Dewettinck, K., Huyghebaert, A. and De Greyt, W. (2001) Gas chromatographic characterization of vegetable oil deodorization distillate. *Journal of Chromatography*, **921**: 277–285.

Wang, H. and Murphy, P.A. (1994) Isoflavone composition of American and Japanese soybeans in Iowa: Effects of variety, crop year, and location. *Journal of Agriculture and Food Chemistry*, **42**: 1674–1677.

Wang, H.L. and Cavins, J.F. (1989) Yield and amino acid composition of fractions obtained during tofu production. *Cereal Chemistry*, **66**: 359–361.

Ware, J.H., Wan, X.S. and Kennedy, A.R. (1999) Bowman-Birk inhibitor suppresses production of superoxide anion radicals in differentiating HL-60 cells. *Nutrition and Cancer*, **33**: 174–177.

Weaver, C.M. and Plawecki, K.L. (1994) Dietary calcium: adequacy of a vegetarian diet. *American Journal of Clinical Nutrition*, **59**(5): 12385–12415.

Weaver, C.M., Heaney R.P., Connor L., Martin B.R., Smith D.L. and Nielsen D.S. (2002) Bioavailability of calcium from tofu as compared with milk in premenopausal women. *Journal of Food Science*, **67**(8): 3144–3147.

Whitten, C., Haddad, E. and Sabate, J. (1997) Developing a vegetarian food guide pyramid: A conceptual framework. *Vegetarian Nutrition*, **1**: 25–29.

Wolfe, B.M., Giovannetti, P.M., Cheng, D.C.H., Roberts, D.C.K. and Carroll, K.K. (1981) Hypolipidemic effect of substituting soybean protein isolate for all meat and dairy protein in the diets of hypercholesterolemic men. *Nutrition Reports International*, **24**: 187–198.

12

Value-Added By-Products from Rice Processing Industries

Kittima Triratanasirichai, Manisha Singh, and Anil Kumar Anal
Asian Institute of Technology, Klong Luang, Pathumthani, Thailand

12.1 Introduction

Rice (*Oryza sativa*) is one of the most common staple foods that people usually consume around the world (Ryan, 2011). Every year, about 600 million tonnes of rice is harvested (Chen *et al.*, 2012). Mostly people consume rice by cooking it (white or brown rice), so a rice milling process is required to remove unwanted material. The main product obtained from the rice milling process is white rice (endosperm), which is about 70%, and other by-products are 20% rice hull, 8% rice bran and 2% rice germ (Figure 12.1) (Wells, 1993; de Deckere and Korver, 1996; Van Hoed *et al.*, 2006). Usually rice hull and rice bran are used as fertilizer or animal feed, with other parts including broken rice, rice germ and rice bran also sometimes applied in beer production. Nowadays, by-products from the rice milling process have started to gain more attention as a functional food because of their high amount of fiber, minerals and vitamins, which might be able to reduce cholesterol levels and have other functional activities (Wilson *et al.*, 2002). Rice is harvested in the form of whole rice kernel. After that, it passes through the drying process for the removal of excess moisture so that its shelf life can be extended. Then the rice milling process is applied. Rice is first exposed to the hull removal process and the outcome is brown rice. Then the outer layer of brown rice is removed and white rice is obtained. The final stage is grading, where broken rice grains are removed (Figure 12.2).

By-products from the rice milling process have high amounts of nutrients when compared to white rice itself. Table 12.1 shows the chemical compositions of various parts of rice by-products. Each rice by-product has different major chemical composition. Rice hull and rice straw both have a very high crude fiber content. But because of their poor digestibility, both are not commonly used in food application, but instead are normally used in livestock feed or crop fertilizer. Rice bran contains the highest amount of fat among all rice by-products, therefore it is commonly used for oil extraction.

Food Processing By-Products and their Utilization, First Edition. Edited by Anil Kumar Anal.

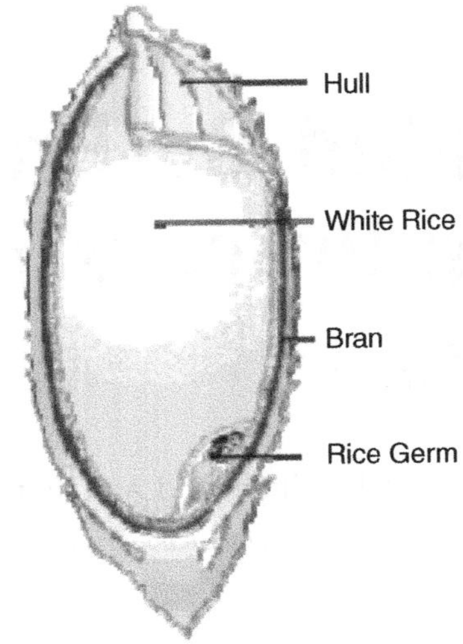

Figure 12.1 Rice kernel compositions.

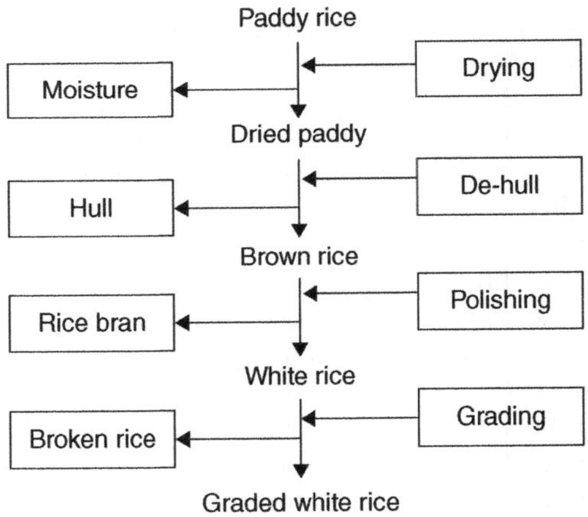

Figure 12.2 Schematic diagram showing rice milling process.

Table 12.1 Chemical compositions of rice by-products (% of dry matter) (adapted from Esa *et al.*, 2013)

Nutrients	Broken rice	Rice husk	Rice bran	Rice polishings	Rice straw
Crude protein (N * 6.25)	6.7–9.8	2.1–3.4	10.6–16.9	11.2–13.4	1.2–7.5
Carbohydrate	–	22.4–35.3	–	51.1–55.0	39.1–47.3
Crude fat	0.5–1.9	0.3–0.9	5.1–19.7	10.1–13.9	0.8–2.1
Crude fiber	0.6	30.0–53.4	7.0–18.9	2.3–3.6	33.5–68.9
Minerals	5.0	13.2–24.4	8.8–28.8	5.2–8.3	12.2–21.4

Broken rice, rice hull and rice straw are the by-products that show the least amount of chemical composition. Broken rice is used in animal feed and the brewing beer industry.

12.2 Rice Bran

Rice bran is rich in protein, fat, fiber, vitamins and minerals. It also contain phenolic compounds and lipophilic antioxidants such as tocopherols, tocotrienols and γ-oryzanol (Min *et al.*, 2011). It was stated in Butsat and Siriamornpun (2010) that rice bran has the highest amount of γ-oryzanol than other parts of rice kernel. Rice bran is mostly used for oil extraction or animal feed. So these compounds can be extracted and utilized in many products with value addition. Rice bran oil is becoming potentially interesting to the food industry, particularly in snack preparation, because rice bran oil has good stability in frying. Nowadays people are becoming more concerned about health and their diets and so households are changing their ways of cooking. Instead of using animal fat or butter, they are using oil that is extracted from plant sources such as soy bean, canola or palm, and rice bran can be another option for people. Meanwhile, rice bran fiber is used as a functional and nutritional ingredient in some snacks, bread or baked food (Sharif, 2009). It not only adds fiber to food product but it also increases protein and lysine content. By adding rice bran to such products, it can help to improve the properties like emulsifying, water absorption and foaming ability (Barber *et al.*, 1981).

12.2.1 Protein and Peptide

Rice bran consists of four types of proteins; albumin, globulin, glutelin and prolamin. They are classified according to their solubility and are water soluble, salt solution soluble, alcohol soluble and alkaline solution soluble, respectively. The additional properties of these proteins are shown in Table 12.2. The protein can be extracted by many methods, such as chemical extraction (alkaline solution), enzymatic extraction

Table 12.2 Protein types found in rice bran

Type	Molecular weight	Solubility	Details
Albumin	10–200 kDa	Water/slightly acidic solution	Precipitates immediately when exposed to heat
Globulin	150 kDa	light *and* medium salt concentration	Hardly dissolved in water and does not dissolved in high salt concentration solution
Prolamin	10–16 kDa	Ethanol (60–80%)	Is a storage protein and highly hydrophobic at the end of the chain
Glutelin	100 kDa	Alkaline solution/ low salt concentration	Storage protein that is the most important in rice, that does not dissolved in neutral medium, alcohol, water and low concentration solution

and physical extraction (high-speed homogenization, high-pressure press, ultrasonic and sonification). Each extraction method has both pros and cons (Fabian and Ju, 2011). Sometimes using a combination of methods can give better extraction. For example, Chittapalo and Noomhorm (2009) studied the extraction of rice bran protein using ultrasonic assisted extraction with alkaline. They found that the time used for extraction decreased and the reaction rate constant increased as the power used in ultrasonic was increased. However, the functional properties like foam and emulsion stability were not significantly different in conventional and ultrasonic extraction.

12.2.2 Protein Extraction Method

Protein extraction using enzymatic method is an effective way to improve functional properties and also to increase the possibility in food application. The proteins and peptides produced by enzymatic hydrolysis have smaller molecular weight than other extraction method. However, due to its high cost, the alkaline extraction method is more commonly used. On the other hand, for a long time fermentation have been known for its ability to produce enzyme, such as in natto, tempeh and other fermented products in many regions worldwide. Therefore, fermentation is another way to extract protein and peptide at much lower costs.

Rice bran protein can be used in many food industries such as in bread (Jiamyangyuen *et al.*, 2005), breakfast cereal, protein supplement, beverages and ingredients in meat and sausages (Prakash and Ramaswamy, 1995). It can be used as an ingredient for infant or child food, because it has essential amino acids that are required by them and it has also been utilized as hypoallergic protein products (Shih, 2003). Moreover, rice bran proteins also have antioxidant and anti-cancer properties (Kawamura and Muramoto, 1993).

The main source of bioactive peptides is protein extracted from animal source such as whey protein (left from the precipitation of casein with acid or enzyme rennin in cheese production), chicken protein and fish muscle protein. Another source is protein from plant sources such as wheat gluten protein, soy protein, sunflower protein and mungbean protein (Megias *et al.*, 2004; Das Neves *et al.*, 2006). All these sources of protein are also the source of bioactive peptides, having functions such as lowering high blood pressure and antioxidant agents. They can inhibit oxidant agents like reactive oxygen species, interact with metal that accelerates the oxidation reaction and act as electron donors in reduction oxidation reactions (Marambe *et al.*, 2008).

According to numerous researches, the factor that affects the properties of protein hydrolysate are sources of protein, type of enzyme used, optimum conditions during hydrolysis (temperature, time and pH) and the concentration of enzyme to substrate. Moreover, the process used to separate molecular size can also effect the structure and quality of the bioactive peptide obtained (Bamdad *et al.*, 2011).

12.2.2.1 Enzymatic Treatment Enzymatic hydrolysis has been one of the important methods to produce bioactive peptides from wheat, soybean, sunflower, rice, barley, corn and buckwheat. Functional properties of a peptide depends on protein source, enzyme used, and conditions in processing (pH, time and temperature) (Wang and Gonzalez de Mejia, 2005). The commercial enzyme used for protein hydrolysis can add value to industrial by-products such as hydrolyzed protein from soy meal using

Alcalase (Wu and Ding, 2002) and hydrolyzed cooking juice of tuna (4% protein) with Orientase (Hwang, 2010). The advantages of enzyme hydrolysis are ease of control, high yield and mild conditions that will not affect the structure of amino acids as compared to hydrolyzation using acid. The benefits of using acid hydrolysis is that it is an easy and cheap method, but have disadvantages like hard-to-control digestion processes and might cause destruction in the structure of amino acid. Therefore, protein hydrolysis using enzyme is commonly used (Wang and Gonzalez de Mejia., 2005). Much research have been done using only one enzyme and a combination of different enzymes, which have resulted in different outcomes, depending on selected enzyme, conditions of hydrolysis and type of protein sample. Venuste *et al.* (2013) studied the hydrolysis of pumpkin meal left from pumpkin oil processing with Alcalase, Flavourzyme, Protamex and Neutrase. They mentioned that all essential amino acids present after hydrolysis met the requirements of the Food and Agriculture Organization of United Nations/World Health Organization (WHO/FAO). The hydrolysis resulted in better antioxidant activity and better ability to suppress the lipid oxidation. The authors mentioned that Alcalase was the most suitable protease in pumpkin protein hydrolysates production with nutritional properties. However, Flavourzyme was best in production of hydrolysates with antioxidant activity. There is also other research that has used different enzymes and whole cell suspension to hydrolyze rice protein. Piu *et al.* (2014) identified that the smaller molecular weight peptide has a higher antioxidant capacity than protein and can be absorbed in the intestine without further digestion.

Rice protein (as by-product) from the rice starch industry was hydrolyzed with enzymes such as Alcalase, Neutrase and Flavourzyme and also by using a microbial strain of *Bacillus spp.* for the whole cell suspension method. The antioxidant activity measured by the ABTS assay indicated the presence of potential antioxidant compound. They stated that the antioxidant activity of protein hydrolysate depends on the enzymes and bacterial strains used. Neutrase resulted in a 30% higher antioxidant activity than Alcalase (Zhang *et al.*, 2009). They also revealed that by treating protein hydrolysate with whole bacterial cells or fermentation and commercial enzymes, functional properties of rice protein can be increased. The author also mentioned that the antioxidant activities are related to the molecular weight of the hydrolysated product. For example, using alcalase and *B. pumilus* produced a low molecular weight peptide with highest antioxidant activity. They also studied rice endosperm that was digested by 5 types of protease. Neutrase appears to be the most desirable in producing high-quality antioxidant peptides from rice endosperm (Zhang *et al.*, 2010).

Alcalase and Protease are the common enzymes used to hydrolyze the sample. Alcalase is one type of protease that is produced by separation and purification from the *B. licheniformis* product. It has the ability to digest protein within the protein molecule. The conditions normally used for Alcalase are in the range of pH 6.5–8.5 at temperatures of 45–60°C. Alcalase has also been used with rice bran protein. Kannan *et al.* (2008) determined the optimum condition of enzymatic hydrolysis to hydrolyze rice bran protein into peptide by the Alcalase enzyme. The optimum condition was obtained at 50°C and pH 8 for 60 min. They also found that peptide derived from rice bran protein hydrolysate possesses a good ability in preventing cancer. Moreover, Hamada *et al.* (1998) studied peptides from rice bran protein hydrolyzed with Alcalase at pH 8.0. Total peptide obtained from this study has 12 fractions, which contained 37% total protein and 57% of amide in hydrolysate.

Van der Ven (2002) reported that protein hydrolysis using protease can help enhance functional properties of protein that have complex structures, are hard to unwind and with low emulsifying properties. However, when the peptide bond was broken, the water molecule interacted with the peptide bond and carboxyl group and also the amount of free amino acids became higher, which lead to more solubility and caused molecular weight to decrease. The structure of globular protein was destroyed and the hydrophobic group was exposed to the outside environment, which lead to better foaming ability and emulsifying properties.

Thawornchinsombut and Kokkaew (2009) studied the optimization of protein hydrolysate production from jasmine rice bran by using protease G6 (endoprotease enzyme produced from *B. licheniformis*). They found that the optimum conditions for extraction are at pH 7.94 and a temperature of 55°C. Protein hydrolysate obtained from optimum conditions showed antioxidant activity of 27.08%, inhibited the ACE enzyme equal to 87.61% and had a degree of hydrolysis (DH) of 17.36%. Another study by Adebiyi *et al.* (2008) on the hydrolysis of rice bran protein, using 5 types of pretease enzyme (from *Aspergillus oryzae, A. melleus, B. subtilis, B. sp.* and pepsin), have shown that rice bran protein hydrolysate also possesses antioxidant properties. Timachai and Thawornchinsombut (2011) used Protease 6G, Protease GN, Flavourzyme 500 MG and Protamex for hydrolyzing defatted rice bran. They found that hydrolysates using Protease 6G showed the highest ACE inhibitory activity.

Cumby *et al.* (2008) extracted protein hydrolysate from canola meal by using Alcalase and Flavournzyme 500 MG and found that protein hydrolyse with Flavournzyme has higher antioxidant activity than the one by Alcalase. However, the combination of both enzymes showed no different result from using Alcalase only. The use of a combination of enzymes has shown some better results. The production of protein hydrolysate from tuna liver by a combination of Alcalase and Neutrase resulted in higher antioxidant activity than using only one commercial enzyme (Je *et al.*, 2009).

12.2.2.2 Fermentation Process Bioactive peptides can be generated through fermentation processes by using either non-starter or starter bacteria. Much research has been reported on producing bioactive peptides with functional properties such as antimicrobial, ACE inhibitory activity, antioxidative and antihypertensive by fermenting dairy products (Matar *et al.*, 2003; Gobbetti *et al.*, 2004; Korhonen and Pihlanto, 2004). Silva *et al.* (2011) also stated that fermentation processes can promote protein hydrolysis in peptide formation and improve the health benefits properties. The main functional property of peptide that is widely studied is ACE inhibitory activity. Angiotensin converting enzyme (ACE) is an important endogenous enzyme in the human body that can increase blood pressure, which is known as hypertension (Oshima *et al.*, 1979; Coates, 2003).

Many researches have used *Bacillus subtilis* as a culture for the fermentation process in order to hydrolyze protein into peptides, because *B. subtilis* can generate protease when fully grown (Muenai, 2009). Wongputtisin (2008) reported that soybean fermented with *B. subtilis* had more soluble peptide than non-fermented soybean. Gallagher *et al.* (1994) found that the amount of peptide hydrolyzed from casein by protease from *B. subtilis* is more than peptides obtained by using bromelain. *B. subtilis* separated from waste water of the animal skin industry is used for fermentation in aerobic bioreactor at 37°C by using casein and gelatin as the culture medium. The protease obtained is alkaline in nature and has optimum enzyme activity at pH 8.5

(Hameed *et al.*, 1996). Chomsri (2001) reported that the protease obtained from *B. spp*. F63.1 has highest activity at pH 7.0. Not only this bacteria, but sometimes other types of microorganisms, have also been used for the synthesis of protease (Bunruang, 2006). Soybean meal was used as a medium for protease production from *A. oryzae* TISTR3019 and indicated best enzyme activity at pH 10 and a temperature of 70°C. Sometimes combining lactic acid bacteria and Flavourzyme can help in accelerating the production of bioactive peptides at a much higher rate than when using only lactic acid bacteria fermentation (Tsai *et al.*, 2007). In 2004, Gibbs *et al.* studied the production and characterization of bioactive peptide from fermented soy products (tempeh and natto). Numbers of bioactive peptides were obtained from soy fermented product after hydrolysis by proteolytic enzymes. It was mentioned that when soy protein is hydrolysed with only exogenous enzyme, and not with *B. subtilis* and *Rhizopus oligosporus*, the production of peptide is less than the fermented one.

Other than providing protease for protein hydrolysis, fermentation also results in peptides with functional properties like ACE inhibitory activity. The ACE inhibitory activity can be related with the amount of peptide extracted (Okamoto *et al.*, 1995; and Kuba *et al.*, 2003, 2004; Wu and Chou, 2009). Quiros *et al.* (2007) used *Enterococcus faecalis* as the bacteria for fermenting milk and after 24 hour, the ACE inhibitory activity corresponded with the peptide content. Figure 12.3 shows the amount of peptide change during the fermentation process. Jakubczyk *et al.* (2013) fermented pea seeds with *Lactobacillus plantarum* 299v and concluded that fermentation can affect both protein and peptide content. After 3 hours of fermentation, the protein content was found to be increasing when compared to non-fermented content. A similar result was obtained by Shekib (1994) during the fermentation of chick pea, lentil, rice and wheat. During the fermentation of tempeh, 7 days of fermentation gave the highest peptide content (Stodolak and Starzynska-Janiszewska, 2008).

He *et al.* (2012) extracted peptides from rapeseed by fermentation with *B. subtilis*. The result indicated that the peptide content increased dramatically within the first 2 days of fermentation, due to the hydrolysis of protein by proteases. They also concluded that the fermentation can help in providing the suitable condition for

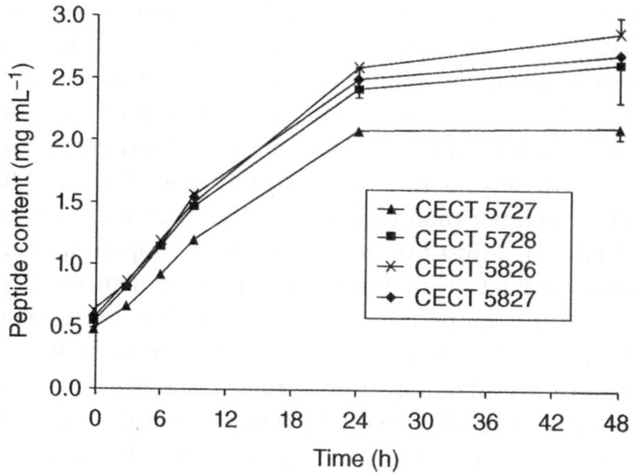

Figure 12.3 Peptide content during milk fermentation with *Enterococcus faecalis* strain CECT 5727, 5728, 5826 and 5827.

production of antioxidant peptides. A similar result was mentioned by Inoue *et al.* (2009), as the total peptide content and ACE inhibition were increased on the third day of miso paste fermentation. Black soybean was fermented with either *B. subtilis* BCRC 14715 or *B. sp.* CN11 at 40°C for 16 hours and the ACE inhibitory activity was found to increase with the increase in fermentation time (Juan *et al.*, 2010). The increasing trend of ACE inhibitory activity during fermentation is also reported by Pyo and Lee (2007). They also stated that *B. spp.* may play a role in the formation of ACE inhibitory substances during black soybean fermentation.

12.2.2.3 Physical Treatment A physical method is generally used as an additional method to enhance the amount of peptide extracted from a sample or to separate peptide fractions from other interference substances or unwanted materials. Sarkar *et al.* (2011) used ion exchange chromatography to purify the extracted peptide by passing it through a cation and anion resin. Once the purification was finished, ultrafiltration was carried out to separate the peptides of different sizes. Liu *et al.* (2010) reported that Porcine plasma (pork blood plasma) hydrolysed by Alcalase has a molecular size of more than 10 kDa and less than 3 kDa. After passing it through ultrafiltration, the protein hydrolysate has a molecular size of less than 3 kDa and exhibits inhibition of DPPH activity. It is said that the smaller molecular weight protein and peptides have antioxidant and antihypertensive activities better than protein and peptide with larger molecular weight (Rajapakse *et al.*, 2005, Ji *et al.*, 2014).

12.2.3 Gamma-Oryzanol (γ-Oryzanol) and Wax

Rice bran has the highest oil content (~12%) among all the chemical compositions. Rice bran oil is mostly saponifiable with a small part of unsaponifiable lipids. The saponifiable part includes mainly triglycerides (68–71%) and the other components such as di- and mono-glycerides, glycolipids, phospholipids, free fatty acids and waxes (Orthoefer, 1996; McCaskill and Zhang, 1999; Patel and Walker, 2004). Rice bran oil also contains various types of vitamins E, such as tocopherols, tocotrienols and γ-oryzanol. The most abundant vitamin E found in rice bran oil is γ-oryzanol. γ-oryzanol content is about 20 times higher than tocopherols and tocotrienols (Bergman and Xu, 2003). γ-oryzanol has been reported to help reduce the level of animal serum-cholesterol and inhibit oxidation of cholesterol (Rong *et al.*, 1997; Akihisa *et al.*, 2000; Xu *et al.*, 2001). The word 'oryzanol' was originally named from *Oryza sativa* by Kaneko and Tsuchiya in 1954, which means rice as it was first discovered and present only in rice bran oil. Oryzanol has very high antioxidant activity, almost 6 times higher than vitamin E. Moreover, it has the ability to reduce bad cholesterol (LDL-C), adjust hormones level in menopausal woman by reducing hot-flush symptoms, protect from ultraviolet light, moisturize the skin, used as an anti-inflammation and more importantly, Japan has used oryzanol for many applications in pharmaceutical, functional food and cosmetic industries.

Gamma-oryzanol consists of mainly trans-ferulic acid esters (trans-hydroxycinnamic acid) and phytosterols (sterols and triterpenic alcohols) such as campesterol, β-sitosterol, cycloartenol and 24-methylenecycloartenol. There are several methods for extraction of γ-oryzanol, such as solid phase extraction, direct solvent extraction, liquid–liquid phase extraction and supercritical fluid extraction. The most commonly-used method is the direct solvent extraction because it is simple, cheap and does not required specific instrumentation. All of these methods require a process where rice bran oil is extracted with hexane many times. Hexane is a very popular solvent for extracting oil from a food source. However, Hu *et al.* (1996) found that the extraction with the mixture of hexane and isopropanol (1:1 v/v) at high temperatures resulted in higher amounts of γ-oryzanol than using only hexane. A similar result was found in the study of Xu and Godber (2000) and Saenjum *et al.* (2012). The most commonly-used methods for determination of γ-oryzanol are high-performance liquid chromatography (HPLC) and ultraviolet (UV) spectrophotometry. The HPLC method is very efficient and accurate, but it is very expensive. However, the spectrophotometry method is cheaper and simpler, but this technique is not as accurate as HPLC, because the absorbance can be affected by many interferences occurring in the sample mixture (Heidtmann-Bemvenuti *et al.*, 2012). The extraction and determination of oryzanol in various varieties of rice bran were studied by Azrina *et al.* (2008). Chloroform and methanol mixture were used to extract oil from the rice bran source. They concluded that the amount of oryzanol obtained would be affected by factors such as variety of rice, solvent used for extraction, the ratio of rice bran to extracting solvent, extraction method applied, temperature of solvent and process that was carried out.

Waxes are the esters that are composed of long-chain high molecular weight carboxylic acids and monohydroxy alcohols. Waxes are the by-products from the oil industry, which is mostly used for cosmetic preparations and partly for pharmaceutical formulations. Recent research indicated that rice bran wax is mainly composed of lignoceric acid and myricyl alcohol. Rice bran wax is an important by-product of the rice bran oil industry, which can be extracted in a cost-effective way.

The amount of rice bran wax extracted depends on the solvent used, the extraction method, temperature and the variety of rice bran. During the oil refining process, several steps are required to remove other impurities or unwanted components from crude oil. These steps include dewaxing, degumming, deacidification, bleaching and deodorization. Dewaxing is the process whereby wax is removed from crude oil by reducing the temperature oil, so that it will solidify and separate by centrifugation or filtration. Rice bran wax can be separated into two parts, hard wax and soft wax. Kim and Godber (2014) compared the extraction of rice bran oil and rice bran wax by soxhlet extraction and microwave-assisted extraction. They concluded that the microwave-assisted extraction method with hexane as the extracting solvent was more efficient in obtaining a higher amount of rice bran wax (Figure 12.4).

Rice bran waxes from rice bran oil have been studied for various applications. Sabale *et al.* (2009) studied the utilization of rice bran wax as an ointment base. Rice bran wax was compared with standard bases in various analyses like washability, spreadability, diffusibility and appearance. The researchers have reported that rice bran wax appears to be shiny with a yellowish-white color. The spreadability of rice bran wax base showed that it is good when compared to the standard base. It has better diffusibility, but less water absorbing capacity and less washability than the standard base. When rice bran wax was applied to pharmaceutical and cosmetic applications, it demonstrated comparable properties with carnauba and other types of waxes. The composition of rice bran wax has not yet been identified and only a small amount of research has been reported (Vali *et al.*, 2004). However, there has been much research on the properties and applications of rice bran wax, and the research should now focus on its composition. It also lacks an efficient purification method, suitable analytical method and has limitations in commercial applications due to the lack of standard

Figure 12.4 Rice bran wax (Ashian Oils Pvt Ltd).

procedure for rice bran wax purification. So better and efficient methods for removing residues, interference and unwanted materials from rice bran wax need to be explored to improve and develop a better-quality of rice bran wax.

12.3 Rice Hull and Rice Bran Fiber

Rice by-products not only contain various types of functional components, but also contain dietary fiber. The fiber can be mostly found in rice hull and the types of fiber present include cellulose, hemicellulose, lignin and hydrated silica (Salanti *et al.*, 2010). These types of fiber can also be found in rice bran and rice straw. However, rice hull, rice bran and rice straw contain different amounts of each fiber (Table 12.3).

It is believed that the outer layer of the rice kernel or rice hull both contain very high amounts of antioxidant compounds, mainly polyphenolic compounds (Asamarai *et al.*, 1996; Lee *et al.*, 2003; Butsat *et al.*, 2009). These polyphenolic compounds can act as antioxidant compounds that can help to protect the rice kernel (endosperm) from oxidation. Various antioxidants are present in the rice husk such as vanillin, phytic acid and flavonoid, but due to its poor digestibility in the human digestive tract, it is used as a fertilizer or in animal feed (Saha and Cotta, 2008). Jeon *et al.* (2006) reported that phenolic compound extracted from rice hull demonstrated very high antioxidant activity on hydrogen peroxide inhibition. Kim *et al.* (2011) also supported this result,

Table 12.3 Fiber compositions of rice hull, rice bran fiber and rice straw

Fiber types (%)	Rice hull	Rice bran fiber	Rice straw
Cellulose	38	30	32.0
Hemicellulose	20	20	35.7
Lignin	22	20	22.3

as they stated that the extracts from Nokmi rice hull showed an effective antioxidant activity.

Since rice bran contains fiber, it is a very good source of dietary fiber (20–51%). Rice bran fiber usually has a laxative effect, so it can help in bowel movement and to loosen the stool. The soluble rice bran fiber is normally applied in food applications to increase the fiber content. It can also decrease blood sugar levels after meals (postprandial glycemia) and in diabetes patient. The soluble fiber will absorb water in the intestine by acting like a sponge. Once it is mixed with food, it will turn food into a gel-like consistency resulting in a slower rate of digestion and absorption (Abdul-Hamid and Luan, 2000). Zumbado *et al.* (1997) used rice bran as a replacement for white flour to develop a traditional bakery product of Cuban Torticas de Moron. They found that replacing 25% of white flour with rice bran resulted in a low fat and high fiber bakery product with acceptable chemical composition, sensory properties and improved shelf life.

12.4 Conclusions

Rice straw, rice hull, broken rice, rice germ, rice bran, rice bran oil and wax are the by-products from the rice industry. These by-products usually have basic applications in their original form, but now can be used as raw materials for different value-added research or in food applications with functional properties. Each by-product can be used in different kinds of applications, depending on their composition, appearance and properties (physical, chemical and functional). Because of the high fiber content in rice hull and rice bran, they are used as ingredients by the bakery industries to increase the fiber content and improve the nutrition of bakery products. Rice bran oil is famous in the household for cooking and frying. It contains several beneficial compounds like vitamin E and γ-oryzanol. Defatted rice bran is also known to have a high protein content. Many methods have been used to extract protein from this source, but fermentation stands out as the better method, as it is cheaper than other methods. And the quality of protein obtained by fermentation is comparable with the other techniques. The major application of broken rice and rice germ is in the brewing or alcohol industries, and not commonly used in other applications. Therefore, more research is required on the other by-products, except for rice bran oil and rice bran, for them to be used with maximum benefit.

References

Abdul-Hamid, A. and Luan, Y.S. (2000) Functional properties of dietary fibre prepared from defatted rice bran. *Food Chemistry*, **68**: 15–19.

Adebiyi, A.P., Adebiyi, A.O., Ogawa, T. and Muramoto, K. (2008) Purification and characterization of antioxidative peptides from unfractionated rice bran protein hydrolysates. *International Journal of Food Science and Technology*, **43**(1): 35–43.

Akihisa, T., Yasukawa, K., Yamaura, M., Ukiya, M., Kimura, Y., *et al.* (2000). Triterpene alcohol and sterol ferulates from rice bran and their anti-inflammatory effects. *Journal of Agricultural and Food Chemistry*, **48**(6): 2313–2319.

Asamarai, A.M., Addis, P.B., Epley, R.J. and Krick, T.P. (1996) Wild rice hull antioxidants. *Journal of Agricultural and Food Chemistry*, **44**; 126–130.

Ashian Oils Pvt. Ltd. Retrieved from: http://www.ashiangroup.co.in/hot-products.htm Last accessed 1 February 2015.

Azrina, A., Maznah, I. and Azizah, A.H. (2008) Extraction and determination of oryzanol in rice bran of mixed herbarium UKMB; AZ 6807: MR 185, AZ 6808: MR 211, AZ6809: MR 29. *ASEAN Food Journal*, **15**(1): 89–96.

Bamdad, F., Wu, J. and Chen, L. (2011) Effects of enzymatic hydrolysis on molecular structure and antioxidant activity of barley hordein. *Journal of Cereal Science*, **54**(1): 20–28.

Barber, S., Benedito-de, B.C. and Martenz, J. (1981) Rice bran proteins: p=Potential value of rice bran fractions as protein food ingredients. *Rev Agroquim Technol Aliment*, **21**: 247–256.

Bergman, C.J. and Xu, Z. (2003) Genotype and environment effects on tocopherols, tocotrienols, and gamma-oryzanol contents of Southern US rice. *Cereal Chemistry*, **80**: 446–449.

Bunruang, R. (2006) Lowering capital cost for protease production from *Aspergillus oryzae* TIST319 by using soy meal as culture media. Faculty of Industrial Technology, King Mongkut's University of Technology North Bangkok, Thailand, Bangkok, p. 72.

Butsat, S. and Siriamornpun, S. (2010) Antioxidant capacities and phenolic compounds of the husk, bran and endosperm of Thai rice. *Food Chemistry*, **119**: 606–613.

Butsat, S., Weerapreeyakul, N. and Siriamornpun, S. (2009) Changes in phenolic acids and antioxidant activity in Thai rice husk at five growth stages during grain development. *Journal of Agricultural and Food Chemistry*, **57**: 4566–4571.

Chen, M.H, Choi, S.H, Kozukue, N., Kim, H.J. and Friedman, M. (2012) Growth-inhibitory effects of pigmented rice bran extracts and three red bran fractions against human cancer cells: relationships with composition and antioxidative activities. *Journal of Agricultural and Food Chemistry*, **60**: 9151–9161.

Chittapalo, T. and Noomhorm, A. (2009) Ultrasonic assisted alkali extraction of protein from defatted rice bran and properties of the protein concentrates. *International Journal of Food Science and Technology*, **44**: 1843–1849.

Chomsri, N. (2001) Thermo-stable protease enzyme. Masters Thesis. Chiangmai University, Thailand, Chiangmai, p. 102.

Coates, D. (2003) The angiotensin converting enzyme (ACE). *The International Journal of Biochemistry & Cell Biology*, **35**: 769–773.

Cumby, N., Zhong, Y., Naczk, M. and Shahidi, F. (2008) Antioxidant activity and water-holding capacity of canola protein hydrolysates. *Food Chemistry*, **109**(1): 144–148.

Das Neves, R.A.M., Campos, T. and Marques, U.M.L. (2006) Modulation of arterial pressure by protein hydrolysates. *Brazilian Journal of Food Technology*, **111(JIPCA)**: 81–86.

de Deckere, E.A. and Korver, O. (1996) Minor constituents of rice bran oil as functional foods. *Nutrition Reviews*, **54**: S120–S126.

Esa, N.M., Ling, T.B. and Peng, L.S. (2013) By-products of rice processing: An overview of health benefits and applications. *Rice Research*, **1**(1): 1–11.

Fabian, C. and Ju, H.Y. (2011) A review on rice bran protein: Its properties and extraction methods. *Critical Reviews in Food Science and Nutrition*, **51**(9): 816–827.

Gallagher, J., Kanekanian, A.D. and Evans, E.P. (1994) Hydrolysis of casein: A comparative study of two proteases and their peptide maps. *International Journal of Food Science and Technology*, **29**: 279–285.

Gibbs, B.F., Zougman, A., Masse, R. and Mulligan, C. (2004) Production and characterization of bioactive peptides from soy hydrolysate and soy-fermented food. *Food Research International*, **37**: 123–131.

Gobbetti, M., Minervini, F. and Rizzello, C.G. (2004) Angiotensin I converting-enzyme-inhibitory and antimicrobial bioactive peptides. *International Journal of Dairy Technology*, **57**: 172–188.

Hamada, J.S., Spanier, A.M., Bland, J.M. and Diack, M. (1998) Preparative separation of value-added peptides from rice bran proteins by high-performance liquid chromatography. *Journal of Chromatography A*, **827**: 319–327.

Hameed, A., Natt, M.A. and Evans, C.S. (1996) Short communication: Production of Alkaline protease by a new *Bacillus subtilis* isolate for use as abating enzyme in leather treatment. *World Journal of Microbiology & Biotechnology*, **12**: 289–291.

He, R., Ju, X., Yuan, J., Wang, L., Girgih, A.T. and Aluko, R.E. (2012) Antioxidant activities of rapeseed peptides produced by solid-state fermentation. *Food Research International*, **49**: 432–438.

Healthy-living.org. Retrieved from: http://healthy-living.org Last accessed 1 February 2015.

Heidtmann-Bemvenuti, R., Nora, N.S. and Badiale-Furlong, E. (2012) Extraction of γ-Oryzanol from rice bran. *Ciência e Agrotecnologia*, **36**(6): 665–673.

Hu, W., Wells, J.H., Shin, T.S. and Godber, J.S. (1996) Comparison of isopropanol and hexane for extraction of vitamin E and oryzanols from stabilized rice bran. *Journal of the American Oil Chemists Society*, **73**: 1653–1656.

Hwang, J.S. (2010) Impact of processing on stability of angiotensin I-Converting enzyme (ACE) inhibitory peptides obtained from tuna cooking juice. *Journal of Food Research International*, **43**(3): 902–906.

Inoue, K., Gotou, T., Kitajima, H., Mizuno, S., Nakazawa, T. and Yamamoto, N. (2009) Release of anti-hypertensive peptides in miso paste during its fermentation by the addition of casein. *Journal of Bioscience and Bioengineering*, **108**: 111–115.

Jakubczyk, A., Karas, M., Baraniak, B. and Pietrzak, M. (2013) The impact of fermentation and in vitro digestion on formation angiotensin converting enzyme (ACE) inhibitory peptides from pea proteins. *Food Chemistry*, **141**: 3774–3780.

Je, J., Lee, K., Lee, M. and Ahn, C. (2009) Antioxidant and antihypertensive protein hydrolysates produced from tuna liver by enzymatic hydrolysis. *Food Research International*, **42**(9): 1266–1272.

Jeon, K.I., Park, E., Park, H.R., Jeon, Y.J., Cha, S.H. and Lee, S.C. (2006) Antioxidant activity of far-infrared radiated rice hull extracts on reactive oxygen species scavenging and oxidative DNA damage in human lymphocytes. *Journal of Medicinal Food*, **9**: 42–48.

Ji, N., Sun, C., Zhao, Y., Xiong, L. and Sun, Q. (2014) Purification and identification of antioxidant peptides from peanut protein isolate hydrolysates using UHR-Q-TOF mass spectrometer. *Food Chemistry*, **161**: 148–154.

Jiamyangyuen, S., Srijesdaruk, V. and Harper, J.W. (2005) Extraction of rice bran protein concentrate and its application in bread. *Songklanakarin Journal of Science and Technology*, **27**(1): 55–62.

Juan, M.Y., Wu, C.H. and Chou, C.C. (2010) Fermentation with *Bacillus* spp. as a bioprocess to enhance anthocyanin content, the angiotensin converting enzyme inhibitory effect, and the reducing activity of black soybeans. *Food Microbiology*, **27**: 918–923.

Kaneko, R. and Tsuchiya, T. (1954) New compound in rice bran and germ oils. *Journal of Chemical Society of Japan*, **57**: 526.

Kannan, A., Hettiarachchy, N., Johnson, M.G. and Nannapaneni, R. (2008) Human colon and liver cancer cell proliferation inhibition by peptide hydrolysates derived from heat-stabilized defatted rice bran. *Journal of Agricultural and Food Chemistry*, **56**: 11643–11647.

Kawamura, Y. and Muramoto, M. (1993) Anti-tumorigenic and immunoactive protein and peptide factors in food stuff. In: *Food and Cancer Prevention Chemical and Biological Aspects* (eds K.W. Waldron, I.T. Johnson and L.R. Fenwick), The Royal Society of Chemistry, Cambridge, pp. 331–401.

Kim, D.J., Oh, S.K., Chun, A., Yoon, M.R., Hong, H.C., *et al.* (2011) Evaluation of biological activities of rice husk extracts. *Journal of Food Science and Nutritional*, **16**: 179–183.

Kim, J. and Godber, J.S. (2014) Comparison of rice bran oil and wax yields: A systematic approach to wax ester analysis. *International Journal of Food Science and Technology*, **49**: 410–415.

Korhonen, H. and Pihlanto-Leppälä, A. (2004) Milk-derived bioactive peptides: Formation and prospects for health promotion. In: *Handbook of Functional Dairy Products. Functional Foods and Nutraceuticals Series 6.0* (eds C. Short and J. O'Brien), CRC Press, Boca Raton, FL, pp. 109–124.

Kuba, M., Shinjo, S. and Yasuda, M. (2004) Antihypertensive and hypocholesterolemic effects of tofuyo in spontaneously hypertensive rats. *Journal of Health Science*, **60**: 670–673.

Kuba, M., Tanaka, K., Tawata, S., Takeda, Y. and Yasuda, M. (2003) Angiotensin I-converting enzyme inhibitory peptides isolated from tofuyo fermented soybean food. *Bioscience, Biotechnology, and Biochemistry*, **67**: 1278–1283.

Lee, S.C., Kim, J.H., Jeong, S.M., Kim, D.R., Ha, J.U., *et al.* (2003) Effect of far-infrared radiation on the antioxidant activity of rice hulls. *Journal of Agricultural and Food Chemistry*, **51**(15): 4400–4403.

Liu, Q., Kong, B., Xiong, Y.L. and Xia, X. (2010) Antioxidant activity and functional properties of porcine plasma protein hydrolysate as influenced by the degree of hydrolysis. *Food Chemistry*, **118**(2): 403–410.

Marambe, P., Shand, P.J. and Wanasundara, J.P.D. (2008) An *in vitro* investigation of selected biological activities of hydrolyzed flaxseed proteins (*Linum usitatissimum L.*) proteins. *Journal of American Oil Chemists Society*, **85**: 115–1164.

Matar, C., LeBlanc, J.G., Martin, L. and Perdigon, G. (2003) Biologically active peptides released in fermented milk: Role and functions. In: *Handbook of Fermented Functional Foods* (ed. E.R. Farnworth). CRC Press, Boca Raton, FL, pp. 177–201.

Megías, C., del Mar Yust, M., Pedroche, J., Lquari, H., Girón-Calle, J., *et al.* (2004) Purification of an ACE inhibitory peptide after hydrolysis of sunflower (*Helianthus annuus L.*) protein isolates. *Journal of Agricultural and Food Chemistry*, **52**(7): 1928–1932.

McCaskill, D.R. and Zhang, F. (1999) Use of rice bran oil in foods. *Journal of Food Technology*, **53**: 50–53.

Min, B., McClung, A.M. and Chen, M.H. (2011) Phytochemicals and antioxidant capacities in rice brans of different color. *Journal of Food Science*, **76**: C117–C126.

Muenai, N. (2009) *Production of dried starter culture from* Bacillus subtilis *for fermented soybean production: Special problem.* Faculty of Science and Agriculture Technology, Rajamangala University of Technology Lama Lampang, Thailand, Lampang, p. 46.

Okamoto, A., Hanagata, H., Matsumoto, E., Kawamura, Y., Koizumi, Y. and Yanagida, F. (1995) Angiotensin I-converting enzyme inhibitory activities of various fermented foods. *Bioscience, Biotechnology and Biochemistry*, **59**(6): 1147–1149.

Orthoefer, F.T. (1996) Rice bran oil: Healthy lipid source. *Food Technology*, **50**: 62–64.

Oshima, G., Shimabukuro, H. and Nagasawa, K. (1979) Peptide inhibitors of angiotensin I-converting enzyme in digests of gelatin by bacterial collagenase. *Biochimica et Biophysica Acta*, **566**: 128–137.

Patel, P. and Walker, T. (2004) Effect of extraction method on the level of important antioxidants in rice bran oil. ASAE/CSAE Meeting Paper No. 046140. ASAE/CSAE Fairmont Chateau Laurier, Government Centre Ottawa, Ontario, Canada.

Piu, L.D., Tassoni, A., Serrazanetti, D.I., Ferri, M., Babini, E., *et al.* (2014) Exploitation of starch industry liquid by-product to produce bioactive peptides from rice hydrolyzed proteins. *Food Chemistry*, **155**: 199–206.

Prakash, J. and Ramanatham, G. (1995) Effect of stabilization treatment of rice bran on functional properties of protein concentrates. *Journal of the Science of Food and Agriculture*, **67**(2): 181–187.

Pyo, Y.H. and Lee, T.C. (2007) The potential antioxidant capacity and angiotensin I-converting enzyme inhibitory activity of Monascus-fermented soybean extracts: Evaluation of Monascus-fermented soybean extracts as multifunctional food additives. *Journal of Food Science*, **72**: s218–s223.

Quiros, A., Ramos, M., Muguerza, B., Delgado, M.A., Miguel, M., *et al.* (2007) Identification of novel antihypertensive peptides in milk fermented with *Enterococcus faecalis*. *International Dairy Journal*, **17**: 33–41.

Rajapakse, N., Mendis, E., Jung, W.K., Je, J.Y. and Kim, S.K. (2005) Purification of a radical scavenging peptide from fermented mussel sauce and its antioxidant properties. *Food Research International*, **38**: 175–182.

Rong, N., Ausman, L.M. and Nicolosi, R.J. (1997) Oryzanol decreases cholesterol absorption and aortic fatty streaks in hamsters. *Lipids*, **32**: 303–309.

Ryan, P.E. (2011) Bioactive food components and health properties of rice bran. *Journal of the American Veterinary Medical Association*, **238**(5): 593–600.

Sabale, V., Sabale, P.M. and Lakhotiya, C.L. (2009) Comparative evaluation of rice bran wax as an ointment base with standard base. *Indian Journal of Pharmaceutical Sciences*, **71**(1): 77–79.

Saenjum, C., Chaiyasut, C., Chansakaow, S., Suttajit, M. and Sirithunyalug, B. (2012) Antioxidant and anti-inflammatory activities of gamma-oryzanol rich extracts from Thai purple rice bran. *Journal of Medicinal Plants Research*, **6**(6): 1070–1077.

Saha, B.C. and Cotta, M.A. (2008) Lime pretreatment, enzymatic saccharification and fermentation of rice hulls to ethanol. *Biomass Bioenergy*, **32**: 971–977.

Salanti, A., Zoia, L., Orlandi, M., Zanini, F. and Elegir, G. (2010) Structural characterization and antioxidant activity evaluation of lignins from rice husk. *Journal of Agricultural and Food Chemistry*, **58**: 10049–10055.

Sarkar, A., Rakwal, R., Agrawal, G.K., Shibato, J. and Sircar, P.K. (2011) Towards the mungbean 'Peptidome': Assessing the bioactivity of low molecular weight peptides isolated from germinating seeds. *Journal of Phytology*, **3**(7): 57–63.

Sharif, M.K. (2009) Rice industrial by-products management for oil extraction and value added products. Doctor of Philosophy Thesis in Food Technology, National Institute of Food Science and Technology, University of Agriculture, Faisalabad, Pakistan, pp. 53–237.

Shekib, L.A. (1994) Nutritional improvement of lentils, chick pea, rice and wheat by natural fermentation. *Plant Foods for Human Nutrition*, **46**: 201–205.

Shih, F.F. (2003) An update on the processing of high-protein rice product. *Nahrung*, **47**(6): 420–424.

Silva, L.H., Celeghini, R.M.S. and Chang, Y.K. (2011) Effect of the fermentation of whole soybean flour on the conversion of isoflavones from glycosides to aglycones. *Food Chemistry*, **128**: 640–644.

Stodolak, B. and Starzynska-Janiszewska, A. (2008) The influence of tempeh fermentation and conventional cooking on anti-nutrient level and protein bioavailability (*in-vitro* test) of grass-pea seeds. *Journal of the Science of Food and Agriculture*, **88**: 2265–2270.

Thawornchinsombut, S. and Kokkaew, H. (2009) *Purification and characterization of rice bran protein hydrolysate on lowering blood pressure*. Khon Kaen University, Thailand, Khon Kaen.

Timachai, S. and Thawornchinsombut, S. (2011) Effect of proteases on bioactive properties of rice bran protein hydrolysate. Master of Science Thesis in Food Technology, Graduate School, Khon Kaen University, Thailand, Khon Kaen.

Tsai, J.S., Chen, T.J., Pan, B.S., Gong, S.D. and Chung, M.Y. (2007) Identification of novel antihypertensive peptides in milk fermented with *Enterococcus faecalis*. *International Dairy Journal*, **17**: 33–41.

Valia, S.R., Jua, Y., Kaimalb, T.N.B. and Chern, Y. (2004) A process for the preparation of food-grade rice bran wax and the determination of its composition. *Journal of the American Oil Chemists' Society*, **82**: 57–64.

Van der Ven, C. (2002). Biochemical and functional characterization of casein and whey protein hydrolysate: A study on the correlations between biochemical and functional properties using multivariate data analysis. Doctoral Thesis, Wageningen Agricultural University, Netherlands, p. 170.

Van Hoed, V., Depaemelaere, G., Vila, A.J., Santiwattana, P., Verhe, R. and De Greyt, W. (2006) Influence of chemical refining on the major and minor components of rice bran oil. *Journal of the American Oil Chemists Society*, **83**: 315–321.

Venuste, M., Zhang, X., Shoemaker, C.F., Karangwa, E., Abbasa, S. and Kamdema, P.E. (2013) Influence of enzymatic hydrolysis and enzyme type on the nutritional and antioxidant properties of pumpkin meal hydrolysates. *Food & Function*, **4**: 811–820.

Wang, W. and Gonzalez de Mejia, E. (2005) A new frontier in soy bioactive peptides that may prevent age-related chronic diseases. *Comprehensive Reviews in Food Science and Food Safety Institute of Food Technology*, **4**: 63–78.

Wells, J.H. (1993) Utilization of rice bran and oil in human diets. *Louisiana Agriculture*, **36**(3): 4–8.

Wilson, T.A., Idreis, H.M., Taylor, C.M. and Nicolosi, R.J. (2002) Whole fat rice bran reduces the development of early aortic atherosclerosis in hypercholesterolemic hamsters compared with wheat bran. *Nutrition Research*, **22**: 1319–1332.

Wongputtisin, P. (2008) Fermented corticated soybean meal production and quality test for animal feed. Doctoral Thesis. Chiangmai University, Thailand, Chiangmai, p. 223.

Wu, C.H. and Chou, C.C. (2009) Enhancement of aglycone, vitamin K_2 and superoxide dismutase activity of black soybean through fermentation with *Bacillus subtilis* BCRC 14715 at different temperatures. *Journal of Agricultural and Food Chemistry*, **57**: 10695–10700.

Wu, J. and Ding, X. (2002) Characterization of inhibition and stability of soy-protein-derived angiotensin I-converting enzyme inhibitory peptides. *Food Research International*, **35**: 367–375.

Xu, Z. and Godber, S. (2000) Comparison of supercritical fluid and solvent extraction methods in extracting γ-oryzanol from rice bran. *Journal of the American Oil Chemists Society*, **77**(5): 547–551.

Xu, Z., Hua, N. and Godber, J.S. (2001) Antioxidant activity of tocopherols, tocotrienols, and gamma-oryzanol components from rice bran against cholesterol oxidation accelerated by 2,2'-azobis (2-methylpropionamidine) dihydrochloride. *Journal of Agricultural and Food Chemistry*, **49**: 2077–2081.

Zumbado, H., Ledesma, L., Fuertes, F. and Ventura, J. (1997) Manufacture of a bakery product with incorporation of high levels of precooked rice bran. *Alimentaria*, **280**: 21–23.

Zhang, J., Zhang, H., Wang, L., Guo, X., Wang, X. and Yao, H. (2009) Antioxidant activities of the rice endosperm protein hydrolysate: Identification of the active peptide. *European Food Research and Technologies*, **229**: 709–719.

Zhang, J., Zhang, H., Wanga, L., Guo, X., Wang, X. and Yao, H. (2010) Isolation and identification of antioxidative peptides from rice endosperm protein enzymatic hydrolysate by consecutive chromatography and MALDI-TOF/TOF MS/MS. *Food Chemistry*, **119**: 226–234.

13
Bioprocessing of Beverage Industry Waste for Value Addition

Surangna Jain and Anil Kumar Anal
Asian Institute of Technology, Klong Luang, Pathumthani, Thailand

13.1 Introduction

Beverages comprise of drinks, both alcoholic and non-alcoholic, which are prepared for the purpose of human consumption. They have a huge consumption worldwide due to increasing consumer demands. However, their increased processing results in the production of primary waste streams in the form of wastewater and solid wastes. Due to the extensive production of huge amounts of waste materials, they have a potential impact on the environment. These wastes generate huge amounts of organic substances, resulting in serious environmental hazards arising from the unsafe disposal of wastes that lead to the pollution of both water and land. Currently, a lot of political and social pressure is being put on the industries to reduce their wastes. It has become mandatory for the industries to take strict measures for waste management, as strong regulations now have to be followed for environmental conservation and sustainability.

The most widely consumed drink worldwide is water, followed by coffee and tea (Kumar and Chandrasekaran, 2013). Beverages also include juices, soft drinks and alcoholic drinks such as beer, wine etc. At each step in the processing cycle of beverages, wastes and by-products are generated, which can be subjected to valorization. This will not only minimize their disposal costs and environmental hazards, but also will allow value addition for the development of new value-added products.

13.2 Coffee

13.2.1 Coffee Processing

Coffee is the most widely consumed beverage in the world and an important trade commodity of many countries due to it being a huge global business. It is a drink

Food Processing By-Products and their Utilization, First Edition. Edited by Anil Kumar Anal.

prepared from the coffee beans following their roasting, which are the seeds of the coffee plants *Coffea arabica* and *Coffea cannephora* (Murugan and Sohaibani, 2013). It has been reported by the Food and Agricultural Organization (FAO) that the world production of coffee in 2010 was 138.3 million bags, which are likely to increase in the coming years. Processing is an important activity for the coffee industries for the conversion of raw coffee beans to liquid coffee. Two methods are generally used for coffee processing, that is the wet method and the dry method (Murthy and Naidu, 2012), which differ in their procedure and quality of the final coffee product.

In wet processing, the fruit pulp which covers the coffee beans is removed, washed and sun-dried. The fruit pulp is also subjected to fermentation for breaking down of the mucilage layer present on the parchment to simple substances that are non-sticky in nature. The coffee obtained after wet processing is referred to as washed or dried parchment coffee (Murthy and Naidu, 2012) and its hulling leads to the production of green coffee. During the dry processing, the whole fruits that are freshly harvested (pulp, parchment and bean) are dried either by sun-drying or in furnaces. The dried products are known as the husks, which are removed to obtain the green coffee. The green coffee obtained is subjected to roasting and brewing to achieve the characteristic flavour and coffee aroma (Castillo *et al.*, 2002). During the roasting process, the coffee bean color changes to yellow along with the changes in their properties and transformation to a mixture of Maillard reaction products (Sacchetti *et al.*, 2009). The coffee is then brewed to obtain high-quality coffee. Depending on the processing method of the coffee beans, by-products and wastes are generated such as husks, peels, pulp and spent coffee beans, which are good sources of nutrients and hence can be used for various purposes.

13.2.2 By-Products and Wastes from Coffee Processing

Huge amounts of by-products are generated during the processing of coffee. These by-products and wastes are responsible for serious environmental hazards in the coffee-producing countries, due to the unsafe disposal of husks and pulp resulting in pollution. The pulp is the major by-product of coffee industries, which have huge disposal problems due to their high moisture content that lead to putrefaction causing serious environmental problems. Both the coffee pulp and husks comprise of carbohydrates, proteins, fibres, pectin and phenolic compounds (Table 13.1). The husks are also characterized by the presence of a high number of volatile compounds. On the other hand, the spent coffee and the peels comprise of caffeine, tannins and polyphenols. The release of caffeine and tannins into the environment results in toxicity to various life processes on disposal (Fan *et al.*, 2003). However, these components can be treated to reduce their toxicity and also components such as tannins and phenolic compounds can be used in various applications due to their anti-nutritional properties. Currently, the coffee-processing wastes are being used only in fertilizers and animal feed, but there is a need to search for more alternative uses for these wastes.

13.2.3 Utilization of Coffee By-Products and Wastes

Traditional methods of waste utilization as fertilizers or in animal feed use only a small portion of the wastes and are also not very efficient. Hence, attempts are being made for detoxification of the by-products and wastes and their utilization for creating

Table 13.1 Production of various by-products during different coffee processing steps (Murthy and Naidu, 2012)

Processing Step	By-product generated	Percentage (%)
	Coffee pulp	29–34%
Coffee Wet Processing	Coffee husk	12–14%
Coffee Dry Processing	Coffee husk	12–17%
Coffee Roasting	Coffee skin	1–3%
Coffee Brewing	Spent coffee waste	45–55%

value-added products such as organic acids, enzymes, mushrooms, flavour and aroma compounds and bioactive compounds.

The coffee husks are used for the production of organic acids such as citric acid by solid-state fermentation using the fungi *Aspergillus niger* (Shankaranad and Lonsane, 1999). It has been observed that the coffee husks lead to higher production of citric acid as compared to wheat bran and rice bran (Pandey *et al.*, 2000). Gibberillic acid is also produced by solid-state fermentation, which can be used as plant growth regulators for the development purpose of plants.

Edible mushrooms can also be cultivated from the coffee wastes, which is a good alternative as the wastes are being directly converted to food for human consumption. The first cultivation of mushrooms using the different coffee by-products such as husks, leaves and spent coffee was made by Fan *et al.* (2001). Cultivation of *Lentinus edodes*, *Pleurotus* sp. and *Volvariella* sp was carried out for mushroom production. It was found that the coffee husks were suitable for mushroom cultivation purposes in an economical way.

Coffee by-products can also be utilized as substrates for the production of industrial enzymes. Mainly, the coffee pulp and husks are used as potential substrates for enzyme production such as pectinase, tannase and caffeise (Murthy and Naidu, 2012). Other enzymes can be produced using the other coffee by-products as well are amylase, xylanase and protease using various fungal organisms (Murthy and Naidu, 2012) by solid-state fermentation due to their low operation cost and effective production.

Another approach for value addition of the coffee wastes, mainly the husks, are to utilize them as substrates for producing flavour and aroma compounds by fermentation using yeast and fungi (Soares *et al.*, 2000). Currently, the demand of consumers is towards natural flavour and aroma compounds as compared to the synthetic ones for consumption purposes. Soares *et al.* (2000) found that yeast fermentation of coffee husks by *Ceratocystis fimbriata* generates an alcoholic and pineapple aroma and a fruity flavour. The fermentation process leads to the production of various compounds such as ethanol, ethyl acetate, acetaldehyde, isopropanol etc. in the cultures that contribute to the characteristic flavour and aroma.

Bioactive compounds are nutritional components present in food that have health benefits. The coffee by-products are good sources of phenolic compounds. Chlorogenic acid, a major phenolic compound, can be found in the spent coffee beans (Petraco, 2005), which can be used as food supplements to prevent various diseases. Coffee pulps are also a good source of polyphenols. The coffee skin has also been evaluated for phenolic compounds by solid state fermentation using different fungal strains (Machado, 2009). It has also been observed that the coffee skins possess antioxidant properties that can be used as potential ingredients in functional foods. The coffee by-products are also a good source of dietary fibres, having a number of health benefits. Coffee skins have a high content of dietary fibres (50–60%) of which 15% includes the soluble dietary fibres and 85% the insoluble dietary fibres (Narita and Inouye, 2014). This content of dietary fibres i much higher than that reported in apples, wheat bran, cabbages, broccoli, potato and carrots (Narita and Inouye, 2014). Anthocyanins, which are the flavonoid compounds responsible for color, can also be obtained from the fresh coffee husks and pulp, which can be used as colorants or ingredients in different food formulations, as they possess many good biological effects.

13.3 Tea

13.3.1 Processing and Production of Tea

Tea is also a popular beverage worldwide. Due to the increasing awareness regarding nutrition and management of health, consumption of tea, mainly green tea, has increased significantly, mostly in Southeast and East Asia. World tea production has been reported as 4.73 million tonnes in 2010 (FAO, 2010), which is increasing continuously. Processing of tea involves various steps which start with the plucking of the young tea leaves whose moisture content is reduced by 50–60% (Murugan and Sohaibani, 2013). The leaves are rich in polyphenols, mainly the polyphenol epigallocatechin gallate. This is followed by fermentation where there is the breakdown of leaf cells to release polyphenol oxidase for polyphenol oxidation. The final step is the extraction of tea leaves by hot water to obtain the finished product, i.e. the tea drink. The processing of green tea is different from that of black tea. For green tea, the leaves are just dried where the enzyme polyphenol oxidase remains inactive, whereas during black tea processing, the leaves are ground and left for 60 minutes. During this step, the polyphenol oxidase converts the catechin polyphenol to other polyphenols that are characteristics of black tea, such as theaflavin and thearubigin (Murugan and Sohaibani, 2013). Hence, the presence of catechins is higher in green tea than in black tea.

13.3.2 Tea By-Products and Wastes and their Utilization

During the processing of tea, a certain portion is accumulated as wastes in the form of tea leaves and tea grounds. These waste forms are disposed of and mainly buried and their under-utilization results in huge environmental pollution. However, due to their high functional and nutritive value, their efficient utilization needs to be explored for solving the economic as well as the environmental problem. The tea leaves are good sources of amino acids, proteins, polyphenols, tannins and vitamins (Yamamoto *et al.*, 2007). The tea grounds are produced from the tea leaves after hot water extraction, which contain high amounts of proteins, polyphenols, tannins, β-carotene and vitamins. Polyphenols derived from both these wastes include flavonoids that demonstrate antioxidant, anticarcinogenic, antimutagenic and cardio protective effects (Wang and Xu, 2013). These are mainly derived from the tea grounds. Due to the presence of components with multiple biological activities, the tea by-products can be possibly used for creation of value-added products with nutraceutical and medicinal value, which will help reduce amount of wastes disposed into the environment. This will also help to solve the problem regarding the limited supply of resources.

13.4 Fruit Juice and Soft Drinks

Soft drinks and fruit juices are also very popular beverages. Soft drinks are non-alcoholic beverages comprising of sugar, carbonated water and various different flavours, whereas fruit juices are drinks prepared by extraction from whole fruits and fruit pulp. Many different fruits can be used for its preparation such as mango, orange, pineapple etc. (Murugan and Sohaibani, 2013). The fruit juices are also an important beverage segment.

Processing of soft drinks does not generate many by-products and solid waste as such can be valorized, but their processing generates large quantities of wastewater (Fillaudeau *et al.*, 2006). This wastewater comprises of moderate amounts of pollutants coming from wasted soft drinks, water after washing of cans and bottles and detergents which includes total suspended solids, biochemical oxygen demand (BOD), chemical oxygen demand (COD), nitrates and phosphates (Nweke *et al.*, 2015). For the prevention of environmental hazards, it is important that these wastewaters are treated before they are discharged into the environment. These wastewaters have no scope for further utilization for value addition purposes.

During the manufacture and processing of fruit juices, the mature fruits used for juice preparation are peeled along with the removal of their seeds, which comprise the by-products and wastes resulting from the juice processing industries. This also includes the rotten and damaged fruits that are not used in the juice production. These wastes are rich in bioactive compounds and comprise of vitamins, dietary fibres, phenolic compounds, carotenoids and phytochemicals, having high content of antioxidant, antimicrobial, antitumor and antimutagenic compounds. These can be further exploited for the production of food supplements and additives with good nutraceutical and medicinal values and the recovery of these compounds will also prove to be economical as these wastes are cheap, renewable, safe to use and available in bulk quantities (Ahmed and Theydan, 2012).

13.5 Alcoholic Beverages

13.5.1 Beer Production

Beer is the most widely-consumed beverage worldwide, after water, coffee and tea. The world beer production was reported to be more than 1.34 hectolitres in 2002, with an average consumption of 23 litres per person per year (Fillaudeau *et al.*, 2006). It has a low alcoholic content in comparison with the other alcoholic beverages. Beer is a yeast-fermented product obtained from malted cereal, i.e. barley in addition with other cereals such as wheat and maize and sugar sources called adjuncts and hops (Mathias *et al.*, 2014). Beer production involves the use of the yeast strain *Sacchromyces* and is categorized into four different processes: malting (steeping of grains such as barley or other cereals in water for their germination followed by drying), mashing (extraction on ground), wort production (producing wort liquid and boiling it with hops) and fermentation (addition of yeast strain to the cooled wort that ferments it).

Beer brewing generates huge amounts of by-products and waste in the form of spent grains, spent hops and yeasts (Mussatto *et al.*, 2006). These wastes comprise of a rich composition of nutritional organic compounds such as fibres, carbohydrates, proteins, ash, vitamins, phenolic compounds and fatty acids (Mathias *et al.*, 2014). Most of these wastes are agricultural products, so they can be easily recycled and re-used. Hence these wastes are more environmental-friendly than waste from other industries. But their proper utilization, can lead to the generation of value-added products with more profits.

13.5.2 By-Products and Wastes from the Brewing Industry and their Utilization

13.5.2.1 Brewers spent grain Brewer's spent grain (BSG) is the first and most abundant by-product of the brewing industries, which accounts for 85% of the total by-products generated by them (Reinold, 1997). They make up 31% of the original malt weight. They are basically the by-products of the mashing step, where malt and the other cereal grains are solubilized for the extraction of wort. During beer production, malted barley is milled and mixed with water to produce a mash. Then the temperature is increased for the hydrolysis of malt constituents, which are mainly starch and proteins followed by the solubilization of the products obtained after breakdown. This enzymatic conversion step produces a liquid known as wort, which is filtered. The filtered wort is subjected to fermentation for the production of beer, and the insoluble residual fraction is referred to as the brewer's spent grain. The brewer's spent grain basically comprises of the husk, pericarp and seed coat layer that covers the original barley grains. They are rich in cellulose and lignin and may also contain certain amounts of proteins, lipids and phenolic compounds. In addition, they also comprise of mineral components such as calcium and different vitamins such as biotin, folic acid, riboflavin, thiamine and vitamin B_6 and different amino acids (Priest and Stewart, 2006).

The BSG is the main by-product of the brewing industries but has been little utilized. Its disposal also poses certain environmental problems. Currently, they are used for the production of animal fertilizers which is good, but they can be further explored for value-added applications. Due to their high protein and fibre content (~20% and 70%, respectively), they can be used in human nutrition (Mussatto *et al.*, 2006), and can also be used as a raw material in various biotechnological processes such as mushrooms and actinobacteria cultivation, creation of value-added products, such as ferulic and p-coumaric acids, xylose, arabinose, r-xylitol and arabitol production (Kumar and Chandrasekaran, 2013).

Until now, the BSG has been used as animal feed due to its high protein and fibre content. However, due to its low cost and high nutritional value, it is now being focused on use in human nutrition for the manufacturing of bread and biscuits. But the BSG cannot be directly incorporated into food. They have to be converted to flour first, but there are limitations in the use of flour as protein additives due to their color and flavour. As the BSG is brownish in color, they are preferred to be used in products with an off-white color such as cookies, cakes and breads, but also in small quantities due to alterations in flavour and texture (Miranda *et al.*, 1994).

The BSG is rich in proteins and polysaccharides, which make it an ideal substrate for many biotechnological processes, as it makes them susceptible to microbial growth and degradation. BSG has been used successfully for cultivating microorganisms such as *Lactobacillus, Penicillium and Streptomyces* spp. Hence, it is considered a good medium component for the isolation of unknown strains and for the fast production of new bioactive components and spores (Szponar *et al.*, 2003). Not only microorganism growth, but the high protein and moisture content of BSG also favor the growth of mushrooms.

Many cereal brans with similar compositions and structures as BSG have been used for the production of commercial enzymes in solid-state fermentations (Sangeetha *et al.*, 2004). This has led to the evaluation of BSG as an alternative substrate for enzyme production. It is considered a good source of nitrogen for the production of xylanase from *Aspergillus awamori* or from a *Streptomyces* isolate. In addition, they help produce a number of enzymes such as β-xylosidases, α-arabinofuranosidases, esterases and α-amylase. The activity of these enzymes and their type is determined by the composition of the substrate and strain used. In addition, BSG can be a source for the creation of value-added products, as they comprise of an abundant amounts of phenolic acids such as hydroxycinnamic acids (ferulic and p-coumaric), which can be used potentially in the food industries (Mussatto *et al.*, 2006). The ferulic acid has a number of bioactive properties such as antioxidant, antimicrobial and anti-inflammatory activities, whereas the p-coumaric demonstrates chemo-protectant and antioxidant properties (Mussatto *et al.*, 2006). BSG can also be used for the production of xylitol, a rare sugar. It is found in low amounts in nature, but is considered a good alternative for sucrose, to be used as a sweetener in various food applications. They can be used for the treatment of diabetes and lipid metabolic disorders. Various agro-industrial wastes can be used for the production of xylitol, but BSG is preferred due to their high initial xylose concentration and due to no initial detoxification steps required.

13.5.2.2 Spent Hops Hops are added in the beer as they are rich in bitter substances such as α-acids (humulones), β-acids (lupulones) and soft and hard resins, which lead to the bitterness and aroma of beer. But only 15% of these hops are utilized in beer production and the remaining 85% comprise the spent hop wastes (Huige, 2006). Following the process of mashing, the wort has to be boiled with the addition of the hops for extraction of aroma and bitter components. During this stage, the hop compounds are precipitated out due to their low solubilization, resulting in the generation of spent hops and this accounts for 10–20% of the precipitates (Priest and Stewart, 2006). The spent hops comprise of huge amounts of fibres and proteins; however, these hops have not been utilized as substrates in various biotechnological processes. These hops cannot be directly used in animal feed as the spent grains comprise of bitter substances which the animals would refuse to eat. Hence, these bitter components need to be removed or degraded prior to their utilization in animal feed.

Many value-added compounds can be recovered from the spent hops, such as organic acids, saccharides and flavour compounds. These compounds can be obtained either by the hydrolysis or the oxidation of spent hops (Oosterveld *et al.*, 2002). The organic acids such as the hop acids are of great potential, as they are considered natural antibacterial agents for alcohol fermentation and can easily replace antibiotics in the production of ethanol (Ruckle and Senn, 2006). Pectins can also be recovered from the spent hops, which can be utilized as gelling and thickening agents in the food industries. These include homogalacturonans and arabinogalactan proteins.

In addition, the spent hops can increase the yield as well as the fermentation process of *Sacchromyces cerevisiae*.

13.5.2.3 Residual Yeast The residual yeast is the second largest by-product of the brewing industries after the spent grains, which are produced in abundant quantities and have a good chemical composition. The most abundant components found in the yeast cells are carbohydrates, proteins, amino acids and vitamins, mainly niacin (Lewis and Young, 1995). In comparison with the spent grains, the residual yeasts have a higher content of proteins, amino acids and vitamins due to which they are currently being utilized in animal feed and nutritional supplements (Chae *et al.*, 2001). But their uses are limited in the production of food products, as the ribonucleic acid (RNA) present can be metabolized to uric acid in the human body, contributing to gout (Huige, 2006). However, due to their rich composition, various attempts are being made to use these residual yeasts in various biotechnological processes.

Various compounds of industrial interest can be recovered from the yeast cells, such as proteins, amino acids, vitamins, enzymes and the purine components present in DNA and RNA (Huige, 2006), by breaking down the yeast cells with the help of endogenous and exogenous enzymes. These components have great potential for use in a wide range of applications. β-glucan, a hydrocolloid, can also be isolated from the brewer's yeast. This compound is of huge interest to the pharmaceutical and functional food industries, because it can improve the functional properties of the food products and can be used as a thickening, water-holding, oil-binding, emulsifying and foaming agent (Thammakiti *et al.*, 2004). It can also restrict the retrogradation of starch (Satrapai and Suphantharika, 2007), leading to great value addition to the starch-based food products.

13.5.3 Wine Production

Wine is an ancient drink that has been consumed by human societies for thousands of years. It originated in ancient Greece, following which it spread throughout other countries such as the ancient Mediterranean, Europe and China. Today, wine is consumed in every continent in the world, mainly in Europe, the Americas, South Africa, Australia and New Zealand. The process of wine-making has evolved throughout the centuries and has changed little throughout the ages. Wine production mainly involves the use of grapes, mainly the *Vitis Vinifera* grapes species (Kumar and Chandrasekaran, 2013). A blend of other grape varieties is also used. Wine production involves a basic fermentation process where the sugars present in the grapes are converted to alcohol. The grapes and the juices derived from it comprise of organic acids and phenol compounds, which contribute to flavour and various vitamins and minerals that are required for yeast growth and fermentation to occur. The basic steps that are involved in wine-making are:

1. *Harvesting*: This is the first step in the wine-making process. The grapes are harvested at the precise time when the sugars, phenol and aroma compounds in them are suitable for wine production.
2. *Destemming and crushing*: This involves the removal of stems and crushing of the grapes by mechanical presses. In addition, certain enzymes such as sulfur dioxides are introduced into the grapes to prevent oxidation and inhibit microbial growth.

The enzymes help to break down the cell walls so that the juice can be released more easily.
3. *Pressing*: The grapes are squeezed properly for juice extraction. which is followed by filtration and centrifugation to separate out the solids.
4. *Fermentation*: The juices obtained are inoculated with the live yeast cells for carrying out the fermentation process, where the sugars are broken down to produce alcohol.
5. *Purification and bottling*: A variety of processes are carried out for the purpose of removing the unwanted salts, solids and microorganisms from the wine, after which the wine is held in barrels for aging and then bottled to be sold.

Recently there has been an increase in the generation of waste materials by wine industries. The wastes produced by these industries consist of both solid and liquid wastes, but mainly comprise of organic materials such as the grape pomace (12%), lees (14%), stalk (12%) and dewatered sludge (12%) (Kumar and Chandrasekaran, 2013). These wastes and by-products can be characterized by the presence of high amounts of suspended solids and biodegradable compounds. The pressing and purification steps are the main steps when major amounts of winery wastes are generated. Almost 25% of the wastes are produced at these steps (Arvanitoyannis *et al.*, 2006).

13.5.3.1 Utilization of Winery Wastes and By-Products The three main by-products that are generated from wine processing include the grape stalks, pomace and wine lees (Bustamante *et al.*, 2008). The stalks and the pomace are produced during the process of crushing and pressing, whereas the wine lees are the sludge-like material remaining after the wine production that comprise of the dead yeast, yeast residues and particles precipitated out at the bottom of the wine tanks and barrels (Hwang *et al.*, 2009). These wastes contain high amounts of organic components such as sugars, phenolic compounds, lipids and pectins, with high chemical and biological oxygen demand due to which it is considered an environmental pollutant (Lafka *et al.*, 2007). Their utilization by their potential conversion to biofuels, animal feed, food ingredients and nutritional supplements is being considered by the industries.

The major organic wastes produced by the wine industries comprise of the grape pomace that accounts for 11–15% of the total grapes crushed. A number of value-added components can be derived from this grape pomace. Grape pomace is generated during the pressing of the grapes and comprises the pressed skins, disrupted cells from the grape pulp, peels and the seeds. Their cell wall is made up of polysaccharides such as cellulose and hence an extra fermentation step is required when they are used in food and feed matrices for preventing gastrointestinal disturbances (Teixeira *et al.*, 2014). They can also be used as fermentation media for the production of single-cell proteins and other valuable metabolites (Tataridis *et al.*, 2006). In addition, grape pomace can be utilized in various processes, including the extraction of grape seed oil, polyphenolic compounds (such as anthocyanins, flavonoids, phenolic compounds and reservatrol), citric acid, methanol, ethanol and xanthan (Tataridis *et al.*, 2006). Major components remaining in the grape pomace are the phenolic compounds that are present in the pulp (1%), the skin (5%) and the seeds (62%). They mainly include the catechins, epicatechins and procyanidins (Yilmaz and Toledo, 2006). Various studies have reported a high antioxidant capacity of these grape pomaces, which makes them a good source of natural antioxidants that can be used in the food, pharmaceutical and cosmetic industries (Rockenbach *et al.*, 2011).

The grape stalks, which are the grape skeletons, are other major by-products of the wine processing industries. They comprise of high amounts of fibres such as lignin and cellulose and huge amounts of nutritive elements such as nitrogen and potassium (Bertran *et al.*, 2004). They are also rich in dietary fibres that make up 77% of its dry matter (Llobera and Canellas, 2007). The grape stalks are generally used for the production of compost and high-quality fertilizers by incorporating them in the winery sludges that are anaerobically digested and centrifuged. They can also be utililzed in the formulation of ruminant and animal feed (Nicolini *et al.*, 1993) and for soils in the vineyards that comprise of low organic matter content due to which they are of high agronomic value. Certain phenolic compounds, such as flavanols, hydroxycinnamic acids and stilbenes, with multiple biological effects can be isolated from the grape stalks for functional and nutraceutical applications (Karvela *et al.*, 2009).

Yeast lees are also the by-products of wine industries, which are generated in small amounts. They can be used as probiotic supplements and in the production of animal feed, as they are good sources of carbohydrates, proteins and vitamins. They can also be used for recovering tartrate for the potential production of calcium tartrate. In addition, β-1,3-glucans, which are present in abundant amounts in the yeast cell walls, can be isolated. They are non-toxic and help to stimulate immune responses against infectious diseases, cancer etc. by activating the macrophages (Kogan, 2000).

13.5.4 Brandy

Various types of brandies are produced by different countries, which are either made from wine or fermented fruit juices or other liquids that contain sugar. Most of them are prepared by heating the wine, which helps to concentrate the alcohol present in wine. As the boiling point of alcohol is less than water, it can be easily boiled off, whereas the watery potion remains as it is in the wine. Many products can be used for the production of brandy, which can be fermented and distilled, such as grapes, apples, blackberries, honey, rice, milk, corn and potato (Kumar and Chandrasekaran, 2013). However, in the production and processing of distilled spirits such as brandies, no separate by-products or wastes are generated. Hence, there does not remain much scope for further utilization and valorization from their processing.

13.6 Conclusion

The beverage industries have huge scope for potentially utilizing the wastes and by-products generated during the processing of coffee, tea, fruit juices and alcoholic beverages. These include the coffee husks, pulp and skin, tea leaves, fruit peels, pomaces, brewer's yeast and brewer's spent grains. These huge quantities of by-products and wastes are however under-utilized. These by-products can be valorized by intensive research and development for identification and the derivation of important bioactive molecules, which will help to create new value-added products of more economical value and health-promoting properties. Few applications of these by-products are known; however, their valorization requires the development of novel extraction, isolation, purification and recovery processes to achieve high amounts of healthy bioactive components. Efforts also need to be made on the isolation and structural elucidation of novel bioactive compounds using chromatographic and spectroscopic methods. This will provide successful achievements for the recovery of important bioactive molecules and the development of innovative products, which can lead to the reduction of environmental pollution from beverage by-products, and provide for the application of isolated bioactive compounds in feed, food, pharmaceutical and cosmetic industries.

References

Ahmed, M.J. and Theydan, S.K. (2012) Equilibrium isotherms, kinetics and thermodynamics studies of phenolic compounds adsorption on palm-tree fruit stones. *Ecotoxicology and Environmental Safety*, **84**: 39–45.

Arvanitoyannis, I.S., Demetrios, L. and Athanasios, M. (2006) Wine waste treatment methodology. *International Journal of Food Science and Technology*, **41**(10): 1117–1151.

Bertran, E., Sort, X., Soliva, M. and Trillas, I. (2004) Composting winery waste: Sludges and grape stalks. *Bioresource Technology*, **95**(2): 203–208.

Bustamante, M.A., Moral, R., Paredes, C., Pérez-Espinosa, A., Moreno-Caselles, J. and Perez-Murcia, M.D. (2008) Agrochemical characterization of the solid by-products and residues from the winery and distillery industry. *Waste Management*, **28**: 372– 380.

Castillo, M.D., Ames, J.M. and Gordon, M.H. (2002) Effect of roasting on the antioxidant activity of coffee brews. *Journal of Agricultural Food Chemistry*, **50**: 3698–3703.

Chae, H.J., Joo, H. and In, M-J. (2001) Utilization of brewer's yeast cells for the production of food-grade yeast extract. Part 1: Effects of different enzymatic treatments on solid and protein recovery and flavor characteristics. *Bioresource Technology*, **76**: 253–258.

Fan, L., Pandey, A. and Soccol, C.R. (2001) Production of *Flammulina velutipes* on coffee husk and coffee spent-ground. *Brazilian Archives of Biology and Technology*, **44**(2): 205–212.

Fan, L., Soccol, A.T., Pandey, A. and Soccol, C.R. (2003) Cultivation of Pleurotus mushrooms on Brazilian coffee husk and effects of caffeine and tannic acid. *Micologia Aplicada Int.*, **15**: 15–21.

FAOSTAT-FAO (2010) Statistical Database, 2010.

Fillaudeau, L., Blanpain-Avet, P. and Daufin, G. (2006) Water, wastewater and waste management in brewing industries. *Journal of Cleaner Production*, **14**(5): 463–471.

Huige, N.J. (2006) Brewery by-products and effluents. In: *Handbook of Brewing*, 2nd edition (eds F.G. Priest and G.G. Stewart). CRC Press, Boca Raton, FL.

Hwang, J.Y., Shyu, Y.S. and Hsu, C.K. (2009) Grape wine lees improve the rheological and add antioxidant properties to ice cream. *LWT-Food Science and Technology*, **42**(1): 312–318.

Karvela, E., Makris, D.P., Kalogeropoulos, N. and Karathanos, V.T. (2009) Deployment of response surface methodology to optimize recovery of grape (*Vitis vinifera*) stems polyphenols. *Talanta*, **79**(5): 1311–1321.

Kogan, G. (2000) 1-3, 1-6-β-D-Glucans of yeasts and fungi and their biological activity. *Studies in Natural Products Chemistry*, **23**: 107–152.

Kumar, R.S. and Chandrasekaran, M. (2013) *Beverages: Valorization of food processing by-products*. Taylor and Francis Group, Boca Raton, FL, pp. 589–614.

Lafka, T.I., Sinanoglou, V. and Lazos, E.S. (2007) On the extraction and antioxidant activity of phenolic compounds from winery wastes. *Food Chemistry*, **104**(3): 1206–1214.

Lewis, M.J. and Young, T.W. (1995) *Brewing*. Chapman & Hall, London.

Llobera, A. and Canellas, J. (2007) Dietary fibre content and antioxidant activity of Manto Negro red grape (*Vitis vinifera*): Pomace and stem. *Food Chemistry*, **101**: 659–666.

Machado, E.S.M. (2009) Reaproveitamento de resíduos da indústria do café como matéria-prima para a produção de etanol. MSc thesis, Department of Biological Engineering, University of Minho, Braga, Portugal.

Mathias, T.R., de Mello, P.P.M. and Servulo, E.F.C. (2014) Solid wastes in brewing process: A review. *Journal of Brewing and Distilling*, **5**(1): 1–9.

Miranda, M.Z., Grossmann, M.V.E., Prudencioferreira, S.H. and Nabeshima, E.H. (1994) Utilization of brewer spent grain (BSG) for production of snacks with fiber. II: Sensory analysis of snacks. *Brazilian Archives of Biology and Technology*, **37**: 9–21.

Murthy, P.S. and Naidu, M.M. (2012) Sustainable management of coffee industry by-products and value addition: A review. *Resources, Conservation and Recycling*, **66**: 45–58.

Murugan, K. and Sohaibani, S.A. (2013) *Coffee, Tea and Cocoa: Valorization of food processing by-products*. Taylor and Francis Group, Boca Raton, FL, pp. 455–487.

Mussatto, S.I., Dragone, G. and Roberto, I.C. (2006) Brewers' spent grain: Generation, characteristics and potential applications. *Journal of Cereal Science*, **43**(1): 1–14.

Narita, Y. and Inouye, K. (2014) Decrease in the acrylamide content in canned coffee by heat treatment with the addition of cysteine. *Journal of Agricultural and Food Chemistry*, **62**(50): 12218–12222.

Nicolini, L., Volpe, C., Pezzotti, A. and Carilli, A. (1993) Changes in *in-vitro* digestibility of orange peels and distillery grape stalks after solid-state fermentation by higher fungi. *Bioresource Technology*, **45**(1): 17–20.

Nweke, C.N., Nwabanne, J.T. and Igbokwe, P.K. (2015) Anaerobic digestion treatment of soft drink wastewater. *Journal of Environment and Human*, **2**(1): 25–35.

Oosterveld, A., Voragen, A.G.J. and Schols, H.A. (2002) Characterization of hop pectins shows the presence of an arabinogalactan-protein. *Carbohydrate Polymer*, **49**: 407–413.

Pandey, A., Soccol, C.R., Nigam, P., Brand, D., Mohan, R. and Roussos, S. (2000) Biotechnological potential of coffee pulp and coffee husk for bioprocesses. *Biochemical Engineering Journal*, **6**(2): 153–162.

Petraco, M. (2005) Our everyday cup of coffee: The chemistry behind its magic. *Journal of Chemical Education*, **82**(8): 1161.

Priest, F.G. and Stewart, G. G. (eds) (2006) *Handbook of Brewing*. CRC Press, Boca Raton, FL.

Reinold, M.R. (1997) *Manual prático de cervejaria*, 1st edition. Aden Editora e Comunicações Ltda, São Paulo. p. 214.

Rockenbach, I.I., Rodrigues, E., Gonzaga, L.V., Caliari, V., Genovese, M.I., *et al.* (2011) Phenolic compounds content and antioxidant activity in pomace from selected red grapes (*Vitis vinifera L.* and *Vitis labrusca L.*) widely produced in Brazil. *Food Chemistry*, **127**: 174–179.

Ruckle, L. and Senn, T. (2006) Hop acids can efficiently replace antibiotics in ethanol production. *International Sugar Journal*, **108**: 139–147.

Sacchetti, G., Mattia, D., Pittia, C. and Mastrocola, D. (2009) Effect of roasting degree, equivalent thermal effect and coffee type on the radical scavenging activity of coffee brews and their phenolic fraction. *Journal of Food Engineering*, **90**(1): 74–80.

Sangeetha, P.T., Ramesh, M.N. and Prapulla, S.G., (2004) Production of fructosyl transferase by *Aspergillus oryzae* CFR 202 in solid-state fermentation using agricultural by-products. *Applied Microbiology and Biotechnology*, **65**: 530–537.

Satrapai, S. and Suphantharika, M. (2007) Influence of spent brewer's yeast-glucan on gelatinization and retrogradation of rice starch. *Carbohydrate Polymer*, **67**: 500–510.

Shankaranand, V.S. and Lonsane, B.K. (1994) Coffee husk: An inexpensive substrate for production of citric acid by *Aspergillus niger* in a solid-state fermentation system. *World Journal of Microbiology and Biotechnology*, **10**(2): 165–168.

Soares, M., Christen P., Pandey, A. and Soccol, C.R. (2000) Fruity flavour production by *Ceratocystis fimbriata* grown on coffee husk in solid-state fermentation. *Process Biochemistry*, **35**(8): 857–861.

Szponar, B., Pawlik, K. J., Gamian, A. and Dey, E.S. (2003) Protein fraction of barley spent grain as a new simple medium for growth and sporulation of soil Actinobacteria. *Biotechnology Letters.* **25**: 1717–1721.

Tataridis, P., Apostolopoulos, K., Moraitaki, E., Spanou, D., Nerantzis, E.T. and Papakonstantinou, S. (2006) *Utilization of grape pomace for the production of bacterial cellulose. Pan-Hellenic Union of Bioscientists Congress 'Biosciences in the 21st Century', 13–15 April*, Athens, Greece, pp. 114 (in Greek).

Teixeira, A., Baenas, N., Dominguez-Perles, R., Barros, A., Rosa, E., *et al.* (2014) Natural bioactive compounds from winery by-products as health promoters: A review. *International journal of molecular sciences*, **15**(9): 15638–15678.

Thammakiti, S., Suphantharika, M. and Phaesuwan, T. (2004) Preparation of spent brewer's yeast-glucans for potential applications in the food industry. *International Journal of Food Science Technology*, **39**: 21–29.

Wang, H. and Xu, C. (2013) Utilization of tea grounds as feedstuff for ruminant. *Journal of animal science and biotechnology*, **4**(1): 54.

Yamamoto, M., Ema, K. and Shibuichi, I. (2007) *In vitro* and *in vivo* anti-allergic effects of 'benifuuki' green tea containing O-methylated catechin and ginger extract enhancement. *Cytotechnology*, **55**(2–3): 135–142.

Yilmaz, Y. and R.T., Toledo. (2006) Oxygen radical absorbance capacities of grape/wine industry by-products and effect of solvent type on extraction of grape seed polyphenols. *Journal of Food Composition and Analysis*, **19**: 41–48.

14

Bioactive Compounds and their Health Effects from Honey Processing Industries

Zjahra Vianita Nugraheni and Taslim Ersam

Department of Chemistry, Faculty of Mathematics and Natural Sciences, Institut Teknologi Sepuluh Nopember, Surabaya, Indonesia

14.1 Introduction

Honey has had a long history in human consumption, and is used in various foods and beverages as a sweetener. Individual flavors of honey depend on the nectar source. Honey has also been used in various traditional medicines to treat many ailments.

Nowadays, honey is graded to determine its quality and further use. It is graded based upon a number of factors, including water content, flavor and aroma, absence of defects and clarity. Honey is also classified by color, though it is not a factor in the grading scale. Table 14.1 presents an example of the honey grade scale of one country, the USA.

Other countries may have differing standards on the grading of honey. India, for example, certifies honey grades based on additional factors, such as the Fiehe's test, and other empirical measurements.

Natural honey has been reported to contain about 200 substances, which consist of not only highly concentrated solutions of sugars, but also a complex mixture of other saccharides, amino acids, peptides, enzymes, proteins, organic acids, polyphenols, and carotenoid-like substances, vitamins and minerals (Gheldof *et al.*, 2002; Sato and Miyata, 2000; White, 1976).

Sugars are the main constituents of honey, comprising about 95% of its dry weight (Alvarez-Suarez *et al.*, 2010). Honey obtains its sweetness from the monosaccharides, fructose and glucose, and has approximately the same relative sweetness as granulated sugar. While glucose and fructose are the dominant constituents, about 25 different sugars have been detected (Doner, 1977; Siddiqui, 1970). It has attractive chemical

Food Processing By-Products and their Utilization, First Edition. Edited by Anil Kumar Anal.

Table 14.1 Honey Grade Scale in the USA

Grade	Water content	Flavor and aroma	Absence of defects	Clarity
A	<18.6%	Good – has a good, normal flavor and aroma for the predominant floral source and is free from caramelization, smoke, fermentation, chemicals and other odor causes	Practically free from defects that affect appearance or edibility	Clear – may contain air bubbles that do not materially affect the appearance; may contain a trace of pollen grains or other finely divided particles of suspended material that do not affect appearance
B	>18.6% and <20.0%	Reasonably good – practically free from caramelization; free from smoke, fermentation, chemicals, and other causes	Reasonably free – do not materially affect appearance or edibility	Reasonably clear – may contain air bubbles, pollen grains, or other finely divided particles of suspended material that do not materially affect appearance
C	<20.0%	Fairly good – reasonably free from caramelization; free from smoke, fermentation, chemicals, and other causes	Fairly free – do not seriously affect the appearance or edibility	Fairly clear – may contain air bubbles, pollen grains, or other finely divided particles of suspended material that do not seriously affect appearance
Sub-standard	>20.0%	Fails Grade C	Fails Grade C	Fails Grade C

(USDA, 2015)

properties for baking and a distinctive flavor that leads some people to prefer it over sugar and other sweeteners. Most microorganisms do not grow in honey because of its low water activity of 0.6. However, honey sometimes contains dormant endospores of the bacterium *Clostridium botulinum*, which can be dangerous to infants, as the endospores can transform into toxin-producing bacteria in infants' immature intestinal tracts, leading to illness and even death.

The view by White (1976) has demonstrated that proteins in honey are mainly enzymes. Honey contains roughly 0.5% proteins (Alvarez-Suarez *et al.*, 2010) and the protein contents in some honeys can be over 1000 μg/g (Azeredo *et al.*, 2003). Although the content of amino acids in honey is relatively small, it has been found that almost all of physiologically essential amino acids are present in honey (Cotte *et al.*, 2004; Hermosín *et al.*, 2003). The primary amino acid is proline, contributing 50–85% of the total amino acids (Hermosín *et al.*, 2003).

Investigations have shown that a wide range of trace elements are present in honey, including Al, Ba, Bi, Co, Cr, Mo, Ni, Pb, Sn and Ti, as well as minerals (Ca, Cu, Fe, K, Na, Mg, Mn and Zn) (Conti, 2000; Stocker *et al.*, 2005); among them, the main mineral element is potassium, while copper presents the lowest amount (Nanda *et al.*, 2003). Vitamins such as thiamin (B_1), riboflavin (B_2), pyridoxine (B_6) and ascorbic acid (C) have also been reported, but their amount is very small in honey (Ball, 2007; Namda *et al.*, 2003). When honey is treated with mild heat or prolonged storage, a compositional change can occur due to caramelization of the carbohydrates, the Maillard reaction, and decomposition of fructose in the acid medium of the honey (Villamiel *et al.*, 2001).

Phytochemicals are chemical substances naturally occurring in plants and many of them are now recognized as having health-promoting activity (Apostolidis *et al.*, 2006). Phenolic substances are the largest group of phytochemicals (King and Young, 1999). The plants containing phytochemicals might be used as a supply for the bees, thereby bioactive compounds can be transferred to honey. Studies have shown that honey contains great variation in contents of different phytochemicals according to floral sources and climatic conditions, which contribute to different characteristic colors, flavors, aromas and bioactivities (Molan, 1996). As herbal medicines are derived from different plants, which can produce different therapeutic properties (Villegas *et al.*, 1997), some honey derived from these specific plants may provide added value for health promotion. Honey produced by bees fed herbal extracts has shown greater antioxidant activity than normal honey (Rosenblat *et al.*, 1997).

The explanations above show that honey is a very useful natural source. It contains many constituents that give many benefits for humans. As a consequence, honey can be used in food, pharmaceutical, cosmetic etc. because of its composition. The nutritional value of honey is listed in Table 14.2.

Honey is also a rich source of organic acid. For many years, formic acid had been considered the principal acid in honey. Several reports before the First World War contradicted this view; it was shown by Farnsteiner in 1908 and confirmed by Merl (1908), Heiduschka and Kaufmann (1911) and Fincke (1912), that only about 10% of the acidity of honey was due to this acid. Farnsteiner believed the acidity was due to malic acid. Heiduschka and Kaufmann (1911) reported the presence of the volatile acids such as butyric, valerie, caproic and capric acids, and also lactic, malic, tartaric, oxalic and succinic acids. The acids were not isolated and identified. Nearly 20 years later, Nelson and Mottern (1931) identified citric, malic and succinic acids in honey. These compounds were isolated and identified by derivative formation; acetic acid was also reported. Shortly afterwards, Angeletti (1932) reported d-gluconic acid in a sample of 'musty' honey; it was separated and identified as the calcium salt and its presence was ascribed to mould activity.

Early determinations of vitamin content in honey by feeding tests were generally negative (Kifer *et al.*, 1929). Later examinations using chemical and microbiological methods (Vivino *et al.*, 1943) gave small but definite values for the various vitamins in honey. There is no doubt that honey contains variable amounts of thiamine, riboflavin, pantothenic acid, nicotinic acid, pyridoxine, ascorbic acid, vitamin K, folic acid and biotin. There is also no doubt that the amounts of these factors commonly found in honey are nutritionally insignificant to humans. A recent controversy regarding the amount of vitamin C in honey has been summarized by Hansson (1949). Early animal assays for vitamin C were negative. Ulrich reported to the *7th International Congress of Entomology* in 1938 that something like vitamin C occurred in honey, and several

Table 14.2 Nutritional value of honey

Nutritional value per 100 g (3.5 oz)	
Energy	1,272 kJ (304 kcal)
Carbohydrates	82.4 g
Sugars	82.12 g
Dietary fiber	0.2 g
Fat	0 g
Protein	0.3 g
Vitamins	
Riboflavin (B_2)	0.038 mg (3%)
Niacin (B_3)	0.121 mg (1%)
Pantothenic acid (B_5)	0.068 mg (1%)
Vitamin B_6	0.024 mg (2%)
Folate (B_9)	2 μg (1%)
Vitamin C	0.5 mg (1%)
Trace metals	
Calcium	6 mg (1%)
Iron	0.42 mg (3%)
Magnesium	2 mg (1%)
Phosphorus	4 mg (1%)
Potassium	52 mg (1%)
Sodium	4 mg (0%)
Zinc	0.22 mg (2%)
Other constituents	
Water	17.10 g

(USDA, 2015)

investigators (Griebel and Hess, 1939) reported rather impressive amounts of so-called vitamin C in honey, as shown by chemical methods –20–300 mg. per 100 g. However, Becker and Kardos (1939) declared, on the basis of animal feeding tests, that the reducing substance present was not vitamin C. Haydak has published a series of vitamin C values, determined chemically; these vary between 0.5 and 6.5 mg per 100 g. honey. These values are very low and Haydak (1942) developed symptoms of vitamin C deficiency while subsisting on a diet of milk and honey (350 g honey a day). The report of Gontarski (1948) that honey contains an enzyme which destroys vitamin C seems to confirm the practical absence of this vitamin in honey.

The enzymes of honey are constituents of great interest and importance. Present only in the merest traces, they have a profound effect on the nature and characteristics of the honey. They may arise from the nectar, or be added by the bee. Some enzyme activity might also be traced to the pollen content, and some might come from microorganisms present in nectar or honey. The early literature on honey enzymes was reviewed by Auzinger (1910). The statement has been made (Gauss, 1932) that the best honey should show all enzyme reactions of the living cell. Of all these enzymes, amylase probably received the greatest amount of attention, because of the importance attached to the diastatic activity of honey in Europe, which will not be reviewed here. It is sufficient to state that it has been thought that since the diastatic activity is decreased by heat, it can serve as an indicator of the degree of heating to which a honey has been subjected. Only unheated or mildly heated honey is acceptable in many European countries. Since some honeys appear to be naturally low in diastatic activity, there has been some controversy, and difficulty has been encountered in exporting honey to these countries. Diastase as such is valuable, that important physiological or nutritive properties of honey are damaged or destroyed by heat, so diastase can be used as an indicator of heat treatment.

Phenolic compounds commonly found in honey include phenolic acids, flavonoids and polyphenols. Honey phenolic acids can be protocatequic, phydroxibenzonic, caffeic, chlorogenic, vanillic, p-coumaric, benzoic, ellagic and cinnamic acids (Estevinho *et al.*, 2008), and flavonoids in honey consist of naringenin, kaempferol, apigenin, pinocembrin, chrysin, galangin, luteolin etc. (Estevinho *et al.*, 2008). The large and complex flavonoids greatly contribute to honey color, flavor, anti-fungal and antibacterial activities (Movileanu *et al.*, 2000). Phenolic compounds or polyphenols are one of the most important groups originating from plants as secondary products (Bravo, 1998). The most important classes of polyphenols are flavonoids and phenolic acids, with more than 5000 compounds already demonstrated (Bravo, 1998). They have been regarded to have effective antioxidant and radical scavenging activities (Takahama, 1998), based on acting on different mechanisms, such as free radical scavenging, hydrogen-donating and metal ion chelating.

Flavonoids are compounds of low molecular weight that commonly occur bound to sugar molecules and can be categorized as flavonols (the most widely-distributed flavonoids, including quercetin, kaempferol and myricetin), flavanones, flavones, anthocyanidins and isoflavones (genistein and daidzein) (King and Young, 1999). Flavonoids can act as antioxidants in various ways, including direct trapping of reactive oxygen species, inhibition of enzymes responsible for superoxide anions formation, chelation of transition metals involved in the processes of free radical formation, and prevention of the peroxidation process (Rice-Evans *et al.*, 1996). Phenolic acids include hydroxybenzoic (the 2 major acids: ellagic and gallic acids) and hydroxycinnamic acids (King and Young, 1999). The hydroxybenzoic acids are mainly found in berries and nuts (Maas *et al.*, 1991). Hydroxycinnamic acids consist mainly of coumaric, caffeic and ferulic acids, which are rarely found in the free form (D'Archivio *et al.*, 2007).

14.2 Biological Applications of Honey

14.2.1 Antibacterial Effects

Antibiotic resistance has increased substantially in recent years and is posing an ever-increasing therapeutic problem. Honey is a natural, nontoxic and inexpensive product for the need of novel therapies against bacterial infections. The clinical use of honey has an enormous potential, especially in the fight against antibiotic-resistant strains. Several bioactive compounds have been identified in honey, which contributed to its antibacterial activity. The main factor is the high osmotic activity of honey, which does not allow bacterial growth. Recent experimental findings indicated that the amylase present in honey increases the osmotic effect in the media by increasing the amount of sugars and consequently increasing the antibacterial activity. Honey contains small amounts of different enzymes, notably diastase (α-and β-amylase), invertase (α-glucosidase), glucose-oxidase, catalase and acid phosphatase, which derive from nectar sources, salivary fluids and the pharyngeal gland secretions of the honeybee. Diastase is a natural enzyme of honey. Alpha-amylase degrades starch to a mixture of the disaccharide maltose, the trisaccharide maltotriose and oligosaccharides known as dextrins. Also phenolic compounds may contribute to antimicrobial activity. The inhibitory effects of polyphenols for α-amylases have attracted great interest among researchers.

Ahmed *et al.* (2013) had studied *in vitro* antibacterial activity of honey alone against *Klebsiella pneumoniae* and in combination with potato starch in the same bacteria, determined by the agar incorporation method. Without starch, the MICs of the six varieties ranged between 14% (v/v) and 24% (v/v). When starch was incubated with honey and added to the media, the MIACs of the six varieties ranged between 8% (w/v) and 20% (w/v), which represents an MIC drop between 65.55% and 50%.

The antimicrobial activity of honey against several pathogens and its dependence on the floral origin has been widely reported. Several activities of honeys that contribute to their antibacterial effect are now well known. A recent study by Ahmed *et al.* (2013) has demonstrated that osmosis determined the bactericidal effects of starch and honey towards *Escherichia coli* and *Pseudomonas aeruginosa*. The use of sugar for the treatment of infected wounds was investigated with *in vitro* experiments with bacteria pathogenic to humans, such as *E. coli, P. aeruginosa, S. aureus* and *K. pneumoniae*. Amylases present in honey were expected to split starch chains to randomly produce dextrin and maltose and probably increase the osmotic effect in the media by increasing the amount of sugars and consequently increase the antibacterial activity.

The mechanism of honey antibacterial activity is complex and might be attributed to the synergistic activity between its various potent biological ingredients such as polyphenols and flavonoids. Flavonoids are the most common polyphenols and are widely studied as α-amylase inhibitors. However, in the present study, no significant correlation was established between α-amylases activity and bioactive compounds.

Honey contains sugars, other carbohydrates, minerals (0.169%), proteins (169 mg/100 g) with a water content of about 17.2%. The combinations of these factors suppress the growth of spoilage bacteria and contribute to the stability of the product without the necessity of particular storage conditions. However, the presence of a small number of bacteria species is anticipated in unpasteurized honeys, comprising their natural microflora. On the other hand, variable amounts of hydrogen peroxide, free radical production, increased osmolarity, acidity, water activity, volatiles, organic acids, phenolic compounds and beeswax, among others, are proposed by many researches as the contributing factors for its antimicrobial activity.

14.2.2 Antioxidant Effects

The phenolic compounds of honey have been known to pose significantly antioxidant activity, including iron-binding and free radical scavenging activity (Wang, 2011). Honey contains a significantly high level of antioxidants, both enzymatic and non-enzymatic, including catalase, phenolic acids, flavonoids, carotenoids, organic acids, ascorbic acid, amino acids, proteins and Maillard reaction products (Aljadi and Kamaruddin, 2004; Gheldof and Engeseth, 2002; Nasuti, *et al.*, 2006; Schramm *et al.*, 2003; Vela, 2007).

The ability of honey to neutralize free radicals has been demonstrated. A clinical trial of honey dressings on burns has indicated that the way in which honey initiates healing in burns is the control of free radicals by the antioxidant activity of honey

(Subrahmanyam, 2003). This antioxidant activity may be partly responsible for the anti-inflammatory action of honey, as oxygen free radicals are involved in various aspects of inflammation. Even if the honey antioxidants do not directly suppress the inflammation, it can at least reduce the amount of damage caused from ROS by scavenging free radicals (Molan, 2001). In addition to removing free radicals via scavenging, honey has the potential to inhibit the generation of free radicals formed from hydrogen peroxide in the first place through a different mechanism of antioxidants (Molan, 2001). While the superoxide produced in inflammation is relatively unreactive and can easily to be converted into less reactive hydrogen peroxide, the generation of peroxide radicals from hydrogen peroxide catalyzed by metal ions, such as iron and copper, can be extremely damaging (Cross *et al.*, 1987). Antioxidants, however, like flavonoids or some polyphenols, are able to sequester these metal irons in a non-catalytic form (Halliwell and Cross, 1994). Although it is important that antioxidants neutralize free radicals when the inflammation has been established, inhibition of the formation of hydroxide radicals would greatly help break the vicious inflammatory cycle.

The antioxidant capacity of different honeys depends on the floral sources used by the bees to collect the nectar, and seasonal and environmental factors, as well as processing methods (Gheldof and Engeseth, 2002). Although the total antioxidant activity of honey is a combination of a wide range of active substances, the content of phenolic compounds can significantly reflect the total antioxidant activity of honey to some extent (Beretta *et al.*, 2005). However, the level of phenolic compounds present in honey is not always positively proportional to its antioxidant activity (Küçük *et al.*, 2007). The explanation for this activity may be due to the presence of variable types of polyphenols, thereby providing variable scavenging activity (Kücük *et al.*, 2007). Darker honey is likely to have higher antioxidant contents than light colored honeys (Gheldof *et al.*, 2002). Also, the antioxidant content is higher in honey with a higher water content (Frankel *et al.*, 1998).

In humans, after honey is consumed, an increase in plasma antioxidants has been reported, and the antioxidants give protection in the bloodstream and within cells (Schramm *et al.*, 2003), demonstrating that the bioavailability and bioactivity of honey gives a high efficiency antioxidant transfer from honey to plasma.

Chemically, a free radical is any atom such as oxygen or nitrogen with at least one unpaired electron present, and is able to exist independently Free radicals can easily be formed in three ways:

1. by the homolytic cleavage of a covalent bond, generally incurring by high energy input;
2. by the loss of a single electron from a normal molecule; and
3. by addition of a single electron to a normal molecule (Cheeseman and Slater, 1993).

These free radicals that are highly reactive molecules can be extremely damaging to the lipids, proteins and cellular DNA (Cochrane, 1991), which may lead to many biological complications, including carcinogenesis, mutagenesis, aging and atherosclerosis (Halliwell and Gutteridge, 1989). The oxygen-derived free radical is an important group formed during metabolism. One of these reactions found in biological pathways is the respiratory burst process, which results in free radical products termed reactive oxygen species (ROS) (Halliwell and Gutteridge, 1989; Winrow *et al.*, 1993).

Examples of ROS include superoxide, hydroxyl radicals and non-oxygen free radical hypochlorites (Conner and Grisham, 1996).

14.2.3 Antiviral Effects

Honey is a natural product that can exhibit anti-inflammatory, wound healing, antioxidant and anti-neoplastic activities. In addition, honey has been used as a traditional remedy for bacterial infections, colds and coughs since ancient times, suggesting that it is effective against many infectious diseases. Natural honey shows a broad spectrum activity against Gram-positive and Gram-negative pathogenic bacteria and fungi, including methicillin-resistant *Staphylococcus aureus* (MRSA) (Jenkins *et al.*, 2014), *Shigella sonnei* (Lusby *et al.*, 2005), *Helicobacter pylori* (Manyi *et al.*, 2010) and *Candida albicans* (Ansari and Usmani, 2013). The antibacterial and antifungal activities of honey are attributable to its osmolarity, low pH, hydrogen peroxide (H_2O_2), phenolic acids and flavonoids. Moreover, natural honey has been reported to exhibit antiviral activities against rubella virus (Zeina, 1996) and varicella-zoster virus (VZV) (Shahzad, 2012) and is used to treat recurrent herpes simplex lesions. From these observations about antiviral activities, traditional use and phytochemicals in honey, it was presumed that honey is effective against the influenza virus. Therefore, in the present study, the anti-influenza viral activities of honey were investigated.

14.2.4 Anti-inflammatory Effects

Inflammation is part of the non-specific immune reaction occurring in response to any type injury in body (Ferrero-Miliani *et al.*, 2007). A mild short-lived inflammation is essentially required to initiate the healing process, but when the inflammation becomes prolonged, it can slow or prevent the healing (Molan, 2009). Once inflammatory stimulus is extended, the superoxide and hydrogen peroxide are continuously generated as they act to recruit more neutrophils (Flohé *et al.*, 1985; Klyubin, 1996). Hydrogen peroxide activates more neutrophils via the activation of the nuclear transcription factor NF-kB to produce specific cytokines, which amplify the inflammatory response by activating leukocytes, and activated neutrophils in turn generate more hydrogen peroxide, which sets up a vicious cycle; hydrogen peroxide from other sources can also trigger this cycle (Molan, 2009). It has been found that the oxidative species produced from hydrogen peroxide, instead of hydrogen peroxide, are responsible for activating NF-kB, and antioxidants prevent this activation (Grimble, 1994).

The most harmful consequence of inflammation is the production of reactive oxygen species in the tissue. These reactive oxygen species (free radicals) can be very damaging as they are very reactive and are able to break down the proteins, lipids and nuclei acids (Molan, 2001). The continued production of free radicals can lead to localized erosion of body tissue. The free radicals are also involved in stimulating the activity of fibroblasts, which is the basis of the repair process (Molan, 2009). The fibroblasts are responsible for synthesis of collagen fibers and other connective tissue components, and if inflammation is continued, the over-stimulation can lead to fibrosis and excessive production of collagen fibers (Molan, 2001). Thus, there are significant benefits for therapeutic use of anti-inflammatory substances. However, pharmaceutical anti-inflammatory medicines have serious limitations:

corticosteroids suppress tissue growth and the immune system (Bucknall, 1984), and the non-corticosteroids are harmful to cells (Brooks, 1985). Honey has been confirmed to possess anti-inflammatory activity without adverse side effects (Molan, 2001).

The anti-inflammation properties of honey have been known well (Molan, 2001). Honey has been found to involve scavenging activity of reactive oxygen species responsible for induction of inflammation (Greten *et al.*, 2004). When honey is applied to wounds, it effectively reduces the inflammation (Burlando, 1978; Subrahmanyam, 1998), as well as reducing oedema around wounds (Dumronglert, 1983; Efem, 1993) and exudation from wounds (Burlando, 1978; Hejase *et al.*, 1996).

Honey has also been observed to relieve the pain that is a feature of inflammation (Burlando, 1978; Subrahmanyam, 1993). Studies of healing animal tissues have indicated that the leucocyte numbers associated with inflammation have been less when the wounds have been treated with honey (Subrahmanyam, 1998). Similar results observed in animal study models have confirmed that the anti-inflammation action of honey cannot be due to removal of bacteria alone (Oryan and Zaker, 1998; Postmes, 1997). Furthermore, honey has been shown to decrease the stiffness of inflamed wrist joints of guinea pigs (Church, 1954). One of the anti-inflammatory effects of honey can be attributed to its antibacterial activity, since components of the bacterial cell wall are potent stimulators of the inflammatory response (Molan, 2009). The presence of slough in wounds also acts as an inflammatory stimulus, and slough removal by honey application to wounds has been shown to help decrease inflammation (Efem, 1988).

The anti-inflammatory action of honey is potentially very important for therapeutic application, as the inflammation can cause major consequences. During the course of inflammation, some mediators called prostaglandins produced by the leucocytes to regulate the activity of surrounding cells can cause the painful symptoms of inflammation, and others can cause blood vessels to dilate and the capillary walls to open up (Molan, 2001). Plasma flows out to cause swelling and to increase the diffusion distance from the capillaries to the cells (Molan, 2001). The opening up of capillaries can cause exudation of serum. If it is prolonged, it could lead to malnutrition (Molan, 2001).

14.3 Conclusion

Bees, the highly beneficial insect species from the family of Apidae, are important for pollinating crops, producing beneficial honey and other medicinal products, and sustaining the world's natural life cycle. The honey bees are bees of the single genus Apis from the tribe of Apini, where the most popular subspecies is *Apis mellifera*, while other subspecies such as *Apis dorsata*, *Apis cerana* and *Apis florea* are found in Southern and Southeastern Asia.

The primary product from bee is honey. Bees produce honey from nectar of blossoms or from honeydew (White, 1976). It has a sweet taste and contains a complex mixture of carbohydrates as major constituents (Mendes *et al.*, 1998). Other constituents in honey include organic acids, lactones, amino acids, minerals, vitamins, enzymes, pollen, wax and pigments (Fallico *et al.*, 2004). Because of its composition, honey can exhibit many biological activities. Many applications of honey have been studied, such as in pharmaceuticals, cosmetics and foods.

References

Ahmed, M., Khiati, B., Aissat, Djebli, S.A.N., Meslem, A. and Bacha, S. (2013) The relationship between bioactive compounds with diastase activity and antibacterial synergy of honey and potato starch combinations against Klebsiella pneumonia. *Clinicali Microbial*, **2**: 6.

Aljadi, A., and Kamaruddin, M. (2004) Evaluation of the phenolic contents and antioxidant capacities of two Malaysian floral honeys. *Food Chemistry*, **85**(4): 513–518.

Alvarez-Suarez, J.M., *et al.* (2010) Antioxidant and antimicrobial capacity of several monofloral Cuban honeys and their correlation with color, polyphenol content and other chemical compounds. *Food and Chemical Toxicology*. **48**: 2490–2499.

Angeleiti, A. (1932) The presence of d-gluconic acid in musty honey. *G. Farm. chim.* **81**: 533–536.

Ansari, M.J., Al-Ghamdi, A., Usmani, S., *et al.* (2013) Effect of jujube honey on *Candida albicans* growth and biofilm formation. *Arch Med Res.*, **44**: 352–360.

Apostolidis, E., Kwon, Y., and Shetty, K. (2006) Potential of cranberry-based herbal synergies for diabetes and hypertension management. *Asia Pacific Journal of Clinical Nutrition*, **15**(3): 433–441.

Auzinger, A. (1910) Ueber Fermente im Honig und den Wert ihres Nachweises fiir die Honigbeurteiling Z. Untersuch. Nahr.- u. *Genussm.* **19**: 65–83.

Azeredo, L., Azeredo, M., Souza, S., and Dutra, V. (2003) Protein contents and physicochemical properties in honey samples of *Apis mellifera* of different floral origins. *Food Chemistry*, **80**: 249–254.

Ball, D.W. (2007) The chemical composition of honey. *Journal of Chemical Education*, **84**: 1643–1646.

Becker, J. and Kardos, R.F. (1939) Ueber den Vitamin-C-Gehalt von Honig Z. *Untersuch. Lebensmitt*, **78**: 305–308.

Beretta, G., Granata, P., Ferrero, M., Orioli, M. and Facino, R.M. (2005) Standardization of antioxidant properties of honey by a combination of spectrophotometric/fluorimetric assays and chemometrics. *Analytica Chimica Acta*, **533**: 185–191.

Bravo, L. (1998) Polyphenol: Chemistry, dietary sources, metabolism and nutritional significance. *Nutrition Reviews*, **56**: 317–333.

Brooks, F.P. (1985) The pathophysiology of peptic ulcer disease. *Digestive Diseases and Sciences*, **30**(11): 15S–29S.

Bucknall, T.E. (1984) Factors affecting healing. In: *Wound Healing for Surgeon* (eds T.E. Bucknall and H. Ellis), Bailliére Tindall, London, pp. 42–74.

Burlando, F. (1978) Sull'azione terapeutica del miele nelle ustioni. *Minerva Dermatologica*, **113**: 699–706.

Cheeseman, K.H. and Slater, T.F. (1993) An introduction to free radical biochemistry. *British Medical Bulletin*, **49**: 481–493.

Church, J. (1954) Honey as a source of the anti-stiffness factor. *Federation Proceedings of the American Physiology Society*, **13**(1): 26.

Cochrane, C.G. (1991) Cellular injury by oxidants. *American Journal of Medicine*, **91**(Suppl. 3c): 23S–30S.

Conner, E.M. and Grisham, M.B. (1996) Inflammation, free radicals, and antioxidants. *Nutrition*, **12**(4): 274–277.

Conti, M. (2000) Lazio region (central Italy) honeys: A survey of mineral content and typical quality parameters. *Food Control*, **11**: 459–463.

Cotte, J.F., Casabianca, H., Giroud, B., Albert, M., Lheritier, J. and Grenierloustalot, M.F. (2004) Characterization of honey amino acid profiles using high-pressure liquid chromatography to control authenticity. *Analytical and Bioanalytical Chemistry*, **378**(5): 134–1350.

Cross, C E., Halliwell, B., Borish, E.T., Pryor, W.A., Ames, B.N., Saul, R.L., *et al.* (1987) Oxygen radicals and human disease. *Annals of Internal Medicine*, **107**: 526–545.

D'Archivio, M., Filesi, C., Benedetto, R., Gargiulo, R., Giovannini, C. and Masella, R. (2007) Polyphenols, dietary sources and bioavailability. *Ann Ist Super Sanità*, **43**: 348–361.

Doner, L.W. (1977) The sugars of honey: A review. *Journal of the Science of Food and Agriculture*, **28**: 443–456.

Dumronglert, E. (1983) A follow-up study of chronic wound healing dressing with pure natural honey. *Journal of the National Research Council of Thailand*, **15**(2): 39–66.

Efem, S.E.E. (1988) Clinical observations on the wound healing properties of honey. *British Journal of Surgery*, **75**: 679–681.

Efem, S.E.E. (1993) Recent advances in the management of Fournier's gangrene: Preliminary observations. *Surgery*, **113**(2): 200–204.

Estevinho, L., Pereira, A.P., Moreira, L., Dias, L.G. and Pereira, E. (2008) Antioxidant and antimicrobial effects of phenolic compounds extracts of Northeast Portugal honey. *Food and Chemical Toxicology*, **46**: 3774–3779.

Fallico, B., Zappalà, M., Arena, E. and Verzera, A. (2004) Effects of conditioning on HMF content in unifloral honeys. *Food Chemistry*, **85**(2): 305–313.

Ferrero-Miliani, L., Nielsen, O.H., Andersen, P.S. and Girardin, S.E. (2007) Chronic inflammation: importance of NOD2 and NALP3 in interleukin-1β generation. *Clinical and Experimental Immunology*, **147**(2): 227–235.

Fincke, H. (1912) Beitrage zur Bestimmung der Arneisensaure in Nahrungsmitteln III Z. Intersuch. Nahr.- u. *Genussm*, **23**: 255–267.

Flohé, L., Beckmann, R., Giertz, H. And Loschen, G. (1985) Oxygen-centred free radicals as mediators of inflammation. In: *Oxidative Stress* (ed. H. Sies), Academic Press, London, pp. 403–435.

Frankel, S., Robinson, G.E. and Berenbaum, M.R. (1998) Antioxidant capacity and correlated characteristics of 14 unifloral honeys. *Journal of Apicultural Research*, **37**(1), 27–31.

Gauss, J. (1932) Eine Methode zur Erganzung der Honiganalyse Z. *Untersuch, Lebensmitt*, **64**: 376–383.

Gheldof, N. and Engeseth, N.J. (2002) Antioxidant capacity of honeys from various floral sources based on the determination of oxygen radical absorbance capacity and inhibition of *in vitro* lipoprotein oxidation in human serum samples. *Journal of Agricultural and Food Chemistry*, **50**: 3050–3055.

Gheldof, N., Wang, X.-H. and Engeseth, N. J. (2002) Identification and quantification of antioxidant components of honeys from various floral sources. *Journal of Agricultural and Food Chemistry*, **50**: 5870–5877.

Gontarski, H. (1948) Ein Vitamin C oxydierendes Ferment der Honigbiene. *Z. Naturf.* **3b** (7/8): 245–249.

Greten, F., Eckmann, L., Greten, T., Park, J., Li, Z., Egan, L., *et al.* (2004) IKK-beta links inflammation and tumorigenesis in a mouse model of colitis-associated cancer. *Cell*, **118**(3): 285–296.

Griebel, C. and Hess, G. (1939) Vitamin C-enthaltende Honige Z. *Untersuch. Lebensmitt.* **78**: 308–314.

Grimble, G.F. (1994) Nutritional antioxidants and the modulation of inflammation: Theory and practice. *New Horizons*, **2**(2): 175–185.

Halliwell, B. and Gutteridge, J. (1989*) Free Radicals in Biology and Medicine*. Clarendon Press, Oxford.

Halliwell, B. and Cross, C.E. (1994) Oxygen-derived species: Their relation to human disease and environmental stress. *Environmental Health Perspectives*, **102**(Suppl 10): 5–12.

Hansson (1949) Innehaller honungen C-vitamin. *Nord. Bitidskr.* **1**(4): 99–101

Haydak, M.H. *et al.* (1942) Vitamin content of honeys. *Journal of Nutrition*, **23**(6): 581–588.

Heiduschka, A. and Kaufmann, G. (1911) Ueber die flilchtigen Siiuren im Honig Z. Untersuch. Nahr.- u. *Genussm.* **21**: 375–378.

Hejase, M.J., Simonin, J.E., *et al.* (1996) Genital Fournier's gangrene: Experience with 38 patients. *Urology*, **47**(5): 734–739.

Hermosín, I., Chicón, R.M., and Cabezudo, D. (2003) Free amino acid composition and botanical origin of honey. *Food Chemistry*, **83**: 263–268.

Jenkins, R., Burton, N. and Cooper R. (2014) Proteomic and genomic analysis of methicillin-resistant *Staphylococcus aureus* (MRSA) exposed to manuka honey *in vitro* demonstrated down-regulation of virulence markers. *Journal of Antimicrobial Chemotherapy.* **69**: 603–615.

Kifer, H.B. and Munsell, H.E. (1929) Vitamin content of honey and honeycomb. *Journal of Agricultural Research*, **39**(8): 355–366.

King, A. and Young, G. (1999) Characteristics and occurrence of phenolic phytochemicals. *Journal of the American Dietetic Association*, **99**: 213–218.

Klyubin, I., Kirpichnikova, K., and Gamaley, I. (1996) Hydrogen peroxide-induced chemotaxis of mouse peritoneal neutrophils. *European Journal of Cell Biology*, **70**(4): 42–44.

Küçük, M., Kolaylı, S., Karaoğlu, Ş., Ulusoy, E., Baltacı, C. and Candan, F. (2007) Biological activities and chemical composition of three honeys of different types from Anatolia. *Food Chemistry*, **100**(2): 526–534.

Lusby, P.E., Coombes, A.L. and Wilkinson, J.M. (2005) Bactericidal activity of different honeys against pathogenic bacteria. *Arch Med Res.*, **36**: 464–467.

Maas, J. and Galletta, G. (1991) Ellagic acid, an anticarcinogen in fruits, especially in strawberries: A review. *Horticultural Science*, **26**(1): 10–13.

Manyi-Loh, C.E., Clarke, A.M., Munzhelele, Tl, *et al.* (2010) Selected South African honeys and their extracts possess *in vitro* anti-*Helicobacter pylori* activity. *Arch Med Res.*, **41**: 324–331.

Mendes, E., Brojo Proença, E., Ferreira, I.. and Ferreira, M.A. (1998) Quality evaluation of Portuguese honey. *Carbohydrate Polymers*, **37**(3): 219–223.

Merl, T. (1908) Zum Nachweis der Ameisensiiure im Bienenhonig Z. Untersuch, Nahr u. *Genussm*, **16**: 385–389.

Molan, P.C. (1996) Authenticity of honey. In: Food Authentication (eds P.R. Ashurst and M.J. Dennis). Blackie Academic & Professional, London, pp. 259–303.

Molan, P.C. (2001a) Potential of honey in the treatment of wounds and burns. *American Journal of Clinical Dermatology*, **2**(1): 13–19.

Molan, P.C. (2001b) Why honey is effective as a medicine. 2: The scientific explanation of its effects. *Bee World*, **82**(1): 22–40.

Molan, P.C. (2009) Why honey works. In: *Honey in Modern Wound Management* (eds R. Cooper, P. Molan and R. White), Wounds UK, Ltd, Aberdeen.

Movileanu, L., Neagoe, I. and Flonta, M.L. (2000) Interaction of the antioxidant flavonoid quercetin with planar lipid bilayers. *International Journal of Pharmaceutics*, **205**: 135–146.

Nanda, V., Sarkar, B.C., Sharma, H.K. and Bawa, A.S. (2003) Physico-chemical properties and estimation of mineral content in honey produced from different plants in Northern India. *Journal of Food Composition and Analysis*, **16**: 613–619.

Nasuti, C., Gabbianelli, R., Falcioni, G. and Cantalamess, F. (2006) Antioxidative and gastroprotective activities of anti-inflammatory formulations derived from chestnut honey in rats. *Nutrition Research,* **26**(3): 130–137.

Nelson, E.K. and Mottern, H.H. (1931) Some organic acids in honey. *Industr. Engng. Chem,* **23**: 335–336.

Oryan, A., and Zaker, S.R. (1998) Effects of topical application of honey on cutaneous wound healing in rabbits. *Journal of Veterinary Medicine Series A*, **45**(3): 181–188.

Postmes, T.J., Bosch, M.M.C., Dutrieux, R., van Baare, J., and Hoekstra, M.J. (1997) Speeding up the healing of burns with honey. An experimental study with histological assessment of wound biopsies. In: *Bee Products: Properties, Applications and Apitherapy* (eds A. Mizrahi and Y. Lensky), Plenum Press, New York, pp. 27–37.

Rice-Evans, C.A., Miller, N.J. and Paganga, G. (1996) Structure-antioxidant activity relationships of flavonoids and phenolic acids. *Free Radical Biology and Medicine*, **20**: 933–956.

Rosenblat, G., Angonet, S., Goroshit, A., Tabak, M. And Neeman, I. (1997) Antioxidant properties of honey produced by bees fed with medical plant extracts. In: *Proceedings of the International Conference on Bee Products: Properties, Applications and Apitherapy*. Plenum Press, New York, pp. 49–55

Sato, T. And Miyata, G. (2000) The nutraceutical benefit, Part III: Honey. *Nutrition*, **16**: 468–469.

Schramm, D.D., Karim, M., Schrader, H.R., Holt, R.R., Cardetti, M. and Keen, C.L. (2003) Honey with high level of antioxidants can provide protection to healthy human subjects. *Journal of Agricultural and Food Chemistry*, **51**: 1732–1735.

Shahzad, A. and Cohrs, R.J. (2012) *In vitro* antiviral activity of honey against *Varicella zoster* virus (VZV): A translational medicine study for potential remedy for shingles. *Transl Biomed.* **3**: 1–7.

Siddiqui, I.R. (1970) The sugars of honey. *Advances in Carbohydrate Chemistry and Biochemistry*, **25**: 285–309.

Stocker, A., Schramel, P., Kettrup, A. and Bengsch, E. (2005) Trace and mineral elements in royal jelly and homeostatic effects. *Journal of Trace Elements in Medicine and Biology*, **19**: 183–189.

Subrahmanyam, M. (1998) A prospective randomised clinical and histological study of superficial burn wound healing with honey and silver sulfadiazine. *Burns*, **24**(2): 157–161.

Subrahmanyam, M., Sahapure, A.G., Nagane, N.S., Bhagwat, V.R. and Ganu, J.V. (2003) Free radical control – the main mechanism of the action of honey in burns. *Annals of Burns and Fire Disasters*, **16**(3): 135–138.

Takahama, U. (1998) Flavonoid. *Protein, Nucleic Acid, and Enzyme*, **33**: 2994–2999.

USDA (2015) United States Department of Agriculture, Agricultural Research Service, National Nutrient Database for Standard Reference Release 28, Full Report (All Nutrients): 19296, Honey.

Vela, L., de Lorenzo, C. And Pérez, R.A. (2007) Antioxidant capacity of Spanish honeys and its correlation with polyphenol content and other physicochemical properties. *Journal of the Science of Food and Agriculture*, **87**(6): 1069–1075.

Villamiel, M., Dolores del Castillo, M., Corzo, N. and Olano, A. (2001) Presence of furosine in honeys. *Journal of the Science of Food and Agriculture*, **81**: 790–793.

Villegas, L., Fernández, I., Maldonado, H., Torres, R., Zavaleta, A., Vaisberg, A., *et al.* (1997) Evaluation of the wound-healing activity of selected traditional medicinal plants from Peru. *Journal of Ethnopharmacology*, **55**(3): 193–200.

Vivino, A.E. *et al.* (1943) Antihemorrhagic vitamin effect of honey. *Proceedings of the Society expo Biol.*, New York. **53**(1): 9–11.

Wang, H. (2011) The Study of the Antioxidant Activity of Phenolic Components of Manuka Honey. University thesis. The University of Waikato.

White, J.W. (1976) Composition of honey. In: *Honey: A comprehensive survey* (ed. E. Crane), Heinemann Newnes, Oxford, pp. 157–206.

Winrow, V.R., Winyard, P.G., Morris, C.J. and Blake, D.R. (1993) Free radicals in inflammation: Second messengers and mediators of tissue destruction. *British Medical Bulletin*, **49**: 506–522.

Zeina, B., Othman, O. and al-Assad, S. (1996) Effect of honey versus thyme on Rubella virus survival *in vitro*. *Journal of Alternative Complementary Medicine*, **2**: 345–348.

15
Advances in Milk Fractionation for Value Addition

Juan M. Gonzalez[1], Deepak Bhopatkar[2], and Dattatreya Banavara[3]

[1]*Global Research and Development, PepsiCo, Barrington, IL, USA*
[2]*Global Research and Development, Mead Johnson Nutrition, Evansville, IN, USA*
[3]*Global Innovation, Firmenich Inc, Plainsboro, NJ, USA*

15.1 Dairy Ingredient Development

Milk is a complete nutritional food designed by nature to support the healthy growth of infant mammals. In broad terms, it is a fluid composed mostly of water where a set of nutritionally complete proteins are dispersed, an emulsion of rich fats is stabilized, and a unique disaccharide and a complex set of oligosaccharides are in a solution shared by mineral salts. The future of dairy food ingredient development will rely on a deeper understanding of the functionality of these components, from the physical-chemical as well as the nutritional and physiological perspectives. The application of novel technologies to fractionate unique milk components, to combine them in unique ratios with themselves or with other food ingredients, may provide a broader spectrum of functional ingredients to develop new products to satisfy the needs of the diverse global population.

The development and approval for use of novel dairy ingredients and dairy processing technologies may be followed by a path of inter-related drivers related to science, technology and society concerns. Science and technology drivers are influenced by health and nutrition, sensory and physical-chemical attributes of the ingredient and product, safety requirements, and novel process and packaging technologies (Figure 15.1) (Gonzalez, 2014a). Societal drivers are influenced by multiple elements that relate to laws and regulations, market, consumer and pricing strategies, import and export policies, agricultural practices, deforestation, and sustainability attitudes towards water, energy consumption and contamination.

These inter-related drivers may be used to trace the development of dairy innovations throughout history. As an example, Marco Polo documented the use of sun-and-wind dried milk for sustenance by early Mongolians (Shephard, 2006). The Mongol horse nomads were responding to 'Health and Nutrition drivers' for the need of a sustainable food source through their journeys far from their villages. The 'Enabler drivers' regarding processes were a) the rudimentary technique for the crude removal of fat, and b) a minimal heat treatment coupled with water removal via sun-and-wind exposure, to control microbial degradation and spoilage. The

Food Processing By-Products and their Utilization, First Edition. Edited by Anil Kumar Anal.

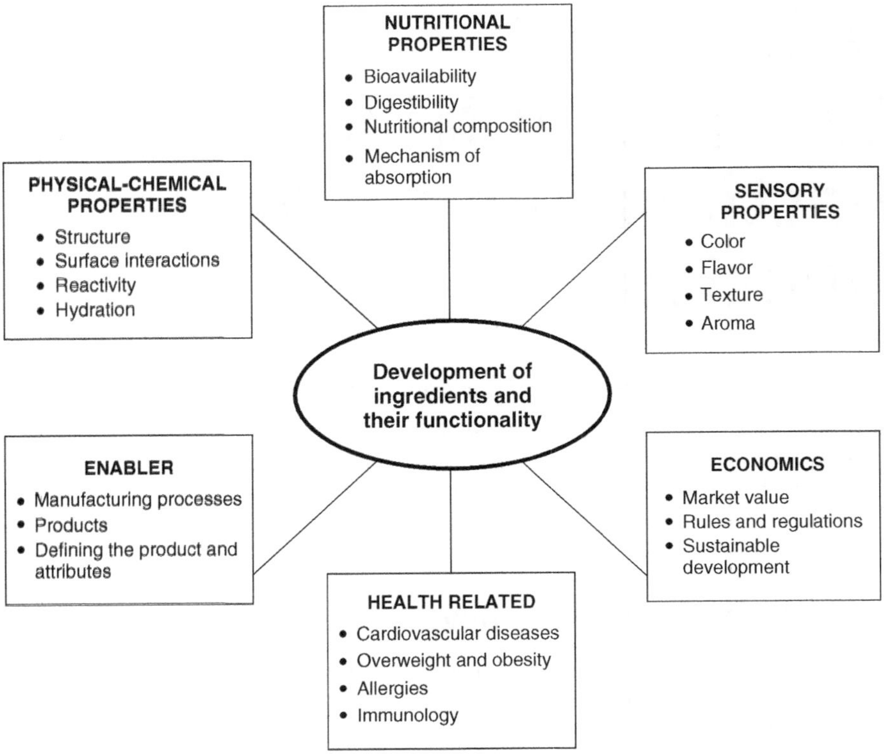

Figure 15.1 Model of functional drivers for the development of dairy food ingredients through their path from industry to the consumer (adapted from Gonzalez, 2014a).

resulting thick paste material addressed the 'Physical-Chemical drivers' of low water activity and fast hydration. These in turn contributed to Enabler drivers regarding convenience of longer shelf life, transportation and consumption away from the source. Further technological advances achieved during the 19th century, such as production of powder milk and condensed milk, were responding to the inevitable call for progress and invention that supports the Enabler driver.

15.2 Milk Proteins

Milk proteins contain all essential amino acids in appreciable amounts. Most milk proteins, such as caseins, are synthesized by the mammary gland with few exceptions that filter from the blood serum, such as serum albumin and immunoglobulins (Fox, 2003). The protein content in milk has a broad degree of variation, with changes occurring between and within mammal species, as well as with the period of lactation (Campbell and Marshall, 1975; Kunz and Lönnerdal, 1992) (Figure 15.2). Milk from Holstein cows may have an average of 3.07 g protein/100 ml, while milk from Jersey cows may deliver 33% more protein with an average of 4.07 g/100 ml (Cerbulis and Farrell, 1975). Milk protein content decreases with lactation period: Human milk varies from colostrum to late lactation in a range of 15.8 to 6.9 g/liter (Lönnerdal and Atkinson, 1995). Bovine milk shows a broader range from 14.9 to 3.2 g/100 ml in colostrum to late lactation,

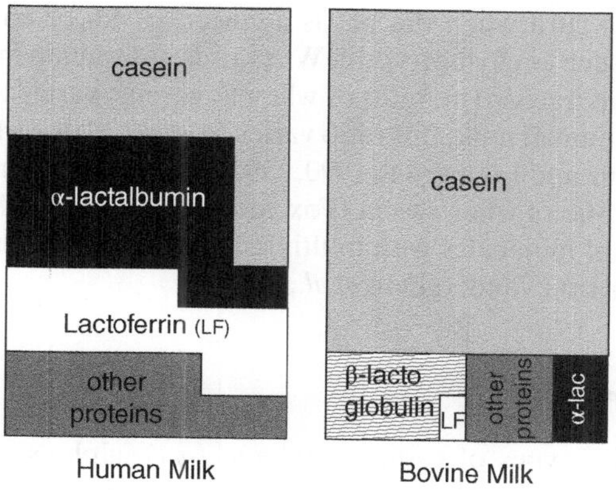

Figure 15.2 Protein composition of mature human and bovine milks. Casein constitutes the major protein portion; however, α-lactalbumin and lactoferrin are predominant in human milk, while β-lactoglobulin predominates in bovine milk.

respectively. This variation is explained by the growth needs of the infant mammal (Campbell and Marshall, 1975).

15.3 Milk Proteins Classification

Milk proteins may be classified into three major fractions: caseins, whey proteins, and milk fat globule membrane (MFGM) proteins (Table 15.1). Caseins are the portion

Table 15.1 Composition of milk proteins in mature bovine milk (adapted from Bylund, 1995)

	Bovine Milk Proteins	
	Concentration, g/100 g	**Ratio of Total Protein, g/100 g**
Casein		
α_{s1}-casein	1.0	30.6
α_{s2}-casein	0.26	8.0
β-casein	1.01	30.8
κ-casein	0.33	10.1
Total Casein	**2.6**	**79.5**
Whey Proteins		
β-lactoglobulin	0.32	9.8
α-lactalbumin	0.12	3.7
Blood Serum Albumin	0.04	1.2
Immunoglobulins	0.07	2.1
Other minor proteins	0.08	2.4
Total Whey Proteins	**0.63**	**19.3**
Fat Globule Membrane Proteins	0.04	1.2
Total Proteins	**32.7**	**100**

that separates into a curd when the pH is adjusted to 4.6 or coagulates under the action of proteases, such as chymotrypsin. Whey proteins remain in the aqueous phase after caseins have separated. The ratio of whey to caseins varies between and within mammal species. In human milk, this ratio varies from 80:20 in early lactation to 60:40 in late lactation (Kunz and Lönnerdal, 1990, 1992). In late lactation, bovine milk has a profile with a 20:80 ratio of whey:casein (Fox and McSweeney, 1998). MFGM proteins are a diverse group of molecules with multiple functions and constitute about 1–4% of the total milk protein (Vanderghem *et al.*, 2010).

15.3.1 Caseins

There are four major types of caseins: alpha-s$_1$ (α_{s1}), alpha-s$_2$ (α_{s2}), beta (β) and kappa (κ). κ- casein (23 kDa, pI 4.1) is a glycopeptide that contains a galactose-based trisaccharide segment. α-caseins (23 kDa) and β-casein (24 kDa, pI 4.1) are mostly hydrophobic molecules that cluster to form thermodynamically stable porous spheres known as submicelles (Fox and McSweeney, 1998). The hydrophylic portions alpha and beta caseins contain phosphate groups that bind calcium ions adding to the integrity of the core (McMahon and Oomen, 2008). These submicelles aggregate into larger clusters known as micelles (~50–600 nm). κ-caseins protrude into the aqueous media via the hydrophilic glycopeptide segments and maintain the micelle in dispersion (McMahon and Oomen, 2008; Dalgleish, 2011).

15.3.2 Whey Proteins

Whey proteins are also referred to as serum proteins or non-casein proteins. The major whey proteins in mature bovine milk are β-lactoglobulin, α-lactalbumin and blood serum albumins. The proportion of these components changes with species and with the stage of lactation (Fox and McSweeney, 1998). In human milk, α-lactalbumin and lactoferrin are the two most abundant whey proteins, constituting about 47% of the total protein, while β-lactoglobulin is almost absent (Kunz and Lönnerdal, 1989), whereas in bovine milk these proteins constitute about 4.5% of the total protein (Boland, 2011).

15.3.2.1 Whey Proteins – Minor Components There are a variety of minor proteins in whey that have a biological role to support the growth of the infant mammal. A study on human milk proteomics identified the presence of over 1300 proteins in whey, and further quantified 615 of those (Zhang *et al.*, 2013). The study showed changes in abundance for the majority of glycoproteins over lactation. Some of these glycoproteins are related to immunity roles. The presence of many of these glycoproteins is high in colostrum and early lactation. Among these glycoproteins are immunoglobulins, lactoferrin, lactoperoxidase, haptocorrin and osteopontin (Recio *et al.*, 2009).

Lactoferrin (80 kDa, pI 8.9) is the second-most prevalent protein in human milk whey, with a 10-fold higher presence in colostrum than in mature milk. Its level in mature bovine milk is 20-fold smaller than in human milk. Lactoferrin tightly binds iron in milk and its functions are related to iron metabolism and antimicrobial (Lönnerdal and Atkinson, 1995). Osteopontin is a highly acidic phosphorylated glycoprotein with about 50 calcium binding sites per molecule; it is involved in formation of mineralized tissue and has antimicrobial properties (Denhardt and Guo, 1993). Immunoglobulins (Ig) are present in high concentration in colostrum and their function is to provide various types of immunity. IgA is most abundant in human milk, whereas IgG is more prevalent in bovine milk (Fox and Sweeney, 1998). Lactoperoxidase is a glycosylated enzyme, abundant in colostrum, and catalyzes the inactivation of a wide range of microorganisms in the lactoperoxidase system (Kussendrager and van Hooijdonk, 2000). Haptocorrin is highly glycosylated and has a role in vitamin B_{12} absorption, as it tightly binds cobalamin (Marchaj *et al.*, 1995; Furger, 2012). Lysozyme is a glycosylated enzyme with antimicrobial properties; it is more abundant in human milk than in bovine milk (Johnson and Larson, 2005).

15.3.3 Milk Fat Globule Membrane Proteins

Milk fat globule membrane (MFGM) is the protective layer that surrounds and stabilizes fat globules. MFGM is composed of about 25–60% proteins (w/w, dry), and represent about 1–2% of the total milk proteins (Singh, 2006; Vanderghem *et al.*, 2010; Le *et al.*, 2013). Over 200 distinct proteins were identified. The major ones are mucin1, xanthine dehydrogenase/oxidase (XDH/XO), periodic acid Schiff III (PASIII), cluster differentiation (CD36), butyrophilin, adipophilin, lactadherin (also known as PAS 6/7) and the fatty acid binding protein (FABP) (Cavaletto *et al.*, 2004). Butyrophilin and XDH/XO constitute about 40% and 12% of total MFGM proteins, respectively. Mucins are highly sialylated proteins, with about half of their weight made up of carbohydrates.

15.3.4 Milk Protein Fractionation Technologies

Several technologies have been successfully used for the development of commercial milk protein fractions. Among those are precipitations, such as isoelectric, salting out, complexation, or heat denaturation, and size exclusion or binding affinity via membranes or resin columns are the most common ones. Recent advances using charged membranes and the use of non-traditional expanded bed adsorption chromatography have shown promising results.

15.3.4.1 Ultrafiltration This is one of the several membrane separation technologies that rely on the use of a pressure-assisted tangential cross-flow filtration through porous membranes. The selection of the membrane pore size would define a molecular weight cut-off that would separate higher molecular weight molecules in the retentate and lower molecular weight ones in the permeate.

The addition of charges to the size exclusion membrane technology provides improvement to a traditional fractionation process. The presence of membrane surface charges create a repulsion effect on charged proteins, restricting their permeation through the membrane. These hybrid membranes allow a combined effect of molecular size exclusion and charge repulsion that yield an improvement in permeability through the selection of wide pores, reduced fouling and increased separation selectivity. Research has shown that negatively-charged 100 kDa ultrafiltration membranes deliver similar protein recovery as traditional 10 kDa membranes but use a two-fold higher flux. This technology has been used to fractionate α-lactalbumin (MW 14.2 kDa, pI 4.2) and β-lactoglobulin (MW 18.4, pI 5.2) from milk serum permeate using a 300 kDa membrane at a 87% and 83% purity, respectively (Bhushan and Etzel, 2009; Kumar and Ulbricht, 2013; Arunkumar and Etzel, 2014, 2015).

15.3.4.2 Chromatography Resin permeation has been used commercially for the separation of high-value commercial milk proteins where high resolution and minimal cross-contamination is needed. The traditional gel permeation technology allows for fractionation of molecules based on their size when the mobile phase flows through the stationary phase, that is, the column bed packed with porous gel. The mobile phase molecules interact with the stationary phase based on their hydrodynamic volume. Smaller molecules are retained longer than larger molecules, as they can penetrate the porous channels of the gel matrix. Different from a membrane process, there is no physical separation of retentate and permeate fractions. Fractions are collected as they elute, based on their retention times and volumes.

A different approach to the use of tightly packed gel beads is the use of stabilized fluid bed adsorption, also known as expanded bed adsorption (EBA). The EBA technology flows the mobile phase upwards, fluidizing a high-density adsorbent phase, where the bed's sedimentation velocity equals the upward liquid flow velocity. This technology eliminates the back-pressure and bed compression issues of packed bed chromatography. The gel beads are manufactured with a high density mineral core, such as tungsten or zirconia. It also allows for larger columns of adsorbent gel as the material is maintained loosely suspended in the mobile phase. Non-clarified crude streams could be eluted without the risk of column clogging (Lihme *et al.*, 1998, 2010; Willis, 2000; Clark, 2013; Du *et al.*, 2014: Banavara *et al.*, 2015).

15.3.5 Milk Protein Ingredients

Today, the development of milk proteins into food ingredients focuses on adding value to dairy streams (Figure 15.3). The market has been dominated by powder dairy ingredients with spray drying as the dominant technology for water reduction, which allows for longer shelf life and low-cost transportation without affecting the physical chemical and nutritional properties (Gonzalez, 2014b).

Commodity ingredients, such as whole milk and skimmed milk powders, are the major dairy protein ingredients marketed globally for applications that range from pediatric nutrition formulations to a variety of beverages. There are other dairy protein ingredients that provide diverse physical properties to the toolbox of food product developers (O'Kennedy, 2011, Boland, 2011):

- *Milk proteins concentrate (MPC):* Obtained by membrane filtration of skimmed milk to remove most lactose and minerals. Caseins and whey proteins remain in the same proportion. Caseins remain as micelles and retain the colloidal calcium phosphate.
- *Casein-whey protein co-precipitate*: Obtained by heating skimmed milk (90°C, 10–20 min) to denature whey proteins onto casein micelles, followed by precipitation of proteins by addition of mineral acids.
- *Phosphocasein/Micellar casein*: Obtained by membrane filtration of skimmed milk to separate whey proteins, lactose and minerals. Caseins remain as micelles and retain the colloidal calcium phosphate.
- *Acid casein*: Obtained by precipitating caseins from skimmed milk by addition of mineral acids, such as hydrochloric or sulfuric acid, to pH 4.6. It has poor solubility.
- *Lactic casein*: Obtained by acidifying skimmed milk to the isoelectric point (pH 4.6) of caseins using the lactic acid produced by *lactobacillus*. Lactic casein contains less mineral salts than acid casein.

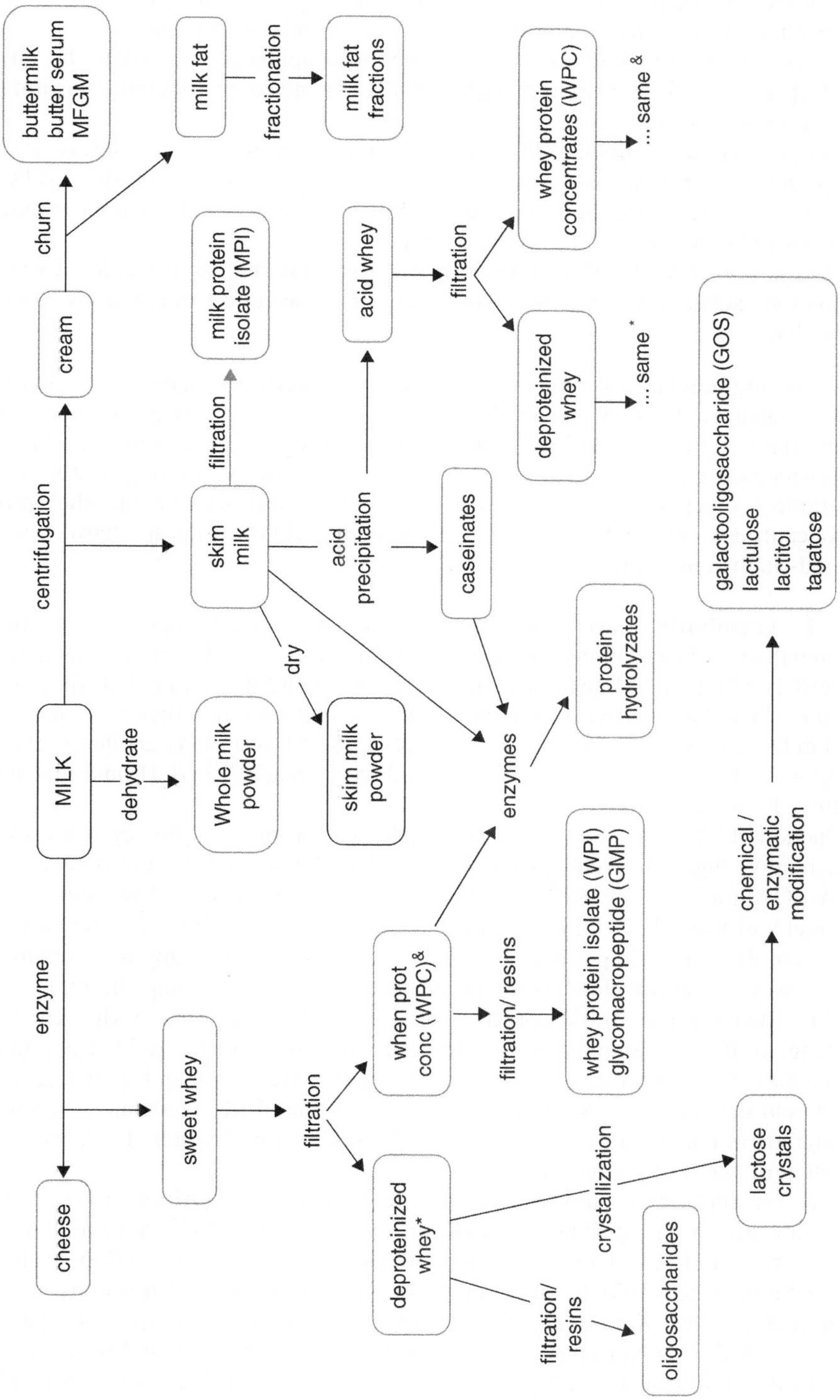

Figure 15.3 Simplified schematic diagram for milk value addition processes (Gonzalez, 2014b).

- *Caseinate*: Acid casein is re-dispersed and re-neutralized using sodium, calcium or ammonium hydroxides, or sodium carbonate. Sodium caseinates yield translucent dispersions, while calcium caseinates are white colloidal dispersions.
- *Rennin casein*: Obtained from the enzymatic coagulation of caseins by chymotrypsin and pepsin. Casein precipitates after the glycomacropeptide portion is cleaved from κ-casein.
- *Whey protein concentrates (WPC):* Obtained after separating the caseins. A range of protein concentrations are commercially available (35–85% w/w, dry). WPCs are manufactured by ultrafiltration, under mild conditions of pH and temperature, to remove lactose and salts to desired levels.
- *Whey protein isolates (WPI):* Obtained by processing the casein-depleted whey using ion exchange resins to increase the final protein concentration above 90%, w/w, dry.

Specialty ingredients are those developed with specific functional properties aimed to satisfy nutritional, physical-chemical or sensory attributes. A variety of specialty dairy ingredients find use in formulations designed for infants, the elderly or sport-focused nutrition. Examples of those are glycoproteins (e.g. lactoferrin, osteopontin), α-lactalbumin, protein hydrolysates with improved flavor, whey protein concentrates enriched with bioactive fractions, MFGM, oligosaccharides and branched-chain amino acids.

15.3.5.1 Lactoferrin Lactoferrin (LF) is a glycoprotein mainly present in the whey of mammalian milk. Its abundance varies between species and period of lactation (Lönnerdal and Iyer, 1995). In human milk, LF is the second-most abundant whey protein after α-lactalbumin. The overall mean LF concentration measured from mature human milk was 1.5–2 g/L. In mature bovine milk, the LF content is smaller, ranging from 0.1–0.3 g/L in milk. Bovine colostrum levels could reach 2–5 g/L (Lönnerdal and Iyer, 1995; Lien *et al.*, 2004).

Production of LF is based on the use ion-exchange chromatography resins leveraging the relative higher isolectric point (pH 8) of lactoferrin compared to other whey proteins (Okonogi *et al.*, 1988; Almécija *et al.*, 2007; Lu *et al.*, 2007; Perraudin, 2011; Voswinkel and Kulozik, 2014; Banavara *et al.*, 2015; Brown, 2015). Lactoperoxidase shares a similar isolectric point as LF and is co-isolated when using ion exchange resins. However, they are eluted from the chromatography resins using different ionic strengths. The low concentration of LF in bovine skim milk and cheese whey creates a challenge in the development of an affordable ingredient for broad incorporation into nutritional food products. Assuming a 50 mg LF/liter whey and a 90% recovery, it would take about 20 K tonnes to process 1 tonne of LF. The major specialty dairy ingredient manufacturers have commercialized high purity ingredients with an LF content of about 95% protein.

Lactoferrin applications mostly support the Health and Nutrition driver, with major relevance in pediatric nutrition (Kaur and Gathwala, 2015). Several clinical studies were conducted, demonstrating its benefits in diarrhea prevention (Ochoa *et al.*, 2015), prevention of late-onset sepsis and necrotizing enterocolitis in neonates (Manzoni *et al.*, 2009; Kaur *et al.*, 2015; Pammi and Abrams, 2015). LF iron affinity has been related to enhancement of the body's immune response, and to a role as an antiviral, antifungal and antiparasitic agent (Jenssen and Hancock, 2009). LF has also been shown to be prebiotic (Kim *et al.*, 2004), and involved in cognition and neural development (Chen *et al.*, 2015). The LF iron affinity has also been leveraged to reduce off-flavor formation in fish oil added-powdered milk (Ueno *et al.*, 2012). A challenge in formulating LF in food products is its high susceptibility to heat denaturation. The most common process is to formulate lactoferrin into pediatric nutritional products via dry blending (Wittke *et al.*, 2012). A method has been patented to produce shelf-stable beverages containing lactoferrin by aseptically combining a heat treated bulk liquid stream with a sterile-filtered lactoferrin stream (Banavara *et al.*, 2012; Gonzalez *et al.*, 2014).

15.3.5.2 Alpha-Lactalbumin Alpha-lactalbumin (α-LA) constitutes about 28% and 4% of the total protein of mature human and bovine milk, respectively (Kunz and Lönnerdal, 1989; Boland, 2011). Its structure is characterized by a high content of tryptophane, lysine and cysteine, and it also has a hydrophobic pocket with strong affinity for calcium, but it can also bind iron and zinc with lower affinities (Lönnerdal, 2014). This pocket is also a binding site for oleic and palmitic acids, as well as vitamin D_3 (Delavari *et al.*, 2015). Alpha-lactalbumin is a heat-labile protein susceptible to thermal denaturation above 65°C that, different from β-lactoglobulin, does not form aggregates and has low ability to form gels (McGuffey *et al.*, 2005). This makes α-LA suitable for beverage applications.

Alpha-lactalbumin is fractionated from whey streams where it may be found in about 25% (w/w, dry) of the total protein present (Marella *et al.*, 2011). Membrane filtration is used to obtain commercial preparations with up to 60% purity. A higher level of purity is achieved by using ion exchange chromatography that leverages the isoelectric point differences between β-lactoglobulin (pI 5.4) and α-LA (pI 4.4). A different method produces α-LA enriched fractions by heating pH-adjusted whey (pH 4–7.5, 85°C) to induce β-lactoglobulin aggregation molecules, and further fractionate the treated whey with ultrafiltration or microfiltration to obtain a α-lactalbumin: β-lactoglobulin ratio greater than 6 to 1 (Uchida *et al.*, 1996). Another process describes the manufacture an infant formula enriched with α-LA by means of selective hydrolysis of beta-lactoglobulin (Murphy *et al.*, 2015). Alpha-lactalbumin has also been used as an effective emulsifier in pediatric nutritional formulations (Alvey *et al.*, 2011).

Alpha-lactalbumin applications mostly support the Health and Nutrition driver, with major relevance on nutritional formulations targeted to infants, sports active consumers and the elderly, and also for stress reduction products (Gonzalez, 2015). A clinical study conducted with infants consuming formulas enriched with α-lactalbumin and glycomacropeptide showed an increase in plasma tryptophan concentration (Sandström *et al.*, 2008). A study on the consumption of α-LA as an evening supplement has been shown to increase blood levels of tryptophan, a serotonin precursor (Markus *et al.*, 2005). Study results revealed improved sleep, alertness and performance on the morning after of patients with sleep complaints. Similarly, Markus *et al.* (2002) showed that a diet protein rich in alpha-lactalbumin improves cognitive performance in stress-vulnerable subjects via increased brain tryptophan and serotonin activities. Alpha-lactalbumin has also been related to appetite control, as serotonin also influences appetite and food intake (Nieuwenhuizen *et al.*, 2010).

15.3.6 Milk Protein Hydrolysates

Milk protein hydrolysates are commercially produced by proteolysis using enzymes of animal, plant or microorganism origin. The substrates most commonly used are caseinates, whey concentrates or a combination of both streams. The reaction may be carried towards a partial hydrolysis, suitable for the development of target peptides and improved absorption, or it may be carried towards a high degree of hydrolysis, suitable for formulating hypoallergenic products.

A challenge in adopting hydrolyzed protein ingredients is the presence of a bitter taste. The bitter notes are attributed to the formation of low molecular weight peptides composed mainly of hydrophobic residues. It is expected that proteins with large hydrophobic domain content, such as caseins, soybean protein and gluten, tend to yield bitter peptides upon hydrolysis. Several processes developed for de-bittering milk protein hydrolysates use incubation with activated charcoal, alcohol extraction, isoelectric precipitation, filtration membrane and hydrophobic adsorption columns (Helbig *et al.*, 1980, Lin *et al.*, 1997; FitzGerald and Ócuinn, 2006). A concern with some of these methods is the loss of peptides with essential amino acids, such as Trp, Phe and Arg. Another processing strategy uses exopeptidases to hydrolyze bitter peptide fractions to produce smaller peptides with reduced bitterness and free amino acids (Saha and Hayashi, 2001; Raksakulthai and Haar, 2003; Khatib *et al.*, 2009).

Milk protein hydrolysates are used as food ingredients with specific value for infant, geriatric and sport nutrition. The use of hydrolyzed protein ingredients also provides an alternative for the development of foods for individuals who have limitations with the dietary intake of intact proteins; in particular, those that present an allergic reaction to proteins, as well as those that have limited luminal hydrolysis or a reduced intestinal membrane absorptive capacity. Multiple studies have demonstrated the efficacy and safety of milk hydrolysates to manage protein allergy in children (Sampson *et al.*, 1991; Halken *et al.*, 1993; Caffarelli *et al.*, 2002; Terheggen-Lagro *et al.*, 2002; Von Berg *et al.*, 2007). A study comparing infant formulas formulated using intact and extensively hydrolysed proteins suggested that the form of the protein used may impact the cognitive development of infants and that the presence of free amino acids may be a contributing factor (Mennella *et al.*, 2015a, b). The study showed that between the ages of 1.5 and 8.5 months, infants fed on the hydrolysed protein formula averaged higher scores in visual reception and general intelligence than those on intact protein formulas.

Additional research on the value of protein peptides has focused on cardiovascular health, obesity, muscle growth, immunity, diabetes and antioxidant properties (Shah, 2000; Clare and Swaisgood, 2000). Angiotensin-converting enzyme (ACE) inhibitor peptides were identified in hydrolyzed casein and whey proteins (Martin *et al.*, 2008; Ortiz-Chao *et al.*, 2009). Antioxidant hydrolysates were prepared from caseins, α-lactalbumin and β-lactoglobulin using commercial proteases (Diaz and Decker, 2014; Hernandez-Ledesma *et al.*, 2005; Rival *et al.*, 2001). Partial β-lactoglobulin hydrolyzates were shown to be effective to micro-encapsulate fish oil and protect it from oxidation (Tamm *et al.*, 2015).

15.3.6.1 Glycomacropeptides The dairy industry generates two major types of whey: acid and sweet whey. Acid whey results from the precipitation of casein after adjusting the milk pH to 4.6. Sweet whey, also known as rennet or cheese whey, results from the separation of caseins by action of proteases, such as chymotrypsin. Sweet whey differs from acid whey in the presence of a glycomacropeptide fraction (GMP) (6.8 kDa. pI <3.8) a hydrophilic portion of κ-casein release by the action of proteases and that is responsible for stabilizing the casein micelles.

Commercial production of GMP is done via ion exchange chromatography and ultrafiltration of sweet whey. Commercial applications of GMP are based on its health benefits. GMP is rich in branched chain amino acids, such as leucine and isoleucine, and has been shown to encourage muscle development, thus finding applications in infant, geriatric and sports formulations. GMP is one of the few proteins that lack phenylalanine, therefore it is a suitable ingredient for development of foods targeted to manage phenylketonuria (LaClair *et al.*, 2009). GMP is also low in methionine content, hence it is suitable for designing foods for hepatic disease patients. GMP carbohydrate portion is rich in galactose, N-acetyl-galactosamine and sialic acid, and constitutes about half of its dry weight. Studies have associated sialic acid with brain development and learning ability. Several pediatric nutritional foods have leveraged this benefit, included GMP in their formulations (LaClair *et al.*, 2009; Bonnaillie *et al.*, 2014).

15.3.6.2 Milk Protein Complexes Caseins and casein micelles are mostly hydrophobic and this property has been leveraged to design effective delivery vehicles for bioactive hydrophobic molecules. Several procedures have been developed with possible commercial expectations.

Plant-derived bioactives have been stabilized by dairy protein matrices. The self-associating behavior of beta-caseins and casein micelles has been used to entrap curcumin, a natural phenolic pigment associated with antioxidant, anti-inflammatory, anti-bacterial and anti-cancer properties (Esmaili *et al.*, 2011; Benzaria *et al.*, 2013; Pan *et al.*, 2013). Curcumin has also been complexed with β-lactoglobulin improving its antioxidant properties (Li *et al.*, 2013). Green tea polyphenols, such as epigallocatechin gallate (EGCG) binds whey proteins, and preferentially caseins, to become more stable to oxidation through the gastrointestinal tract (Hasni *et al.*, 2011; Kanakis *et al.*, 2011; Ferruzi *et al.*, 2012; Moser *et al.*, 2014; Haratifar and Corredig, 2014; Sabouri *et al.*, 2015).

Fat soluble vitamins have increased their stability through interaction with dairy proteins. Casein micelles entrap fat-soluble vitamin A and carotenes via hydrophobic interactions between the hydrophobic domains of the micelles (Mohan *et al.*, 2013; Saiz-Abajo *et al.*, 2013). A high pressure-assisted process was developed to re-assemble casein micelles and enhance their ability to entrap vitamin D_2 (Menendez-Aguirre *et al.*, 2011). Whey protein aggregates provide heat and light protection to α-tocopherols (Relkin and Shukat, 2012). Alpha-lactalbumin binds vitamin D_3 in its hydrophobic pocket, forming a complex that provides added stability and solubility to the molecule (Delavari *et al.*, 2015). The hydrophobic pocket of α-lactalbumin also binds oleic acid to create a complex, known as BAMLET (bovine α-lactalbumin made lethal to tumor cells), that has been demonstrated to cause apoptosis in cancer tumor cells (Håkansson *et al.*, 1995; Lišková *et al.*, 2010).

Maillard reaction products of whey proteins with hydrocolloids, such as dextran, have produced complexes with increased denaturation temperature and lowered isolectric point (Liu and Zhong, 2013, 2015). These complexes have improved emulsifying properties (Zhu *et al.*, 2010) and produced gels with reduced stiffness (Sun *et al.*, 2011). The *in-vitro* digestibility studies using whey protein isolate-dextran complexes showed lower digestion rate constants as compared to WPI (Böttger *et al.*, 2013), hence opening opportunities to developing hypoallergenic foods.

15.4 Milk Fats

Milk fat triglycerides are formed in the endoplasmic reticulum and released into the cytosol as pre-globules, or micro-lipid droplets (<0.5 microns) surrounded by a layer of phospholipids and proteins. The micro-droplets grow in size in the cytoplasm by collapsing with other micro-droplets as they move towards the inner face of the plasma membrane surrounding the milk-secreting cells. Upon contact with this phospholipid double layer membrane, they are surrounded by additional proteins and are pinched-out into the serum as membrane-encapsulated globules (Heid and Keenan, 2005). Milk fat globules form a colloidal suspension in milk with an average distribution size of about 0.02–20 microns in human milk (Michalski and Januel, 2006) and 0.2–15 microns in bovine milk (Singh, 2006).

15.4.1 Milk Fat Classification

In bovine milk, the core of the milk fat globule lipids are mostly triglycerides (~98%), with a small amount of free fatty acids and mono/di-glycerides, whereas MFGM is composed of phospholipids (0.5–1% of total lipids in milk) and sterols (0.2–0.5% of total lipids) with cholesterol as the major sterol (10–20 mg/dl of milk) (Jensen *et al.*, 1991). In general, there are over 400 different fatty acids in milk fat (Fox and McSweeney, 1998). Milk fatty acids with 4 to 14 carbons are synthesized in the mammary gland, while higher chain fatty acids originate from dietary lipids (Månsson, 2008; Huppertz *et al.*, 2009). The higher content of butyric acid (C4) in bovine milk fat over that of human milk fat is the result of the ruminal fermentation process. Butyric acid in bovine milk fat is about 2–5 g/100 g total fatty acids, whereas human milk fat has a content that is about 20-times smaller (Givens and Shingfield, 2006).

The content of saturated and unsaturated fatty acids in milk fat varies with species, breed, type of feed and nutritional status, season, health and age, stage of lactation and intervals between milking (Jensen *et al.*, 1978; Hamosh *et al.*, 1985; Fox and Sweeney, 1998; Singh, 2006). In human milk, about 50% of the fatty acids are unsaturated, whereas in bovine milk, about 80% are saturated. In human milk, about 73% of the fatty acids are palmitic (C16), oleic (C18:1) and linoleic (C18:2), whereas in bovine fat, about 75% are myristic (C14), palmitic, oleic and linoleic (Table 15.2; Precht and Molkentin, 1999; Givens and Shingfield, 2006).

Compound lipids in milk are diverse and include molecules such as phosphates, choline, sphingosine, sugars and sialic acid. They are mostly associated with the MFGM, such as phospholipids, gangliosides, cerebrosides and sphingomyelins. Other derived lipids in milk are not fatty acid related molecules, but exhibit a hydrophobic nature, such as steroids, carotenoids and fat soluble vitamins. Cholesterol is the most well-known derived milk lipid, with a content of about 204–382 mg/100 g of bovine or human milk fat (Nakayama, 1981; Jensen *et al.*, 1978; 1991, Precht, 2001). There are traces of vitamins A, D, E and K with higher concentrations in colostrum than mature milk (Jansson *et al.*, 1981; Mohan *et al.*, 2013).

15.4.2 Milk Fat Ingredients

Milk fat products are manufactured via a wide range of technologies. Some processes do not alter the milk fat composition, but change the overall fat content and the

Table 15.2 Fatty acid composition of bovine and human milk fat (g/100 g total fatty acids). Saturated fatty acids are more abundant in bovine milk fat than in human milk fat. Butyric and caproic acids (C4:0 and C6:0) are common in ruminants milk fat, but are almost absent in human milk fat (adapted from Precht and Molkentin, 1999: Givens and Shingfield, 2006).

Concentration of various fatty acids (g/100g total fatty acids)			
Fatty Acids		**Human Milk**	**Bovine Milk**
Butyric	C4:0	0.2	**3.9**
Caproic	C6:0	0.2	**2.5**
Caprilic	C8:0	0.1	1.4
Capric	C10:0	0.6	3.1
Lauric	C12:0	3.1	4.2
Myristic	C14:0	6.4	**11.6**
Palmitic	C16:0	**25.3**	**29.4**
Stearic	C18:0	7.1	**11.4**
Others		trace	2.8
Total Saturated		***43.0***	***70.1***
Myristoleic	C14:1 *cis* ϖ 9	0.3	1.1
Palmitoleic	C16:1 *cis* ϖ 9	2.8	1.9
Oleic	C18:1 *cis* ϖ 9	**33.7**	**21.9**
Linoleic	C18:2 ϖ 6	**10.6**	1.9
Linolenic	C18:3 ϖ 3	0.9	0.6
Arachidonic	C20:4 ϖ 6	0.6	
Others		0.4	1.7
Total Unsaturated		***49.3***	***29.1***

physical properties of the product or ingredient. For example:, centrifugation is a basic unit operation applied to whole milk to separate skim milk and cream. Churning converts cream into butter and separates buttermilk. Other processes alter the milk fat composition and may also change the physical and chemical profile. For example, temperature and solvent fractionation methods, cholesterol removal techniques and fermentation processes, such as those that produce cultured cream and buttermilk products.

15.4.2.1 Cream Cream, also known as sweet cream, is the low-density fat containing portion separated from whole milk by centrifugation. It is an oil-in-water (O/W) emulsion that is enriched in milk fat globules and their corresponding MFGM components. The cream is heat processed to inactivate viable microorganisms and to inactivate enzymes, such as lipases, that may affect flavor and reduce shelf life. It is also homogenized to provide better shelf stability. There are different types of commercial creams available: *Half cream*, or coffee cream, is characterized by a 10–18% cream content. *Light cream*, also known as single, contains 18–30% milk fat and finds use in beverages, soups and desserts. *Whipping cream* contains no less than 35% milk fat and is used for aeration in cakes and pastry fillings. *Whipped cream* is cream that has been already aerated. *Heavy cream*, or double, contains no less than 36% milk fat and is used in pouring or spooning for desserts, and can also be whipped for cakes and pastry fillings. *Sour cream* is produced by the souring of cream via inoculation and incubation of lactic acid bacteria. It contains not less than 18% fat, and if optional ingredients are added, such as sweeteners and flavoring agents, its fat content should not be less than 14.4%. *Acidified* sour cream is produced by the addition of suitable acidulants, with or without the action of lactic acid bacteria (USCFR, 1993).

15.4.2.2 Buttermilk The composition of buttermilk obtained from sweet cream is similar to skim milk. It contains lactose, oligosaccharides, minerals and proteins. It also has higher lipid content and MFGM fragments compared to skim milk which makes it a good starting material for MFGM enrichment. A two-step process was developed to increase the buttermilk concentration of polar MFGM lipids by microfiltration and supercritical fluid extraction (Astaire *et al.*, 2003; Costa *et al.*, 2010). The membrane filtration removed most of the lactose, oligosaccharides and minerals, while the supercritical extraction removed exclusively non-polar lipids. This process does not rely on the use of organic solvents, and yields a material with about 73% protein and 21% lipids, of which 60% are polar MFGM lipids (Astaire *et al.*, 2003). Another process concentrates the MFGM components in buttermilk by combining membrane filtration with the addition of citrates (Corredig *et al.*, 2003). The citrate causes the dissociation of casein micelles and allows their permeation through the membrane.

15.4.2.3 Milk Fat Globule Membrane Milk fat globule membrane (MFGM) layer is easily ruptured during milk processing and is carried with the cream during the fat separation process (Le *et al.*, 2013). The MFGM has a complex structure with three distinct layers (Figure 15.4): a single polar lipid layer adjacent to the milk fat, a layer of diverse cytoplasmic proteins, and an outer bi-layer of polar lipids with embedded proteins (Cavaletto *et al.*, 2004; Lopez *et al.*, 2008; Dewettinck *et al.*, 2008).

The phospholipid monolayer, mostly composed of phosphatidyl-ethanolamine (PE), phosphatidyl-serine (PS) and phosphatidyl-inositol (PI), surrounds the triglyceride. The phospholipid bilayer that originates from the secretory cell

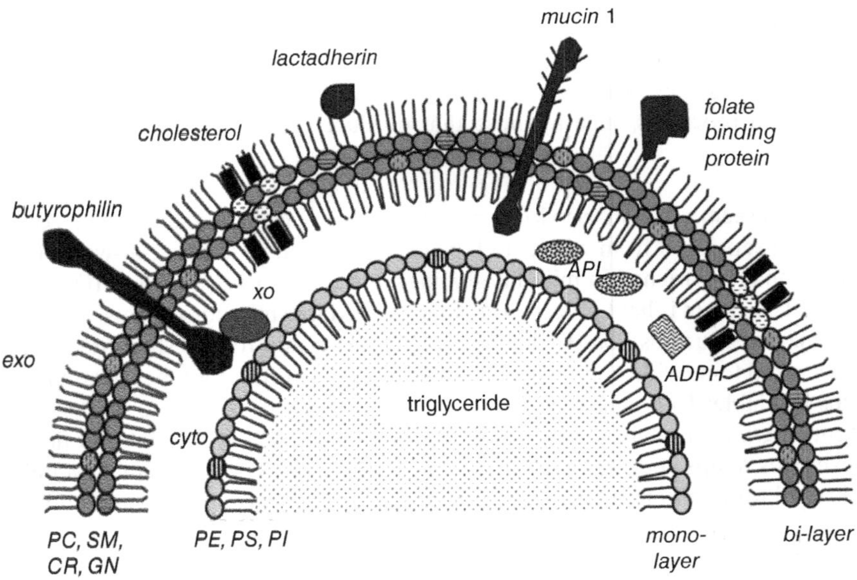

Figure 15.4 Structure of the milk fat globule membrane and associated lipids and proteins.

membrane, mostly composed of phosphatidyl-choline (PC), sphingomyelins (SM), cerebrosides (CR) and gangliosides (GN), enrobes the monolayer with some cytoplasm proteins, such as apolipoproteins (APL), xanthine dehydrogenase/oxidase (XO), adipophilin (ADPH) and others. The bilayer also contains trans-membrane proteins, such as butyrophilin, mucins, polymeric Ig receptors and others. The cholesterol is shown as part of the sphingomyelin-rich domains, also known as lipid rafts (Cavaletto *et al.* 2004; Lopez *et al.* 2008; Dewettinck *et al.* 2008).

There are similarities between human and bovine milk MFGM structures and components and these may be responsible of delivering intact bioactive molecules to the intestine, while facilitating efficient absorption and processing of milk fats, without metabolic stress (Garcia and Innis, 2013). These structures are nowhere else replicated in nature; however, it may be possible to replicate them via processing using MFGM-rich ingredients for functional health benefits. Milk handling processes, such as aging, agitation, air inclusion, temperature changes and heat treatments, affect the composition of MFGM (Michalski and Januel, 2006; Singh, 2006). Homogenization ruptures the fat globule, reduces droplet size and releases MFGM fragments that remain suspended in the milk aqueous serum. The two main fractions, buttermilk and butter serum, both by-products of butter manufacturing (Figure 15.3), are rich in MFGM proteins, phospholipids, sphingomyelins and gangliosides (Rombault and Dewettinck, 2006). MFGM ingredients are derived from buttermilk and butter serum and commercialized as enriched WPC fractions.

15.4.2.3.1 Phospholipids Phospholipids constitute less than 0.1% of milk; however, they are found in higher levels in the dairy fractions rich in MFGM fragments, such as cream and buttermilk. MFGM phospholipids (PL) are diverse, with phosphatidyl choline and phosphatidyl ethanolamine constituting about 63% of the total MFGM PL. Phosphatidyl inositol, phosphatidyl serine and lysophosphatidyl choline represent most of the remaining MFGM PL portion (Singh, 2006). The flow chart shown in Figure 15.5 shows the approximate values of fat and phospholipids of the different streams during processing. Whey or buttermilk is commonly used as a starting stream to develop phospholipid-rich ingredients. Whey is membrane filtered to remove fat prior to further process into protein concentrates and isolates. The defatted whey protein ingredients have improved shelf life. The removed lipid fraction is known as whey phospholipid concentrate, and represents about 14–18 % of the whey processed and contains about 4–8% of phospholipids (Burrington *et al.*, 2011). Another method proposed to remove MFGM fragments from whey is by forming a membrane-metal complex using zinc (Damodaran, 2010). However, no study has been conducted to assess the value of this MFGM-metal complex.

15.4.2.3.2 Benefits of Milk Fat Globule Membrane The value of buttermilk is mostly driven by the presence of MFGM fragment components. The potential of MFGM components to address Health and Nutrition benefit drivers are a current topic of research. Benefits have been demonstrated related to cancer cell growth inhibition (Zanabria *et al.*, 2013), immunity and rotavirus inhibition (Fuller *et al.*, 2013), as well as gut and heart health (Spitsberg, 2005; Contarini and Povolo, 2013). Pediatric nutritional formulations prepared using MFGM-enriched ingredients support healthy growth (Billeaud *et al.*, 2011), enhance brain and cognitive development (McJarrow *et al.*, 2009; Gurnida *et al.*, 2012; Timby *et al.*, 2014), have antipathogenic and antidiarrheal effect (Rueda *et al.*, 2008; Zavaleta *et al.*, 2011) and produce

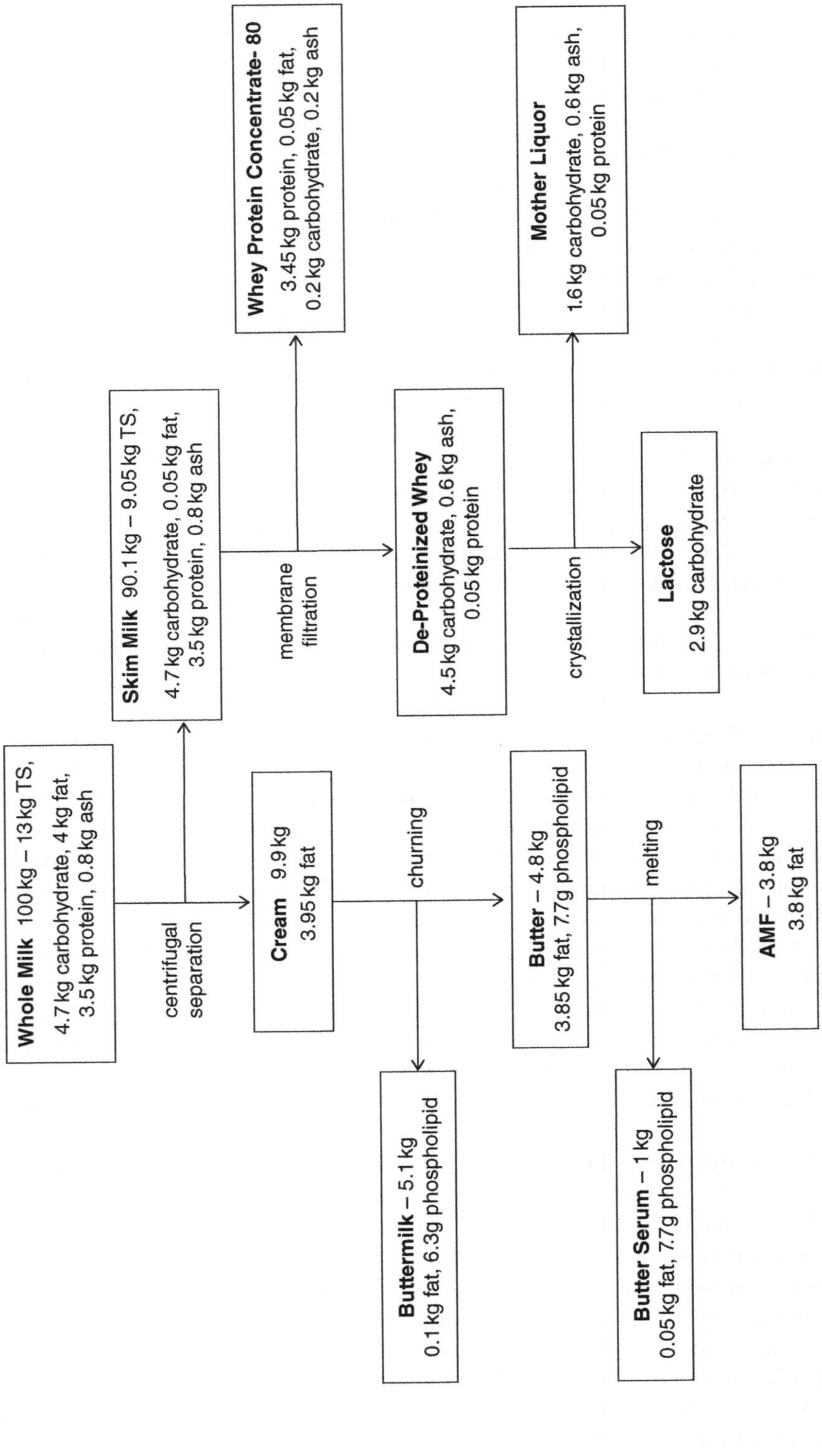

Figure 15.5 Whole milk process flow chart for fat and carbohydrate separation, showing approximate yield values for 100 kg of liquid milk.

a microbiota profile closer to breast-fed infants (Rueda *et al.*, 2008). A patented pediatric nutritional formulation developed using at least 300 mg/l of milk phospholipids claims benefits of reducing the risk of developing metabolic syndrome, increased weight gain, increased fat deposition, overweight, obesity, insulin resistance, glucose intolerance or diabetes mellitus later in the said infant's life (Huber-Haag *et al.*, 2015). A study conducted in adults (30–51 years old) demonstrated that consumption of a phospholipid-rich beverage provided stress protective effects by normalizing their cortisol levels, as well as improve memory performance (Schubert *et al.*, 2011).

The value of buttermilk is also driven by the Physical-Chemical drivers. In standard buttermilk, the phospholipid content is about seven times higher than in whole milk. Phospholipid-enriched buttermilk ingredients have excellent emulsifying and whipping properties useful in a variety of food applications (Corredig and Dalgleish, 1997; Roesch *et al.*, 2004; Miocinovic *et al.*, 2014, Phan *et al.*, 2014a, b). Pediatric nutritional formulations were processed using MFGM-rich ingredients to produce infant milk formulas with fat globules with a particle size distribution and interface structures closer to human breast milk (Banavara *et al.*, 2014a, b; Gallier *et al.*, 2015). In these formulations the phospholipids became the major component stabilizing the fat droplets, compared to traditional infant formulas where denatured proteins are the dominant molecules at the droplet interface. The chemical composition, size and structure of the fat globules in these formulas may enhance lipid digestion in infants by providing structured fat globules that are more accessible to lipases (Banavara *et al.*, 2014a)

15.4.2.4 Butter Butter is produced by churning, or tumbling, cream or cultured cream in vats at low temperatures, where the oil-in-water emulsion is inverted into a water-in-oil (W/O) emulsion. Butter grains are formed and pressed to remove the aqueous fraction, or buttermilk, and the final milk fat content is increased to not less than 80%. Salted and unsalted butter have multiple uses in bakery and in the preparation of creams and sauces.

15.4.2.4.1 Cholesterol Reduction Hypercholesterolemia is a modifiable risk factor for cardiovascular disease, since elevated cholesterol concentrations have been associated with the development of endothelial dysfunction and atherosclerosis (Oggioni *et al.*, 2015). The role of dietary cholesterol is controversial. It has been challenging to conduct human intervention studies assessing dietary cholesterol intake on the occurrence of cardiovascular events, and existing studies have provided conflicting results (Hansel and Giral, 2015). Total body cholesterol is regulated by the interaction of three factors: absorption, synthesis and excretion. Absorption of dietary cholesterol was found to increase with intake; however, this increase in absorption was compensated for by an increase in excretion and a decrease in body synthesis (Quintao *et al.*, 1971). It is this uncertainty and lack of conclusive information that confers a negative nutritional image to milk fat and presents a challenge for the adoption of butter as a main stream food ingredient.

Several commercial technologies are available for the removal of cholesterol from milk fat (Boudreau and Arul, 1993). Short path molecular distillation may reach up to 90% of cholesterol reduction by vacuum evaporation of fat components based on molecular weight. Supercritical carbon dioxide fluid extraction, described in the section on anhydrous milk fat, may also reach about a 90% reduction. Vacuum steam distillation relies on the higher volatility of cholesterol, compared to the major triglycerides in milk fat, to reduce its content to about 75% of the initial level.

Melt fractionation concentrates cholesterol, as well as flavor and pigments, into the olein liquid fraction of milk fat. Other methods leverage the ability of cyclodextrins (Oakenfull *et al.*, 1991; Kwak *et al.*, 2000; Soni, 2012) and saponins (Sundfeld *et al.*, 1994) to form complexes with cholesterol, which can then be further removed from the dairy stream via centrifugation. A simple method extracts cholesterol from milk by sharing a mixture of vegetable oil and milk, followed by centrifugation to separate the distinct phases (Merchant *et al.*, 1995). Another process preferentially removes cholesterol from milk fat globule membranes using an organic polar solvent, recovering a solvent-free, low-cholesterol material (Johnson and Conte, 1991). These enabling technologies may allow for the commercialization of cholesterol-reduced milk fat and milk fat fractions, with improved physical-chemical properties and defined health benefits, which can change the consumer's perception of milk fat and expand its food applications (Elling and Duncan, 1996).

15.4.2.5 Anhydrous Milk Fat The processing of butter, or homogenized cream, via centrifugation and heating under vacuum, yields an ingredient with over 99% of milk fat. This high fat ingredient may be washed with water to remove proteins, or it may be treated with alkali to neutralize free fatty acids. Cholesterol may be removed by additional processing with complex carbohydrates. The ingredient is known as anhydrous milk fat (AMF) if it contains at least 99.8% milk fat and has no traces of neutralizer additives, whereas if it has similar fat levels but traces of alkali, it is known as anhydrous butter oil (ABO). Both AMF and ABO have a broad melting point range that limits their use in certain foods. However, it is this broad melting point range that facilitates the fractionation of milk fat without the use of solvents. At room temperature (20°C), milk fat is a mixture of liquid, semi-hard and hard fats. Its melting temperature range starts at –30°C and is complete at 40°C (Boudreau and Arul, 1993).

15.4.2.5.1 Anhydrous Milk Fat Fractionation The development of milk fat fractions allows for ingredients with different physical-chemical and nutritional properties. These fractions have renewed commercial relevance to milk fat. Technologies suitable for milk fat fractionation are crystallization, supercritical carbon dioxide extraction, steam vacuum distillation, molecular distillation and crystallization from solvents (Gibon, 2006).

15.4.2.5.2 Supercritical Carbon Dioxide Extraction The supercritical extraction process relies on using carbon dioxide (CO_2), liquefied under pressure, as a solvent to leverage differences in solubility of different triglycerides. The fractions are obtained by de-pressurizing the separated liquid portion that releases the CO_2 from the fat fraction. The solubility of triglycerides in the liquid CO_2 increases with temperature and degree of unsaturation of the carbon chains, while it decreases with an increase in the carbon chain length. The polarity of CO_2 may be altered by changes in pressure and temperature. Hence, it may be possible to create a variety of milk fat fractions by modifying process conditions. The same technique can also be used to separate carotenoids and flavor compounds naturally present in milk fat, as well as to reduce cholesterol from milk fat (Shukla *et al.*, 1994).

A study comparing fat fractionation through supercritical CO_2 extraction and melt fractionation showed that the bovine milk fat fractions obtained had different physical and chemical properties (Bhaskar *et al.*, 1998). The supercritical CO_2 method separated milk fat into four fractions: Super-Stearin, Stearin, Olein and Super-Olein. The

Table 15.3 Fatty acid composition of anhydrous bovine milk fat (AMF) fractions separated via melt crystallization with cooling stops at 45°, 30°, 20° and 10°C (adapted from Bhaskara *et al.*, 1998)

Melt Crystallization Fractions	Saturated Fatty Acids	Unsaturated Fatty Acids	C 4	C 16:0	C 18:0	C18:1 cis
Milk Fat	63.9	28.6	3.4	28.7	10.3	20.9
AMF 45	72.7	20.2	3.1	34.4	11.8	15.1
AMF 30	66.1	26.3	3.4	29.1	9.9	19.5
AMF 20	66.4	26.2	4.5	29.6	8.6	20.2
AMF 10	58.6	33.3	4.7	22.2	6.1	25.3

Table 15.4 Fatty acid composition of anhydrous bovine milk fat fractions separated via supercritical carbon dioxide extraction (adapted from Bhaskara *et al.*, 1998)

Supercritical CO_2 Extraction Fractions	Saturated	Unsaturated	C 4	C 16:0	C 18:0	C 18:1 cis
Milk Fat	63.9	28.6	3.4	28.7	10.3	20.9
Super Stearin	51.4	39.0	0.5	25.0	15.6	29.8
Stearin	63.1	29.5	2.9	28.9	10.5	21.5
Olein	67.1	26.0	4.6	29.5	9.2	18.7
Super Olein	70.7	23.1	6.0	29.6	8.2	16.5

Super-Olein fraction had a C4 content about a 13-fold higher, and a C18 and C18:1 content about half the level of Super-Stearin (Table 15.3). The melt fractionation process also separated anhydrous milk fat (AMF) into four fractions within a temperature range of 45–10°C. Compared with the AMF-45, the AMF-10 fraction showed a C4 and C18:1 content that was a 1.5–1.7-fold higher; however, the C18 content was a 0.5-fold lower (Table 15.4).

15.4.2.5.3 Melt Fractionation Milk fat melt fractionation, also known as dry fractionation, takes place by cooling a batch of melted fat to a set temperature and removing, via centrifugation or filtration, the crystallized fraction. The first fraction removed at the highest cooling temperature is stearin, whereas the one that remains liquid at the lowest cooling temperature is olein. Hence, compared to the non-fractionated milk fat, oleins have a lower melting point and may have a higher content of unsaturated fatty acids. Stearins have a higher melting point and may be richer in saturated fatty acids (Amer *et al.*, 1985). Milk fat has a melting range that varies between 4°C and 40°C. This broad melting point range is due to the presence of fatty acids with diverse chain lengths (C4 to C20) and their position on the glycerol backbone. It is possible to separate them into 3–5 fractions, depending on the desired melting point range (Bhaskara *et al.*, 1998).

Milk fat fractions of different melting points find application to modify texture, flavor and shelf life of food products. High melting milk fat fractions (>40°C) are used in chocolate and confectionary, and medium melting fat (37°C) are used in baked goods like pastries to provide a layered structure (Kaylegian *et al.*, 1993; Metin and Hartel, 2012). Low melting milk fats are used in ice-cream. recombined dairy foods and other nutrition products. Butters of different spreadability have been produced; however, they have not been very successful in the market place. Melt crystallization is a lower-cost process compared to supercritical CO_2 fractionation, as it relies simply on melting temperature differences between triglycerides and does not require the use of pressurized vessels. Melt fractionation does have limitations, since milk fatty acids are randomly distributed in triglycerides of various melting points. Hence researchers have tried to take support from solvents which utilize the polarity of triglycerides along with melting temperatures (Bhaskara *et al.*, 1998).

15.5 Milk Carbohydrates

Carbohydrates are the most abundant component of mammalian milk on a dry basis. The carbohydrate content varies with the species and the stage of lactation. There are small amounts of glucose and galactose, significant amounts of lactose, and a diverse set of oligosaccharides. Lactose is a unique disaccharide solely found in milk and constitutes its primary carbohydrate. Lactose content in mature human and bovine milk is about 54% and 41% (w/w, dry), respectively. In human milk, oligosaccharides are the third-most abundant component after water, lactose and fat (Kunz *et al.*, 2000). Mature human milk contains about 15 g oligosaccharides/liter compared to about 9 g protein/liter (Coppa *et al.*, 1993). In comparison, mature bovine milk only has traces of oligosaccharides. A young mammal digests mono- and di-saccharides mainly for energy purposes; however, oligosaccharides remain undigested by the innate enzyme system, serving as a valuable substrate for the intestinal microbiota (Chichlowski *et al.*, 2011).

15.5.1 Lactose

Lactose is a disaccharide composed of β1-4 linked D-galactose and D-glucose. The synthesis of lactose in the mammary gland requires the formation of a complex between the enzyme beta-1,4-galactosyltransferase and alpha-lactalbumin to form the lactose synthetase system (Ramakrishnan *et al.*, 2002). This complex catalyzes the formation of lactose at physiological levels of glucose and galactose (Brew *et al.*, 1972). Lactose abundance varies between species; in human milk it is about 7%, whereas in cow and goat milk, lactose content is smaller to about 4.6% (Schaafsma, 2008). Lactose has two structural configurations (α and β) that co-exist in equilibrium at room temperature. The anhydrous β-isomer is soluble under ambient conditions, whereas the monohydrated α-isomer, the most common commercial form, is soluble at higher temperatures. Mutarotation between both configurations is affected by temperature, pH, presence of salts and solvents (Gänzle *et al.*, 2008; Wong and Hartel, 2014).

The dairy industry faces challenges regarding lactose production, as it is considered a low return product due to inconsistent yields, cyclic price variations, and product quality of edible lactose. The ingredient finds application in pediatric nutritional formulations, as a substrate for fermentation processes, or in chemical and enzymatic reactions to produce valuable components such as galactooligosaccharides, lactulose, lactitol and tagatose. The use of lactose in product development is also limited by the almost generalized prevalence of lactose intolerance, also known as lactase non-persistence, and lactose malabsorption in the global adult population. It may be estimated that about 80% of the world population has difficulties digesting lactose (Savaiano and Levitt, 1987; Schaafsma, 2008). This popular rejection of lactose is unjustified, because lactase deficient and malabsorption subjects may consume up to 11 g per day, 1–2 servings of milk, without side effects when this amount is distributed over the day and ingested with meals (Schaafsma, 2008).

15.5.1.1 Lactose Process Improvements Amorphous lactose is produced by fast cooling of concentrated demineralized whey permeate, while crystalline lactose is achieved through controlled cooling of permeate. Crystalline lactose may be further dried and milled to attain desired particle size. Concerns with both processes are low yields of about 65–70% (Wong and Hartel, 2014) and the mother liquor by-product that contains about 25–30% of solids, of which about 50% is lactose (Priyanka, 2012). The mother liquor has a market as animal feed or as a fermentation culture medium for the production of organic acids, such as lactic acid.

The industrial production of crystalline lactose follows three main steps: whey permeate concentration, crystallization and purification. Improvements in the whey permeate quality and controlling the lactose crystallization process may increase the quality of the ingredient for food and pharmaceutical uses.

15.5.1.2 Permeate Treatment The whey permeate is concentrated by means of vacuum evaporators. Both ultra-filtrated cheese and lactic whey permeate contain high amounts of calcium and phosphorus that leads to formation of hydrated amorphous calcium phosphate. The presence of permeate salts tend to foul evaporators, reducing process efficiency and increasing cleaning frequency. Previously, pretreatment of whey permeates, by means of pH and temperature adjustments, provided a meaningful reduction in calcium phosphate impurities (Hayes *et al.*, 1974; Hickey and Hill, 1980; Pederson, 1980; Matthews, 1984). Ion exchange demineralization is used for softening of permeates by using sequential columns filled with cationic resin and anionic resins (Lifran *et al.*, 2011). To achieve maximum decalcification, resins are frequently regenerated with salt solution. The presence of sodium and potassium salts in permeate streams creates a competitive regeneration effect that reduce the demineralization efficiency. In addition, the presence of peptides negatively affects the softening ability of resins. Other technology uses an EDTA–polyphosphate mixture added to the concentrated permeate to chelate calcium ions (Evans *et al.*, 1982a, b).

Nanofiltration technology has been used to further purify permeates and to increase lactose yields (Guu and Zall, 1992). Calcium phosphate is removed prior to nanofiltration to avoid fouling of the membranes. This technology allows the use of higher solids due to a reduction in slurry viscosity (Mikkonen *et al.*, 2001). A patented process for lactose purification combines ion exchange and nano-filtration to treat permeated whey (Durham *et al.*, 2002). The filtration retentate is further purified by ion exclusion chromatography, while the permeate passes through reverse osmosis process and is then used to regenerate the ion exchange column. Ion exclusion chromatography is used to prepare purified pharmaceutical grade lactose.

The color of edible or crude lactose has an impact on its market value. Riboflavin, a yellow-colored soluble vitamin, is present in whey concentrate and co-crystallizes with lactose. Riboflavin is challenging to remove, even after repeated washes. Activated charcoal has been successfully used to remove this and other impurities from lactose solution. Montilla *et al.* (2005) showed that as low as 5 mg/L of activate charcoal could reduce as much as 92% of the color. However, significantly increase in the operation costs, complexities around regeneration processes, microbial growth risks, and health hazards associated with activated charcoal limits its application. Pragono *et al.* (2008) used adsorptive ion-exchange resin to remove colored compounds, showing that about 90% of riboflavin could be removed from 15% lactose solution without attaining the saturation level. This method offers a more economically feasible solution compared to activated charcoal, since resins are much easier to regenerate.

15.5.1.3 Crystallization Improvements in the lactose crystallization process may yield more uniform shaped and sized crystals that may improve ingredient properties, such as bulk density and flow properties, as well as increased process yield and reduced costs. In the traditional industrial process, lactose crystal nucleation occurs when super saturated concentrated whey permeate is cooled from 75°C to 25°C. The low solubility of lactose at 25°C facilitates lactose nuclei growth. During crystallization, mutarotation maintains an equilibrium between the α-lactose monohydrate and anhydrous β-lactose forms. The size distribution of lactose crystals varies significantly due to the kinetics of nucleation, mutarotation and crystal growth. The rate of mutarotation is temperature dependent, equilibrium is reached within few minutes at 70°, while it takes several hours at 20°C. Hence, a low cooling rate during crystal growth would induce a uniform crystal growth development. A patented process describes a rapid cooling of concentrated permeate from 80°C to 50°C, to facilitate nucleation, followed by gradually cooling to 30°C to achieve crystal growth with a narrow size distribution (Shi *et al.*, 2006). Other processes, such as microfluidic crystallization and ultra-sonication were used to achieve a uniform lactose crystal distribution (Vu *et al.*, 2003, 2004; Dombrowkski *et al.*, 2007; Dhumal *et al.*, 2008; Patel and Murthy, 2009; Kougoulos *et al.*, 2010; Dincer *et al.*, 2014). Sonication facilitates crystallization by accelerating primary nucleation and supporting a secondary nucleation to produce small and purer crystals. In a sonication-assisted process, crystallization occurs at lower lactose concentrations than observed in traditional processes. It is also possible to control crystal size and yield through sonication time and frequency (Dincer *et al.*, 2014).

15.5.2 Enzymatic and Chemical Modification

15.5.2.1 Galactooligosaccharides Galactooligosaccharides (GOS) are linear oligosaccharides resulting from the attachment of galactose units to a lactose. GOS is produced industrially using microbial β-galactosidases via the transgalactosylation reaction of lactose. β-galactosidase creates a covalent bond with the galactose moiety of lactose, releasing a glucose and transferring a galactose to a nucleophile receptor (Otieno, 2010). In reactions where there is a low lactose concentration, such as 4–6%, water is the most likely nucleophile receptor and the reaction outcome is lactose hydrolysis. This hydrolysis reaction is useful in producing low-lactose beverages with increased sweetness targeted to lactose intolerant consumers. In reactions where lactose concentrations are higher than 30%, there is a higher probability for lactose or a galactosyl derivative, to become the nucleophile receptor and the reaction outcome is a transgalactosylation with GOS as the outcome (Guerrero *et al.*, 2015). GOS has a structure Glu-[Gal]*n*, where *n* is the degree of polymerization, and the associated glycosidic linkages are β 1-4, β 1-6, and β 1-3. The most predominant GOS species include 2 to 4 lactose molecules (Figure 15.6).

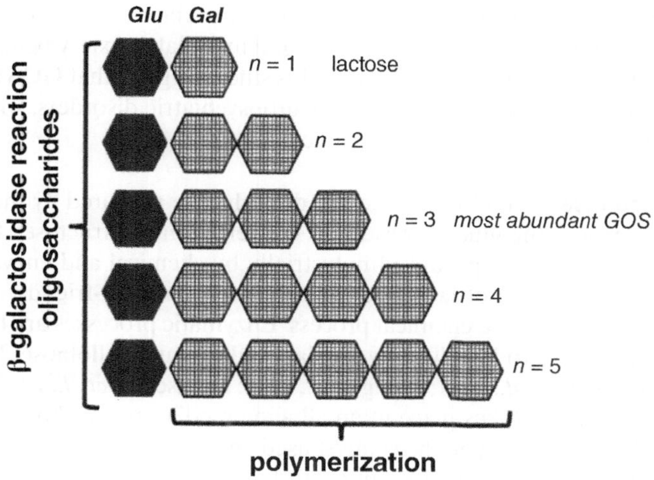

Figure 15.6 Schematic representation of sugar structures in GOS. A transgalactosylation of a galactose to a lactose (n=2) yields a tri-saccharide. Steric hindrances limit the polymerization significantly beyond n=5.

Several methods were patented for the production of GOS using β-galactosidase by free enzyme addition (Matsumoto *et al.*, 1988; Kan and Kobayashi, 1990; Sawatzki, 2005; Gibson *et al.*, 2007; Jeong *et al.*, 2011), by enzyme immobilization (Jovanovic-Malinovska *et al.*, 2012) or using whole microbial cells (Avalakki *et al*., 2011). Concentrated de-proteinized whey permeates are used commercially as low-cost substrates. However, membrane or resin filtration is needed to remove some of the salts, peptides and free amino acids present in the whey stream. The amino groups from residual peptides produce dark Maillard reaction compounds that yield a brownish tone to the GOS ingredient. Pure lactose solutions are also used to eliminate the step of removal of impurities (Kobayashi *et al.*, 1990). GOS may be further purified to increase purity to about 85% using ion exchange resins (Matsumoto *et al.*, 1988). Nano-filtration using 800–1000 Da cut-off membranes has been suggested as a low-cost alternative to remove about 90% of monosaccharides and 52% of disaccharides, obtaining a GOS yield of about 70% (Feng *et al.*, 2009).

GOS is used mostly as a prebiotic supporting the Health and Nutrition drivers to encourage the growth of lactobacilli and bifidobacteria (Otieno, 2010). GOS is used as a prebiotic in pediatric nutritional formulations as an attempt to derive similar benefits as those provided by the complex mixture of naturally present milk oligosaccharides. There is evidence that GOS and other added prebiotics alter the infant gastrointestinal microbiota stimulating the growth of bifidobacteria (Vandenplas *et al.*, 2014). This prebiotic activity is observed through a change in the metabolic activity of the infant microbiota, resulting in softening stool consistency, decreasing stool pH, increasing short-chain fatty acids and bringing defecation frequency closer to those of breast-fed infants. However, there is limited evidence that these changes in gastrointestinal microbiota induce a significant clinical benefit for the immune system (Vandenplas *et al.*, 2015). A study using a combination of GOS and maltodextrins as dietary supplements in adults showed enhanced bifidogenic properties and inhibited the growth of undesirable bacteria. Also, as was observed in infants, there was an increase in short-chain fatty acids and a decrease in stool pH (Musilova *et al.*, 2015).

Changes in gastrointestinal microbiota using specific prebiotics have also reduced the inflammatory response, alter brain chemistry, and modulate anxiety behavior in both rodents and humans (Savignac *et al.*, 2015). This study suggests that GOS has a potential role as a prebiotic in the treatment of neuropsychiatric disorders, where anxiety and neuro-inflammation are prominent clinical features.

15.5.2.2 Lactulose Lactulose is a ketose disaccharide prepared by the isomerization of lactose, where the glucose residue is converted into a fructose (Panesar and Kumari, 2011). Lactulose is produced industrially by chemical and enzymatic methods. Enzymatic production of lactulose is a safer, environment-friendly and simpler method in comparison to the chemical process. Enzymatic processes are based on the conversion of lactose into lactulose by β-galactosidases and cellobiose 2-epimerases (Lee *et al.*, 2004; Tang *et al.*, 2011; Wang *et al.*, 2013; Rentschler *et al.*, 2015). Chemical isomerization of lactose takes place in an alkaline media, such as borax, with yields of about 20–85%. depending on the catalyst used and process conditions (Krumbholz and Dorscheid, 1991; Panesar and Kumari, 2011). Concerns associated with chemical processes are the catalyst removal cost, presence of colored by-products and product safety. Lactulose is highly soluble in water, and is difficult to obtain as a stable powder form, thus it is usually commercialized as a syrup. A method to produce stable trihydrate lactulose crystals from a lactulose syrup (70–90% solids, 10% lactulose) has been developed (Tomita *et al.*, 1994). The process requires cooling the concentrate to about 2°–20°C, seeding with trihydrate crystals of lactulose, and inducing the trihydrate crystallization by stirring.

Lactulose has also been used as a substrate for the production of prebiotic oligosaccharides via enzymatic polymerization using β-galactosidase (Padilla *et al.*, 2015). Corzo-Martinez *et al* (2013) developed a method to produce GOS from a lactulose-enriched whey permeate. The galactose in the permeate had been isomerized into lactulose with a 16% w/w yield, then it was further polymerized using β-galactosidase to obtain a mixture composed of 50% w/w of prebiotic carbohydrates such as GOS and lactulose.

Lactulose Health driver benefits are associated with its role as a prebiotic promoting the growth of bifidobacteria and lactobacilli in the gastrointestinal tract (Schumann, 2002; Mao *et al.*, 2014), and as inhibitor for the growth of pathogenic bacteria such as Salmonella (Bovee-Oudenhoven and Van der Meer, 1997). It also maintains blood glucose and insulin levels (Bianchi *et al.*, 1994), treats constipation (Sanders, 1978), prevents recurrence of hepatic encephalopathy in patients with cirrhosis (Sharma *et al.*, 2009) and lowers the recurrence of colonic adenomas (Roncucci *et al.*, 1993). Lactulose has also been used in food product development as a sugar substitute in beverages, yoghurt, dairy products, infant formula and foods for diabetics.

15.5.2.3 Lactitol Lactitol is a sugar alcohol containing galactose and sorbitol. It is derived from the catalytic hydrogenation of lactose (Zacharis, 2012). Lactitol is industrially-produced following a traditional method of catalytic hydrogenation of d-lactose under high pressure (Wolfrom *et al.*, 1938). The conditions use a temperature of about 70–130°C, a Raney nickel catalyst (nickel-aluminum alloy) under about 30–74 bar hydrogen pressure, and low lactose concentrations, such as 30–40% (Van Velthuijsen, 1979; Wilson, 2000). The hydrogenated material is purified using resins and activated charcoal followed by crystallization. Yields of about 98% have been obtained using ruthenium-on-carbon as a catalyst that allows operation at lower hydrogen pressures (40–60 bar) (Kuusisto *et al.*, 2008). However, catalyst deactivation and formation of lactobionic acid as by-product poses a challenge for its industrial implementation. Crystalline lactitol occurs in anhydrous, monohydrate, dihydrate and trihydrate forms. Anhydrous and monohydrate lactitol are of commercial interest due to their low hygroscopicity. High-purity lactitol monohydrate is produced by controlling the lactitol solution concentration, temperature and agitation during crystallization (Heikkilae *et al.*, 1998), or by atomizing a pure lactitol solution over a fluid-bed of lactitol particles (Caboche, 1999).

Lactitol has a relative sweetness of about 0.3–0.4 compared to sucrose (Gänzle *et al.*, 2008). It is used in confectionary for diabetics due to its insulin independent response. Lactitol also has applications based on its ability to prevent dental caries, as a low-calorie sweetener, as a laxative and as a prebiotic (Patil *et al.*, 1987; Chen *et al.*, 2007; Faruqui and Chandrakant, 2012; Miller *et al.*, 2014). Lactitol is better tolerated than lactulose as a stool consistency regulator in adults and in paediatric populations (Pitzalis *et al.*, 1995; Copé *et al.*, 1996).

15.5.2.4 Tagatose Tagatose is a hexose monosaccharide, stereoisomer of fructose. Several methods were patented for the commercial production of d-tagatose via isomerization of d-galactose. The reaction is aided by metal oxides catalysts, such as aluminum hydroxide, barium hydroxide, calcium hydroxide, magnesium hydroxide and strontium hydroxide (Beadle *et al.*, 1991; Xu, 2013; Oroskar, 2014). The resulting tagatose-rich stream is purified using ion exchange resins and crystallized. Enzymatic methods isomerize d-galactose via L-arabinose isomerase. The most favored d-lactose source is low-cost deproteinized whey permeate or whey permeate that may be treated with β-galactosidase (Bertelsen *et al.*, 2006) or fermented to obtain d-galactose (Ibrahim and Spradlin, 2000). Enzymatic processes may provide a cleaner, lower cost, and more efficient option than the current chemical methods.

Compared to sucrose, it is about 90% sweeter, with 38% of its calories (1.5 kcal/g) (Beadle *et al.*, 1991). D-Tagatose is used as a sweetener in beverages, dietary supplements and health foods. It is generally recognized as safe (GRAS) by the US and the FAO/WHO since 2001 (Beadle *et al.*, 1991). D-Tagatose also has prebiotic health benefits (Laerke *et al.*, 2000; Levin, 2002; Venema *et al.*, 2009).

15.6 Milk Oligosaccharides

Milk oligosaccharides are a group of over 200 molecules of complex composition with more than 3 sugar monomers linked by 13 possible glycosidic linkages (Kunz *et al.*, 2000; Ninonuevo *et al.*, 2006). They are synthesized in the mammary gland, and their composition and abundance varies with lactation (Chichlowski *et al.*, 2011). Oligosaccharides have linear or branched structures with a lactose as the reducing end followed by β-glycosidic linkages of galactose and *N*-acetylgucosamine (Bode, 2006). This general pattern may have additional attachments of fucose and sialic acid. The smallest oligosaccharides are 2′-fucosyllactose, a neutral trisaccharide formed by a fucose linked to a lactose, and 3′- or 6′- sialyllactose, an acidic trisaccharide formed by a sialic acid linked to a lactose (Figure 15.7). However, these smaller structures are less abundant than the larger more complex ones (Chichlowski *et al.*, 2011).

Oligosaccharides may be classified based on their sialic acid content. Acidic ones contain sialic acid in their structure, neutral ones do not (Boehm and Stahl, 2007;

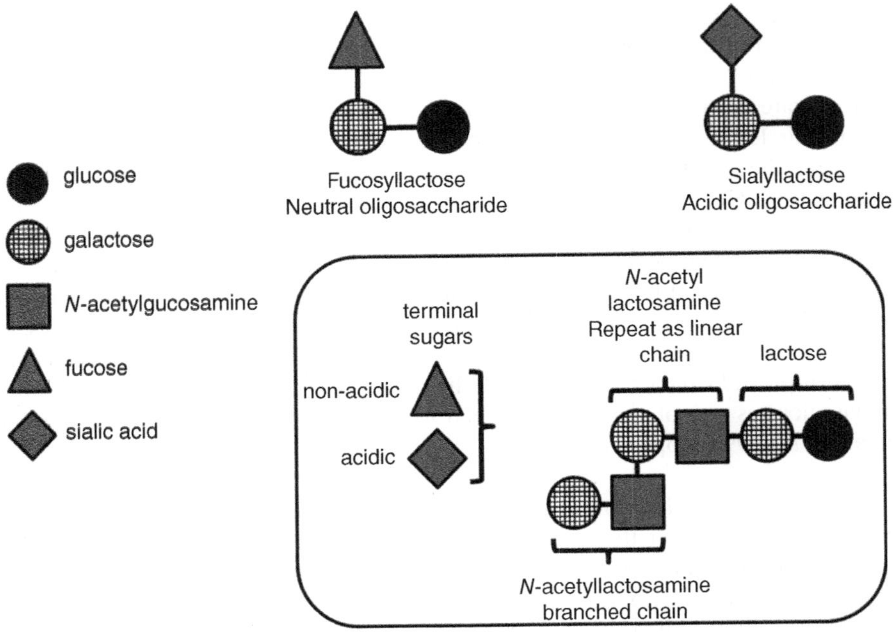

Figure 15.7 Milk oligosaccharides have a lactose at the reducing end, and most have a backbone formed by repeating units of N-acetyllactosamine. Bovine milk is mostly acidic, with 3′- and 6′-sialyllactose the most abundant. Human milk oligosaccharides are mostly neutral with a fucosyl terminal (adapted from Newburg and Grave, 2014).

Newburg and Grave, 2014). Oligosaccharides fractionation methods based on ion exchange chromatography leverage this difference in charges. The sialyllactoses are the most abundant acidic oligosaccharides in bovine and human milk. In bovine milk colostrum, these sialylated molecules may reach about 600 mg/liter, and reduce to about 60–200 mg/liter in mature milk. In contrast, human milk may reach about 850 and 600 mg/liter in colostrum and mature milk, respectively (Martin-Sosa *et al.*, 2003; Nakamura *et al.*, 2003; McJarrow and van Amelsfort-Schoonbeek, 2004).

The function of milk oligosaccharides has been closely associated with their structural complexity and diversity. Most of the oligosaccharides resist gastric and intestinal enzymatic digestion, thus reaching the colon intact and serving as substrate for colonic microbiota. Neutral oligosaccharides are relevant for the development of the intestinal microbiota in infants supporting the growth of Bifidobacterium and lactobacilli, enhancing the immune system (Zivkovic and Barile, 2011). Acidic oligosaccharides have a role in preventing adhesion of pathogenic virus and bacteria to the gastrointestinal wall. There is evidence that some oligosaccharides, such as 2′-fucosyllactose, are absorbed through the intestinal wall, showing positive impacts on cognitive domains and improving learning and memory in rodents (Vazquez *et al.*, 2015). There are several patents related to the use of oligosaccharides in nutritional formulations, with claims of benefits towards modulating the immune system (Joost van *et al.*, 2011a), benefiting growth of beneficial bifidobacteria (Buck *et al.*, 2012a, b; Frantz, 2014), reducing the risk of development of infant obesity (Joost van *et al.*, 2011b), reducing oxidative stress and inflammation (Davis *et al.*, 2012; Buck *et al.*, 2012b), improving airway respiratory health (Thomas and Brooks, 2004), treating enteric viral infection, preventing injury and improving the healing of the gastrointestinal tract (Davis and Chow, 2014), and decreasing the incidence of necrotizing enterocolitis in infants (Chow *et al.*, 2012).

15.6.1 Oligosaccharide Processing

A number of processes have been developed to isolate siallyllactose-related molecules, such as 3′-sialyllactose, 6′-sialyllactose, 6′-sialyllactosamine, 3′-sialyllactosamine and disialyllactose from delactosed whey permeate using membranes and/or ion-exchange resins. A membrane method using filtration (2.5 KDa cut-off) to prepare a concentrate from whey permeate or lactose mother liquor, is preferable from that which is enriched to about 5–10% (w/w, dry) of sialyllactose (Holst *et al.*, 2013). Another membrane method recovers sialyllactose from whey permeate after treatment with β-galactosidase to hydrolyze lactose and phytase to de-phosphorylate compounds (Spade *et al.*, 2003). The hydrolyzed stream is nano-filtered (1 KDa cut-off) to remove enzymatic end-products and yields a fraction containing up to 30% (w/w, dry) of sialyloligosaccharides. Ionic-exchange resin methods bind siallylactose to strong cationic resins after previously desalting the whey permeate stream using weak anion and cation exchange resins. Bonte *et al.* (2011) used strong cationic resins to produce an 8–15% purity phosphorus-free siallylactose ingredient. Scott and Krishnapillai (2015) combined several weak and strong anionic and cationic resins to desalt and produce a 25–30% purity siallylactose ingredient. The major challenge in commercializing an oligosaccharide-rich dairy ingredient relates to the low abundance of these carbohydrates in bovine milk and the large volumes that are needed to be processed to produce scale-up commercial volumes. It may be required to process over 400 tons (MT) of delactosed whey permeate to produce 1 MT of a 10% (w/w, dry) sialyllactose-rich ingredient, assuming a 75% recovery.

15.7 Future Outlook

Milk and dairy-derived products continue to be prime nutritional choices for infant and adults. Progress has been made in the development of Enabler drivers to fractionate, increase yields and purify the major and minor milk components and their derived materials. Clinical research has expanded the knowledge on benefits that continue expanding Health and Nutrition drivers. Furthermore, industry has continued investing in technologies to enhance the traditional physical-chemical and sensory attributes of dairy ingredients and their applications in product development. The increasing demand for nutritious and functional food products will continue driving the consumption of milk and dairy products. There is an optimistic outlook regarding the identification of new health and nutrition benefits, as well as on the development of novel technologies that will make dairy components more available and affordable for novel and exciting food products.

References

Almécija, M.C., Ibáñez, R., Guadix, A. and Guadix, E.M. (2007) Effect of pH on the fractionation of whey proteins with a ceramic ultrafiltration membrane. *Journal of Membrane Science*, **288**(1–2): 28–35.

Alvey, J., Gonzalez, J.M., Taylor, B.J., Morris, K.L., Anthony, J.C., *et al.* (2011) Formulations and methods for nutrient delivery. World Patent 2011106482 A1, 1 September.

Amer, M.A., Kupranycz, D.B. and Baker, B.E. (1985) Physical and chemical characteristics of butterfat fractions obtained by crystallization from molten fat. *Journal of the American Oil Chemists Society*, **62**(11): 1551–1557.

Arunkumar, A. and Etzel, M.R. (2014) Fractionation of α-lactalbumin and β-lactoglobulin from bovine milk serum using staged, positively charged, tangential flow ultrafiltration membranes. *Journal of Membrane Science*, **454**: 488–495.

Arunkumar, A. and Etzel, M.R. (2015) Negatively-charged tangential flow ultrafiltration membranes for whey protein concentration. *Journal of Membrane Science*, **475**: 340–348.

Astaire, J.C., Ward, R., German, J.B. and Jiménez-Flores, R. (2003) Concentration of polar MFGM lipids from buttermilk by microfiltration and supercritical fluid extraction. *Journal of Dairy Science*, **86**(7): 2297–2307.

Avalakki, U.K., Maheswaran, P. and Saravanan, R. (2011) Process for production of galactooligosaccharides (GOS). US Patent 9139856 B2, 17 March.

Banavara, D., Alvey, J.D. and Gonzalez, J.M. (2012) Composition comprising heat labile milk proteins and process for preparing same. US Patent 20120171328 A1, 5 July.

Banavara, D., Alvey, J. and Gonzalez, J.M. (2014a) Nutritional compositions containing structured fat globules and uses thereof. US Patent 20140255538 A1. 11 September.

Banavara, D., Alvey, J. and Gonzalez, J.M. (2014b) Nutritional compositions containing structured fat globules and uses thereof. US Patent 20140255539 A1. 11 September.

Banavara, D., Alvey, J.D., Peters, J.A and Gonzalez, J.M. (2015) Expanded bed adsorption methods for isolation of basic milk proteins including lactoferrin. US Patent 20150024462 A1, 22 January.

Beadle, J.R., Saunders, J.P. and Wajda, T.J. (1991) Process for manufacturing tagatose. US Patent 5002612 A, 26 March.

Benzaria, A., Maresca, M., Taieb, N. and Dumay, E. (2013) Interaction of curcumin with phosphocasein micelles processed or not by dynamic high-pressure. *Food Chemistry*, **138**(4): 2327–2337.

Bertelsen, H., Eriknauer, K., Bøttcher, K., Christensen, H.J.S., Stougaard, P., *et al.* (2006) Process for manufacturing of tagatose. US Patent 6991923 B2, 31 January.

Bhaskar, A.R., Rizvi, S.S.H., Bertoli, C., Fay, L.B. and Hug, B. (1998) A comparison of physical and chemical properties of milk fat fractions obtained by two processing technologies. *Journal of the American Oil Chemists Society*, **75**(10): 1249–1264.

Bhushan, S. and Etzel, M. (2009) Charged ultrafiltration membranes increase the selectivity of whey protein separations. *Journal of Food Science*, **74**(3): E131–E139.

Bianchi, G.P., De Mitri, M.S., Bugianesi, E., Abbiati, R., Fabbri, A. and Marchessini, G. (1994) Lowering effects of a preparation containing fibres and lactulose on glucose and insulin levels in obesity. *Italian Journal of Gastroenterology*, **26**(4): 174–178.

Billeaud, C., Puccio, G., Saliba, E., Guillois, B., Pecquet, S. and Steenhout, P. (2011) Growth of infant fed starter formula containing milk fat globule membranes. *Pediatric Research*, **70**: 782–782.

Bode, L. (2006) Recent advances on structure, metabolism, and function of human milk oligosaccharides. *Journal of Nutrition*, **136**(8): 2127–2130.

Boehm, G. and Stahl, B. (2007) Oligosaccharides from milk. *The Journal of Nutrition*, **137**(3): 847S–849S.

Boland, M. (2011) Whey proteins. In: *Handbook of Food Proteins* (eds G.O. Phillips and P.A. Williams). Woodhead Publishing Ltd, Cambridge, pp, 30–55.

Bonnaillie, L.M., Qi, P., Wickham, E. and Tomasula, P.M. (2014) Enrichment and purification of casein glycomacropeptide from whey protein isolate using supercritical carbon dioxide processing and membrane ultrafiltration. *Foods*, **3**(1): 94–109.

Bonte, A.W., Klarenbeek, G. and Becker, M.J.A. (2011) Process for isolating sialic acid containing oligosaccharides, and the compositions containing sialic acid containing oligosaccharides obtainable thereby. US Patent 20110098244 A1, 28 April.

Böttger, F.H., Etzel, M.R. and Lucey, J.A. (2013) *In vitro* infant digestion of whey protein–dextran glycates. *Food Digestion*, **4**(2–3): 76–84.

Boudreau, A. and Arul, J. (1993) Cholesterol reduction and fat fractionation technologies for milk fat: An overview. *Journal of Dairy Science*, **76**(6): 1772–1781.

Bovee-Oudenhoven, I. and Van Der Meer, R. (1997) Protective effects of dietary lactulose and calcium phosphate against Salmonella infection. *Scandinavian Journal of Gastroenterology, Supplement*, **32**(222): 112–114.

Brew, K., Findlay, J.B. and Khatra, B.S. (1972) The structure and regulation of the lactose synthetase system. *Biochemical Journal*, **130**(1): 2P–3P.

Brown, A. (2015) Process for purifying growth factors from milk and products thereof. US Patent 20150232823 A1, 20 August.

Buck, R., Chow, J.M. and Davis, S.R. (2012a) Human milk oligosaccharides to promote growth of beneficial bacteria. US Patent 20120171165 A1, 5 July.

Buck, R., Duska-McEwen, G.O. and Schaller, J.P. (2012b) Human milk oligosaccharides for modulating inflammation. US Patent 20120172330 A1, 5 July.

Burrington, K., Schoenfuss, T. and Patel, S. (2011) *Technical Report: Coproducts of Milk and Whey Processing*, Dairy Research Institute and US Dairy Export Council.

Bylund, G. (1995) The chemistry of milk. In: *Dairy Processing Handbook* (ed. G.S. Bylund). Tetra Pak Processing Systems AB Lund, Sweden, pp. 13–37.

Caboche, J.J. (1999) Lactitol composition and process for the preparation thereof. US Patent 5989352 A, 23 November.

Caffarelli, C., Plebani, A., Poiesi, C., Petroccione, T., Spattini, A. and Cavagni, G. (2002) Determination of allergenicity to three cow's milk hydrolysates and an amino acid-derived formula in children with cow's milk allergy. *Clinical and Experimental Allergy*, **32**(1): 74–79.

Campbell, J.R. and Marshall, R.T. (1975) Nutritional changes in milk composition. In: *The Science of Providing Milk to Man* (eds J.R. Campbell and R.T. Marshall). McGraw-Hill Ryerson, Ltd, UK, pp. 2–24.

Cavaletto, M., Giuffrida, M.G. and Conti, A. (2004) The proteomic approach to analysis of human milk fat globule membrane. *Clinica Chimia Acta*, **347**: 41–48

Cerbulis, J. and Farrell, H.M. (1975) Composition of milks of dairy Cattle. I: Protein, lactose, and fat contents and distribution of protein fraction. *Journal Dairy Science*, **58**(6): 817–827.

Chen, C., Li, L., Wu, Z., Chen, H. and Fu, S. (2007) Effects of lactitol on intestinal microflora and plasma endotoxin in patients with chronic viral hepatitis. *Journal of Infection*, **54**(1): 98–102.

Chen, Y., Zheng, Z., Zhu, X., Shi, Y., Tian, D., *et al.* (2015) Lactoferrin promotes early neurodevelopment and cognition in postnatal piglets by upregulating the BDNF signaling pathway and polysialylation. *Molecular Neurobiology*, **52**(1): 256–269.

Chichlowski, M., German, J.B., Lebrilla, C.B. and Mills, D.A. (2011) The influence of milk oligosaccharides on microbiota of infants: Opportunities for formulas. *Annual Review of Food Science and Technology*, **2**: 331–351.

Chow, J.M., Davis, S.R., Buck, R., Duska-McEwen, G.O. and Linke, H.K. (2012) Methods for decreasing the incidence of necrotizing enterocolitis in infants, toddlers, or children using human milk oligosaccharides. US Patent 20120172319 A1, 5 July.

Clare, D. and Swaisgood, H. (2000) Bioactive milk peptides: A prospectus. *Journal of Dairy Science*, **83**(6): 1187–1195.

Clark, N. (2013) Novel process separates whey proteins into high-value ingredients. ScientistLive. 1 April.

Contarini, G. and Povolo, M. (2013) Phospholipids in milk fat: Composition, biological and technological significance, and analytical strategies. *International Journal of Molecular Sciences*, **14**(2): 2808–2831.

Copé, R., Debou, J., Cohen, S. and Callens, A. (1995) Study of early postoperative lactitol versus paraffin following anal surgery in 110 adult patients. *Annales de chirurgie*, **50**(7): 570–576.

Coppa, G.V., Gabrielli, O., Pierani, P., Catassi, C., Carlucci, A. and Giorgi, P.L. (1993) Changes in carbohydrate composition in human milk over 4 months of lactation. *Pediatrics*, **91**(3): 637–641.

Corredig, M. and Dalgleish, D.G. (1997) Isolates from industrial buttermilk: Emulsifying properties of materials derived from the milk fat globule membrane. *Journal of Agricultural and Food Chemistry*, **45**(12): 4595–4600.

Corredig, M., Roesch, R.R. and Dalgleish, D.G. (2003) Production of a novel ingredient from buttermilk. *Journal of Dairy Science*, **86**(9): 2744–2750.

Corzo-Martínez, M., Copoví, P., Olano, A., Moreno, F.J. and Montilla, A. (2013) Synthesis of prebiotic carbohydrates derived from cheese whey permeates by a combined process of isomerisation and transgalactosylation. *Journal of the Science of Food and Agriculture*, **93**(7): 1591–1597.

Costa, M.R., Elias-Argote, X.E., Jiménez-Flores, R. and Gigante, M.L (2010) Use of ultrafiltration and supercritical fluid extraction to obtain a whey buttermilk powder enriched in milk fat globule membrane phospholipids. *International Dairy Journal*, **20**(9): 598–602.

Dalgleish, D.G. (2011) On the structural models of bovine casein micelles – Review and possible improvements. *Soft Matter*, **7**(6): 2265–2272.

Damodaran, S. (2010) Zinc-induced precipitation of milk fat globule membranes: A simple method for the preparation of fat-free whey protein isolate. *Journal of Agricultural and Food Chemistry*, **58**(20): 11052–11057.

Davis, S.R. and Chow, J.M. (2014) Human milk oligosaccharides for preventing injury and/or promoting healing of the gastrointestinal tract. US Patent 20140335065 A1, 13 November.

Davis, S.R., Buck, R. and Duska-McEwen, G.O. (2012) Methods for reducing the incidence of oxidative stress using human milk oligosaccharides, vitamin C and anti-inflammatory agents. US Patent 20120172307 A1, 5 July.

Delavari, B., Saboury, A.A., Atri, M.S., Ghasemi, A., Bigdeli, B., *et al.* (2015) Alpha-lactalbumin: A new carrier for vitamin D 3 food enrichment. *Food Hydrocolloids*, **45**: 124–131.

Denhardt, D.T. and Guo, X. (1993) Osteopontin: A protein with diverse functions. *The FASEB Journal*, **7**(15): 1475–1482.

Dewettinck, K., Rombaut, R., Thienpont, N., Le, T.T., Messens, K. and Van Camp, J. (2008) Nutritional and technological aspects of milk fat globule membrane material. *International Dairy Journal*, **18**(5): 436–457.

Dhumal, R.S., Biradar, S.V., Paradkar, A.R. and York, P. (2008) Ultrasound-assisted engineering of lactose crystals. *Pharmaceutical Research*, **25**(12): 2835–2844.

Díaz, M. and Decker, E.A. (2004) Antioxidant mechanisms of caseinophosphopeptides and casein hydrolysates and their application in ground beef. *Journal of Agricultural and Food Chemistry*, **52**(26): 8208–8213.

Dincer, T., Zisu, B., Vallet, C., Jayasena, V., Palmer, M. and Weeks, M. (2014) Sonocrystallisation of lactose in an aqueous system. *International Dairy Journal*, **35**(1): 43–48.

Dombrowski, R.D., Litster, J.D., Wagner, N.J. and He, Y. (2007) Crystallization of alpha-lactose monohydrate in a drop-based microfluidic crystallizer. *Chemical Engineering Science*, **62**(17): 4802–4810.

Du, Q.-Y., Lin, D.-Q., Zhang, Q.-L. and Yao, S.-J. (2014) An integrated expanded bed adsorption process for lactoferrin and immunoglobulin G purification from crude sweet whey. *Journal of Chromatography B*, **947**: 201–207.

Durham, R.J., Hourigan, J.A., Sleigh, R.W. and Johnson, R.L. (2002) Process for the purification of nutrients from food process streams. US Patent 6475390 B1, 5 November.

Elling, J.L. and Duncan, S.E. (1996) Physical properties of 20% milk fat reformulated creams manufactured from cholesterol-reduced butteroil. *Journal of Food Science*, **61**(2): 375–378.

Esmaili, M., Ghaffari, S.M., Moosavi-Movahedi, Z., Atri, M.S., Sharifizadeh, A., *et al.* (2011) Beta casein-micelle as a nano vehicle for solubility enhancement of curcumin, food industry application. *LWT – Food Science and Technology*, **44**(10): 2166–2172.

Evans, J.W. and Young, G.C. (1982a) Production of USP quality lactose. US Patent 4316749 A. 23 February.

Evans, J.W., Young, G.C. and Stager, C.W. (1982b) Production of a stable lactose product. US Patent 4342604 A. 3 August.

Faruqui, A.A. and Chandrakant, J. (2012) Lactitol: A review of its use in the treatment of constipation. *International Journal of Recent Advances in Pharmaceutical Research*, **2**(1): 1–5.

Feng, Y.M., Chang, X.L., Wang, W.H. and Ma, R.Y. (2009) Separation of galacto-oligosaccharides mixture by nanofiltration. *Journal of the Taiwan Institute of Chemical Engineers*, **40**(3): 326–332.

Ferruzzi, M.G., Bordenave, N. and Hamaker, B.R. (2012) Does flavor impact function? Potential consequences of polyphenol–protein interactions in delivery and bioactivity of flavan-3-ols from foods. *Physiology and Behavior*, **107**(4): 591–597.

FitzGerald, R. and Ócuinn, G. (2006) Enzymatic debittering of food protein hydrolysates. *Biotechnology Advances*, **24**(2): 234–237.

Fox, P.F. (2003) The major constituents of milk. In: *Dairy Processing – Improving Quality* (ed. G. Smit). Woodhead Publishing, Cambridge, pp. 2–38.

Fox, P.F. and McSweeney, P.L.H. (1998) Milk proteins. In: *Dairy Chemistry and Biochemistry* (eds P.F. Fox and P.L.H. McSweeney), Springer, Cambridge, pp. 146–238.

Frantz, D.C. (2014) Modulation of growth of bifidobacteria using a combination of oligosaccharides found in human milk. US Patent 20140371173 A1, 18 December.

Fuller, K.L., Kuhlenschmidt, T.B., Kuhlenschmidt, M.S., Jiménez-Flores, R. and Donovan, S.M. (2013) Milk fat globule membrane isolated from buttermilk or whey cream and their lipid components inhibit infectivity of rotavirus *in vitro*. *Journal of Dairy Science*, **96**(6): 3488–3497.

Furger, E., Fedosov, S.N., Launholt Lildballe, D., Waibel, R., Schibli, R., *et al*. (2012) Comparison of recombinant human haptocorrin expressed in human embryonic kidney cells and native haptocorrin. *PLoS ONE*, **7**(5): e37421.

Gallier, S., Vocking, K., Post, J.A., Van De Heijning, B., Acton, D., *et al*. (2015) A novel infant milk formula concept: Mimicking the human milk fat globule structure. *Colloids and Surfaces B: Biointerfaces*, **136**: 329–339.

Gänzle, M.G., Haase, G. and Jelen, P. (2008) Lactose: Crystallization, hydrolysis and value-added derivatives. *International Dairy Journal*, **18**(7): 685–694.

Garcia, C. and Innis, S. (2013) Structure of the human milk fat globule. *Lipid Technology*, **25**(10): 223–226.

Gibon, V. (2006) Fractionation of lipids for use in food. In: *Modifying Lipids for Use in Food* (ed. F.D. Gunstone). Woodhead Publishing, Cambridge, pp. 201–233.

Gibson, G., Slupinski, J.W., Tzortzis, G. and Wynne, A.G. (2007) Novel galactooligosaccharide composition and the preparation thereof. US Patent 20070274955 A1, 29 November.

Givens, D.I. and Shingfield, K.J. (2006) Optimising dairy milk fatty acid composition. In: *Improving the Fat Content of Foods*, (eds C. Williams, and J. Buttriss), Woodhead Publishing, Cambridge, pp. 25–280.

Gonzalez, J.M. (2014a) The role of Food Science in ingredient functionality and food systems. In: *Consumer Behavior and Food Science Innovations for Optimal Nutrition*. Sackler Institute for Nutrition Sciences, New York Academy of Sciences: New York.

Gonzalez, J.M. (2014b) Milk protein ingredients for infant and clinical nutrition products. In: *South Dakota State University-NIZO joint symposium on High Quality Dairy Ingredients Spray Drying*, Brookings, SD.

Gonzalez, J.M. (2015) Product development drivers to support alpha-lactalbumin formulations. In: *Symposium on Alpha-Lactalbumin by Davisco-Agropur*, Twin Falls, ID.

Gonzalez, J.M., Banavara, D. and Alvey, J. (2014) Oral electrolyte solution containing lactoferrin and uses thereof. US Patent 20140255365 A1, 11 September.

Guerrero, C., Vera, C., Conejeros, R. and Illanes, A. (2015) Transgalactosylation and hydrolytic activities of commercial preparations of β-galactosidase for the synthesis of prebiotic carbohydrates. *Enzyme and Microbial Technology*, **70**: 9–17.

Gurnida, D.A., Rowan, A.M., Idjradinata, P., Muchtadi, D. and Sekarwana, N. (2012) Association of complex lipids containing gangliosides with cognitive development of 6-month-old infants. *Early Human Development*, **88**(8): 595–601.

Guu, Y. and Zall, R. (1992) Nanofiltration concentration effect on the efficacy of lactose crystallization. *Journal of Food Science*, **57**(3): 735–739.

Håkansson, A., Zhivotovsky, B., Orrenius, S., Sabharwal, H. and Svanborg, C. (1995) Apoptosis induced by a human milk protein. *Proceedings of the National Academy of Sciences*, **92**(17): 8064–8068.

Halken, S., Host, A., Hansen, L. and Østerballe, O. (1993) Safety of a new, ultrafiltrated whey hydrolysate formula in children with cow milk allergy: a clinical investigation. *Pediatric Allergy and Immunology*, **4**(2): 53–59.

Hamosh, M., Bitman, J., Wood, D.L. and Mehta, N.R. (1985) Lipids in milk and the first steps in their digestion. *Pediatrics*, **75**(1 II Suppl.) : 146–150.

Hansel, B. and Giral, P. (2015) Cholestérol alimentaire et morbi/mortalité cardiovasculaire. *Cahiers de Nutrition et de Diététique*, **50**(4) : 202–208.

Haratifar, S. and Corredig, M. (2014) Interactions between tea catechins and casein micelles and their impact on renneting functionality. *Food Chemistry*, **143**: 27–32.

Hasni, I., Bourassa, P., Hamdani, S., Samson, G., Carpentier, R. and Tajmir-Riahi, H.-A. (2011) Interaction of milk α-and β-caseins with tea polyphenols. *Food Chemistry*, **126**(2): 630–639.

Hayes, J., Dunkerley, J., Muller, L. and Griffin, A. (1974) Studies on whey processing by ultrafiltration. II: Improving permeation rates by preventing fouling. *Australian Journal of Dairy Technology*, **29**(3): 132–140.

Heid, H.W. and Keenan, T.W. (2005) Intracellular origin and secretion of milk fat globules. *European Journal of Cell Biology*, **84**: 245–258

Heikkila, H.O., Nurmi, J.V. and Pepper, T. (1998) Crystalline anhydrous lactitol and a process for the preparation thereof as well as use thereof. US Patent 5779806 A, 14 July.

Helbig, N., Ho, L., Christy, G. and Nakai, S. (1980) Debittering of skim milk hydrolysate by adsorption into acidic beverages. *Journal of Food Science*, **45**(2): 331–335.

Hernandez-Ledesma, B., Dávalos, A., Bartolomé, B. and Amigo, L. (2005) Preparation of antioxidant enzymatic hydrolysates from α-lactalbumin and β-lactoglobulln. Identification of active peptides by HPLC-MS/MS. *Journal of Agricultural and Food Chemistry*, **53**(3): 588–593.

Hickey, M. and Hill, R. (1980) Investigations into the ultrafiltration and reverse osmosis of wheys. II: The effects of some minor whey constituents. *New Zealand Journal of Dairy Science and Technology*, **15**(2): 123–130.

Holst, H.H., Gunther, W.S., Mogensen, M.T. and Jorgensen, A.S. (2013) Concentrate derived from a milk product enriched in naturally occurring sialyllactose and a process for preparation thereof. US Patent 8445053 B2, 21 May.

Huber-Haag, K.J., Mace, C., Steenhout, P. and Tran, L.A. (2015) Nutritional compositions with phospholipids. World Patent 2015078506 A1, 4 June.

Huppertz, T., Kelly, A.L. and Fox, P.F. (2009) Milk lipids – Composition, origin and properties. In: *Dairy Fats and Related Products* (ed. A.Y. Tamime), Wiley-Blackwell, Oxford, pp 1–27.

Ibrahim, O.O. and Spradlin, J.E. (2000) Process for manufacturing D-tagatose. US Patent 6057135 A, 2 May.

Jansson, L., Akesson, B. and Holmberg, L. (1981) Vitamin E and fatty acid composition of human milk. *American Journal of Clinical Nutrition*, **34**(1): 8–13.

Jensen, R G., Hagerty, M.M. and McMahon, K.E. (1978) Lipids of human milk and infant formulas: A review. *The American Journal of Clinical Nutrition*, **31**(6): 990–1016.

Jensen, R.G., Ferris, A.M. and Lammi-Keefe, C.J. (1991) The composition of milk fat. *Journal of Dairy Science*, **74**(9): 3228–3243.

Jenssen, H. and Hancock, R.E.W. (2009) Antimicrobial properties of lactoferrin. *Biochimie*, **91**(1), 19–29.

Jeong, H., Kwon, H. and Choi, J.S. (2011) High-purity galactooligosaccharides and uses thereof. US Patent 20110189342 A1, 4 August.

Johnson, B.R. and Conte, J.A. (1991) Production of low-cholesterol milk fat by solvent extraction. US Patent 4997668 A, 5 March.

Johnson, E. and Larson, A. (2005) Lysozyme antimicrobials in food. In: *Antimicrobial in Food* (eds P.M. Davidson, J.N. Sofos and A. Branen). CRC Press, Taylor and Francis, Boca Raton, FL, pp. 361–371.

Joost Van, N.R.J., Schoterman, H.C. and Nauta, A. (2011a) Use of sialyl oligosaccharides to modulate the immune system. World Patent 2011096809 A1, 11 August.

Joost Van, N.R.J., Schoterman, H.C. and Nauta, A. (2011b) Use of sialyl oligosaccharides in weight management. World Patent 2011096808 A1, 11 August.

Jovanovic-Malinovska, R., Fernandes, P., Winkelhausen, E. and Fonseca, L. (2012) Galacto-oligosaccharides synthesis from lactose and whey by β-galactosidase immobilized in PVA. *Applied Biochemistry and Biotechnology*, **168**(5): 1197–1211.

Kan, T. and Kobayashi, Y. (1990) Method for producing oligosaccharide. US Patent 4957860A, 18 Sepember.

Kanakis, C., Hasni, I., Bourassa, P., Tarantilis, P., Polissiou, M. and Tajmir-Riahi, H.-A. (2011) Milk β-lactoglobulin complexes with tea polyphenols. *Food Chemistry*, **127**(3): 1046–1055.

Kaur, G. and Gathwala, G. (2015) Efficacy of bovine lactoferrin supplementation in preventing late-onset sepsis in low birth weight neonates: A randomized placebo-controlled clinical trial. *Journal of Tropical Pediatrics*, **61**(5): 370–376.

Kaylegian, K.E., Hartel, R. and Lindsay, R.C. (1993) Applications of modified milk fat in food products. *Journal of Dairy Science*, **76**(6): 1782–1796.

Khatib, K.A., Chiang, W.C. and Ehninger, J. (2009) Method for decreasing bitterness and improving taste of protein-free and hydrolyzed infant formulas. US Patent 20090117256 A1, 7 May.

Kim, W.S., Ohashi, M., Tanaka, T., Kumura, H., Kim, G.Y., *et al.* (2004) Growth-promoting effects of lactoferrin on *L. acidophilus* and *Bifidobacterium* spp. *BioMetals*, **17**(3): 279–283.

Kobayashi, Y., Kan, T. and Terashima, T. (1990) Method for producing processed milk containing galactooligosaccharide. US Patent 4944952 A, 31 July.

Kougoulos, E., Marziano, I. and Miller, P. (2010) Lactose particle engineering: Influence of ultrasound and anti-solvent on crystal habit and particle size. *Journal of Crystal Growth*, **312**(23): 3509–3520.

Krumbholz, R.E. and Dorscheid, M.G. (1991) Method of manufacturing lactulose. US Patent 5071530 A, 10 December.

Kumar, M. and Ulbricht, M. (2013) Novel antifouling positively charged hybrid ultrafiltration membranes for protein separation based on blends of carboxylated carbon nanotubes and aminated poly (arylene ether sulfone). *Journal of Membrane Science*, **448**: 62–73.

Kunz, C. and Lönnerdal, B. (1989) Human milk protein: separation of whey proteins and their analysis by polyacrylamide gradient gel electrophoresis, fast protein liquid chromatography (FPLC), gel filtration and anion-exchange chromatography. *American Journal of Clinical Nutrition*, **49**: 464–470.

Kunz, C. and Lönnerdal, B. (1990) Human-milk proteins: analysis of casein and casein subunits by anion-exchange chromatography, gel electrophoresis, and specific staining methods. *American Journal of Clinical Nutrition*, **51**: 37–46.

Kunz, C. and Lönnerdal, B. (1992) Re-evaluation of the whey protein/casein ratio of human milk. *Acta Pædiatrica*, **81**(2): 107–112.

Kunz, C., Rudloff, S., Baier, W., Klein, N. and Strobel, S. (2000) Oligosaccharides in human milk: structural, functional, and metabolic aspects. *Annual Review of Nutrition*, **20**(1): 699–722.

Kussendrager, K.D. and Van Hooijdonk, A.C.M. (2000) Lactoperoxidase: Physico-chemical properties, occurrence, mechanism of action and applications. *British Journal of Nutrition*, **84**(Suppl. 1): S19–S25.

Kuusisto, J., Mikkola, J.-P., Sparv, M., Wärnå, J., Karhu, H. and Salmi, T. (2008) Kinetics of the catalytic hydrogenation of d-lactose on a carbon supported ruthenium catalyst. *Chemical Engineering Journal*, **139**(1): 69–77.

Kwak, H.S., Ahn, J.J. and Lee, D.K. (2000) Method for removing cholesterol from milk and cream. US Patent 6110517 A, 29 August.

LaClair, C.E., Ney, D.M., MacLeod, E.L. and Etzel, M.R. (2009) Purification and use of glycomacropeptide for nutritional management of phenylketonuria. *Journal of Food Science*, **74**(4): E199–E206.

Laerke, H.N., Jensen, B.B. and Hojsgaard, S. (2000) *In vitro* fermentation pattern of D-tagatose is affected by adaptation of the microbiota from the gastrointestinal tract of pigs. *Journal of Nutrition*, **130**(7): 1772–1779.

Le, T.T., Debyser, G., Gilbert, W., Struijs, K., Van Camp, J., *et al.* (2013) Distribution and isolation of milk fat globule membrane proteins during dairy processing as revealed by proteomic analysis. *International Dairy Journal*, **32**(2): 110–120.

Lee, Y.J., Kim, C.S. and Oh, D.K. (2004) Lactulose production by β-galactosidase in permeabilized cells of *Kluyveromyces lactis*. *Applied Microbiology and Biotechnology*, **64**(6): 787–793.

Levin, G.V. (2002) Tagatose, the new GRAS sweetener and health product. *Journal of Medicinal Food*, **5**(1): 23–36.

Li, M., Ma, Y. and Ngadi, M.O. (2013) Binding of curcumin to β-lactoglobulin and its effect on antioxidant characteristics of curcumin. *Food Chemistry*, **141**(2): 1504–1511.

Lien, E., Jackson, J., Kuhlman, C., Pramuk, K., Lönnerdal, B. and Janszen, D. (2004) Variations in concentrations of lactoferrin in human milk: A nine country survey. In: *Advances in Experimental Medicine and Biology*, **554**: 423–426.

Lifran, E.V., Sleigh, R.W., Johnson, R.L., Steele, R.J., Hourigan, J.A. and Dalziel, S.M. (2011) Method for purification of lactose. US Patent 20110034685 A1. 10 February.

Lihme, A.O.F., Hansen, T.C.B. and Nielsen, C.S. (1998) Permeable hollow particles having an outer shell of mechanically rigid porous material. US Patent 5770416 A, 23 June.

Lihme, A.O.F., Hansen, M.B., Andersen, I.V. and Burnouf, T. (2010) A novel core fractionation process of human plasma by expanded bed adsorption chromatography. *Analytical Biochemistry*, **399**(1): 102–109.

Lin, S.-B., Nelles, L.P., Cordle, C.T. and Thomas, R.L. (1997) Debittering casein hydrolysates with octadecyl-siloxane (C18) columns. *Journal of Food Science*, **62**(4): 665–670.

Lišková, K., Kelly, A.L., O'Brien, N. and Brodkorb, A. (2010) Effect of denaturation of α-lactalbumin on the formation of BAMLET (bovine α-lactalbumin made lethal to tumor cells). *Journal of Agricultural and Food Chemistry*, **58**(7): 4421–4427.

Liu, G. and Zhong, Q. (2013) Thermal aggregation properties of whey protein glycated with various saccharides. *Food Hydrocolloids*, **32**(1): 87–96.

Liu, G. and Zhong, Q. (2015) High temperature-short time glycation to improve heat stability of whey protein and reduce color formation. *Food Hydrocolloids*, **44**: 453–460.

Lönnerdal, B. (2014) Infant formula and infant nutrition: bioactive proteins of human milk and implications for composition of infant formulas. *The American Journal of Clinical Nutrition*, **99**(Suppl.): 712S–717S.

Lönnerdal, B. (2003) Nutritional and physiologic significance of human milk proteins. *The American Journal of Clinical Nutrition*, **77**(6): 1537S–1543S.

Lönnerdal, B.O. and Atkinson, S. (1995) Nitrogenous components of milk. A: Human milk proteins. In: *Handbook of Milk Composition* (ed. R.G. Jensen). Academic Press: San Diego, pp. 351–368.

Lönnerdal, B. and Iyer, S. (1995) Lactoferrin: Molecular structure and biological function. *Annual Review of Nutrition*, **15**: 93–110.

Lopez, C., Briard-Bion, V., Menard, O., Rousseau, F., Pradel, P. and Besle, J.M. (2008) Phospholipid, sphingolipid, and fatty acid composition of the milk fat ft globule membrane are modified by diet. *Journal of Agricultrue Food Chemistry*, **56**(13): 5226–5236.

Lu, R.R., Xu, S.Y., Wang, Z. and Yang, R.J. (2007) Isolation of lactoferrin from bovine colostrum by ultrafiltration coupled with strong cation exchange chromatography on a production scale. *Journal of Membrane Science*, **297**(1–2): 152–161.

Månsson, H.L. (2008) Fatty acids in bovine milk fat. *Food and Nutrition Research*, 52.

Manzoni, P., Rinaldi, M., Cattani, S., Pugni, L., Romeo, M.G., *et al.* (2009) Bovine lactoferrin supplementation for prevention of late-onset sepsis in very low-birth-weight neonates: A randomized trial. *JAMA – Journal of the American Medical Association*, **302**(13): 1421–1428.

Mao, B., Li, D., Zhao, J., Liu, X., Gu, Z., *et al.* (2014) *In vitro* fermentation of lactulose by human gut bacteria. *Journal of agricultural and Food Chemistry*, **62**(45): 10970–10977.

Marchaj, A., Jacobsen, D.W., Savon, S.R. and Brown, K.L. (1995) Kinetics and thermodynamics of the interaction of cyanocobalamin (vitamin B_{12}) with haptocorrin: Measurement of the highest protein-ligand binding constant yet reported. *Journal of the American Chemical Society*, **117**(47): 11640–11646.

Marella, C., Muthukumarappan, K. and Metzger, L.E. (2011) Evaluation of commercially available, wide-pore ultrafiltration membranes for production of α-lactalbumin-enriched whey protein concentrate. *Journal of Dairy Science*, **94**(3): 1165–1175.

Markus, C.R., Olivier, B. and De Haan, E.H.F. (2002) Whey protein rich in α-lactalbumin increases the ratio of plasma tryptophan to the sum of the other large neutral amino acids and improves cognitive performance in stress-vulnerable subjects. *American Journal of Clinical Nutrition*, **75**(6): 1051–1056.

Markus, C.R., Jonkman, L.M., Lammers, J.H.C.M., Deutz, N.E.P., Messer, M.H. and Rigtering, N. (2005) Evening intake of α-lactalbumin increases plasma tryptophan availability and improves morning alertness and brain measures of attention. *American Journal of Clinical Nutrition*, **81**(5): 1026–1033.

Martin, M., Wellner, A., Ossowski, I. and Henle, T. (2008) Identification and quantification of inhibitors for angiotensin-converting enzyme in hypoallergenic infant milk formulas. *Journal of Agricultural and Food Chemistry*, **56**(15): 6333–6338.

Martin-Sosa, S., Martín, M.-J., García-Pardo, L.-A. and Hueso, P. (2003) Sialyloligosaccharides in human and bovine milk and in infant formulas: variations with the progression of lactation. *Journal of Dairy Science*, **86**(1): 52–59.

Matsumoto, K., Natsuko, T. and Tsunekazu, W. (1988) Method for producing galactooligosaccharide. European patent 0272095A2. 22 June.

Matthews, M.E. (1984) Whey protein recovery processes and products. *Journal of Dairy Science*, **67**(11): 2680–2692.

McGuffey, M.K., Epting, K.L., Kelly, R.M. and Foegeding, E.A. (2005) Denaturation and aggregation of three α-lactalbumin preparations at neutral pH. *Journal of Agricultural and Food Chemistry*, **53**(8): 3182–3190.

McJarrow, P. and van Amelsfort-Schoonbeek, J. (2004) Bovine sialyl oligosaccharides: seasonal variations in their concentrations in milk, and a comparison of the colostrums of Jersey and Friesian cows. *International Dairy Journal*, **14**(7): 571–579.

McJarrow, P., Schnell, N., Jumpsen, J. and Clandinin, T. (2009) Influence of dietary gangliosides on neonatal brain development. *Nutrition Reviews*, **67**(8): 451–463.

McMahon, D.J. and Oommen, B.S. (2008) Supramolecular structure of the casein micelle. *Journal of Dairy Science*, **91**(5): 1709–1721.

Menéndez-Aguirre, O., Stuetz, W., Grune, T., Kessler, A., Weiss, J. and Hinrichs, J. (2011) High pressure-assisted encapsulation of vitamin D_2 in reassembled casein micelles. *High Pressure Research*, **31**(1): 265–274.

Mennella, J.A., Trabulsi, J. and Papas, M. (2015a) Effects of cow milk versus extensive protein hydrolysate formulas on infant cognitive development. *The FASEB Journal*, **29**(1 Suppl.).

Mennella, J.A., Trabulsi, J.C. and Papas, M.A. (2015b) Effects of cow milk versus extensive protein hydrolysate formulas on infant cognitive development. *Amino Acids*, **47**(11): 1–9.

Merchant, Z.M., Wrezel, P.W., Spurlock, L.L. and Carpenter, D.E. (1995) Oil extraction of cholesterol from milk products. US Patent 5378487 A, 3 January.

Metin, S. and Hartel, R.W. (2012) Milk fat and cocoa butter. In: *Cocoa Butter and Related Compounds* (eds N. Garti and N. Widlak), Elsevier, US, pp. 365–392.

Michalski, M.-C. and Januel, C. (2006) Does homogenization affect the human health properties of cow's milk? *Trends in Food Science and Technology*, **17**(8): 423–437.

Mikkonen, H., Helakorpi, P., Myllykoski, L. and Keiski, R.L. (2001) Effect of nanofiltration on lactose crystallisation. *Milchwissenschaft*, **56**(6): 307–310.

Miller, L.E., Tennila, J. and Ouwehand, A.C. (2014) Efficacy and tolerance of lactitol supplementation for adult constipation: A systematic review and meta-analysis. *Clinical and Experimental Gastroenterology*, **7**: 241–248.

Miocinovic, J., Le Trung, T., Fredrick, E., Van Der Meeren, P., Pudja, P. and Dewettinck, K. (2014) A comparison of composition and emulsifying properties of MFGM materials prepared from different dairy sources by microfiltration. *Food Science and Technology International*, **20**(6): 441–451.

Mohan, M.S., Jurat-Fuentes, J.L. and Harte, F. (2013) Binding of vitamin A by casein micelles in commercial skim milk. *Journal of Dairy Science*, **96**(2): 790–798.

Montilla, A., Del Castillo, M.D., Sanz, M.L. and Olano, A. (2005) Egg shell as catalyst of lactose isomerisation to lactulose. *Food Chemistry*, **90**(4): 883–890.

Moser, S., Chegeni, M., Jones, O. G., Liceaga, A. and Ferruzzi, M.G. (2014) The effect of milk proteins on the bioaccessibility of green tea flavan-3-ols. *Food Research International*, **66**: 297–305.

Murphy, E.G., Roos, Y.H., Hogan, S.A., Maher, P.G., Flynn, C.G. and Fenelon, M.A. (2015) Physical stability of infant milk formula made with selectively hydrolysed whey proteins. *International Dairy Journal*, **40**: 39–46.

Musilova, S., Rada, V., Marounek, M., Nevoral, J., Dušková, D., Bunesova, V., *et al.* (2015) Prebiotic effects of a novel combination of galactooligosaccharides and maltodextrins. *Journal of Medicinal Food*, **18**(6): 685–689.

Nakamura, T., Kawase, H., Kimura, K., Watanabe, Y., Ohtani, M., *et al.* (2003) Concentrations of sialyloligosaccharides in bovine colostrum and milk during the prepartum and early lactation. *Journal of Dairy Science*, **86**(4): 1315–1320.

Nakayama, T. (1981) Changing values of lipid, total-cholesterol and phospholipids in human milk in North Japan. *Pediatrics International*, **23**(3): 403–403.

Newburg, D.S. and Grave, G. (2014) Recent advances in human milk glycobiology. *Pediatric Research*, **75**(5): 675–679.

Nieuwenhuizen, A.G., Hochstenbach-Waelen, A. and Westerterp-Plantenga, M. (2010) α-lactalbumin in the regulation of appetite and food intake In: *Weight Control and Slimming Ingredients in Food Technology* (ed. S.S. Cho). Wiley-Blackwell, Oxford, pp. 109–120.

Ninonuevo, M.R., Park, Y., Yin, H., Zhang, J., Ward, R.E., *et al.* (2006) A strategy for annotating the human milk glycome. *Journal of Agricultural and Food Chemistry*, **54**(20): 7471–7480.

O'Kennedy, B.T. (2011) Caseins. In: *Handbook of Food Proteins* (eds G.O. Phillips and P.A. Williams), Woodhead Publishing, Cambridge, pp. 13–29.

Oakenfull, D.G., Sidhu, G.S. and Rooney, M.L. (1991) Cholesterol removal. World Patent 1991016824 A1, 14 November.

Ochoa, T.J., Zegarra, J., Cam, L., Llanos, R., Pezo, A., *et al.* (2015) Randomized controlled trial of lactoferrin for prevention of sepsis in Peruvian neonates less than 2500 g. *Pediatric Infectious Disease Journal*, **34**(6): 571–576.

Oggioni, C., Cena, H., Wells, J.C.K., Lara, J., Celis-Morales, C. and Siervo, M. (2015) Association between worldwide dietary and lifestyle patterns with total cholesterol concentrations and DALYs for infectious and cardiovascular diseases: An ecological analysis. *Journal of Epidemiology and Global Health*, **5**(4): 315–325.

Okonogi, S., Tomita, M., Tomimura, T., Tamura, Y. and Mizota, T. (1988) Process for producing bovine lactoferrin in high purity. US Patent 4791193 A, 13 December.

Oroskar, A.R. (2014) Tagatose production from deproteinized whey and purification by continuous chromatography. US Patent 20140366869 A1, 18 December.

Ortiz-Chao, P., Gómez-Ruiz, J.A., Rastall, R., Mills, D., Cramer, R., *et al.* (2009) Production of novel ACE inhibitory peptides from β-lactoglobulin using Protease N Amano. *International Dairy Journal*, **19**(2): 69–76.

Otieno, D.O. (2010) Synthesis of β-Galactooligosaccharides from lactose using microbial β-galactosidases. *Comprehensive Reviews in Food Science and Food Safety*, **9**(5): 471–482.

Padilla, B., Frau, F., Ruiz-Matute, A.I., Montilla, A., Belloch, C., *et al.* (2015) Production of lactulose oligosaccharides by isomerisation of transgalactosylated cheese whey permeate obtained by β-galactosidases from dairy Kluyveromyces. *Journal of Dairy Research*, **82**(3): 356–364.

Pammi, M. and Abrams, S.A. (2015) Oral lactoferrin for the prevention of sepsis and necrotizing enterocolitis in preterm infants. *The Cochrane database of Systematic Reviews*, **2**, CD007137.

Pan, K., Zhong, Q. and Baek, S.J. (2013) Enhanced dispersibility and bioactivity of curcumin by encapsulation in casein nanocapsules. *Journal of Agricultural and Food Chemistry*, **61**(25): 6036–6043.

Panesar, P.S. and Kumari, S. (2011) Lactulose: Production, purification and potential applications. *Biotechnology Advances*, **29**(6): 940–948.

Patel, S.R. and Murthy, Z. (2009) Ultrasound assisted crystallization for the recovery of lactose in an anti-solvent acetone. *Crystal Research and Technology*, **44**(8),: 889–896.

Patil, D.H., Grimble, G.K. and Silk, D.B. (1987) Lactitol, a new hydrogenated lactose derivative: intestinal absorption and laxative threshold in normal human subjects. *The British Journal of Nutrition*, **57**(2): 195–199.

Pederson, H.T. (1980) Treatment of whey. US Patent 4202909 A, 13 May.

Perraudin, J.P. (2011) Method for production of lactoferrin. US Patent 20110301077 A1, 8 December.

Phan, T.T.Q., Le, T.T., Van der Meeren, P. and Dewettinck, K. (2014a) Comparison of emulsifying properties of milk fat globule membrane materials isolated from different dairy by-products. *Journal of Dairy Science*, **97**(8): 4799–4810.

Phan, T.T.Q., Moens, K., Le, T.T., Van der Meeren, P. and Dewettinck, K. (2014b) Potential of milk fat globule membrane enriched materials to improve the whipping properties of recombined cream. *International Dairy Journal*, **39**(1): 16–23.

Pitzalis, G., Deganello, F., Mariani, P., Chiarini-Testa, M., Virgilii, F., *et al.* (1995) Lactitol in chronic idiopathic constipation in children. *La Pediatria medica e chirurgica: Medical and surgical pediatrics*, **17**(3): 223–226.

Pragono, R., Vu, L.T.T., Hourigan, J.A., Durham, R.J. and Sleigh, R.W. (2008) Sorption kinetics of riboflavin from diluted lactose solution in batch systems [online]. In: *Chemeca 2008: Towards a Sustainable Australasia.* [Barton, ACT]: Engineers Australia, pp. 523–532.

Precht, D. (2001) Cholesterol content in European bovine milk fats. *Nahrung – Food*, **45**(1): 2–8.

Precht, D. and Molkentin, J. (1999) C18: 1, C18: 2 and C18: 3 trans and cis fatty acid isomers including conjugated cis Δ9, trans Δ11 linoleic acid (CLA) as well as total fat composition of German human milk lipids. *Nahrung – Food*, **43**(4): 233–244.

Priyanka, R. (2012) Environmental sound management of lactose mother liquor through bacterial fermentation. *International Journal of Research in Engineering and Applied Sciences*, **2**(2): 1616–1622.

Quintão, E., Grundy, S.M. and Ahrens, E.H. (1971) Effects of dietary cholesterol on the regulation of total body cholesterol in man. *Journal of Lipid Research*, **12**(2): 233–247.

Raksakulthai, R. and Haard, N.F. (2003) Exopeptidases and their application to reduce bitterness in food: A review. *Critical Reviews in Food Science and Nutrition*, **43**(4): 401–445.

Ramakrishnan, B., Boeggeman, E. and Qasba, P.K. (2002) β-1, 4-galactosyltransferase and lactose synthase: Molecular mechanical devices. *Biochemical and Biophysical Research Communications*, **291**(5): 1113–1118.

Recio, I., Moreno, F.J. and López-Fandiño, R. (2009) Glycosylated dairy components: Their roles in nature and ways to make use of their biofunctionality in dairy products. In: *Dairy-Derived Ingredients* (ed. M. Corredig). Woodhead Publishing, Cambridge, pp. 170–211.

Relkin, P. and Shukat, R. (2012) Food protein aggregates as vitamin-matrix carriers: Impact of processing conditions. *Food Chemistry*, **134**(4): 2141–2148.

Rentschler, E., Schuh, K., Krewinkel, M., Baur, C., Claaßen, W., *et al.* (2015) Enzymatic production of lactulose and epilactose in milk. *Journal of Dairy Science*, **98**(10): 6767–6775.

Rival, S.G., Boeriu, C.G. and Wichers, H.J. (2001) Caseins and casein hydrolysates. 2. Antioxidative properties and relevance to lipoxygenase inhibition. *Journal of Agricultural and Food Chemistry*, **49**(1): 295–302.

Roesch, R.R., Rincon, A. and Corredig, M. (2004) Emulsifying properties of fractions prepared from commercial buttermilk by microfiltration. *Journal of Dairy Science*, **87**(12): 4080–4087.

Rombaut, R. and Dewettinck, K. (2006) Properties, analysis and purification of milk polar lipids. *International Dairy Journal*, **16**(11): 1362–1373.

Roncucci, L., Di Donato, P., Carati, L., Ferrari, A., Perini, M., *et al.* (1993) Antioxidant vitamins or lactulose for the prevention of the recurrence of colorectal adenomas. *Diseases of the Colon and Rectum.* **36**(3): 227–234.

Rueda, R., Barranco, A., Ramirez, M., Vazquez, E., Valverde, E., *et al.* (2008) Enriched infant formulas. US Patent 20080057178 A1, 6 March.

Sabouri, S., Geng, J. and Corredig, M. (2015) Tea polyphenols association to caseinate-stabilized oil–water interfaces. *Food Hydrocolloids*, **51**: 95–100.

Saha, B.C. and Hayashi, K. (2001) Debittering of protein hydrolyzates. *Biotechnology Advances*, **19**(5): 355–370.

Sáiz-Abajo, M.J., González-Ferrero, C., Moreno-Ruiz, A., Romo-Hualde, A. and González-Navarro, C.J. (2013) Thermal protection of β-carotene in re-assembled casein micelles during different processing technologies applied in food industry. *Food Chemistry*, **138**(2–3): 1581–1587.

Sampson, H.A., Bernhisel-Broadbent, J., Yang, E. and Scanlon, S.M. (1991) Safety of casein hydrolysate formula in children with cow milk allergy. *The Journal of Pediatrics*, **118**(4, Part 1): 520–525.

Sanders, J.F. (1978) Lactulose syrup assessed in a double-blind study of elderly constipated patients. *Journal of the American Geriatrics Society*, **26**(5), 236–239.

Sandström, O., Lönnerdal, B., Graverholt, G. and Hernell, O. (2008) Effects of α-lactalbumin–enriched formula containing different concentrations of glycomacropeptide on infant nutrition. *The American Journal of Clinical Nutrition*, **87**(4): 921–928.

Savaiano, D.A. and Levitt, M.D. (1987) Milk intolerance and microbe-containing dairy foods. *Journal of Dairy Science*, **70**(2): 397–406.

Savignac, H.M., Couch, Y., Stratford, M., Bannerman, D.M., Tzortzis, G., *et al.* (2015) Prebiotic administration normalizes lipopolysaccharide (LPS)-induced anxiety and cortical 5-HT2A receptor and IL1-*β* levels in male mice. *Brain Behavior and Immunity*, **52**: 120–131.

Sawatzki, G. (2005) Process for continuous production of galacto-oligosaccharides. European Patent 1352967 B1, 31 August.

Schaafsma, G. (2008) Lactose and lactose derivatives as bioactive ingredients in human nutrition. *International Dairy Journal*, **18**(5): 458–465.

Schubert, M., Contreras, C., Franz, N. and Hellhammer, J. (2011) Milk-based phospholipids increase morning cortisol availability and improve memory in chronically stressed men. *Nutrition Research*, **31**(6): 413–420.

Schumann, C. (2002) Medical, nutritional and technological properties of lactulose. *An update. European Journal of Nutrition*, **41**(Suppl. 1): 17–25.

Scott, S.N. and Krishnapillai, A. (2015) Isolation and purification of components of whey. US Patent 8932659 B2, 13 January.

Shah, N.P. (2000) Effects of milk-derived bioactives: An overview. *British Journal of Nutrition*, **84**(S1): 3–10.

Sharma, B.C., Sharma, P., Agrawal, A. and Sarin, S.K. (2009) Secondary prophylaxis of hepatic encephalopathy: An open-label randomized controlled trial of lactulose versus placebo. *Gastroenterology*, **137**(3): 885–891.

Shephard, S. (2006) Milk products. In: *Pickled, potted, and canned: How the art of science of food preserving changed the world* (ed. S. Shephard). Simon and Schuster Paperback, New York, pp. 142–162.

Shi, Y., Liang, B. and Hartel, R. (2006) Crystal refining technologies by controlled crystallization. US Patent 20060128953 A1, 15 June.

Shukla, A., Bhaskara, A.R., Rizvi, S.S.H. and Mulvaney, S.J. (1994) Physicochemical and rheological properties of butter made from supercritically fractionated milk fat. *Journal of Dairy Science*, **77**(1): 45–54.

Singh, H. (2006) The milk fat globule membrane: A biophysical system for food applications. *Current Opinion in Colloid and Interface Science*, **11**(2–3): 154–163.

Soni, K.N. (2012) Low cholesterol butter and process of preparation. World Patent 2012025931 A1, 1 March.

Spade, M.E., Weil, J.K. and McHale, M.S. (2003) Process for removal of phosphorous from a dairy stream. US Patent 6623954 B1, 23 September.

Spitsberg, V.L. (2005) Invited review: Bovine milk fat globule membrane as a potential nutraceutical. *Journal of Dairy Science*, **88**(7): 2289–2294.

Sun, W.-W., Yu, S.-J., Yang, X.-Q., Wang, J.-M., Zhang, J.-B., *et al.* (2011) Study on the rheological properties of heat-induced whey protein isolate–dextran conjugate gel. *Food Research International*, **44**(10): 3259–3263.

Sundfeld, E., Krochta, J.M. and Richardson, T. (1994) Aqueous process to remove cholesterol from food products. US Patent 5370890 A, 6 December.

Tamm, F., Gies, K., Diekmann, S., Serfert, Y., Strunskus, T., Brodkorb, A., *et al.*, (2015) Whey protein hydrolysates reduce autoxidation in microencapsulated long chain polyunsaturated fatty acids. *European Journal of Lipid Science and Technology*, **117**: 1960–1970.

Tang, L., Li, Z. A., Dong, X.X., Yang, R.J., Zhang, J.H. and Mao, Z.G. (2011) Lactulose biosynthesis by β-galactosidase from a newly isolated Arthrobacter sp. *Journal of Industrial Microbiology and Biotechnology*, **38**(3): 471–476.

Terheggen-Lagro, S.W.J., Khouw, I.M.S.L., Schaafsma, A. and Wauters, E.A.K. (2002) Safety of a new extensively hydrolysed formula in children with cow's milk protein allergy: A double blind crossover study. *BMC Pediatrics*, **2**: 10.

Thomas, R. and Brooks, T. (2004) Common oligosaccharide moieties inhibit the adherence of typical and atypical respiratory pathogens. *Journal of Medical Microbiology*, **53**(9): 833–840.

Timby, N., Domellöf, E., Hernell, O., Lönnerdal, B. and Domellöf, M. (2014) Neurodevelopment, nutrition, and growth until 12 mo of age in infants fed a low-energy, low-protein formula supplemented with bovine milk fat globule membranes: a randomized controlled trial. *The American Journal of Clinical Nutrition*, **99**(4): 860–868.

Tomita, M., Shimamura, S., Tamura, Y., Mizota, T., Nakano, S. and Suzawa, I. (1994) Crystalline lactulose trihydrate and a method for its manufacture. US Patent 5304251 A, 19 April.

Uchida, Y., Shimatani, M., Mitsuhashi, T. and Koutake, M. (1996) Process for preparing a fraction having a high content of α-lactalbumin from whey and nutritional compositions containing such fractions. US Patent 5503864 A. 2 April.

Ueno, H.M., Shiota, M., Ueda, N., Isogai, T. and Kobayashi, T. (2012) Iron-lactoferrin complex reduces iron-catalyzed off-flavor formation in powdered milk with added fish oil. *Journal of Food Science*, **77**(8): C853–C858.

United States Code of Federal Regulations (USCFR) (1993) Standard of Identity. *Milk and Cream Sections* **131**.150–131.160.

Van Velthuijsen, J.A. (1979) Food additives derived from lactose: Lactitol and lactitol palmitate. *Journal of Agricultural and Food Chemistry*, **27**(4): 680–686.

Vandenplas, Y., Greef, E.D. and Veereman, G. (2014) Prebiotics in infant formula. *Gut Microbes*, **5**(6): 681–687.

Vandenplas, Y., Zakharova, I. and Dmitrieva, Y. (2015) Oligosaccharides in infant formula: More evidence to validate the role of prebiotics. *British Journal of Nutrition*, **113**(9): 1339–1344.

Vanderghem, C., Bodson, P., Danthine, S., Paquot, M., Deroanne, C. and Blecker, C. (2010) Milk fat globule membrane and buttermilks: From composition to valorization. *Biotechnology, Agronomy and Society and Environment*, **14**(3): 485–500.

Vázquez, E., Barranco, A., Ramírez, M., Gruart, A., Delgado-García, J.M., *et al.* (2015) Effects of a human milk oligosaccharide, 2′-fucosyllactose, on hippocampal long-term potentiation and learning capabilities in rodents. *The Journal of Nutritional Biochemistry*, **26**(5): 455–465.

Venema, K., Vermunt, S.H.F. and Brink, E.J. (2009) D-Tagatose increases butyrate production by the colonic microbiota in healthy men and women. *Microbial Ecology in Health and Disease*, **17**(1): 47–57.

von Berg, A., Koletzko, S., Filipiak-Pittroff, B., Laubereau, B., Grübl, A., *et al.* (2007) Certain hydrolyzed formulas reduce the incidence of atopic dermatitis but not that of asthma: three-year results of the German Infant Nutritional Intervention Study. *Journal of Allergy and Clinical Immunology*, **119**(3): 718–725.

Voswinkel, L. and Kulozik, U. (2014) Fractionation of all major and minor whey proteins with radial flow membrane adsorption chromatography at lab and pilot scale. *International Dairy Journal*, **39**(1): 209–214.

Vu, I.T.T., Hourigan, J.A., Sleigh, R.W., Ang, M.H. and Tade, M.O. (2003) Metastable control of cooling crystallisation. In: *Computer Aided Chemical Engineering*, **14**: 527–532.

Vu, L.T.T., Durham, R.J., Hourigan, J.A. and Sleigh, R.W. (2004) Dynamic modelling and optimisation of α-lactose monohydrate seeded batch cooling crystallisation. In: *Computer Aided Chemical Engineering*, **18**: 847–852.

Wang, H., Yang, R., Hua, X., Zhao, W. and Zhang, W. (2013) Enzymatic production of lactulose and 1-lactulose: Current state and perspectives. *Applied Microbiology and Biotechnology*, **97**(14): 6167–6180.

Willis, R. (2000) Expanded-bed adsorption. *Todays Chemist at Work*, **9**(11): 23–28.

Wilson, L.L. (2000) Process for the manufacture of a lactitol syrup. US Patent 6090429 A, 18 July.

Wittke, A., Banavara, D. and Munoz, C. (2012) Use of nutritional compositions including lactoferrin and one or more prebiotics in inhibiting adhesion of pathogens in the gastrointestinal tract. US Patent 20120171176 A1, 5 July.

Wolfrom, M.L., Burke, W.J., Brown, K.R. and Rose, R.S. (1938) Crystalline lactositol. *Journal of the American Chemical Society*, **60**(3): 571–573.

Wong, S.Y. and Hartel, R.W. (2014) Crystallization in lactose refining: A review. *Journal of Food Science*, **79**(3): R257–R272.

Xu, Y. (2013) Process for manufacturing tagatose and glucose. US Patent 20130081613 A1, 4 April.

Zacharis, C. (2012) Lactitol. In: *Sweeteners and Sugar Alternatives in Food Technology*, vol. **2** (eds K. O'Donnell and M. Kearsley). Wiley-Blackwell, UK, pp. 275–293.

Zanabria, R., Tellez, A.M., Griffiths, M. and Corredig, M. (2013) Milk fat globule membrane isolate induces apoptosis in HT-29 human colon cancer cells. *Food and Function*, **4**(2): 222–230.

Zavaleta, N., Kvistgaard, A.S., Graverholt, G., Respicio, G., Guija, H., *et al.* (2011) Efficacy of an MFGM-enriched complementary food in diarrhea, anemia, and micronutrient status in infants. *Journal of Pediatric Gastroenterology and Nutrition*, **53**(5): 561–568.

Zhang, Q., Cundiff, J.K., Maria, S.D., McMahon, R.J., Woo, J.G., *et al.* (2013) Quantitative analysis of the human milk whey proteome reveals developing milk and mammary-gland functions across the first year of lactation. *Proteomes*, **1**(2): 128–158.

Zhu, D., Damodaran, S. and Lucey, J.A. (2010) Physicochemical and emulsifying properties of whey protein isolate (WPI) – Dextran conjugates produced in aqueous solution. *Journal of Agricultural and Food Chemistry*, **58**(5): 2988–2994.

Zivkovic, A.M. and Barile, D. (2011) Bovine milk as a source of functional oligosaccharides for improving human health. *Advances in Nutrition*, **2**(3): 284–289.

16

Bioprocessing of Chicken Meat and Egg Processing Industries' Waste to Value-Added Proteins and Peptides

Surangna Jain, Damodar Dhakal, and Anil Kumar Anal
Asian Institute of Technology, Klong Luang, Pathumthani, Thailand

16.1 Introduction

Waste is obvious where food is processed. With industries such as dairy, alcoholic and non-alcoholic beverage plants, meat and fish processing, fruits and vegetables, sugar and jiggery etc., a variety of waste are generated right from the farm to the fork. It has now become mandatory for each food processor and its stakeholders to take prominent steps towards waste management, since rigorous regulation are in force regarding the environment conservation. At present, global pork consumption is 114 million tonnes per annum, whereas for poultry it is 106 million tonnes; 2.5% annual growth rate of chicken meat will overtake pork, having 1.5% growth rates before 2020 (UNFAO). The European Union reported that 37 million MT/year of biowaste is produced by the food industries. Rational utilization of waste from these industries is a matter of necessity and challenges. The biological instability, pathogenic and oxidizing nature, high moisture content and being highly prone to enzymatic activity are the reasons behind the difficulty in the valorization of the waste (Jayathilakan *et al.*, 2012a).

Slaughterhouse waste from chicken processing industries includes skin, tendon, cartilage, bones and other connective tissues that are the source materials for collagen extraction. Burning, incineration, composting and using for animal feed supplements are the different traditional applications of those waste. From food to medical, collagen has wider applications (Rodziewicz-Motowidło *et al.*, 2008). Higher quality of food product can be achieved by the use of collagen as a supplement. There is an abundant possibility of collagen extraction from chicken feet by using different

Food Processing By-Products and their Utilization, First Edition. Edited by Anil Kumar Anal.

acids (Liu *et al.*, 2001). A lot of work has been done on the extraction of collagen from chicken waste. The acid and pepsin solubilized method yields 43.9% and 88.5% collagen on the dry weight basis method respectively. Poultry feet have an abundant amount of collagen whose extraction can be used for collagen film formation and also for the powdered collagen for the control of wound exudation (Pachence, 1992). Among the use of two enzymes, papain and pepsin, chicken feet-extracted collagen was found to be higher quality, so can be recommended as an alternative to commercial collagen production, with a yield of 18.16% by the dry weight method (Hashim *et al.*, 2012).

The frequency of chicken consumption worldwide has been ever increasing, whereas the waste products remain in the shade (Table 16.1). Due to this, the meat processing industry has to think of waste valorization, not only for the cost minimization but also for the rational distribution of resources and for the protection of the environment. Over the past few decades, researchers have been trying to convert the waste into feed and food applications, but now new doors of advanced research has opened to widen the scope of waste into more valued products for medical and industrial applications. Furthermore, it has been found that the waste from chicken industries requires multiple processing steps for its conversion and extraction, which implies a huge capital for research and development. So, there is always a brighter side in finding out the economic technology for the waste recovery system; on the other hand, environmental impacts should not be undermined when taking initiative for transformation of waste into new valued products like proteins and other bioactive peptides.

Chicken eggs are widely consumed worldwide, as they are cheap with a very high nutritional value (Table 16.2). The consumption pattern of chicken is increasing as per the increase in the population growth. Global egg production increased by almost double from 35 million tonnes in 1990 to 61.2 million tonnes in 2011 (Windhorst, 2011).

Table 16.1 World poultry meat and chicken meat production by region (million tonnes) (adopted from website: www.thepoultrysite.com)

Region	2000	2005	2006	2007	2008	2009	2010	2011	2012
Africa	3	3.6	3.6	4	4.2	4.4	4.6	4.8	4.9
America	30.1	35.9	37	38.9	41.1	40.1	41.8	42.8	43
Asia	22.9	27.3	28.3	30.1	31.8	32.9	34.5	36.1	37.4
Europe	11.9	13.2	13.1	14	14.4	15.7	16.1	16.6	16.9
Oceania	0.8	1	1	1	1	1	1.1	1.3	1.3
World	68.6	80.9	83	87.9	94.2	92.5	98.1	101.6	103.5

Table 16.2 World egg production (million tonnes) (adopted from website: www.thepoultrysite.com)

Region	2000	2005	2006	2007	2008	2009	2010	2012
Africa	1.9	2.2	2.4	2.5	2.6	2.6	2.8	2.9
America	10.4	11.7	12.3	12.3	12.5	12.6	12.8	13.1
Asia	29	32.6	33.0	34.5	36.2	37.2	37.4	38.1
Europe	9.5	9.9	10.1	10	10.2	10.3	10.5	10.6
Oceania	0.2	0.2	0.2	0.2	0.2	0.2	0.3	0.3
World	51	56.6	57.9	59.6	61.8	62.8	63.8	65

Egg shell and its membrane in the form of waste is generated while processing, consuming and during of the hatching of chickens.

16.2 By-Products and Wastes Generated During Chicken Meat and Egg Processing

Poultry processing generates a number of organic by-products such viscera, blood, feathers, head etc. (Hashim *et al.*, 2012). The majority of these by-products are produced during the slaughtering process. The slaughterhouse waste comprises of the portion that cannot be utilized or sold as meat. This includes bones, skin, blood and internal organs (Hashim *et al.*, 2012). On the other hand, egg breaking generates almost 1.5 million tonnes of shell waste annually. Currently, due to their unalluring appearance and poor quality, these by-products are underutilized and generally discarded and disposed off as wastes. However, they must be dealt with more efficiently, as the growth of these industries mainly depends on the management of their by-products.

Waste utilization is a great challenge for the growth of chicken and egg processing industries (Jayathilakan *et al.*, 2012). A waste management center in the USA is involved in converting the value products from poultry and animal waste. Excreta, mortalities, feather, skin, bones, blood, feet etc. are the potential source for different variable biological components, which come from the slaughterhouse. Non-utilization of these by-products and the extra cost for proper disposal can also sometimes cause serious environmental problems (Jayathilakan *et al.*, 2012). Hence, it is essential to find applications for these wastes, as they are becoming a serious environmental issue. By-products from poultry and egg processing also have high nutritive value (Figure 16.1). They are good sources of bioactive compounds like proteins and several bioactive peptides, which have various beneficial effects on the human health.

Previously, the poultry wastes were converted to feed and organic fertilizers using the processes of rendering, composting, chemical and microbial treatments. Also waste management techniques like land filling and dumping were carried out. But due to legislations in certain countries, indiscriminate dumping and land-filling of animal wastes is not permitted (Hashim *et al.*, 2012). Hence, better options are needed for proper waste utilization. Waste valorization is an interesting new concept that offers a range of alternatives for management of waste other than disposal or land filling. Valorization allows us to explore the possibility of reusing nutrients in the production of main products, and thus highlights the potential gains that can be achieved. It allows for processing of poultry wastes into useful products (Luque and Clark, 2013). Valorization of by-products from the poultry industry is not fully exploited as compared to the by-products from other industries like dairy, fruits and vegetables etc. Valorizing the poultry waste components can lead to the recovery of important bioactive compounds that can be used for the creation of new value-added products and functional ingredients that can be utilized in food, nutrition and many non-food applications.

Various techniques have been developed for the extraction, isolation and purification of these bioactive compounds that are trapped in the by-products. These include enzymatic extraction and hydrolysis, microbial fermentation, chromatographic techniques etc. This will allow for efficient and less expensive utilization of the by-products that are currently under-utilized and will ultimately lead to high profits in the poultry and egg processing industries. Hence, processing of the by-products can convert a

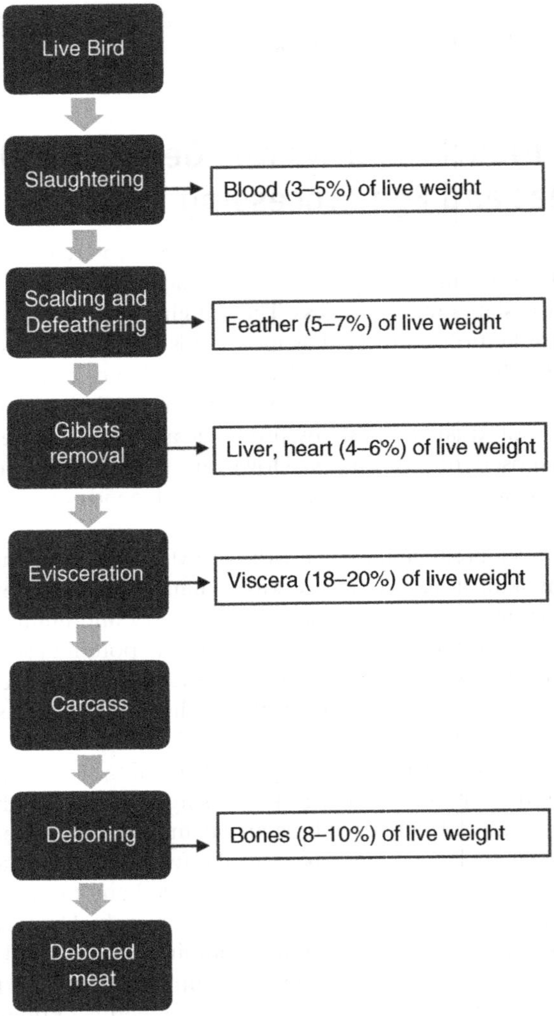

Figure 16.1 By-products generated during poultry processing (adapted from Rathinaraj and Sachindra, 2013)

low-value product into a value-added product that covers all the disposal costs as well as reducing any environmental damage.

16.2.1 Feather

Feather comprises 5–7% of total body weight of the fowl, which comprises 91% protein (keratin), 1% lipids and 8% water (Thyagarajan and Sakthivadivu, 2013). Feather acts as an insulating, locomotory and structural functions in the birds, which constitutes a rich amount of keratinous materials (Onifade *et al.*, 1998). It contains useful amino acids that can be used as a cheap source of proteins for feed stuffs. The limitation behind the feather utilization is highly variable amino acids composition (Wang and Parsons, 1997) and also deficiencies of methionine, lysine, histidine and tryptophan. Keratinolytic protease can be produced by fermentation using various soil bacteria (Paul *et al.*, 2014) and used for many industrial applications. The strong covalent bonds present in the protein of feather are very difficult to break (Saravanan and Dhurai, 2012).

16.2.2 Skin

Collagen and gelatin are the two prominent components that can be extracted from chicken skin. Since there is only mammalian collagen and gelatin available commercially, chicken can be used as a best medium in this regard. So chicken skin as waste from poultry processing industries can be converted into value-added products with similar chemical composition to bovine extract and also with better physicochemical properties than fish extract gelatins (Mhd Sarbon *et al.*, 2013).

16.2.3 Bones

Collagen is abundant in chicken bone, which can be extracted as per the method described by a study carried out by Zhang *et al.* (2011). If we can come up with this technology, then not only waste disposal problems are solved but also many functional components can be obtained from this, which have both social and economic values (Zhang *et al.*, 2011).

16.2.4 Trachea

Tons of viscera are generated as waste from the poultry industry every day. The main component of trachea is cartilage – a kind of connective tissue which comprises of the type || collagen and glycosaminoglycans. A significant amount of a chrondroitin sulphate (CS) and collagen are found in the trachea. CS and collagen can be used as neutraceuticals for joint problems. There is an increasing demand for these products, but because of limited sources such as from shark cartilage, this makes it expensive and from bovine sources there is always a religious obstacle. So the chicken trachea can be used as an important substitute for chrondroitin sulphate and collagen (Liu *et al.*, 2001).

16.2.5 Blood

Blood is the most important slaughterhouse waste, which is a rich source of quality dietary protein. Chicken contains 40 ml blood/kg live weight that has 60–70% plasma and 20–40% red blood corpuscles. The same blood can be used to extract the bioactive peptides and also as meal cake for feeding purpose. The proteins that can be recovered from the poultry blood include thrombin, fibrinogen, immunoglobulins, Bovine serum Albumin (BSA), etc. (Warhadpande *et al.*, 2010).

16.2.6 Feet

Chicken feet are often treated as waste, whereas in some places they are considered a delicacy. When considered as waste, there is also a huge scope for collagen and gelatin extraction. Acid extraction of collagen can be carried out from the chicken feet (Lin and Liu, 2006). Poultry feet contain abundant collagen whose extraction is used for the formation of film and collagen-based powder for wound healing (Pachence, 1992).

16.2.7 Eggshell and Eggshell Membrane

The shell membrane and the shell containing calcium are the main by-products from the egg waste. These untreated forms are not edible and do not have any saleable value. Furthermore, they contain a lot of bioactive peptides and other active compounds, which has a lot of application in food and medicine. Egg shell waste consists of 94% calcium carbonate, 1% magnesium carbonate, 45 organic matter and 1% calcium phosphate (Wei *et al.*, 2009).

16.3 Proteins and Peptides derived from Chicken Processing By-Products and Waste

By-products from the chicken meat industries have high nutritional value, comprising of high-quality proteins and peptides. The utilization of these by-products for the extraction of proteins and peptides is more beneficial than discarding them as waste. Non-utilization or the inappropriate disposal of these by-products in landfills not only causes loss of potential revenues and increased cost of disposal, but also serious environmental pollution and related hazards (Jayathilakan *et al.*, 2012). Hence it is essential to find applications for these wastes that are now becoming a serious environmental issue. Waste valorization is an interesting new concept that offers a range of alternatives for management of waste other than disposal and land-filling. They provide a viable option for dealing with these wastes by increasing the economic value of the by-products (Luque and Clark, 2013). Since these by-products contain organic matter such as proteins, peptides and amino acids, they can be utilized for the creation of value-added products that find a wide range of applications in the cosmetic, food and medical sector (Nakano *et al.*, 2003).

16.3.1 Collagen

Collagen is one of the major structural proteins of the connective tissue that makes up 25% to 30% of the total protein in animals (Kiew and Don, 2013). Collagen has a native structure comprising of coiled polypeptides wound around each other to form a triple helix (Wang *et al.*, 2008). The chains comprise of repeating structures of (Gly-X-Y) (Parenteau-Bareil *et al.*, 2010) and the chains and subunits are held together by the hydrogen bonds that also link together the glycine residues of the neighboring polypeptides. Any amino acids can be present at the X and Y position, but they are generally occupied by proline and hydroxyproline. The glycine residue is the main amino acid for the formation of the triple helix. The structure of collagen also comprises of telopeptides, which is made up of 15–26 amino acids at the C and N terminal regions. These telopeptides cannot form the triple helical structure due to their aldehyde derivatives and intra- and inter-molecular covalent links, but they are important for the cross-linking and surface assembly of collagen (Pati *et al.*, 2010).

Currently, there are 29 different types of collagen of different polymeric forms (Parenteau-Bareil *et al.*, 2010). However, they all exhibit the triple helical structure, which is a characteristic of collagen. The most prominent one is type 1 collagen, which has three polypeptide chains (Wang *et al.*, 2008).

16.3.1.1 Extraction and Characterization of Collagen Previously, the common sources for collagen extraction were pigs and cattle. But after the outbreak of Bovine Spongiform Encephalopathy (BSE), there became a need for an alternative source of collagen due to the possibility of human transmission and due to aesthetic and religious issues (Kiruthiga *et al.*, 2013). More efforts are now being made to find alternative sources of collagen that includes chicken by-products. The common chicken by-products that can be used for extraction of collagen are mainly skin, bones and tendons. As collagen is an abundant protein, it can be easily isolated from different parts of the animal. The oldest method for collagen extraction from animal tissues involves boiling them in water. However, this method denatures some of the collagen. To isolate collagen in its native and un-denatured form, collagen can be extracted from the chicken wastes by using two different methods. One is acid treatment followed by solubilization and the other is solubilization by the enzyme treatment (Kiruthiga *et al.*, 2013).

Collagen extraction is generally carried out using organic acids like acetic acid and lactic acid. Inorganic acids like hydrochloric acid can also be used for collagen extraction, but their efficiency is less compared to the organic acids (Wang *et al.*, 2009). Acetic acid is considered the most promising solvent for collagen extraction. Solubilization of collagen in the acid solution plays an important role in the extraction efficiency and it depends on the concentration of acid used. Enzymatic treatment using pepsin is also a common method for collagen extraction (Wang *et al.*, 2009). The enzyme pepsin removes the non-helical ends of collagen and increases the collagen solubility after the treatment. However, pepsin treatment can degrade collagen as the enzyme cleaves the terminal non-helical regions that contain the intermolecular cross-linkages. Papain, a proteolytic enzyme from papaya can also be used to obtain better-quality collagen with low immunogenicity (Gunasekaran, 2003).

16.3.1.2 Functional and Bioactive Properties of Collagen Collagen is a major component of the extracellular matrix (ESM) that provides support to skin, bones, tendons and connective tissues. They play an important role in tissue and organ formation and the functional expression of cells (Lee *et al.*, 2001). Functional properties of collagen include properties relating to their gel behavior, such as thickening, gel formation and water-binding capacity, and surface properties such as emulsion formation, stabilization and film-forming ability (Gómez-Guillén *et al.*, 2011). Collagen also has been thought of as a good source of bioactive peptides (antioxidant, antihypertensive and antimicrobial), with promising applications in nutrition and medicine. These bioactive peptides can be obtained from the collagenous wastes by the enzymatic hydrolysis method (Gómez-Guillén *et al.*, 2011).

16.3.1.3 Applications of Collagen The applications of collagen are growing tremendously in the nutraceutical, biomedical, pharmaceutical and cosmetics field, due to their desirable properties. Also, it has been reported that the monetary value of collagen is very high ($1000/g) (MacNeil, 2001).

Collagen is found to have an important function in improving textural and sensory properties such as stability, consistency, juiciness and mouth-feel in the different food products such as meat products and ready-to-eat meals (Gómez-Guillén *et al.*, 2011). Hence they are being used as food additives. In addition, they are used for increasing the protein content in the food products.

Collagen has been identified as an important biomaterial due to their special characteristics such as biocompatibility, biodegradability and weak antigenicity (Zhang *et al.*, 2006). Also, collagen can form extra strong and stable fibers by cross-linking and self-aggregation. They can be used in drug delivery systems and tissue engineering. Using collagen as a drug delivery system is quite diverse, where they can be used in different forms of delivery systems such as sponges for burns/wounds, collagen shields in ophthalmology, tablet form for protein delivery, nanoparticles for gene delivery, transdermal delivery and gel formulation combined with liposomes for sustained drug delivery (Lee *et al.*, 2001). In tissue engineering, main applications of collagen are in wound dressings, heart valves, implants, scaffolds, bone grafts and replacements for artificial blood vessels (Lee *et al.*, 2001).

Collagen is also an important component of many cosmetic formulations that helps reduce signs of aging and it can also be used as a base material for encapsulation and biofilm for packaging (Benjakul and Visessanguan, 2000).

16.3.2 Gelatin

Gelatin is a soluble protein obtained from collagen, whose yield depends upon the age of animal and type of collagen used (Sukkwai *et al.*, 2011). Global demand for gelatin has increased, as it has a wide range of applications in food and pharmaceutical products. It is obtained by the thermal denaturation of collagen with a dilute acid and consists of high amounts of certain amino acids like glycine, proline and hydroxyproline (Nik Aisyah *et al.*, 2014). Glycine is the prominent amino acid in gelatin, having a typical Gly-X-Y sequence and proline and hydroxyproline are mainly responsible for stabilization of its structure (Russell *et al.*, 2007). However, gelatin is low in cysteine, methionine and tyrosine, as they are degraded during the hydrolysis process.

The main raw materials for gelatin extraction are skin, tendons, bones and cartilage. Porcine skin was first used for gelatin extraction and is still widely used; however, alternative sources such as fish and chicken processing by-products have received great attention in the recent years (Cheow *et al.*, 2007) (Table 16.3).

16.3.2.1 Extraction of Gelatin from Chicken Waste Collagen containing tissues are the initial raw materials for gelatin extraction. Gelatin can be extracted from collagen by their denaturation, which breaks the non-covalent bonds between the

Table 16.3 Sources of gelatin from chicken waste

Source	Process	Gelatin Yield (%)	Reference
Chicken feet	Acid-alkali process	4.42	(Lim *et al.*, 2001)
Chicken skin	Acid-alkali process	6.67	(Mhd Sarbon *et al.*, 2013)
Chicken meat residue	Acid-alkali process	16.03	(Rammaya *et al.*, 2012)
Silky fowl feet	Enzymatic process	7.30	(Cheng *et al.*, 2009)

polypeptide chains and some intra- and inter-molecular covalent bonds. This results in the triple helical structure of collagen to change to an amorphous form of gelatin (Karnjanapratum and Benjakul, 2015).

Acid alkaline and the enzymatic processes are used for conversion of collagenous materials to gelatin. Gelatin prepared by the acidic process is referred to as type A gelatin and that prepared by the alkaline process is referred as type B gelatin (Zhou and Regenstein, 2005). The enzymatic process involves the hydrolysis of covalent cross-links in collagen- using proteolytic enzymes followed by heat treatment for their deactivation (Zhang *et al.*, 2011). The gelatin extraction process comprises of three steps that includes pretreatment for removing non-collagenous materials from the initial material, followed by extraction that allows for controlled hydrolysis of collagen to gelatin (Ockerman and Hansen, 1999). The optimum conditions for gelatin extraction depends on the type of collagen used and extraction conditions like pH, time and temperature also affect the properties of the gelatin extracted. The final step is purification to obtain the finished final product.

16.3.2.2 Functional Properties of Gelatin The functional properties of gelatin are associated with their chemical attributes, such as their unique structure, amino acid content and molecular weight distribution (Mariod and Adam, 2013). These include surface, rheological and film-forming properties.

Due to its surface active properties, such as their amphoteric characteristics and the presence of hydrophobic groups in the peptide chain, gelatin can be used as a foaming and emulsifying agent. It can also be used as an emulsifier in oil–water emulsions such as in margarine, salad dressings and milk cream (Alfaro *et al.*, 2015). However, they are weaker emulsifiers when compared to other surface active agents such as gum arabic and other globular proteins. To improve its emulsifying ability, gelatin has to be used along with anionic surfactants or modified by adding non-polar side groups (Surh *et al.*, 2005). Gelatin is also involved in foam formation, which is affected by the source of protein, protein conformation, composition and their properties.

Gelatin is a hydrocolloid, which when mixed with water results in its thickening and gelling activity. Gelatins are soluble in water and capable of forming thermo-reversible gels (Alfaro *et al.*, 2015). The gel strength of gelatin is an important property, which is expressed as the Bloom value that is directly related to its resistance to degradation. The Bloom value is defined as weight, in grams, that is required for depressing the surface of the gel to a particular depth under standard conditions (Schreiber and Garris, 2007). In addition to gel strength, viscosity is another important property of gelatin, which depends on molecular weight of the proteins. The lower the molecular weight, the lower will be the viscosity and the gel strength of the protein solution (Alfaro *et al.*, 2015).

Gelatin has been studied widely for its film-forming properties. By using gelatin, it is possible to obtain transparent, strong, water soluble and colorless films. The characteristics of these films can be improved further by mixing with other proteins and polysaccharides and addition of cross-linking reagents as they reduce its hydrophilic properties and strengthen the covalent bonds in the protein network of the films (Alfaro *et al.*, 2015).

16.3.2.3 Applications of Gelatin Gelatin is the partially-hydrolyzed form of collagen that possesses a unique gel-forming nature. Hence, gelatins are being used as food additives to improve the water-holding capacity, texture and stability of many products such as meat products, baked goods, confectionary etc. (Mariod and Adam, 2013). In the dairy industry, they help to improve and stabilize the texture of yoghurt and provide a good mouth-feel and smooth texture to cheese products. Gelatin also has a melt-in-the-mouth property, due to which it is used for the release of flavor and aroma. Gelatin is used in the food industries, not only for its functional properties, but also to increase the protein content in the food products. Furthermore, for people suffering from obesity, it can be used as a dietetic agent for its low calorie content.

Gelatin is also used in the pharmaceutical industry for the manufacture of hard-gel and soft-gel capsules (Mariod and Adam, 2013). Capsules, tablet coatings, homeostatic sponges, blood plasma substitutes and vitamin and nutritional supplements encapsulation are the ranges of usage of gelatin in pharmaceuticals (Nik Aisyah *et al*, 2014). In addition, gelatin is used in the manufacture of edible films and packaging materials that protects the food products from light and oxygen. This helps to improve the quality and the shelf life of the different products.

16.3.3 Keratin

Keratin is a fibrous protein that is found mainly in feathers, nails, hairs etc. (Shi and Dumont, 2014). A great amount of chicken feathers (~1.8 million tons) are generated every year in the form of wastes. Some portion of the feathers is used as animal feed, whereas the rest are discarded. These feathers are a valuable and renewable protein source, as they are composed of 90% proteins, the main one being keratin (Wang and Cao, 2012). They are the major structural proteins found in feathers and are characterized by high amounts of cysteine and hydroxyl amino acids like serine. Keratin has two conformations – alpha-keratin and beta-keratin. The alpha-keratins are mainly found in soft tissue fibers, such as hair and skin, where they are twisted to form a rope-like structure (Gupta *et al.*, 2012). On the other hand, the beta-keratins are found in the hard tissues, like feathers and nails, and their role is protecting the keratin fibers from any physical or chemical damage. Alpha-keratins are rich in cysteine and poor in proline and hydroxyproline sequences. and the beta-keratins are rich in glycine and alanine and poor in cysteine and hydroxyproline (Gupta *et al.*, 2012). Keratins have a high sulphur content, which is due to the presence of cysteine residues in the protein sequence.

16.3.3.1 Keratin Extraction The keratin proteins comprise of various covalent and non-covalent interactions such as disulfide bonds, hydrophobic forces and hydrogen bonds. A native form of keratin is difficult to extract, as it is strongly linked with the disulfide bonds (Yin *et al.*, 2013). Different reducing agents can be used for keratin extraction, as they can reduce the stability of solid keratin and break down the disulphide and hydrogen bonds of the keratin fibers. Reducing agents such as dithiothreitol and 2-mercaptoethanol, sodium sulfide, thioglycate and potassium cyanide can increase keratin solubility by disrupting the disulphide bonds (Yin *et al.*, 2013; Gupta *et al.*, 2012). These reducing agents are fast and do not affect the protein yield as well as their chemical structure. However, the disadvantage of using reducing agents for keratin extraction is the consumption of huge amounts of reagents, which cannot be recycled (Wang and Cao, 2012). So research is now focusing on developing simple and eco-friendly methods for keratin extraction. Extraction of keratin using ionic liquids is one green and efficient method that has gained prominence (Yang *et al.*, 2006). Ionic liquids are inflammable and non-volatile and possess good thermal stability and solubility. The Shindai method for extraction of keratin is also being used widely, which comprises of three steps: pretreatment with ethanol, hydrochloric acid pretreatment and deoxidization using 2-mercaptoethanol (Khosa and Ullah, 2013).

In addition, the keratinous wastes, such as chicken feathers, can be hydrolysed further by feather degrading microorganisms that possess keratinolytic activity. These microorganisms include fungi, actinomycetes and certain *Bacillus* species (Fakhfakh *et al.*, 2011). This allows for efficient bioconversion and enhancement of their nutritional value.

16.3.3.2 Applications of Keratin Ninety percent of crude proteins in the form of keratin are present in feathers, which are used as alternative protein supplements in animal feed (Onifade *et al.*, 1998). However, the feather meal has limited use in animal feed, as it is deficient in histidine and methionine. Hence, the nutritional enhancement of feathers can be carried out using feather degrading microorganisms, so as to increase their nutritional applications (Fakhfakh *et al.*, 2011).

Keratin can be used for developing many keratin-based biomaterials that can be used in various biomedical applications. The keratin proteins possess the ability to self assemble into three-dimensional structures and polymerize to form fibrous scaffolds and hydrogels that have good biocompatibility (Rouse and Dyke, 2010). These scaffolds and hydrogels can be used for wound healing purposes, bone regeneration during injury, trauma or chronic bone defects, nerve repair, vascular grafts and medical devices (Balaji *et al.*, 2012). In addition, the presence of cell adhesion sequences in keratin, such as arginine-glycine-aspartic acid (RGD) and leucine-aspartic acid-valine (LGD), help the keratin biomaterials to provide cell attachment and growth (Vasconcelos and Cavaco-Paulo, 2013). These cell adhesion sequences are also found in various extracellular matrix proteins.

Bio-based films can also be developed from the keratin proteins, due to their availability and biodegradability. These films can be used as packaging materials and composites. Keratin protein possesses excellent film-forming properties for biodegradable packaging that helps to reduce the use of harmful synthetic polymers for food packaging (Song *et al.*, 2014). These polymers degrade the environment and produce various toxic components, hence have limited applications. Biodegradable films produced from keratin have good potential for food preservation, as they protect the food components from oxygen, moisture and volatile substances (Kim and Ustunol, 2001). Keratin can also be used for developing films for controlled drug release purposes in drug delivery systems (Li *et al.*, 2011). Polymeric network structures made from keratin are used to incorporate the therapeutic agents in a manner that allows them to be released slowly. Keratins are better than synthetic polymers used, due to the toxicity issues associated with these polymers. Current research is now focused on enhancing and optimizing the strength and flexibility of the keratin films and also maintaining their good biological activities.

Keratin can also be applied in the formulation of anti-wrinkle cream, hair straightener, conditioning shampoo etc. (Kelly *et. al*, 2006). They help to protect the hair cortex from any damage caused due to heat and chemicals and help restore the hair care processes (Villa *et al.*, 2013).

16.3.4 Plasma Proteins

Animal blood is composed of plasma and red blood cells or haemoglobin (Figure 16.2). The plasma is a yellow-colored watery fluid portion of the blood, which is responsible for transporting nutrients and wastes throughout the body. They make up 60% of the blood and comprise of functional proteins with commercial potential value (Bah *et al.*, 2013).

Proteins are important macromolecules that play a very important role in the growth and development of the human body, along with the repair and maintenance of the damaged tissues. The proteins in the plasma are a complex mixture that is composed not only of simple proteins but conjugated proteins, as well such as glycoproteins and some types of lipoproteins. The plasma derived from blood has 7–8% proteins, whereas the red blood cells have 34–38% protein content, respectively (Rathinaraj and Sachindra, 2013). The main proteins present in the plasma are albumin (60%), globulins and clotting proteins like fibrinogen. These proteins possess various functionalities such as gelation, foam formation etc. (Chandrasekaran, 2013). Other minor proteins, such as fibronectin and lipoprotein, constitute less than 1% of total plasma protein content (Saroat *et al.*, 2004). All the plasma proteins are synthesized mainly in the liver, except for immunoglobulins and each plasma protein have a characteristic half-life in the blood circulation. Essential amino acids make up 48% of the total amino acids in the blood plasma, with high amounts of lysine, valine, isoleucine, leucine, threonine and phenylalanine (Okanovic *et al.*, 2009).

16.3.4.1 Extraction and Characterization Chicken by-products, such as the blood, can be utilized as a valuable raw material for proteins and peptides. Blood can be used as a source for the extraction and isolation of plasma proteins from which peptides can be derived. Extraction is a process carried out for collecting beneficial compounds from various plant and animal sources. It includes a separation process that results in the separation of substances from the matrix (Se-Kwon Kim, 2013). Cohn's method has been used extensively for plasma protein fractionation, where the proteins are precipitated using ethanol (Cohn *et al.*, 1946). However, the use of alcohol denatures

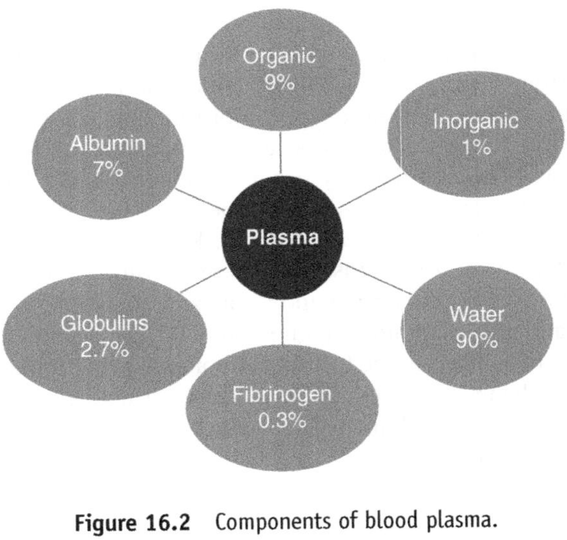

Figure 16.2 Components of blood plasma.

the proteins during the process, due to which other methods are being considered for use. Use of salts and polyethylene glycol (PEG) are alternative methods for plasma protein fractionation. They lead to lesser protein denaturation and the use of PEG also prevents a removal step, as they are retained in the supernatant with no protein damage (Sharma and Kalonia, 2004). Hence, use of PEG is now being done in the food and therapeutic industries for protein isolation and purification. Figure 16.3 shows one study done on fractionation of plasma proteins from chicken blood using polyethylene glycol (Saroat *et al.*, 2004).

16.3.4.2 Value-Added Applications of Plasma Proteins The plasma proteins possess good functional and nutritional properties, due to which they are of great interest for the food industries. They enhance the nutritional value of the food products, due to their essential amino acid content and good digestibility (Furlan *et al.*, 2012). Due to their functional properties, the plasma proteins can be used in processed food for human consumption, such as in cakes as a replacement of egg whites and in burgers as fat replacers (Viana *et al.*, 2005). Various other food applications of the plasma proteins have been reported. The dried plasma protein powder can be utilized in meat products as protein supplements and in fish surimi products as inhibitors of endogenous proteases that cause a decrease in the gel strength (Rawdkuen *et al.*, 2007). Proteinase inhibitors are found in the plasma proteins, which include α-trypsin inhibitors and α-chymotrypsin (Saroat *et al.*, 2004). Since the plasma protein shows a proteinase inhibition nature and also gel-strengthening abilities, it is used during heat-induced gelation in surimi production (Benjakul and Visessanguan, 2000). Fresh plasma proteins are used as functional ingredients in textured meat products, due to their gelling and emulsification behaviour.

Plasma proteins also have several therapeutic applications, as they contain thrombin, plasmin, fibrinogen, serotonin, immunoglobulins etc. (Young and Lawrie, 2007). Purified albumin can be used to replenish the fluid loss in animals and maintain the volume of blood following surgery or traumatic injury. It can also be used in antibiotic sensitivity tests, nutrient for tissue culture, microbial growth medium etc. (Kurbanoglu and Kurbanoglu, 2004).

The plasma proteins can also be used for preparing protein hydrolysates and bioactive peptides that have many valuable uses (Wanasundara *et. al*, 2003). These include

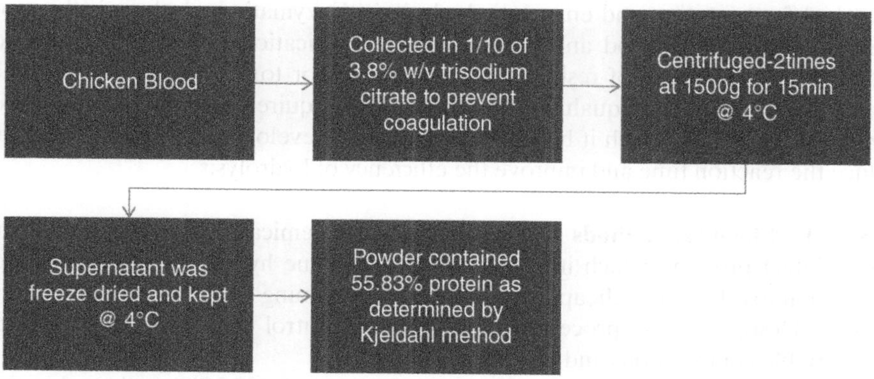

Figure 16.3 Extraction of plasma proteins (adapted from Saroat *et al.*, 2004).

functional enhancers for flavor development and composite foods, nutritional supplements and bioactivities such as antihypertensive, antioxidant and antimicrobial.

16.3.5 Bioactive Peptides

The chicken by-products are a good source of valuable bioactive peptides that can be incorporated in the food products and therapeutics to make them functional. These peptides are gaining a lot of attention as potential components of health promoting foods, nutraceuticals and pharmaceuticals, due to their good biological properties.

Bioactive peptides are peptides in the range of 3–20 amino acids that are found to be encrypted within the primary structure of the proteins (Korhonen and Pihlanto, 2003). Within the parent protein sequence, they are inactive and must be released in order to exert an effect. They can be released during food processing, fermentation, gastrointestinal digestion or *in-vitro* enzymatic hydrolysis. Following their release, the bioactive peptides can be absorbed in the intestine, they can enter the blood circulation and exert systematic effects, or they can produce local effects in the gastro-intestinal tract (Korhonen and Pihlanto, 2003) and bioavailability of the peptides *in vivo* is facilitated leading to physiological responses at the target site (Erdmann *et al.*, 2008).

Their potential for carrying out biological responses in the body is based on their ability to cross intestinal epithelium and bind to sites on the epithelial cell surface receptors or enter the blood circulation (Korhonen and Pihlanto, 2003). The effects in the gastro-intestinal tract include enzyme activity, enzyme inhibition, enhancing nutrient absorption etc. The absorption of the bioactive peptides is more efficient as compared to the large intact proteins. Various bioactivities demonstrated by bioactive peptides from various food sources and by-products include antimicrobial activity, antioxidant properties, immunomodulatory effects, ACE-inhibitory and blood pressure-lowering abilities and opioid activities (Bah *et al.*, 2013). Certain peptides can carry out more than one effect and hence are multifunctional.

16.3.5.1 Extraction of Bioactive Peptides A number of bioactive proteins and peptides have been obtained from various fish and animal wastes, but those from chicken wastes have yet to be explored. Various methods can be used for isolation of proteins and peptides depending on the nature, amino acid composition and sequence of the protein (Se-Kwon Kim, 2013). They include solvent extraction, microbial fermentation and enzymatic hydrolysis. Enzymatic hydrolysis is the most preferred method for food and pharmaceutical applications, because they do not result in the production of residual organic solvents or toxic chemicals in the final products (Agyei and Danquah, 2012). However, it requires a long reaction time of several hours, due to which it becomes necessary to develop a new method that can reduce the reaction time and improve the efficiency of hydrolysis.

16.3.5.1.1 Chemical Methods There are two main chemical methods for the hydrolysis of intact proteins, which includes acid and alkaline hydrolysis. These methods are comparatively easier, cheaper and less time-consuming and are generally used in industry. However, these processes are difficult to control and can lead to products with variable compositions and functionalities.

Acid hydrolysis is carried out at high temperatures (110–120 °C) for a long duration (18–96 hours). Generally, hydrochloric acid is used for hydrolysis, as it increases the cleavage rate of the peptide bonds as compared to the other acids used at the same concentration (Se-Kwon Kim, 2013). Using hydrochloric acid also results in faster hydrolysis and excessive hydrochloric acid can be removed by the process of evaporation. Sulphuric acid is another acid that is used frequently for hydrolysis. However, the use of strong acids is ecologically unacceptable and results in the reduction of the nutritional qualities of the product (Se-Kwon, 2013).

Alkaline hydrolysis utilizes a strong alkali at temperatures of 180°C for hydrolyzing the proteins. Sodium hydroxide and potassium hydroxide are the main alkalis used for breaking down the proteins into peptides (Se-Kwon, 2013). However, this method has certain drawbacks, such as the formation of undesirable components and destruction of certain amino acids like serine and threonine (Pasupuleti and Braun, 2010). Many deleterious reactions can also occur during hydrolysis, which includes the raceamization of L-amino acids to produce D-amino acids that are not absorbed by humans.

16.3.5.1.2 Microbial Fermentation Fermentation is an eco-friendly method for protein hydrolysis that utilizes the microorganism, Lactic Acid Bacteria (LAB). The microbial strains of LAB that are commonly used are *Lactobacillus Helveticus, Lactobacillus Plantarum, Lactococcus Lactis ssp. diacetylactis, Lactobacillus delbreucki ssp. Bulgaris* and *Lactobacillus acidophilus* (Agyei and Danquah, 2012). These microorganisms are either present in the wastes as natural flora or they are added as inoculum. As the by-products are poor in carbohydrates, external sources such as glucose and sucrose are added that can be metabolized by the microorganisms to produce organic acids such as lactic acid that results in the desired pH and also enhances preservation. Also, breakdown of proteins to peptides takes place that possess good bioactivity.

These microorganisms have a proteolytic system that comprises of the cell wall bound proteinases and a variety of peptidases like endopeptidases, dipeptidases etc. (Agyei and Danquah, 2012), which degrade the proteins to form oligopeptides, dipeptides and tripeptides that serve as good sources for production of bioactive peptides. Production of bioactive peptides by microbial fermentation is affected by the various growth parameters like the inocula conditions, as well as the peptide content of the media used (Korhonen and Pihlanto, 2003). Controlling these parameters is essential for enhancing the proteolytic potential of the microorganisms. The choice of the strains influences the release of the effective bioactive peptides. When choosing the strain, it should not be highly proteolytic, otherwise the products will be damaged, but it must be rightly proteolytic for the production of high concentrations of bioactive peptides.

16.3.5.1.3 Enzymatic Hydrolysis Enzymes can be used for the hydrolysis and degradation of proteins into peptides and amino acids. Enzymatic hydrolysis of the proteins is defined as the hydrolysis of proteins to produce protein hydrolysates using isolated enzymes, in order to use them for food or medical applications. Using enzymes, i.e. proteases, is comparatively a mild application. The amino acids in proteins are linked together by peptide bonds, which are cleaved during the process of hydrolysis (Damodaran, 1997). The catalytic action of the protease enzymes is determined by the degree of hydrolysis (DH), which gives an idea about the number of peptide bonds that are cleaved.

Enzymatic hydrolysis is the most commonly used method for the production of protein hydrolysates, with the desired functional and biological properties. It is preferred over the acid and alkaline hydrolysis of the proteins, as it minimizes side effects or production of undesirable by-products due to the mild processing conditions and the specificity of the enzymes. The use of enzymatic hydrolysis is also preferred over the microbial fermentation process. due to the lack of residual organic solvents or toxic chemicals in the products (Agyei and Danquah, 2012).

The enzymes used for the modification of proteins can be categorized based on their origin (plant, animal or microbial), catalytic activity (endo- or exo-peptidases) and the number of catalytic sites. Endo-peptidases are the enzymes generally employed during protein hydrolysis that can be further classified into cysteine proteases, serine proteases, aspartic proteases and metallo-proteases (Whitaker and Scheffner. 2003).

The parent proteins isolated from chicken waste can be hydrolysed either by using endogenous proteolytic enzymes present in the chicken by-products or by using exogenous enzymes from various sources (Guerard, 2007). Many different exogenous enzymes, such as alcalase, pepsin, pancreatin and trypsin etc., are used to produce hydrolysates from the different chicken by-products (Agyei and Danquah, 2012). The enzyme alcalase has been used extensively in many studies due to its ability to produce hydrolysates in a shorter time period and due to its broad specificity (Lasekan *et al.*, 2013).

Recent studies have shown improvement in the efficiency of the enzymatic hydrolysis process using ultrasound or microwave technology as pretreatment methods. The ultrasound technology involves the use of ultrasonic waves with a frequency of 20 kHz and greater. It produces a cavitation effect where the bubbles obtained from cavitation burst due to crushing, resulting in high amounts of mechanical shearing force. This leads to alterations in the protein conformation, in turn leading to exposure of more hydrophilic groups that increases the protein solubility and allows the enzyme to easily bind to the substrate (Yu *et al.*, 2012). This helps to increase the hydrolysis efficiency. The microwave technology utilizes electromagnetic waves with a frequency of 0.3–300 GHz. Microwaves produces heat within the material that increases the heating rate and shortens the reaction time, leading to an increase in the protein digestion speed (Li *et al.*, 2010).

16.3.5.2 Purification and Characterization of Bioactive Peptides Fractionation and purification of bioactive peptides is required, mainly if the peptides are to be used as neutraceuticals and therapeutics. High-performance liquid chromatography (HPLC) is the most commonly-used technique for the separation, identification and purification of bioactive peptides. Other methods that are used for peptide fractionation and extraction include ion-exchange chromatography, gel filtration technologies and ultrafiltration (Agyei and Danquah, 2012). This step is important for identifying the physiochemical properties of the peptides and evaluating their respective bioactivities.

High-performance liquid chromatography is efficient and is an important method for separating and purifying the peptides. The different peptides in the sample will move through the column at different velocities, which will depend on their chemical properties, their interactions with the stationery phase and the mobile phase composition and will separate out. Due to their high resolution and recovery, it is possible to separate the closely-related peptides as well. Also it has good reproducibility due to the stability of the sorbent materials (Agyei and Danquah, 2012).

Ultra-filtration is a popular type of membrane filtration that helps to remove or concentrate the bioactive peptides. It separates the peptides on the basis of size and shape.

Here the membrane acts as a barrier that allows small molecules to pass through while retaining the large molecules. So the peptides can be separated by choosing membranes with proper pore size. As compared to the other methods used for purification, the ultra-filtration method is not only effective at the laboratory scale, but it is the most suitable technology for commercial scale-up. However, the membranes used in ultrafiltration are not effective for the separation of peptides having molecular weight of less than 1 kDa (Agyei and Danquah, 2012).

Size exclusion chromatography is a chromatographic technique where the peptides are separated by their size and not their molecular weight. This method allows a good separation of the large molecules from the smaller ones. This method has good sensitivity, due to the formation of narrow bands and well-defined separation times.

Ion-exchange chromatography allows the separation of peptides with similar molecular weight but different surface charges. It is very frequently used for the fractionation and purification of peptides due to its simplicity and high capacity and resolving power. The principle of this method is that by controlling the pH and the ionic strength, the peptides with different charges can be separated out (Agyei and Danquah, 2012).

The gel filtration technique is widely used for the separation and purification of bioactive peptides. It is based on the molecular size and shape of the peptides. The large proteins are unable to diffuse through the pores and pass through the void volume column due to which they are eluted first. On the other hand, the small peptides are able to diffuse through the pores and are eluted later. This method can be used to separate the peptides that have significant differences in their molecular weights. Since the separation of the peptides is size-based, it becomes ineffective for small peptides or the peptides having similar molecular sizes. In such cases, other methods are needed (Agyei and Danquah, 2012).

Following purification of the bioactive peptides with any one of the separation techniques, the peptides can be characterized by mass spectrometry (MS) analysis. Mass spectrometry is an important tool that is used for the determination of the molecular mass and the protein sequences and also for the identification of the protein conformation. LC-MS/MS is commonly used to determine the peptide sequence and MALDI-TOF helps to generate the peptide profiles. SDS-PAGE and ultracentrifugation are methods used in addition to MS for determining the amino acid composition of the peptides for their structural characterization. The peptide sequences obtained can then be screened against a database to identify the possible peptide and their bioactivity can be confirmed using various cellular and animal assays (Singh *et al.*, 2014).

16.3.5.3 Functionalities of Bioactive Peptides

16.3.5.3.1 Antihypertensive Activity Hypertension, or high blood pressure, is a serious health issue that affects 20% of the world adult population and is considered a risk for cardiovascular diseases such as coronary heart disease, peripheral artery disease and stroke. Angiotensin converting enzyme (ACE) is one main regulator of blood pressure. They are a part of the rennin-angiotensin system that converts angiotensin I to angiotensin II, which is a potent vasoconstrictor. Various ACE inhibitors are available for treating hypertension, but they have many undesirable effects (Huang and Liu, 2010). Currently, one promising approach to address this problem is increasing the consumption of proteins with biologically active peptide sequences that can reduce the risk of hypertension. Antihypertensive peptides are a group of bioactive peptides that inhibit the angiotensin converting enzyme *in vitro* (Onuh *et al.*, 2013). They are a

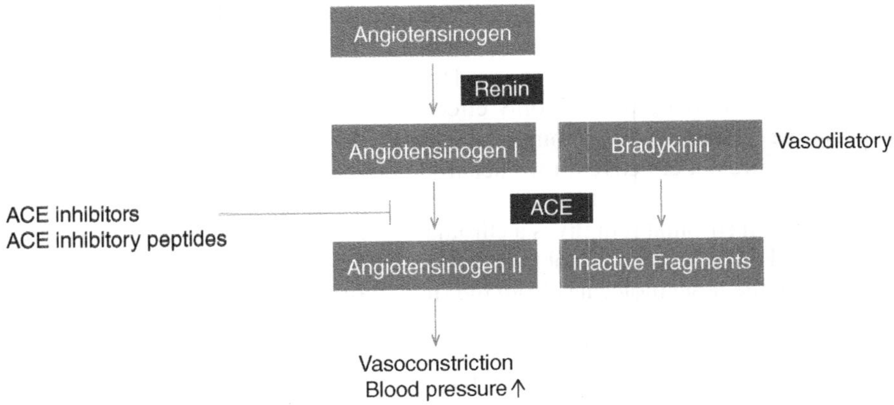

Figure 16.4 ACE inhibition of the renin angiotensin system (Belovic *et al.*, 2011).

good alternative, as they are relatively cheaper to synthesise and have very little or no side effects that are associated with the ACE inhibitors (Onuh *et al.*, 2013). These peptides can be used as treatment options in individuals with minor hypertension, or as supplemental treatment by developing functional foods that can prevent hypertension (Erdmann *et al.*, 2008).

Various studies indicate the effect of peptide structure on the peptides ACE-inhibitory activity (Figure 16.4). According to the structure-activity correlation studies, the tripeptides residues of tryptophan, alanine and proline at the carboxyl terminal plays an important role in competitively binding to the ACE active site and inhibiting the ACE activity (Cheung *et al.*, 1980). The presence of the proline residues also resists the peptides from enzymatic degradation, while they are passing through the gastrointestinal tract (Agyei and Danquah, 2012).

16.3.5.3.2 Antioxidative Activity Oxidation is an important process that occurs in all living organisms. But, it leads to various side effects, such as the production of reactive oxygen species (ROS) such as hydroxyl radicals, peroxyl radicals, superoxide anions etc. The reactive oxygen species produced can lead to extensive damage to the biological macromolecules like DNA, RNA, proteins and lipids (Erdmann *et al.*, 2008). Also, they cause diseases such as hypertension, diabetes, cancer etc. Hence, the inhibition of such free radicals is required to reduce the risk of these diseases. Many endogenous antioxidants are present in the body that protects the tissues and organs from oxidative damage due to these ROS. They include enzymes such as superoxide dismutase, catalase, glutathione peroxidase and non-enzymes such as vitamin C, selenium and α-tocopherols (Decker and Xu, 1998). In addition, various amino acids and peptides present naturally, lead to antioxidative property in the cells and help in maintaining the balance of the biological tissues. Dietary consumption of antioxidants helps in providing additional benefits to the endogenous antioxidant defence system in order to fight against oxidative stress. Well-known dietary antioxidants are vitamin C, vitamin E, polyphenols and carotenoids. In addition to these, many other dietary components can provide defences against oxidative damage. But we still require more.

The balance between the antioxidants and ROS in the human body can be disrupted due to age progression and other factors such as high fat diets, fatigue and environmental pollution. These factors can weaken the body's immune system, due to which the body becomes prone to oxidative attack. Dietary antioxidants have increased exponentially and include α-tocopherols, vitamin C, phytochemicals, lycopene etc. However, the use of antioxidant peptides is a new trend, where recent research is being conducted. Antioxidant peptides are the peptides that help in protecting the body, by scavenging the free radicals and reactive oxygen species. This is important for the survival of the cells in an organism and also for preventing undesirable oxidative processes in food (Hartmann *et al.*, 2007). In food products, the oxidation of lipids and fatty acids results in the development of unwanted chemical compounds, rancid flavors and unacceptable taste. Their further oxidation affects the food quality and results in low shelf life. Toxic products may also form, whose consumption may cause certain diseases.

Antioxidant activity of these peptides is attributed to the presence of certain amino acids. The antioxidant effect is related to high amounts of histidine and hydrophobic amino acids. Antioxidant activity of histidine-containing peptides is related to the hydrogen donating ability, lipid peroxy radical trapping and the metal ion chelating ability of the imadizole group. Addition of some leucine or proline residues at the N-terminus further enhances the antioxidant activity (Erdmann *et al.*, 2008). Hydrophobic amino acids such as valine and leucine increases interactions between the peptides and fatty acids, which help to scavenge the free radicals in the lipid phase (Ranathunga *et al.*, 2006). The molecular weight of most of the antioxidative peptides ranges from 500–1800 Da (Kim *et al.*, 2007).

16.3.5.3.3 Antimicrobial Activity Over recent years, there has been a huge increase in the number of antibiotic resistant bacteria. Antibiotic resistance is the ability of the bacteria to grow, even in the presence of antibiotics. Emergence of antibiotic resistance can seriously affect human health and economics. One solution to this problem is to look into a new class of antibiotics, referred to as antimicrobial peptides (Hancock, 2001).

Antimicrobial peptides have now gained a lot of attention as potential alternatives to the currently-used antimicrobial compounds. These peptides carry out their action either by disrupting the integrity of the cell membranes or by membrane translocation and acting on internal targets. They can further inhibit pathogen growth by withholding the nutrients that are essential for microbial proliferation and they also protect the gastrointestinal tract from pathogenic bacteria and viruses (Agyei and Danquah, 2012). The antimicrobial peptides exhibit selectivity against different microorganisms. Some of the peptides display a broad spectrum activity against Gram-negative bacteria, while others preferentially kill the Gram-positive bacteria (Agyei and Danquah, 2012).

The antimicrobial peptides exhibit many common features. This includes their size being less than 50 amino acids, of which 50% are hydrophobic in nature. Hence they form amphipathic three-dimensional structures. The antimicrobial peptides possess great potential for their use in food preservation and for therapeutics in healthcare. The traditional methods of food preservation have less efficiency to inhibit their growth. So the need is for novel antimicrobial agents having a natural origin. Bioactive peptides having antimicrobial properties are potential candidates to be used as food additives. The advantages of these peptides over the chemical preservatives include lesser adverse effects, minimal processing required, and the ability to retain the nutritional and organoleptic properties of food. In therapeutics, they are better than the conventional antibiotics, as they are more efficient in killing the bacteria and are unperturbed by the antibiotic-resistance mechanisms that are faced by some of the antibiotics (Shahidi and Zhong, 2008).

Antimicrobial peptides from animal by products have been studied extensively, but little from chicken wastes. They are yet to be explored for the generation of antimicrobial peptides that are effective against various bacteria.

16.3.5.4 Applications of Bioactive Peptides The idea of using bioactive peptides as functional food ingredients is highly attractive. They have gained interest as promising candidates in health-promoting functional foods. New emerging technologies, such as microencapsulation and nano-encapsulation, allow for the improved stability of bioactive peptides in numerous food products and during digestion. However, the biosafety of these peptides has to be considered before they can be used in the food products formulation. In addition, they have applications in food preservation, where they can either be used as food additives or be incorporated in the food packaging materials.

Functional foods are a rapidly growing segment of the food market and they comprise of many biologically active compounds that provide health-promoting abilities. The field of functional foods is very new and still there exists many gaps in their development. Bioactive peptides are health-promoting compounds that are potential ingredients for designing the functional foods. This is because they play an important role in disease risk management. Of all the peptides isolated from the chicken wastes, antihypertensive peptides would be excellent ingredients for improving health and decreasing healthcare costs. For incorporating the peptides in food, they must be effective and must not negatively affect the food quality, so that consumer needs are met and high levels of compliance are achieved (Decker and Park, 2010).

The bioactive peptides can be introduced in the food products in two ways. The first method is to add proteins containing bioactive peptides into food, which would generate the peptides on hydrolysis in the gastrointestinal tract. But this method is difficult, as very high levels of proteins have to be included in the food for reaching a sufficient dose for them to carry out their respective bioactivities. An alternative method is to generate peptides using enzymatic hydrolysis of the food proteins and then concentrating the peptide fractions rich in bioactive peptides and introducing them as nutraceuticals ingredients in foods at low levels. This method is more feasible, but the bioactive peptides being incorporated should not alter food quality and should maintain the bioactive forms of the peptides. One potential method is microencapsulation or nano-encapsulation, where the peptides can be encapsulated in the emulsions or the lipid carrier systems like liposomes. The efficiency of these peptides depends on their interactions with the proteases and acids occurring naturally in the stomach and the intestine. Encapsulation of the bioactive peptides in a lipophilic matrix is able to inhibit such interactions and when they enter the intestine, the lipids are digested and the peptides are released so that they can enter the blood flow, after which they can be carried to the different target organs, so as to carry out their respective activities (Decker and Park, 2010).

Foodborne diseases are of major concern and mainly meat contamination with pathogenic bacteria leads to serious foodborne outbreaks. In addition, the meat industries have to face huge economic losses. Another serious concern in the meat industry is the sensory quality of the product. Oxidation reaction is one major factor that reduces the shelf life of meat products. Hence, novel preservation methods are required. Due to the increasing demand for safe and preservative-free products, many innovations are being made for the development of active packaging, which not only acts as a barrier to external conditions, but also plays its part in the desired functions in food preservation. Also, consumers now want foods that contain natural additives rather than synthetic ones. Hence, there is a need to find natural compounds that can be used and incorporated into the active packaging.

Antimicrobial and antioxidant packaging are gaining a lot of attention in the food industries for the preservation of minimally-processed food products such as meat products (Coma, 2012). Addition of antimicrobial and antioxidant peptides in the packaging materials is one major area of development in the packaging technology. This can be done either to increase the shelf life of the product or to reduce the microbial load of the packaging before they can be used. This will allow low levels of peptides to be utilized effectively and also increase the range of microorganisms whose growth can be inhibited. Also the peptides can be used in combination with other non-thermal methods of preservation, such as high hydrostatic pressure, ultrasonication and irradiation for preventing pathogen growth in the various food products. This approach can reduce the processing costs and improve the product quality (Touch *et al.*, 2009).

With emerging infectious diseases and development of bacterial resistance, there is a search for alternatives for the conventional antibiotics. Antimicrobial peptides are a promising approach in the antibiotic research and have the potential to be used as new antimicrobial substances, as they have low MIC (Minimum Inhibitory Concentration) *in vitro* and have broad spectrum activity. Over the past few years, a number of antimicrobial peptides have been identified from animal-processing wastes, although antimicrobial peptides from chicken waste are yet to be explored. These peptides kill the bacteria by disrupting the membrane integrity, therefore are likely less to induce resistance in pathogenic microorganisms (Yang and Blecha, 2008). Hence, they are being evaluated as novel antimicrobial drugs in place of antibiotics. Also, they have a broad range of activity and are mainly bactericidal as compared to many antibiotics that are bacteriostatic. Chicken by-products are currently under-utilized sources of such antimicrobial peptides. They can be isolated and developed into high-value products such as biopreservatives and pharmaceuticals. However, further research and development needs to be done before they can be commercialized and their full advantage taken.

16.4 Valorization of Egg Waste

Egg wastes are generated in huge amounts every year, such as those discarded by hatcheries and egg processing industries, as well as households that consume the albumen and yolk. They mainly comprise of the shell and the shell membranes that represents 11% of the total weight of the egg (Stadelman, 2000). These by-products are not edible and are not disposed of properly, thus leading to environmental pollution. These wastes are mainly disposed off in landfills without any pretreatment (Yoo *et al.*, 2009), which leads to production of strong odors on biodegradation. However, the eggshells, mainly the eggshell membrane, comprise of organic matter such as proteins, peptides and amino acids (Nakano *et al.*, 2003) that can be utilized for the creation of value-added products from waste material. They find a wide range of applications in the cosmetic, food and medical sectors.

The eggshell membrane is made up of 70% proteins, of which collagen (type I, V and X) constitutes 10% of the total proteins (Nys and Gautron, 2007). The presence of hydroxyproline in the membranes suggests the presence of collagen. However, the chemical composition of the outer and inner membranes is different from outer ESM, comprising of mainly type I collagen and the inner one consisting of both type I and type V collagen (Balaz, 2014). Collagen X is present in both the membranes. The matrix also comprises of 11% of polysaccharides, such as chrondroitin sulphate, hyaluronic acid and uronic acids (Nys and Gautron, 2007), which are important contents responsible for water retention due to which they are of important interest in the field of pharmaceuticals, food and cosmetics. Other compounds such as lipids and carbohydrates are also present, but in smaller quantities.

Various bio-proteins can be identified from the eggshells, which include ovocleidin-17, ovocalxyin-32 and osteopontin (Abdou *et al.*, 2013). Even egg white proteins, such as ovalbumin, lysozyme and ovotransferrin proteins, can be found in the eggshells. The eggshell membranes are also a good source of functional protein hydrolysates and peptides, with various applications in the food industries (Jain and Anal, 2016). In addition, the eggshells are a good source of calcium, which helps increase bone mineral density in humans with osteoporosis (Abdou *et al.*, 2013).

16.5 Conclusion

Substantial research and development has been done on the utilization of animal by-products for beneficial purposes. Chicken waste and by-products have attracted a lot of attention, due to their abundance and their potential biological activities. They are under-utilized resources that have the potential to be used in various industrial applications. The main focus of the chicken processing industry should be to minimize waste generation so as to protect the environment and the efficient utilization of the by-products. High biological value proteins and bioactive peptides can be derived from these wastes. Chicken waste derived proteins play an important role in promoting health and nutrition. Also, it has been suggested that they are promising for the exploration of bioactive peptides. However, before they can be utilized, various issues must be addressed, such as their large-scale production, bioavailability and long-term stability. Suitable environmentally-friendly methods are needed for waste disposal and utilization. Biotechnological methods, such as enzymatic hydrolysis and fermentation, are found to be better than chemical ones. In addition, a more detailed understanding is required regarding the mechanisms by which the proteins and bioactive peptides carry out their physiological effects, for which metabolomics studies have to be done. One bioprocess challenge in the area of their production is their isolation and purification. Several isolation and purification methods are out there, but they have to be selected based on their advantages and disadvantages.

References

Abdou, A.M., Kim, M. and Sato, K. (2013) *Functional Proteins and Peptides of Hen's Egg Origin: Bioactive food peptides in health and disease*. InTech, Rijeka, Croatia, pp. 115–144.

Agyei, D. and Danquah, M.K. (2012) Pharmaceutical applications of bioactive peptides. *OA Biotechnology*, **1**(2): 5.

Alfaro, A., Balbinot, E., Weber, C.I., Tonial, I.B. and Machado-Lunkes, A. (2015) Fish gelatin: Characteristics, functional properties, applications and future potentials. *Food Engineering Reviews*, **7**(1): 1–12.

Bah, C.S.F., Din A., Carne, A. and McConnell, M.A. (2013) Slaughterhouse blood: An emerging source of bioactive compounds. *Comprehensive Reviews in Food Science and Food Safety*, **12**: 314–331.

Balaji, S., Kumar, R., Sripriya, R., Kakkar, P., Ramesh, D.V., *et al.* (2012) Preparation and comparative characterization of keratin–chitosan and keratin–gelatin composite scaffolds for tissue engineering applications. *Materials Science and Engineering: C*, **32**(4): 975–982.

Balaz, M. (2014) Eggshell membrane biomaterial as a platform for applications in materials science. *Acta Biomaterialia*, **10**: 3827–3843.

Belovic, M., Mastilovic, J.S., Torbica, A.M., Tomic, J.M., Stanic, D.R. and Dzinic, N.R. (2011) Potential of bioactive proteins and peptides for prevention and treatment of mass non-communicable diseases. *Food and Feed Research*, **38**(2): 51–61.

Benjakul, S., and Visessanguan, W. (2000) Pig plasma protein: potential use as proteinase inhibitor for surimi manufacture; inhibitory activity and the active components. *Journal of the Science of Food and Agriculture*, **80**(9): 1351–1356.

Chandrasekaran, M. (2013) Valorization of food processing by-products. Taylor & Francis Group, CRC Press, Boca Raton, FL.

Cheng, F.Y., Hsu, F.W., Chang, H.S., Lin, L.C. and Sakata, R. (2009) Effect of different acids on the extraction of pepsin-solubilized collagen containing melanin from silky fowl feet. *Food Chemistry*, **113**(2): 563–567.

Cheow, C., Norizah, M., Kyaw, Z. and Howell, N. (2007) Preparation and characterisation of gelatins from the skins of sin croaker (*Johnius dussumieri*) and shortfin scad (*Decapterus macrosoma*). *Food Chemistry*, **101**(1): 386–391.

Cheung, H.S., Wang, F.L., Ondetti, M.A., Sabo, E.F. and Cushman D.W. (1980) Binding of peptide substrates and inhibitors of angiotensin-converting enzyme: Importance of the COOH-terminal dipeptide sequence. *Journal of Biological Chemistry*, **255**(2): 401–407.

Cohn, E.J., Oncley, J.L., Strong, L.E., Hughes, W.L. and Armstrong, S.H. (1946) Chemical, clinical and immunological studies on the products of human plasma fractionation. I: The characterization of the protein fractions of human plasma. *Journal of Clinical Investigation*, **23**: 417–432.

Coma, V. (2012) Antimicrobial and antioxidative packaging for meat and poultry. In: *Advances in Meat, Poultry and Seafood Packaging*. (ed. J.P. Kerry). Woodhead Publishing Ltd, Cambridge.

Damodaran, S. (1997) Food proteins: An overview. In: *Food Proteins and their Applications*. Marcel Dekker Inc., New York, pp. 21–24.

Decker, E.A. and Park, Y. (2010) Healthier meat products as functional foods. *Meat Science*, **86**: 49–55.

Erdmann, K., Belinda, C. and Schroder, H. (2008) The possible roles of food-derived bioactive peptides in reducing the risk of cardiovascular disease. *Journal of Nutritional Biochemistry*, **19**(10): 643–654.

Fakhfakh, N., Ktari, N., Haddar, A., Mnif, I.H., Dahmen, I. and Nasri, M. (2011) Total solubilisation of the chicken feathers by fermentation with a keratinolytic bacterium, *Bacillus pumilus* A1, and the production of protein hydrolysates with high antioxidative activity. *Process Biochemistry*, **46**: 1731–1737.

Furlan, L.R., Lecot, J., Padilla, A.P., Campderrós, M.E. and Zaritzky, N. (2012) Stabilizing effect of saccharides on bovine plasma protein: A calorimetric study. *Meat Science*, **91**(4): 478–485.

Gómez-Guillén, M., Giménez, B., López-Caballero, M.A. and Montero, M. (2011) Functional and bioactive properties of collagen and gelatin from alternative sources: A review. *Food Hydrocolloids*, **25**(8): 1813–1827.

Guerard, F. (2007) Enzymatic methods for marine by-products recovery. In: Shahidi, F., ed. *Maximizing the value of marine by-products*. Woodhead Publishing Ltd, Cambridge, pp. 107–143.

Gunasekaran, S. (2003) Purifying type I collagen using two papain treatments and reducing and delipidation agents: Google Patents.

Gupta, A., Kamarudin, N.B., Kee, C.Y.G. and Yunus, R.B.M. (2012) Extraction of keratin protein from chicken feather. *Journal of Chemical Engineering*, **6**: 732–737.

Hancock, R. (2001) Cationic peptides: Effectors in innate immunity and novel antimicrobials. *The Lancet Infectious Diseases*, **1**: 156–164.

Hartmann, R. and Meisel, H. (2007) Food-derived peptides with biological activity: From research to food applications. *Current Opinions in Biotechnology*, **18**(2): 163–169.

Hashim, P., Ridzwan, M.M. and Bakar, J. (2012) Isolation and characterization of collagen from chicken feet. *International Journal of Biological, Veterinary, Agricultural and Food Engineering*, **8**(3): 238–242.

Huang, S.C. and Liu, P.J. (2010) Inhibition of angiotensin I-converting enzymes by enzymatic hydrolysates from chicken blood. *Journal of Food and Drug Analysis*, **18**(6): 458–463.

Jain, S. and Anal, A.K. (2016) Optimization of extraction of functional protein hydrolysates from the chicken eggshell membrane (ESM) by ultrasonic assisted extraction (UAE) and enzymatic hydrolysis. *LWT-Food Science and Technology*, **69**: 295–302.

Jayathilakan, K., Sultana, K., Radhakrishna, K. and Bawa, A.S. (2012) Utilization of byproducts and waste materials from meat, chicken and fish processing industries: a review, *Journal of Food Science and Technology,* **49**(3): 278–293.

Karnjanapratum, S. and Benjakul, S. (2015) Characteristics and antioxidative activity of gelatin hydrolysates from unicorn leatherjacket skin as affected by autolysis assisted process. *Journal of Food Processing and Preservation*, **30**(6): 915–926.

Kelly, G., Agbogbo, F.K. and Holtzapple, M.T. (2006) Lime treatment of keratinous materials for the generation of highly digestible animal feed. II: Animal hair. *Bioresource Technology,* **97**(11): 1344–1352.

Khosa, M. and Ullah, A. (2013) A sustainable role of keratin biopolymer in green chemistry: A Review. *Journal of Food Processing and Beverages,* **1**(1): 8.

Kiew, P.L. and Don, M.M. (2013) The influence of acetic acid concentration on the extractability of collagen from the skin of hybrid *Clarias* sp. and its physicochemical properties: A preliminary study. *Focusing on Modern Food Industry,* **2**(3): 123–128.

Kim, S.J. and Ustunol, Z. (2001) Solubility and moisture sorption isotherms of whey protein-based edible films as influenced by lipid and plasticizer incorporation. *Journal of Agricultural and Food Chemistry*, **49**(9): 4388–4391.

Kim, Y.C., Ludovice, P.J. and Prausnitz, M.R. (2007) Transdermal delivery enhanced by magaininpore-forming peptide. *Journal of Control Release,* **122**: 375–383.

Kiruthiga, N., Sujithra, S. and Kumaresan, R. (2013) Extraction and characterization of type I collagen from broiler chicken waste. *International Journal of Institutional Pharmacy and Life Sciences,* **3**(6): 53–60.

Korhonen H. and Pihlanto A. (2003) Food-derived bioactive peptides: Opportunities for designing future foods. *Current Pharmaceutical Design,* **9**(16): 1297–1308.

Kurbanoglu, E. and Kurbanoglu, N. (2004) Utilization as peptone for glycerol production of ram horn waste with a new process. *Energy Covers Manage,* **45**(2): 225–234.

Lasekan A., Bakar F.A. and Hashim D. (2013) Potential of chicken by-products as sources of useful biological resources. *Waste Management,* **33**: 552–565.

Lee, C.H., Singla, A. and Lee, Y. (2001) Biomedical applications of collagen. *International Journal of Pharmaceutics,* **221**(1): 1–22.

Li, F.Y., Wang, R.M., He, Y.F., Li, X.X., Song, P.F., *et al.* (2011) Keratin films from chicken feathers for controlled drug release. *Journal of Controlled Release,* **152**: 92–93.

Li, J.F., Wei, F., Dong, X.Y., Guo, L.L., Yuan, G.Y., *et al.* (2010) Microwave-assisted approach for the rapid enzymatic digestion of rapeseed meal. *Food Science Biotechnology,* **19**(2): 463–469.

Lim, J., Oh, S. and Kim, K.O. (2001) The effects of processing conditions on the properties of chicken feet gelatin. *Food Science and Biotechnology,* **10**(6): 56–63.

Lin, Y.K. and Liu, D C. (2006) Effects of pepsin digestion at different temperatures and times on properties of telopeptide-poor collagen from bird feet. *Food Chemistry,* **94**(4): 621–625.

Liu, D.C., Lin, Y.K. and Chen, M.T. (2001) Optimum condition of extracting collagen from chicken feet and its characteristics. *Asian-Australasian Journal of Animal Sciences,* **14**(11): 1638–1644.

Luque, R. and Clark, J.H. (2013) Valorisation of food residues: Waste to wealth using green chemical technologies. *Sustainable Chemical Processes,* **1**(10): 1–3.

MacNeil, J.H. (2001) Method and apparatus for separating a protein membrane and shell material in waste egg shells: Google Patents.

Mariod, A.A. and Adam, H.F. (2013) Review: Gelatin, source, extraction and industrial applications. *ACTA Scientiarum Polonorum Technologia Alimentaria,* **12**(2): 135–147.

Mhd Sarbon, N., Badii, F. and Howell, N.K. (2013) Preparation and characterisation of chicken skin gelatin as an alternative to mammalian gelatin. *Food Hydrocolloids,* **30**(1): 143–151.

Nakano, T., Ikawa, N. and Ozimek, L. (2003) Chemical composition of chicken eggshell and shell membranes. *Chicken Science,* **82**(3): 510–514.

Nik Aisyah, N.M., Nurul, H., Azhar, M.E. and Fazilah, A. (2014) Poultry as an alternative source of gelatin. *Health and the Environmental Journal,* **5**(1): 37–49.

Nys, Y. and Gautron, J. (2007) Structure and formation of the eggshell. In: *Bioactive Egg Compounds* (eds R. Huopalahti, R. Lopez-Fandino, M. Anton and R. Schade), Springer, Berlin, p. 99.

Ockerman, H.W. and Hansen, C.L. (1999) *Animal By-Product Processing and Utilization.* CRC Press, Taylor & Francis Group, Boca Raton, FL.

Okanovic, D., Ristic, M., Kormanjos, S. and Njezic, Z. (2009) Chemical and nutritional quality of slaughter pigs by-products. *Quality of Life,* **1**(1): 55–60.

Onifade, A A., A1-Sane, N., Al-Musallam, A. and Al-Zarban, S. (1998) A review: Potentials for biotechnological applications of keratin-degrading microorganisms and their enzymes for nutritional improvement of feathers and other keratins as livestock feed resources. *Bioresources Technology* **66**: 1–11.

Onuh, J.O., Girgih, A.T., Aluko, R.E. and Aliani, M. (2013) Inhibitions of renin and angiotensin converting enzyme activities by enzymatic chicken skin protein hydrolysates. *Food Research International,* **53**: 260–267.

Pachence, J.M. (1992) Process for extracting type I collagen form an avian source, and applications thereof: Google Patents.

Parenteau-Bareil, R., Gauvin, R. and Berthod, F. (2010) Collagen-based biomaterials for tissue engineering applications. *Materials*, **3**(3): 1863–1887.

Pasupuleti, V.K. and Braun S. (2010) *Protein Hydrolysates in Biotechnology*. Springer, New York.

Pati, F., Dhara, S. and Adhikari, B. (2010) Fish collagen: A potential material for biomedical application. Paper presented at the Students' Technology Symposium (TechSym), 2010 IEEE.

Paul, T., Jana, A., Das, A., Mandal, A., Halder, S.K., *et al.* (2014) Smart cleaning-in-place process through crude keratinase: An eco-friendly cleaning techniques towards dairy industries. *Journal of Cleaner Production*, **76**: 140–153.

Rammaya, K., Ying, V.Q. and Babji, A.S. (2012) Physicochemical analysis of gelatin extracted from mechanically deboned chicken meat (MDCM) residue. *International Journal of Food, Nutrition and Public Health*, **5**: 147–168.

Ranathunga, S., Rajapakse, N. and Kim, S.K. (2006) Purification and characterization of antioxidative peptide derived from muscle of conger eel (*Conger myriaster*). *European Food Research Technology*, **222**: 310–315.

Rathinaraj, K. and Sachindra, M. (2013) *Meat, Chicken and Eggs. Valorization of Food Processing By-Products* (pp 649-684). CRC Press, Taylor & Francis Group, Boca Raton, FL.

Rawdkuen, S., Benjakul, S., Visessanguan, W. and Lanier, T.C. (2007) Cysteine proteinase inhibitor from chicken plasma: Fractionation, characterization and autolysis inhibition of fish myofibrillar proteins. *Food Chemistry*, **101**(4): 1647–1657.

Rodziewicz-Motowidło, S., Śladewska, A., Mulkiewicz, E., Kołodziejczyk, A., Aleksandrowicz, A., *et al.* (2008) Isolation and characterization of a thermally stable collagen preparation from the outer skin of the silver carp *Hypophthalmichthys molitrix*. *Aquaculture*, **285**(1): 130–134.

Rouse, J.G. and Van Dyke, M.E. (2010) A review of keratin-based biomaterials for biomedical applications. *Materials*, **3**(2): 999–1014.

Russell, J.D., Dolphin, J.M. and Koppang, M.D. (2007) Selective analysis of secondary amino acids in gelatin using pulsed electrochemical detection. *Analytical Chemistry*, **79**(17): 6615–6621.

Saroat, R., Soottawat, B., Wonnop, V. and Tyre, L.C. (2004) Chicken plasma protein: Proteinase inhibitory activity and its effect on surimi gel properties. *Food Research International*, **37**: 156–165.

Saravanan, K. and Dhurai, B. (2012) Exploration on the amino acid content and morphological structure in chicken feather fiber. *Journal of Textile and Apparel, Technology and Management*, **7**(3): 1–6.

Schreiber, R. and Gareis, H. (2007). *Gelatine Handbook: Theory and industrial practice*. John Wiley & Sons, Germany, FL.

Se-Kwon, K. (2013) *Marine Proteins and Peptides: Biological activities and applications*. Wiley-Blackwell.

Shahidi, F. and Zhong, Y. (2008) Bioactive peptides. *Journal of OAOC International*, **91**(4): 914–929.

Sharma, V.K. and Kalonia, D.S. (2004) Polyethylene-glycol induced precipitation of interferon alpha-2a followed by vacuum drying: Development of a novel process for obtaining a dry, stable powder. *American Association of Pharmaceutical Sciences*, **6**: 1–14.

Shi, W. and Dumont, M.J. (2014) Review: Bio-based films from zein, keratin, pea, and rapeseed protein feedstocks. *Journal of Materials Science*, **49**(5): 1915–1930.

Singh, B.P., Vij, S. and Hati, S. (2014) Functional significance of bioactive peptides derived from soybean. *Peptides*, **54**: 171–179.

Song, N.-B., Lee, J.-H., Al Mijan, M. and Song, K. B. (2014) Development of a chicken feather protein film containing clove oil and its application in smoked salmon packaging. *LWT-Food Science and Technology*, **57**(2): 453–460.

Stadelman, W.J. (2000) Eggs and egg products. In Francis, F.J. (ed.). *Encylopaedia of Food Science and Technology*. John Wiley & Sons, New York. pp. 593–599.

Sukkwai, S., Kijroongrojana, K. and Benjakul, S. (2011) Extraction of gelatin from bigeye snapper (*Priacanthus tayenus*) skin for gelatin hydrolysate production. *International Food Research Journal*, **18**(3): 1129–1134.

Surh, J., Gu, Y.S., Decker, E.A. and Mcclements, D J. (2005) Influence of environmental stresses on stability of O/W emulsions containing cationic droplets stabilized by SDS-fish gelatin membranes. *Journal of Agricultural and Food Chemistry*, **53**(10): 4236–4244.

Thyagarajan, M.B., Barathi, M. and Sakthivadivu, R. (2013) Scope of poultry waste utilization. *IOSR Journal of Agriculture and Veterinary Science (IOSR-JAVS)*, **6**(5): 29–35.

Touch, V., Hayakawa, S. and Commins, T. (2009) Natural antimicrobial proteins: A review of current challenges and solutions for food applications. *Asian Journal of Food and Agro Industry*, **2**(1): 1–16.

Vasconcelos, A. and Cavaco-Paulo, A. (2013) The use of keratin in biomedical applications. *Current Drug Targets*, **14**(5): 612–619.

Viana, F.R., Silva, V.D.M., Delvivo, F.M., Bizzotto, C.S. and Silvestre, M.P.C. (2005) Quality of ham patte containing bovine globin and plasma as fat replacers. *Meat Science*, **70**: 153–160.

Villa, A.L.V., Aragão, M.R.S., dos Santos, E.P., Mazotto, A.M., Zingali, R.B., *et al.* (2013) Feather keratin hydrolysates obtained from microbial keratinases: Effect on hair fiber. *BMC Biotechnology*, **13**(1): 15.

Wanasundara, J.P.D., Pegg, R.B. and Shand, P.J. (2003) Value added applications for plasma proteins from the beef processing industry. *CMCA News*, pp. 10–15.

Wang, L., Yang, B., Du, X., Yang, Y. and Liu, J. (2008) Optimization of conditions for extraction of acid-soluble collagen from grass carp (*Ctenopharyngodon idella*) by response surface methodology. *Innovative Food Science and Emerging Technologies*, **9**(4): 604–607.

Wang, L., Yang, B. and Du, X. (2009) Extraction of acid soluble collagen from grass carp (*Ctenopharyngodon idella*) skin. *Journal of Food Process Engineering*, **32**(5): 743–751.

Wang, X. and Parsons, C. (1997) Effect of processing systems on protein quality of feather meals and hog hair meals. *Poultry Science*, **76**(3): 491–496.

Wang, Y.X. and Cao, X.J. (2012) Extracting keratin from chicken feathers by using a hydrphobic ionic liquid. *Process Biochemistry*, **47**: 896–899.

Warhadpande, R.M., Dutta, K.K., Mahanta, J.D. and Hazarika, M. (2010) Effect of incorporation of chicken blood plasma on physico-chemical properties of cakes. *Journal of Food Science Technology*, **47**(6): 693–696.

Wei, Z., Xu, C. and Li, B. (2009) Application of waste eggshell as low-cost solid catalyst for biodiesel production. *Bioresource Technology*, **100**(11): 2883–2885.

Whitaker, N.J. and Scheffner, M. (2003) Human papillomavirus-induced carcinogenesis and the ubiquitin-proteasome system. *Seminar Cancer Biology*, **13**(1): 59–67.

Windhorst, H.-W. (2011) The changing global egg industry. *Lohmann Information*, **46**(2): 3–7.

Yang, J.Z., Zhang, Q.G., Wang, B. and Tong, J. (2006) Study on the properties of amino acid ionic liquid. *Journal of Physical Chemistry*, **110**: 22521–22524.

Yang, S. and Blecha, F. (2008) Antimicrobial peptides and bacteriocins: Alternatives to traditional antibiotics. *Animal Health Research Reviews*, **9**(2): 227–235.

Yin, X.C., Li, F.Y., He, Y.F., Wang, Y. and Wang, R.M. (2013) Study on effective extraction of chicken feather keratins and their films for controlling drug release. *Biomaterials Science*, **1**: 528–536.

Yoo, S., Hsieh, J.S., Zou, P. and Kokoszka, J. (2009) Utilization of calcium carbonate particles from eggshell waste as coating pigments for ink-jet printing paper. *Bioresource Technology*, **24**: 6416–6421.

Young, R. and Lawrie, R. (2007) Utilization of edible protein from meat industry by-products and waste. *International Journal of Food Science Technology*, **9**(2): 171–177.

Yu, L., Sun, J. and Yang, Q. (2012) Ultrasonic-assisted enzymolysis to improve the antioxidant activities of peanut antioxidant hydrolysate. *International Journal of Molecular Science*, **13**(7): 9051–9068.

Zhang, F., Xu, S. and Wang, Z. (2011) Pre-treatment optimization and properties of gelatin from freshwater fish scales. *Food and Byproducts Processing*, **89**(3): 185–193.

Zhang, Z., Li, G. and Shi, B. (2006) Physicochemical properties of collagen, gelatin and collagen hydrolysate derived from bovine limed split wastes. *Journal Society of Leather Technologists and Chemists*, **90**(1): 23.

Zhou, P. and Regenstein, J.M. (2005) Effects of alkaline and acid pretreatments on Alaska pollock skin gelatin extraction. *Journal of Food Science*, **70**(6): 392–396.

17

Bioprocessing of Beef and Pork Meat Processing Industries, 'Waste to Value-Add'

Damodar Dhakal, Sajal Man Shrestha, and Anil Kumar Anal

Asian Institute of Technology, Klong Luang, Pathumthani, Thailand

17.1 Introduction

Food wastes are generated from every type of food processing industry, which cannot be avoided. However, we can utilize and dispose of the waste in such a way that it cannot only reduce environmental issues but also increase the returns of any company in terms of profits. Although it is not that easy to valorize the waste, because of its difficulty to easily decompose, pathogenic nature, highly prone to auto-oxidation, biologically instable in nature and also highly affected by the enzyme actions. All this things are directly related to sustainable development and also production.

The leading problem that the processors have been facing is that these millions of tons of processing wastes are produced every day, but the disposal of such wastes remains an issue. This is because processing and disposal of these by-products and wastes experience significant cost to processors for their treatment, which have to comply with strict regulations. Even though varies technologies has been introduced for the treatment and reduction of these by-products, it is believed that most often they are under-utilized and represent a treasured resource if treated in an appropriate manner. Due to its strong economic potential and a wide range of importance, it is no longer advisable to discard these by-products and wastes. It is desirable to process all by-products into valuable products associated to human foods, animal feeds, pharmaceuticals or fertilizers and even biodiesel generation. Newer technologies and industries have been developed for by-product utilization, yet they face a larger problem during their application due to complications in combining the use of innovative technology, processing methods and suitable marketing of the product.

Food Processing By-Products and their Utilization, First Edition. Edited by Anil Kumar Anal.

Table 17.1 Examples of traditional consumption of different edible meat by-products (adapted from Toldrá *et al.*, 2012)

By-Products	Traditional Consumption	Examples of Countries
Liver	Splinantero, Sheep liver, Paté Almondega and Faggots	Greece, Turkey, Iran, Portugal, UK
Spleen	Pani ca meusa (bread with spleen)	Italy
Heart	Cooked and diced	South America
Kidney	Kidney pie	UK
Brain	Sesos	South America, Spain
Tongue	Boiled or marinated Ingredient in meat products	South America, Spain
Testicles	Criadillas	Spain
Tripe	Blóomör, slátur, Dinuguan, Korouch, Callos, Cold appetizer, Haggis and Pieds et paquets	Iceland, Philippines, Spain, Lebanon, China, Scotland, France
Intestines	Casings for sausages	Mediterranean
Blood	Morcilla sausage, Black pudding and Thuringer blood sausage	Spain, UK, Germany
Feet	Pork's feet and Khash	Spain, Armenia
Ear	Pork's ears and Khash	Spain, Armenia
Tails	Rabo de toro (Bull's tail) and Tail pork	Spain, Mediterranean
Bones	Gelatin soups and Osso bucco	Mediterranean, Italy
Lard, tallow	Lard for cooking and covering hams	Northern Europe, Mediterranean

According to the United Nation Environment Programme's 2000 report, China and the USA are the world's leading producers of beef and pork, whereas Brazil, Mexico, the Russian Federation and a number of Western European countries are also large producers (Roupas *et al.*, 2007). Although information on volumes of generated waste is variable, it depends on the type of meat being processed. Table 17.1 discusses the various traditional forms of meat consumption of different edible by-products around the globe.

17.2 Different By-Products and Waste coming from Beef and Pork Meat Processing Industries

In the meat processing industries there are various fragments which are considered as inedible chunks, such as the skin, blood, bones, fatty tissues, meat trimmings, horns, hoofs, feet and other internal organs. Slaughterhouses, as well as other meat processing industries including retailers and renderers, produce a substantial amount of solid and liquid by-products. These unwanted wastes can be used as animal feed, fertilizer or fuel and even in pharmaceutical and cosmetic activities (Figure 17.1).

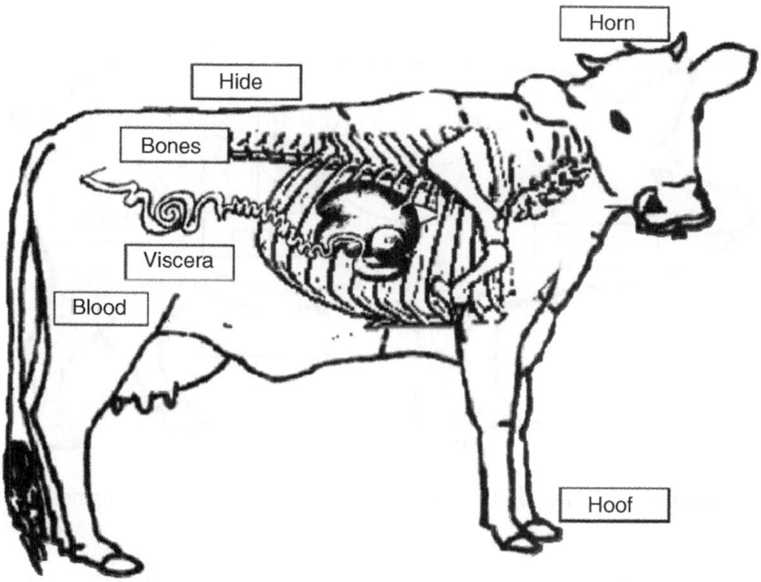

Figure 17.1 The waste by-products from beef processing industries.

These by-products are the potential for the generation for large quantities of solid wastes and wastewater with a biochemical oxygen demand (BOD). The amounts of wastewater generated and the contaminants depends on the type of meat being processed. For example, the processing of gut has a significant impact on the quantity and quality of wastewater generated (as measured by levels of BOD and chemical oxygen demand, COD). The wastewater from a slaughterhouse usually comprises blood, manure, fat, hair and bones. The wastewater may be at a high temperature and may perhaps contain organic material and nitrogen, while some pathogens such as bacteria, parasite eggs and amoebic lumps might also be present. The pesticides present residues from the treatment of animals or their feed (Figure 17.2).

17.2.1 Skin

Gelatin and collagen can be extracted from pork skin. It is believed that pork skin was the only source of gelatin during the past. Now gelatin, an animal protein, can be obtained by a controlled hydrolysis of the fibrous insoluble collagen that is present in the bones and skin, which is usually generated as waste during animal slaughtering and processing. Hence the technologists have come up with newer ideas for the outsourcing of these wastes, which helps in its production and in obtaining a moderately low cost and excellent functional and filmogenic properties (Hanani *et al.,* 2012).

The animal skin containing gelatin is composed of an exceptional sequence of amino acids and characteristics. Its key feature is the repeating sequences of glycine, proline and hydroxyproline amino acids, which is the main reason for its ability to form gels. Also, in the UK, pork skin is used to make a popular snack food called 'pork scratchings', made by immersing, boiling, drying and finally frying the pork skin. Collagen

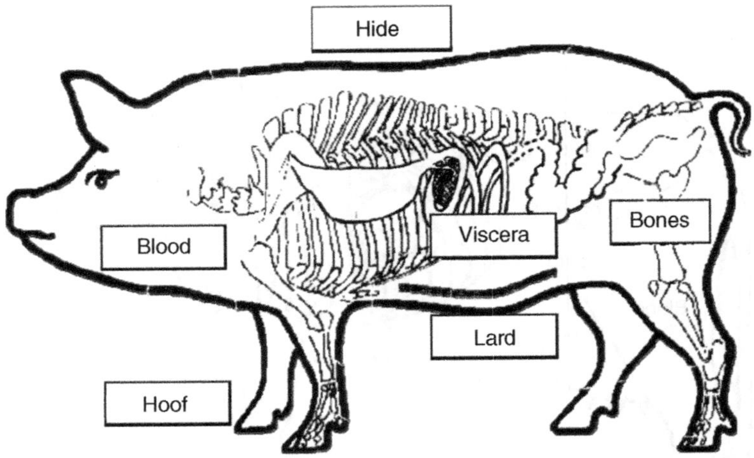

Figure 17.2 The waste by-products from pork processing industries.

can be used as a filler in meat products. Because of the 'melt in mouth' properties, the gelatin is used in dessert products. Also, both the gelatin and collagen extracted from pork skin has a lot of applications in pharmaceuticals and surgery (Hanani *et al.*, 2012).

17.2.2 Bones

Bone is one of the most cherished by-products that is found in the meat processing industries. The bone marrow inside the bone is often used as a food delicacy. The bone is well known for its use in soup and gelatin. Hence, newer technologies are being introduced in order to extract more meat from the bones, therefore beef, pork or lamb meats are produced by mechanical deboning produces of the tissues, which are known as 'mechanically separated', 'mechanically deboned' or 'mechanically removed'. Now such meats are popular in many developed countries (Jayathilakan *et al.*, 2012).

In recent years, meat and bone meal (MBM) has been widely recommended and also used as animal nutrition, as a rich source of protein in terms of proteinaceous feeds due to its high value contents and the availability of essential amino acids, minerals and vitamin B_{12}. Therefore, MBM and its related rendered protein possessions are potential factors in other applications other than animal feed, including use as a fuel or a phosphorus fertilizer (Jayathilakan *et al.*, 2012). Proper utilization and processing of this by-product is essential in terms of nutrition and economic benefits. Newer technologies are being introduced and innovative ideas are also essential in order to gain the maximum utilization from these by-products.

17.2.3 Hides and Hooves

Leather is one such by-product, which is obtained from the animals which have been slaughtered in meat processing industries or found dead. Pig, cow and buffalo are the most common domestic animals preferred for leather manufacture. We should understand that these hides and hooves from the animal provide an economic rout between the factory farming and the leather business, because the farmers sell these by-products in order to minimize the waste and maximize the revenue and profits. A direct trade with these two parties may reduce the waste elements, but yet it is still considered that the whole process has been underutilized.

Hides signify a higher weight content of the live animal, where the percentage may vary from 4% to 11% (e.g. cattle: 5.1– 8.5%, average: 7.0%; sheep: 11.0–11.7%; swine: 3.0– 8.0%). Due to their significant values, hides and skins are considered one of the most valuable by-products. Some of finished products that can be obtained from hides of cattle and pigs are leather shoes and bags, rawhide, athletic equipment, reformed sausage casings and cosmetic products, sausage skins, edible gelatine and glue (Benjakul *et al.*, 2009). Biopolymers such as collagen, which has high added value and can generate economic value, are unused and unprocessed. Collagen proteins are enormously important structural proteins present in connective tissues. Collagen protein has numerous benefits such as a wide raw material resource, it is renewable, degradable and biocompatible and also its structure is adjustable. It has a high demand in healthcare products, cosmetics, medical ingredients, feedstuff, animal feed, and even paper.

Proper processing of waste by-product such as hides and hooves may complete the demands of the leather production and increase the collagen protein supply to various other industries.

17.2.4 Horn

Horn is one by-product that is mostly underutilized, even though it is of high value. Horn occurs in animals belonging to the bovid family, which includes cattle, sheep and waterbuck. The horn consists of a keratinous sheath overlapping a bony core. The keratin comprises of hard α-keratin and has a crystalline fiber phase along with an amorphous matrix phase. The crystalline phase present in horns comprises microfibrils with an α-helical structure, while the amorphous phase is made up of microfibrils with a non-helical structure and other morphological components. The keratin fibers are significantly parallel to the growth direction and are stacked against each other to form a lamellar structure. Keratin fibers are also embedded in an amorphous and non-fibrous protein matrix through many bonds, such as disulfide bonds, hydrogen bonds, van der Waals forces and ionic interaction. Therefore, horn keratin is an excellent biological model having a hierarchical structure that ranges from the nanometer to micrometer scale. As a result, it has good performance, such as high toughness, stiffness and strength (Zhang *et al.*, 2015).

Due to the stable three-dimensional structure of horn, it forms keratin with a higher chemical stability in physiological environments that results in its non-degradability, which delays its processing activity. We have come up with various extraction methods for soluble keratin, including the oxidation and reduction methods. The soluble keratins are the considered to be advantageous in several fields, such as wound care, tissue reconstruction, cell seeding and diffusion, and drug delivery. Since it does not have a three-dimensional structure, it has poor mechanical properties and processing abilities. Therefore, the practical applications of soluble keratin are forced to be restricted.

Hence, the goal is to prepare a keratin with both good mechanical properties and degradation (Zhang *et al.*, 2015).

17.2.5 Blood

One of the most generated waste materials in a slaughterhouse is animal blood, which is usually washed away or underutilized. Blood not only consists of water but also protein, lipid, carbohydrate and minerals, which are obtained from bovine and porcine sources and are made up of the cellular and plasma fractions (Toldra *et al.*, 2012).

Blood-derived products have vital ingredients, which provides value addition in the food industry and as dietary supplements. Gelation, foaming agents and emulsification are some of the qualities of blood proteins. For instance, fractionated plasma proteins such as immunoglobulins, fibrinogen and serum albumin, are recommended as an additional factor in food and feed ingredients, because of their good gelation and emulsification properties. Also, some show a good cross-linking ability between proteins and protease linkage and hence can be used in the manufacturing of pastas to improve protein addition. The meat industry can also benefit from the blood's cellular fraction, as it may be used as a color enhancer in sausages. It is a better alternative to avoid sensory adverse effects by removing the hemoglobin to produce globin, also known as decolorized blood that exhibits relevant technological functions for replacing fat in meat products (Toldra *et al.*, 2012).

Various types of devices have been developed to carry out blood hemoglobin hydrolysis with simultaneous extraction of the hydrolyzate. We have also seen the membrane-ultrafiltration reactors that are made for such purposes and more recently, modern-day technologies have successfully built immobilized reactors with pepsin immobilized as a mineral support like alumina (Toldra *et al.*, 2012).

17.2.6 Lard

Lard is considered one of the chief by-products of the pork processing industry, which is simply the fat which has been rendered from the pork. The most common method of gaining lard is either by the dry or wet rendering method. The reason why lard is so valued is because of its role in the deep-frying industries, since the frying medium is also a concern for various factors. Previously, lard and other animal fats were used as frying mediums for a number food products. However, we must not forget the fact that it comes with a higher content of saturated fatty acids and a lower percentage of polyunsaturated fatty acids, which is of great health concern and therefore this by-product has lost its popularity as a frying medium. Also, the cholesterol percentage in these kinds of animal fats is considered bad by the public, as consuming high cholesterol is a major risk factor for coronary heart disease. This does not mean that we no longer use this by-product, as we can make some changes to still make use of it.

Due to health implications and firm religious prescriptions in many countries, the lard detection methods have been regulated very strictly. The primary detection techniques were based on GLC and HPLC, and focused mainly on raw or fresh products and some processed products (Marikkar *et al.*, 2003). Hence the need of detection and the utilization of the by-product is a necessity and there is a must for newer and modern technologies to do so.

17.2.7 Viscera

Viscera are the animal organs which are present in the cavities of the body, especially present in the abdominal cavity. They are considered a good source of protein. But the problem that we have to deal with is the high moisture content and issues regarding handling and storage. Viscera are usually recovered as a consumable product such as from heart or liver, while it can also be separated for inedible rendering or processing from parts like the lungs.

The animal's stomach containing digested feeds is estimated to range from 27 to 40 kg. The usual practices that are carried out regarding the waste management of these paunches are:

1. *Total dumping*: Entire paunch contents recovery is flushed away into the sewer.
2. *Wet dumping*: The paunch contents are washed out and the wet slurry is partitioned in the presence of gross solids, which are then removed.
3. *Dry dumping*: The paunch contents are dumped for subsequent rendering or for disposal as solid waste without needless water flushing.
4. *Whole paunch handling*: The entire paunch may be removed, intact, for rendering or disposal as solid waste (Verheijen *et al.*, 1996).

Intestines may be rendered directly, or hashed and washed prior to rendering. For the processing of intestines, de-sliming prior to thorough washing is necessary.

17.3 Valorization of Beef and Pork Meat Processing Waste

17.3.1 Collagen

The by-product of pork, skin and bones can be used to extract the collagen, having higher values which help to increase the profit of the pig industry. Collagen is a fibrous protein with branched chains which are found in the skin, tooth, blood vessel, ligament, cartilage, tendon and bone of the animal. Different organs are supported and protected by the collagen protein, which also helps to bind tissues together. Being renewable, biocompatible and degradable, collagen can be used as a raw material for wider applications.

Collagen protein is the most widely present proteins in mammals, which constituents 25–30% of total body protein and 6% of live body weight. Because of its wider applications from feed to food, from cosmetics to surgery, its research is always worthwhile. Collagen exists in the body as an insoluble macromolecular structure with proteoglycan and glycoprotein (Hua Yang, 2014).

17.3.1.1 Extraction and Characterization For collagen protein preparation, we use the selection, preparation of sample, extraction, and separation and purification processes. Extraction can be done by either chemical means, rDNA synthesis methods or enzymatic synthesis methods (Hua Yang, 2014). By the salting method, collagen is separated by the use of different concentrations of salt solution of Nacl, Tris-HCL, phosphate or citrates. Another way to extract the hydrolysed collagen is the alkali method. For the general alkali method, the extracted collagen fraction consist of

25.9%, 10.3% of ash and 23.2 mg/kg of chromium (Hua Yang, 2014). Also, acetic acid, citric acid and hydrochloric acid can be used in the extraction process. All the above methods, including enzymatic methods (pepsin, papain, trypsinase), have better results with less collagen destruction. Pepsin was found to be best for maintaining the structure of collagen, with around 100 KDa molecular weight and whitish in color or lyophilized, while the rate of reaction was slow. Reaction conditions should be maintained in such a way that collagen fiber is not damaged during the process of extraction and purification (Hua Yang, 2014).

17.3.1.2 Functional Properties Swelling and solubility deals with the hydration properties of collagen. Two groups, such as gelling behavior, thickening, water-binding capacity, gel formation and texture deal with identical types of properties, whereas emulsion, foam formation, adhesion, cohesion, stabilization, colloidal and film-forming capacity deal with the other types of properties of collagen (Schrieber and Gareis, 2007).

17.3.1.3 Applications of Collagen Collagen has enormous applications in the food, cosmetic and biomedical sectors. Collagen can be a wider raw materials resource, which is renewable, biodegradable and also biocompatible. With having such robust characteristics, it can be employed in health-oriented products, food, feed, cosmetics and surgery (Hua Yang, 2014). It is used as one of the prominent constituents of soaps, body lotions, shampoos, cosmetics, and food grade proteins supplements. In the medical field, it can be used in surgery, neurology, orthopedics, ophthalmology and urology. Collagen proteins, as antibodies, can be used as powerful tools for highly sensitive and specific assays in the laboratory. Homostatic agents such as sutures, replacements of tissue and casings for sausages, are the various applications of insoluble collagen in the form of dispersion. Moreover, it can be used as a micro-carrier in the form of solid support in enzyme production (Meena *et al.*, 1999).

17.3.2 Gelatin

Traditionally, gelatins are made from pork and beef. Forty-six percent derives from pig skin, 29.4% from hides of bovines and 23.1% from bones of cattle and pig (with only 1.5% from the fish) are the major raw materials for gelatin production (Gómez-Guillén *et al.*, 2009). From the waste of pork and beef, such as skin, hides, bones etc., by the process of controlled hydrolysis of collagen, gelatin can be obtained. Gelling biopolymer-gelatin is used to provide the elasticity, stability and viscosity in foods. More economic, excellent functional and filmographic properties are produced by this method (Eastoe and Leach, 1977).

Gelatin possesses a unique sequence of amino acids with distinct characteristics. The repeating sequence of glycine, proline and hydroxyproline amino acids are responsible for the formation of gel. Collagen and gelatin have almost the same composition of amino acids, which can be characterized by Gly-X-Y triplets, where X indicates proline and Y indicates hydroxyproline (Eastoe and Leach, 1977). Gelatins, having different molecular masses ranging from 15 to 300 kDa, are extracted during the extraction process, where the melting point and the strength of gel are the important physical properties (Zhou *et al.*, 2006).

17.3.2.1 Extraction and Characterization Pork skin is the significant raw material source for gelatin production and also the first one raw material source for gelatin extraction up to the 1930s. During the extraction process, transformation of collagen into gelatin leads to some changes in the composition of amino acids. Moreover, the fibrous collagen helixes will lose their conformation during the process of heating, whereas during cooling partial recover takes place (Duconseille *et al.*, 2015).

Gelatin is produced by the hydrolysis of collagen under controlled conditions. The fresh raw waste from pork (hides/bones/skin) is taken and then is followed by three major steps. Primarily, the non-collagenous materials are eliminated from the raw material, then it is subjected towards the hydrolysis of collagen and finally recovery takes place (Jayathilakan *et al.*, 2012).

17.3.2.2 Functional Properties The melt-in-mouth property of gelatin is used in food products. Gelatin mixtures help to make smoother desserts with enhanced flavor (Gómez-Guillén *et al.*, 2011). Introduction of the gas phase into gelatin-based products improves the texture and appearance of food products such as fruit jellies. Moreover, energy consumption can be reduced by the application of less calories of air-type particles in food, whereas bubbles of air, nitrogen or helium weakens the gel structure with opaque and distinct mechanical characteristics (Zúñiga and Aguilera, 2009).

Gelatin hydrogels are formed by the cross-linking of covalent matrices through swelling. Swelling, while maintaining the shape, is governed by the two hydrophilic and hydrophobic properties of gelatin. Gelatin hydrogels are inexpensive, biodegradable and biocompatible with different combinations. The balance in charge related to the pH, influencing the ionic and electrostatic interactions in gel formation (Farris *et al.*, 2009).

17.3.2.3 Applications Gelatin is a main constituent of jellies, meat pies and aspic, because of its melt-in-the-mouth properties. It is also used as a stabilizer in ice-cream and other desserts. As a protective colloid in ice-cream, cream pies and yoghurt, gelatin is used because of its high blooming characteristics. Moreover, gelatin inhibits the formation of ice crystals in ice-cream and also recrystallization of lactose during storage (Gómez-Guillén *et al.*, 2011).

Approximately 5% of the total gelatin is used in the pharmaceutical industry (Hidaka and Liu, 2003), for making outer coverings of capsules, an ingredient for protective ointment, sterile sponges, etc. Also, from immortal times, it has been using for food and cosmetic applications. The water absorbing ability in the form of gelatin casings has widened the application of gelatin in different fields of pharmacy, medicine, food and agriculture (Gómez-Guillén *et al.*, 2011).

17.3.3 Blood Products

Blood is sterile when it is inside the body of healthy animal. The protein content of blood is high, with a high content of amino acids; 2.4–8% of the live animal body weight is composed of blood; 3–4% of blood can be recovered from the pig and beef. Blood has a dark color with less palatability, so its plasma has a higher value because of its more functional properties (Jayathilakan *et al.*, 2012).

Blood from the slaughterhouse, as a waste from the meat animal, has a huge potential to obtain the various range of good ingredients useful for meat products.

Emulsification, texture and coloration effect can be best obtained from blood. The main focus should be on more improved techniques that provide maximum ability to recover and economically exploit the blood (Parés *et al.*, 2014).

17.3.3.1 Extraction and Characterization Blood is composed of a cellular fraction of red blood corpuscles, white blood corpuscles and platelets and also the plasma fraction (Hsieh, 2011). Blood proteins found in the plasma contain the relevant important properties, with gellifying, foaming and emulsifying having boosted blood products in food and dietary supplements (Hsieh, 2011). The thrombin and fibrinogen are found together on the meat surface, then enzyme of thrombin converts the soluble fibrinogen into insoluble polymer of fibrin, forming a half-staggered structure, namely protofibril that aggregates into fibers giving rise to a 3-dimensional fibrin cloth network (Kenny *et al.*, 2008). Thus, the final network will give modified physicochemical and textural characteristics in meat emulsions (Herrero *et al.*, 2007). The cellular fraction can be used to enhance the color of sausages (Hsieh, 2011). Different types of reactors, such as membrane ultrafiltration (Kapel *et al.*, 2003) and immobilized reactors with pepsin immobilized on mineral supports (Ticu *et al.*, 2005), have been designed for the hydrolysis of blood hemoglobin.

17.3.3.2 Applications Blood of the animal is one of the major edible by-products that contains high protein and also the iron-heme (Wan *et al.*, 2002). Blood can be used as an emulsifier, clarifier, nutritional component in blood meal-like puddings, cakes, curds etc. Blood sausages, biscuits, blood puddings and bread etc. are made from the blood of animals in Europe, whereas blood puddings, blood cake and blood curd are taken in Asia as a meal (Ghosh, 2001). Other applications, such as fertilizers, binders and feedstuffs are used as non-food items. Plasma helps to increase the gel strength of protein (Silva and Silvestre, 2003) and has several therapeutic applications as it contains fibrinogen, serotonin, immunoglobulins etc. (Young and Lawrie, 1974). The purified albumin can be used to replenish fluid loss in animals. It can be used in antibiotic sensitivity tests, nutrient for tissue cultures and microbial growth medium etc. (Kurbanoglu and Kurbanoglu, 2004).

17.3.4 Bioactive Peptides

Large amounts of peptides and free fatty acids are generated during meat processing, due to the proteolytic activity (Calpains/Cathepsins). Furthermore, peptides can be obtained by the hydrolysis of trimmings, organs, collagen and haemoglobins etc., by the action of bromalain, papain, pronase etc. (Vercruysse *et al.*, 2005).

17.3.4.1 Extraction and Characterization Among hundreds of peptides released, our interest is in those having the activity of antihypertensive and antioxidant. The inhibition of angiotension I-converting Enzyme (ACE), which takes part in renin-angiotension system where angiotension I is converted into angiotension II, helps to constrict the arteries and increases blood pressure (Ahhmed and Muguruma, 2010). The thermolysis hydrolysis of pork meat proteins contains two pentapeptides and six tripeptides with prominent anti-hypertensive activity (Arihara *et al.*, 2001). The papain hydrolysis of pork myofibrillar protein shows the proper antioxidant and free-radical scavenging peptides (Arihara *et al.*, 2001) (Figure 17.3).

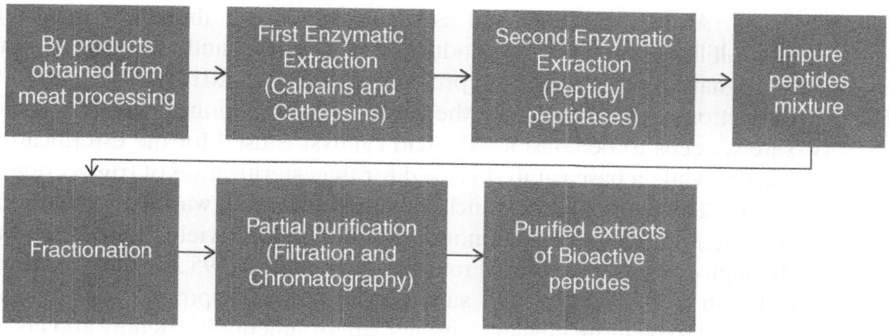

Figure 17.3 Extraction of bioactive peptides (Toldrá *et al.*, 2012).

17.3.4.2 Functional Properties Conjugated linoleic acid, taurine, L-carnitine, carnosine, glutathione, anserine and creatine are the different bioactive substances present in meat that have various physiological properties (Arihara *et al.*, 2001). Hygienic meat products can be designed through the emphasis on several physiological activities. Manipulating the animal feed helps to improve the animal products. Conjugated linoleic acid and carnitine were found to increase the modification of animal feed (Mir *et al.*, 2004). As a result, meat and meat products having different characteristics are produced.

17.3.4.3 Applications Antioxidant, antimicrobial and anti-proliferative activities are shown by the bioactive peptides derived from the animal meat under *in vitro* conditions, whereas the anti-hypertensive effects are exhibited under *in vivo* conditions. The oligopeptide which is presents in the capsule for lowering blood pressure is converted into the active form with the help of a digestive enzyme through the process of ingestion (Hartmann and Meisel, 2007). Peptide bioavailability will determine the physiological properties *in vivo*. The overall mechanism depends on the factor such as peptide resistance towards the hydrolysis governed by peptidases of serum and intestinal tract, and also absorption towards the epithelium of the intestine (Vermeirssen *et al.*, 2004).

17.3.5 Biodiesel

Biodiesel is gaining popularity as an alternative source of fuel for petroleum diesel, as it is less polluting and also can reduce price-hiking of petroleum products. Biodiesel has been found to produce less emissions of carbon monoxide, sulphur and hydrocarbons than those of fossil-based diesel (Nelson and Schrock, 2006). Having different advantages such as low prices, energy value, economic importance and the environmentally-friendly nature of tallow and lard, it can be widely used as a feedstuff for the biodiesel production. Also, less risk of bovine spongiform encephalopathy will exist if we go for biodiesel production rather than other traditional end users, which will lead to restraints in the trading of fat of animals (Nelson and Schrock, 2006).

Fatty acid methyl esters are a major component, for the biodiesel can be easily produced from animal fats and vegetable oils. Comparisons between vegetable oil and animal fats found that the vegetable oil is more costly than animal fat and used-cooking oil. Animal waste fat and used-cooking oil are abundantly available and can easily fulfill the need for the production of biodiesel and also help to solve environmentally-harmful waste disposal problems (Shin *et al.*, 2011).

For the production of bio-diesel from the waste oil catalyzation process, the inclusion of two steps needs to be considered. Acid catalyst is used for the esterification of free fatty acids, while a base catalyst is used for the esterification of triglyceride in animal waste oils. The overall process including neutralization, washing, purification and separation etc., generates a huge amount of water and low yield of biodiesel, but comprises the high cost of processing. From the same method, 95.3% pure biodiesel was produced with a 66.4% yield. The supercritical methanol process for biodiesel production is found to be more efficient, simpler, environmentally-friendly and also is not affected by the presence of water and FFA (Shin *et al.*, 2011).

17.3.5.1 Extraction and Characterization Waste from the pork and beef processing industries, such as slaughterhouse animal fat (SHAF), pork lard, beef tallow and oil from flesh, are taken as a raw materials for biodiesel production. Temperatures of 100–110°C are required to melt the fats and then solid residues are further separated after removal of water (Chakraborty *et al.*, 2014). Before the transesterification reaction, a 110°C temperature treatment is required to dry the sample. For pork skin, a high pressure cooking technique is employed to extract the fat.

Among the solvent extraction methods, the supercritical CO_2 technique is an efficient process where continuous fat is extracted from animal fat (Taher *et al.*, 2011). From the freeze-dried meat sample, 87.4% fat is extracted under the optimized conditions of 45°C, 3 ml/min of CO_2 at 500 bar pressure (Al-Zuhair *et al.*, 2012). In that study, it was found that increasing the supercritical CO_2 from 3 to 5 ml caused a linear decrease in the yield of extraction. However, fat yield was increased while increasing the pressure from 300 to 500 bar at 55°C (Al-Zuhair *et al.*, 2012).

Transesterification with a suitable catalyst has a very crucial role to play in the conversion of biodiesel from feedstock. For the feedstock with FFA in low amounts, NaOH and KOH are taken as catalysts from a homologous base, whereas the higher quantity of FFA results in the saponification reaction. Although the acid catalyst has a 4000 times slower reaction rate, it can help to prevent soap formation (Chakraborty *et al.*, 2014). On the other hand, a higher practicability and lower process complexity can be achieved in the acid catalyzed method. However, the base catalyzed process was found to be less costly in terms of fixed capital, but the overall cost estimation has proven that the acid catalyzed method is much more cost-effective, as it consists of only one step in the transesterification process (Chakraborty *et al.*, 2014).

Modification of fuel and engine helps to mitigate the emissions of NO, and as a result biodiesel can be employed as an alternative to petroleum diesel. Qualitative and quantitative analysis can be achieved through infrared techniques, which is not only rapid but also low-cost and reproducible. Different properties of fuel and parametric sensitivity, analytical tools and operating conditions can be assessed by the same technique (Meher *et al.*, 2006).

In comparison with the acid or base catalytic processes, the enzymatic catalytic method is found to be more eco-friendly. The lipases from *Mocor miehei*, *Bacillus subtilis* and *Pseudomonas cepcia* were found to more effective in the transesterification of feedstocks having high amounts of FFA (Akoh *et al.*, 2007). The specificity with minimum side reactions are the advantages of this method, whereas the higher cost of enzymes and the presence of residual enzyme are the disadvantages. It was found that the transesterification process catalyzed by lipases is a time-consuming process, with less yield but is more environmentally-friendly (Al-Zuhair *et al.*, 2012).

17.3.5.2 Applications The frequent increase in the price, emissions of the pollutants from combustion and continual diminishing of fossil fuels makes biomass an alternative source for bio-diesel generation. The best way to replace the diesel fuel is bio-diesel in the diesel engine, which has better efficiency than other petroleum diesel. It has a huge potential for the engine with compression-ignition system. Also, the relatively high heating value of diesel makes it a promising fuel, which we can now consider. The higher heating values (HHV) of gasoline is 46 MJ/kg, petroleum 42 MJ/kg and petrodiesel-43 MJ/kg and for biodiesel it is 39 MJ/kg, which is higher than the 32–37 MJ/kg of coal. Biodiesel, having 100% fuel, is also termed as a neat fuel or B100. BXX is a term used to define the blends of biodiesel with petro-diesel (Atadashi *et al.*, 2010).

The entire combustion process is facilitated by the presence of oxygen, which helps to decrease the oxidation potential. The homogeneity from the structural oxygen helps to enhance the combustion efficiency. Also, it has higher efficiency of combustion than petro-diesel and the overall injector is found to be considerably low (Atadashi *et al.*, 2010). No difference was found when visual inspection was carried out between petro-diesel and biodiesel. Carbon-dioxide emission from petro-diesel is higher than the biodiesel fuel (Atadashi *et al.*, 2010).

17.3.6 Keratin

Keratin is a fibrous protein present in by-products of animal meat, such as hair, bristle, hooves, horns, hides, skin etc. It is not suitable for human consumption, but has a wide application for cosmetic purposes. It is insoluble and also resistance against protease enzymes. The abundant disulsulphide bonds present in α-keratin make keratin processing a difficult task (Kornillowicz-Kowalska and Bohacz, 2011). Keratinolytic microorganisms, such as prokaryotes and fungi having a keratinophilic nature, are responsible for the biodegradation of keratin from animal meat-processing waste. These microorganisms colonize the different waste, such as hides, hooves, hairs etc., which are both parasitic and saprophytic in nature. Generally, they are present in soil and other habitants, which are keratin-rich like nests of birds, waste from animals etc. (Kornillowicz-Kowalska and Bohacz, 2011).

Keratins are also known as scleroproteins, produced by the epithelial cells of aves, reptiles and mammals, which helps to give mechanical strength and protects cells. The intercellular system of both fibrous protein and microtubules governed by the cytoskeleton is also composed by keratin, cytoskeleton and filaments (Kornillowicz-Kowalska and Bohacz, 2011). The filaments keratin is composed of 15 acidic, 15 neutral and 15 alkaline polypeptides. Hair, horns, hooves, hides, beaks, feather etc. providing the mechanical strength, are the sources of keratin that are present in the keratinized epidermal cells and also in corneous skin (Kornillowicz-Kowalska and Bohacz, 2011).

Keratin proteins are very resistant to physic-chemical factors such as insolubility in water, light acid and bases, organic solvents and also even sensitivity to some proteolytic enzymes. Among the amino-acids, high cysteine in keratin is the most distinguishable characteristic, which differentiates keratin from collagen and elastin

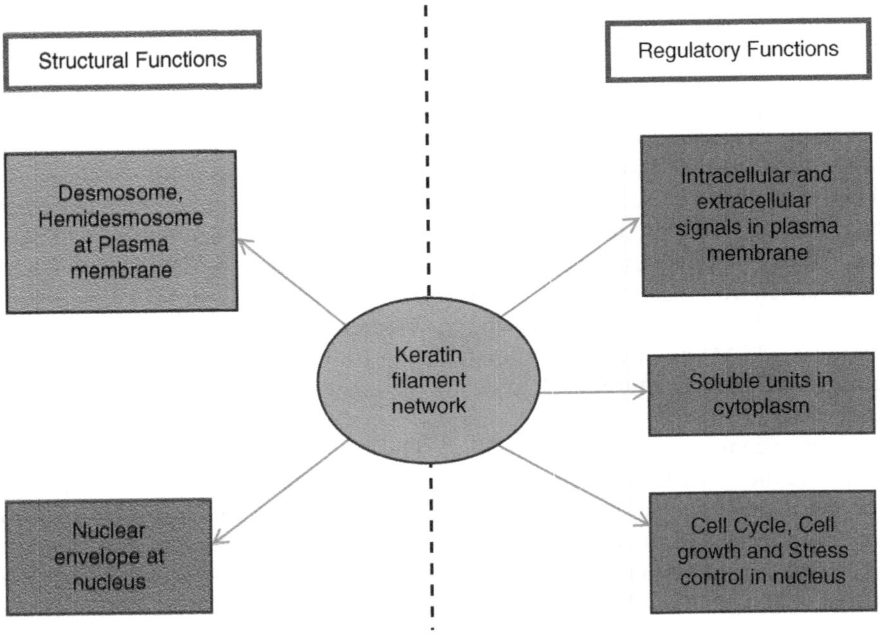

Figure 17.4 A schematic of a wool fiber drawn by Bruce Fraser and Tom MacRae (Copyright CSIRO Australia 1996. Reproduced with permission from The Lennox Legacy. Published by CSIRO Publishing, Melbourne, Australia) (Hill *et al.*, 2010).

(Figure 17.4). Enzymatic degradation is also resisted by the numbers of disulphide bonds binding the peptide chains. Keratin-polypeptide chains occur as a pleated sheet with a β-sheet and α-helix (Fraser *et al.*, 1972). Keratin can also be described as soft, having less than 2% cysteine and hard, having up to 22%. Soft keratin is found in the epidermis and hair core with weak cross-linking and also less resistance to different chemical factors (Fraser *et al.*, 1972). Hard keratin is present in horns, hides, hooves, beaks and skin, and has a diversified morphological structure and also high resistance to physic-chemical factors (Bruce Fraser and Parry, 2003; Fraser *et al.*, 1972).

17.3.6.1 Extraction and Characterization The easiest and cheapest method for the hydrolysis of feather, hides, hooves and horns into soluble proteins like keratin, is of utmost importance, as it can help to recover the rare serine, cysteine and proline amino-acids. Keratinase is a kind of enzyme that helps to hydrolyze the substrate for keratin extraction (Lateef *et al.*, 2010). Similarly, *Bacillus cereus* has a high keratinolytic activity, which can be used as important tools in the biodegradation of keratin substrate (Lateef *et al.*, 2010).

The bovine hooves and horns were taken into account for keratin extraction after removal of debris and properly washing, drying and subjected to chopping and milling into 60 mesh particle sizes. The sample was sterilized at 121°C for 15 min by keeping it in different concentrations of $NaNO_3$, NaCl, KH_2PO_4, $MgSO_4$, $FeSO_4.7H_2O$, $CaCO_3$ and agar-agar, where 0.05 g/l of sterile nystatin was added to inhibit the growth of fungi (Lateef *et al.*, 2010). The substrate degrading bacterial colony was isolated from the soil from where the previously the keratin substrate was taken. The overall activity of keratinase was determined by the modified Cheng *et al.* (1995) method, where 0.5 ml crude enzyme was incubated in a 1.5 g feather powder with 2 ml of phosphate buffer of 7.5 pH. Similarly, azocasein assay was used to determine the proteolytic activity (Sarath *et al.*, 1989). The keratinolytic activity of enzyme for horn substrate was found to be in the range of 10.9–63.1 U/ml for the cultivation for 120 h, whereas at 72 h, its activity was found to be at a maximum (Lateef *et al.*, 2010).

For the extraction of keratin, 20 g of agricultural horn-hoof was taken with 14% of commercial nitrogen and then washed using sodium dodecyl sulphate aqueous solution, followed by distilled water rinsing, washing with acetone, again distilled water rinsing and further conditioned for 24 h at RH of 65% and temperature of 20°C. Three percent yield of keratin was gained from horn-hoof with 8 M-Urea, 0.1 M-DTT and 5 mM-Tris, with pH 8.6 using 5 N-HCL and shaking for 4 h under nitrogen atmosphere. Whereas, 10% yield of keratin was gained from horn-hoof when 8 M-urea, 0.5 M-metabisulphite and 0.05 M-SDS, with pH 6.5 using 5 N NaOH and shaking at 65°C for 5 h (Zoccola *et al.*, 2009). The sulphitolysis method at 65°C was found to be the more effective sample (Zoccola *et al.*, 2009).

The dried protein in keratin materials contains the amino acids, having cysteine and a small amount of methionine (Table 17.2). Maximum amounts of amino acids such as glutamic acid, aspartic acid, arginine lysine and leucine, contribute to α-helix assembling in low sulphur proteins of horn and hoof (Zoccola *et al.*, 2009).

α-Keratin of hide and hoof, having a low sulphur group, is responsible for the varying molecular weight bands in the patterns of electrophoretical separation. It also resembles the hardness of tissue with respect to the function of original tissue (Zoccola *et al.*, 2009). The lesser intensity of peak for water evaporation, NIR analysis and moisture content determination confirms that horn-hoof keratins are

Table 17.2 Amino-acid composition in horn and hoof (wt %)

Amino Acid	Horn-hoof (wt %)
CYA	0
ASP	8.7
SER	7.2
GLU	16.3
GLY	4
HIS	1.3
ARG	10.8
THR	5.1
ALA	3.7
PRO	4.4
1/2CYS	7.1
TYR	4.5
VAL	5.3
MET	1
LYS	4.6
ILE	4.1
LEU	8.7
PHE	3.2

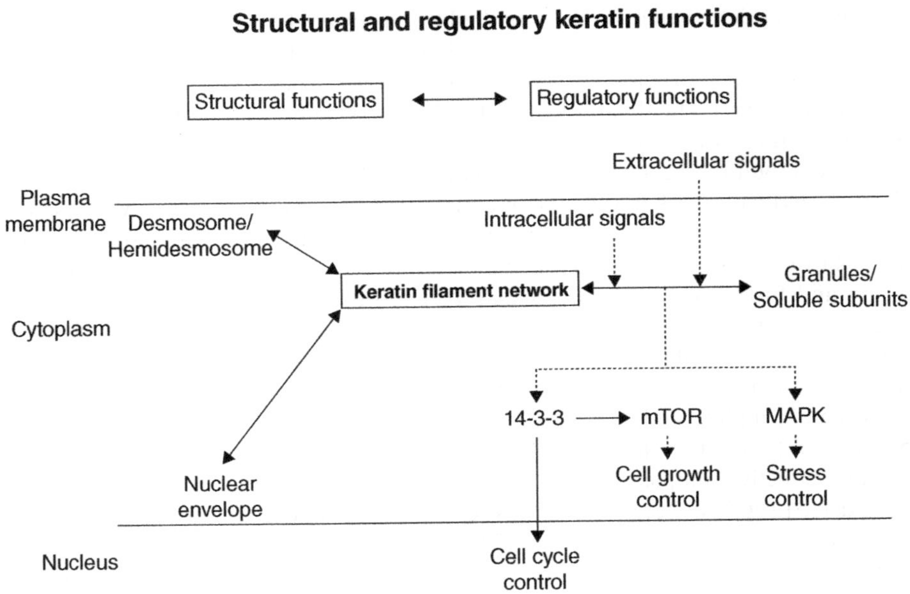

Figure 17.5 Structural versus regulatory functions of keratins (Magin *et al.*, 2007).

lower hygroscopically. Futhermore, at temperatures of more than 210°C, different molecular weights of crystal line fraction become denatured, which shows the different endothermal peaks (Zoccola *et al.*, 2009).

17.3.6.2 Functional Properties Keratin is a building block for epithelial cells, having the largest group of filamentous proteins. Epithelial cells are protected by the keratin from different mechanical and other physicochemical stresses, which final results in cell death. Cell signally, apoptosis and the response towards stress are the different specific functions of keratin (Figure 17.5). There is also a complex modification of keratin, which involves post-translational and complex interactions with various associated proteins (Coulombe and Omary, 2002).

Also, dynamic scaffolding materials can emerge from keratin that can help in the determination of cell size, control of translation, proliferation, transformation of malignant cells, cellular transportation in specific organelles, and other responses towards stress. Moreover, these sorts of functional attributes come from the intense complex trends of phosphorylation and also different associations at the molecular level (Magin *et al.*, 2007).

17.3.6.3 Applications The main structural components of horn, feather, hooves, hair etc. are formed by the fibrous protein called keratin. Keratin is a major by-product, which arises from the animal and bird meat processing industries. As we know, these by-products have a great potential in food, cosmetic, pharmaceutical and fertilizer industries, with a significant amount being frequently wasted. Furthermore, being difficult to degrade by itself is one of its environmental problems. Also, it has a very good application in the tanning industry for making quality leathers (Karthikeyan *et al.*, 2007).

The keratinolytic proteases obtained from advancements in enzyme technology help to convert the polluting waste into a nutritionally-rich protein feedstuff for cattle and poultry, through economic and energy efficient means (Onifade *et al.,* 1998). The product, having a nutritional and higher biological digestability, can be more achievable from the naturally present enzymes in the microorganisms responsible for keratin degradation, which is many times better than conventional processing using hydrothermal means (Onifade *et al.*, 1998).

The phase of keratin-based biomolecules is in the development stage, because of the plentiful advances in the different steps of extraction, purification and characterization from animal and poultry processing waste. The keratin thus obtained has most promising intrinsic biological functions and is also compatible with biological objects. The self-assembly nature of keratin contributed to different behaviors for the recognition of cellular regulation (Rouse and Van Dyke, 2010). Moreover, the qualities thus obtained are also suitable for biomaterial developments based on keratin, which can be used for the healing of wounds, scaffolds for tissue engineering, biomedical application, delivery of drugs, trauma etc. Due to its intrinsic compatibility against biomaterial, biodegradability, durabilty against mechanical stress and also being highly present in nature, has produced more frontiers for keratin-based biomaterial development than for other proteins (Rouse and Van Dyke, 2010).

17.4 Conclusion

From the excess review of literature about waste valorization, to date it is found that the majority of studies have been concentrated only in the lab and the research scale needs to widen towards commercialization. Proper technical studies, dealing with the advanced processes for the exploitation of waste and the economics behind them, need to be analyzed extensively. The plethora of applications from chicken and meat waste can be achieved with minimal justified investment. Furthermore, along with the usual applications for food, feed, rendered fat in cosmetics, and hides and hooves for leather, there are many other such areas of use, such as the recent innovations including animal and avian proteins for gaining nutritional properties, generation of different bioactive and antimicrobial peptides, and also fats as an alternative source for diesel synthesis. Regardless of the final products, it is of utmost importance to search for effective tools and techniques to exploit these waste-based resources, which cannot only benefit mankind but also make our planet a better place to live in. Hence, it is always imperative that cost-effective methods have no critical effects on environment, and be developed to make meat waste truly viable.

References

Ahhmed, A.M. and Muguruma, M. (2010) A review of meat protein hydrolysates and hypertension. *Meat Science*, **86**(1): 110–118.

Akoh, C.C., Chang, S.-W., Lee, G.-C. and Shaw, J.-F. (2007) Enzymatic approach to biodiesel production. *Journal of Agricultural and Food Chemistry*, **55**(22): 8995–9005.

Al-Zuhair, S., Hussein, A., Al-Marzouqi, A.H., and Hashim, I. (2012) Continuous production of biodiesel from fat extracted from lamb meat in supercritical CO_2 media. *Biochemical Engineering Journal*, **60**: 106–110.

Arihara, K., Nakashima, Y., Mukai, T., Ishikawa, S. and Itoh, M. (2001) Peptide inhibitors for angiotensin I-converting enzyme from enzymatic hydrolysates of porcine skeletal muscle proteins. *Meat Science*, **57**(3): 319–324.

Atadashi, I.M., Aroua, M.K. and Aziz, A.A. (2010) High quality biodiesel and its diesel engine application: A review. *Renewable and Sustainable Energy Reviews*, **14**(7): 1999–2008.

Benjakul, S., Oungbho, K., Visessanguan, W., Thiansilakul, Y. and Roytrakul, S. (2009) Characteristics of gelatin from the skins of bigeye snapper, *Priacanthus tayenus* and *Priacanthus macracanthus*. *Food Chemistry*, **116**(2): 445–451.

Bruce Fraser, R. and Parry, D.A. (2003) Macrofibril assembly in trichocyte (hard α-) keratins. *Journal of structural Biology*, **142**(2): 319–325.

Chakraborty, R., Gupta, A.K., and Chowdhury, R. (2014) Conversion of slaughterhouse and poultry farm animal fats and wastes to biodiesel: Parametric sensitivity and fuel quality assessment. *Renewable and Sustainable Energy Reviews*, **29**: 120–134.

Cheng, S.W., Hu, H.M., Shen, S.W., Takagi, H., Asano, M. and Tsai, Y.C. (1995) Production and characterization of keratinase of a feather-degrading *Bacillus licheniformis* PWD-1. *Bioscience, Biotechnology and Biochemistry*, **59**(12): 2239–2243.

Coulombe, P.A. and Omary, M.B. (2002) 'Hard'and 'soft' principles defining the structure, function and regulation of keratin intermediate filaments. *Current Opinions in Cell Biology*, **14**(1): 110–122.

Duconseille, A., Astruc, T., Quintana, N., Meersman, F. and Sante-Lhoutellier, V. (2015) Gelatin structure and composition linked to hard capsule dissolution: A review. *Food Hydrocolloids*, **43**: 360–376.

Eastoe, J. and Leach, A. (1977) Chemical constitution of gelatin. In: *The Science and Technology of Gelatin* (eds A.G. Ward and A. Courts), Academic Press, London, pp. 73–108.

Farris, S., Schaich, K.M., Liu, L., Piergiovanni, L. and Yam, K.L. (2009) Development of polyion-complex hydrogels as an alternative approach for the production of bio-based polymers for food packaging applications: A review. *Trends in Food Science & Technology*, **20**(8): 316–332.

Fraser, R., MacRae, T. and Rogers, G.E. (1972) *Keratins: Their composition, structure, and biosynthesis*. Charles C. Thomas.

Ghosh, R. (2001) Fractionation of biological macromolecules using carrier phase ultrafiltration. *Biotechnology & Bioengineering*, **74**(1): 1–11.

Gómez-Guillén, M.C., Pérez-Mateos, M., Gómez-Estaca, J., López-Caballero, E., Giménez, B., and Montero, P. (2009) Fish gelatin: A renewable material for developing active biodegradable films. *Trends in Food Science & Technology*, **20**(1): 3–16.

Gómez-Guillén, M.C., Giménez, B., López-Caballero, M.E. and Montero, M.P. (2011) 11Functional and bioactive properties of collagen and gelatin from alternative sources: A review. *Food Hydrocolloids*, **25**(8): 1813–1827.

Hanani, Z.N., Beatty, E., Roos, Y.H., Morris, M A. and Kerry, J.P. (2012) Manufacture and characterization of gelatin films derived from beef, pork and fish sources using twin screw extrusion. *Journal of Food Engineering*, **113**(4): 606–614.

Hartmann, R. and Meisel, H. (2007) Food-derived peptides with biological activity: From research to food applications. *Current Opinions in Biotechnology*, **18**(2): 163–169.

Herrero, A., Cambero, M., Ordóñez, J., Castejón, D., Romero de Avila, M. and De La Hoz, L. (2007) Magnetic resonance imaging, rheological properties, and physicochemical characteristics of meat systems with fibrinogen and thrombin. *Journal of Agricultural and Food Chemistry*, **55**(23): 9357–9364.

Hidaka, S. and Liu, S. (2003) Effects of gelatins on calcium phosphate precipitation: a possible application for distinguishing bovine bone gelatin from porcine skin gelatin. *Journal of Food Composition and Analysis*, **16**(4): 477–483.

Hill, P., Brantley, H. and Van Dyke, M. (2010) Some properties of keratin biomaterials: kerateines. *Biomaterials*, **31**(4): 585–593.

Hsieh, J.A.O. (2011) Blood-derived products for human consumption. *Revelation and Science*, **01**(1): 14–21.

Hua Yang, Z.S. (2014) The extraction of collagen protein from pigskin. *Journal of Chemical and Pharmaceutical Research*, **6**(2): 683–687.

Jayathilakan, K., Sultana, K., Radhakrishna, K. and Bawa, A.S. (2012) Utilization of by-products and waste materials from meat, poultry and fish processing industries: A review. *Journal of Food Science Technology*, **49**(3): 278–293.

Kapel, R., Froidevaux, R., Nedjar-Arroume, N., Fertin-Bazus, A., Dhulster, P. and Guillochon, D. (2003). Continuous production of a peptidic fraction containing the intermediate opioid peptide LVV-haemorphin-7 (LVVh-7) by peptic hydrolysis of bovine haemoglobin in a continuous membrane reactor. *Biotechnology and Applied Biochemistry*, **37**(3): 317–324.

Karthikeyan, R., Balaji, S. and Sehgal, P.K. (2007) Industrial applications of keratins: A review. *Journal of Scientific & Industrial Research*, **66**: 710–715.

Kenny, T., Lennon, A., Ward, P., Sullivan, P., McDonald, K. and K. O'Neill (2008) Adding value to beef forequarter muscles. Ashtown Food Research Centre Research Report No. 98, Teagasc.

Kornillowicz-Kowalska, T. and Bohacz, J. (2011) Biodegradation of keratin waste: Theory and practical aspects. *Waste Management*, **31**(8): 1689–1701.

Kurbanoglu, E.B. and Kurbanoglu, N.I. (2004) Utilization as peptone for glycerol production of ram horn waste with a new process. *Energy Conversion and Management*, **45**(2): 225–234.

Lateef, A., Oloke, J.K., Gueguim Kana, E.B., Sobowale, B.O., Ajao, S.O. and Bello, B.Y. (2010) Keratinolytic activities of a new feather-degrading isolate of *Bacillus cereus* LAU 08 isolated from Nigerian soil. *International Biodeterioration & Biodegradation*, **64**(2): 162–165.

Magin, T.M., Vijayaraj, P. and Leube, R.E. (2007) Structural and regulatory functions of keratins. *Experimental Cell Research*, **313**(10): 2021–2032.

Marikkar, J.M.N., Ghazali, H.M., Long, K. and Lai, O.M. (2003) Lard uptake and its detection in selected food products deep-fried in lard. *Food Research International*, **36**(9–10): 1047–1060.

Meena, C., Mengi, S.A. and Deshpande, S.G. (1999) Biomedical and industrial applications of collagen. *Proceedings of the Indian Academy of Sciences – Chemical Scences,* **111**(2): 319–329

Meher, L., Vidyasagar, D. and Naik, S. (2006) Technical aspects of biodiesel production by transesterification: A review. *Renewable and Sustainable Energy Reviews*, **10**(3): 248–268.

Mir, P.S., McAllister, T.A., Scott, S., Aalhus, J., Baron, V., *et al.* (2004) Conjugated linoleic acid–enriched beef production. *The American Journal of Clinical Nutrition*, **79**(6): 1207S–1211S.

Nelson, R. G. and Schrock, M.D. (2006) Energetic and economic feasibility associated with the production, processing, and conversion of beef tallow to a substitute diesel fuel. *Biomass and Bioenergy*, **30**(6): 584–591.

Onifade, A.A., Al-Sane, N.A., Al-Musallam, A.A. and Al-Zarban, S. (1998) A review: Potentials for biotechnological applications of keratin-degrading microorganisms and their enzymes for nutritional improvement of feathers and other keratins as livestock feed resources. *Bioresources Technology*, **66**(1): 1–11.

Parés, D., Toldrà, M., Saguer, E. and Carretero, C. (2014) Scale-up of the process to obtain functional ingredients based in plasma protein concentrates from porcine blood. *Meat Science*, **96**(1): 304–310.

Roupas, P., Silva, K, Smithers, G., Ferguson, A., and Waldron, K. (2007) Waste management and co-product recovery in red and white meat processing. In: *Handbook of Waste Management and Co-Product Recovery in Food Processing.* vol. **1**, pp. 305–331.

Rouse, J.G. and Van Dyke, M.E. (2010) A review of keratin-based biomaterials for biomedical applications. *Materials*, **3**(2): 999–1014.

Sarath, G., Motte, R. and Wagner, F. (1989) Commercially available protease inhibitors. In: *Proteolytic Enzymes – A Practical Approach*, IRL Press, Oxford, pp. 22–25.

Schrieber, R. and Gareis, H. (2007) *Gelatine Handbook: Theory and industrial practice*, John Wiley & Sons.

Shin, H.-Y., Lee, S.-H., Ryu, J.-H. and Bae, S.-Y. (2011) Biodiesel production from waste lard using supercritical methanol. *The Journal of Supercritical Fluids*. doi: 10.1016/j.supflu.2011.09.009

Silva, V.D. and Silvestre, M.P. (2003) Functional properties of bovine blood plasma intended for use as a functional ingredient in human food. *LWT – Food Science and Technology*, **36**(7): 709–718.

Taher, H., Al-Zuhair, S., Al Marzouqui, A. and Hashim, I. (2011) Extracted fat from lamb meat by supercritical CO_2 as feedstock for biodiesel production. *Biochemical Engineering Journal*, **55**(1): 23–31.

Ticu, E.-L., Vercaigne-Marko, D., Froidevaux, R., Huma, A., Artenie, V. and Guillochon, D. (2005) Use of a protease-modified-alumina complex to design a continuous stirred tank reactor for producing bioactive hydrolysates. *Process Biochemistry*, **40**(8): 2841–2848.

Toldrá, F., Aristoy, M.C., Mora, L. and Reig, M. (2012) Innovations in value-addition of edible meat by-products. *Meat Science*, **92**(3): 290–296.

Vercruysse, L., Van Camp, J. and Smagghe, G. (2005) ACE inhibitory peptides derived from enzymatic hydrolysates of animal muscle protein: A review. *Journal of Agricultural & Food Chemistry*, **53**(21): 8106–8115.

Verheijen, L.A.H.M., Wiersema, D. and Hulshoff Pol, L.W. (1996) Livestock and the Environment – Finding a balance. *Management of Waste from Animal Product Processing*. International Agriculture Centre. Coordinated by the FAO, USAID and World Bank.

Vermeirssen, V., Camp, J. V. and Verstraete, W. (2004) Bioavailability of angiotensin I converting enzyme inhibitory peptides. *British Journal of Nutrition*, **92**(03): 357–366.

Wan, Y., Ghosh, R. and Cui, Z. (2002) High-resolution plasma protein fractionation using ultrafiltration. *Desalination,* **144**(1): 301–306.

Young, R. and Lawrie, R. (1974) Utilization of edible protein from meat industry by-products and waste. *International Journal of Food Science andTechnology*, **9**(2): 171–177.

Zhang, Q., Shan, G., Cao, P., He, J., Lin, Z., *et al.* (2015) Mechanical and biological properties of oxidized horn keratin. *Materials Science and Engineering: C*, **47**: 123–134.

Zhou, P., Mulvaney, S.J. and Regenstein, J.M. (2006) Properties of Alaska pollock skin gelatin: A Comparison with tilapia and pork skin gelatins. *Journal of Food Science*, **71**(6): C313–C321.

Zoccola, M., Aluigi, A. and Tonin, C. (2009) Characterisation of keratin biomass from butchery and wool industry wastes. *Journal of Molecular Structure*, **938**(1–3): 35–40.

Zúñiga, R. and Aguilera, J. (2009) Structure–fracture relationships in gas-filled gelatin gels. *Food Hydrocolloids*, **23**(5): 1351–1357.

18

Aquaculture and Marine Products Contribution for Healthcare Application

Maushmi S. Kumar
Shobhaben Pratapbhai Patel School of Pharmacy and Technology Management, SVKM'S NMIMS, Vile Parle West, Mumbai, India

18.1 Introduction

Aquaculture is the practice of cultivating, growing and harvesting aquatic organisms such as fish, crustaceans, molluscs and aquatic plants. It also involves seawater populations under controlled conditions, and is different from commercial fishing. The freshwater and marine water product industry accounts for up to 75% of the catch, depending on postharvest or industrial preparation processes (Rustad *et al.*, 2011). Besides seafood, marine bioresources produce a great variety of biomolecules, including natural organic compounds such as fatty acids, polysaccharides, polyethers, peptides, proteins, enzymes, lectins, vitamins and minerals responsible for specific bioactivity. These products contribute significantly to pharmaceutical and biotechnology industries from a vast array of sources including marine plants, microorganisms and sponges, all of which contain unique chemical structured biomolecules responsible for therapeutic activity. However, this resource is relatively untapped for the discovery of new biomaterials. The biodiversity of the marine environment and the associated chemical diversity constitute an unlimited resource of novel active substances for the development of bioactive products (Giri and Ohshima, 2012).

Recently, a great deal of interest has been expressed worldwide in aquaculture and marine-derived bioactive products due to their numerous health benefits and applications in the food industry. Fishery products are the major contributing source from the oceans, which constitutes about 13 000 species of fish, 50 000 species of mollusks, and 1000 species of cephalopods. There is great potential for these marine organism and their bioprocessed products to be developed as functional ingredients. This chapter focuses on the freshwater and marine organism's products, for their suitability as functional ingredients in healthcare applications for the prevention and treatment of chronic diseases. Major classes are shown in Figure 18.1, which is discussed for its application in the following sections.

Food Processing By-Products and their Utilization, First Edition. Edited by Anil Kumar Anal.

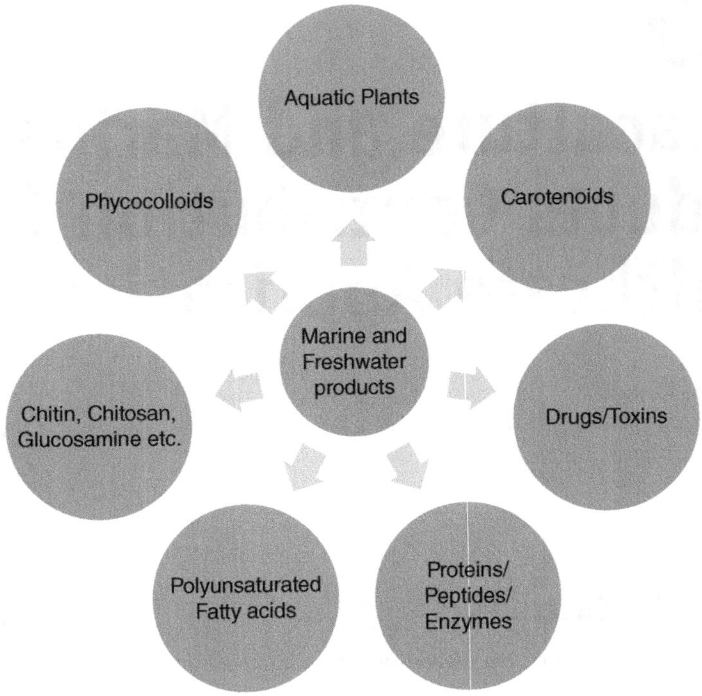

Figure 18.1 Major classes of marine and freshwater products.

18.2 Various Classes of Freshwater and Marine Products and their Healthcare Application

18.2.1 Proteins and Peptides

According *to Scientific American*: the 'Blue Food Revolution' aims to provide a sustainable feeding of protein food source to a future population of 9 billion people, which is only possible by open ocean marine aquaculture (Simpson, 2011). If this revolution is to be well executed, fish farming can provide much needed protein for the world, while minimizing the expansion of land-based farming and the environmental costs. Currently, the world's population obtains about 25% of its protein from fish (Khora, 2013). They provide about 55% of all the animal protein consumed in Asia (WHO, 2013). It is a rich source of easily digestible protein, with a digestibility of more than 90%. It contains all the essential amino acids and has a very high biological value. Fish proteins are classified into three broad classes: structural (actin, myosin, tropomyosin and actomyosin) – 70–80% protein; sarcoplasmic (myoalbumin, globulin and enzymes) – 25–30% protein; and connective tissue (collagen) – 3% protein. Seafood serves as an excellent source of protein because of its low fat content and omega-3 fatty acids. Research shows that eating just 6 ounces of fish a week may lower the risk of dying from heart disease by more than 36%. The enhanced digestibility is mainly due to the absence of strong collagenous fibers and tendons in fish muscle, which are common in land animals (Friedman, 2006; Khora, 2013). Fish protein is available as fish protein concentrate, fish protein supplements and fish protein hydrolysates. Other important protein sources explored are macro and microalgae (*Spirulina, C. vulgaris, S. platensis* and *Isochrisis galbana* and *Aphanizomenon* species) for its bio-functional proteins.

Generally, the protein content of seaweeds/macroalgae is low compared to microalgae. The protein content varies among species according to the seasonal conditions of their growth. According to Fleurence (1999), brown algae have shown less protein content (3–15% of the dry weight) with high phenolic content, with respect to the green and red seaweeds (10–47% of the dry weight). Dried Spirulina biomass protein content is reported to be 68% and it contains all the essential amino acids; microalga Chlorella contains about 50–60% of proteins of quality comparable to the yeast proteins, soy flour and milk powder (Blazencic, 2007). Algal proteins have shown antioxidant, antihypertensive, anticoagulant, antiproliferation and immune-stimulant activity. Their mode of action includes mimicking hormones and drug-like activities. Additionally, they could bind to specific receptors on target cells or inhibition by enzyme actions (FitzGerald and Murray, 2007; Kitts and Weiler, 2003). Another study was used to determine the nutritive value of sea urchins, *Paracentrotus lividus* and *Echinus esculentus*. The effect of this diet was studied on leucine aminopeptidase and gamma-glutamyltranspeptidase, and intestinal and hepatic enzymes in rat models. Nutritive values of these marine species proved to be a good protein source and similar to casein (Gonazalez *et al.*, 2001). The pharmacological effects of fish peptides show antihypertensive, immunomodulatory, antioxidant, antitumor and antimicrobial activities. Thus fish being highly nutritious can also be considered as a valuable source of unique peptides with a broad spectrum of biological activities (Urakova *et al.*, 2012). Recently, research has focused on peptides from marine animal sources. They are mostly considered for biological activity and structural leads, since they are found as secondary metabolites from sponges, ascidians, tunicates and mollusks. Most of them have shown anti-oxidant and cytotoxic effects on human cancer cells, which is promising for the development of drugs in the future.

Interestingly, the by-products of the fish processing industry (skins, scales and bones), like some collagen-derived peptides also exhibit antioxidant, potent antihypertensive and antimicrobial activity. In addition, they show protective effects on cartilage by stimulating bone formation. Several hydrolysates exhibiting antioxidant activity have been obtained from collagen or gelatin (a hydrolyzed collagen) of different marine sources, as Alaska Pollock skin, hoki fish skin, cobia skin, brown-stripe red snapper skin, tuna backbones, sole skin, jellyfish umbrella, cod, or squid skin. They are also mild or moderate ACE inhibitors to target hypertension. They are resistant to protein digestion, and thus can be considered as potential ingredients of health promoting foods. ACE-inhibitory peptides have been isolated and identified in gelatin hydrolysates derived from fish skins, scales, chum salmon cartilage, sea cucumbers and squid tunics (Aleman and Alvarez, 2013). These peptides are also understood to exert antibacterial activity due to a high proportion of select residues particularly Pro, Arg or Gly, which are abundant in collagen (Brogden *et al.*, 2003).

In a recent study, Gomez-Guillen *et al.* (2010) obtained antimicrobial tuna and squid skin hydrolysates. These gelatin hydrolysates were highly active against different strains of Gram-positive and Gram-negative bacteria. Fish gelatin is a common ingredient in pharmaceutical applications for the preparation of hard and soft capsules, for encapsulation of drugs and bioactive molecules, as an antioxidative hydrolysate, as a carrier in controlled drug delivery, as a component in dietary health supplements and in wound care. It has also been reported that live attenuated viral vaccines used for immunization against measles, mumps, rubella, Japanese encephalitis, rabies, diphtheria and tetanus contain gelatin as a stabilizer (Burke *et al.*, 1999).

18.2.2 Marine Enzymes

Marine animals and microorganism have been accessed for the search of novel enzymes for their unique biocatalytic functions. They also have been an important product, mostly developed for food and feed products like fish protein hydrolysate and fish sauces. Enzymes like transglutaminase production from microorganisms make it affordable for use in food products. Fish enzymes are also used in fish processing, starting with skin and scale removal. Researchers have isolated enzymes like proteases and carbohydrases from deep-sea hydrothermal locations. Psychrophilic marine microorganisms have been the source for cold active metabolic enzymes. Proteases, carbohydrases and peroxidases are noteworthy among the most widely-studied enzymes. Enzymes from marine animals and plants have been primarily studied for their metabolic functions, although peroxidases and proteases have found industrial applications. Novel techniques of molecular biology are used to assess the diversity of chitinases, nitrate, nitrite, and ammonia-metabolizing and pollutant-degrading enzymes. Genes encoding for chitinases, proteases and carbohydrases from microbial and animal sources have been cloned and characterized (Debashish *et al.*, 2005). A variety of fish proteases (trypsin, alcalase etc.) display potentially new biochemical properties in industrial applications, such as chitin extraction. Alkaline protease extracts from the viscera of goby (*Zosterisessor ophiocephalus*), thornback ray (*Raja clavata*) and scorpion fish (*Scorpaena scrofa*) have been studied (Miled *et al.*, 2011). Two detergent-stable alkaline serine-proteases (BM1 and BM2) purified from *Bacillus mojavensis* A21 were isolated from seawater by Haddar *et al.* (2009). Both proteases were highly stable towards non-ionic surfactants. In addition, both of them showed excellent compatibility with a wide range of commercial liquid and solid detergents. A novel extracellular phospholipase C was purified from a marine *Streptomycete*, selected from approximately 400 marine bacteria (Mo *et al.*, 2009). It had optimum enzyme activity at 45°C and pH 8.0, and it hydrolyzed only phosphatidylcholine.

On the other hand for polysaccharide degrading enzymes, a variety of chitinase genes have been cloned from marine bacteria and fungi. In one of their works, Suolow and Jones (1988) inserted two chitinase genes (ChiA, ChiB) into *E. coli*, and subsequently these genes ware transferred into *Pseudomonas*, which acquired four high-yielding chitinase strains. Meanwhile, in another study the chitinase gene was put into tobacco plant cells by Roberts and Cabib (1982), from which they were able to develop a new tobacco plant with a strong disease resistance to the pathogen *Alternaria longipes*. Alginate lyases, characterized as either mannuronate or guluronate lyases, have been isolated from a wide range of marine algae, invertebrates and terrestrial microorganisms. Agar-degrading microorganisms produced agarase enzyme. In 1902, Gran isolated agar-degrading *Pseudomonas galatica* from seawater. Until now, researchers have found the presence of agarase from species within the genus *Cytophaga, Bacillus, Vibrio, Alteromonas, Pseudoalteromonas, Streptomyces* (Aoki *et al.*, 1990; Leon *et al.*, 1992; Hosoda *et al.*, 2003). *Pseudomonas, Cytophaga,*

Alteromonas atlantica, *Alteromonas carrageenovora*, and some unidentified strains have been found to produce carrageenan-degrading enzymes.

A distinct λ-Carrageenan-degrading *Pseudoalteromonas* bacterium (CL19) was reported by Yukari and Yuji (2006); the molecular mass of this purified enzyme was approximately 100 kDa (Ohta *et al.*, 2006). The oligosaccharides obtained from carrageenan degradation show a variety of specific physiological activities, such as anti-viral, anti-tumor, anti-coagulation etc., indicating the importance of these enzymes (Roberts *et al.*, 2007). Furukawa *et al.* (1992) purified fucoidanase from marine *Vibrio* and obtained three enzymes, which could hydrolyze substrate fucoidan to small molecule-oligosaccharides. In recent years, research programs investigating extremophilic microorganisms have attracted a new area of interest in healthcare and other industrial applications. Apart from many thermophiles already reported, a novel thermostable non-specific nuclease from thermophilic bacteriophage GBSV1 was isolated, which could degrade various nucleic acids, including RNA, single-stranded DNA and double-stranded DNA (Song and Zhang, 2008). Marine microbial enzymes can offer novel biocatalysts with extraordinary properties in the future, while continuously investigating the occurrence and bioprocessing of marine microbial enzymes.

18.2.3 Polyunsaturated Fatty Acids

Fish oils contain triacylglycerides and variable amounts of phospholipids, glycerol ethers and wax esters. They contain a wide range of highly unsaturated long-chain fatty acids, with the number of carbon atoms ranging from 14–22 (Schmidtsdorff, 1995). Fish oil studies have been shown to protect against coronary heart disease due to the beneficial effects of constituents such as long-chain n-3 polyunsaturated fatty acids, eicosapentenoic acid and docosahexaenoic acid, also known as omega-3 fatty acids. Long chain n-3 polyunsaturated fatty acids show an effect by decreasing the plasma level of fasting and postprandial triacylglycerol (Roos *et al.*, 2012). In another very interesting recent report by Zheng *et al.* (2013), the results showed lowering of the risk of breast cancer with a higher daily consumption of dietary marine n-3 PUFA. It was speculated that the PUFAs might have an important role in regulating the activities of molecules involved in the growth and spread of cancerous tissues. Alternatively, they can also act by their powerful anti-inflammatory properties. There are also reports of PUFA from marine macro and microalgae having potential for nutritional and pharmaceutical applications (Pereira *et al.*, 2012). Microalgae, such as cyanobacteria *Spirulina* and *Chlorella minutissima*, is suggested to be a good source for the purification of gamma-linolenic acid and PUFA-rich nutraceutical supplement (Mahajan and Kamat 1995; Seto *et al.*, 1984). n-3 PUFAs, especially eicosapentaenoic acid (EPA) and docosahexaenoic acid (DHA), helps in the prevention and treatment of several diseases such as atherosclerosis and arrhythmias in modulation of platelet aggregation and decreased synthesis of pro-inflammatory agents. These fatty acids have indicated insulin-sensitizing effects in metabolic disorders, which might be the result of antagonizing n-6 PUFA (arachidonic acid)-induced pro-inflammatory prostaglandin E_2 (PGE_2) formation. They can also impact their anti-inflammatory effects via reduction of nuclear factor-κB activation. Finally, beneficial effects of EPA and DHA on insulin resistance include their ability to increase secretion of adiponectin, an anti-inflammatory adipokine (Siriwardhana *et al.*, 2012). This summarizes the n-3 PUFAs multiple health benefits; thus their consumption from food sources should be encouraged.

18.2.4 Seafood Processing By-Products

18.2.4.1 Chitin and Chitosan Chitin is the second-most important polymer after cellulose. It is found in the outer skeleton of crabs, shrimps and lobsters and in the internal structures of other invertebrates. It is composed of ß(1-4) units of the amino sugar N-acetyl glucosamine and is the main source of production of chitosan, which is a deacetylated derivative of chitin. Polysaccharide chitin is abundantly found in nature, making its derivative chitosan, a relatively inexpensive product. They are non-toxic, weakly antigenic, antibacterial and biodegradable and used in a number of applications such as a flocculating agent, wound-healing agent, sizing and strengthening agent for paper and a delivery vehicle for pharmaceuticals and genes. Other biomedical applications include tissue engineering scaffolds, drug delivery, and stent coatings and wound dressings. Unlike chitin, chitosan is water soluble in acidic media, or under precisely specified conditions at neutral pH, allowing much development in the domains of solutions and hydrogels (Rinaudo, 2006). Chitosan has served as an excellent excipient. It has been used in direct compression and binding, controlled release formulations, gel formulations, and ophthalmic, intestinal, buccal and transdermal delivery. It also finds its place in immobilization of enzymes and the food industry. Various derivatives of chitosan are being developed and tested as drug carriers. Chitosan micro-carriers are produced for drugs in six different classes, interacting with anions, thermal cross-linking with citric acid, solvent evaporation, interfacial acylation, coating on preformed micro-particles and cross-linking with chemicals for cell culture. They have been used in hepatocyte culture, where apart from their biocompatibility and biodegradability, they are a suitable choice due to their structure being similar to glycosaminoglycans, which is a part of the liver extracellular matrix (Gameil *et al.*, 2014).

18.2.4.2 Glucosamine Glucosamine is one of the most abundant monosaccharaides, produced commercially by the hydrolysis of crustacean exoskeletons or outer shells, and also sometimes by fermentation of grains like corn or wheat. It is one of the most common non-vitamin, non-mineral, dietary supplements used by adults. Edible crustaceans such as shrimp, lobster, crab and crawfish can provide trace amounts of glucosamine in their shells and tails. However, this quantity consumed is not sufficient to achieve its medicinal property. Vegetarian glucosamine is fermented from corn and serves as an alternative used by vegetarians or those who are allergic to shellfish. Glucosamine is sold in combination with other supplements, such as chondroitin sulfate for treating osteoarthritis. However, there is no evidence that the consumption of glucosamine by sports persons prevents or limits joint damage after injury (Hespel *et al.*, 2006). Reports are divided about its clinical studies, with some reporting relief from arthritic pain and stiffness, but larger group studies have reported no benefit above placebo (Barrette *et al.*, 2010). Glucosamine has earlier shown suppressed synovial hyperplasia, cartilage destruction and inflammatory cell infiltration in rat adjuvant arthritis. It also reduced the production of nitric oxide and prostaglandin E(2) in plasma ($p < 0.05$) (Hua *et al.*, 2005). In a recent study by Haleagrahara *et al.*, 2014, when collagen-induced arthritis in female Dark Agouti rats was treated with 300 mg/kg of glucosamine hydrochloride, it reversed the arthritic changes, hence suggesting that glucosamine has a therapeutic effect against collagen-induced arthritis.

18.2.4.3 Phycocolloids Phycocolloids is used to refer to three main starch-like products, alginate, carrageenan, propylene glycol alginate and agar, which are extracted from brown and red seaweeds respectively. They are used in cosmetics, food and medicine. Alginates is a generic term, meaning the various derivatives of alginic acid that either occur naturally in certain brown seaweeds, or are produced from natural derivatives. Alginates are essential components as thickening, gelling or stabilizing agents in food applications (McHugh 1987; Perez *et al.*, 1992), and also as an immuno-stimulatory agent (Gopinathan and Panigrahy, 1983). They have important use in wound dressing materials for the treatment of acute or chronic wounds. They are used in cystic fibrosis, where the bacterial biofilms are formed from alginate gels (Hatch and Schiller, 1998; Pawar and Edgar, 2012). They are also used widely in the drug delivery applications of formulations (Badwan *et al.*, 1985; Tonnesen and Karlsen, 2002). New research has also affirmed the weight-loss potential of the dietary supplement sodium alginate, which also aids in the burning of excess fat. When taken daily before meals, it helps overweight people to shed the extra fat, and can improve weight loss in obese subjects who complete a 12-week dietary intervention (Jensen *et al.*, 2012).

Propylene glycol alginate is an additive, used mostly as thickening agents in food. It helps in thickening and stabilizing certain types of food materials.

Another phycocolloid, carrageenan, is made from various red algae or seaweeds. It is a high-molecular-weight (food grade >100 000 Da) strongly anionic polymer that is used for the textural stabilization of foods. It is used to treat coughs, bronchitis, tuberculosis and intestinal problems. The French used a carrageenan form to treat peptic ulcers, and as a bulk laxative. Chemicals present in carrageenan decreases stomach and intestinal secretions. Large amounts of carrageenan seem to pull water into the intestine, for a laxative effect. However, more evidence is needed to rate the effectiveness of carrageenan for these therapeutic uses. Ongoing research of carrageenan use in food has been linked to growing health concerns, including gastrointestinal inflammation, inflammatory bowel disease, ulcers, gastrointestinal bleeding, tumors, malignancies, insulin resistance and systemic inflammation. When ingested, it can cause an immune and symptomatic response similar to a pathogen invasion (Tobacman, 2001).

Agar is another important polymer widely used worldwide. It consists of sugar galactose subunits. It is a very good source of calcium, iron and fiber. It does not contain sugar, fat or carbohydrates. It is a component of the cell walls of several species of red-purple marine algae (Class Rhodophyceae) including Gelidium, Pterocladia and Gracilaria. It helps in digestion as well as weight loss, by carrying toxic waste out of the body. Agar can also be used as a nutritional supplement, both as a bulking agent as well as lubricating the stool for treating hemorrhoids. Other benefits include its ability to reduce inflammation, calm the liver, and relief to the lungs. In Japan, agar is also used to treat diabetes and constipation. In pharmaceutical preparations, agar is used as an excipient in emulsions, suspensions, gels and certain suppositories.

18.2.4.4 Carotenoids Carotenoids are the most common pigments and considered to be a key molecule of life. In nature, they are synthesized by all photosynthetic organisms and fungi. Marine animals contain various carotenoids that show structural diversity. In general, animals do not synthesize carotenoids *de novo*, and so those found in animals are either directly accumulated from food or partly modified through metabolic reactions. Marine animals accumulate carotenoids from foods such as algae and other animals in their environment and modify them through metabolic reactions. An interesting way of using microalgae as a carotenoid carrier in the food industry and using many other antioxidant compounds in the micro-algae biomass, shows positive impact on human health, sometimes acting synergistically (Belay, 1993). Carotenoid metabolites of β-carotene, fucoxanthin, peridinin, diatoxanthin, alloxanthin and astaxanthin etc. are present in these animals such as crustaceans, tunicates, sponges etc. Many fish accumulate carotenoids in their integuments and gonads.

Astaxanthin is widely distributed in both marine and freshwater fish from dietary crustacean zooplankton. The carotenoids such as lutein, β-carotene, lycopene etc. have shown antioxidant bioactivity in immunoresponse modulation, and positive effects in signalling transduction between cells and in anti-inflammatory response mechanisms (Giovannucci *et al.*, 1995; Vilchez *et al.*, 2011). The most important function of carotenoids in the human body is their ability to convert into retinol (provitamin A), which is converted into vitamin A. β-carotene has been assigned a provitamin A activity of 100% (Graham and Rosser, 2000). β-carotene, astaxanthin, cantaxanthin and zeaxanthin have also shown promoting reduction in size and number of liver neoplasias *in vivo* (Nishino *et al.*, 2002). They also protect LDL against oxidation, and daily dietary β-carotene supplementation in mammal's results in decreased plasma levels of total lipids, cholesterol and triglycerides (El-Baky *et al.*, 2007). Lutein and zeaxanthin are the two main pigments responsible for both the yellowing and maintenance of normal visual function of the human eye (Le and Xiao-Ming, 2010). Possessing such benefits, carotenoids have been the most sought after products and are being commercially produced from tomatoes, algae and fungi.

18.2.4.5 Aquatic Plants Aquatic plants predominantly grow in fresh and salt water, having the greatest number of adaptations to life in a water environment. They fall into one of four common class types: algae, floating plants, submerged plants and emerged plants. Algae are the most common type of aquatic plants, devoid of roots, stems and leaves. They are primarily found in the ocean and are microscopic, like phytoplankton; and others are macroalgae, visible to the naked eye. They form the base for the marine food chain. Some important types of algae are *Cyanophyta lyngbya*, microcystis, *Chlorophyta spirogyra*, cladophora, *Xanthophyta vaucheria* and *Charophyta chara*, nitella. Blue-green algae (BGA) contains high amounts of protein, iron and other mineral content, absorbed when taken orally. It also contains bioactive components such as phycocyanin, carotenoids, γ-linolenic acid, fibers and plant sterols, which can promote optimal health in humans. It is being researched for its effects on the immune system, swelling (inflammation) and viral infections. Studies have also demonstrated their plasma cholesterol and triglyceride-lowering properties due to their modulation of intestinal cholesterol absorption and hepatic lipogenic gene expression. It can also reduce inflammation by inhibiting the nuclear factor κB activity, resulting in reduction of the production of pro-inflammatory cytokines. Furthermore, BGA also inhibits lipid peroxidation and shows free radical scavenging activity, beneficial in oxidative stress (Ku *et al.*, 2013). Kelp, another marine algae, is known as a rich source of iodine, important and critical for thyroid health. Other important nutrients present are essential fatty acids, potassium, sodium, protein and fiber. Spirulina and chlorella are two super foods derived from algae. Alphanizomenon Flos Aquae, or AFA, is a recent addition as a powerful brain food. It is a rich source of essential amino acids, minerals, vitamins and low molecular weight neuropeptides, which are capable of penetrating readily across the blood-brain barrier. It helps in fighting inflammation owing to a compound phycocyanin, a natural COX-2 inhibitor. This compound relieves pain caused by inflammation without any side effects (Cousens, 1985).

Traditionally, a variety of aquatic plants from fresh water are also used in curative therapy by some communities. For example, *Bacopa monnieri* is used for stomach disorders and coughs in small children. It is also used for purifying blood disorders, and curing some skin diseases. Aponogeton, commonly known as Kekatiya, has a therapeutic value for diabetes. It is used in hyperdipsia, cholera, diarrhea, helminthiasis, hemorrhage, menorrhagia and bronchitis in Ayurveda, Siddha and Unani medicines. *Acanthus ilicifolius* is used in asthma, paralysis, leucorrhoea, neuralgia, rheumatism and an expectorant in folk and Siddha system of medicines (Munasinghe *et al.*, 2009; Jyothi and Sureshkumar, 2013). The therapeutic and curative properties of these aquatic plants are found to be consistent with studies of Jayaweera (1982).

18.2.4.6 Marine Toxins Vibrantly colored organisms from the depths of the ocean harbor toxins that can potentially act as powerful anti-cancer drugs. Several organic compounds containing nitrogen as part of a heterocyclic system have been isolated from marine animals. These compounds are mostly toxic. Tetrodotoxin is a toxin known for many years; several derivatives of tetrodotoxin have been isolated from crabs, an octopus, a goby, molluscs, flatworms and even a terrestrial amphibian, all suggesting its origin from a microbial source. It is considered as a good hypotensive agent, potent respiratory inhibitor, and pain relieving agent in the neurogenic form of Hansen's disease. Saxitoxin has been isolated from mussels, crabs and dinoflagellates. It blocks nerve conductions by specifically interfering with the initial increase in sodium permeability of the membrane. Saxitoxin also produces effects on the cardiovascular system, and has a marked hypotensive effect (Nishikawa *et al.*, 2004; Daly 2004). In porifera-marine sponges, the toxins produced exist symbiotically and act by disrupting the function of actin. Actin serves as major protein part of eukaryotic cell structure. It forms long chains, essential for cellular division, growth and locomotion.

Recent research on structural and functional studies has revealed the mechanism of toxic marine macrolides common mechanism of interacting with actin in cancer cells, constraining their development as a pharmacological agent. However, it is proved that the marine toxins affect several forms of cancer. The findings enable the synthesis of toxins instead of biological sources from the ocean (Saito, 2009). A few compounds derived from marine sources that are currently in clinical trials for relieving pain associated with cancer are Tetrodotoxin, XEN-2174, Leconotide and Keyhole Limpet Hemocyanin. Some other marine derived drugs are approved; still in clinical trials are Cytarabine, ET743, PM-10450, Lurbinectedin, Eribulin, Brentuximab Vedotin, Glembatumumab Vedotin, ABT-414, PSMA-ADC, DCDT-2980S, DCDS-4501A, *Enfortumab vedotin*, *Vorsetuzumab mafdotin* etc. (Newman and Cragg, 2014).

18.2.4.7 Marine Drugs Numerous metabolites of unique structure and exhibiting biological activities have been reported from marine animals. Some of these bioactive metabolites have biomedical potential. The bioactive metabolites that are of human interest have been mainly isolated from marine sponges, jelly fish, sea anemones,

Table 18.1 FDA and EU approved marine drugs

Compound name	Trademark	Marine organism source	Compound class	Application	Company/ Institution
Brentuximab vedotin (SGN-35)	Adcetris®	Mollusk/cyano-bacterium	Antibody Drug Conjugate	Cancer	Seattle Genetics
Cytarabine (Ara-C)	Cytosar-U®	Sponge	Nucleoside	Cancer	Bedford Laboratories
Trabectedin (ET-743) (EU Registered only)	Yondelis®	Tunicate	Alkaloid	Cancer	Pharmamar
Vidarabine (Ara-A)	Vira-A®	Sponge	Nucleoside	Antiviral	King Pharma-ceuticals
Ziconotide	Prialt®	Cone snail	Peptide	Pain	Jazz Pharma-ceuticals
Omega-3-acid ethyl esters	Lovaza®	Fish	Omega-3 fatty acids	Hypertri-glyceridemia	GlaxoSmithKline
Eribulin Mesylate (E7389)	Halaven®	Sponge	Macrolide	Cancer	Eisai Inc.

corals, bryozoans, molluscs, echinoderms, tunicates and crustaceans (Bhakuni and Rawat, 2005). These bioactive metabolites can be divided into different chemical classes of steroids, terpenoids, isoprenoids, prostaglandins, quinones, brominated compounds etc. Cyanobacteria in marine forms in particular are one of the richest sources of known and novel bioactive compounds including toxins. Several metabolites from seaweeds, sponges, cnidarians, bryozoans, molluscs, tunicates and echinoderms have been studied extensively. Metabolites extracted from fish, sea snakes and aquatic mammals are meager. Sponges (37%), coelenterates (21%) and microorganisms (18%) are the major sources of bioactive compounds followed by algae (9%), echinoderms (6%), tunicates (6%), molluscs (2%), bryozoans (1%) etc. (Blunt *et al.*, 2004). Table 18.1 lists the marine-derived compounds, which had been FDA-approved up to June 2014 (Mayer *et al.*, 2010). There are 25 marine-derived compounds that are currently in clinical trials. The potential of marine natural products is significant and will contribute more in future. New techniques and industrial-academia collaborations are essential to ensure the future success of marine natural products as new and novel therapeutic drug entities.

18.3 Recent Patents in Healthcare Applications

18.3.1 Chitin and Chitosan

Konno *et al.* (2001) disclosed chitin oligosaccharides as useful for preventing or treating common colds and pain. The active ingredient in these treatment was a water soluble mixture available in oral form and selected from the range of chitin oligomers: di-N-acetyl chitobiose, tri-N-acetyl chitotriose, tetra-N-acetyl chitotetraose, penta-N-acetyl chitopentaose and hexa-N-acetyl chitohexaose, with the water soluble oral chitosan oligomers selected from chitobiose, chitotriose, chitotetraose, chitopentaose, chitohexaose and chitoheptaose. Hwa *et al.* (2010) provided a solid supplement additive containing mineral, germanium, zeolite, selenium, natural methyl sulfonyl methane, chitosan and calcium responsible for accelerating the immunity and growth of organisms. Lin (2012) disclosed that the chitosan oligosaccharide marine organism anti-inflammation anus suppository can be used for anus disinfection in people of different ages and is capable of treating increasing of anus secreta, vaginitis secretion stimulation, and pruritus caused by fissure in ano, anal fissure, hemorrhoids, anal sinusitis, anal papilla, rectal prolapsed, and anal incontinence; it possesses excellent bactericidal effects on venereal disease pathogenic bacteria; and is capable of diminishing inflammation and removing unpleasant odours. In the year 2013, Wu took out several patents contributing to marine products invention and applications. He reported a mouthwash for dental diagnosis and treatment, which is also applicable for diagnosis and treatment in dental clinics (Wu 2013a).

The mouthwash for dental diagnosis and treatment containing a marine organism antibacterial component disclosed by the invention had the advantages of:

1. being remarkable in antibacterial effect;
2. being free from irritation to mucous membrane of mouth; and
3. being environment-friendly and non-toxic due to the marine organism component.

In another invention, Wu (2013b) disclosed an oral decayed-tooth spray for children, containing the marine ingredients that gave antibacterial and anti-inflammatory effects and was convenient for use by being uniformly sprayed on the surface of teeth and capable of eliminating malignant bacteria of oral inflammation, retaining oral flora balance and treating ulcer and inflammation rapidly and moderately. Later Wu (2013c) mentioned an antibacterial anti-inflammatory gargle containing marine biological ingredients. It comprised of ingredients (% w/w): 5.0–8.0 chitosan oligosaccharide, 4.0–7.0 algal polysaccharides, 2.5–3.5 radix angelicae, 1.5–2.5 rheum officinale, 0.01–0.1 almond extracting solution, 5.0–7.0 proteolytic enzyme, and the rest as purified water. The gargle mildly removed harmful bacteria of oral cavity inflammation and maintained oral flora balance.

Several dietary foods for diabetes, cardiovascular diseases and tumor are patented by Wu (2013). A marine dietary food for diabetics made of chitosan oligosaccharide, chitosan, astaxanthin, beta-carotene, coenzyme Q_{10}, vitamin C, vitamin D, vitamin E, vitamin B_1, vitamin B_2, vitamin B_3, vitamin B_6, folic acid, vitamin B_{12}, calcium and magnesium as macro-elements as well as iron, zinc, iodine, copper, manganese, selenium and chromium as trace elements, was also formulated. The chitosan oligosaccharide and other natural active ingredients present in this interacted to beneficially control blood glucose levels in diabetics, relieved the symptoms and reduced the occurrence rate of diabetic complications, so that the food was suitable for long-term use by the diabetics. A special diet food for cardiovascular patients containing oligochitosan as one of the actives was also made available. The combined action of the oligochitosan and other natural active ingredients improved the symptoms of cardiovascular diseases and reduced the occurrence rate of coronary heart disease complications.

Another composite marine special dietary food was made for tumor patients. The synergy effect of the chitosan oligosaccharide and other natural active components made it suitable for long-time uses in tumor patients, particularly patients who were on radio- and chemotherapy (Wu 2013a,b,c). A special diet formula containing chitosan oligosaccharide and oligopeptide was also provided for hepatopaths and related to the technical field of healthcare food (Huan and Nianqiang, 2013). A composition for reducing sclerotin loss was disclosed in another patent by Lee *et al.* (2013). The composition contained calcium compound, marine fish bone collagen oligopeptide powder, glucosamine hydrochloride, casein phosphoric acid peptide, natural vitamin E, vitamin C, vitamin K_1, vitamin B_{12}, vitamin D_3 and dietary fiber. Animal and human tests showed its effect of increasing bone density, enhancing bone intensity, reducing bone loss, delaying senility and enhancing the immunity of the organism; it was also suggested to be used as a healthcare food for middle-aged and elderly people.

18.3.2 Phycocolloids

Alamuru *et al.* (2007) patented a nutritious, tasty and affordable drink made from the sap of *Kappaphycus alvarezii* seaweed, which can be readily cultivated. The process of refining sap enabled the seaweed drink to be palatable and widely accepted, besides bestowing adequate shelf life. The process of preparation was integrated with preparation of carrageenan and plant nutrients, making optimum use of the seaweed and minimizing waste. Bhavanath *et al.* (2006) worked on a halophilic marine *Pseudomonas* sp. isolated from decayed algae, capable of hydrolyzing k-carrageenan. To maximize k-carrageen's production in a cost-effective manner, a novel medium with minimum components was defined along with their optimum concentration by statistical methods. A recent patent of Athwal (2014) was related to cosmetic compositions for the skin. The compositions include any amount or form of fucoidan, any amount or form of beta glucan and any amount of a marine extract. When the cosmetic composition was applied to the skin, the appearance of the skin improved.

Eswaran *et al.* (2005) developed an integrated method for maximum extent of the fresh biomass of seaweeds, such as *Kappaphycus alvarezii,* that can be crushed to release sap, which is useful as a potent liquid fertilizer after suitable treatment with additives and dilution, while the residue was a superior raw material for extraction of k-carrageenan, thereby enhancing the value of the seaweed. Other advantages of the invention included reduced drying time, reduced drying area for the raw material for k-carrageenan production, reduced cost of transporting and storing bulk, easier handling, and direct use in gel preparation. Another work by Ceccoli *et al.* (2009) disclosed an extract of marine algae, *Sarcodiotheca gaudichaudii*, for use in cosmetic compositions. The marine bioferment of *S. gaudichaudii* was useful in anti-aging cosmetic applications, and particularly useful in combination with extract of red algae, such as *Chondrus crispus*.

18.3.3 Carotenoids

The invention by Aki *et al.* (2001) provides a new microorganism with high productivity of a carotenoid substance and a higher unsaturated fatty acid, used as nutritional supplements. It is a method for manufacturing the carotenoid substance and the higher unsaturated fatty acid in high yields. Gross *et al.* (2012) disclosed a method for promoting oral health in animals by a composition comprising an oral health-promoting effect. Olson (2008) invented a composition comprising of cartilage extract, grape seed extract and tomato extract, having remarkable anti-oxidant effects and free-radical inhibition. The composition comprising of a hydrophilic antioxidant, a lipophilic (hydrophobic) antioxidant, and a cartilage extract dramatically increased collagen synthesis in the dermis. Furthermore, the composition lowered collagenase activity and levels of advanced glycation end-products (AGE). Compositions of the invention are intended for the treatment of ageing skin and for delaying the onset of the signs of ageing in healthy skin.

Inventors Montoya-Olvera and Torres-Quiroga (2014) disclosed micro-micelle forms of carotenoids and/or xanthophylls, with improved absorption and bioavailability, providing higher levels of carotenoids in the blood stream. Consequently, it was deposited at the target tissues at a faster rate and more efficiently than crystalline carotenoids. The micro-micelle forms did not contain any crystal forms of carotenoids, and were obtained as micro-micelles after melting Lutein diacetate/Lutein dipropionate in their natural original lipid vegetable matrix, in the presence of lipids, phospholipids, fatty acids, emulsifiers and moisture. The micro-micelle forms can be ingested at a dosage between 2 mg and 50 mg to provide an absorption of the carotenoid solubilizate at least 20% above that of crystalline carotenoids. It provided a macular pigment deposit that exceeded at least 10% of Macular Pigment Optical Density, than the deposition obtained by ingesting crystalline lutein. Dillon and Wolfson (2009) provided oral skin care supplements and nutraceutical compositions and methods of use for improving the health and appearance of skin.

Nutraceutical compositions were also provided, which when orally consumed provided nourishment and delivered essential skin health nutrition to the skin. These nutraceutical compositions included omega-3 rich oils, vitamins, lutein, marine collagen and elastin. It also provided a unique microalgae oral delivery system targeting the health and appearance of skin. Inventor Grune (2008) described a sunblock composition providing ultraviolet radiation protective agents that can be used in various compositions or formulations, specifically those of a higher SPF value that utilized micronized zinc dioxide, micronized titanium oxide, glass microspheres, glycerin and/or aloe, a carrier oil and an emulsifier. The emulsifier was selected from the group comprising fatty acid salts, fatty acid sucrose esters, phospholipids, borate salts, cocoate esters, essential oils, non-petroleum derived waxes, chitosan, water and/or ethanol, a thickening agent, a pigment and a carotenoid or any single or multiple combination of this group of emulsifiers. Miyashita *et al.* (2013) disclosed fucoxanthin extracted from wakame seaweed (*Undaria pinnatifida*) to be effective as an anti-obesity agent.

However, fucoxanthin was contained in mature (adult) brown algae in a small amounts, and accordingly, it took a long time and high cost to obtain fucoxanthin in an amount sufficient for practical use. However, an anti-obesity active agent showing a stronger effect is required and therefore it is desirable to specify a carotenoid exhibiting anti-obesity activity stronger than that of fucoxanthin. Intensive and extensive studies explored the carotenoids as having an anti-obesity activity stronger than that of fucoxanthin and material to be used in order to economically obtain such carotenoid named fucoxanthinol. In the present invention, fucoxanthinol was extracted particularly from internal organs and hard skins of sea squirt (ascidian), and revealed strong anti-obesity activity. Martin (2014) describes a composition which can be oral or topical for treating the cutaneous signs of ageing, comprising collagen and a combination of Reactive Oxygen Species (ROS) scavengers/antioxidants such vitamin C, disulfides, tocopherols/tocotrienols, carotenoids and coenzyme Q10 and collagen precursors and use of such compositions for treating cosmetic ageing. Benade and Opperman (2014) disclosed an omega-3 fatty acid emulsion comprising of omega-3 fatty acids, carotenoids, tocopherols, tocotrienols, zinc, selenium, manganese and water.

18.4 Conclusion

Our diet plays a key role in shaping our overall health. With the recent growing interest in marine natural products and marine foods, it is a challenge for food and pharmaceutical industries to supply healthy food and lifesaving drugs. The fundamental applications of marine products are industrial products, health care/pharmaceutical biotechnology, consumer products, public services and infrastructure. Marine nutraceuticals are an attractive option for the food industry, as there are a multitude of functional food ingredients that can be derived from marine sources. It is also believed that, as people become increasingly aware of the association between diet and good health, the consumption of fishery products will most likely increase (Kadam and Prabhasankar, 2010). The marine environment has historically been used as a source of ingredients such as the marine polysaccharides carrageenan, alginate and agar. Recently, bioactive substances, particularly marine oils and glucosamine, have been added to foods to provide additional health benefits. Functional foods containing omega-3 fats in particular are widely–consumed, with several hundred new products being introduced worldwide each year. Other marine functional foods being developed are bioactive peptides derived from fish muscle, such as ACE inhibitors, gelatin and collagen compounds, and material from shellfish, particularly chitin, chitosan and oligo-chitosan. The marine-derived products playing a diverse role in various areas are compiled in Figure 18.2.

With continued efforts, marine products have a great scope and potential in the future, with the growing number of patents and intellectual property rights in this field. The continued research will help us to understand and fill the gaps in our very

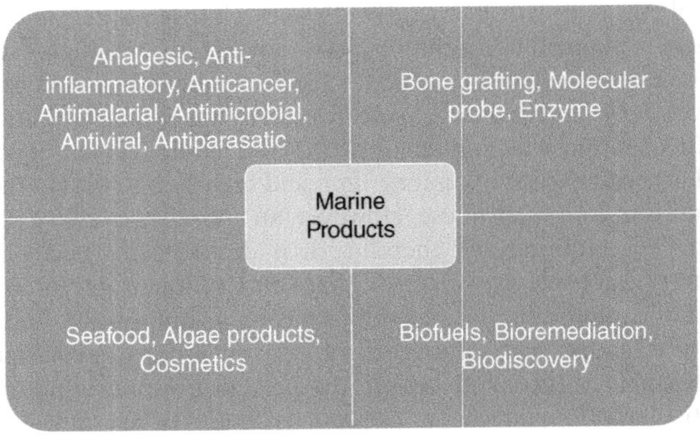

Figure 18.2 Major uses of marine and freshwater products.

little knowledge about the role of these products as biomedical products, biopolymers, bioremediation, and food production and to further its processing. The growing usage of marine products in food, cosmetics and agriculture industries is driving the market for marine biotechnology. Advances in drug discovery, fisheries and aquaculture are also important in driving the market. Lack of exploration in oceans is constraining the growth, as many unknown marine resources (animals and plants) become extinct, even before their proper useful characteristics are discovered.

References

Aki, Y., Kawamoto, M., Ono, K., Shigeta, M. and Suzuki, O. (2001) Marine microorganism, and method for manufacturing carotenoid pigment and/or higher unsaturated fatty acid by using the micro-organism, Patent No. JP2003052357.

Alamuru, V.R.R., Bandyopadhyay, S., Devmurari, C.V., Eswaran, K., Ghosh, P.K., *et al.* (2007) A soft drink and its manufacturing method, Patent No. JP05312325B2.

Aleman, A. and Martinez-Alvarez, Oscar. (2013) Marine collagen as a source of bioactive molecules: A review. *The Natural Products Journal*, **3**(2), 105–114.

Aoki, T., Araki, T. and Kitamikado, M. (1990) Purification and characterization of a novel β-agarase from *Vibrio* sp. AP-2. *European Journal of Biochemistry*, **187**; 461–465.

Athwal, G. (2014) Seaweed derived cosmetic compositions, Patent No. US20140242130A1.

Badwan, A.A., Abumalooh, A., Sallam, E., Abukalaf, A. and Jawan, O. (1985) A sustained release drug delivery system using calcium alginate beads. *Drug Development and Industrial Pharmacy,* **11**(2 & 3): 239–256.

Barrett, S. (2010) Glucosamine and Chondroitin for Arthritis: Benefit is unlikely. Quackwatch. Last accessed 13 July 2014.

Belay, A. (1993) Current knowledge on potential health benefits of *Spirulina platensis*. *Journal of Applied Phycology*, **5**: 235–240.

Benade, A.J.S. and Opperman, A. (2014) Omega 3-fatty acid emulsion, Patent No. CA2830275A1.

Bhakuni, D.S. and Rawat, D.S. (2005) *Bioactive Metabolites of Marine Invertebrates in Bioactive Marine Natural Products*, 1st edition. Anamaya Publishers, New Delhi, pp. 26–54.

Bhavanath, J., Haresh, M.K., Jha, B., Khambhaty, Y.N., Mody, K.H. and Najmuddin, K.Y. (2006) A method for the preparation of K-carrageenase, Patent No. KR1358770B1.

Blazencic, J. (2007) *Sistematika Algi*. Beograd, NNK International, Beograd.

Blunt, J.W., Copp, B.R., Munro, M.H.G., Northcote, P.T. and Prinsep, M.R. (2004) Marine natural products. *Natural Product Reports*, **21**: 1–49.

Brogden, K.A., Ackermann, M., McCray, P.B. and Tack, B.F. (2003) Antimicrobial peptides in animals and their role in host defences. *International Journal of Antimicrobial Agents*, **22**: 465–478.

Burke, C.J., Hsu, T.A. and Volkin, D.B. (1999) Formulation, stability, and delivery of live attenuated vaccines for human use. *Critical Reviews in Therapeutic Drug Carrier Systems*, **16**: 1–83.

Ceccoli, J.D., Costello, B. and Pillai, S. (2009) Marine extracts and bio-fermentations for use in cosmetics, Patent No. US8202519.

Cousens, G. (1985) Report of Treatment of Alzheimer's disease with *Aphanizomenon flosaquae. Orthomedicine*, (Winter/Spring), **8**: 1–2.

Daly, J.W. (2004) Marine toxins and non-marine toxins: Convergence or symbiotic organisms? *Journal of Natural Products*, **67**: 1211.

Debashish, G., Malay, S., Barindra, S. and Joydeep, M. (2005) Marine enzymes. *Advances in Biochemical Engineering/Biotechnology*, **96**: 189–218.

Dillon H.F. and Wolfson J.S. (2009) Oral carotenoid supplementation methods for improving the health and appearance of skin, US Patent No. 20090285850 A1.

El-Baky, H., El-Baz, F. and El-Baroty, G.S. (2007) Production of carotenoids from marine microalgae and its evaluation as safe food colorant and lowering cholesterol agents. *American-Eurasian Journal of Agricultural and Environmental Sciences*, **2**(6): 792–800.

Eswaran, K., Ghosh, P.K., Mehta, A.S., Mody, K.H., Pandya, J.B., *et al.* (2005). Method for production of carrageenan and liquid fertilizer from fresh seaweeds, Patent No. KR689982B1.

FitzGerald, J.R. and Murray, A.B. (2007) Bioactive peptides and lactic fermentations. *International Journal Diary Technology*, **59**: 118–125.

Fleurence, J. (1999) Seaweed proteins: Biochemical, nutritional aspects and potential uses. *Trends in Food Science and Technology*, **10**: 25–28.

Friedman, M. (2006) Nutrition. In: *Food Proteins; Properties and Characterization* (eds S. Nakai and H.W. Modler). VCH, Cambridge, p. 281.9.

Furukawa, S.I., Fujikawa, T. and Koga, D. (1992) Purification and some properties of exo-type fucoidanases from *Vibrio* sp. N-5. *Bioscience, Biotechnology and Biochemistry*, **56**: 1829–1834.

Gameil, A.H.M., Hashim, Y.Z. and Salleh, H.M. (2014) Chitosan micro-carriers in mammalian cell culture. *International Conference on Biological, Chemical and Environmental Sciences* (BCES-2014), 14–15 June, Penang (Malaysia).

Giovannucci, E., Ascherio, A., Rimm, E.B., Stampfer, M.J., Colditz, G.A. and Willett, W.C. (1995) Intake of carotenoids and retinol in relation to risk of prostate cancer. *Journal of the National Cancer Institute*, **87**: 1767–1776.

Giri, A. and Ohshima, T. (2012) Bioactive marine peptides: Nutraceutical value and novel approaches. *International Journal of Food Science and Technology*, **46**: 2001–2014.

Gomez-Guillon, M.C., Lopez-Caballero, M.E., Lopez de Lacey, A., Alemon, A., Gimonez, B. and Montero, P. (2010) Antioxidant and antimicrobial peptide fractions from squid and tuna skin gelatin. In: *Sea By-Products as a Real Material: New ways of application*, Chapter 7 (eds E. Le Bihan and N. Koueta). Transworld Research Network Signpost, Kerala (India), Chapter 7, pp. 89–115.

Gonzalez, M., Caride, B., Lamas, A. and Taboada, C. (2001) Nutritive value of protein from sea urchin, and its effects on intestinal leucine aminopeptidase and intestinal and hepatic gamma-glutamyltranspeptidase. *International Journal of Food Sciences and Nutrition*, **52**(3): 219–224.

Gopinathan, C.P. and Panigrahy, R. (1983) Seaweed resources. *CMFRI Bulletin*, **34**: 47–51.

Graham, R.D. and Rosser, J.M. (2000) Carotenoids in staple foods: Their potential to improve human nutrition. *Food and Nutrition Bulletin*, **21**: 405–409.

Gross, K., Gross K.L., Logan, E., Logan, E.I., Logang, E.I., *et al.* (2012) Method to promote oral health in companion animals, US8168161B2.

Grune, G.L. (2008) Sunblock formulations, US20080219939A1.

Haddar, A., Agrebi, R., Bougatef, A., Hmidet, N., Sellami-Kamoun, A. and Nasri, M. (2009) Two detergent stable alkaline serine-proteases from *Bacillus mojavensis* A21: Purification, characterization and potential application as a laundry detergent additive. *Bioresource Technology*, **100**: 3366–3373.

Hatch, R.A. and Schiller, N.L. (1998) Alginate lyase promotes diffusion of aminoglycosides through the extracellular polysaccharide of mucoid *Pseudomonas aeruginosa*. *Antimicrobial Agents and Chemotherapy*, **42**(4): 974–977.

Haleagrahara, N., Tudawe, D., Chakravarthi, S. and Radhakrishnan, A.K. (2013) Amelioration of collagen-induced arthritis in female dark agouti rats by glucosamine treatment. *ISRN Pharmacology*, **2013**: Article ID 562905, 7 pp.

Hespel, P., Maughan, R.J. and Greenhaff, P.L. (2006) Dietary supplements for football. *Journal of Sports Sciences*, **24**(7): 749–761.

Hosoda, A., Sakai, M. and Kanazawa, S. (2003) Isolation and characterization of agar-degrading *Paenibacillus* spp. associated with the Rhizosphere of spinach. *Bioscience Biotechnology and Biochemistry*, **67**: 1048–1055.

Hua, J., Suguro, S., Hirano, S., Sakamoto, K. and Nagaoka, I. (2005) Preventive actions of a high dose of glucosamine on adjuvant arthritis in rats. *Inflammation Research*, **54**(3): 127–132.

Huan, L. and Nianqiang, J. (2013) Special diet formula containing chitosan oligosaccharide and oligopeptide for hepatopaths, Patent No. CN102715396B.

Hwa, S.S. (2010) Solid supplement additive capable of being efficient to health of human and improving the using convenience of the additive, Patent No. KR1160801B1.

Jayaweera, D.M.A. (1982) *Medicinal Plants (Indigenous and Exotic) used in Sri Lanka*, 1st edition. The National Science Council, Sri Lanka, pp. 1–4.

Jensen, M., Kristensen, M., Belza, A., Knudsen, J.C. and Astrup, A. (2012) Acute effect of alginate-based preload on satiety feelings, energy intake, and gastric emptying rate in healthy subjects. *Obesity* (Silver Spring), **20**(9): 1851–1858.

Jyothi, P.V. and Kumar S.S (2013) Herbal medicines from freshwater macrophytes in Ponnani kole wetlands, India and their usage in different systems of medicine. *Reviews of Progress*, **1**(10): 1–10.

Kadam, S. and Prabhasankar, P. (2010) Marine foods as functional ingredients in bakery and pasta products. *Food Research International*, **43**: 1975–1980.

Khora, S.S. (2013). Marine fish-derived bioactive peptides and proteins for human therapeutics. *International Journal of Pharmacy and Pharmaceutical Sciences*, **5**(3): 31–37.

Kitts, D.D. and Weiler, K. (2003) Bioactive proteins and peptides from food sources. Applications of bioprocesses used in isolation and recovery. *Current Pharmaceutical Design*, **9**: 1309–1323.

Konno, A.I. (2001) Administering a water soluble mixture of Chitin and Chitosans, US Patent No. 6492350.

Ku, C.S., Yang, Y., Park, Y. and Lee, J. (2013) Health benefits of blue-green algae: prevention of cardiovascular disease and non-alcoholic fatty liver disease. *Journal of Medicinal Food*, **16**(2): 103–111.

Le, M. and Xiao-Ming, L. (2010) Effects of lutein and zeaxanthin on aspects of eye health. *Journal of the Science of Food and Agriculture*, **90**: 2–12.

Lee, A., Lixia, K., Zhihui, L., Liying, Y., Xuan, T., *et al.* (2013) Composition for reducing bone loss and preparation method thereof, Patent No. CN102318835.

Leon, O., Quintana, L., Peruzzo, G. and Slebe, J.C. (1992) Purification and properties of an extracellular agarase from *Alteromonas* sp. strain C-1. *Applied Environmental Microbiology*, **58**: 4060–4063.

Lin, D. (2012) Chitosan oligosaccharide marine organism anti-inflammation anus suppository, Patent No. CN103720715A.

Mahajan, G. and Kamat, M. (1995) γ-Linolenic production from *Spirulina platensis*. *Applied Microbiology and Biotechnology*, **43**(3): 466–469.

Martin, R.M. (2014) A composition, Patent No. WO2014041528A2.

Mayer, A.M.S., Glaser, K.B., Cuevas, C., Jacobs, R.S., Kem, W., *et al.* (2010). The odyssey of marine pharmaceuticals: A current pipeline perspective. *Trends in Pharmacological Sciences.* **31**(6): 255–265.

McHugh, D.J. (1987) Production, properties and uses of alginates. *FAO Fish Technology*, **288**: 58–115.

Miled, N., Nasri, M., Kishimura, H. and Rebah, F.B. (2011) Proteins and enzymes from marine resources. *Journal of Amino Acids*, Article ID 594646, 1 page.

Miyashita, K., Hosokawa, M. and Sashima, T. (2013) Fucoxanthinol as anti-obesity agent. Patent No. EP 1884238 B1.

Mo, S.J., Kim, J.H. and Cho, K.W. (2009) Enzymatic properties of an extracellular Phospholipase C purified from a marine Streptomycetes. *Bioscience Biotechnology and Biochemistry*, **73**: 2136–2137.

Montoya-Olvera, R. and Torres-Quiroga J.O. (2014) Compositions and applications of carotenoids of improved absorption and bioavailability, Patent No. US20140099386A1.

Munasinghe, J.U., Dilhan, M.A.A.B. and Sundaramurthy, T.V. (2009) Utilization of aquatic plants: A method to enhance the productivity of water in seasonal tanks in the Anuradhapura District. In: *Proceedings of the National Conference on Water, Food Security and Climate Change in Sri Lanka*, BMICH, Colombo, Sri Lanka, **1**: 23–32.

Nishikawa, T., Urabe, D. and Isobe, M. (2004) An efficient total synthesis of optically active tetrodotoxin. *Angewandte Chemie International Edition*, **43**(36): 4782–4785.

Nishino, H., Murakoshi, M., Li, T., Takemura, M., Kuchide, M., *et al.* (2002) Carotenoids in cancer chemoprevention. *Cancer Metastatis Review*, **21**: 257–264.

Newman, D.J. and Cragg G. M. (2014) Marine-sourced anti-cancer and cancer pain control agents in clinical and late preclinical development. *Marine Drugs*, **12**(1): 255–278.

Ohta, Y. and Hatada, Y. (2006) A novel enzyme, lambda-Carrageenase, isolated from a deep-sea bacterium. *Journal of Biochemistry*, **140**: 475–481.

Olson, B.K. (2008) Combined marine and plant extract compositions, Patent No. US7435432B2.

Pawar, S.N. and Edgar, K.J. (2012) Alginate derivatization: A review of chemistry, properties and applications. *Biomaterials*, **33**: 3279–3305.

Pereira, H., Barreira, L., Figueiredo, F., Custodio, L., Vizetto-Duarte, C., *et al.* (2012) Polyunsaturated fatty acids of marine macroalgae: Potential for nutritional and pharmaceutical applications. *Marine Drugs*, **10**(9): 1920-1935.

Perez, R., Kaas, R., Campello, F., Arbault, S. and Barbaroux, O. (1992) The cultivation of Seaweed in the world, Ifremer. http://archimer.ifremer.fr/doc/00000/4402/.

Rinaudo, M. (2006) Chitin and chitosan: Properties and applications. *Progress in Polymer Science*, **31**: 603–632.

Roberts, J.N., Christopher, B.B., Cynthia, D.T., Rhonda, K., Marcelino, B., *et al.* (2007). Genital transmission of HPV in a mouse model is potentiated by nonoxynol-9 and inhibited by carrageenan. *Nature Medicine*, **13**: 857–861.

Roberts, R.L. and Cabib, E. (1982) *Serratia marcescens* Chitinase: One-step purification and use for the determination of chitin. *Analytical Biochemistry*, **127**: 402–412.

Roos, B., Sneddon, A. and Macdonald, H. (2012) Fish as a dietary source of healthy long chain n-3 polyunsaturated fatty acids (LC n-3 PUFA) and vitamin D. *A Review of Current Literature*, Food and Health Innovation, June.

Rustad, T., Storro, I. and Slizyte, R. (2011) Possibilities for the utilization of marine by-products, *Advances in Food and Nutrition Research*, **65**: 73–105.

Saito, S. (2009) Toxins affecting actin filaments and microtubules. *Progress in Molecular and Subcellular Biology*, **46**: 187–219.

Seto, A., Wang, H.L. and Hesseltine, C.W. (1984) Culture conditions affect eicosapentenoic acid content of *Chlorella minutissima*. *Journal of the American Oil Chemists Society*, **61**(5): 892–894.

Schmidtsdorff, W. (1995) Fish meal and fish oil – not only by-products. In: *Fish and Fishery Products – Composition, Nutritive Properties and Stability* (ed. A Ruiter). CAB International, Wallingford, pp. 347–376.

Simpson, S. (2011) The Blue food revolution: Making aquaculture a sustainable food source. *Scientific American*, **304**: 54–61.

Siriwardhana, N., Kalupahana, N.S. and Moustaid-Moussa, N. (2012) Health benefits of n-3 polyunsaturated fatty acids: Eicosapentaenoic acid and docosahexaenoic acid. *Advances in Food and Nutrition Research*, **65**: 211–222.

Song, Q. and Zhang, X.B. (2008) Characterization of a novel non-specific nuclease from thermophilic bacteriophage GBSV1. *BMC Biotechnology*, **8**: 43.

Suolow, T.V. and Jones, J. (1988) Chitinase-producing bacteria, Patent No. US4751081.

Tobacman, J.K. (2001) Review of harmful gastrointestinal effects of carrageenan in animal experiments. *Environmental Health Perspectives*, **109**(10): 983–984.

Tonnesen, H.H. and Karlsen, J. (2002) Alginate in drug delivery systems. *Drug Development and Industrial Pharmacy*, **28**: 621–630.

Urakova, I N., Pozharitskaya, O.N., Demchenko, D.V., Shikov, A.N. and Makarov, V.G. (2012) The biological activities of fish peptides and methods of their isolation. *Russian Journal of Marine Biology*, **38**(6): 417–422.

Vilchez, C., Forjan, E., Cuaresma, M., Bedmar, F., Garbayo, I. and Vega J.M. (2011) Marine carotenoids: Biological functions and commercial applications. *Marine Drugs*, **9**(3): 319–333.

WHO (2013) Availability and consumption of fish. *Global and regional food consumption patterns and trends*, **3**.5: 3.

Wu, Q. (2013a) Composite marine food for special dietary uses for diabetics, Patent No. CN103431392A.

Wu, Q. (2013b) Complex ocean special-diet food eaten by patients with cardiovascular disease, Patent No. CN103404875A.

Wu, Q. (2013c) Composite marine special dietary food for consumption of tumor patients, CN103404765A.

Yukari O. and Yuji H. (2006) A novel enzyme, λ-carrageenase, isolated from a deep-sea bacterium. *Journal of Biochemistry*, **140**(4): 475–481.

Zheng, J.S., Hu, X.J., Zhao, Y.M., Yang, J. and Li, D. (2013) Intake of fish and marine n-3 polyunsaturated fatty acids and risk of breast cancer: Meta-analysis of data from 21 independent prospective cohort studies. *British Medical Journal*, **346**: f3706.

19

Seafood By-Products in Applications of Biomedicine and Cosmeticuals

Ngo Dang Nghia

Institute of Biotechnology and Environment, Nha Trang University, Vietnam

19.1 Introduction

The oceans hide in their depths a huge number of diverse organisms ranging from primary life forms to highly-structural animals and plants that have provided seafood for humans since the earliest of times. The seafood, with its general nutritional composition including protein with essential amino acids, lipids rich in polyunsaturated fatty acids, carotenoids expressing high activity, and diverse minerals and vitamins, has proved beneficial to health and is consumed more and more. World fish production increased from 71.9 MT (million tonnes) in 1980 to 158 MT in 2012, in which marine fish has been augmented from 64.5 MT to 104.4 MT, respectively (FAO, 2014). In seafood processing, after removing the edible part such as fillets, the rest may constitute 20–60% of the raw material and become the input for production of low-value products like fish meal and fish oil, mostly used in animal feed, or even as fertilizers (Ferraro *et al.*, 2010). Under the pressure of global population increase, especially in the developing countries, the scientific community has reviewed the value of fish by-products and discovered many precious bioactive compounds that are still hidden within the huge quantity of this marine resource (Kim, 2014; Ferraro *et al.*, 2010; Olsen *et al.*, 2014).

Besides the use of marine life as food, the by-products contain many bioactive compounds that possess antioxidative, antimicrobial, anticancer, antihypertension, anti-inflammable and anti-aging properties, preventing cardiovascular diseases, reduction of risk from the environment such as radiation, smoke and oxidative agents. Along with health benefits, these compounds also take part in the anti-aging process, especially in skin care and hair care and have become the preferred cosmeticuals (Ferraro *et al.*, 2010; Kim and Mendis, 2006).

Food Processing By-Products and their Utilization, First Edition. Edited by Anil Kumar Anal.

In this chapter, seafood by-products will be reviewed to provide the panorama from resource to bioactive substances and their application in biomedicine and cosmetics, the areas attracting increasing attention in research and commercialization.

19.1.1 Global Fishery Production

Global total production in 2012 of fish, crustaceans, molluscs and other aquatic animals, more than doubled since 1980, with 71.9 MT and now 158 MT, in which 91.3 MT is from capture and the rest, 66.6 MT, associated with aquaculture. While marine production obtains the main part from capture of 79.7 MT and from inland waters of 11.6 MT, marine aquaculture obtains 24.7 MT, with inland water production being larger at 41.9 MT. Total marine production, including capture and aquaculture, was 104.4 MT in 2012, accounting for 66.1% of the global total production (FAO, 2014).

The main fishing countries are China, followed by Indonesia, the USA, India and Peru. Although there are almost 1600 harvested marine species reported, only 23 major species represent about 40% of the global total capture. Among them, anchovies are the highest production species, followed by Alaska pollock, skipjack tuna, Atlantic herring and chub mackerel. In aquaculture, China is also the leading country with 41.1 MT, followed by India with 4.2 MT and Vietnam with 3.1 MT (FAO, 2014).

The human consumption of fish worldwide is 136.2 MT and the non-food use, mainly fishmeal and fish oil, was 21.7 MT in 2012. The global per capita consumption of fish has increased regularly, at about 1.6% every year, from 11.5 kg in 1980 to 19.2 kg in 2012 (FAO, 2014),

19.1.2 Important Species

In data from 2012 (FAO, 2015), herrings, sardines and anchovies belong to the leading group in capture, with production up to 17.5 MT. Following is the group of cod, hake and haddock at 7.7 MT. One of the important species with a high-value by-product is the salmon, both from capture and aquaculture. The capture production of salmon, trout and smelts was 973 000 T. The fish oil industry is very important in salmon-producing countries such as Norway, the USA and Chile. The most important commercial salmon species is Atlantic salmon (*Salmo salar*), with an aquaculture production of over 2 MT in 2012, higher in 2011 (1.7 MT) and 2010 (1.4 MT) (FAO, 2014). Norway is the most important country in salmon aquaculture, with 40 years of experience. This country has been successful in biological and technological aspects, such as smolt rearing and dry feed, which has laid the foundation for advanced salmon aquaculture (Liu *et al.*, 2011). The Atlantic salmon is now the dominant species, with production of 1 232 095 tonnes in 2012 (Fisheries, Norway).

One of the main resources of fishmeal and fish oil is the by-product from the group of tuna, bonitos and billfishes. In 2012, the capture production of this group was 7.2 MT (FAO, 2014). With tuna only, the by-products can range from 50–70% of raw material (Saadi *et al.*, 2014). The tuna and tuna-like species are one of the main commodities of fisheries. Although there are more than 48 species of tuna distributed throughout the Atlantic, Indian and Pacific oceans and the Mediterranean Sea, they are divided by market into the tropical and temperate tunas. The first category is represented by the bigeye (*Thunnus obesus*), skipjack (*Katsuwonus pelamis*) and yellowfin (*Thunnus albasares*), the latter including the albacore (*T. alalunga*), Atlantic bluefin (*T. thynnus*), Pacific bluefin (*T. orienalis*) and southern bluefin (*T. maccoyii*) (Herpandi *et al.*, 2011). Among the tunas, skipjack tuna is the most important, with a global production of nearly 2.8 MT in 2012 (FAO, 2014).

In crustaceans, shrimp and crab are species where their shells have a broad range of applications from chitin, a polysaccharide in the exoskeleton. This compound and its derivative known as chitosan reveal diverse properties, depending on molecular structures. This group has a capture production in 2012 of about 4.8 MT (FAO, 2014). Shrimp production in Thailand was reduced dramatically by EMS (Early Mortality Syndrome) disease and decreased to 200 000 T in 2014, compared with 500 000 T in 2012 (www.globefish.org, 2014).

19.1.3 Seafood By-Products

The purpose of seafood processing is separation of the edible parts and preserving under different conditions such as salt, low temperature, low water activity or canning. The pre-treatment of the fish includes mechanical operations such as blooding, heading, gutting, filleting, skinning and scaling. After the edible parts are removed, the remains of the fish consist of head, frame, skin, fin, scale, viscera and blood. Depending on the species and the method of handle, the quantity and composition of the by-product will vary. In the case of shrimp/crab processing, the remaining part is the shell.

When the processing is carried out with fishing gear, the by-products are often discarded directly into the sea, due to the limited space on the fishing vessels, which means it is not brought to land. These by-products are only retained on larger ships, which include a large cold room (Olsen *et al.*, 2014).

The main technology to preserve the fish is refrigeration and most of the fishery plants have the facilities for freezing. In these factories, the fish are bled, gutted, filleted, skinned and then the final products, the fillets, are frozen and packaged before distribution. These kinds of the factories have suitable facilities for keeping the by-products in good condition by using cold storage. In developed countries, the by-products can be preserved carefully so that the quality can meet the requirement for human uses.

The situation is not so good in developing countries, where by-products are left outside and spoil quickly, so most of them can only be used as animal feeds or fertilizers. The main method for preservation of these by-products in these areas is sun drying in the open air, with high risk of contamination from insects and bacteria.

In some Asian countries, such as Thailand, Vietnam, the Philippines and Myanmar, an important quantity of small fish, like the anchovy, are used to produce fish sauce. In this technology, the fish is not pre-treated but is fermented as a whole fish (Eong and Hui, 2010). After fermentation, the by-products comprise mainly bone with very small amounts of protein retained. This by-product is mostly used in fertilizers. Some main species with high production will be discussed below, to better estimate the extent of by-products.

19.1.3.1 Salmon Atlantic salmon is cultured very successfully in Norway. The concentration of the large number of farms and neighbouring processing factories allows receiving and treating of the by-products in a short time and so retaining the good quality. This is why by-products from salmon are used to produce high-value products like rich omega-3 fish oil or fishmeal for human consumption. The cold weather in Scandinavia also helps in retaining the quality of the by-products. After removing the fillet, the remainder includes viscera, 9%; head, 18%; backbone, 8.6%; and skin, 2% (Ramirez, 2007). With 40% of raw materials, the total by-products of 2 MT Atlantic salmon every year are estimated at about 800 000 T, resulting in an impressive quantity for fish oil, astaxanthin and fish hydrolysate.

19.1.3.2 Shrimp The waste from shrimp is composed of the head and the exoskeleton (shell and tail) (Cao *et al.*, 2009). Depending on the species, the head, the shell and the tail contribute in different percentages. In the Northern pink shrimp, the head is 38.9%, the shell is 10.7% and the tail is 2.3%, while in the spotted shrimp, these percentages are 37.9%, 11.5% and 2.3%, respectively (Heu *et al.*, 2003). With the red-spotted shrimp in Brazil, the composition of the waste includes the head of 28%, the shell of 8% and the tail of 4% (Sánchez-Camargo *et al.*, 2011). *Penaeus notialis* and *Parapenaeopsis atlantica* have the portions of head at about 33% and 34%, and the shell and tail at about 14% and 17%, respectively. In general, the by-products from shrimp are about 40–50% of the raw materials (Balogun and Akegbejo-Samsons, 1992).

The main chemical composition of the shrimp head is protein, chitin, lipids and carotenoids (Trung and Pham, 2012). The data in Table 19.1 show the high content of protein in the shrimp by-products, from 9–11% (wet basic), but the lipid content is small, from 0.6–0.7% (wet basic), but with a high content of minerals, from 7–8%. When considering only the shrimp head, the protein content is up to 15% (wet basic) or 54% (dry basic), and the carotenoids are up to 20.6 mg/kg. Therefore, in by-product processing, to reduce the chemical consummation and recover higher-value components such as carotenoids and protein, there is a tendency to separate the head and shell by mechanical means. Then the head and the shell are treated with different technologies to obtain the carotenoids, protein and chitin.

19.1.3.3 Tuna Tuna products consist of fresh, frozen and canned commodities. Especially, the sashimi, a traditional meal in Japan, is made from fresh tuna. This product requires the high quality and large-sized tunas such as the bigeye and bluefin. The tuna loins and fillets in general constitute 37.1% and 17.9% of a headless tuna, respectively, while the bone and dark meat obtain 17.9% (Herpandi *et al.*, 2011). The skipjack tuna has a waste portion of about 30%, in which the head accounts for 22.98% and viscera 7% (Balogun and Talabi, 1985). The by-products from tuna include head, backbone, trimmings or cutoffs, skin and intestines.

The tuna has the higher proportion of red muscle (dark meat) than any other fish species. The yellowfin tuna dark muscle is composed of 26.92% protein, 4.87% lipids and 0.97% ash (Sánchez-Zapata *et al.*, 2011). The compositions of amino acids of white meat and red meat in tuna vary, and in the case of the skipjack tuna, the percentages of histidines are 11.371% and 7.940%, lysine are 7.776% and 2.614%, and arginine are 0.398% and 6.729%, respectively (James and Kumar, 2013). The roes of skipjack, tongol and bonito contain 18–20% protein, 3.3–5.7% lipid and 1.8–2.1% ash (Intarasirisawat *et al.*, 2011).

Table 19.1 Proximate chemical composition of shrimp by-products in different species

Species	Moisture	Crude protein	Total lipid	Crude ash	Carotenoids	Chitin	References
White shrimp (*Penaeus vanamei*) (shrimp head)	77.6% (ww)	13.8% (ww)	2.82% (ww)	4.69% (ww)	–	–	(Cao *et al.*, 2009)
Penaeus vannamei (shrimp head)	–	54.4% (dw)	11.9% (dw)	21.2% (dw)	20.6 mg/100g (dw)	9.3% (dw)	(Trung and Pham, 2012)
Parapenaeopsis atlantica (shrimp head)	75.7%	15% (ww)	0.8% (ww)	8.49% (ww)	–	–	(Akegbejo-Samsons and Yemi, 1992)
Pacific white (shrimp shell)	–	43.98% (dw)	1.53% (dw)	18.77% (dw)	1.42% (dw)	34.92% (dw)	(Senphan *et al.*, 2014)
Redspotted shrimp (*Farfantepenaeus*) (shrimp by-products)	–	49% (dw)	4.9% (dw)	27% (dw)	5.3 mg/100g (dw)	–	(Sánchez-Camargo *et al.*, 2011)
Northern pink shrimp (*Pandalus borealis*) (shrimp by-products)	79.1%	9.3% (ww)	0.6%(ww)	8.2% (ww)	–	-	(Heu *et al.*, 2003)
Spotted shrimp (*Trachypena curvirostris*) (shrimp by-products)	78.5%	11.6% (ww)	0.7% (ww)	7.0% (ww)	–	–	(Heu *et al.*, 2003)

www: wet weight basic, dw: dry weight basic

Tuna oil is produced from the head, meat and bone. This oil is rich in polyunsaturated fatty acids (PUFAs), especially eicosapentaenoic acid (EPA; 20:5, n-3) and docosahexaenoic acid (DHA; 22:6, n-3). In Atlantic bluefin tuna, the oil content is in the muscle, liver and adipose tissue at 8%, 12.6% and 73%, respectively. The total n-3 PUFA content in tuna oil depends on the organs used, about 20% in muscle and liver, 24% in adipose tissue, but up to 35% in the ovary (Sprague *et al.*, 2012).

19.1.3.4 Basa Fish Basa fish and Tra fish are the commercial names of fish originated from the Mekong Delta in Vietnam, and now are cultured largely in this location, with an annual production of about 1 MT. In Vietnam, there are these two main species: Tra fish (*Pangasius hypophthalmus*) and Basa fish (*Pangasius bocourti*). The proportion of fillet is about 33–36%, so the total by-products from Pangasius can rise by up to 600 000 MT annually (Nguyen Thi *et al.*, 2007). The by-products from Pangasius are the materials for fishmeal and fish oil. The protein concentration of fishmeal from head and frame of catfish ranges from 37–43%, ash from 17.3–33.8%, with calcium from 8–13%, and lipids from 10.7–21% (Nguyen Thi *et al.*, 2007).

With the global fishery production of 158 MT in 2012 and a main part of this as by-products, the seafood industry provides a huge and valuable resource for marine biomaterials. Ranging from protein, polysaccharide, lipids, to carotenoids, minerals, vitamin and enzymes, the seafood by-products still require more trials to discover new applications. For a long time, this resource was discarded or used in low-value products such as feed or fertilizers. Currently, with further progress in physiology, chemical synthetics, biotechnology, bioengineering and pharmaceuticals, an increasing number of interesting bioactive substances from seafood by-products have been identified, characterized and applied in functional food, pharmaceuticals, biomedicine and cosmetics.

19.2 Seafood By-Products and Biomedicine

The impressive development of science and technology during the last century has built up modern life with more help from machines and transportation. People now spend less energy in working and more time on sedentary activities. In addition, agriculture has achieved many successes, with more food for developed countries. Eating more, but with less exercise, means that most big cities has been introduced to many modern-age chronic diseases such as diabetes, cardiovascular, stroke, cancer, obesity, osteoporosis and stroke (Wildman and Kelley, 2007).

Pharmaceuticals during the last two decades have discovered many effective drugs that can be used to treat chronic diseases. Nevertheless, the drugs based on the chemicals often have many side effects, because the chemicals accumulate in the patient's body, are difficult to degrade and consequently cause many uncomfortable symptoms. In this situation, the new trend to develop new drugs with natural compounds has received more attention. In the past, many drugs based on the compounds extracted from plants have increasingly proved their effectiveness, and the biodiversity of marine organisms has revealed the rich resources for biomedicine (Dias *et al.*, 2012). In addition, with the development of aquaculture, fishery production increases and supports the huge amount of by-products, which can help in reducing the costs involved in drugs manufacture.

19.2.1 Fish Protein Hydrolysate

Fish protein hydrolysates are one of the most interesting products made from the waste generated by fishery processing. The head, frame, bone and viscera still contain important proteins and lipids that are hard to isolate by mechanical methods. Although there are many methods to obtain the protein hydrolysate, such as chemical hydrolysis, enzymatic hydrolysis and microbial fermentation, the enzymatic hydrolysis is the most adaptive (Najafian and Babji, 2012). Enzyme hydrolysis requires a small amount of enzyme, is easily controlled and stopped by deactivating the enzyme by heat. By using different enzymes, diverse fragments with a wide range of properties can be obtained. In addition, enzymes cleave the peptide bonds, but do not destroy the amino acids and neither do they generate chloropropanols (Herpandi *et al.*, 2011). The main enzymes used to generate the biopeptides are alcalase, pepsin, trypsin, chymotrypsin and thermolysin. Every enzyme attaches to a specific target and breaks the parent protein molecules into different fragments.

Protein contents of hydrolysate in most species are 60–90%, with a fat content of below 5%, and improved amino acid composition makes fish protein hydrolysate one of the most important resources for functional food and biopeptides (Chalamaiah *et al.*, 2012). The products obtained after hydrolysis (hydrolysates) are a complicated mixture of polypeptides, oligopeptides and amino acids. Depending on the protein in the raw materials, the methods of hydrolysis, and the conditions of the degradation of protein, the hydrolysate has a distinct composition and so the distribution of molecular weight will have different patterns. The Tilapia by-product hydrolysate is composed mainly of polypeptides of 46.9% and oligopeptides of 35.8%, with a high diversity of 1374 identified peptides (Robert *et al.*, 2014). The protein hydrolysates, therefore, express diverse bioactivities that benefit health.

19.2.1.1 Antioxidative Properties A large number of studies published on protein hydrolysis with antioxidative properties, which are identified in raw materials (species of fish, organs like skin, viscera, frame, head, liver), in conditions of hydrolysis, in enzymes and methods of isolation.

Some authors use only one specific enzyme and concentration to find out the best conditions for hydrolysis. The backbone of the Atlantic cod (*Gadus morhua*), fresh and frozen, was minced and hydrolyzed by Protamex 0.1% by weight of raw material and kept at 55°C over different times periods. The results show that the hydrolysate from fresh backbone gives antioxidative properties better than from a frozen sample. The DPPH radical scavenging activity increased with the time of hydrolysis (Šližytė *et al.*, 2009).

The tuna dark muscle by-products were hydrolyzed by Alcalase PROLYVE 1000 under optimal conditions with temperature of 55°C, time of 60 min and pH of 8.5. These conditions lead to a degree of hydrolysis of about 20%, with a yield of protein solubilization of 89%. To obtain bioactive peptides with molecular weights between 1 and 4 kDa, the combination of ultrafiltration and nanofiltration membranes was applied (Saadi *et al.*, 2014).

Different enzymes were used to obtain diverse fractions of peptides from the same raw materials. The liver of skipjack tuna was hydrolyzed by different proteases, such as Alcalase 2.4L FG, Flavourzyme 500MG, Neutrase 0.8L and Protamex. The process was carried out at a temperature of 50°C and pH of 7 and separated into two steps. In the first step, the liver was hydrolyzed with each of Alcalase, Neutrase, Protamex and Flavourzyme. In the second step, every product (by Alcalase, Neutrase and Protamex) from the first step was hydrolyzed with Flavourzyme. The reason of the combination is that the three enzymes Alcalase, Neutrase and Protamex, are endopeptidase, but Flavourzyme is exopeptidase. The results show that when using the low concentration of enzyme, below 1.0 mg/mL, the DPPH scavenging activity of the second-step hydrolysate is higher than that of the first step. When increasing the concentration of enzyme, the second step has little effect. The two-step hydrolysate has more influence on hydrogen peroxide scavenging activity, hydroxyl radical scavenging and reducing power compared to using only first step (Je *et al.*, 2009).

The viscera of fish contain many enzymes that can be extracted and used for autolysis. The Nile tilapia processing waste was hydrolyzed by enzymes from intestine in comparison to Alcalase. The activity of Alcalase is lower than that of intestine enzymes. This study shows that the waste can be used as an important resource, not only for protein but also for enzymes (Silva *et al.*, 2014).

In general, when breaking the protein into smaller fragments, their bioactivities are expressed more clearly, with every fragment revealing a different activity. The protein hydrolysate from cod (*Gadus morhua*) was fractionated into three parts of >5 kDa, 3–5 kDa and <3 kDa. The <3 kDa fractions had very good radical scavenging activity, Fe^{2+} chelating activity and reducing power, while the fraction 3–5 kDa presented stronger protection against oxidation in a liposome model system (Farvin *et al.*, 2014). However, when breaking down so far, some bioactivities may be reduced. The protein hydrolysate from cod with molecular weight below 1 kDa (through a tangential flow filtration system) was separated into four fractions by reversed-phase high performance liquid chromatography (RP-HPLC). When testing the antioxidant properties, the unfractionated cod protein hydrolysate (<1 kDa) displayed improved capacity to scavenge superoxide, hydroxyl radical, ferric reduction and chelation of iron than the RP-HPLC fractions (Girgih *et al.*, 2015).

19.2.1.2 Anticancer Cancer is one of the fatal diseases of humans. In 2012, an estimated 14 million new cases of cancer per year were reported, and this number is expected to rise to 22 million annually within the next two decades. At the same time, cancer deaths are predicted to rise from an estimated 8.2 million annually to 13 million per year. The highest cancer cases were lung (1.8 million cases, 13.0% of the total), following by breast (1.7 million, 11.9%), large bowel (1.4 million, 9.7%), liver (0.8 million, 9.1%) and stomach (0.7 million, 8.8%) (Stewart and Wild, 2014). Finding the effective drugs is the first priority in global research. There are many different approaches to resolving the problem, from biological cells to molecular techniques. One of the goals in the treatment of the cancer is how to limit the proliferative activity of cancer cells. A number of fishery peptides were proved to inhibit the growth and division of the cancer cells. The hydrolysates from Atlantic salmon, Atlantic cod, plaice, blue whiting, Atlantic emperor, pollock and Portuguese dogfish were tested for anticancer activity in the two human breast carcinoma cell lines, MCF-7/6 and MDA-MB-231. The results have shown that anticancer peptides exist in fish hydrolysates. Among 18 fish protein hydrolysates, 10 samples from blue whiting, cod, plaice and salmon present significant growth inhibitory activity in these two cancer cell lines (Picot *et al.*, 2006). The anticancer activities of biopeptides extracted from oyster (*Saccostrea cucullata*) protein hydrolysate were evaluated *in vitro*. The peptide SCAP 1 had anticancer activities against the human colon carcinoma HT-29 cell line through cytotoxic activity, apoptotic morphological changes and DNA damage (Umayaparvathi *et al*, 2014).

19.2.1.3 Angiotensin I-Converting Enzyme (ACE) Inhibition Fish protein hydrolysates, rich in bioactive peptides, can inhibit ACE and thus reduce blood pressure. Tuna cooking juice with 4% protein was hydrolyzed by orientase enzyme and fractionated to obtain peptides by gel-filtration. The fractions show inhibitory activity in ACE and they reversed the properties under various temperature, levels of pressure and pH (Hwang, 2010). The by-product protein hydrolysates from sardine (*Sardinella aurita*) exhibited the inhibitory activity in ACE with an IC50 value of 1.2 mg/ml (Bougatef *et al.*, 2008). Frame of Alaska pollock (*Theragra chalcogramma*) after hydrolysis also exhibited the inhibition of ACE with the strong active peptide isolated (Je *et al.*, 2004).

Many companies have successfully developed the fish protein hydrolysates with brand names such as Amizate, LIQUAMEN, MOLVAL, Nutripeptin, Seacure, Stabilium and SEAGEST (Chalamaiah *et al.*, 2012).

19.2.2 Carotenoprotein

Carotenoids are the family of organic pigments that are found in plant, algae, bacteria and fungi. The animals and human cannot synthesize these compounds so must ingest them through diet. There are now over 600 carotenoids reported and classified into two types, carotenes and xanthophylls. Lycopene and β-carotene belong to carotene carotenoids, while astaxanthin, canthaxanthin, capsorubin, lutein, violaxanthin and zeaxanthin are xanthophyll carotenoids (Venugopal, 2006). The effects of carotenoids affect animals and humans based on their polarity and the way they configure with membranes (Pashkow *et al.*, 2008).

Among numerous carotenoid compounds, astaxanthin is the most studied because of its wide range of health benefits. It has antioxidative and anti-inflammatory effects and is a high potential agent in quenching the singlet oxygen. The activities of carotenoids are very strong in quenching the singlet oxygen compared to α-tocopherol, from 35–540 times, with astaxanthin as the highest (Shimidzu *et al.*, 1996).

Astaxanthin has three stereoisomers: (3S,3'S), (3R,3'S) and (3R,3'R). The compositions of these isomers vary from species to species. The percentages of (3S,3'S), (3R,3'S) and (3R,3'R) in microalgae and red yeast *Phaffia rhodozyma* are 80%, 11.7% and 8.3% and 14.4%, 10.7% and 74.9%, respectively (Moretti *et al.*, 2006). The bioavailability of every isomer is different. The (3S,3'S) is more highly absorbed than other isomers in the human body, comprising 54% of total astaxanthin in the blood, while only 31% of total astaxanthin is in the administered dose (Coral-Hinostroza *et al.*, 2004).

The role of astaxanthin in cardiovascular health and disease has received much attention from scientists. The medical effects of astaxanthin were researched in the cell model, animal studies and clinical trials (Fassett and Combes, 2012). The studies on rats, rabbits and dogs show the positive effects of astaxanthin on the cardiovascular system, as a reduction in incidence of thrombosis (Lauver *et al.*, 2008), significant reduction of myocardial infarct size (Gross and Lockwood, 2004), reduction in blood pressure (Hussein *et al*, 2005), trend in lowering of lipid peroxidation products (Gross *et al.*, 2006) and reduction in nitric oxide end-products (Hussein *et al.*, 2006). In order to confirm the effects of astaxanthin in humans, clinical studies were organized to investigate the safety, bioavailability and effects on oxidative stress. The volunteers who used the diet with salmon (wild and cultivated) had the plasma concentration of astaxanthin (3S,3'S) isomer higher than its proportionate concentration in the salmon. These results may be explained by the different bioavailability of astaxanthin isomers (Rüfer *et al.*, 2008). Astaxanthin represented the positive effects in reducing the LDL oxidation in healthy volunteers (Iwamoto *et al.*, 2000), significantly decreasing triglycerides and increasing HDL cholesterol in hyperglyceridemic participants (Yoshida *et al.*, 2010). These studies supported the suggestions that the bioactivities of astaxanthin protect the cardiovascular system.

Besides fish protein waste, the by-products from shrimp receive more interest because of the chemical composition of the head and the shell. The data in Table 19.1 show the contents of protein, chitin, lipid, carotenoids and minerals in shrimp by-products. In crustacean like shrimp and crab, the carotenoids are bonded to a high-density lipoprotein and build up the stable complexes referred to as carotenoprotein (Lakshman and Okoh, 1993). Among the carotenoids, the astaxanthin (3,3′-dihydroxy-β, β-carotene-4,4′-dione) is the pigment giving the blue to green colour of crustaceans. When cooked, the astaxanthin is separated partly or completely from the protein moiety to produce the red to orange colour (Schiedt *et al.*, Vecchi and Glinz, 1986). Astaxanthin is used to provide the red-orange colour in salmon meat, because this fish cannot biosynthesize astaxanthin. Under natural conditions, salmonids have to eat the rich astaxanthin-containing microalgae. In aquaculture, people add the astaxanthin into feed to obtain the salmon with red meat (Storebakken *et al.*, 1987).

Composition of major carotenoids (astaxanthin, β-carotene, canthaxanthin, lutein, zeaxanthin and crustacyanin) in different shrimp head waste (*Penaeus monodon* wild and culture, *P. indicus*, *Metapenaeus monocerous*) was determined. The results show that astaxanthin dominated the carotenoids with percentages from 53.3% (culture *P. monodon*) to 67.4% (wild *P. monodon*), with the other carotenoids taking the low levels below 10%. The total carotenoids were highest in the head waste of wild *P. monodon*, with 50.2 μg/g, while in the cultured *P. monodon*, this content was 24.3% (Babu *et al*, 2008).

The carotenoids can be extracted from shrimp waste by ensilage, fermentation or enzymatic hydrolysis, in which the fermentation seems more feasible. Shrimp waste of *Penaeus indicus* was fermented with glucose as carbohydrate substrate and *Lactobacillus plantarum* B 4496 (10^6 cells/ml, 24 h old). In order to compare, the same raw material was ensiled with 15% acid propionic. After 75 days of storage, the oil extracted from carotenoids in fermented silage was 31.3 μg/g, higher than the one in acid silage of 26.18 2μg/g. The results indicate that the fermentation is useful in recovery of carotenoids in shrimp waste (Sachindra *et al*, 2007). In another work, the lactic acid fermentation of *Litopenaeus vannamei* was carried out with *Pediococcus pentosaceus* to obtain the freeze-dried carotenoprotein. This carotenoprotein were hydrolyzed in two treatments:

1. using the commercial protease Savinase; and
2. combining Savinase and Lipolase.

The carotenoprotein obtained after hydrolyzing for 24 h with Savinase rose up to 852 mg/g and total carotenoids of 48.18 mg/g. When combining the two enzymes, the results were higher, 896 mg/g and 66.24 mg/g, respectively. The carotenoprotein and total carotenoids increased when using the two enzymes protease and lipase, probably because of the complexes of protein-lipid-carotenoids. The lipase broke the linkage of lipid and that facilitated the protease to achieve the apoprotein subunits in the carotenoproteins (Armenta and Guerrero-Legarreta, 2009). In shrimp processing, the hepatopancreas is removed and can be the major source of protease, mainly trypsin and chymotrypsin. These proteases can be used to treat the shell in order to obtain carotenoids. From the shell of Pacific white shrimp, carotenoids were extracted by hydrolysis with protease from hepatopancreas proteases of the same species. These carotenoids comprised of astaxanthin in three forms: free, mono- and diester, and had strong antioxidant activity (Senphan *et al.*, 2014).

19.2.3 Bioactive Peptides

To obtain more effectiveness and specific activities, the hydrolysates were separated into individual peptides with identified sequences and tested bioactivities. The enzymatic hydrolysis seems the most appropriate way to derive the peptides from intact proteins in order to reveal the bioactivities. After selection for specific targets, these peptides will be the template for chemical synthesis with high mass production, which is the new trend of pharmaceuticals. The number of identified peptides was reported increasingly but most of them were from the muscle protein hydrolysates and fewer in the by-products. Nevertheless, from the same composition of muscle in fillet/meat and by-products, we can predict the properties of biopeptides in head, broken meat or frame, except for viscera, the skin and exoskeleton.

The bioactive peptides represented many health benefits such as scavenging oxidative stress, antihypertensive agents, modulating immune disorders, anticoagulant activities, opiate-like activity, antimicrobial, anticancer and anti-tumor activities (Saadi *et al.*, 2014). Related to antioxidative property, there are many articles that report the peptide sequences and activities. The two peptides with sequence Leu-Pro-Thr-Ser-Glu-Ala-Lys-Tyr and Pro-Met-Asp-Tyr-Met-Val-Thr extracted from tuna dark muscle after hydrolysis by Orientase and Protease XXIII revealed antioxidative *in vitro* (Hsu, 2010). In tuna cooking juice hydrolysate, three peptides with sequences Pro-Gly-Val-Gly-Pro-Leu-Gly-Pro-Ile-Gly-Pro-Cys-Tyr-Glu, Cys-Ala-Tyr-Gln-Try-Gln-Arg-Pro-Val-Asp-Arg-Ile-Arg and Pro-Ala-Cys-Gly-Gly-Phe-Try-Ile-Ser-Gly-Arg-Pro-Gly showed the inhibition effect of dipeptidyl-peptidase IV with IC50 values of 116.1, 78.0 and 96.4 μM, respectively (Huan *et al.*, 2012). From the sardine (*Sardinella aurita*) by-products, seven short fragments of peptides were isolated including P1: Leu-His-Tyr, P2: Leu-Ala-Arg-Leu, P3: Gly-Gly-Glu, P4: Gly-Ala-His, P5:Gly-Ala-Try-Ala, P6: Pro-His-Tyr-Leu and P7: Gly-Ala-Leu-Ala-Ala-His, in which the first peptide revealed the highest DPPH radical-scavenging activities (Bougatef *et al.*, 2010).

The antihypertensive activity of biopeptides has gained increasing attention from scientists. The peptide Gly-Asp-Leu-Gly-Lys-Thr-Thr-Thr-Val-Ser-Asn-Trp-Ser-Pro-Pro-Lys-Try-Lys-Asp-Thr-Pro derived from tuna frame protein hydrolysate was evaluated for ACE-inhibitory activity. This peptide was shown to decrease systolic blood pressure in the Spontaneously Hypertensive Rats (SHR) model without cytotoxicity (Lee *et al.*, 2010). Other peptides extracted from protein hydrolysate of Alaska pollock frame (Je *et al.*, 2004) and yellowfin sole frame (Jung *et al.*, 2006) also expressed ACE inhibitory activity. The sequences and bioactivities of peptides obtained from some seafood by-products are shown in Table 19.2.

In comparing papers focused on antioxidative and ACE inhibition, the antimicrobial property of biopeptides from by-products is very little. From the liver of skipjack tuna, a 2.3 kDa peptide revealed high homology with the C-terminus of hemoglobin β-chain and showed potent antimicrobial activity against Gram-positive bacteria such as *Bacillus subtilis*, *Staphylococcus aureus* and *Streptococcus iniae* and Gram-negative bacteria such as *Escherichia coli* D31, *Pseudomonas aeruginosa*, *Salmonella enterica* and *Shigella sonnei* (Seo *et al.*, 2014).

Although the list of peptides with sequence, molecular mass and bioactivity is increasingly elongating, the real effects in the human body still need more evidence. While the tests *in vitro* and in animal models of bioactivities are available, clinical trials are still far from making clear the role of bioactive peptides in physiological and health benefits. Owing to expressing the activities in human organisms, the bioavailability of the biopeptides is the important factor. This notion means that the peptides have to retain their structure in the GI tract and be suitable for different absorption routes such as the paracellular route, the passive diffusion, and the invention of a specific transporter, via endocytosis and via the lymphatic system (Saadi *et al.*, 2014). There are some strategies to protect the peptides from enzyme biodegradation, such as eliminating enzyme inhibitors, improving the fragile structures, enhancing capacity of absorption, and designing a genus complex system, for example emulsion, liposomes and muco-adhesion models (Lee, 2002). Among them, microencapsulation is revealed to be an excellent technique to protect the activities and to enhance the bioavailability of peptides in the enterocyte domain. This technique is widely applied in nutraceuticals, controlled drug release in pharmaceuticals and immobilization in biomedicine. The biomaterials used for support are often biopolymers like alginate, chitosan and gelatin (Bickerstaff, 1997; Friess, 1998; Higuera-Ciapara *et al.*, 2004; Lee *et al.*, 2011).

19.2.4 Glycosaminoglycans (GAGs)

Glycosaminoglycans are heteropolysaccharides composed of a repeating disaccharide unit in which one of the two monosaccharides is always an amino sugar (N-acetylgalactosamine or N-acetylglucosamine) and the other is a uronic acid. The GAGs exist in the extracellular matrix and in the animal cell surfaces, with functions to bind and regulate different proteins such as enzymes, cytokines or growth factors. Chondroitin sulfate (CS) and hyaluronic acid (HA) are the most interesting GAGs, because they have many functions and bioactivities that make them useful in biomedicine and cosmetics. The GAGs are often obtained from rooster combs, cartilage and umbilical cords due to their availability. However, many marine animals contain GAGs, such as sponges, sea cucumbers, squids, mollusks and invertebrates. The by-products from fish (i.e. shark, ray, salmon, tuna) that consist of cartilaginous material seem to be potential materials for GAGs production.

Table 19.2 Biopeptides from different marine resources

Resources	Isolation methods	Peptides	Properties	References
Myofibillar protein of Patin (*Pangasius sutchi*)	Papain, alcalase, flavourzyme	VPKNYFHDIV LVMFLDNQHRVIRH FVNQPYLLYSVHMK	Antioxidant	Najafian and Babji (2015)
Skin mucus of yellow catfish (*Pelteobagrus fulvidraco*)	purification	GKLNLFLSRLEILKLFVGAL	Antimicrobial	Su (2011)
Skin of Pacific cod (*Gadus macrocephalus*)	Alcalase, Neutrase, papain, trypsin, pepsin, α-chymotrypsin	TCSP TGGGNV	Antioxidant ACE inhibitory	Ngo *et al.* (2015)
Liver of skipjack tuna (*Katsuwonus pelamis*)	Acidification		Antimicrobial	Seo *et al.* (2014)
Roe of skipjack tuna (*Katsuwonus pelamis*)	Alcalase	DWMKGQ MLVFAV MCYPAST FVSACSVAG LADGVAAPA YVNDAATLLPR DLDLRKDLYAN	Antioxidant ACE inhibitory	(Intarasirisawat *et al.* (2013)
Skin of chum salmon (*Oncorhynchus keta*)	Trypsin	GLPLNLP	ACE inhibitory	(Lee *et al.* (2014)
Cooking juice of tuna	Protease Orientase	PGVGGPLGPIGPCYE CAYQWQRPVDRIR PACGGFYISGRPG	Dipeptidyl-peptidase IV inhibitory	Huang *et al.* (2012)
Frame of Alaska Pollack (*Theragra chalcogramma*)	Pepsin	FGASTRGA	ACE inhibitory	Je *et al.* (2004)

19.2.4.1 Chondroitin Sulfate (CS) Chondroitin sulfate (CS) is the chain of repeating disaccharide units of glucuronic acid (GlcA) and N-acetylgalactosamine (GalNAc) linked by β(1→3) glycosidic bonds and sulfated in different carbon positions. Corresponding to the position of sulfated carbon in the pyranose ring, we have CS-O (non-sulfated), CS-A (carbon 4), CS-C (carbon 6), CS-E (both carbon 4 and 6), CS-D (carbon 6 in GalNAc and carbon 2 in GlcA) and CS-B (carbon 4 of GalNAc and carbon 2 in GlcA) (Malavaki *et al.*, 2008). The composition and concentration of CS are different in organism and tissue, for example CS-D in shark, CS-A and CS-D in dogfish, and CS-E in squid and salmon (Garnjanagoonchorn *et al.*, 2007; Lauder, 2009).

CS is present in the extracellular matrix of connective tissue and is responsible for different functions like elasticity of the articular cartilage, hemostasis and inflammation, regulation of cell development, and proliferation and differentiation (Schiraldi *et al.*, 2010). CS has many applications in tissue engineering, such as repair of bone, cartilage and cutaneous wounds. CS combined with other biopolymers was used to make the scaffolds with controlled biodegradability. CS also revealed many activities such as antiviral activity, and inhibition in different types of cancer. The most successful applications of CS are in cartilage regeneration, anti-inflammation and osteoarthritis (Pothacharoen *et al.*, 2006; Volpi, 2009).

19.2.4.2 Hyaluronic Acid (HA) Hyaluronic acid, also called hyaluronan, HA, discovered by Meyer and Palmer in 1934 in the vitreous humor of cattle eye, is a high molecular weight polysaccharide (Meyer and Palmer, 1934). HA is a linear and unbranched biopolymer, composed of a simple repeating disaccharide: N-acetyl-D-glucosamine (GlcNAc) and D-glucuronic acid (GlcA) linked through a β(1→4) glycosidic bond. The disaccharides are linked through a β(1→3) glycosidic bond to form the polysaccharide chain. The average molecular weight of HA is about 105 to 107 (Vázquez *et al.*, 2013). HA is found in the intercellular matrix of connective tissues, but is mostly extracted from bovine vitreous humor, rooster combs, umbilical cord, synovial fluid and cell walls of Streptococci bacteria. In seafood by-products, HA is found in the vitreous humor of fish and in cartilage of chondrichthyes such as ray and shark (Murado *et al.*, 2011). The concentration of HA in rooster combs is 8–45 g/kg (Nakano *et al.*, 1994), in bovine synovial liquid 15–40 g/kg (Cullis-Hill, 1989), in vitreous humor of swordfish 0.055 g/kg, and in vitreous humor of shark 0.3 g/kg (Murado *et al.*, 2011). Compared to animals, the content of HA in seafood is lower and only some high-production fish species may have commercial potential.

HA can link with a large number of water molecules and form a huge hydrodynamic space in solution. The water-holding capacity and other characteristics create the diversity in physicochemical and biological properties for HA, such as lubricity, viscoelasticity, biocompatibility, angiogenic and immunostimulatory. Based on these bioactivities, HA is used in biotechnology, regenerative medicine and cosmetics, such as arthritis treatment, joint injection and intra-ocular surgery and plastic surgery (Vázquez *et al.*, 2013).

19.2.5 Polyunsaturated Fatty Acids

Fatty acids are carboxylic acids with hydrocarbon chains, fully saturated or containing one or more double bonds. The hydrocarbon chains, ranging from 4 to 36 carbons long, have a common pattern in the location of the double bonds. Most monounsaturated fatty acids have the double bond between C9 and C10, signed Δ9, with the numbering from the carboxyl end, while polyunsaturated ones often present the other double bonds in Δ12 and Δ15. When counting from the methyl end to the position of the first double bond, the fatty acids are signed ω-n, for example ω3 (n-3) or ω6 (n-6). The double bonds in polyunsaturated fatty acids (PUFA) are almost not conjugated and are separated by a methylene group.

In nature, nearly all fatty acids possess the double bonds in cis configuration, but the trans fatty acids are produced in the rumen of dairy animals or in the hydrogenation of fish oil or vegetable oil. The trans fatty acids with high amounts in the diet may relate to an increase the LDL (bad cholesterol) and decrease the HDL (good cholesterol) in the blood (Nelson and Cox, 2008). The fatty acids are normally bonded with glycerol to form mono-, di- or tri-acylglycerols, referred to as acylglycerides, fat or neutral fats. The triacylglycerols with the same kind of fatty acids linked in three positions of glycerol are called simple triacylglycerols. When two or more fatty acids are different, we have the mixed triacylglycerols. Most natural fats contain a mixture of different simple and mixed triacylglycerols.

The human body can synthesize many fatty acids, except the group of PUFA: ω-3 (n-3) and ω-6 (n-6) fatty acids that refer to essential fatty acids. Therefore, this fatty acid group must be taken in the diet and is necessary for health. The two important fatty acids are EPA (Eicosapentaenoic, 20:5(n-3)) and DHA (Docosahexaenoic, 22:6(n-3)). The n-6:n-3 fatty acid ratio recommended by nutrition experts should be less then 5:1 (FAO, 1994). Nowadays, fast food often contains fatty acids with the high ratios of n-6, n-3, and up to 25:1 and may put consumers at risk (Rubio-Rodríguez *et al.*, 2010). Therefore, the omega-3 fatty acids gain more attention and become the important ingredients in many functional foods.

Fish oil, rich of omega-3 fatty acids, is the main resource for pharmaceuticals. The concentrations of lipid, EPA and DHA of some seafood are presented in Table 19.3. The percentage total lipid, EPA and DHA of Atlantic bluefine tuna by-products were

Table 19.3 The lipid, EPA and DHA of some seafood

Species	Lipid (g/100 g wet basic)	EPA (g/100 g total lipid)	DHA (g/100 g total lipid)	n-6/n-3	References
Atlantic salmon	9.92	61.12	302.63	–	Ho and Paul, (2009)
Asian seabass	0.67	1.55	12.90	–	Ho amd Paul, (2009)
Tra catfish	2.55	0.76	10.00	–	Ho amd Paul, (2009)
Black tiger shrimp	1.23	8.58	14.9	0.77	Sriket *et al.*, (2007)
White shrimp	1.30	9.46	9.99	1.0	Sriket *et al.*, (2007)
Northern pink shrimp (by-products)	0.6%	8.9	10.7	–	Heu *et al.* (2003)
Spotted shrimp (by-products)	0.7%	10.7	10.9	–	Heu *et al.* (2003)
Red-spotted shrimp	–	11.69	12.24	0.3	Sánchez-Camargo *et al.* (2011)

5%, 12.4% in liver; 6.7%, 25% in ovary; and 6%, 14.1% in adipose, respectively. In addition, the n-6:n-3 ratio of the above-mentioned by-products range from 0.11–0.18 (Sprague *et al.*, 2012). Ho and Paul (2009) compared the fatty acid profile of three species: Tra catfish (*Pangasius hypophthalmus*), Atlantic salmon (*Salmo solar*) and Asian seabass (*Lates calcarifer*). The crude fats (g/100 g wet basic) were 2.55, 9.92 and 0.67, respectively. The EPA (mg/100 g wet basic) was highest in Atlantic salmon, with 61.12, while very low in Asian seabass and Tra fish, 1.55 and 0.76, respectively. The DHA in Atlantic salmon is very high, 302.6 (g/100 g wet basic), but in Asian seabass and Tra fish, only 12.9 and 10, respectively. Therefore, the salmon by-products should be treated as the main resource for humans in term of omega-3 supplying. Shrimps, although having low lipid concentrations, present an improved profile of fatty acids. The contents of EPA and DHA in red-spotted shrimp was 11.69 (g/100 g total fatty acids) and 12.24 (g/100 g), respectively, with the n-6:n-3 of 0.3 (Sánchez-Camargo *et al.*, 2011). In black tiger shrimp and white tiger shrimp, the EPA was 8.58 (g/100g) and 9.46 (g/100 g), respectively, and DHA was 14.9 (g/100 g) and 9.99 (g/100 g), respectively (Table 19.3).

19.2.6 Chitin/Chitosan

Chitin is a linear polysaccharide composed of β(1→4) linked N-acetyl-D-glucosamine residues, but the acetylation is not filled in all monomers of the polymer chain. Chitin plays the role as a major component in supporting tissues in exoskeletons of crustacean like shrimp, crab, krill, in mollusk like squid pen, and in insects and fungi. Chitin in nature exists in three forms α, β and γ, depending on the orientation of the adjacent chains. In the α form, the chains are arranged in anti-parallel orientation with strong intra- and inter-molecular hydrogen bonds and this is the most abandon form in nature. The β form comprises of parallel chains linked by the weak intra-hydrogen bonds and found in squid pen. The last one, the γ form, is the mixture of the antiparallel and parallel chains, although the existence of this form is still debated (Rinaudo, 2006).

Chitin in the α form, owing to strong hydrogen bonds between the chains, is insoluble in water, most organic acids and diluted acid and alkaline solution. It dissolves in concentrated hydrochloric, sulfuric and phosphoric, dichloroacetic, trichloroacetic and formic acids. In the chitin molecules, the ratio of 2-acetamido-2-deoxy-D-glucopyranose to 2-amino-2-deoxy-D-glucopyranose residues is called the degree of acetylation (DA). The degree of acetylation imposes important effects on the properties of chitin. When the degree of N-acetylation in the polymer chain is less than 0.50, the solution is better and referred to as chitosan (Rinaudo, 2006). However, the properties of chitin and chitosan not only depend on the molecular weight and the DA, but the distribution patterns of the acetyl group along the chain contribute to delicate features of the chitin/chitosan. This characteristic is very important in application of chitin/chitosan, but very hard to control in production. The acetylation (FA) of intact chitin depends on the source, for example in shrimp *P. granulosa* and *P. vularis*, the FA is 0.87 and 0.77, respectively (Galed *et al.*, 2005). When dissolving the chitosan in concentrated acids, the two reactions occur, namely the hydrolysis of the glycosidic linkage (depolymerization) and the N-acetyl linkage (de-N-acetylation). The rate of hydrolysis of a glycosidic linkage following an N-acetylated unit was 54 times higher than the rate of de-N-acetylation and 115 times higher than rate of hydrolysis of a glycosidic linkage following a de-N-acetylated unit. When the hydrolysis is complete, we obtain the monomer glucosamine, the most important product in dietary supplements in the USA (Einbu and Vårum, 2008).

The waste in the processing of shrimp head and shell (including the tail) is the main material for chitin/chitosan processing. The chitin processing consists of two steps: first, removal of the minerals, mostly calcium carbonate by acidic solution and second, obtaining the protein by alkali solution. The remainder of the shell is mainly chitin. This method is still used today owing to its economic effect, is the cause of pollution by the rich-protein waste and so does not recover the value protein, mainly in the head of the shrimp. Some studies used mechanical and enzymatic treatments to obtain protein before isolation of chitin by chemicals. This integrated technology produced more protein, carotenoids, mainly astaxanthin and reduced the chemicals and so is environmentally-friendly. The fermentation of shrimp waste with lactic acid bacteria is also a good alternate for chitin production (Vázquez *et al.*, 2013).

Chitin possesses many properties suitable for use in medicine, such as non-toxicity, biodegradability, physiological inertness, hydrophilicity and gel-forming. Related to gel-forming capacity, chitin is widely-used to immobilize enzymes and cells, to make chitin film and fiber in wound-dressing materials and controlled drug release. Chitin is also used as an excipient in drugs (Rinaudo, 2006).

As the most important derivative of chitin, chitosan has been discovered as useful in biomaterials, pharmaceuticals and cosmetics. Chitosan can dissolve in dilute acidic solution and can be used to make fiber, film, micro-morphologies and nanofibers (Geng *et al.*, 2005). Chitosan is a unique cationic polymer and can combine with other anionic polymers such as alginate to form alginate-chitosan gel that can be applied in immobilization and drug release. Chitosan and its derivatives support the alternative routes of administration of insulin, such as oral, nasal, transdermal and buccal delivery systems (Wong, 2009). Chitosan can protect the insulin from enzymatic degradation and prolong the retention time of insulin. Chitosan and its derivatives have excellent muco-adhesive and absorption-enhancing properties that make them useful in delivery of therapeutic protein and antigens through mucosal routes. Chitosan can interact with mucus and epithelial cells and induces the opening of cellular tight junctions and increases the paracellular permeability of the epithelium (Artursson *et al.*, 1994). Chitosan nanoparticles can be used in controlling DNA release and extend its action (Erbacher *et al.*, 1998). Attaching the drug to the chitosan and introducing this combination to the specific target site in the body, then releasing the drug in a controlled fashion and exhibiting the bioactivity of the drug, are the advanced technologies in pharmaceuticals. Chitosan can be prepared into micro/nanoparticles loaded with the drug for nasal and pulmonary delivery, into injectable hydrogels in the case of cancer therapy (Swatantra *et al.*, 2010).

In tissue engineering, the scaffold materials need to meet requirements to be biocompatible, bio-absorbable, biodegradable and exhibit predictable porosity. The scaffold is not only the framework on which the new tissues grow, but it must facilitate the proliferation and the differentiation of the cells. Chitosan, based on its properties, satisfies many severe requirements in preparation of the scaffold and gains more interest from scientists. The scaffold made from chitosan has been applied in cartilage tissue, intervertebral disc and bone tissue engineering (Di Martino *et al.*, 2005). TGF-ß1 controlled released from collagen/chitosan/GAG and promoted cartilage regeneration. Chitosan also improved the mechanical properties and stability of the collagen network in the scaffold through the inhibition of the collagenases (Lee *et al.*, 2004; Taravel and Domard, 1996). In spin tissue engineering, chitosan can be used in three different fields, such as spine fusion, gene therapy and nucleus pulposus regeneration. Chitosan could be used to develop a composite graft material with predictable macroporous structure and degradation that link the growth factors to support osteogenic cell in spin fusion progress. In intervertebral disc engineering, a chitosan solution was mixed with the cells (nucleus pulposus or mesenchymal stem cells), then was injected into the disc space when the gel was formed at body temperature (37°C). In another approach, the cells were seeded onto the fibrous scaffold made from chitosan and collagen and left to mature *in vitro*. After the cells reached the desired maturation degree, the scaffold was transplanted *in vivo* to repair annulus fibrosis (Di Martino *et al.*, 2005).

Owing to its cationic nature, chitosan can interact with anionic glycosaminoglycans (GAG), proteoglycans. This feature has great meaning in manufacture, because many growth factors are linked to GAG and therefore the scaffold based on chitosan-GAG can retain and concentrate growth factors secreted by colonizing cells (Madihally and Matthew, 1999).

19.2.7 Collagen, Gelatin

Collagen is the common name of a specific abundant protein category, consisting of 29 types identified and constituting about one-third by weight of total protein in the bodies of mammals (Söderhäll *et al.*, 2007). Collagen presents mostly as an extracellular matrix of connective tissues like skin, tendon, cartilage and bone, although it is found in many other organs like blood vessels and even muscle. Collagen has a complex hierarchical conformation, including four structural levels. The primary structure of collagen has the two features:

1. the strict repeat of triplet of amino acid, -(Gly-X-Y)n- along the entire length of the polypeptide, with the total of about 1000 amino acids; and
2. there is the high proportion of proline (X) and hydroxyproline (Y) in the tripeptide sequences.

Hydroxyproline does not commonly appear in other proteins, but in the case of collagen, it constitutes more than 50% of total amino acid content. In the secondary structure, the polypeptide has the α-helix conformation in which glycine, the smallest amino acid, is essential for the formation of the three left-handed helices. Every polypeptide chain possesses two ends with non-helical domains, the C-terminus is involved in the initiation of triplex-helix formation and the N-terminus is related to the regulation of primary fibril diameters. The short non-helical telopeptides of collagen are domains to link with collagen or other molecules. The tertiary structure, triple helix, is formed by the three α-helixes into a right-handed coil, in which all glycines are located inside the coil, while other amino acids remain on the outer surface. The triple helix is a long, rode-shaped conformation, about 1.5 nm in diameter and 300 nm long. Depending on the chain, the triplex helix can be homo- or hetero-trimers. In collagen type I, the triple helix consists of the two α1(I) chains and one α2(I) chain, signed (α1(I))2 α1(I). Collagen type I is found in skin, tendon, bone, cornea, dentin, fibrocartilage, large vessels, intestine, uterus, dentin, dermis and tendon. Collagen type II is homo-trimer with three chains of α1(II), signed (α1(II))3, located in hyaline cartilage, vitreous, nucleus pulposus and notochord (Friess, 1998).

To assembly the quaternary structure of collagen fibrils, five triplex helix molecules are arranged into a quarter-stagger package pattern, highly-oriented with D-periodic banding spaces, with D at about 67 nm. The non-helical regions, called telopeptides, composed of about 20 amino acids at the ends of chains, have an important role in fibrillogenesis and stabilization of the mature collagen molecules through cross-link formation (Gelse, 2003).

The collagen molecules in different tissues have different structures, depending on their functions. The raw materials for production of collagen are often the connecting tissues like skin, tendon, cartilage, bone and fish scale, which give the body its strength. Collagen in these tissues is insoluble and has many covalent cross-links in the structure. The procedure for extraction of collagen from different raw materials (organs and species) therefore requires different solvents, conditions and techniques. The general principles of collagen extraction are to remove lipid, non-collagen protein, and then to extract collagen (dissolve) in salt, acidic, alkali or enzymatic solutions. Because of the thermal sensitivity of linkages in the collagen structure, the processing is carried out under low temperatures of under 4°C. The solution is centrifuged to obtain the supernatant, then precipitated by adjusting the pH, salt concentration or temperature to receive the purified collagen. The neutral salt often used to extract collagen is NaCl, with concentrations of about 0.15 to 2 M (Liu *et al.*, 2012; Minh Thuy *et al.*, 2014; Sadowska and Kołodziejska, 2005).

The dilute acidic solvent of 0.5 M acetic acid or hydrochloric acid pH 2–3 is more efficient then neutral salt solutions. However, the dilute acids cannot dissolve the less labile cross-links such as keto-imime bonds, which are found in bone, cartilage and other tissues from older animals (Bazin and Delaumay, 1976). The salt and dilute acid solutions only extract about 2% of tissue collagen. The dissolved collagen can be reconstituted into large fibrils with similar properties to native fibrils at suitable pH and temperature. The remaining 98% is referred to as insoluble collagen that needs further treatment by strong alkali or enzymes to cleave the cross-links for dissolution. In alkali extraction, the remaining collagen after acid dissolution is treated in a solution of 10% sodium hydroxide and 10% sodium sulfate for 48 h (Cioca, 1981). The protease such as proase, ficin, pepsin and chymotrypsin can give a higher yield of extraction through the selection cleavage in the non-helical regions, breaking the peptide bonds near cross-links and releasing the fragments into solution (Piez, 1984).

After extraction, the soluble collagen can be precipitated by salt or biopolymer such as glycosaminoglycans (GAGs) and polysaccharides. Sadowska and Kolodziejska (2005) have used kappa carrageenan in precipitation of collagen extracted from Baltic cod (*Gadus morhua*). The yield of precipitated collagen increased with the ratio of carrageenan:collagen and completed at 0.4:1.

Gelatin refers to the warm water-soluble mixture of polypeptides obtained from part hydrolysis of collagen. Because the hierarchy structure of collagen is at four structural levels, hydrolysis will first cleave the non-covalent bonds then the covalent bonds, releasing the tertiary structural fragments (triple helix) and secondary structural polypeptides (polypeptide chains). If hydrolysis is carried out more deeply, we will obtain the short fragments of polypeptide and oligopeptides, which are soluble in cold water, called hydrolysate gelatin or, by its more popular commercial name, hydrolysate collagen, mainly used in cosmetics. The hydrolysis of collagen, which can proceed through chemical (acid or alkali) or enzymatic denaturation, is known as the conditioning process in the gelatin industry. Depending on the hydrolysis agents, alkali or acid in the conditioning process, type B (alkali) or type A (acid) gelatin is obtained. When hydrolyzing by alkali, both asparagine and glutamine are almost completely converted into aspartic and glutamic acid. The change in amino acid composition leads to the difference of iso-electric points (IEP), about 8–9 for type A and 4.8–5.5 for type B gelatin (Schrieber and Gareis, 2007). Beside the amino acid composition, the properties of gelatin change over a large range, depending on the molecular weight distribution and therefore, how to control the parameters in conditioning processes to obtain gelatin to satisfy the specific application, which is the main goal of the manufacturer. From a technical aspect, the gelatin is characterized by its gel-forming capacity, viscosity, gelling and melting temperature.

Besides the traditional raw materials such as pig skin, fish by-products have received considerable attention. More studies published relate to the conditions and characterization of collagen and gelatin extracted from skin, scale and fins in different species (Sadowska *et al.*, 2003; Pati *et al.*, 2010; Minh Thuy *et al.*, 2014; Veeruraj *et al.*, 2015; Singh *et al.*, 2011). In terms of technical properties, the raw materials are divided into coldwater fish like cod, Atlantic salmon, Alaska pollock, haddock and hake, and warmwater (the tropical and sub-tropical species) fish represented by tilapia, Nile perch, yellow tuna, channel fish, grass carp and giant squid (Wang *et al.*, 2014). Coldwater fish gelatin has low gelling (4–12°C) and melting (<17°C) temperatures, while in warmwater species, these properties are higher with gelling at 18–19°C and melting at 24–29°C (Aberoumand, 2010). The gel strength and the viscosity of gelatin from coldwater fish are also significantly lower than those of the warmwater species (Gómez-Guillén *et al.*, 2011). There are some studies published on optimization of collagen extraction to obtain the high gel strength value. The gelatin from fish skin of rainbow trout (*Onchorhynchus mykiss*) under optimal conditions had the gel strength of 459 g with the melting point of 20.4°C (Tabarestani *et al.*, 2010). The dorsal skin of yellow tuna showed high gel strength with a Bloom of 426 g, gelling temperature of 18.7°C and melting point of 24.3°C, compared to those of porcine gelatin with 295, 25.6 and 36.5, respectively (Cho., 2005; Karim and Bhat, 2009). The low gelling temperature of fish gelatin can introduce the new application in micro-encapsulation of vitamins, pharmaceutical additives and colorants. The similar features of collagen from warmwater animals can extend the application of fish collagen in biomedicine. In clinics, the collagens have a large spectrum in neural repair, bladder repair, skin substitute, ligament and tendon repair, and also in dental applications (Abou Neel *et al.*, 2013).

One of the important applications of collagen is forming the scaffold in tissue engineering. The selective materials for this technique have to satisfy a series of requirements such as biocompatibility, biodegradability, mechanical stability and more specific features that provide the appropriate signals for directing the cellular processes in tissue formation (Yang *et al.*, 2001). Collagen is a suitable biomaterial for making the scaffold in bone tissue engineering in four groups:

1. injectable hydrogels;
2. membranes and films;
3. sponges and scaffolds; and
4. micro- and nanospheres (Ferreira *et al.*, 2012).

Jellyfish (*Stomolophus nomurai meleagris*) collagen exhibited higher cell viability than bovine collagen, gelatin, hyaluronic acid and glucan. The jellyfish collagen scaffold had a high porosity and interconnected pore structure that was very useful for high-density cell seeding. It also induced an immune response (Song *et al.*, 2006).

Similar to fish protein hydrolysate, the collagen and gelatin from fish skin, scales and squid tunics can be the potential material for bioactive peptides by enzymatic digestion. These peptides showed antimicrobial, antihypertensive, antioxidants, antithrombotic, anticoagulant, antidiabetic, hypocholesterolemic and anticancer activity (Kim, 2014). Skate (*Okamejei kenojei*) skin gelatin hydrolysate below 1 kDa exhibited the highest angiotensin-I converting enzyme (ACE) when digested by alcalase. The two identified peptides, LGPLGHQ, 720 Da and MVGSAPGVL, 829 Da, showed potent ACE inhibition with IC50 values of 4.22 and 3.09 μM, respectively (Ngo *et al.*, 2015). An increasing number of bioactive peptide sequences had been reported from different species, such as fish skin of Alaska pollock with activity in inhibition of lipid peroxidation (Kim *et al.*, 2001) and being antihypertensive (Byun and Kim, 2001); squid skin and tunic of *Dosidicus gigas* with radical scavenging (Mendis *et al.*, 2005) and antihypertensive (Alemán *et al.*, 2011); fish skin of *Jonius belengerii* with inhibition of lipid peroxidation and radical scavenging (Mendis *et al.*, 2005).

19.3 Marine Cosmeticuals

19.3.1 Cosmetics and Cosmeceuticals

Since prehistoric times, the appearance of human beings, particularly the need to look more beautiful, has become an important part of their culture. Although cosmetics now refer to products used to enhance the beauty of skin, eyes, hair and other parts of the body, the early purpose of these products related more to spiritual aspects, seen as protection from evil powers. The ingredients were used in and on the body for religious, medical and cosmetic purposes (Murube, 2013). The Egyptians developed cosmetics very early, with coloring materials for skin, hair and eyes. The ingredients made from botanical sources, such as henna, myrrh, incense, gum arabic and olives; from animals, such as fat, honey and milk; or from minerals such as copper sulfate, zinc oxide, plumbic sulfate and antimony sulfide (Murube, 2013; El-Seedi *et al.*, 2013).

At first, the cosmetic materials just covered the skin with the chosen colours and did not have any long-term effects. The powders, lipsticks, oils and essences helped the skin to achieve an even colour, covering blemishes and wrinkles. Increasingly, people knew how to condition the skin with a mask of nutritional compounds from vegetables, herbs, milk, honey and minerals. The skin, when supported by this nutrition, had a better appearance.

Today, with the knowledge of the structure, biological functions and the aging process of the skin, people can contribute to the physiological functions and slow the aging progress that makes the skin appear degraded with unexpected signs such as wrinkles, dryness, less elasticity and uneven colour. Appearance now tends to be perceived in terms of youth, based on well-being, not only the artificial outside mask (Giacomoni, 2008).

The integration of the two concepts, cosmetics and pharmaceuticals, has introduced a new one, namely cosmeceuticals that enhance beauty based on the health benefits through the ingredients possessing bioactive functions. This concept was coined by Raymond Reed in 1961 and was reactivated by Albert Klingman in 1971 (Preetha and Karthika, 2009). The cosmeceuticals have many effects as activating a receptor, enhancing the barrier function, cellular repair, inhibiting oxidation, modulating pigmentation and protection from solar radiation. These cosmeceuticals now seem be divided into eight categories: retinoids, sunscreen, moisturizers, antioxidants, hydroxyacids, depigmentation, proteins/peptide and growth factors. However, cosmeceutical ingredients can also be classified into anti-inflammatory, depigmenting agents, barrier enhancing agents, antioxidants and skin renewal agents (Abdullah and Nasreen, 2012).

Although the concept of cosmeceuticals is still controversial and not yet approved by the FDA, the market of cosmeceuticals is developing quickly. Following the report Cosmeceuticals: Products and Global Markets 2013, the global sales of topical and injectable cosmeceutical products was valued at $35 billion in 2012 and is expected to rise to about $45.6 billion in 2017, with a 5-year compound annual growth rate (CAGR) of 5.4% in the period from 2012 to 2017 (Reportlinker.com).

Besides supporting the precursor for skin matrix like collagen, hyaluronic acid, chondroitin and minerals, the antioxidants that are believed to retard the aging process attract more interest. At this early stage, the antioxidants extracted from vegetables and herbs have focused on vitamins A, C and E, carotenoids, tocopherols and many pigments that give leaves and fruits attractive colours. Currently, marine biomaterials express strong activities in cosmetics in terms of supporting skin nutrition, protecting the skin from oxidative stress and stimulating the proliferation of cells in skin structure.

19.3.2 Skin Care

In cosmetics, skin care plays the most important role, because it expresses beauty and youth. But the skin is not only the simple outside cover, but also performs many vital functions such as protection of the body from external physical, chemical and biological actions, serves for thermoregulation, and prevents water loss (Kanitakis, 2002). Skin is also a sensitive organ and is related to the brain. Modern skincare strategy is based on an understanding of anatomy, physiology, function and aging of the skin, so as to design an integrated approach including nutrition, protection from stress, and delay the aging process, with the aim to retain the youthful condition of the skin.

The skin is composed of three layers: the epidermis, the dermis and the hypodermis (subcutaneous tissue). The epidermis contains mainly the keratinocytes and is divided into four layers: the basal cell layer (*Stratum germinativum*), the squamous cell layer (*Stratum spinosum*), the granular cell layer (*Stratum granulosum*) and the cornified or horny cell layer (*Stratum corneum*) (James *et al.*, 2006). The keratinocyte is a type of cell whose function is to synthesize keratin, a fibrous intermediate filament that serves as part of the cell's cytoskeleton. The colour of the skin is believed to be the function of the melanocytes, a dendritic, pigment-synthesizing cell found mainly in the basal layer. The melanocytes produce pigment melamine and transfer it to keratinocytes (Chu, 2008). The epidermis is a dynamic tissue and continually undergoing renewal.

The dermis, directly below the epidermis, is integrated into a system of fibrous, filamentous and amorphous connective tissue containing nerves, vascular networks, fibroblasts, macrophages and mast cells. The main component of the dermis is collagen, a major stress-resistant material of the skin, representing 70% of the skin's dry weight (Kolarsick *et al.*, 2011). Combined with collagen, elastin protein is responsible for maintaining the elasticity to the skin. The base of the dermis is a complex system of glycosaminoglycans, glycoproteins, hyaluronic acid and chondroitin sulphate.

The bottom layer of the skin is the hypodermis that contains the fat cell (lipocytes). Considered an endocrine organ, the subcutaneous tissue provides the body with buoyancy and storage of energy. Lipocytes produce leptin, a hormone that regulates the weight of body through the hypothalamus (Kolarsick *et al.*, 2011).

In addition to the natural aging process, there are many factors that influence the premature aging of skin such as diet, smoking, exposure to harsh environments and life style. A regime rich in carbohydrate and saturated fat, an imbalance ratio between omega 6 and omega 3, and high salt and calorie intake, induce the speed of skin aging. The North American diet seems not to support healthy skin (Cosgrove *et al.*, 2007), while the Mediterranean dietary pattern based on more fruits, vegetables and unsaturated fatty acids represents the spectrum of nutrients and bioactive compounds necessary for health and skin (Panagiotakos *et al.*, 2011).

Having a tanned skin is the preference for many European people and this tan is caused by over-exposure to sunlight with its dangerous ultra-violet ray (UV). The long wave UV (UV-A) representing 90–95% in the sunlight reaching the Earth's surface, can penetrate deeper into the epidermis and dermis of the skin, where it induces the oxidative stress that stimulates post-UV inflammation and hyperpigmentation. Although only accounting for 5% in sunlight, the short wave UV (UV-B) causes higher damage to DNA and keratinocytes and is believed to be responsible for non-melanoma skin cancer (Miyamura *et al.*, 2011).

The oxidative stress is one of important reasons for skin aging, therefore scavenging the oxidative compounds is one of the main goals in designing cosmetic products in skincare. The marine biomaterials from seafood by-products, such as collagen hydrolysate, hyaluronic acid, astaxanthin, bioactive peptides, oligopolysaccharides, polyunsaturated fatty acids and chondroitin, represent high potential cosmeticuals for this purpose. These compounds have been described above with their different function and bioactivities. We will now discuss further the mechanism of the most important compounds related to physiology of the skin.

19.3.3 Bioactive Compounds from Seafood By-Products for Skin Care

19.3.3.1 Collagen Hydrolysate The use of collagen hydrolysate is now popular in cosmetic products focusing on skin care, in terms of increasing the moisture and recovery of the collagen system. In structure, it comprises of the small fragment of peptides with the molecular weights ranging from 2–6 kDa. With short chains, these peptides dissolve easily, even in cold water, and are very suitable for preparation. Collagen after hydrolyzation reveals the short fragments of amino acids with side groups exposed, which make the collagen hydrolysate more active in antioxidants, antimicrobials and anti-aging, thus reducing the risk of free radicals in causing chronic diseases. Collagen hydrolysate can be introduced into many food, diary, drink and skin-care products. In seafood by-products processing, collagen hydrolysate is obtained from fish skin, bone and scale. Under endogenous and environmental conditions, the skin collagenous extracellular matrix is degraded by the matrix metalloproteinase (MMPs). This enzyme family can attack all components of extracellular matrix such as collagen, elastin, fibronectin, proteoglycans and laminin. MMP1 and MMP13 initiate the degradation of the type I collagen, MMP9 further breaks up the collagen fragments produced, while MMP2 degrades the type IV collagen, reducing lines and wrinkles (Sárdy, 2009). The oral administration of collagen hydrolysate proved to be beneficial through significantly increasing the type I collagen in the skin of rats with four-fold, type IV collagen, with three-fold compared to the control sample.

While collagen hydrolysate decreased the MMP2 activity; however, the MMP9 activity was not affected by collagen hydrolysate (Zague *et al.*, 2011). The effect in mice of collagen hydrolysate from jellyfish umbrella was tested under the irradiation of UVA and UVB. The results showed that the collagen hydrolysate protected the skin from moisture loss and inhibited wrinkle formation. More importantly, the UV reduces the type I collagen and increases the type III collagen, which means increasing the ratio of type III to type I collagen. In this experiment, with high concentrations of collagen hydrolysate, this ratio was restored (Fan *et al.*, 2013). The fish collagen hydrolysate extracted from skin of Nile tilapia (*Oreochromis niloticus*) expressed the capacity to protect the human reconstructed skin model from damage of UVA. The collagen hydrolysate could strengthen the collagen fibrils and preserved the keratinocytes (Katoet *et al.*, 2011). There are more new collagen hydrolysate-based products developed by cosmetics companies in oral supplement, topical and injectable administration.

19.3.3.2 Hyaluronic Acid Hyaluronic acid is present in the intracellular matrix of the dermis and serves for bulk, lubrication and shock absorption. By binding with water molecules, hyaluronic acid increases the hydration of the skin and provides elasticity and viscosity. In cosmetics, hyaluronic acid can be used through oral, topical or injectable administrations. Its derivatives were first approved by the FDA for soft-tissue augmentation in 2003, under the trade name of Restylan, and now there are 12 dermal filler products based on hyaluronic acid that have been accepted. The aim of this product category, referred to as hyaluronic acid fillers, is to improve the stabilization of the skin via increasing the elasticity, viscosity and residency of the tissue. The hyaluronic acid derivatives can improve the appearance of facial skin in terms of reducing lines and wrinkles, increasing hydration and enhancing elastic properties. The injection sites in the facial area are mental groove, infraorbital and supraorbital hollows, soft acne or scars, temporal hollow, malar region and philtral columns. Depending on the treatment area, the longevity of hyaluronic acid can be 3–4 months in lips, and more than 6 months in tear trough (Rohrich *et al.*, 2007). Beside the role of a skin constituent, hyaluronic acid is also used as a topical vehicle for delivery of drugs to the skin. More attention has been paid to hyaluronic acid because this matrix can satisfy the requirements in this field, such as the permeation through stratum corneum, localization of drugs to target tissue, but avoidance of diffusion all the way through the skin into the blood system. One example in this application is the product Solaraze®, a mixture of 3% diclofenac in 2.5% hyaluronic acid gel, which expressed effective results in treatment of actinic keratoses (Brown and Jones, 2005). For enhancing the permeation of drug, a hyaluronic acid-based nano-emulsion has been developed. With α-tocopherol as the model ingredient to assess skin permeability and bioavailability, the nano-emulsion based on hyaluronic acid exhibited a high capacity to penetrate across stratum corneum and diffuse deeper into the dermis in rat dorsal skin. These results suggest hyaluronic acid nano-emulsion can be used as a transdermal carrier for active lipophilic ingredients in skin care for drug and cosmetic delivery (Kong *et al.*, 2011).

19.3.3.3 Astaxanthin With a structure including the two polar ends and the nonpolar middle part, astaxanthin fits to the cell membranes and plays the role of a bridge for conduction of electrons through the membranes. Therefore, astaxanthin provides the cell membranes with the capacity of protection against free radicals or other types of oxidative stress (Kidd, 2011). Astaxanthin also reveals the potent protection of DNA from oxidative damage When receiving 2 mg/day or 8 mg/day, both astaxanthin dosages lowered the plasma 8-hydroxy-2′-deoxyguanosine (8-OHdG), an indicator of oxidative DNA breakdown, in healthy women (Park *et al.*, 2010). In heavy smokers, after administration of astaxanthin for 3 weeks, the two oxidative biomarkers, plasma malondialdehyde (MDA) and isoprostanes (ISP) levels decreased (Kim *et al.*, 2011). Astaxanthin from *Haematococcus pluvialis* was confirmed in its safety for the skin, increasing moisture retention, and improving moistness, smoothness and elasticity (Seki *et al.*, 2001). When combining the oral supplementation and topical treatment, astaxanthin showed improvements in skin wrinkles, age spot size, elasticity, skin texture, moisture content of the corneocyte layer and corneocyte condition. The mechanism of action of astaxanthin in wrinkle reduction may be explained by the recovery of the collagen system in the dermis through scavenging the singlet oxygen and therefore protecting the skin from oxidative damage (Tominaga *et al.*, 2012). Astaxanthin also inhibited the melanogenesis in age spots by suppressing the oxidative polymerization in melanocytes and inflammation in the epidermis (Yamashita, 1995).

19.4 Conclusions

With the diverse and high bioactive compounds reported, seafood by-products processing is developing and showing impressive successes, going beyond the initial opinion as the waste from seafood processing and now achieving high value-added products. In the situation of rising side effects from chemicals in medicine and cosmetics, the trend of using natural extracts, especially marine materials, has been receiving more attention. The results of numerous works, from laboratory to industrial scale, from physiology and anatomy to commercial products, have showed the potentials and perspectives of bioactive substances extracted from seafood by-products for humans in biomedicine and cosmetics.

References

Abdullah, B. and Nasreen, R. (2012) Cosmeceuticals: A revolution in cosmetic market. *International Journal of Pharmacy and Technology*, **4**(1): 3925–3942.

Aberoumand, A. (2010) Isolation and characteristics of collagen from fish waste material. *World Journal of Fish and Marine Science*, **2**(5): 471–474.

Abou Neel, E.A, Bozec, L., Knowles, J.C., Syed, O., Mudera, V., *et al.* (2013) Collagen-emerging collagen based therapies hit the patient. *Advanced Drug Delivery Reviews*, **65**(4): 429–456.

Balogun, A.M. and Akegbejo-Samsons, Y. (1992) Waste yield, proximate and mineral composition of shrimp resources of nigeria's coastal waters. *Bioresource Technology*, **40**: 157–161.

Alemán, A., Giménez, B., Pérez-Santin, E., Gómez-Guillén, M.C. and Montero, P. (2011) Contribution of Leu and Hyp residues to antioxidant and ACE-inhibitory activities of peptide sequences isolated from squid gelatin hydrolysate. *Food Chemistry*, **125**(2): 334–341

Armenta, R. E. and Guerrero-Legarreta, I. (2009) Amino acid profile and enhancement of the enzymatic hydrolysis of fermented shrimp carotenoproteins. *Food Chemistry*, **112**(2): 310–315.

Artursson, P., Lindmark, T., Davis, S. and Illum, L. (1994) Effect of chitosan on the permeability of monolayers of intestinal epithelial cells (Caco-2). *Pharmaceutical Research*, **11**: 1358–1361.

Babu, C.M., Chakrabarti, R. and Surya Sambasivarao, K.R. (2008) Enzymatic isolation of carotenoid-protein complex from shrimp head waste and its use as a source of carotenoids. *LWT – Food Science and Technology*, **41**(2): 227–235.

Balogun, A.M. and Talabi, S.O. (1985) Proximate analysis of the flesh and anatomical weight composition of skipjack tuna (*Katsuwonus pelamis*). *Food Chemistry*, **17**(2): 117–123.

Bazin, S., AND Delaumay, A. (1976). Preparation of acid and citrate soluble collagen. In: *The Methodology of Connective Tissue Research* (ed. D.A. Hall). Joynson–Bruvvers, Oxford, pp. 13–18.

Bickerstaff, G.F. (ed.) (1997) *Immobilisation of Enzymes and Cells*. Humana Press, New Jersey.

Bougatef, A., Nedjar-Arroume, N., Ravallec-Plé, R., Leroy, Y., Guillochon, D., *et al.* (2008) Angiotensin I-converting enzyme (ACE) inhibitory activities of Sardinelle (*Sardinella aurita*) by-products protein hydrolysates obtained by treatment with microbial and visceral fish serine proteases. *Food Chemistry*, **111**(2): 350–356.

Bougatef, A., Nedjar-Arroume, Naïma Manni, L.,Ravallec-plé, R., Barkia, A., *et al.* (2010) Purification and identification of novel antioxidant peptides from enzymatic hydrolysates of sardinelle (*Sardinella aurita*) by-products proteins. *Food Chemistry*, **118**: 559–565.

Brown, M.B. and Jones, S A. (2005) Hyaluronic acid: A unique topical vehicle for the localized delivery of drugs to the skin. *Journal of the European Academy of Dermatology and Venereology*, **19**: 308–318.

Byun, H.-G. and Kim, S.-K. (2001) Purification and characterization of angiotensin I converting enzyme (ACE) inhibitory peptides from Alaska pollock (*Theragra chalcogramma*) skin. *Process Biochemistry*, **36**(12): 1155–1162.

Cao, W., Zhang, C., Hong, P., Ji, H., Hao, J. and Zhang, J. (2009) Autolysis of shrimp head by gradual temperature and nutritional quality of the resulting hydrolysate. *LWT – Food Science and Technology*, **42**(1): 244–249.

Chalamaiah, M., Dinesh Kumar, B., Hemalatha, R. and Jyothirmayi, T. (2012) Fish protein hydrolysates: Proximate composition, amino acid composition, antioxidant activities and applications: A review. *Food Chemistry*, **135**(4): 3020–3038.

Cho, S.M., Gu, Y.S. and Kim, S.B. (2005) Extracting optimization and physical properties of yellowfin tuna (*Thunnus albacares*) skin gelatin compared to mammalian gelatins. *Food Hydrocolloids*, **19**(2): 221–229.

Cioca, G. (1981) Process for preparing macromolecular biologically active collagen, US Patent 4279812.

Chu, D.H. (2008) Development and structure of skin. In: *Fitzpatrick's Dermatology in General Medicine* (eds L.A. Goldsmith, S.L. Katz, B.A. Gilchrest, A.S. Paller, D.J. Leffell and K. Wolff), McGraw Hill Education, US, pp. 57–91.

Coral-Hinostroza, G.N., Ytrestøyl, T., Ruyter, B. and Bjerkeng, B. (2004) Plasma appearance of unesterified astaxanthin geometrical E/Z and optical R/S isomers in men given single doses of a mixture of optical 3 and 3'R/S isomers of astaxanthin fatty acyl diesters. *Comparative Biochemistry and Physiology. Toxicology and Pharmacology□: CBP*, **139**(1–3): 99–110.

Cosgrove, M.C., Franco, O.H., Granger, S.P., Murray, P.G. and Mayes, A E. (2007) Dietary nutrient intakes and skin-aging appearance among middle-aged American women. *The American Journal of Clinical Nutrition*, **86**: 1225–1231.

Cullis-Hill, D. (1989) Preparation of hyaluronic acid from synovial fluid, US Patent 4879375A.

Dias, D.A., Urban, S. and Roessner, U. (2012) A historical overview of natural products in drug discovery. *Metabolites*, **2**: 303–336.

Di Martino, A., Sittinger, M. and Risbud, M.V. (2005) Chitosan: a versatile biopolymer for orthopaedic tissue-engineering. *Biomaterials*, **26**(30): 5983–5990.

Einbu, A. and Vårum, K.M. (2008) Characterization of chitin and its hydrolysis to GlcNAc and GlcN. *Biomacromolecules*, **9**: 1870–1875.

El-Seedi, H.R., Burman, R., Mansour, A., Turki, Z., Boulos, L., *et al.* (2013).The traditional medical uses and cytotoxic activities of sixty-one Egyptian plants: Discovery of an active cardiac glycoside from *Urginea maritima*. *Journal of Ethnopharmacology*, **145**(3): 746–757.

Eong, Y.S., and Hui, D.T.B. (2010) *Good Manufacturing Practice (GMP) for Small and Medium-sized Traditional Fish Products Processing Establishments in ASEAN*. Marine Fisheries Research Department, Southeast Asian Fisheries Development Center, Singapore.

Erbacher, P., Zou, S., Bettinger, T., Steffan, A. and Remy, J. (1998) Chitosan-based vector/DNA complexes for gene delivery: Biophysical characteristics and transfection ability. *Pharmaceutical Research*, **15**: 1332–1339.

Fan, J., Zhuang, Y. and Li, B. (2013) Effects of collagen and collagen hydrolysate from jellyfish umbrella on histological and immunity changes of mice photoaging. *Nutrients*, **5**: 223–233.

FAO (1994) *Fats and Oils in Human Nutrition*. FAO, Rome.

FAO (2014) *FAO Year Book: Fishery and Aquaculture Statistics 2012*. FAO, Rome.

Farvin, K.H.S., Lystbæk, L., Hauch, H., Jacobsen, C., Jakobsen, G., *et al.* (2014).Antioxidant activity of Cod (*Gadus morhua*) protein hydrolysates: *In vitro* assays and evaluation in 5% fish oil-in-water emulsion. *Food Chemistry*, **149**(0): 326–334.

Fassett, R.G. and Coombes, J.S. (2012) Astaxanthin in cardiovascular health and disease. *Molecules (Basel, Switzerland)*, **17**(2): 2030–2048.

Ferraro, V., Cruz, I.B., Jorge, R.F., Malcata, F.X., Pintado, M.E. and Castro, P.M.L. (2010) Valorisation of natural extracts from marine source focused on marine by-products: A review. *Food Research International*, **43**(9): 2221–2233.

Ferreira, A.M., Gentile, P., Chiono, V. and Ciardelli, G. (2012) Collagen for bone tissue regeneration. *Acta Biomaterialia*, **8**(9): 3191–3200.

Friess, W. (1998) Collagen – Biomaterial for drug delivery 1. *European Journal of Pharmaceutics and Biopharmaceutics*, **45**: 113–136.

Galed, G., Miralles, B., Paños, I., Santiago, A. and Heras, Á. (2005) N-Deacetylation and depolymerization reactions of chitin/chitosan: Influence of the source of chitin. *Carbohydrate Polymers*, **62**: 316–320.

Garnjanagoonchorn, W., Wongekalak, L. and Engkagul, A. (2007) Determination of chondroitin sulfate from different sources of cartilage. *Chemical Engineering and Processing*, **46**: 465–471.

Gelse, K. (2003) Collagens – Structure, function, and biosynthesis. *Advanced Drug Delivery Reviews*, **55**(12): 1531–1546.

Geng, X., Kwon, O. and Jang, J. (2005) Electrospinning of chitosan dissolved in concentrated acetic acid solution. *Biomaterials*, **26**: 5427–5432.

Giacomoni, P.U. (2008) Advancement in skin aging: *The future cosmeceuticals*, **26**(4): 364–366.

Girgih, A.T., He, R., Hasan, F.M., Udenigwe, C.C., Gill, T.A. and Aluko, R.E. (2015) Evaluation of the *in vitro* antioxidant properties of a cod (*Gadus morhua*) protein hydrolysate and peptide fractions. *Food Chemistry*, **173**: 652–659.

Gómez-Guillén, M.C., Giménez, B., López-Caballero, M.E. and Montero, M.P. (2011) Functional and bioactive properties of collagen and gelatin from alternative sources: A review. *Food Hydrocolloids*, **25**(8): 1813–1827.

Gross, G.J. and Lockwood, S.F. (2004) Cardioprotection and myocardial salvage by a disodium disuccinate astaxanthin derivative (Cardax). *Life Sciences*, **75**(2): 215–224.

Gross, G.J., Hazen, S. and Lockwood, S. (2006) Seven day oral supplementation with Cardax (disodium disuccinate astaxanthin) provides significant cardioprotection and reduces oxidative stress in rats. *Molecular and Cellular Biochemistry*, **283**: 23–30.

Herpandi, N.H., Rosma, A. and Wan Nadiah, W.A. (2011) The Tuna fishing industry: A new outlook on fish protein hydrolysates. *Comprehensive Reviews in Food Science and Food Safety*, **10**(4): 195–207.

Heu, M.-S., Kim, J.-S. and Shahidi, F. (2003) Components and nutritional quality of shrimp processing by-products. *Food Chemistry*, **82**(2): 235–242.

Higuera-Ciapara, I., Felix-Valenzuela, L., Goycoolea, F.M. and Argüelles-Monal, W. (2004) Microencapsulation of astaxanthin in a chitosan matrix. *Carbohydrate Polymers*, **56**(1): 41–45.

Ho, B. and Paul, D. (2009) Fatty acid profile of Tra Catfish (*Pangasius hypophthalmus*) compared to Atlantic Salmon (*Salmo solar*) and Asian Seabass (*Lates calcarifer*). *International Food Research Journal*, **16**: 501–506.

Hsu, K. (2010) Purification of antioxidative peptides prepared from enzymatic hydrolysates of tuna dark muscle by-product. *Food Chemistry*, **122**: 42–48.

Huang, S.-L., Jao, C.-L., Ho, K.-P. and Hsu, K.-C. (2012) Dipeptidyl-peptidase IV inhibitory activity of peptides derived from tuna cooking juice hydrolysates. *Peptides*, **35**(1): 114–121.

Hussein, G., Sankawa, U., Goto, H., Matsumoto, K. and Watanabe, H. (2006) Astaxanthin, a carotenoid with potential in human health and nutrition. *Journal of Natural Products*, **69**: 443–449.

Hwang, J.-S. (2010) Impact of processing on stability of angiotensin I-converting enzyme (ACE) inhibitory peptides obtained from tuna cooking juice. *Food Research International*, **43**(3): 902–906.

Intarasirisawat, R., Benjakul, S. and Visessanguan, W. (2011) Chemical compositions of the roes from skipjack, tongol and bonito. *Food Chemistry*, **124**(4): 1328–1334.

Intarasirisawat, R., Benjakul, S., Wu, J. and Visessanguan, W. (2013) Isolation of antioxidative and ACE inhibitory peptides from protein hydrolysate of skipjack (*Katsuwana pelamis*) roe. *Journal of Functional Foods*, **5**(4): 1854–1862.

Iwamoto, T., Hosoda, K., Hirano, R., Kurata, H., Matsumoto, A., *et al.* (2000) Inhibition of low-density lipoprotein oxidation by astaxanthin. *Journal of Atheroscler Thromb*. **7**: 216–222.

James, A., Pople, J., Parish, W., Moore, A. and Dunbar, N. (2006) Histological evaluation of hyperpigmentation on female Filipino axillary skin. *International Journal of Cosmetic Science*, **28**(4): 247–253.

James, R. and Kumar, V.T.V. (2013) Variation of amino acids in white and red meat of *Skipjack tuna (Katsuwonus pelamis) caught from Arabian Sea*. **2**(7): 2843–2846.

Je, J.-Y., Park, P.-J., Kwon, J.Y. and Kim, S.-K. (2004) A novel angiotensin I converting enzyme inhibitory peptide from Alaska Pollack (*Theragra chalcogramma*) frame protein hydrolysate. *Journal of Agricultural and Food Chemistry*, **52**(26): 7842–7845.

Je, J.-Y., Lee, K.-H., Lee, M.H. and Ahn, C.-B. (2009) Antioxidant and antihypertensive protein hydrolysates produced from tuna liver by enzymatic hydrolysis. *Food Research International*, **42**(9): 1266–1272.

Jung, W.-K., Mendis, E., Je, J.-Y., Park, P.-J., Son, B.W., *et al.* (2006) Angiotensin I-converting enzyme inhibitory peptide from yellowfin sole (*Limanda aspera*) frame protein and its antihypertensive effect in spontaneously hypertensive rats. *Food Chemistry*, **94**(1): 26–32.

Kanitakis, J. (2002) Anatomy, histology and immunohistochemistry of normal human skin. *European Journal of Dermatology*, **12**(4): 390–399.

Karim, A. and Bhat, R. (2009) Fish gelatin: properties, challenges, and prospects as an alternative to mammalian gelatins. *Food Hydrocolloids*, **23**(3): 563–576.

Katherine, E., Christina-Maria, K., Panagiotakos B,D. and Giugliano, D. (2011) Mediterranean diet and weight loss: Meta-analysis of randomized controlled trials. *Metabolic Syndrome and Related Disorders*., **9**(1): 1–12.

Kato, S., Matsui, H., Saitoh, Y. and Miwa, N. (2011) Fish collagen-containing drink is subcutaneously absorbed and attenuates the UVA-induced tissue-integrity destruction and DNA damages in 3D-human skin tissue model. *Journal of Functional Foods*, **3**(1): 50–55.

Kidd, P. (2011) Astaxanthin, cell membrane nutrient with diverse clinical benefits and anti-aging potential. *Alternative Medicine Review*, **16**(4): 355–364.

Kim, J.H., Chang, M.J., Choi, H.D., Youn, Y.-K., Kim, J.T., *et al.* (2011) Protective effects of haematococcus astaxanthin on oxidative stress in healthy smokers. *Journal of Medicinal Food*, **14**(11): 1469–1475.

Kim, S. (ed.) (2014) *Seafood Processing By-Products: Trends and Applications*. Springer, New York.

Kim, S.-K. and Mendis, E. (2006) Bioactive compounds from marine processing byproducts – A review, **39**: 383–393.

Kim, S., Kim, Y., Byun, H., Nam, K., Joo, D. and Shahidi, F. (2001) Isolation and characterization of antioxidative peptides from gelatin hydrolysate of Alaska pollock skin. *Journal of Agricuture and Food Chemistry*, **49**(4): 1984–1989.

Kong, M., Chen, X.G., Kweon, D.K. and Park, H.J. (2011) Investigations on skin permeation of hyaluronic acid based nanoemulsion as transdermal carrier. *Carbohydrate Polymers*, **86**(2): 837–843.

Lakshman, M.R. and Okoh, C. (1993) *Carotenoids Part B: Metabolism, Genetics, and Biosynthesis. Methods in Enzymology*. **214**: 74–86.

Lauder, R. (2009) Chondroitin sulphate: A complex molecule with potential impacts on a wide range of biological systems. *Complementary Therapies in Medicine*, **17**: 56–62.

Lauver, D.A., Driscoll, E.M. and Lucchesi, B.R. (2008) Disodium disuccinate astaxanthin prevents carotid artery rethrombosis and *ex vivo* platelet activation. *Pharmacology*, **82**: 67–73.

Lee, H. (2002) Protein drug oral delivery: The recent progress. *Archives of Pharmacal Research*, **25**(5): 572–584.

Lee, J., Kim, S., Kwon, I., Ahn, H., Cho, H., *et al.* (2004) Effects of a chitosan scaffold containing TGF-beta1 encapsulated chitosan microspheres on *in vitro* chondrocyte culture. *Artificial Organs*, **28**(9): 829–839.

Lee, J.K., Jeon, J.-K, and Byun, H.-G. (2014) Antihypertensive effect of novel angiotensin I converting enzyme inhibitory peptide from chum salmon (*Oncorhynchus keta*) skin in spontaneously hypertensive rats. *Journal of Functional Foods*, **7**: 381–389.

Lee, J.-S., Park, S.-A., Chung, D. and Lee, H.G. (2011) Encapsulation of astaxanthin-rich Xanthophyllomyces dendrorhous for antioxidant delivery. *International Journal of Biological Macromolecules*, **49**(3): 268–73.

Lee, S., Qian, Z. and Kim, S. (2010) A novel angiotensin I converting enzyme inhibitory peptide from tuna frame protein hydrolysate and its antihypertensive effect in spontaneously hypertensive rat. *Food Chemistry*, **118**: 96–102.

Liu, D., Liang, L., Regenstein, J.M. and Zhou, P. (2012) Extraction and characterisation of pepsin-solubilised collagen from fins, scales, skins, bones and swim bladders of bighead carp (*Hypophthalmichthys nobilis*). *Food Chemistry*, **133**(4): 1441–1448.

Liu, Y., Olaf Olaussen, J. and Skonhoft, A. (2011) Wild and farmed salmon in Norway – A review. *Marine Policy*, **35**(3): 413–418.

Madihally, S. and Matthew, H. (1999) Porous chitosan scaffolds for tissue engineering. *Biomaterials*, **20**(12): 1133–1142.

Malavaki, C., Mizumoto, S., Karamanos, N. and Sugahara, K. (2008) Recent advances in the structural study of functional chondroitin sulfate and dermatan sulfate in health and disease. *Connective Tissue Research*, **49**: 133–139.

Mendis, E., Rajapakse, N. and Kim, S. (2005a) Antioxidant properties of a radicals scavenging peptide purified from enzymatically prepared fish skin gelatin hydrolysate. *Journal of Agriculture and Food Chemistry*, **53**: 581–587.

Mendis, E., Rajapakse, N., Byun, H. and Kim, S. (2005b) Investigation of jumbo squid (*Dosidicus gigas*) skin gelatin peptides for their *in vitro* antioxidant effects. *Life Science*, **77**(17): 2166–2178.

Meyer, K. and Palmer, J.W. (1934) The polysaccharide of the vitreous humor. *Journal of Biological Chemistry*, **107**: 629–634. Available at: http://www.jbc.org/cgi/content/long/107/3/629

Minh Thuy, L.T., Okazaki, E. and Osako, K. (2014) Isolation and characterization of acid-soluble collagen from the scales of marine fishes from Japan and Vietnam. *Food Chemistry*, **149**: 264–270.

Miyamura, Y., Coelho, S. G., Schlenz, K., Batzer, J., Smuda, C., *et al.* (2011) The deceptive nature of UVA-tanning versus the modest protective effects of UVB-tanning on human skin. *Pigment Cell Melanoma Research*, **24**(1): 136–147.

Moretti, V.M., Mentasti, T., Bellagamba, F., Luzzana, U., Caprino, F., *et al.* (2006) Determination of astaxanthin stereoisomers and colour attributes in flesh of rainbow trout (*Oncorhynchus mykiss*) as a tool to distinguish the dietary pigmentation source. *Food Additives and Contaminants*, **23**(11): 1056–1063.

Murado, M.A., Montemayor, M.I., Cabo, M.L., Vázquez, J.A. and González, M.P. (2011) Optimization of extraction and purification process of hyaluronic acid from fish eyeball. *Food and Bioproducts Processing*, **90**(3): 491–498.

Murube, J. (2013) Ocular cosmetics in ancient times. *The Ocular Surface*, **11**(1): 2–7.

Najafian, L. and Babji, A.S. (2012) A review of fish-derived antioxidant and antimicrobial peptides: Their production, assessment, and applications. *Peptides*, **33**(1): 178–185.

Najafian, L. and Babji, A.S. (2015) Isolation, purification and identification of three novel antioxidative peptides from patin (*Pangasius sutchi*) myofibrillar protein hydrolysates. *LWT - Food Science and Technology*, **60**(1): 452–461.

Nakano, T., Nakano, K. and Sim, J.S. (1994) A simple rapid method to estimate hyaluronic acid concentrations in rooster comb and wattle using cellulose acetate electrophoresis. *Journal of Agricultural and Food Chemistry*, **42**: 2766–2768.

Nelson, D.L. and Cox, M.M. (2008) *Lehninger Principle of Biochemistry*, W.H. Freeman and Company, New York, p. 1119.

Ngo, D., Kang, K., Ryu, B., Vo, T., Jung, W., *et al.* (2015) Angiotensin-I converting enzyme inhibitory peptides from antihypertensive skate (*Okamejei kenojei*) skin gelatin hydrolysate in spontaneously hypertensive rats. *Food Chemistry*, **174**: 37–43.

Nguyen Thi, T., Nguyen Tan, L. and Lindberg, J. (2007) Survey of the production, processing and nutritive value of catfish by-products meals in the Mekong Delta of Vietnam. *Livestock Research for Rural Development*, **19**: Available at: http://www.lrrd.org/lrrd19/9/thuycit.htm

Olsen, R.L., Toppe, J. and Karunasagar, I. (2014) Challenges and realistic opportunities in the use of by-products from processing of fish and shellfish. *Trends in Food Science AND Technology*, **36**(2): 144–151.

Park, J., Chyun, J., Kim, Y., Line, L. and Chew, B. (2010) Astaxanthin decreased oxidative stress and inflammation and enhanced immune response in humans. *Nutrition and Metabolism*, **5**: 7–18.

Pashkow, F.J., Watumull, D.G. and Campbell, C.L. (2008) Astaxanthin: A novel potential treatment for oxidative stress and inflammation in cardiovascular disease. *The American Journal of Cardiology*, **101**(10A): 58D–68D.

Pati, F., Adhikari, B. and Dhara, S. (2010) Isolation and characterization of fish scale collagen of higher thermal stability. *Bioresource Technology*, **101**(10): 3737–3742.

Paul, A.J., Kolarsick, B., Kolarsick, M.A. and Goodwin, C. (2011) Anatomy and physiology of the skin. *Journal of the Dermatology Nurses' Association*, **3**(4): 203–213.

Picot, L., Bordenave, S., Didelot, S., Fruitier-Arnaudin, I., Sannier, F., *et al.* (2006) Antiproliferative activity of fish protein hydrolysates on human breast cancer cell lines. *Process Biochemistry*, **41**: 1217–1222.

Piez, K.A. (1984) Molecular and aggregates structures of the collagen. In: *Extracelluolar Matrix Biochemistry* (ed K.A. Piez and A.H. Reddi). Elsevier, New York, pp. 1–40.

Pothacharoen, P., Teekachunhatean, S., Louthrenoo, W., Yingsung, W., Ong-Chai, S., *et al.*, (2006).Raised chondroitin sulfate epitopes and hyaluronan in serum from rheumatoid arthritis and osteoarthritis patients. *Osteoarthritis Cartilage*, **14**: 299–301.

Preetha, J.P. and Karthika, K. (2009) Cosmeceuticals – An evolution. *International Journal of ChemTech Research*, **1**(4): 1217–1223.

Ramirez, A. (2007) *FAO Fisheries Circular No. 1027 Salmon By-Product Proteins*, vol. **1027**. FAO, Rome.

Rinaudo, M. (2006) Chitin and chitosan: Properties and applications. *Progress in Polymer Science*, **31**: 603–632.

Robert, M., Zatylny-Gaudin, C., Fournier, V., Corre, E., Le Corguillé, G., *et al.* (2014). Molecular characterization of peptide fractions of a Tilapia (*Oreochromis niloticus*)

by-product hydrolysate and *in vitro* evaluation of antibacterial activity. *Process Biochemistry*. doi:10.1016/j.procbio.2014.12.022

Rohrich, R.J., Ghavami, A. and Crosby, M.A. (2007) The role of hyaluronic acid fillers (Restylane) in facial cosmetic surgery: Review and technical considerations. *Plastic and Reconstructive Surgery*, **120**: 41S–54S.

Rubio-rodríguez, N., Beltrán, S., Jaime, I., Diego, S.M. De, Sanz, M.T. and Carballido, J.R. (2010) Production of omega-3 polyunsaturated fatty acid concentrates: A review. *Innovative Food Science and Emerging Technologies*, **11**(1): 1–12.

Rüfer, C.E., Moeseneder, J., Briviba, K., Rechkemmer, G. and Bub, A. (2008) Bioavailability of astaxanthin stereoisomers from wild (*Oncorhynchus* spp.) and aquacultured (*Salmo salar*) salmon in healthy men: A randomised, double-blind study. *The British Journal of Nutrition*, **99**(5): 1048–1054.

Saadi, S., Deratani, A., Belleville, M.-P. and Amar, R.B. (2014a) Production and fractionation of tuna by-product protein hydrolysate by ultrafiltration and nanofiltration: Impact on interesting peptides fractions and nutritional properties. *Food Research International*, **65**: 453–461.

Saadi, S., Saari, N., Anwar, F., Abdul Hamid, A. and Ghazali, H.M. (2014b) Recent advances in food biopeptides: Production, biological functionalities and therapeutic applications. *Biotechnology Advances*. doi:10.1016/j.biotechadv.2014.12.003

Sachindra, N.M., Bhaskar, N., Siddegowda, G.S., Sathisha, A.D. and Suresh, P.V. (2007) Recovery of carotenoids from ensilaged shrimp waste. *Bioresource Technology*, **98**: 1642–1646.

Sadowska, M. and Kołodziejska, I. (2005) Optimisation of conditions for precipitation of collagen from solution using k-carrageenan. Studies on collagen from the skin of Baltic cod (*Gadus morhua*). *Food Chemistry*, **91**: 45–49.

Sadowska, M., Kolodziejska, I. and Niecikowska, C. (2003) Isolation of collagen from the skins of Baltic cod (*Gadus morhua*). *Food Chemistry*, **81**: 257–262.

Sánchez-Camargo, A., Almeida Meireles, M.Â., Lopes, B.L.F. and Cabral, F.A. (2011) Proximate composition and extraction of carotenoids and lipids from Brazilian redspotted shrimp waste (*Farfantepenaeus paulensis*). *Journal of Food Engineering*, **102**(1): 87–93.

Sánchez-Zapata, E., Amensour, M., Oliver, R., Fuentes-Zaragoza, E., Navarro, C., *et al.* (2011) Quality characteristics of dark muscle from yellowfin tuna *Thunnus albacares* to its potential application in the food industry. *Food and Nutrition Sciences*, **02**(01): 22–30.

Sárdy, M. (2009) Role of Matrix metalloproteinases in skin ageing. *Connective Tissue Research*, **50**(2): 132–138.

Schiedt, K., Vecchi, M. and Glinz, E. (1986) Astaxanthin and its metabolites in wild rainbow trout (*Salmo gairdneri* R.). *Comparative Biochemistry and Physiology Part B: Comparative Biochemistry*, **83**(1): 9–12.

Schiraldi, C., Cimini, D. and Rosa, M. De (2010) Production of chondroitin sulfate and chondroitin. *Applied Microbiology and Biotechnology*, **87**: 1209–1220.

Schrieber, R. and Gareis, H. (2007) *Gelatine Handbook Theory and Industrial Practice*. Weinheim, Wiley-VCH Verlag GmbH & Co.KGaA.

Seki, T., Sueki, H., Kohno, H., Suganuma, K. and Yamashita, E. (2001) Effects of astaxanthin from *Haematococcus pluvialis* on human skin. *Fragrance Journal*, **12**: 98–103.

Senphan, T., Benjakul, S. and Kishimura, H. (2014) Characteristics and antioxidative activity of carotenoprotein from shells of Pacific white shrimp extracted using hepatopancreas proteases. *Food Bioscience*, **5**: 54–63.

Seo, J.-K., Lee, M.J., Jung, H.-G., Go, H.-J., Kim, Y.J. and Park, N G. (2014) Antimicrobial function of SHβAP, a novel hemoglobin β chain-related antimicrobial peptide, isolated from the liver of skipjack tuna, *Katsuwonus pelamis*. *Fish and Shellfish Immunology*, **37**(1): 173–83.

Shimidzu, N., Goto, M. and Miki, W. (1996) Carotenoids as singlet oxygen quenchers in marine organisms. *Fisheries Science*, **62**(1): 134–137.

Silva, J.F.X., Ribeiro, K., Silva, J.F., Cahú, T.B. and Bezerra, R.S. (2014) Utilization of tilapia processing waste for the production of fish protein hydrolysate. *Animal Feed Science and Technology*, **196**: 96–106.

Singh, P., Benjakul, S., Maqsood, S. and Kishimura, H. (2011) Isolation and characterisation of collagen extracted from the skin of striped catfish (*Pangasianodon hypophthalmus*). *Food Chemistry*, **124**(1): 97–105.

Šližytė, R., Mozuraitytė, R., Martínez-Alvarez, O., Falch, E., Fouchereau-Peron, M. and Rustad, T. (2009) Functional, bioactive and antioxidative properties of hydrolysates obtained from cod (*Gadus morhua*) backbones. *Process Biochemistry*, **44**(6): 668–677.

Söderhäll, C., Marenholz, I., Kerscher, T., Rüschendorf, F., Esparza-Gordillo, J., *et al.* (2007) Variants in a novel epidermal collagen gene (COL29A1) are associated with atopic dermatitis. *PLoS Biology*, **5**(9): e242.

Song, E., Yeon Kim, S., Chun, T., Byun, H.-J. and Lee, Y.M. (2006) Collagen scaffolds derived from a marine source and their biocompatibility. *Biomaterials*, **27**(15): 2951–61.

Sprague, M., Dick, J.R., Medina, A., Tocher, D.R., Bell, J.G., *et al.* (2012) Lipid and fatty acid composition, and persistent organic pollutant levels in tissues of migrating Atlantic bluefin tuna (*Thunnus thynnus*, L.) broodstock. *Environmental Pollution (Barking, Essex: 1987)*, **171**: 61–71.

Sriket, P., Benjakul, S., Visessanguan, W. and Kijroongrojana, K. (2007) Comparative studies on chemical composition and thermal properties of black tiger shrimp (*Penaeus monodon*) and white shrimp (*Penaeus vannamei*) meats. *Food Chemistry*, **103**(4): 1199–1207.

Stewart, B. and Wild, C. (2014) *World Cancer Report 2014*. International Agency for Research on Cancer, World Health Organization.

Storebakken, T., Foss, P., Schiedt, K., Austreng, E., Liaaen-Jensen, S. and Manz, U. (1987) Carotenoids in diets for salmonids. *Aquaculture*, **65**(3–4): 279–292.

Su, Y. (2011) Isolation and identification of pelteobagrin, a novel antimicrobial peptide from the skin mucus of yellow catfish (*Pelteobagrus fulvidraco*). *Comparative Biochemistry and Physiology. Part B, Biochemistry & Molecular Biology*, **158**(2): 149–154.

Swatantra, K.K., Awani, R.K. and Satyawan, S. (2010) Chitosan: A platform for Targeted drug delivery. *Pharmaceutical Technology*, **2**(4): 2271–2282.

Tabarestani, H.S., Maghsoudlou, Y., Motamedzadegan, A. and Mahoonak, A.S. (2010) Optimization of physico-chemical properties of gelatin extracted from fish skin of rainbow trout (*Onchorhynchus mykiss*). *Bioresource Technology*, **101**(15): 6207–6214.

Taravel, M. and Domard, A. (1996) Collagen and its interactions with chitosan. III: Some biological and mechanical properties. *Biomaterials*, **17**(4): 451–455.

Tominaga, K., Hongo, N., Karato, M. and Yamashita, E. (2012) Cosmetic benefits of astaxanthin on humans subjects. *Acta Biochimica Polonica*, **59**(1): 43–47.

Trung, T.S. and Pham, D.P. (2012) Bioactive compounds from by-products of shrimp processing industry in vietnam, **20**: 194–197.

Umayaparvathi, S., Meenakshi, S., Vimalraj, V. and Arumugam, M. (2014) Antioxidant activity and anticancer effect of bioactive peptide from enzymatic hydrolysate of oyster (*Saccostrea cucullata*). *Biomedicine & Preventive Nutrition*, **4**(3): 343–353.

Vázquez, J.A., Rodríguez-Amado, I., Montemayor, M.I., Fraguas, J., González, M. del P. and Murado, M.A. (2013) Chondroitin Sulfate, hyaluronic acid and chitin/chitosan production using marine waste source: Characteristics, applications and eco-friendly processes: A review. *Marine Drugs*, **11**: 747–774.

Veeruraj, A., Arumugam, M., Ajithkumar, T. and Balasubramanian, T. (2015) Isolation and characterization of collagen from the outer skin of squid (*Doryteuthis singhalensis*). *Food Hydrocolloids*, **43**: 708–716.

Venugopal, V. (2006) *Marine products for healthcare*. CRC Press, New York.

Volpi, N. (2009) Quality of different chondroitin sulfate preparations in relation to their therapeutic activity. *Journal of Pharmacy and Pharmacology*, **61**: 1271–80.

Wang, C., Chang, T., Shi, L., Yang, H., Cui, M. and Tambalu, L. (2014) Seafood processing by-products. In: *Seafood processing By-Products: Trends and applications* (ed. S.-K. Kim). Springer, New York.

Wildman, R.E. and Kelley, M. (2007) Nutraceuticals and functional foods. In: *Handbook of Nutraceuticals and Functional Foods*, 2nd edition. (ed. R.E. Wildman). CRC Press, New York.

Wong, T. (2009) Chitosan and its use in design of insulin delivery system. *Recent Patents on Drug Delivery AND Formulation*, **3**: 8–25.

Yamashita, E. (1995) Suppression of post-UVB hyperpigmentation by topical astaxanthin from krill. *Fragrance Journal*, **14**: 180–185.

Yang, S., Leong, K., Du, Z. and Chua, C. (2001). The design of scaffolds for use in tissue engineering. Part I: Traditional factors. *Tissue Engineering*, **7**(6): 679–689.

Yoshida, H., Yanai, H., Ito, K., Tomono, Y., Koikeda, T., *et al.* Tsukahara, H. and Tada, N. (2010) Administration of natural astaxanthin increases serum HDL-cholesterol and adiponectin in subjects with mild hyperlipidemia. *Atherosclerosis*, **209**: 520–523.

Zague, V., Freitas, V. De, Rosa, C. and Jaeger, R.G. (2011) Collagen hydrolysate Intake Increases Skin collagen expression and suppresses matrix metalloproteinase 2 activity. *Journal of Medicinal Food*, **14**(6): 618–624.

20

Food Industry By-Products as Protein Replacement in Aquaculture Diets of Tilapia and Catfish

Gabriel Arome Ataguba[1,2], Manoj Tukaram Kamble[1], and Krishna R. Salin[1]

[1]*Asian Institute of Technology, Klong Luang, Pathumthani, Thailand*
[2]*University of Agriculture, Makurdi, Nigeria*

20.1 Introduction

20.1.1 Overview of Aquaculture

Aquaculture production throughout the world has grown by about 33.5% from the production level of 49.9 million tonnes in 2007 to 66.6 million tonnes in 2012, contributing to 42% of the total fish production worldwide (FAO, 2014a). Aquaculture, like animal husbandry, requires feed to ensure good production, especially under the intensive culture system, which use a great deal of input. Aquaculture is more efficient in production of animal protein compared to terrestrial animal production, because aquaculture consumes between 2.5 and 4.5 kg dry feed to yield a kilogram of consumable fish, compared to between 3.0 and 17.4 for the terrestrial meat production sector (Costa-Pierce *et al.*, 2012).

The increase in aquaculture production is expected to be sustained via increased production using semi-intensive small-scale systems from developing countries in two continents, Africa and Asia, and this increase will be supported by robust nutrition and feeding (Hasan, 2001). Aquaculture in the developing world relies mostly on the use of both semi-intensive and extensive systems with relatively low technology. Commercial production in developed countries often evolves from high-priced, niche marketing to commodity status, when complete diets have been formulated and produced for the target species (Trushenski *et al.*, 2006). Therefore, the majority of the freshwater or saltwater fish that are secondary or tertiary consumers but are highly demanded as

Food Processing By-Products and their Utilization, First Edition. Edited by Anil Kumar Anal.

human food in the developed nations, are farmed intensively and maintained using feeds that are produced from high-value ingredients to give full nutritional complement (Hasan 2001). In Canada, the use of fishmeal and fish oil in the production of salmon has been reported to be efficient, hence production of high-value aquaculture products from reduction fisheries (Sarker *et al.*, 2013). In the developing countries, cultured fish are fed either complete rations that supply all nutrients or supplementary feeds that supply part of the required nutrients while the rest is obtained from natural food in the pond (Eyo, 2003). The use of supplementary feed is also critical in the semi-intensive system, as it determines the growth of the aquatic organisms and effectively differentiates it from an extensive system. In China, the advantages of polyculture as a means of conserving the use of fishmeal in fish diets is overshadowed by the high demand occasioned by a growing economy and increased income, hence use of aquaculture feeds made with fishmeal is expected to increase (Chiu *et al.*, 2013). Economic viability and sustainability are driving factors for modern aquaculture, hence there are a plethora of studies that support the minimal use of fishmeal and fish oil in aquaculture diets (Bureau, 2006).

20.1.2 Use of Fishmeal

The quality of protein in fish meal has made it an ingredient of choice for livestock and aquaculture feed manufacturing, due to its amino acid profile. It may make up to 60% or more of the ingredients in an aquaculture feed (Goddard, 1995). Feed generally accounts for between 60 and 70% of variable cost in a production cycle (Fagbenro 1998; Isyagi *et al.*, 2009). The number of ingredients used in commercial catfish feed production is limited, principally due to the limited number of ingredients that contain the required nutrient density and physical characteristics that will satisfy the nutritional requirements of fish (Gatlin, 1995). Fish meal is in high demand because its amino acid profile is adequate for aquaculture species, while fish oil also supplies a blend of poly-unsaturated fatty acids (Ω-3 fatty acids) as the essential nutrient for fish prior to consumption by consumers to attain health benefits (Jackson, 2006). According to Rana *et al.* (2009), the demand for fish meal has led to an increase in its value in the international market, with an increase of 73% in its cost between 2005 and 2008, while the average price of other aquaculture feed ingredients rose by 20–92% between mid-2007 and mid-2008. The production of aquaculture feeds as an industry does not exist in isolation from the other animal feed industries, or even the human food industry. It draws its raw materials from the same pool of ingredients as that of the terrestrial animal agriculture feed industry and even the human food industry (Rana *et al.*, 2009), considering ingredients such as cereal grains, oil seeds and by-products, as well as fish meal and fish oil.

Fishmeal and fish oil are two ingredients currently being used in aquacultural diets, but controversy trails their use as arguments ensue between producers, scientists and conservationists. Tacon and Metian (2008) calculated the conversion rates between fishmeal as a percentage of diets and biomass fish flesh produced via a ratio named as the fish-in–fish-out ratio (unit weight of fishmeal converted into unit weight of fish meat) and reported a range from 0.2 for milkfish and carps to 4.9 for salmon, a carnivorous species. However, a swift response from the International Fishmeal and Fish Oil Organization (IFFO) shows that this ratio lies between 0.1 for milkfish and 3.1 for eel (Jackson, 2009). It is estimated that 3.1 million tonnes of fish meal was used to produce 31.6 million tonnes of feed in 2015. with 756 000 tonnes of fish oil being used in the process, but efficiency as well as use of other protein sources is expected to cause a reduction in fishmeal and fish oil use by 2020 to 2.4 million tonnes and 712 000 tonnes, respectively (Tacon and Metian, 2008).

Tacon (1995) advocated the use of sustainable farming methods to enable aquaculture assist in subverting food insecurity in developing countries by providing cost-effective animal protein. The increased use of fishmeal in aquaculture diets has been used as a weapon against aquaculture by its antagonists. However, Tacon and Forster (2003) have advocated the control of off-farm effluents as against regulating on-farm feed input as a means to reduce the effects of aquaculture feeds on the environment. A search for alternatives to replace fishmeal in the aquaculture feed industry is an imperative task. However, the use of alternatives must consider bio-safety of products (Kamble *et al.*, 2014). A global framework that ensures that producers show evidence of origin to elicit traceability of products will go a long way to show support for the ecosystem (Deutsch *et al.*, 2007). Considering climate change, the nature of response to impact is critical to sustainability of wild fisheries as a supply source for fishmeal as a feed ingredient for aquaculture (Merino *et al.*, 2010).

20.1.3 Siluridae

Catfish are a group of fish represented by many species and discoveries are still ongoing (Sullivan *et al.*, 2006) with culture being practiced across five continents. Three species are commonly cultured including *Ictalurus punctatus* (channel catfish), *Clarias gariepinus* (African catfish) and *Pangasius hypophthalmus* (Tra catfish). These are carnivorous species that require feed with nutrients that have high quality and cost, especially animal protein feed ingredients, and particularly fishmeal and fish oil (Goddard, 1995). Catfish feed production worldwide utilizes between 2 and 5% of fishmeal and between 1 and 4% of fish oil (Jackson, 2007; Tacon and Metian, 2008), considering the total world aquaculture feed production.

20.1.4 Cichlidae

According to Nelson (2006), the family Cichlidae comprises of about 3000 species, making it one of the most diverse families in the sub-phylum vertebrata. Tilapia is a group of cichlids native to Africa that have been introduced into various parts of the world for culture. Tilapias have several characteristics that have made them a choice species for aquaculture, such as environmental tolerance, good growth, disease and stress resistance, prolific nature and excellent feed conversion (Yi *et al.*, 1996; El-Sayed, 2006). As at 2011, the global production of tilapia was about 3.95 million tonnes (FAO, 2014a).

Tilapia nutrition under culture conditions depends on the scale, for example extensive, semi-intensive and intensive culture scales. The protein requirements of tilapia, which is a function of the amount of energy ingredients, origin of protein and age/size of the fish tends to fall with increase in size as a result of growth (El-Sayed, 2006). Depending on species and size, tilapia require between 28 and 56% protein levels in their diets (Winfree and Stickney, 1981; Twibell and Brown, 1998; El-Sayed, 2006; Bahnasawy, 2009; Larumbe-Moran *et al.*, 2010). However, variation in salinity also affects dietary protein requirements, with increasing salinity leading to increased protein requirements (Larumbe-Moran *et al.*, 2010).

In 2008, global aquafeed production was about 29.3 million tonnes and studies on tilapia and catfish feed production utilizes seven ingredient classes (fishery products, terrestrial livestock products, protein-rich cereal products, plant oil seed products, pulse and grain seed products, terrestrial plant oils and low-protein terrestrial plant feedstuff and by-products) (Cocker, 2014). With the drive for sustainability, the aquaculture feed industry must join the drive to save wild fisheries from collapse, as a result of reduction in fishmeal and fish oil to serve the industry and avoid any detrimental effects on the environment and even the profitability of aquaculture, hence sustainable alternatives need to be investigated and utilized. However, the quality of such alternatives may be limited by protein content and amino acid profile.

20.1.5 Food Industry By-Products

Processing of food produces by-products are usually of less value due to the cost of assembly and recovery for further use. On the other hand, if these products are discarded continuously, environmental problems are bound to occur, hence the need for reuse. Agricultural products are processed using different methods with varying yields of desired products as well as by-products. A summary of the yield of various by-products of interest to the aquaculture industry, as well as other industries, are presented in Table 20.1. It has been indicated that feeding dairy cattle with food processing industry by-products can lower costs of milk production and also fight climate change by reducing emission of greenhouse gases (Vasta *et al.*, 2008; Vellinga *et al.*, 2011).

Trials with alternative feed ingredients that are mostly unconventional appear in a plethora of literature on fish nutrition research. Various ingredients of plant, animal and other origins have been tested to determine suitability in catfish diets. Food industry waste or by-products can be utilized in two broad ways in line with their agricultural roots, either as animal feed or as fertilizers, with examples including spent grains, distillers wash, oil seed cakes, filtration sludge and carbonation sludge (Russ and Schnappinger, 2007).

Table 20.1 Percentage yield of by-products from processing of raw food crops

Crop (Grain/Seed)	By-Product	Percentage Yield	Source
Rice	Bran	5 - 12	Kahlon, 2009; Sharif *et al.*, 2014; Ozdestan *et al.*, 2014
	Polishing	2 - 3	Sharif *et al.*, 2014
Wheat	Bran	15	Cornell and Hoveling, 1998
	Germ	0.3	
Groundnut	Groundnut Cake	50	Head *et al.*, 1995
Cotton	Cotton seed meal	45	Jones and Wedegaertner, 1986
Corn	Germ meal	25 - 30	Davis, 2001
	Gluten		
	Bran		
Soybean	Soybean meal	35	CRB, 2007

20.2 Alternatives to Fishmeal in Catfish Diets

20.2.1 Ingredients of Plant Origin

20.2.1.1 Oil Seed Cakes World oilseed meal production stood at 269 million tonnes in 2013 and rose by 9.3% in 2015 (USDA, 2015), with an increase of 70 million tonnes being expected by 2022 (OECD-FAO, 2013). Utilization of oilseed meals and cakes was 274 million tonnes in 2013, with greater utilization in Asia and Europe (Figure 20.1) compared to other continents.

Oilseed meals that are being used heavily in livestock feeds include soybean meal, cotton seed meal (CSM), groundnut cake, sesame seed meal, canola and sunflower seed meals as protein supplements with interchangeable use being possible. Webster *et al.* (1995) reported the possibility of a complete elimination of fishmeal from the diet of the blue catfish (*Ictalurus furcatus*) by substituting with soybean meal (SBM). These authors totally replaced fishmeal with SBM (70%) coupled with supplementation of L-methionine at 0%, 0.3%, 0.6% and 0.9% levels, using a diet that is similar to commercial catfish diet as a control. They discovered that growth and feed utilization did not differ among treatments and body proximate composition was similar for all treatments. However, the inclusion of fishmeal in feed compounded for the channel catfish was reported by Li *et al.* (2010) to create effectiveness in feed utilization with better weight gain and reduced feed conversion ratio. According to the authors, two vegetable protein diet combinations comprising SBM, corn meal and CSM, as one diet and a combination that included wheat middlings alongside the aforementioned ingredients as the second alternative, performed in a similar manner as those with fishmeal included. Inclusion of soybean meal at the 100% and 75% levels to replace fishmeal had no significant effect on the growth of channel catfish in comparison with diets containing 100% fishmeal (Goda *et al.*, 2007).

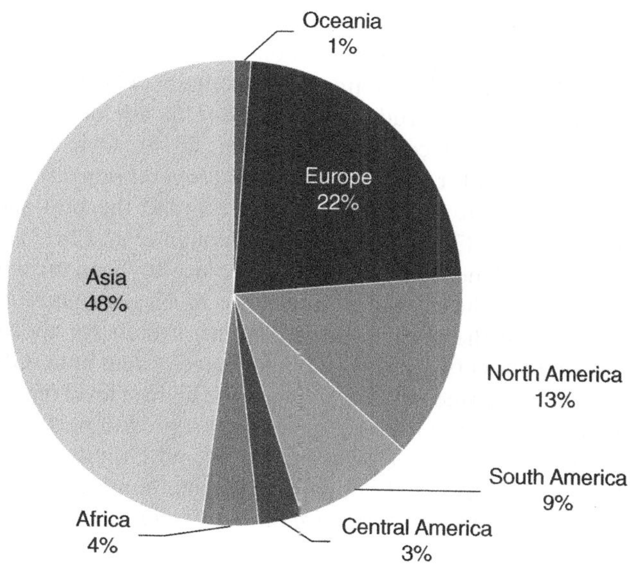

Figure 20.1 World utilization of oilseed meals 2013 (*Source:* FAO, 2014b).

Canola seed meal is a product of oil extraction from the canola seed (*Brassica rapa*), which was genetically selected by Canadian scientists to reduce the erucic acid and glucosinolate content of the seed (Gatlin III *et al.*, 2007), with similar research being conducted in Iran, yielding about 78% glucosinolate reduction and increased production (Enami, 2011). Solvent extracted canola seed meal contains 3.5% residual fat, 6% ash, 12% crude fibre and 35% crude protein on an as-fed basis (10% moisture), with phytic acid being 4% (Gatlin III *et al.*, 2007). Canola meal can spare 36% SBM in diets formulated and tested for juvenile channel catfish (*I. Punctatus*), without affecting the food utilization, body composition and haematological indices of the fish (Lim *et al.*, 1998). Similarly, Slawski *et al.* (2011) reported the substitution of soybean meal by up to 25% with rapeseed protein concentrate in the diet of the wels catfish (*Silurus glanis*), without any untoward effects on growth but with similarities in feed conversion and protein utilization by using 25% levels of inclusion of both SBM and rapeseed meal in the fish diets. This suggests the interchangeable possibility of SBM and rapeseed meal in the diet of *S. glanis*.

Mechanically extracted sesame seed meal contains 44.4% crude protein, 7.8% crude fibre, 11.1% residual crude lipid and 12.4% ash on a dry matter basis, while the solvent extracted meal contains 48.5% crude protein, 10.1% crude fibre, 2.6% residual crude lipid and 12.6% ash (Feedipedia, 2015). Jimoh *et al.* (2014) evaluated the performance of the African catfish (*C. gariepinus*) fed different temporal wet thermally treated (10, 20 and 30 min) and sesame seed meal obtained from mechanical oil press at three substitution levels for soybean meal (15, 30 and 45% levels). Wet thermal treatment carried out at various temporal levels had no apparent consequence on the proximate composition of the ingredient. Growth performance of the fish and feed utilization by the fish did not differ from that recorded for soybean meal. Inclusion of sesame seed meal from the mechanical press at levels from 50% and above, however, had a negative effect on growth and efficiency of protein ingested, but with a positive effect on the feed conversion ratio (FCR) for African catfish (*Clarias gariepinus*) fed sesame seed meal-based diets (Jimoh and Aroyehun, 2011).

Groundnut cake is a food by-product from oil production using groundnut seeds. Groundnut cake (GNC) on a dry matter basis, contains 49–56.2% crude protein, 7.0–14.0% crude fibre, 0.7–10.1% crude lipid, and 5.8–7.1% ash, depending on method of processing and presence of cortex (Feedipedia, 2015). Ovie and Ovie (2007) formulated and tested a diet for the African catfish (*Heterobranchus longifilis*) using GNC as a fishmeal substitute at 10% level and recorded the best specific growth rate for diets containing 44.17% crude protein. Tiamiyu *et al.* (2013) reported that either GNC or cotton seed meal will successfully replace soybean meal at 50% levels of each, without affecting the growth performance of *C. gariepinus*. Adewolu and Benfey (2009) modulated the crude protein content of diets for the bagrid catfish (*Chrysichthys nigrodigitatus*) by varying the percentage of rice bran, GNC, fish meal and yellow maize. They reported that the diet with the highest level of groundnut cake coupled with the highest level of fishmeal gave the best growth performance, with its crude protein content being 35%. However, Davies and Ezenwa (2010) advocated the use of amino acid supplementation, particularly methionine and lysine, in diets with GNC at 0.45% for optimal performance of *C. gariepinus*.

Considering dry matter, de-hulled solvent extracted sunflower seed meal contains 41% crude protein, 0.5% ether extract, 14% crude fibre and 5.9% ash (Hesley, 1994), with an indirect association existing between protein and fibre composition as a result of the processing technique (Moghaddam *et al.*, 2012). According to

Nyina-Wamwiza *et al.* (2010), the processing of sunflower oil cake affected nutrient utilization by juvenile *C. gariepinus* fed diets containing a combination of sunflower oil cake and groundnut cake. Consequently, de-hulling increased dry matter apparent digestibility coefficient (ADC) by 11.6% from 60.5% for the crude cake, while soaking increased ADC for protein and amino acids, dry matter, neutral detergent fibre (NDF) and energy. They reported a maximum inclusion level of 25% of sunflower oil cake in the diet of juveniles of the African catfish.

Cotton seed meal (CSM), as used for animal nutrition, can be obtained from three different processing techniques: mechanically extracted seed meal, solvent extracted seed meal and ammoniated cotton seed meal. The mechanical and solvent extracted meals are used in aquaculture feeds and contain crude protein from 36% and above (AAFCO, 2001). According to NRC (1993), crude protein from solvent extracted cotton seed meal is 41.7%, with lysine being 1.8% and methionine 0.5% of amino acids. CSM contains either 45% or 40% crude protein on a dry matter basis, depending on fibre content (Feedipedia, 2015). However, CSM contains the yellow glandular pigment, gossypol, hence its use in fish feed is limited and will require proper processing to eliminate the compound. This is toxic to fish and other animals (Morgan, 1989; Gadelha *et al.*, 2014), with the primary effect being impairment of reproduction (Randel *et al.*, 1992; Gadelha *et al.*, 2014), as well as reduction in immunity (Gadelha *et al.*, 2014). It causes reduced growth in *I. punctatus* at a dose of 671 mg/kg (Barros *et al.*, 2000, 2002). The reduction of gossypol content of CSM has been achieved using several methods, such as mechanical removal of glands (Buser and Abbas, 2001), solvent extraction (Pelitire *et al.*, 2014), breeding for low content (Bi *et al.*, 1999; Kulkarni *et al.*, 2002; Romano and Scheffler, 2008) and microbial fermentation (Zhang *et al.*, 2006).

A combination of three protein sources: cotton seed cake, blood meal and fish meal in diets of *C. gariepinus* produced the least FCR, highest efficiency of protein and specific growth rates compared to four other diets with two devoid of CSM and the other two containing palm kernel meal, brewer's yeast and groundnut cake (Agbabiaka, 2010). Li and Robinson (2008) suggested the restriction of inclusion percentage of CSM in fish diets to between 10 and 15% as a result of the degree of variation in free gossypol from various types of meals and its toxicity. However, Robinson and Li (2008) reported that the safe level of inclusion can be as high as 50% of the diet, based on tests with *I. punctatus* by switching SBM with CSM, a combination of SMB/CSM, CSM/DDGS (distillers dried grains) and SBM/DDGS. Feed conversion ratio (1.97) of the SBM/CSM combination and SBM/DDGS combination were not significantly different. A similar discovery was also reported by Tiamiyu *et al.* (2013) with an FCR of 1.17 for 50% replacement of CSM with SBM in the diet of *C. gariepinus*, with increasing FCR as replacement percentage increased. Diets of catfish containing CSM in lieu of SBM at 100% substitution and fortified with lysine can be utilized effectively following a report by Robinson (1991).

According to Fagbenro (1998), oilseed meals are highly digested by the African catfish (*Clarias* sp.), with reference to protein, with soybean meal being the most digestible (86.9%), while groundnut cake, sunflower cake and macademia presscake had similar protein digestibilities of 86.6%, 86.5% and 83.5%, respectively. These digestibility values were comparable to 92.8% for the Menhaden fish meal and are therefore reliable substitutes for fishmeal for African catfish diets. Furthermore, Li *et al.* (2013a) reported that soybean meal is better utilized more than other plant protein sources, because digestibility, protein and amino acid availability from other sources such as canola meal, distiller's dry grains, and two corn processing by-products (corn gluten and germ meal), are lower than soybean meal.

20.2.1.2 Cereal By-Products Cereal by-products are rich sources of starch, with content varying from between 62% and 72% as in maize and wheat, while protein is typically low (10–20%) compared to oil seed cakes, low essential amino acid profile, especially lysine, and low mineral content, except for phosphorous, which is bound as phytate (Metailler and Guillaume, 1999).

Maize is processed industrially by wet milling, with resultant products being corn oil and starch, which can be further processed to yield dextrin and glucose. Corn gluten meal, corn germ meal and maize bran are the resultant by-products that are mixed and marketed as maize gluten feed. This product has low lysine content and should be used in tandem with other feed ingredients that are rich in lysine (Teles *et al.*, 2011). The high content of starch in cereal grains implies viscosity of feed and this has been found to increase digesta viscosity with effects on intestinal fermentation, nutrient digestibility and dry matter content of digesta in the African catfish, but this has no antagonistic effects on the performance of the fish in terms of growth (Leenhouwers *et al.*, 2007).

Hu *et al.* (2012) tested two protein levels (28% and 32%) of feed with or without the inclusion of corn gluten feed on channel-blue catfish hybrids. Their results showed that protein levels affected only the hepato-somatic index, while feed type affected intra-peritoneal fat ratio, but on the whole, the protein levels did not interact significantly with the feed type, hence a uniform utilization potential with satisfactory output. Furthermore, Li *et al.* (2013b) suggested the possibility of using corn germ meal alternatively with corn gluten feed as a replacement for corn, wheat and soybean in diets of *I. punctatus*, following results that indicate non-significant effects on feed utilization parameters of up to 35% level of inclusion.

The isolation of starch from wheat flour results in the production of wheat gluten, which is obtained through a process that removes soluble fibres and starch with accompanying recovery of two fractions of insoluble proteins, gliadins and insoluble glutenin. Although wheat gluten has been reported to have good digestibility in other species of fish, effects in catfishes are not common. On the other hand, wheat bran is obtained from the coat of the wheat grain during milling and comprises the cuticle, pericarp and seed coat, which altogether form the outer layer of the grain as well as small quantity of starchy endosperm from the kernel. Wheat bran contains approximately 17% crude protein and 10% fibre on a dry matter basis (Feedipedia, 2015). Its high fibre content limits its use in diets for omnivorous and carnivorous fish, with recommended levels being between 2% and 5% (Hertrampf and Piedad-Pascual, 2000). Usmani and Jafri (2002) reported that apparent digestibility of wheat bran was affected by temperature in the catfish *Heteropneustes fossilis* and by size in the African catfish *C. gariepinus*, with ADC in *H. fossilis* increasing as temperature increases and ADC in *C. gariepinus* decreasing as size increases. Apparent digestibility of protein in wheat bran by jundia catfish (*Rhamdia quelen*) was reported to be next to broken rice, which was better digested, but protein digestibility of wheat bran was not significantly different from soybean hulls, citrus pulp and ground corn (Rodrigues *et al.*, 2012).

Rice bran contains between 11% and 19% crude protein (dry matter basis), depending on fat content (Prakash and Ramaswamy, 1996), with microwaved rice bran containing 19.4% crude protein as against 18.9% and 16.6% for roasted and whole rice bran, respectively (Faria *et al.*, 2012). Rice bran contains oil with high concentrations of polyunsaturated fatty acids that are easily oxidized, hence rancidity, but its high fibre content that can be more than 20% (Feedipedia, 2015) reduces its value for use as feed ingredient in fish diets and makes pellets unstable in water. Apparent digestibility of rice bran by fingerlings of *Pangasianodon hypophthalmus* was reported to be 86.3% in comparison with groundnut cake, which was 94.6% (Da *et al.*, 2013). These authors reported that individual and total essential amino acids were highest in fish and shrimp by-products, with GNC, snail meal and rice bran being least in that order. This is in line with the report by Usmani and Jafri (2002) that rice bran was least digestible compared to fishmeal and soybean meal in diets of the African catfish at various sizes and *H. fossilis* at various temperatures.

20.2.1.3 Distiller's Dried Grains and Brewer's Yeast Slurry Distiller's dry grain is the by-product of ethanol production following fermentation of grain starch by yeast. Different grains are used for this process, with different nutrient composition of the by-products. Wheat distillers grain with solubles (starch level of <7%) contains 37.3% crude protein with lysine and methionine being 2.3% and 1.5%, respectively, of the protein content (Feedipedia, 2015). Corn distiller's dry grain contains 29.4% crude protein (Lim and Yildirim-Aksoy, 2008) with lysine and methionine being 3% and 2%, respectively, of the protein content (Feedipedia, 2015).

Renukdas *et al.* (2014) reported that putative channel catfish and hybrid catfish can utilize up to 20% of solvent extracted distiller's dried grains at 28% crude protein level without negative effects on the feed processing and production process, with lower feed conversion ratios being observed with the use of solvent extracted DDGS. Similarly, Robinson and Li (2008) had earlier reported a reduction in FCR as DDGS is added to channel catfish diet in tandem with cotton seed meal and soybean meal. Furthermore, channel catfish diets containing 10% distiller's solubles resulted in greater feed consumption compared to 30% inclusion, but without differences in growth and feed utilization (Li *et al.*, 2010). A diet with DDGS at the 30% level and one with 1% brewer's yeast in the diet of the channel catfish resulted in feed efficiency ratios that are similar to fishmeal-based diets, while ethanol extracted DDGS at the 30% level in the diet compares to all-plant ingredient diets in terms of feed efficiency ratio (Li *et al.*, 2011). The growth of fish fed this diet in terms of weight gain was lower than 30% DDGS and the authors suggested that if ethanol extraction can be economically optimized, incorporation of the extracted DDGS will be possible at higher percentages in diets for the channel catfish.

The production of beer and ale yields spent yeast, known as brewer's yeast slurry. When it is dried and sterilized without extraction, it is termed brewer's dried yeast. In terms of proximate composition, brewer's yeast (*Saccharomyces cerevisae*) slurry holds about 45% protein, 1% lipid and 2.7% of crude fibre (Raven and Walker 1980). The deficiency of methionine and cystine is a shortcoming of this feed ingredient, even though it has an excellent amino acid profile (Huige, 2006). Feed for the African catfish with 4% graded baker's yeast has been reported to produce an FCR of 0.56 (Aderolu *et al.*, 2011).

Ezenwaji *et al.* (2012) reported that total mortality of *C. gariepinus* occurs when soybean is replaced totally with yeast, but optimal inclusion is the 50% level with acceptability increasing to this point. Increasing apparent net protein utilization and protein efficiency ratio point to the fact that low level inclusion of yeast in the diet of *C. gariepinus* does not seem to affect protein metabolism, as yeast is included in the diet up to the 2% level (Essa *et al.*, 2011).

20.2.2 Ingredients of Animal Origin

The protein quality of fishmeal remains an issue that has made it almost impossible to jettison it for other plant protein ingredients in fish diets. There is a difference of between 19% and 26% in the protein content of fishmeal in comparison with soybean meal (FAO, 1986). Anchovy meal contains about 75% crude protein (Turan *et al.*, 2007; Sankar *et al.*, 2013), while the product from herring will contain about 71% crude protein as against 66% for white fish and its offal (Windsor, 2001). The amino acid profile of fishmeal is balanced with a high percentage of the two limiting essential amino acids, lysine (7.5% of protein) and methionine (2.8% of protein) (Feedipedia, 2015). According to Winsor (2001), herring and white fish meals contain 7.7% and 6.9% lysine, respectively, and methionine content is 2.9% and 2.6% for herring and white fish meals, respectively. Although the quality of protein is high, the cost of fishmeal is also very high, and supplies may be affected by seasonal changes every year, hence there is a push via research to replace fishmeal with alternative ingredients of animal origin (Li *et al.*, 2002). However, issues concerning the quality, market value and supply of animal by-products may limit their use (Li and Robinson, 2008), hence problems can be avoided if processing is done in such a manner that produces high-quality products. Processing of fish generates wastes such as skin, heads and internal organs. According to Lovell (1982), for every 1 tonne of catfish processed, 400 kg comes out as waste, but this can range from 600 to 700 kg (Knuckey *et al.*, 2004), considering the species and methods of processing.

Menhaden fishmeal has been reported to contain a full complement of growth factors for channel catfish and enhancement of Menhaden meal substituted diets with limiting amino acids did not produce the same effect as Menhaden meal diets (Andrews and Page, 1974). Li *et al.* (2002) scrutinized the corollary of using other animal protein sources in lieu of Menhaden fish meal in the diet of the channel catfish. It was observed that feed efficiency (gain/feed), intake of feed and increase in weight were similar among the fish, hence the use of these alternative sources would not affect channel catfish growth. Similarly, Adewolu *et al.* (2010) evaluated the performance of the African catfish using a mixture of three animal protein sources. The mixture comprised two by-products: chicken feather meal that was hydrolyzed and chicken offal meal as well as maggot meal (Magmeal), which is a product of waste decomposition in a ratio of 4:3:2 in that order (Ogunji et al., 2006, 2008; Sing *et al.*, 2014) There was a reduction in carcass protein as the level of substitution for fish meal increased from 25% to 100%. Lipid content of carcass from fish fed between 25% and 50% animal protein mixture as well as 100% fishmeal was higher than carcass from fish fed 75% to 100% of the alternative protein source. Furthermore, Chor *et al.* (2013) used chicken feather meal as a substitute for fish meal in the diet of fingerlings of the African catfish and it was discovered that 20% replacement was ideal since it produced a similar FCR to fishmeal diet and also the best carcass protein content after the 100% fishmeal diet. Fish meal can be replaced by fermented feather meal in feed for the hybrid catfish (*C. macrocephalus* x *C. gariepinus*) at the 25% level with comparable FCR, weight gain and protein retention to diets of fish meal as a protein source (Arunlertaree and Rakyuttithamkul, 2006). In addition, feed intake decreased as feather meal inclusion increased and carcass fat composition reduced markedly from initial values in fish fed all diets, while the protein levels of carcass were similar.

Giri *et al.* (2000), used dried fish and chicken viscera as alternative protein sources in diets for *Clarias batrachus* fingerlings, with comparable performance between the dried fish viscera and fish meal in terms of growth (weight gain), PER and FCR, while dried chicken viscera meal and a vegetable-based diet elicited a connate response from the fish. Either extract of fish fed dried fish viscera was high compared to those fed both fish meal and dried chicken viscera by 4% and from soybean based diet by 4.4%. Inclusion of poultry by-product meal in the diet of *C. gariepinus* at high rates was discovered to impede growth, with negative effects on carcass protein composition, but without differences between fish meal-based diets and poultry by-product meal diets when considering the lipid content of the carcass of fish fed the diets (Abdel-Warith *et al.*, 2001). Although this ingredient was found to alter the liver structure of the fish, it was recommended to be used at the 40% substitution level for fish meal. Similarly, inclusion of 30% and 40% levels of poultry waste protein processed via thermal anaerobic digestion in the diet of channel catfish (*I. punctatus*) was found to reduce growth performance, FCE and protein utilization, with a recommended safe level being 20%, although 10% inclusion gave the best results (Eya *et al.*, 2010). Conversely, an inclusion level of chicken viscera meal at levels between 60 and 100% in exchange for fish meal from marine sources (by-catch) in feeds of *C. batrachus* produced the best growth and nutrient utilization with increasing lipid content in the carcass as chicken viscera meal increased (Giri *et al.*, 2010).

Soltan *et al.* (2008a), replaced fishmeal with fermented fish silage (FFS) made from fish by-products in the diets of *C. gariepinus*. A substitution level up to 50% in the diet was found to have no effect on weight gain, SGR, FCR and PER in comparison with 100% fishmeal protein diet, while feed intake reduced with increased levels of FFS. However, carcass protein composition was affected with inclusion of FFS with reducing carcass protein content as inclusion of FFS increased. Lipid levels, however, were similar between inclusion levels of 25%, 50% and the fishmeal-based diet. Furthermore, Nwanna (2003) reported the use of shrimp head waste meal that was fermented in diets to feed fingerlings of *C. gariepinus* at 0, 5, 10, 20, 30 and 40% inclusion levels without significant differences in protein and energy digestibility, FCR and PER, but with the 40% fermented shrimp head waste meal included diet giving the lowest SGR and weight gain. Carcass protein and lipid composition were not different for all diets. Using economic considerations, the author recommended the inclusion of 30% level of the alternative protein source in the diet of *C. gariepinus*. Furthermore, shrimp head meal has an apparent protein digestibility of 90% (Da *et al.*, 2013).

Protein from catfish processing waste meal is low in methionine + cystine as well as lysine and it is collagenous in nature, hence its quality is below Menhaden fish meal (Lovell, 1982). Early growth of channel catfish seems unaffected by inclusion of crab by-products in place of Menhaden fish meal, but prolonged feeding in cages produced significant differences between fish meal-based diets and crab by-product diets (Dean *et al.*, 1992). Fish waste meal included in diets for the African catfish up to 15% level produced the same growth and feed utilization as fish meal included at the same level (Sotolu, 2009). However, increasing levels of fish waste meal from 50% and above affected the growth and nutrient utilization of the African catfish negatively (Obasa *et al.*, 2011). Tilapia by-product meal and poultry by-product meals were combined as animal protein sources in diets for the African catfish at various ratios, with plant protein sources with the lowest plant protein ratio producing the best growth performance (Yousif *et al.*, 2014). Channel catfish utilized catfish waste meal effectively when fed diets containing 50% catfish waste meal, with weight gain averaging 80% of those fed Menhaden fish meal, with utilization of the waste meal being enhanced by further processing in the form of cooking (Lovell, 1982). The aforementioned is supported by the fact that meals produced from catfish waste has 95% apparent protein digestibility (Da *et al.*, 2013).

20.2.3 Other By-Products and Immuno-Modulation

Katya *et al.* (2014) used a fermented by-product made from the mushroom (*Pleurotus ostreatus*) to feed the Amur catfish (*Silurus asotus*) at 0%, 0.1%, 0.2%, 0.4% and 0.8% levels. They reported that growth was better at 0.1% and 0.2% inclusion levels. Lysozyme activity was found to be highest in fish administered the 0.2% mushroom by-product diet and chemi-luminescence was highest with fish fed the 0.1% by-product diet. Carcass proximate composition followed a staggering pattern, with reduction in crude protein content of between 0.1% and 0.4% inclusions, while an increase was observed at the 0.8% inclusion level. Conversely, lipid content increased in between the afore-mentioned levels and reduced at 0.8% level.

Yeast and its by-products have been used in several studies in fish nutrition. Peterson *et al.* (2012) discovered that the survival of disease (*Edwardsiella ictaluri*) challenged *I. punctatus* fed a yeast-derived protein source was similar among all treatments. The use of a yeast-derived protein (NuPro®) in place of Menhaden fish meal had no inimical effect on channel catfish growth and utilization of the feed by the fish. Carcass protein content was not affected by inclusion levels, but carcass fat content declined from the 75% inclusion level upwards. The authors suggested that Nupro® or yeast-derived proteins can be added to channel catfish diets to replace fish meal by up to 100%.

Conclusively, using alternatives to fish meal in the diet of various catfish can be done with efforts to reduce undesirable effects associated with the various ingredients being used. Plant protein sources in particular present challenges due to anti-nutritional factors and the binding of phosphorous as phytate that inhibits utilization by fish due to inability to synthesize phytase. Moreover, the amino acid profile of most plant ingredients shows limitation in lysine and methionine. These factors coupled with palatability and the carnivorous nature of catfish present problems for their effective utilization and subsequently the valorization of by-products for the catfish feed industry. Animal by-products are better placed to replace fishmeal in diets of the catfish. Processing and standardization for industrial application must be achieved if the potential they hold is to be realized.

20.3 Alternatives to Fishmeal in Tilapia Diets

20.3.1 Plant By-Product Protein Source

20.3.1.1 Soybean Meal (SBM) Soybean meal (SBM) is the most beneficial feedstuff of all plant protein feedstuffs, commonly used as replacement or complement to marine proteins (Olvera-Nova and Olivera-Castillo, 2000). For many reasons, such as its availability, palatability and its inexpensive cost, SBM is a commonly-utilized protein of vegetable origin in commercial fish diets. SBM is also known for its fair amino acid profile and adequate nutritional value (Tacon, 2000; Samocha *et al.*, 2004). However, SBM diets have resulted in reduction of fish growth (Mahmoud *et al.*, 2014) for many reasons, such as use of crude or partially-heated SBM, which causes activity of protease inhibitors (Dabrowski and Kozak, 1979; Wilson and Poe, 1985). SBM lacks amino acid balance (FAO, 1980; Dabrowski *et al.*, 1989), has low energy content in diets for fish compared to fish meal (FM) (Viola *et al.*, 1983; Hilton and Slinger, 1986) and decreased digestibility of phosphorus than fish meal (Liebowitz, 1981).

A plethora of studies have evaluated partial or complete substitution of commercial feed of tilapia, with FM being replaced either singularly or in combination with protein from vegetable sources, without affecting growth parameters and carcass composition. A report by El-saidy and Gaber (2002) shows that using 55% SBM supplemented with 0.5% L-lysine, fish meal can be completely replaced in a diet based on maximum values of protein, fat and energy digestibility coefficients, with no untoward effects on growth parameters and carcass composition of fingerlings of tilapia. The methionine level is the limiting factor in SBM compared to FM. Incorporation of single cell protein (SCP), (marine and freshwater algae) into tilapia diets to make up for methionine, valine and histidine shortages in treated soybeans, resulted in significant outcome on food conversion and protein efficiency ratios related to control groups, hence 50% treated soya and dried algae can partly replace fish meal safely and economically in feeds of Nile tilapia fry (Attalla and Mikhail, 2008). Nguyen *et al.* (2009) reported on the basis of growth parameters that FM in diets of juvenile tilapia can be totally replaced with dehulled solvent-extracted soybean meal (DSESM) and also revealed that using normal levels of cotton seed meal (CSM), DSESM, and meat and bone meal (MBM) as essential protein sources will not effectively make methionine a limiting factor in the diet.

20.3.1.2 Cottonseed Oil Meal Cottonseed meal (CSM) is a by-product of cotton seed oil extraction, rich in protein, hence its universal use as an alternative protein source in animal and fish diets. Mbahinzireki *et al.* (2001) reported that CSM inclusion in practical diets of up to 50% significantly improved growth performance and feed utilization, with CSM as a principal protein ingredient in the diet partly replacing FM in commercial feeds for tilapia farming. However, the authors reported that diets with 25, 50, 75 and 100% CSM produced a proportional increase of fish liver gossypol concentration (32.3, 72.3, 99.4 and 132.1 μg/g ww, respectively). Crude cottonseed oil cake is characterized by high fiber and protein contents. Cottonseed products (oilcake and meal) can serve as cost-effective protein ingredients in fish feeds as a result of their high protein content of up to 40% (Ramachandran *et al.*, 2007). Presence of gossypol has anti-nutritional effects that limit its use as an alternative protein source (Rinchard *et al.*, 2003; Yue and Zhou, 2008). The inclusion of cottonseed oil cake at 15–30% as an FM substitute in commercial diet of tilapia juveniles has been recommended, since this range did not produce any significant toxic effects (Meric *et al.*, 2011).

20.3.1.3 Plant Protein Mixture (PPM) Plant protein mixture (PPM) contains mixture of by-products of the edible oil industry. These can be incorporated into diets as a single protein source, considering their amount of protein with a good EAA profile, although deficiency in some EAAs and the presence of anti-nutritional factors (NRC, 1993) may be a hindrance to their utilization. PPM may provide better results than single ingredients in replacing animal protein, since their combination usually results in better amino acid profiles with possible limitation of anti-nutritional factors.

Soltan *et al.* (2008b) reported maximum growth performance in fingerlings of Nile tilapia fed practical diets with 45% of FM replaced by PPM (cottonseed, sunflower, canola, sesame and linseed meals). In addition, they also revealed that hemoglobin, hematocrit and the activity of aspartate aminotransferase (AST) and alanine aminotransferase (ALT) reduced, when increasing PPM in the test diets compared to the control. PPM comprising SBM (25%), CSM (25%), SFM (25%) and LSM (25%) resulted in no significant difference in growth performance and apparent protein digestibility for PPM replacement until 75% of the protein of FM.

On the basis of carcass composition and cost-benefit analysis, Elsaidy and Gaber (2003) reported that the PPM regime was cost-effective compared to FM and it can totally substitute FM protein in the commercial diets for tilapia. Gonzalez-Felix *et al.* (2010) reported that the use of PPM (soybean meal, alfalfa meal and corn gluten meal) can reduce use of fishmeal in commercial diets for Nile tilapia without affecting growth performance. Obirikorang *et al.* (2015) reported that FM could be partly substituted by using copra and palm kernel meals, considering results for short-term growth, nutrient utilization and digestibility. In addition, a complex mixture of plant ingredients (canola protein concentrate, maize gluten meal, pea protein concentrate, SBM and dehulled flax) could be applied in tilapia diets, with positive effects on SGR, PER, daily weight gain and FCR compared to a simple mixture comprising corn gluten meal and soybean meal (Borgeson *et al.*, 2006). Agbo (2008) reported that Ghanaian oilseed by-products: soybean meal (*Glycine* sp.), cottonseed meal (*Gossypium* sp.), groundnut cake (*Arachis hypogaea* L.) and groundnut husk at 50% inclusion, can replace fish meal with improved growth and feed utilization obtained, compared to the use of unprocessed meals in diets of *O. niloticus*. Borski *et al.* (2011) sought cheaper and more sustainable sources of protein ingredients for tilapia by using copra meal, cassava meal and rice bran and reported that a combination of these protein sources and alternate feeding days can lower feeding cost of tilapia without adversely impacting yield from commercial tilapia production.

20.3.1.4 Groundnut Meal/Cake Groundnut meal/cake is an oilseed by-product obtained after oil extraction from groundnuts, either by mechanical or solvent means. Its protein content is high but, like other plant products, it has sub-optimal levels of certain amino acids such as lysine, methionine and cystine, and is deficient in calcium and vitamin B_{12} (Jauncey, 1998). Groundnut cake (GNC) is used as human food at the village level; however, industrially produced cake is sold for animal feed (Agyenim-Boateng *et al.*, 1990). Groundnut husk (GNH), as it is known in Ghana, is another by-product of groundnut and is actually the *testa*, or skin of the kernels, which is usually removed after roasting. In the commercial diet of tilapia fingerlings, 20 or 50% of FM can be replaced slightly by GNC and GNH, while maintaining good growth, utilization of feed and protein digestibility (Agbo *et al.*, 2011).

20.3.1.5 Distillers Dried Grains with Solubles (DDGS) Distillers dried grains with solubles (DDGS) are fundamental residues from breweries and ethanol distillation plants following yeast fermentation of cereal grains. DDGS is a comparatively cost-effective ingredient in diet formulations. The properties of DDGS offer several potential advantages to animals, including moderate protein and lipid, as well as phosphorus, vitamins and trace minerals. In addition, DDGS has no anti-nutritional factors that are available in diverse feed ingredients of plant origin. Combining a variety of ingredients with DDGS may lower the demand of FM in commercial diet of tilapia, while maintaining good growth performance and feed utilization.

Coyle *et al.* (2004) reported that diets with 30% mixture of DDGS with MBM and SBM and devoid of fish meal produced good growth in tilapia. Chatvijitkul (2013) reported that lysine supplemented LE-DDGS is a promising protein source that can partially spare SBM in formulated feeds containing approximately 36% protein for hybrid tilapia. Wheat DDGS vary considerably in chemical composition from maize DDGS. Wheat DDGS contain more crude protein, crude fiber and lower crude fat than maize DDGS (Spiehs *et al.*, 2002). Li *et al.* (2011) suggested that wheat DDGS could be integrated at a level of 300 g/kg as a replacement in Nile tilapia diets for a mixture of SBM and maize meal (MM) without addition of lysine.

20.3.1.6 Rice Polish and Rice Bran The rice milling industry produces a by-product called rice polish, which is accessible throughout the year in Bangladesh, China, Thailand, Vietnam and other rice-producing and processing countries. Rice bran, maize meal and red wheat flour contain lower amounts of protein than rice polish, as proven by a number of studies (Saunders, 1990; Nyirenda *et al.*, 2000). Rice wine residue is produced as waste in the process of producing rice wine, which occurs in two steps: first, the conversion of rice starch to sugar by fungi through amylolysis and, second, the fermentation of the sugar by yeast (Vechklang *et al.*, 2011).

Khan *et al.* (2013) revealed that rice polish and mustard oil cake can partially replace fish meal without any significant change in growth performance. Liti *et al.* (2006) reported that rice bran is not economical for use as a sole feeding material for tilapia compared to maize bran and wheat bran, and male tilapia are sensitive to the use of the three brans compared to female tilapia.

20.3.1.7 Rapeseed Meal Rapeseed meal (RSM) is obtained from rapeseed, after extraction of oil as a by-product. Global RSM is about 14% of global oilseed production, with the quantity of about 71.5 million tonnes produced between 2013 and 2014 (FAO, 2014b). The approximate composition of RSM shows crude protein of 35%, crude lipid of 3.5%, ash content of 6%, crude fiber of 12% and moisture content of 10%.

Davies *et al.* (1990) suggested that experimental incorporation of RSM in tilapia diet should be around 15%. Increased levels in the diet cause retarded growth, following poor feed utilization. According to Jatta and Sigurgeirsson (2013), low protein diets for tilapia fry should not contain RSM, because this elicits a negative effect on carcass composition, hence substitution of 54% FM with RSM in diets of fish at 3 g size was strongly discouraged by the authors, based on its effect on growth and feed use.

20.3.1.8 Taro Leaf Meal In Vietnam and Cambodia, the use of taro cocoyam (*Colocasia esculenta*) in animal nutrition has been investigated. Fodder leaves and petioles of this plant has been fed to pigs after being cooked (Ty *et al.*, 2007, Toan and Preston, 2010, Hang and Preston, 2010). After heat treatment (Lewu *et al.*, 2009), leaves are used partially in the human diet in most Asian countries, some regions in the Pacific and a few parts of Africa (Onwueme, 1999). Leaves of taro are rich in some nutrients that are essential to human and animal nutrition, such as minerals (iron, calcium and phosphorus) and vitamins (A, C, thiamine, riboflavin and niacin) (Onwueme, 1999; Aregheore and Perera, 2003). In fertilized ponds, 23% of taro leaf protein can adequately substitute shrimp meal protein without negatively affecting growth parameters, survival and apparent food conversion ratio (Mathia and Fotedar, 2012).

20.3.1.9 Tomato/Potato By-Product Meal (T/PBM) Tomato waste is one of the canning wastes and the processing of tomatoes yields several by-products such as seeds and peels, which are mostly classified as pomace, seed meal, seed cake and seed oil of tomato. Potato processing is a very specialized field, which can only be described briefly here. The potato processing industry produces several products and by-products. Potato waste meal (potatoes, potato pulp and peels) is produced by drying and grinding of culls of potatoes. Soltan (2002) reported that substitution by up to 50% SBM by TBM and up to 40% yellow corn meal by PBM did not significantly affect fish growth performance and feed utilization, with 10.93% reduction in feed cost for 50% replacement of SBM by TBM and 7.53% reduction in feed cost with 40% replacement of Yellow Corn (YC) by PBM.

20.3.1.10 Recycled Food Wastes The application of recycled domestic waste as a substitute protein ingredient in fish feed is a possibility, but presently little information is available. Wastes of organic origin can provide a cost-effective and balanced additional source of protein and could be a better diet for larvae of fish. The suitable mixture of recycled food waste can eventually lower the fish meal amount in commercial diets of aquaculture species (Bake *et al.*, 2009). The use of extruded leftover food supplemented with premix has been found to produce a similar SGR to commercial diets, as opposed to feed without premix, although FCR and PER are in favour of the commercial feed, hence economic viability of using extruded and premix supplemented leftover food is questionable, even though it is reported as a possibility (Al-Ruqaie, 2007). The need for environmentally-sensitive feed production has led to the development of eco-friendly feed using various food wastes and the development of safety standards for its production (Sugiura *et al.*, 2009).

20.3.2 Animal By-Product Protein Source

20.3.2.1 Rendered Animal Protein Ingredients The transformation of animal waste into energy and a nutrient source for fish culture is a cost-effective option for recycling, in order to decrease environmental issues and at the same time expand animal production and provide cheap protein of animal origin, particularly for the poor. In the same manner, like fish meal, nutrients from rendered animal products are preferable sources of huge amounts of protein, with an acceptable suite of amino acids, digestible energy and dry matter (Bureau *et al.*, 1999; Allan *et al.*, 2000). They are also used as binders in formulation of fish feed (Jayathilakan *et al.*, 2012). However, on a standalone basis, meals from rendered animal protein have essential amino acid compositions that are either deficient or excessive, hence affecting the output of fish culture, particularly when applied solely as the primary protein source in diets of fish (Hegedius *et al.*, 1990; Davies *et al.*, 1991; Millamena, 2002).

Fasakin *et al.* (2005) reported that a combination of hydrolyzed feather meal with SBM led to improved fish performance in terms of growth and feed utilization, while fish productivity was unaffected regardless of use of enzyme treated feather meal with or without inclusion of dietary SBM. In another study, Ayoola (2010) reported that yeast extract, mechanically deboned meat and poultry by-product meal (PBM) proteins are feasible alternatives for FM, but economically PBM is the best alternative to FM in the commercial diet of Nile tilapia. Proteins from rendered animals have economic advantage and dietary promise in commercial feed of tilapia, but their proper application has been vigorously restricted by the volatile character and shortcomings of amino acids. The possible use of by-product of animal origin (PBM and porcine meal (PM)) due to the opportunity for an enhanced nutrition of cultured animals has been given attention. Hernandez *et al.* (2010) fed four diets containing three protein sources (PM, FM and PBM) and a commercial diet to tilapia and found a statistically comparable growth and nutrient usage among tilapia fed FM and PBM diets, while the

PM diet elicited poor performance by the fish compared to the FM diet, but similar to the commercial diet, hence both PBM and PM are suitable candidates to replace FM in tilapia diets.

20.3.2.2 Poultry By-Product Meal A complete or partial substitution of FM with less costly protein from PBM can lower the price of feed (Bureau, 2007; Pine *et al.*, 2008), without reduced palatability (Allen Davis and Arnold, 2004; Rawles *et al.*, 2009), reduce essential fatty acids in carcasses (Parés-Sierra *et al.*, 2014), reduce essential amino acids content of the diet (Pine *et al.*, 2008; Rawles *et al.*, 2009; Hernandez *et al.*, 2014) and reduce digestibility (Erturk and Sevgili, 2003; Hernandez *et al.*, 2014). Meals from poultry processing waste and bone and meat meal are by-products of terrestrial animal meat processing and have high amounts of protein and relevant EAA profiles (Tacon, 1993). Yildirim *et al.* (2009) suggested that 50% meal from poultry by-product could be applied in a feed of *T. zilli* cultured in brackish water without any negative effect on the growth of the fish. However, according to Suleiman and Ahmed (2011), FM is superior to poultry by-products (offal + intestine) in the diets of Nile tilapia.

20.3.2.3 Feather Meal Keratin, a composite protein, is the main material that makes up feather meal. It is digestible via hydrolysis. Cooking time and pressure directly affects the digestibility of feather. Cystine, threonine and arginine are three essential amino acids abundant in feather meal and it has a high level of pepsin digestible protein (75–87%) (Bureau *et al.*, 2000). The profile of amino acids in feather meal shows that FM and SBM are identical (Sinchermiri *et al.*, 1988; Sarmwatanakul and Bamrongtum, 2000). The improvement of protein quality in feather meal by fermentation with microorganism has been studied for subsequent use as a replacement for FM in diets of fish. From the foregoing, hydrolysed feather meal (HFM) is a good candidate for replacing fishmeal in fish diets, especially due to unavailability of anti-nutritional factors.

Arunlertaree and Moolthongnoi (2008) concluded that up to 50% fermented meal from feather can be applied in the commercial feed of Nile tilapia without compromising growth and nutrient utilization efficiency. One of the latest feed products comprises 18 free amino acids and is called 'hydrolyzed feather meal'. It is a product of a Chinese biotechnology firm, Autoau (Hubei) Biotechnology Co., Ltd. SBM and CSM can be partly substituted by CHFM (Zhang *et al.*, 2014) in the feed of hybrid tilapia, with no negative effects on immunity against *Aeromonas hydrophila*.

20.3.2.4 Blood Meal Blood meal (BM) is an animal waste product that is readily available in abattoirs and is a cheap alternative protein source. The development of a more sustainable aquaculture feed product will depend on identifying and establishing alternative feedstuffs to fish meal (Olukayode and Emmanuel, 2012). According to Aladetohun and Sogbesan (2013), protein in the diet of tilapia can be supplied totally by BM, without any adverse effects on growth and survival. Contrary reports have indicated that BM is not effectively utilized by tilapia (El-Sayed, 1998), but digestibility could be improved by using different preparation techniques such as spray drying as against steam-tube and roto-plate drying methods (Bureau *et al.*, 1999). Furthermore, potential bioavailability of amino acids from BM, especially lysine, is affected by the method of processing and is higher with either spray dried BM or flash dried BM, compared to l-lysine HCL (El-Haroun and Bureau, 2007).

20.3.2.5 Meal of Sand Smelt Sand smelt enter different freshwater bodies by coincidence, but become dominant in these habitats because of its biological features (Küçük *et al.*, 2006). It is generously and accidentally found in natural environments and, as an invasive species, challenges for food, DO and space with native fish. However, Gumus *et al.* (2010) suggested that up to 75% meal of sand smelt could substitute fish meal in Nile tilapia fry diets.

20.3.2.6 By-Product Derived from Local Fisheries/Fish Processing Thailand is a rapidly developing country with a booming small-scale operation of agro-industry. One of the small-scale industries-fermented fish enterprises, is worthy of note. Fermented products can be made from cut and unprocessed edible parts of cultured freshwater fishes. Throughout this process, it always produces an obnoxious odor from the solid waste. There is paucity of information on the likelihood of substituting mainstream fish meal in diets of tilapia with fish meal produced from by-products of fish processing. Chitmanat *et al.* (2009) reported that dietary fish meal protein could be substituted in Nile tilapia fingerling commercial diet by only 25% of the local fish processing by-product.

A considerable amount of waste is generated from the catching and processing of fish. The amount of fish processing waste as well as fishery discards combined surpasses the capacity of fish stocks aimed at every year for the production of fish meal. By-catch and meals from processing waste have higher ash contents and less protein and lipid compared to anchovy meal (Goddard *et al.*, 2008). Fish fed with high ash meals, mixed benthic species meal (MBM), mixed catfish species meal (CAM) and tuna cannery waste (TCW) retained phosphorus more significantly than lower ash meals (commercial anchovy meal (COM) and small pelagic species meal (SPM). It was concluded that fisheries by-catch (MBM, SPM and CAM), and processing waste meals are suitable as juvenile Nile tilapia diet protein ingredients.

20.3.2.7 Shrimp Industry Waste Meal Waste from the shrimp industry can be utilized as a protein ingredient in fish diets with high-quality protein being utilized, since it can be isolated hence it has good nutritional value and quality. When shrimp meals separated from chitin are included in diets, there is an improvement in growth and survival. Shrimp head silage protein hydrolysate (SPH) inclusion at 10% and 15% in tilapia diets resulted in a significantly better growth and other parameters as against diets containing fish meal or higher SPH inclusions in diets, hence a maximum of 15% incorporation of SPH is recommended in the commercial diet of tilapia (Plascencia-Jatomea *et al.*, 2002). Diop *et al.* (2013) reported that meals from shrimp waste are crucial nutrition for tilapia and up to 45% of the diet for tilapia can be made from meals of steamed shrimp waste.

From the heads of Pacific white shrimp, Silva (2004) produced shrimp protein hydrolysate (SPH), an excellent protein material considering its profile of amino acids and small amount of fiber. The level of free amino acids from processing waste of shrimp products makes it possible for use as a feed ingredient and additive in feed formulations (Heu *et al.*, 2003; Ruttanaporn *et al.*, 2005). Leal *et al.* (2009) stated that 6% shrimp protein hydrolysate (SPH) could be a promising protein feedstuff for diets of Nile tilapia, with no negative growth performance and feed utilization.

In another study, Cavalheiro *et al.* (2007) reported that shrimp head silage can totally substitute fish meal as a protein source in the diet of tilapia, with cost-effectiveness and no deleterious effect on the characteristics of the diet.

20.3.2.8 Fish Silage Fish silage is a fluid and to assure an excellent application, its amount of water has to be decreased in diets of animals. The amount of water is reduced by co-drying (Lopez, 1990; Dong *et al.*, 1993), a process that recovers solubilized proteins and reduces moisture through addition of dry feedstuffs to wet fish silage. Stone *et al.* (1989) stated that co-drying enhances the nutrient characteristics of feed blends, alters the physical characteristics and promotes their establishment as traditional food. Soltan and El-Laithy (2008) reported that the utilization of feed as well as growth performance of Nile tilapia were not significantly affected when fish silage (fermented with *Lactobacillus plantarum*) protein was included in the diet to provide as much as 30% of dietary protein.

Hancock (*Pterigoplichthys multiradiatus*) is an invasive fish species in Mexico with excellent meat properties, but is given little consideration (Rueda-Jasso *et al.*, 2013). The fish has a hard skin with less meat, hence making extraction of meat protein for direct consumption very expensive (Rueda-Lopez *et al.*, 2013). These authors used four diets with stepwise increases in levels of acid treated *Pterigoplichthys multiradiatus* silage as a replacement for fish meal in the diet of juvenile tilapia and observed that the feed conversion ratio was least at 15% substitution, with SGR highest at 5% substitution, while cost-effectiveness is achieved at 15% substitution with about 24% cost reduction.

20.3.2.9 Gambusia Fish Meal Gambusia or mosquito fish (*Gambusia affinis*) is rich in protein, with crude protein contents above 55% on a dry matter basis (Abdelghany, 2003). Abdelghany (2003) suggested that 50% Gambusia (*Gambusia affinis*) fish meal (GFM) can replace herring fish meal (HFM) with no adverse effects on Nile tilapia growth. Furthermore, Ahmad (2008) investigated the possible use of GFM in diets of Nile tilapia. Diets containing 35% crude protein were formulated, with replacement of HFM by GFM at 0, 10, 25, 50, 75 and 100% levels. It was observed that fish growth, PER, ANPU and feed intake increased with increasing GFM by up to 75%, while FCR reduced with increasing GFM by up to 75%. Body composition was not affected by GFM inclusion and digestibility coefficient of dry matter was not significantly different for all diets, hence substitution of HFM by GFM in tilapia diets can be feasible up to 75% level.

20.3.2.10 Tuna By-Products Meal (TBM) or Tuna Silage Hydrolysates (TSH) Global tuna catch in 2012 was well over 7 million tonnes (FAO 2014a). According to Hernandez *et al.* (2013), about 52–54% of the 134 500 tonnes of tuna landed in Mexico are scrapped as waste by the canning industry and a portion is transformed to economically-feasible meal of tuna by-products (TBM). Tuna canning factories produce waste from virtually all processes and this waste contains a high content of proteins (Veiga *et al.*, 1994). Researchers have advocated its combination with nutrients of plant proteins in feeds of fish (Goddard *et al.*, 2008). Hernandez *et al.* (2013) reported that tilapia fed TBM as the sole protein component of the feed had greater growth and lower feed utilization than those fed feed consisting of tuna silage hydrolysate (TSH).

20.3.3 Other By-Product Protein Source

20.3.3.1 Yeast Yeast contains a high amount of crude protein comparable to soybean. It is also a rich source of vitamin B (Olvera-Novoa *et al.*, 2002). Nonetheless, it is inadequate in methionine and cystine, but has a high content of lysine (Huige, 2006). Yeast protein contains 10–15% nucleic acids (Reed and Nagodawithana, 1991), which tends to increase nucleic acid concentration in the digestive tract, especially purines leading to accumulation of uric acid as prompted by lactate production (Murray and Pizzorno, 2012), hence protein metabolism is hampered (Reed and Nagodawithana, 1991). Ozorio *et al.* (2012) stated that 15% dried *S. cerevisiae* inclusion could improve feed utilization and growth of tilapia juveniles. The point of origin, as well as medium of growth of the yeast, determines its nutritional value, with alkane growth medium deemed producing better-quality yeast than carbohydrate medium (Sarlin, 2005). Inclusion of yeast (*Candida utilis*) in the diet of tilapia fry of up to 30% of protein source and 65% of animal protein could be substituted with a mixture of plant proteins with no unfavourable effects on growth, yield and profit of fish culture (Olvera-Novoa *et al.*, 2002).

20.3.3.2 Mango Peel Mango peel is an energy-rich feed ingredient that is obtained from the mango fruit after processing. Its use as animal feed has been stepped up over the years (IITA, 1990), mainly due to the need to reduce its disposal as waste and thus protect the environment. Omojowo *et al.* (2010) reported that on the basis of excellent growth performance, mango peel meal could substitute 25% of maize in Nile tilapia commercial diets.

20.3.3.3 Microalgae Algae play a crucial role in aquaculture, because green water aquaculture has produced better survival, growth and feed conversion than clear water aquaculture (Rodolfi *et al.*, 2003; Lio-Po *et al.*, 2005). Ease of culture, non-toxicity, size, nutritional content, digestibility of the cell wall and shape are critical factors that determine the suitability of marine microalgal species for use as a feed ingredient in aquaculture (Brown *et al.*, 1997; 1999, Guedes and Malcata 2012, Renaud *et al.*, 2002). The advantages of using marine microalgae in aquaculture notwithstanding, their use is limited due to difficulty in production, concentration and storage (Becker, 2004). Fermented Cladophora could replace plant-based protein in the commercial diet of tilapia at 10% level of inclusion (Dewi *et al.*, 2014). Spirulina could be used in tandem with phosphorous supplementation without affecting Nile tilapia growth parameters (Olvera-Novoa *et al.*, 1998). Azaza *et al.* (2008) reported that green algae (*Ulva rigida*) can replace 20% of soybean meal in Nile tilapia feeds, with no negative response from body composition and diet utilization. Similarly, Hussein *et al.* (2013) concluded that spirulina nutrients could substitute up to 50% of corn gluten meal in Nile tilapia feeds, with no negative effect on growth and body mineral composition.

20.3.3.4 Melon Shell Meal Husks, nutshells, stalks, residues and manure etc. are generated as wastes from the production and processing of agricultural products. About 140 billion tonnes of these are produced annually worldwide (UNEP, 2009). These can either be efficiently applied as diet constituents for animal and fish feeds or converted into energy. Generally, residues of crop origin have low digestibility given their high fibre content and they have little value as protein ingredients. Orire and

Ricketts (2013) reported up to 75% addition of meal from melon shell in Tilapia diet to replace maize meal.

20.3.3.5 By-Products from Alcoholic Beverage Production Processes Vechklang *et al.* (2011) stated that most of the Asian countries produce a traditional fermented liquor known as rice wine, which is distilled with various nomenclature, such as choujiu (China), cheong (Korea), sake (Japan), santi (India) and sato (Thailand). Rice wine residue (RWS) is usually discarded after separation from the liquid wine produced after amylolytic conversion and fermentation. However, RWS contains a mixture of nutrients from both rice and microorganisms, with the possibility for use as an ingredient in fish feed. Vechklang (2011) reported that RWS (at 22.5%) has the possibility of application in the diets of Nile tilapia with no negative growth response.

20.3.3.6 Brewer's Waste Brewer's waste is spent grains left over after production of beer (Kaur and Saxena, 2004). Although this product contains high dietary fibre, it can be utilized effectively if the lignocelluloses are removed mechanically (Levic *et al.*, 2010). Zerai *et al.* (2008) concluded that since brewer's waste did not elicit a negative response from tilapia regarding body composition and food utilization, it may substitute up to 50% of fishmeal protein.

20.3.3.7 Cheese Processing By-Product Meal (Whey Powder) Whey powder contains 11–14.5% crude protein, while whey protein concentrate contains between 25% and 89% crude protein (USDEC, 2004). Abdel-Tawwab *et al.* (2011) reported that CPBM can replace 75% FM in commercial feed of Nile tilapia with no negative growth performance.

References

AAFCO (2001) Association of American Feed Control Officials, *AFFCO Official Publication*. The Association of Feed Control Officials, Atlanta, GA.

Abdelghany, A.E. (2003) Partial and complete replacement of fish meal with gambusia meal in diets for red tilapia (*Oreochromis niloticus* x *O. mossambicus*). *Aquaculture Nutrition*, **9**(3): 145–154.

Abdel-Tawwab, M., Abbass, F.E. and Seden, M.E.A. (2011) Partial and total replacement of fish meal with cheese processing by-product meal in practical diets for Nile tilapia, *Oreochromis niloticus* (L.): A preliminary study. In: *Ninth International Symposium on Tilapia in Aquaculture* (eds L. Liu and K. Fitzsimmons). Proceedings of the. Aquaculture and Fisheries Collaborative Research Support Program, pp. 283–291.

Abdel-Warith, A.A., Russell, P.M. and Davies, S.J. (2001) Inclusion of a commercial poultry by-product meal as a protein replacement of fish meal in practical diets for African catfish *Clarias gariepinus* (Burchell 1822). *Aquaculture Research*, **32**(Suppl. 1): 296–305.

Aderolu, A.Z., Lawal, M.O., Ali, T.O. and Aarode, O.O. (2011) Utilization of baker's yeast (*Saccharomyces cerevisiae*) in the diet of juvenile African catfish (*Clarias gariepinus*). *Journal of Scientific Research and Development*, **13**: 19–27.

Adewolu, M.A. and Benfey, T.J. (2009) Growth, nutrient utilization and body composition of juvenile bagrid catfish, *Chrysichthys nigrodigitatus* (actinopterygii: siluriformes: claroteidae), fed different dietary crude protein levels. *Acta Ichthyologica Et Piscatoria*, **39**(2): 95–101.

Adewolu, M.A., Ikenweiwe, N.B. and Mulero, S.M. (2010) Evaluation of an animal protein mixture as a replacement for fishmeal in practical diets for fingerlings of *Clarias gariepinus* (Burchell, 1822). *The Israeli Journal of Aquaculture – Bamidgeh*, **62**(4): 237–244.

Agbabiaka, L.A. (2010) Evaluation of some under-utilized protein feedstuffs in diets of *Clarias gariepinus* (Burchell, 1822) fingerlings. *International Journal of Tropical Agriculture and Food Systems*, **4**(1): 10–12.

Agbo, N.W. (2008) Oilseed meals as dietary protein sources for juvenile Nile tilapia (*Oreochromis niloticus* L.). PhD Thesis, University of Stirling, UK, pp. 224.

Agbo, N.W., Adjei-Boateng, D. and Jauncey, K. (2011) The potential of groundnut (*Arachis hypogaea* L.) by-products as alternative protein sources in the diet of Nile tilapia (*Oreochromis niloticus*). *Journal of Applied Aquaculture*, **23**(4): 367–378.

Agyenim-Boateng, C.E., Vincke, M.M.J. and Wijkstrom, N. (1990) Manures and agro-industrial by-products for fish-farming in Ghana. Field working paper 4. FAO, Rome.

Ahmad, M.H. (2008) Evaluation of Gambusia, *Gambusia affinis*, fish meal in practical diets for fry Nile tilapia, *Oreochromis niloticus*. *Journal of the World Aquaculture Society*, **39**(2): 243–250.

Aladetohun, N.F. and Sogbesan, O.A. (2013) Utilization of blood meal as a protein ingredient from animal waste product in the diet of *Oreochromis niloticus*. *International Journal of Fisheries and Aquaculture*, **5**(9): 234–237.

Allan, G.L., Parkinson, S., Booth, M.A., Stone, D.A.J., *et al.* (2000) Replacement of fish meal in diets for Australian silver perch, *Bidyanus bidyanus*: I: Digestibility of alternative ingredients. *Aquaculture*, **186**: 293310.

Al-Ruqaie, I.M. (2007) Extruded Leftover Food as Animal Feed. I: Effect of extruded feed on growth and feed utilization of Tilapia (*Oreochromis niloticus*) in Saudi Arabia. *Pakistan Journal of Biological Sciences*, **10**: 3248–3253.

Andrews, J.W. and Page, J.W. (1974) Growth factors in the fishmeal component of catfish diets. *Journal of Nutrition*, **104**: 1091–1096.

Aregheore, E.M. and Perera, D. (2003) Dry matter, nutrient composition and palatability/acridity of eight exotic cultivars of cocoyams – Taro (*Colocassia esculenta*) in Samoa. *Plant Foods for Human Nutrition*, **58**: 1–8.

Arunlertaree, C. and Rakyuttithamkul, E. (2006) Utilization of fermented feather meal as a replacement of fish meal in the diet of hybrid Clarias catfish. *Kasetsart Journal (Natural Science)*, **40**: 436–448.

Arunlertaree, C. and Moolthongnoi, C. (2008) The use of fermented feather meal for replacement fish meal in the diet of *Oreochromis niloticus*. *Environment and Natural Resources Journal*, **6**(1): 13–24.

Attalla, R.F. and Mikhail, S.K. (2008) Effect of replacement of fish meal protein with boiled full fat soybean seeds and dried algae on growth performance, nutrient utilization and some blood parameters of Nile tilapia (*Oreochromis niloticus*). *Egyptian Journal of Aquatic Biology and Fisheries*, **12**(4): 41–61.

Ayoola, A.A. (2010) Replacement of Fishmeal with Alternative Protein Source in Aquaculture Diets. MSc Thesis, Faculty of North Carolina State University, pp. 145.

Azaza, M.S., Mensi, F., Ksouri, J., Dhraief, M.N., Brini, B., *et al.* (2008). Growth of Nile tilapia (*Oreochromis niloticus* L.) fed with diets containing graded levels of green algae ulva meal (*Ulva rigida*) reared in geothermal waters of southern Tunisia. *Journal of Applied Ichthyology*, **24**: 202–207.

Bahnasawy, M. (2009) Effect of dietary protein levels on growth performance and body composition of Monosex Nile tilapia, *Oreochromis niloticus* L. Reared in fertilized tanks. *Pakistan Journal of Nutrition*, **8**(5): 674–678.

Bake, G.G., Endo, M., Akimoto, A. and Takeuchi, T. (2009) Evaluation of recycled food waste as a partial replacement of fishmeal in diets for the initial feeding of Nile tilapia *Oreochromis niloticus*. *Fisheries Science*, **75**(5): 1275–1283.

Barros, M.M., Lim, C., Evans, J.J. and Klesius, P.H. (2000) Effect of iron supplementation to cottonseed meal diets on the growth performance of channel catfish, *Ictalurus punctatus*. *Journal of Applied Aquaculture*, **10**(1): 65–86.

Barros, M.M., Lim, C. and Klesius, P.H. (2002) Effect of soybean meal replacement by cottonseed meal and iron supplementation on growth, immune response, and disease resistance of channel catfish (*Ictalurus punctatus*) to *Edwardsiella ictaluri* challenge. *Aquaculture*, **207**: 263–279.

Becker, W. (2004) Microalgae for aquaculture. The nutritional value of microalgae for aquaculture. In: *Handbook of Microalgal Culture* (ed. A. Richmond). Blackwell, Oxford, pp. 380–391.

Bi, I.V., Maquet, A., Baudoin, J.P., du-Jardin, P., Jacquemin, J.M. and Mergeai, G. (1999) Breeding for 'low-gossypol seed and high-gossypol plants' in upland cotton. Analysis of tri-species hybrids and backcross progenies using AFLPs and mapped RFLPs. *Theoretical and Applied Genetics*, **99**(7): 1233–1244.

Borgeson, T.L., Racz, V.J., Wilkie, D.C., White, L.J. and Drew, M.D. (2006) Effect of replacing fishmeal and oil with simple or complex mixtures of vegetable ingredients in diets fed to Nile tilapia (*Oreochromis niloticus*). *Aquaculture Nutrition*, **12**(2): 141–149.

Borski, R.J., Bolivar, R.B., Jimenez, E.B.T., Sayco, R.M.V., Arueza, R.L.B., *et al.* (2011) Fishmeal-free diets improve the cost effectiveness of culturing Nile tilapia (*Oreochromis niloticus* L.) in ponds under an alternate day feeding strategy. In: *Proceedings of the Ninth International Symposium on Tilapia in Aquaculture. Aquaculture and Fisheries Collaborative Research Support Program* (eds L. Liu and K. Fitzsimmons), pp. 111–118.

Brown, M.R., Jeffrey, S.W., Volkman, J.K. and Dunstan, G.A. (1997) Nutritional properties of microalgae for mariculture. *Aquaculture*, **151**: 315–331.

Brown, M.R., Mular, M., Miller, I., Farmer, C. and Trenerry, C. (1999) The vitamin content of microalgae used in aquaculture. *Journal of Applied Phycology*, **11**: 247–255.

Bureau, D. (2006) Rendered products in fish aquaculture feeds. In: *Essential Rendering: All About The Animal By-Products Industry* (ed. D.L. Meeker). National Renderers Association, 302 pp.

Bureau, D.P. (2007) Fish meal replacement: Opportunities for rendered products. Render August 2007, pp. 10–12. [Online] Last accessed 29th September, 2015. Available from: https://d10k7k7mywg42z.cloudfront.net/assets/4d90dcdfdabe9d7a9b00004c/fishmealreplacement.pdf

Bureau, D.P., Harris, A.M. and Cho, C.Y. (1999) Apparent digestibility of rendered animal protein ingredients for rainbow trout (*Oncorhynchus mykiss*). *Aquaculture*, **180**: 345–358.

Bureau, D.P., Harris, A.M., Bevan, D.J., Simmons, L.A., Azevedo, P.A. and Cho, C.Y. (2000) Feather meals and meat and bone meals from different origins as protein sources in rainbow trout (*Oncorhynchus mykiss*) diets. *Aquaculture*, **181**: 281–291.

Buser, M.D. and Abbas, H.K. (2001) Mechanically processing cottonseed to reduce gossypol and aflatoxin levels. *Journal of Toxicology – Toxin Reviews*, **20**(3 & 4): 179–208.

Cavalheiro, J.M., Oliveira, E.S. and Bora, P.S. (2007) Utilization of shrimp industry waste in the formulation of tilapia (*Oreochromis niloticus* Linnaeus) feed. *Bioresource Technology*, **98**: 602–606.

Chatvijitkul, S. (2013) Lipid Extracted Distillers Dried Grains with Solubles (LE-DDGS) as a partial replacement for soybean meal in hybrid Tilapia (*Oreochromis niloticus* × *Oreochromis aureus*). MSc Thesis, Aubusrn University, 64 pp.

Chitmanat, C., Tipin, A., Chaibu, P. and Traichaiyaporn, S. (2009) Effects of replacing fishmeal with wastes derived from local fisheries on the growth of juvenile tilapia, *Oreochromis niloticus*. *Songklanakarin Journal of Science and Technology*, **31**(1): 105–110.

Chiu, A., Li, L., Guo, S., Bai, J., Fedor, C. and Naylor, R.L. (2013) Feed and fishmeal use in the production of carp and tilapia in China. *Aquaculture*, **414-415**: 127–134.

Chor, W-K, Lim, L-S. and Shapawi, R. (2013) Evaluation of feather meal as a dietary protein source for african catfish fry, *Clarias gariepinus*. *Journal of Fisheries and Aquatic Science*, **8**: 697–705.

Cocker, L.M. (2014) Strategic review on African aquaculture feeds. Partnership for African Fisheries (PAF) Aquaculture Working Group report [Online] Last accessed 6 October 2015. Available from: http://www.sarnissa.org/dl687

Cornell, H.J. and Hoveling, A.W. (1998) *Wheat: Chemistry and Utilization*. CRC Press, New York, 426 pp.

Costa-Pierce, B.A., Bartley, D.M., Hasan, M., Yusoff, F., Kaushik, S.J., *et al.* (2012) Responsible use of resources for sustainable aquaculture. In: *Farming the Waters for People and Food. Proceedings of the Global Conference on Aquaculture 2010, Phuket, Thailand* (eds R.P. Subasinghe, J.R. Arthur, D.M. Bartley, S.S. De Silva, M. Halwart, *et al.*). 22–25 September 2010. FAO, Rome and NACA, Bangkok, pp. 113–147.

Coyle, S.D., Mengel, G.J., Tidwell, J.H. and Webster, C.D. (2004) Evaluation of growth, feed utilization, and economics of hybrid tilapia, *Oreochromis niloticus* x *Oreochromis aureus*, fed diets containing different protein sources in combination with distillers dried grains with solubles. *Aquaculture Research*, **35**(4): 365–370.

CRB (2007) *The CRB Commodity Yearbook 2007*. Commodity Research Bureau. John Wiley & Sons, New Jersey, 400 pp.

Da, C.T., Lundh, T. and Lindberg, J.E. (2013) Digestibility of dietary components and amino acids in animal and plant protein feed ingredients in striped catfish (*Pangasianodon hypophthalmus*) fingerlings. *Aquaculture Nutrition*, **19**: 741–750.

Dabrowski, K. and Kozak, B. (1979) The use of fish meal and soyabean meal as a protein source in the diet of grass carp fry. *Aquaculture*, **18**: 107–114.

Davies, O.A. and Ezenwa, N.C. (2010) Groundnut cake as alternative protein source in the diet of *Clarias gariepinus* fry. *International Journal of Science and Nature*, **1**(1): 73–76.

Davies, S.J., McConnell, S. and Bateson, R.I. (1990) Potential of rapeseed meal as an alternative protein source in complete diets for tilapia (*Oreochromis mossambicus*, Peters). *Aquaculture*, **87**: 145–154.

Davies, S.J., Williamson, J., Robinson, M. and Bateson, R.I. (1991) Practical inclusion levels of common animal by-products in complete diets for tilapia (*Oreochromis mossambicus* Peters). In: *Fish Nutrition in Practice* (eds S.J. Kaushik and P. Luquet). INRA, Paris, pp. 325–332.

Davis K.S. (2001) Corn milling, processing and generation of co-products. *Minnesota Nutrition Conference Minnesota Corn Growers Association Technical Symposium*, 11 September 2001 [Online] Last accessed 11 September 2015. Available from: http://www.distillersgrains.org/files/grains/k.davis--drywetmillprocessing.pdf

Davis, A,D. and Arnold, C.R. (2004) Red Drum, *Sciaenops ocellatus*, production diets: Replacement of fish meal with soybean meal. *Journal of Applied Aquaculture*, **15**(3–4): 173–181.

Dean, J.C., Nielsen, L.A., Helfrich, L.A. and Garling, D.L. (1992) Replacing fish meal with seafood processing wastes in channel catfish diets. *The Progressive Fish Culturist*, **54**(1): 7–13.

Deutsch, L., Graslund, S., Folke, C., Troell, M., Huitric, M., *et al.* (2007) Feeding aquaculture growth through globalization: Exploitation of marine ecosystems for fishmeal. *Global Environmental Change*, **17**: 238–249.

Dewi, A.P.W.K., Nursyam, H. and Hariati, A.M. (2014) Response of fermented Cladophora containing diet on growth performances and feed efficiency of Tilapia (*Oreochromis* sp.). *International Journal of Agronomy and Agricultural Research,* **5**(6): 78–85.

Diop, D., Fall, J., Sagne, M., Loum, A., Ndong, D., Diouf, M., *et al.* (2013) Use of Biochemically improved shrimp industry waste in fry Tilapia (*Oreochromis niloticus*, Linnaeus 1758) diets: Effects on growth performance and carcass composition. *Journal of Biology and Life Science*, **4**(2): 328–340.

Dong, F.M., Fairgrieve, W.T., Skonberg, D.I. and Rasco, B.A. (1993) Preparation and nutrient analysis of lactic acid bacterial ensiled salmon viscera. *Aquaculture*, **109**: 351–366.

El-Haroun, E.R. and Bureau, D.P. (2007) Comparison of the bioavailability of lysine in blood meals of various origins to that of L-lysine HCL for rainbow trout (*Oncorhynchus mykiss*). *Aquaculture*, **262**: 402–409.

El-Saidy, D.M.S.D. and Gaber, M.M.A. (2002) Complete replacement of fish meal by soybean meal with dietary l-lysine supplementation for Nile tilapia *Oreochromis niloticus* (L.) fingerlings. *Journal of the World Aquaculture Society*, **33**(3): 297–206.

El-Saidy, D.M.S.D. and Gaber, M.M.A. (2003) Replacement of fish meal with a mixture of different plant protein sources in juvenile Nile tilapia, *Oreochromis niloticus* (L.) diets. *Aquaculture Research*, **34**(13): 1119–1127.

El-Sayed, A.F.M. (1998) Total replacement of fish meal with animal protein sources in Nile tilapia (*Oreochromis niloticus* L.) feeds. *Aquaculture Research*, **29**: 275–280.

El-Sayed, A.F.M. (2006) *Tilapia Culture*. CABI Publishing International, Oxford, 277 pp.

Enami, H.R. (2011) A review of using canola/rapeseed meal in aquaculture feeding. *Journal of Fisheries and Aquatic Science*, **6**: 22–36.

Erturk, M.M. and Sevgili, H. (2003) Effects of replacement of fish meal with poultry by-product meals on apparent digestibility, body composition and protein efficiency ratio in a practical diets for rainbow trout, *Onchorynchus mykiss*. *Asian-Australasian Journal of Animal Sciences*, **16**(9): 1355–1359.

Essa, M.A., Mabrouk, H.A., Mohamed, R.A. and Michael, F. (2011) Evaluating different additive levels of yeast, *Saccharomyces cerevisiae*, on the growth and production performances of a hybrid of two populations of Egyptian African catfish, *Clarias gariepinus*. *Aquaculture*, **320**(1): 137–141.

Eya, J.C., Ashame, M.F. and Pomeroy, C. (2010) Nutritive evaluation of protein from anaerobically digested poultry wastes as a dietary ingredient replacer for channel catfish, *Ictalurus punctatus*. *Journal of the World Aquaculture Society*, **41**(S2): 179–190.

Eyo, A.A. (2003) Fundamentals of fish nutrition and diet development. An overview. *National Workshop on Fish Feed Development and Feeding Practices in Aquaculture Organized by Fisheries Society of Nigeria (FISON)*. In Collaboration with National Institute of Fresh Water Fisheries Research (NIFFR) and FAO National Special Programme for Food Security (FAO-NSPFS), pp. 1–33.

Ezenwaji, N.E., Iluno, A., Atama, C., Nwaigwe, C.O. and Nwaigwe, C.U. (2012) Substitution of soyabean meal with bioactive yeast in the diet of *Clarias gariepinus*: Effect on growth rate, haematological and biochemical profile. *African Journal of Biotechnology*, **11**(91): 15802–15810.

Fagbenro, O.A. (1998) Apparent digestibility of various oilseed cakes/meals in African catfish diets. *Aquaculture International*, **6**: 317–322.

FAO (1980) *Fish Feed Technology*. Lectures presented at the FAO/UNDP Training Course in Fish Feed Technology, held at the College of Fisheries, University of Washington, Seattle, Washington, 9 October–15 December 1978. ADCP/REP/80/11. Food and Agriculture Organization of the United Nations, Rome, 1980. http://www.fao.org/docrep/x5738e/x5738e00.htm#Contents

FAO (1986) *The Production of Fish Meal and Oil*. FAO, Fisheries Technical Paper 142. Food and Agriculture Organisation of the United Nations, Rome. [Online] Last accessed 29 March 2015. Available from: http://www.fao.org/docrep/003/x6899e/X6899E00.HTM

FAO (2014a) *The State of World Fisheries and Aquaculture: Opportunities and challenges*. Food and Agriculture Organization of the United Nations, Rome. 223 pp.

FAO (2014b) FAO *Market Summaries: Oil crops*. FAO Trade and Marketing Division. FAO Outlook October 2014 [Online] Last accessed 6 March 2015. Available from: http://www.fao.org/fileadmin/templates/est/COMM_MARKETS_MONITORING/Oilcrops/Documents/Food_outlook_oilseeds/Oilcrops_Foodoutlook_Oct14.pdf

Faria, S.A.S.C., Bassinello, P.Z. and Penteado, M.V.C. (2012) Nutritional composition of rice bran submitted to different stabilization procedures. *Brazilian Journal of Pharmaceutical Sciences*, **48**(4): 651–657.

Fasakin, E.A., Serwata, R.D. and Davies, S.J. (2005) Comparative utilization of rendered animal derived products with or without composite mixture of soybean meal in hybrid tilapia (*Oreochromis niloticus* × *Oreochromis mossambicus*) diets. *Aquaculture*, **249**(1–4): 329–338.

Feedipedia (2015) Animal feed resources information system [Online] Last accessed 9 March. Available from: http://www.feedipedia.org/

Gadelha, I.C., Fonseca, N.B., Oloris, S.C., Melo, M.M. and Soto-Blanco, B. (2014) Gossypol toxicity from cottonseed products. *Scientific World Journal*, **2014**: 231635.

Gatlin, D.M. (1995) Review of red drum nutrition. In: *Nutrition and Utilization Technology in Aquaculture* (eds C. Lim and D.J. Sessa). The American Oil Chemists Society, pp. 41–49

Gatlin, D.M., Barrows, F.T., Brown, P., Dabrowski, K., Gaylord, T.G., *et al.* (2007). Expanding the utilization of sustainable plant products in aquafeeds: A review. *Aquaculture Research*, **38**: 551–579.

Giri, S.S., Sahoo, S.K., Sahu, A.K. and Mukhopadhyay, P.K. (2000) Growth, feed utilization and carcass composition of catfish *Clarias batrachus* (Linn.) fingerlings fed on dried fish and chicken viscera incorporated diets. *Aquaculture Research*, **31**: 767–771.

Giri, S.S., Sahoo, S.K. and Mohanty, S.N. (2010) Replacement of by-catch fishmeal with dried chicken viscera meal in extruded feeds: Effect on growth, nutrient utilisation and carcass composition of catfish *Clarias batrachus* (Linn.) fingerlings. *Aquaculture International*, **18**: 539–544.

Goda, A.M., El-Haroun, E.R. and Chowdhury, M.A.K. (2007) Effect of totally or partially replacing fish meal by alternative protein sources on growth of African catfish *Clarias gariepinus* (Burchell, 1822) reared in concrete tanks. *Aquaculture Research*, **38**: 279–287.

Goddard, S. (1995) *Feed Management in Intensive Aquaculture*. Springer, US.

Goddard, S., Al-Shagaa, G. and Ali, A. (2008) Fisheries by-catch and processing waste meals as ingredients in diets for Nile tilapia, *Oreochromis niloticus*. *Aquaculture Research*, **39**(5): 518–525.

Gonzalez-Felix, M.L., Perez-Velazquez, M., Villalba-Villalba, A.G., Civera-Cerecedo, R., Ezquerra, J. and Goytortua-Bores, E. (2010) Tailoring a diet for Nile tilapia (*Oreochromis niloticus*) culture in Northwest Mexico. *Journal of Marine Science and Technology*, **18**(5): 674–681.

Guedes, A.C. and Malcata, F.X. (2012) Nutritional value and uses of microalgae in aquaculture, In: *Aquaculture, InTech* (ed. Z. Muchlisin). Available from: http://www.intechopen.com/books/aquaculture/nutritional-value-and-uses-of-microalgae-in-aquaculture

Gumus, E., Kaya, Y., Balci, B.A., Aydin, B., Gulle, I. and Gokoglu, M. (2010) Replacement of fishmeal with sand smelt (*Atherina boyeri*) meal in practical diets for Nile tilapia fry (*Oreochromis niloticus*). *The Israeli Journal of Aquaculture – Bamidgeh*, **62**(3): 172–180.

Hasan, M.R. (2001) Aquaculture in the third millennium. In: *Technical Proceedings of the Conference on Aquaculture in the Third Millennium, Bangkok, Thailand* (eds R.P., Subasinghe, P. Bueno, M.J. Phillips, C. Hough, S.E. McGladdery, *et al.*). 20–25 February 2000. NACA, Bangkok and FAO, Rome, 471 pp.

Head, S.W., Swetman, A.A., Hammonds, T.W., Gordon, A., Southwell, K.H. and Harris, R.V. (1995) *Small-Scale Vegetable Oil Extraction*. NRI, 105 pp.

Hegedius, M., Bokori, J., Andri, E. and Kiw, O. (1990) Optimizing protein quality of mixture of blood meal, feather meal and bone meal. *Acta Vet. Ihaw.* **38**: 143–152.

Hernández, C., Olvera-Novoa, M.A., Hardy, R.W., Hermosillo, A., Reyes, C. and Gonza Lez, B. (2010) Complete replacement of fish meal by porcine and poultry by-product meals in practical diets for fingerling Nile tilapia *Oreochromis niloticus*: Digestibility and growth performance. *Aquaculture Nutrition*, **16**(1): 44–53.

Hernández, C., Olvera-Novoa Miguel, A., Voltolina, D., Hardy, R.W., Gonzalez-Rodriguez, B. and Dominguez-Jimenez, P. (2013) Use of tuna industry waste in diets for Nile tilapia, *Oreochromis niloticus*, fingerlings: Effect on digestibility and growth performance. *Latin American Journal of Aquatic Research*, **41**(3): 468–478.

Hernández, C., Osuna-Osuna, L., Benitez-Hernandez, A., Sanchez-Gutierrez, Y., González-Rodríguez, B. and Dominguez-Jimenez, P. (2014) Replacement of fish meal by poultry by-product meal, food grade, in diets for juvenile spotted rose snapper (*Lutjanus guttatus*). *Latin American Journal of Aquatic Research*, **42**(1): 111–120.

Hertrampf, J.W. and Piedad-Pascual, F. (2000) *Handbook on Ingredients for Aquaculture Feeds*. Kluwer Academic Publishers, 624 pp.

Hesley, J. (ed.) (1994) *Sunflower Meal Use in Livestock Rations*. National Sunflower Association, Bismarck, ND.

Heu, M.S., Kim, J.S. and Shahid, F. (2003) Components and nutritional quality of shrimp processing by-products. *Food Chemistry*, **82**: 235–242.

Hilton, J.W. and Slinger, S.J. (1986) Digestibility and utilization of canola meal in practical-type diets for rainbow trout (*Salmo gairdneri*). *Canadian Journal of Fisheries and Aquatic Science*, **43**: 1149–1155.

Hu, B., Ferrell, M., Lim, C.E. and Davis, D.A. (2012) Evaluation of traditional diet and corn gluten feed substituted alternative diet for pond-raised hybrid catfish on production and xanthophyll level. *Aquaculture*, **354–355**: 22–26.

Huige, N.J. (2006) Brewery by-products and effluents. In: *Handbook of Brewing* (eds F.G. Priest and G.G. Stewart). Taylor & Francis, New York, pp. 655–713.

Hussein, E.E., Dabrowski, K., El-Saidy, D.M.S.D. and Lee, B-J. (2013) Enhancing the growth of Nile tilapia larvae/juveniles by replacing plant (gluten) protein with algae protein. *Aquaculture Research*, **44**(6): 937–949.

IITA (1990) *Cassava in Tropical Africa. A reference manual*. International Institute of Tropical Agriculture, Ibadan, Nigeria, 176 pp.

Isyagi, N.A., Veverica, K.L., Asiimwe, R. and Daniels, W.H. (2009) *Manual for the Commercial Pond Production of the African Catfish in Uganda*. USAID-FISH Cooperative Agreement: 617-A-00-05-00003-00, Department of Fisheries and Allied Aquacultures; Auburn University. Walimi Fish Co-op Society Ltd, Kampala, 222 pp.

Jackson, A. (2006) Feed ingredients. The importance of fishmeal and fish oil in aquaculture diets. *International Aquafeed*, **November–December**: 16–19.

Jackson, A. (2007). Challenges and opportunities for the fishmeal and fish oil industry. *Feed Technology Update*, **2**(1): 1–9.

Jackson, A.J. (2009) Fish in-fish out (FIFO) ratios explained. *Aquaculture Europe*, **34**(3): 5–10 [Online] Last accessed 10 November 2014. Available from: www.iffo.net/downloads/100.pdf

Jatta, S. and Sigurgeirsson, O. (2013) The effects of substituting fishmeal with rapeseed meal at three protein levels on growth and body composition of Nile tilapia fingerlings (*Oreochromis Niloticus*). *UNU-Fisheries Training Programme*, 33 pp.

Jauncey, K. (1998) Tilapia feeds and feeding. Pisces Press, Stirling, UK.

Jayathilakan, K., Sultana, K., Radhakrishna, K. and Bawa, A.S. (2012) Utilization of byproducts and waste materials from meat, poultry and fish processing industries: A review. *Journal of Food Science and Technology*, **49**(3): 278–293.

Jimoh, W.A. and Aroyehun, H.T. (2011) Evaluation of cooked and mechanically defatted sesame (*Sesamum indicum*) seed meal as a replacer for soybean meal in the diet of African catfish (*Clarias gariepinus*). *Turkish Journal of Fisheries and Aquatic Sciences*, **11**: 185–190.

Jimoh, W.A., Fagbenro, O.A. and Adeparusi, E.O. (2014) Response of African catfish, *Clarias gariepinus* (Burchell 1822), fingerlings fed diets containing differently timed wet-heat-treated sesame (*Sesamum indicum*) seedmeal. *Agricultural Sciences*, **5**: 1159–1171.

Jones, L.A. and Wedegaertner, T.C. (1986) Advances in feeding cottonseed products. In: *Proceedings of the World Conference on Emerging Technologies in the Fats and Oils Industry* (ed. A.R. Baldwin). *The American Oil Chemists Society*, 431 pp.

Kahlon, T.S. (2009) Rice bran: Production, composition, functionality and food applications, physiological benefits. In: *Fiber Ingredients: Food Applications and Health Benefits* (S.S. Cho and P. Samual). CRC Press, pp. 305–321.

Kamble, M.T., Chavan, B.R., Ataguba G., Azpeitia, T., Medhe, S.V., *et al.* (2014) Application of *Moringa oleifera* for development of sustainable and biosecure aquaculture. *Proceedings of International Conference of Aquaculture Indonesia (ICAI)*, 20–21 June: 86–95.

Katya, K., Yun, Y., Yun, H., Lee, J. and Bai, S.C. (2014) Effects of dietary fermented by-product of mushroom, *Pleurotus ostreatus*, as an additive on growth, serological characteristics and nonspecific immune responses in juvenile Amur catfish, *Silurus asotus*. *Aquaculture Research*, **2014**: 1–9.

Kaur, V.I. and Saxena, P.K. (2004) Incorporation of brewery waste in supplementary feed and its impact on growth in some carps. *Bioresource Technology*, **91**: 101–104.

Khan, M.S.K., Siddique, M.A.M. and Zamal, H. (2013) Replacement of fish meal by plant protein sources in Nile tilapia (*Oreochromis niloticus*) diet: Gowth performance and utilization. *Iranian Journal of Fisheries Sciences,* **12**(4): 860–872.

Knuckey, I., Sinclair, C., Surapaneni, A. and Ashcroft, W. (2004) Utilisation of seafood processing waste – challenges and opportunities. *SuperSoil 2004: 3rd Australian New Zealand Soils Conference*, 5–9 December, University of Sydney, Australia. [Online] Last accessed 18 September 2015. Available from: http://www.regional.org.au/au/asssi/supersoil2004/s7/oral/1662_knuckeyi.htm

Küçük, F., Gülle, İ., Güçlü, S.S., Gümüş, E. and Demir, O. (2006) Effect on fishery and lake ecosystem of nonnative sand smelt (*Atherina boyeri* Risso, 1810) in Eğirdir Lake. I: *Ulusal Balıklandırma ve Rezervuar Yönetimi Sempozyumu*, Antalya (in Turkish).

Kulkarni, V.N., Khadi, B.M. and Sangam, V.S. (2002) Pre-breeding efforts for low gossypol seed and high gossypol plant in *G. herbaceum* L. cotton utilizing *G. australe* Mueller. *Current Science*, **82**(4): 434–439.

Larumbe-Moran, E., Hernandez, M.P., Olvera-Novoa, M.A. and Perez-Rostro, C.I. (2010) Protein requirements of Nile tilapia (*Oreochromis niloticus*) fry cultured at different salinities. *Aquaculture Research*, **41**(8): 1150–1157.

Leal, A.L.G., de Castro, P.F., de Lima, J.P.V., de Souza Correia, E. and de Souza Bezerra, R. (2009) Use of shrimp protein hydrolysate in Nile tilapia (*Oreochromis niloticus*, L.) feeds. *Aquaculture International*, **18**(4): 635–646.

Leenhouwers, J.I., ter Veld, M., Verreth, J.A.J. and Schrama, J.W. (2007) Digest characteristiscs and performance of African catfish (*Clarias gariepinus*) fed cereal grains that differ in viscosity. *Aquaculture*, **264**: 330–341.

Levic, J., Djuragic, O. and Sredanovic, S. (2010) Use of new feed from brewery by-products for breeding layers. *Romanian Biotechnological Letters*, **15**(5): 5559–5565.

Lewu, M.N., Adebola, P.O, and Afolayan, A.J. (2009) Effect of cooking on the proximate composition of seven accessions of *Colocasia esculenta* (L.) Schott tubers growing in Kwazulu-Natal Province of South Africa. *African Journal of Biotechnology*, **8**(8): 1619–1622.

Li, E., Lim, C., Cai, C. and Klesius, P.H. (2011) Growth response and resistance to *Streptococcus iniae* of Nile tilapia, *Oreochromis niloticus*, fed diets containing different levels of wheat distiller's dried grains with solubles with or without lysine supplementation. *Animal Feed Science and Technology*, **170**(3-4): 246–255.

Li, M.H. and Robinson, E.H. (2008) Recent advances in channel catfish nutrition and feeding research. In: *Aquaculture Research Trends* (ed. S.H. Schwartz). Nova Science Publishers, Hauppauge, New York, pp. 263–277,

Li, M.H., Manning, B.B. and Robinson, E.H. (2002) Comparison of various animal protein sources for growth, feed efficiency, and body composition of juvenile channel catfish *Ictalurus punctatus*. *Journal of the World Aquaculture Society*, **33**(4): 489–493.

Li, M.H., Robinson, E.H., Bosworth, B.G., Oberle, D.F. and Lucas, P.M. (2010) Effects of varying dietary compositions using common feed ingredients on growth and feed efficiency of pond-raised channel catfish *Ictalurus punctatus* (Rafinesque). *Aquaculture Research*, **41**: 1133–1139.

Li, M.H., Oberle, D.F. and Lucas, P.M. (2011) Evaluation of corn distillers dried grains with solubles and brewer's yeast in diets for channel catfish *Ictalurus punctatus* (Rafinesque). *Aquaculture Research*, **42**: 1424–1430.

Li, M.H., Oberle, D.F. and Lucas, P.M. (2013a) Apparent digestibility of alternative plant-protein feedstuffs for channel catfish, *Ictalurus punctatus* (Rafinesque). *Aquaculture Research*, **44**: 282–288.

Li, M.H., Robinson, E.H., Oberle, D.F., Lucas, P.M. and Bosworth, B.G. (2013b) Use of corn germ meal in diets for pond-raised channel catfish, *Ictalurus punctatus*. *Journal of the World Aquaculture Society*, **44**(2): 282–287.

Liebowitz, H. E. (1981) Replacing fish meal with soybean meal in practical catfish diets. Master's thesis. Auburn University, Auburn, Al.

Lim, C. and Yildirim-Aksoy, M. (2008) Distillers dried grains with solubles as an alternative protein source in fish feeds. *Proceedings of the 8th International Symposium on Tilapia in Aquaculture*, 12–14 October, Cairo, Egypt [Online]. Last accessed 29 March 2015. Available from: https://ag.arizona.edu/azaqua/ista/ISTA8/ChhornLim.pdf

Lim, C, Klesius, P.H. and Higgs, D.A. (1998) Substitution of canola meal for soybean meal in diets for channel catfish *Ictalurus punctatus*. *Journal of the World Aquaculture Society*, **29**: 161–168.

Lio-Po, G.D., Leaño, E.M., Peñaranda, M.M.D., VillaFranco, A.U., Sombito, C.D. and Guanzon, N.G. (2005) Antiluminous Vibrio factors associated with the 'green water' growout culture of the tiger shrimp *Penaeus monodon*. *Aquaculture*, **250**: 1–7.

Liti, D.M., Mugo, R.M., Munguti, J.M. and Waidbacher, H. (2006) Growth and economic performance of Nile tilapia (*Oreochromis niloticus* L.) fed on three brans (maize, wheat and rice) in fertilized ponds. *Aquaculture Nutrition*, **12**: 239–245.

Lopez, C.S. (1990) Microbial ensilage of trash fish for animal feeds. In: Post-Harvest Technology, Preservation And Quality if Fish in South-East Asia (eds P.J.A. Reilly, R.W.H. Parry and L.E. Barile)., International Foundation of Science (IFS), Stockholm, Sweden, pp. 189–192.

Lovell, R.T. (1982) Utilization of catfish processing waste. *Alabama Agricultural Experiment Station Bulletin*, p. 521.

Mahmoud, M.M.A., Kilany, O.E. and Dessouki, A.A. (2014) Effects of fish meal replacement with soybean meal and use of exogenous enzymes in diets of Nile tilapia (*Oreochromis niloticus*) on growth, feed utilization, histopathological changes and blood parameters. *Life Science Journal*, **11**(2): 6–18.

Mathia, W.M. and Fotedar, R. (2012) Evaluation of boiled taro leaves, *Colocasia esculenta* (L.) Schott, as a freshwater shrimp, *Caridina nilotica* Roux protein replacement, in diets of Nile tilapia, *Oreochromis niloticus* (Linnaeus). *Aquaculture*, **356-357**: 302–309.

Mbahinzireki, G.B., Dabrowski, K., Lee, K-J., El-Saidy, D. and Wisner, E.R. (2001) Growth, feed utilization and body composition of tilapia (*Oreochromis* sp.) fed with cottonseed meal-based diets in a recirculating system. *Aquaculture Nutrition*, **7**(3): 189–200.

Meric, I., Wuertz, S., Kloas, W., Wibbelt, G. and Schulz, C. (2011) Cottonseed oilcake as a protein source in feeds for juvenile Tilapia (*Oreochromis niloticus*): Antinutritional effects and potential detoxification by iron supplementation. *The Israeli Journal of Aquaculture – Bamidgeh*, **63**: 1–8.

Merino, G., Barange, M., Mullon, C. and Rodwell, L. (2010) Impacts of global environmental change and aquaculture expansion on marine ecosystems. *Global Environmental Change*, **20**: 586–596.

Metailler, R. and Guillaume, J. (1999) Raw materials and additives used in fish foods. In: *Nutrition and Feeding of Fish and Crustaceans* (eds J. Guillaume, S. Kaushik, P. Bergot and R. Metailler). Springer, London, pp. 281–296.

Millamena, M.O. (2002) Replacement of fish meal by animal by-product meals in a practical diet for grow-out culture of grouper *Epinephelus coioides*. *Aquaculture*, **204**: 75–84.

Moghaddam, H.N., Salari, S., Arshami, J., Golian, A. and Maleki, M. (2012) Evaluation of the nutritional value of sunflower meal and its effect on performance, digestive enzyme activity, organ weight, and histological alterations of the intestinal villi of broiler chickens. *Journal of Applied Poultry Research*, **21**: 293–304.

Morgan, S.E. (1989). Gossypol as a toxicant in livestock. *Veterinary Clinics of North America, Food Animal Practice*, **5**(2): 251–262.

Murray, M.T. and Pizzorno, J. (2012). *The Encyclopedia of Natural Medicine*. Atria Paperbacks, New York. 1232 pp.

Nelson, J.S. (2006) *Fishes of the World*. John Wiley & Sons, New York, 601 pp.

Nguyen, T.N., Davis, D.A. and Saoud, P.I. (2009) Evaluation of alternative protein sources to replace fish meal in practical diets for juvenile tilapia, *Oreochromis* spp. *Journal of the World Aquaculture Society*, **40**(1): 113–121.

NRC (1993) *Nutrient Requirement of Fish*. National Research Council. National Academic Press, Washington. 140 pp.

Nwanna, L.C. (2003) Nutritional value and digestibility of fermented shrimp head waste meal by African catfish *Clarias gariepinus*. *Pakistan Journal of Nutrition*, **2**(6): 339–345.

Nyina-Wamwiza, L., Wathelet, B., Richir, J., Rollin, X. and Kestemont, P. (2010) Partial or total replacement of fish meal by local agricultural by-products in diets of juvenile African catfish (*Clarias gariepinus*): growth performance, feed efficiency and digestibility. *Aquaculture Nutrition*, **16**: 237–247.

Nyirenda, J., Mwabumba, M., Kaunda, E. and Sales, J. (2000) Effect of substituting animal protein sources with soybean meal in diets of *Oreochromis karongae* (Trewavas, 1941). *Naga, The ICLARM Quarterly*, **23**(4): 13–15.

Obasa, S.O., Akinyemi, A.A., Ogundijo, O.P. and Alade, O.O. (2011) Use of fish waste meal as a replacement for fish meal in the practical diets of African mud catfish *Clarias gariepinus* fingerlings. *Journal of Agricultural Science and Environment*, **11**(1): 68–77.

Obirikorang, K.A., Amisah, S., Fialor, S.C. and Skov, P.V. (2015) Digestibility and postprandial ammonia excretion in Nile tilapia (*Oreochromis niloticus*) fed diets containing different oilseed by-products. *Aquaculture International*, **23**(5): 1249–1260.

OECD-FAO (2013) Oilseeds and oilseed products. 139–282 In: OECD-FAO Agricultural Outlook 2013, OECD Publishing, Rome.

Ogunji, J.O., Kloas, W., Wirth, M., Schulz, C. and Rennert, B. (2006) Housefly maggot meal (Magmeal): An emerging substitute of fishmeal in tilapia diets. *Conference on International Agricultural Research for Development*, Tropentag, Berlin, pp. 1–7.

Ogunji, J.O., Toor, R.U.A.S., Schulz, C. and Kloas, W. (2008) Growth performance, nutrient utilization of Nile tilapia *Oreochromis niloticus* fed housefly maggot meal (Magmeal) diets. *Turkish Journal of Fisheries and Aquatic Sciences*, **8**: 141–147.

Olukayode, A.M. and Emmanuel, B.S. (2012) The potential of two vegetable carried blood meals as protein sources in African catfish (*Clarias gariepinus* Burchell) juvenile diet. *Open Journal of Animal Science*, **2**(1): 15–18.

Olvera-Novoa, M.A. and Olivera-Castillo, L. (2000) Potencialidad del uso de las leguminosas como fuente proteica en alimentos para peces. In: *Avances en Nutricion Acuicola. Memorias del IV simposio Internacional de Nutricion Acuicola* (ed. R. Civera-Cerecedo)m 15–18 November 1998, Universidad de Neuvo Leon, Mexico.

Oliveira Cavalheiro, J.M., Oliveira de Souza, E. and Bora, P.S. (2007) Utilization of shrimp industry waste in the formulation of tilapia (*Oreochromis niloticus* Linnaeus) feed. *Bioresource Technology*, **98**(3): 602–606.

Olvera-Novoa, M.A., Domínguez-Cen, L.J., Olivera-Castillo, L. and Martínez-Palacios, C.A. (1998) Effect of the use of the microalga *Spirulina maxima* as fish meal replacement in diets for tilapia, *Oreochromis mossambicus* (Peters) fry. *Aquaculture Research*, **29**(10): 709–715.

Olvera-Novoa, M.A,, Martinez-Palacios, C.A, and Olivera-Castillo, L. (2002) Utilization of torula yeast (*Candida utilis*) as a protein source in diets for tilapia (*Oreochromis mossambicus* Peters) fry. *Aquaculture Nutrition*, **8**(4): 257–264.

Omojowo, T.M., Omojowo, F.S. and Alatise, P.S. (2010) Growth response and nutritional evaluation of mango peel-based diets on tilapia (*Oreochromis niloticus*) fingerlings. *Researcher*, **2**(6): 44–49.

Onwueme, I. (1999) *Taro Cultivation in Asia and the Pacific*. Food and Agriculture Organization of the United Nation (FAO). Regional office for Asia and the Pacific Bangkok, Thailand. RAP publication: 1999/16.

Orire, A.M. and Ricketts, O.A. (2013) Utilisation of melon shell as dietary energy source in the diet of Nile tilatia (*Oreochromis niloticus*). Research inventory. *International Journal of Engineering & Science*, **2**(4): 5–11.

Ovie, S.O. and Ovie, S.I. (2007) The effect of replacing Fish meal with 10% of groundnut cake in the diets of *Heterobranchus longifilis* on its growth, food conversion and survival. *Journal of Applied Sciences and Environmental Management*, **11**(3): 87–90.

Ozdestan, O., Erol, T. and Acar, B. (2014) Phytosterols in rice bran and usage of rice bran in food industry. In: *9th Baltic Conference on Food Science and Technology 'Food for Consumer Well-Being' FOODBALT 2014 Conference Proceedings*. Jelgava, LLU, pp. 24–27.

Ozório, R.O.A., Portz, L., Borghesi, R. and Cyrino, J.E.P. (2012) Effects of dietary yeast (*Saccharomyces cerevisiae*) supplementation in practical diets of tilapia (*Oreochromis niloticus*). *Animals*, **2**(4): 16–24.

Pares-Sierra, G., Durazo, E., Ponce, M.A., Badillo, D., Correa-Reyes, G. and Viana, M.T. (2014) Partial to total replacement of fishmeal by poultry by-product meal in diets for juvenile rainbow trout (*Oncorhynchus mykiss*) and their effect on fatty acids from muscle tissue and the time required to retrieve the effect. *Aquaculture Research*, **45**(9): 1459–1469.

Pelitire, S.M., Dowd, M.K. and Cheng, H.N. (2014) Acidic solvent extraction of gossypol from cottonseed meal. *Animal Feed Science and Technology*, **195**: 120–128.

Peterson, B.C., Booth, N.J. and Manning, B.B. (2012) Replacement of fish meal in juvenile channel catfish, *Ictalurus punctatus*, diets using a yeast-derived protein source: The effects on weight gain, food conversion ratio, body composition and survival of catfish challenged with *Edwardsiella ictaluri*. *Aquaculture Nutrition*, **18**: 132–137.

Pine, H.J., Daniels, W.H., Allen Davis, D., Jiang, M. and Webster, C.D. (2008) Replacement of fish meal with poultry by-product meal as a protein source in pond-raised sunshine bass, *Morone chrysops* (female) × *M. saxatlis* (male), diets. *Journal of the World Aquaculture Society*, **39**(5): 586–597.

Plascencia-Jatomea, M., Olvera-Novoa, M.A., Arredondo-Figueroa, J.L., Halland, G.M. and Shirai, K. (2002) Feasibility of fishmeal replacement by shrimp head silage protein hydrolysate in Nile tilapia (*Oreochromis niloticus* L) diets. *Journal of the Science of Food and Agriculture*, **82**(7): 753–759.

Prakash, J. and Ramaswamy, H.S. (1996) Rice bran proteins: Properties and food uses. *Critical Reviews in Food Science and Nutrition*, **36**(6): 537–552.

Ramachandran, S., Singh, S.K, Larroche, C., Soccol, C.R. and Pandey, A. (2007) Oil cakes and their biotechnological applications – A review. *Bioresearch Technology*, **98**: 2000–2009.

Rana, K.J., Siriwardena, S. and Hasan, M.R. (2009) Impact of rising feed ingredient prices on aquafeeds and aquaculture production. *FAO Fisheries and Aquaculture Technical Paper. No. 541*. FAO, Rome, 63 pp.

Randel, R.D., Chase, C.C. and Wyse, S.J. (1992) Effects of gossypol and cottonseed products on reproduction of mammals. *Journal of Animal Science*, **70**(5): 1628–1638.

Raven, P. and Walker, G. (1980) Ingredients for fish feed manufacture in the United States. In: *Fish Feed Technology: Lectures presented at the FAO/UNDP Training Course in Fish Feed Technology*, College of Fisheries, University of Washington, Seattle, 9 October–15 December 1978. Food & Agriculture Organization of the United Nations.

Rawles, S.D., Gaylord, T.G., McEntire, M.E. and Freeman, D.W. (2009) Evaluation of poultry by-product meal in commercial diets for hybrid striped bass, *Morone chrysops* (Female) x *Morone saxatilis* (male), in pond production. *Journal of the World Aquaculture Society*, **40**(2): 141–156

Reed, G. and Nagodawithana T.W. (1991) *Yeast Technology*. AVI-Van Nostrand Reinhold, New York. 454 pp.

Renaud, S.M., Thinh, L-V., Lambrinidis, G. and Parry, D.L. (2002) Effect of temperature on growth, chemical composition and fatty acid composition of tropical Australian microalgae grown in batch cultures. *Aquaculture*, **211**: 195–214.

Renukdas, N., Engle, C., Lochmann, R., Li, M.H., Avery, J., *et al.* (2014) Performance of alternative diets containing solvent-extracted distillers dried grains with solubles compared to traditional diets for pond-raised channel catfish, *Ictalurus punctatus*, and hybrid catfish, *Ictalurus punctatus* × *Ictalurus furcatus*. *Journal of the World Aquaculture Society*, **45**(3): 290–300.

Rinchard, J., Lee, K.J., Dabrowski, K., Ciereszko, A., Blom., J.H. and Ottobre, J.S. (2003) Influence of gossypol from dietary cottonseed meal on haematology, reproductive steroids and tissue gossypol enantiomer concentrations in male rainbow trout (*Oncorhynchus mykiss*). *Aquaculture and Nutrition*, **9**: 275–282.

Robinson, E.H. (1991) Improvement of cottonseed meal protein with supplement lysine in feed for channel catfish. *Journal of Applied Aquaculture*, **1**: 11–14. https://doi.org/10.1300/J028v01n02_01

Robinson, E.H. and Li, M.H. (2008) Replacement of soybean meal in channel catfish, *Ictalurus punctatus*, diets with cottonseed meal and distiller's dried grains with solubles. *Journal of the World Aquaculture Society*, **39**(4): 521–527.

Rodolfi, L., Chini Zittelli, G., Barsanti, L., Rosati, G. and Tredici, M.R. (2003) Growth medium recycling in *Nannochloropsis* sp. mass cultivation. *Biomolecular Engineering*, **20**: 243–248.

Rodrigues, A.P.O., Gominho-Rosa, M.D.C., Cargnin-Ferreira, E., De Francisco, A. and Fracalossi, D.M. (2012) Different utilization of plant sources by the omnivores: Jundia catfish (*Rhamdia quelen*) and Nile tilapia (*Oreochromis niloticus*). *Aquaculture Nutrition*, **18**: 65–72.

Romano, G.B. and Scheffler, J.A. (2008) Lowering seed gossypol content in glanded cotton (*Gossypium hirsutum* L.) lines. *Plant Breeding*, **127**: 619–624.

Rueda-Jasso, R.A., Campos-Mendoza, A., Arreguín-Sánchez, F., Díaz-Pardo, E. and Martínez-Palacios, C.A. (2013) The biological and reproductive parameters of the invasive armored catfish *Pterygoplichthys disjunctivus* from Adolfo López Mateos El Infiernillo Reservoir, Michoacán-Guerrero, Mexico. *Revista Mexicana de Biodiversidad*, **84**(1): 318–326.

Rueda-Lopez, S., Teresa Viana, M., Badillo-Zapata, D., Martinez-Palacio, C.A. and Correa-Reyesb, G. (2013) Sustainable tilapia production in a backyard system for rural areas using fish by-products in formulated diets. *Tropical and Subtropical Agroecosystems*, **16**(3): 475–482.

Russ, W. and Schnappinger, M. (2007) Waste related to the food industry: A challenge in material loops. In: *Utilization of By-Products and Treatment of Waste in the Food Industry* (eds V. Oreopoulou and W. Russ). Springer, New York, pp. 1–13.

Ruttanaporn, V.Y., Ikeda, M. and Hara, K. (2005) Effect of shrimp head protein hydrolysates on the state of water and denaturation of fish myofibrils during dehydration. *Fish Science*, **71**: 220–228.

Samocha, T, Davis, D.A, Saoud, I.P. and DeBault, K. (2004) Substitution of fish meal by coextruded soybean poultry by product meal in practical diets for the Pacific white shrimp, *Litopenaeus vannamei*. *Aquaculture*, **231**: 197–203.

Sankar, T.V., Anandan, R., Mathew, S., Asha, K.K., Lakshmanan, P.T., *et al.* (2013) Chemical composition and nutritional value of anchovy (*Stolephorus commersonii*) caught from Kerala coast, India. *European Journal of Experimental Biology*, **3**(1): 85–89.

Sarker, P.K., Bureau, D.P., Hua, K., Drew, M.D., Forster, I., *et al.* (2013) Sustainability issues related to feeding salmonids: A Canadian perspective. *Reviews in Aquaculture*. **5**: 1–21.

Sarlin P.J. (2005) Marine yeasts as source of single cell protein and immunostimulant for application in Penaeid prawn culture systems. Thesis submitted to the Cochin University of Science and Technology in partial fulfillment of the requirements for the degree of Doctor of Philosophy in Microbiology. 250 pp. [Online] Last accessed: 31 August 2015. Available from: http://shodhganga.inflibnet.ac.in/handle/10603/2950

Sarmwatanakul, A. and Bamrongtum, B. (2000) Aquarium fish nutrition. *Extension paper No. 1/2000*. Ornament Fish Research and Public Aquarium. Bangkok.

Saunders, R.M. (1990) The properties of rice bran as a foodstuff. *Cereal Foods World*, **35**: 632–636.

Sharif, M.K., Butt, M.S., Anjum, F.M. and Khan, S.H. (2014) Rice bran: A novel functional ingredient. *Critical Reviews in Food Science and Nutrition*, **54**(6): 807–816.

Silva, C.P. (2004) Composição nutricional e análise microbiológica do hidrolisado protéico de subprodutos de camaraá da espãöcie *Litopenaeus vannamei*. Dissertação (Mestrado), Universidade Federal de Pernambuco.

Sinchermiri, D., Harnbunchon, A. and Boaban, R. (1988) A study on native value of hydrolyzed feather meal product in Thailand. In: *The Proceedings of the 42nd Kasetsart University Annual Conference*. Kasetsart University, Bangkok.

Sing, K.W., Kamarudin, M.S., Wilson, J.J. and Sofian-Azirun, M. (2014) Evaluation of blowfly (*Chrysomya megacephala*) maggot meal as an effective, sustainable replacement for fishmeal in the diet of farmed juvenile red tilapia (*Oreochromis* sp.). *Pakistan Veterinary Journal*, **34**(3): 288–292.

Slawski, H., Adem, H., Tressel, R.P., Wysujack, K., Koops, U., *et al.* (2011) Replacement of fish meal with rapeseed protein concentrate in diets fed to Wels catfish (*Silurus glanis* L.). *Aquaculture Nutrition*, **17**: 605–612.

Soltan, M.A. (2002) Using of tomato and potato by-products as non-conventional ingredients in Nile tilapia, *Oreochromis niloticus* diets. *Annals of Agricultural Science, Moshtohor Journal*, **40**(4): 2081–2096.

Soltan, M.A. and El-Laithy, S.M. (2008) Evaluation of fermented silage made from fish, tomato and potato by-products as a feed ingredient for Nile tilapia, *Oreochromis niloticus*. *Egyptian Journal of Aquatic Biology and Fisheries*, **12**: 25–41.

Soltan, M.A., Hanafy, M.A. and Wafa, M.I.A. (2008a) An evaluation of fermented silage made from fish by-products as a feed ingredient for African catfish (*Clarias gariepinus*). *Global Veterinaria*, **2**(2): 80–86.

Soltan, M.A., Hanafy, M.A. and Wafa, M.I.A. (2008b) Effect of replacing fish meal by a mixture of different plant protein sources in Nile tilapia (*Oreochromis niloticus* L.) diets. *Global Veterinaria,* **2**(4): 157–164.

Sotolu, A.O. (2009) Comparative utilizations of fish waste meal with imported fishmeal by African catfish (*Clarias gariepinus*). *American-Eurasian Journal of Scientific Research,* **4**(4): 285–289.

Spiehs, M.J., Whitney, M.H. and Shurson, G.C. (2002) Nutrient database for distiller's dried grains with solubles produced from new ethanol plants in Minnesota and South Dakota. *Journal of Animal Science*, **80**: 2639–2645.

Stone, F.E., Hardy, R.W., Shearer, K.D. and Scott, T.M. (1989) Utilization of fish silage by rainbow trout (*Salmo gairdneri*). *Aquaculture*, **76**: 109–118.

Sugiura, K., Yamatani, S., Watahara, M. and Onodera, T. (2009) Ecofeed, animal feed produced from recycled food waste. *Veterinaria Italiana*, **45**(3): 397–404.

Sulieman, H.M.A. and Ahmed, F.I.K. (2011) Performance of Nile tilapia (*Oreochromis niloticus*) fed fish meal and poultry by-product. *Online Journal of Animal and Feed Research*, **1**(6): 444–447.

Sullivan, J.P., Lundberg, J.G. and Hardman, M. (2006) A phylogenetic analysis of the major groups of catfishes (Teleostei: Siluriformes) using rag1 and rag2 nuclear gene sequences. *Molecular Phylogenetics and Evolution*, **41**: 636–662.

Tacon, A.G.J. (1993) Feed ingredients for warmwater fish: Fish meal and other processed feedstuffs. *FAO Fisheries Circular No. 856*, Rome, 64 pp.

Tacon, A.G.J. (1995) Aquaculture feeds and feeding in the next millennium: Major challenges and issues. *First FAO Conference on Tropical Feeds and Feeding Systems.* pp. 145–157.

Tacon, A.G.J. (2000) Rendered animal by-products: a necessity in aquafeeds for the new millennium. *Global Aquaculture Advocate*, **3**(4): 18–19.

Tacon, A.G.J. and Forster, I.P. (2003) Aquafeeds and the environment: Policy implications. *Aquaculture*, **226**: 181–189.

Tacon, A.G.J. and Metian, M. (2008) Global overview on the use of fish meal and fish oil in industrially compounded aquafeeds: Trends and future prospects. *Aquaculture*, **285**: 146–158.

Teles, A.O., Lupatsch, I. and Nengas, I. (2011) Nutrition and feeding of Sparidae, In: *Sparidae: Biology and Aquaculture of Gilthead Sea Bream and other Species* (eds M. Pavlidis and C. Mylonas). Wiley-Blackwell, Chichester, UK, pp. 199–232

Tiamiyu, L.O., Okomoda, V.T. and Iber, B. (2013) Growth response of *Clarias gariepinus* fingerlings fed diet substituted groundnut cake meal and cotton seed meal. In: *Livestock Research for Rural Development*, vol. 25, Article No. 76. Last accessed 11 March 2015. Available from: http://www.lrrd.org/lrrd25/5/tiam25076.htm

Toan, N.H. and Preston, T.R. (2010) Taro as a local feed resource for pigs in small scale household condition. In: *Livestock Research for Rural Development*, vol. 22, Article No. 152. Last accessed 26 April 2017, from http://www.lrrd.org/lrrd22/8/toan22152.htm

Trushenski, J.T, Kasper, C.S, and Kohler, C.C. (2006) Challenges and opportunities in finfish nutrition. *North American Journal of Aquaculture*, **68**: 122–140.

Turan, H., Kaya, Y. and Erkoyuncu, I. (2007) Protein and lipid content and fatty acid composition of anchovy meal produced in Turkey. *Turkish Journal of Veterinary and Animal Science*, **31**(2): 113–117.

Twibell, R.G. and Brown, P.B. (1998) Optimal dietary protein concentration for hybrid tilapia (*Oreochromis niloticus* × *Oreochromis aureus*) fed all-plant diets. *Journal of the World Aquaculture Society*, **29**: 9–16.

Ty, C., Borin, K., Preston, T.R. and Sokveasna, M. (2007) Intake, digestibility and N retention by growing pigs fed ensiled or dried Taro (*Colocasia esculenta*) leaves as the protein supplement in basal diets of rice bran/broken rice or rice bran/cassava root meal. In: *Livestock Research for Rural Development*, vol. 19, Article No. 137. Last accessed 26 April 2017, from http://www.lrrd.org/lrrd19/9/chha19137.htm

UNEP (2009) *Converting Waste Agricultural Biomass into a Resource: Compendium of Technologies*. United Nations Environmental Programme Division of Technology, Industry and Economics International Environmental Technology Centre, Osaka/Shiga, Japan, 437 pp.

USDA (2015) *Oilseeds: World Markets and Trade*. United States Department of Agriculture, Foreign Agricultural Service [Online]. Last accessed 6 March. Available from: http://apps.fas.usda.gov/psdonline/circulars/oilseeds.pdf

USDEC (2004) Products definition, composition, functions. *Reference Manual for US Whey and Lactose Products*, pp. 27–40.

Usmani, N. and Jafri, A.K. (2002) Effect of fish size and temperature on the utilization of different protein sources in two catfish species. *Aquaculture Research*, **33**: 959–967.

Vasta, V., Nudda, A., Cannas, A., Lanza, M. and Priolo, A. (2008) Alternative feed resources and their effects on the quality of meat and milk from small ruminants. *Animal Feed Science and Technology*. **147**: 223–246.

Vechklang, K., Boonanuntanasarn, S., Ponchunchoovong, S., Pirarat, N. and Wanapu, C. (2011) The potential for rice wine residual as an alternative protein source in a practical diet for Nile tilapia (*Oreochromis niloticus*) at the juvenile stage. *Aquaculture Nutrition*, **17**(6): 685–694.

Veiga, M.E., Mendez, R. and Lema, J.M. (1994) Waste water treatment for fisheries operations, In: *Fisheries Processing: Biotechnological applications* (ed. A.M. Martin). Springer Science+Business Media, Dordrecht, pp. 344–369.

Vellinga, T.V., de Haan, M.H.A., Schils, R.L.M., Evers, A. and Dasselaar, A.P. (2011) Implementation of GHG mitigation on intensive dairy farms: Farmers' preferences and variation in cost effectiveness. *Livestock Science*, **137**: 185–195.

Viola, S., Mokady, S. and Arielli, Y. (1983) Effects of soybean processing methods on the growth of carp (*Cyprinus carpio*). *Aquaculture*, **32**: 27–38.

Webster, C.D,, Goodgame-Tiu, L.S. and Tidwell, J.H. (1995) Total replacement of fish meal by soy bean meal, with various percentages of supplemental L-methionine, in diets for blue catfish, *Ictalurus furcatus* (Lesueur). *Aquaculture Research*, **26**: 299–306.

Wilson, R.P. and Poe, W.E. (1985) Effects of feeding soybean meal with varying trypsin inhibitor activities on growth of fingerling channel catfish. *Aquaculture*, **46**: 19–25.

Windsor, M.L. (2001) Fish meal. Department of Trade and Industry, *Torry Research Station, Torry Advisory Note No. 49* [Online]. Last accessed 28 March 2015. Available from: http://www.fao.org/wairdocs/tan/x5926e/x5926e01.htm

Winfree, R.A. and Stickney, R.R. (1981) Effects of dietary protein and energy on growth, feed conversion efficiency and body composition of *Tilapia aurea*. *Journal of Nutrition*, **111**: 1001–1012.

Yi, Y., Lin, C.K. and Diana, J.S. (1996) Influence of Nile tilapia (*Oreochromis niloticus*) stocking density on their growth and yield in cages and ponds containing the cages. *Aquaculture*, **146**: 205215.

Yıldırım, O., Turker, A., Sebahattin, E., Murat, Y. and Gulsahin, A. (2009) Growth performance and feed utilization of *Tilapia zillii* (Gervais, 1848) fed partial or total replacement of fish meal with poultry by-product meal. *African Journal of Biotechnology*, **8**(13): 3092–3096.

Yousif, M.O., Krishnankutty, M.K., Abdul-fatah, A. and Abdul-rahman, A. (2014) Growth performance, feed efficiency and carcass composition of African catfish, *Clarias gariepinus* (Picess: Clariidae) fingerlings fed diets composed of agricultural by-products. *International Journal of Biosciences*, **4**(10): 276–284.

Yue, Y.R. and Zhou, Q.C. (2008) Effect of replacing soybean meal with cottonseed meal on growth, feed utilization, and hematological indexes for juvenile hybrid tilapia, *Oreochromis niloticus x O. aureus. Aquaculture*, **284**: 185–189.

Zerai, D.B., Fitzsimmons, K.M., Collier, R.J. and Duff, G.C. (2008) Evaluation of brewer's waste as partial replacement of fish meal protein in Nile tilapia, *Oreochromis niloticus*, diets. *Journal of the World Aquaculture Society*, **39**(4): 556–564.

Zhang, W., Xu, Z., Sun, J. and Yang X. (2006) Effect of selected fungi on the reduction of gossypol levels and nutritional value during solid substrate fermentation of cotton seed meal. *Journal of Zhejiang University SCIENCE B*, **7**(9): 690–695.

Zhang, Z., Xu, L., Liu, W., Yang, Y., Du, Z. and Zhou, Z. (2014) Effects of partially replacing dietary soybean meal or cottonseed meal with completely hydrolyzed feather meal (defatted rice bran as the carrier) on production, cytokines, adhesive gut bacteria, and disease resistance in hybrid tilapia (*Oreochromis niloticus* (female) x *Oreochromis aureus* (male)). *Fish Shellfish Immunology,* **41**(2): 517–525.

21
Value-Added By-Products from Sugar Processing Industries

Ali Akbar[1] and Imran Ali[2,3]

[1]Department of Microbiology, Faculty of Life Sciences, University of Balochistan Quetta, Pakistan
[2]Plant Biomass Utilization Research Unit, Department of Botany, Chulalongkorn University, Bangkok, Thailand
[3]Institute of Biochemistry, Faculty of Life Sciences, University of Balochistan Quetta, Pakistan

21.1 Introduction

Sugar is a source of energy for all living things. Crystalline sugar mainly comes from sugar cane and sugar beet. Approximately 125 to 130 million tonnes of sugar are produced every year worldwide, with about two-thirds from sugar cane, and one-third from sugar beet (Jai *et al.*, 2012). The reported global sugar production in the year 2009 was 160 million tonnes, which was 4.5% higher than in the previous year (Fang *et al.*, 2011). The estimated worldwide production of sugar cane is approximately 1557 tonnes, where Brazil (514 tonnes), India (356 tonnes), China (106 tonnes) and Thailand (64 tonnes) are the major producers (Figure 21.1). The estimated production of dry bagasse from each tonne of cane is 130 kg, which is roughly equal to a total amount of 200 million tonnes of bagasse per year (Stanmore, 2010).

Bagasse and molasses are the major by-products of sugar industries. Molasses contains up to 48% sugar. It is commonly used in feeds for livestock and in other industrial processing. It can also be a major cause of pollution in water bodies, in the case of untreated effluent discharge (Jai *et al.*, 2012). Molasses is one of the most common substrates for fermentation industries, as a feedstock for ethanol and backer's yeast production. Its abundance, high sugar content and low cost availability make it more attractive for industrial fermentation and other similar uses (Kanawade, 2014).

Sugar processing is a complex procedure consisting of several steps for the production of final products (Figure 21.2). A number of physical and chemical procedures are part of industrial processing of sugar. The main chemicals used are calcium hydroxide (milk of lime, $Ca(OH)_2$), carbon dioxide (CO_2), phosphoric acid (H_3PO_4), sulphur dioxide (SO_2), polyelectrolytes, polyacrylamide flocculent, caustic soda

Food Processing By-Products and their Utilization, First Edition. Edited by Anil Kumar Anal.

Figure 21.1 Sugar cane crop in fields, Thailand (left) and Pakistan (right).

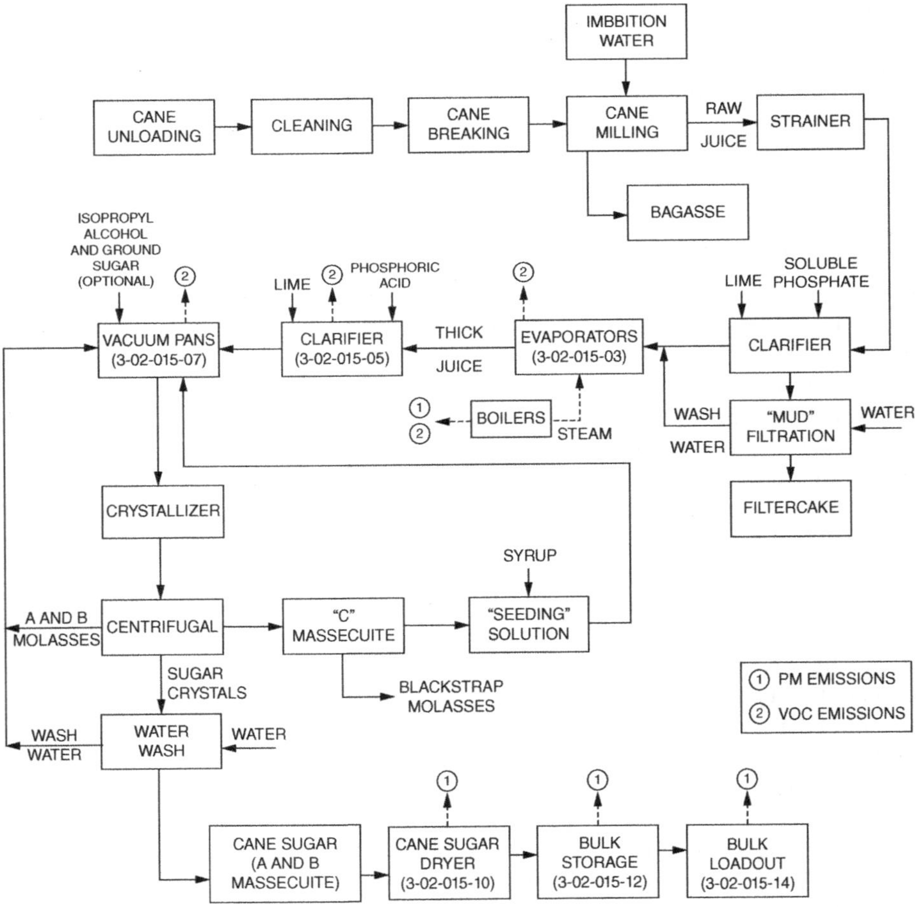

Figure 21.2 Cane sugar production flow chart (adapted from US EPA, 1997).

(NaOH), soda ash (Na_2CO_3), lead sub-acetate and hydrochloric acid (HCl). The major by-products of sugar processing industries are bagasse, mud or filter cake, molasses etc. The presence of all these complex contents in sugar industrial waste effluents are creating environmental pollution on one side and opportunity for valuable by-products isolation and re-use of the contaminating materials on the other (Akbar and Khwaja, 2006).

The sugar industries are considered one of the prominent causes of environmental pollution. The effluent causes eutrophication in water supplies due to its highly nutritious contents released into the environment and water bodies, creating anaerobic conditions, which are toxic for the majority of aquatic fauna (Raghukumar and Rivonkar, 2001). The effluent from sugar industries contains a variety of organic and inorganic contents, which can be utilized for different beneficial purposes as valuable by-products. Such valuable utilization of the by-products from sugar industries is the focus of this chapter. Some of these are mentioned in Figure 21.3.

Figure 21.3 Summary of useful by-products obtained from sugar industries.

21.2 Pulp and Paper Production

Paper or boards are the final products of pulp. The use of bagasse for paper production is almost a century-old in tradition.

21.2.1 Pulp Production

Pulp is the basic material required for the production of paper. It is produced mechanically or chemically from raw materials such as bamboo, wood and bagasse. Pulp is produced from sugar industry by-products via a two-stage process. During the first stage, the loosening of the fiber bundles is acquired by the chemical degradation of hemicelluloses and lignin present in bagasse. In the second stage, the fiber bundles are mechanically separated into single fibers by the process of delignification. Delignification may or may not be obtained by the pre-chemical treatment of fibers. Partial delignification can be obtained by the use of light chemicals and low temperatures, while the higher degree of delignification may require high temperatures. However, delignification at high temperatures may result in low pulp yield production. Semi-chemical pulp is generally used in the manufacture of newsprint, cultural and corrugated paper. This pulp has the capability of producing paper with brightness and high surface qualities (Rainey, 2012; Poopak and Reza, 2012; Delgado and Casanova 2001).

21.2.1.1 Mechanical Pulp Production from Bagasse Bagasse has opacity, brightness and low resistance, similar to soft woods. The pulp from bagasse offers value-added advantages in the printing quality, while showing the lower basic weight. The bagasse is embedded in hot water of approximately 80°C in the mechanical pulping process, prior to the refinement of the pulp. Although the paper produced by mechanical bagasse pulping is of acceptable quality, according to many experts the mechanical processing of the bagasse must be coupled with either chemical or thermal pre-treatment or with both pre-treatments. Chemi-mechanical pulp production was first introduced by the Cuba-9 Research Institute in Cuba. The process involves chemical and thermal pre-treatments prior to the mechanical pre-treatments of the pulp. In this process, a high fiber content bagasse (85%) is mixed in a special impregnating mixer along with hot dilute sodium hydroxide. The pulp moisture content is adjusted to the 20–30% levels, which then becomes reduced to 10–20% during the refining process. The pulp obtained is further refined by the two-stage process, cleaning the fractions by the three cyclones system. The raw pulp fractions pass through the bleaching process by treating them with hydrogen peroxide in a descending bleaching tower, followed by washing with sulphur dioxide to obtain 50–55% brightness (Rainey and Clark, 2004; Galves, 1988).

21.2.1.2 Chemical Pulp Production from Bagasse Production of chemical pulp from bagasse is generally accomplished by soda process. This pulp is mostly utilized in the production of cultural paper (i.e. writing, printing, bonds and forms) and for the manufacture of tissue and Bristol boards. The sulphate and soda pulps are generally known as alkaline pulps. The processing of sulphate pulps includes the blending of sodium sulphide and sodium hydroxide. Sulphate pulps are similar to soda pulps, but their use is somewhat less popular than soda pulps. The production of chemical pulp includes the moist depithing of bagasse to approximately 85% of fiber content, followed by depithing in suspension. The processed bagasse then undergoes digestion for 10–15 min under a steam pressure of 7.5 kg/cm^3. The digested bagasse is washed in the rotary vacuum filters, screened by vibratory screeners and cleaned with hydro-cyclones (Rainey and Clark, 2004; Galves, 1988).

The pulp obtained can be bleached in two to four steps, either individually treating it with chemicals such as peroxide, hypochlorite, chlorine dioxide and chlorine or treating it with a combination of all these chemicals. The process of bleaching is followed by the extraction process, which is attained by using sodium hydroxide. The incorporation of additives such as oxygen and anthraquinone into alkaline soda pulping has been found to enhance both the yield and properties of bagasse pulps. However, few reports suggest that pulping of bagasse with oxygen has not shown the improvement in its properties, as the presence of iron in the pulp has been found to increase its viscosity. The addition of anthraquinone at low levels has been found to increase yields and lower the consumption of chemicals in soda pulping. Chemical pulp produced with anthraquinone has been found to be similar to soda pulp in properties. The pulp is then sent to a storage tower prior to bleaching (Delgado and Casanova, 2001).

21.2.2 Paper Production from Bagasse Pulp

Various kinds of papers with different qualities are produced from the bagasse pulp. Each kind of paper requires different mechanisms and processes for its production, depending on its demands by the paper industry. Approximately 3.4 tonnes of bagasse are required for the production of 1 tonne of paper. The bagasse pulp is blended with chemical and mechanical wood pulp and then transferred to a mixing tank, where it is mixed with paper fragments. The process is then followed by the addition of resins and chemical reagents to enhance the resistance of the mixture in the liquid penetration. In order to control the pH, the resins are added with alum and to increase the opacity, fillers and colorants such as kaolin are added. The blended pulp mixture with additives produces a paste that can be fed into the forming machine. Sheets are formed by removing most of the water in the wet section of this machine, leaving the sheet with approximately 20% of dry matter. Further pressing for dewatering provides 30–40% of dry matter, and finally 90–95% dry matter content is attained by the heat transferred through the rolls (Poopak and Reza, 2012).

21.3 Agglomerated Products Production from Bagasse

Agglomerated products are those consisting of the materials obtained from plant origin, made into a continuous solid medium by an agglutinating agent. The additives such as derived materials, resins and even cement can be incorporated into the agglomerated products. Particle board and fiber board are examples of agglomerated products. The fiber board is mostly produced from waste from the wood industry, but adequate quantities of fiber boards are produced from straw, cereals and bagasse. Particle and fiber boards differ with each other due to the methodology by which they are being formed. The fibrous material used in the production of particle board is treated mechanically, while that of fiber board is treated thermally, chemically or mechanically (Yadav and Solomon, 2006; Gupta and Ahuja, 1989).

21.3.1 Particle Board Production

Particle board production involves depithing, grinding and drying. On the basis of their relative size, the ground particles are dried and classified. During the process of drying, the humidity of these particles is reduced to 1–4%, in order to achieve the conditions required for gluing, formation and pressing. Over-drying must be avoided to prevent dust pollution, electrostatic problems, damage to the machine, risk of fire and other detrimental problems. Gluing is the most costly process in particle board production due to the cost of resins, which are used for binding purposes in the particle board. The quality of particle boards is determined by the properties of resins used in them. Urea formaldehyde, melamine formaldehyde and phenol formaldehyde are the most utilized resins in particle boards. The urea formaldehyde is cheaper than other resins, but the use of phenol formaldehyde is recommended for persistence of particle boards under severe weather conditions. Natural derivatives of wood, such as tannins mixed with phenol resins, are generally employed as the gluing agent (Rainey and Clark, 2004; Delgado and Casanova, 2001; Gupta and Ahuja, 1989).

Most factories employ formaldehyde as the gluing agent, along with other additives such as sodium hydroxide, paraffin emulsions and ammonium chloride. These additives either impregnate the special characteristics into the particle board or speed up or facilitate the chemical reaction occurring in the formation of particle board. Pentachlorophenol provides protection to these particle boards from fungal attack. Similarly, the ammonium salts, borax and inorganic phosphates provide fire-proofing effects. These additives are separately prepared and then later added to the gluing agents. The viscosity of the mixture must be carefully monitored in order to obtain the desired results. In multi-layer board production, the addition of the mixture is carefully dealt with, when separating the inner layers from the outer layers. The outer layers consume larger amounts of mixtures as compared to the inner layers. A homogenous mat of particles is obtained at the formation stage by using mechanical or pneumatic techniques, or both. The mat is then pressed, following which the boards undergo the finishing stage to complete polymerization of resins, and provide final dimensions to the design and polishing of the surface. The polymerization step is accomplished by the incubation of boards to the ambient conditions for approximately seven days (Delgado and Casanova, 2001).

21.3.2 Fiber Board Production

The fibrous material is subjected to chemical, thermal and mechanical treatments during fiber board production. These processes facilitate the softening of lignin, which allows the separation of its constituent fibrous material. These separated fiber bundles then undergo cross-linking whereby they stick together. For the production of medium to dense fiber boards, the bagasse from sugar industries is considered as the best choice, due to its fibrous nature and low density. Bagasse provides lightness and good mechanical behavior in the fiber board. The first step for fiber board production involves the preparation of pulp by using thermal, mechanical, chemical or thermo-mechanical processes. The pulp produced by these mechanisms is refined to obtain the fibers of appropriate dimensions. In order to enhance the mechanical properties of the board, the chemical agents may be added. Asphalt emulsions and phenol resins can improve the quality of the board. Similarly, paraffin can increase the water repellent ability of fiber boards. To avoid bacterial and fungal attacks, other chemical agents can also be incorporated into the production process. The pH of the mixture must be well controlled and monitored in order to guarantee the flocculation of gels around the fibers (Rainey and Clark, 2004; Delgado and Casanova, 2001).

The production of fiber board may involve dry or wet processing. In wet processing, the mat formation is attained with the fibers suspended in water. Wet processing provides the inter-cross-linking of the fibers, resulted in harboring good mechanical properties in the fiber board. The wet processing may require large amounts of water and thus high investment cost. During dry processing, the dried fibers along with the additives are pneumatically led over a travelling mesh, using suction through the fibers in order to achieve mat formation. The mat thus obtained is pre-pressed without heating prior to being transferred to the heating press. Hot pressing of the board is either performed by dry or wet pressing. Dry pressing provides a smooth surface on both sides, while wet processing provides smoothness on only one. Subsequently, for heat pressings, the boards are adjusted in size, by trimming to obtain the desired length and width. The fiber boards are re-humidified to obtain the necessary equilibrium of water. The fiber board quality is improved by heat treatment and painting with a thin film of drying oil (Poopak and Reza, 2012; Rainey, 2012; Galves, 1988).

21.4 Alcohols

Fossil fuel reserves are continuously becoming depleted, while the demand for energy is increasing worldwide, with the increase in population and economic growth (Ali *et al.*, 2011). Global energy consumption is expected to reach 702 quadrillion British thermal units (QBtu) by 2030. Similarly, the demand for electricity by developing countries is expected to rise 40% by 2030 (Ali *et al.*, 2013, 2014a). Bio-renewable fuels are handy choices to fulfill these increasing energy needs. Fuels are the major uses of by-products obtained from the sugar processing industries. The fuels obtained as by-products are alcohols, which are capable of fulfilling these energy requirements. The utilization of sugar cane and sugar beet for alcoholic fuels are on the rise in the USA and the UK. Alcohol is the major by-product produced from the sugar industries. Mostly, the alcohol is obtained either in its anhydrous form having over 99% of alcohol, or as an azeotropic mixture with water having 93–95% alcohol. Beet pulp does not contain rich fiber like that of the bagasse of sugar cane. Similarly, unlike bagasse, the beet pulp cannot be burned directly to produce steam through which electricity can be generated (Delgado and Casanova, 2001).

21.4.1 Production of Alcohol

Alcohol production depends on the types of raw material provided for the process. The basic raw inputs are clarified juice molasses, mixed juice obtained from mills, and a mixture of different clarified juice syrup. Alcohol is produced by the following major steps.

21.4.2 Substrate Preparation

When cane juice is applied for the alcohol production from the mash, the process used is similar to that used for the clarification of juice in sugar production. In order to obtain the final pH of 5.5, the milk of lime is added to the mixture, and to accelerate the coagulation of proteins and the formation of calcium phosphate. the juice is heated to 105°C. Flocculation and settling of materials is attained in the clarifier. The final concentration of the clarified juice is obtained by mixing it with molasses or syrup. When molasses is taken from the mash or main substrate, fermentation yields are improved due to the presence of the probiotic components of molasses. To avoid the undesirable fermentation by microbes, the heating of mash is very important.

21.4.3 Preparation and Inoculation of Yeast

Biotechnologists and microbiologists are continuously searching for yeast strains that can produce high amounts of enzymes such as zymase and invertase and stability in terms of high temperature and osmotic pressure (Ali *et al.*, 2014b). The most widely-used yeast in alcohol production from the sugar industries is *Saccharomyces cerevisiae*. Fresh or recycled yeast can be used in the production of alcohol.

The section of yeast strain is performed aseptically, with controlled temperature, pH, total solid contents, nutrients and cell concentration. The yeast is grown in a metallic grower, under aerobic conditions and then transferred through a pipeline to the processing plant. Cell growth and multiplication in the final stage takes place in the pre-fermenter, while the remaining growth and multiplication of cells takes place in stainless steel tanks. The aerobic conditions required for the respiration and growth of yeast in the tanks is obtained by the presence of spargers located at the bottom end of the tank. An air filter with a glass fiber base enclosed in the metallic cylinder delivers air to these spargers. For further aseptic conditions, hot steam is blown in the filter.

The Melle or MelleBoinot system is an alternative method for the preparation of yeast inoculum for the fermentation process. This process involves the repeated re-use of the selected yeast. Alcohol produced at the end of each fermentation cycle is recovered by the process of centrifugation. By the help of sulphuric acid, the pH of the precipitate obtained during the centrifugation is maintained at 3.0. The amount of recovered yeast, separated in terms of dry basis, is always more than the applied inoculum size. Excessive quantities of produced yeast are mostly utilized as cattle feed (Delgado and Casanova, 2001).

21.4.4 The Process of Fermentation

The process of fermentation is obtained with the help of enzymes, which were produced during the inoculation of the yeast. The disaccharides, such as sucrose, are broken into monosaccharides by the invertase enzyme, while the monosaccharides are converted into carbon dioxide and alcohol by the zymase enzyme. In the initial phase of this process, the production of alcohol is slow due to the presence of high amounts of monosaccharides, while in the later stages there is a subsequent increase in the production of alcohol by zymase enzyme activity (Delgado and Casanova. 2001).

21.4.5 Alcohol Purification

The alcohol produced during the process of fermentation also contains small quantities of sugars and some acids such as butyric acid, acetic acid and propionic acid, esters and some other compounds which are part of sugar cane juice or molasses. Through the process of distillation, the alcoholic mixture is separated and purified. Dehydration is performed to obtain the anhydrous form of alcohol, depending upon the requirement and production need of the particular sugar industry. The process of distillation may require large quantities of water, which is recycled by the process of condensation (Delgado and Casanov, 2001).

21.4.6 Kinds of Alcohols Obtained from Sugar Industries

The major alcoholic fuels obtained from sugar processing industries are ethanol, methanol and butanol. These three alcoholic fuels carry different properties, which makes each of them somehow superior or inferior to each other. Ethanol, a two-carbon fuel from sugar processing, in particular is gaining more attention than methanol and butanol.

21.4.6.1 Ethanol Ethanol is also known as ethyl alcohol. It is one of the most common forms of alcohol, mostly produced from the fermentation of sugars. Ethanol consists of two carbons, six hydrogens and one oxygen atom. It is highly volatile, colorless and carries a light odor. Ethanol is an excellent solvent, soluble in water and in many organic solvents (Haynes, 2011). Its lower boiling point than water and its anti-freezing abilities, make it as an excellent choice to be used as fuel. Brazil produces and utilizes large amounts of bioethanol, obtained mostly from the sugar cane industries. The USA also produces large amounts of bioethanol, which is mostly obtained from the corn processing industry. However, bioethanol from sugar cane has been found to be more efficient than that from corn (Leite *et al.*, 2009). Ethanol can be converted into sub-chemicals, as classified in Figure 21.4.

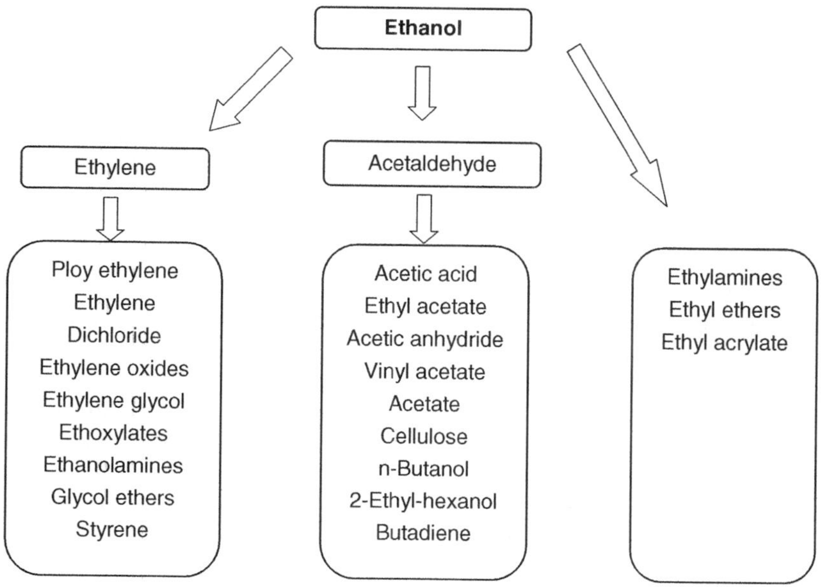

Figure 21.4 Chemicals from ethanol (adapted from Bruscino, 2009).

21.4.6.2 Methanol Methanol is also called single carbon alcohol. It is composed of one carbon, four hydrogens and a single oxygen molecule. The structure of methanol is similar to that of ethanol and methane, which is a natural gas. Methanol is widely applied in several chemical processing industries, which makes it available to different markets (Amend *et al.*, 1993).

Methanol, like ethanol, has a low freezing point, making it a useful anti-gelling and anti-freezing compound, which can be applied in diesel engines and gas lines. Both fuels serve as an oxygenate in the combustion process, cleaning the engine exhaust. The production of methanol from the enzymatic fermentation of the sugar processing industries is relatively cheaper than that of ethanol production. Methanol does not readily absorb water and can be transported easily through pipelines in its concentrated form or mixed with petroleum products. Both methanol and ethanol burn with an invisible flame and can readily be distributed, making it difficult to handle any fire caused by them. Methanol is more toxic than ethanol. Methanol can accumulate in the liver after being absorbed through mucus membranes or the skin. Prolonged exposure to methanol can cause damage to optic nerves, which may cause blindness. Methanol has a lower energy content than ethanol. Methanol can damage aluminum and aluminum-based alloys due to its corrosiveness. By the process of oxidation, aluminum in catalytic converters and newly-made engines. it is stripped and dissolves in aluminum components, producing several toxic heavy metal components in the exhausts of automotives. Formaldehydes have been detected from methanol exhausts. Atlantic Biomass Conversions, Inc. estimates the total cost of production for methanol from sugar beet pulp is approximate $0.50 per gallon, if produced on a commercial scale (Watts and Associates, 2007).

21.4.6.3 Butanol Butanol has higher levels of stored potential energy than methanol or ethanol, due to the presence of four carbon atoms, making it a more energetic and complex molecule. Butanol is less corrosive than ethanol and can tolerate water. This property makes is more feasible for transportation through pipelines. Unlike other alcohols, butanol can be easily separated from the water mixture. Butanol is mostly blended with other fuels such as diesel and gasoline with approximately an 85% strength. The blended fuel containing butanol is more stable, which facilitates its storage and distribution. Under cold conditions, butanol can work better as a starter fuel for cars (Mahalaxmi and Williford, 2012). Unlike methanol and ethanol, butanol can be used in cars without any changes to the engine and produces almost the same amount of power as that of gasoline (Smith and Workman, 2004). In industry, butanol is mostly used as solvent. However, it has several drawbacks in comparison to its counterpart alcohols. Unlike methanol and ethanol, butanol is dense which makes it more difficult to flow through fuel lines as compared to lighter fuels. Butanol has a low boiling point, which makes it vaporize quickly at room temperature if available in unblended form. The octane value in butanol is lower than methanol and ethanol. In 2006, some UK companies, namely British Sugar, DuPont Chemicals and British Petroleum, made a partnership for the formation of butanol by utilizing sugar beet. The butanol was later decided to be made in the blended form with ethanol from sugar industry by-products (Watts and Associates, 2007).

21.5 Animal Feed

The by-products from sugar industries have been extensively utilized as animal feed. The by-products obtained from the beet sugar industry and the cane sugar industry are discussed below.

21.5.1 Animal Feed from Beet Sugar Industries

For decades the beet pulps, as both wet or dried forms, have been used traditionally as substitutes for hormones for animals. Highly digestible characteristics, being rich sources of proteins and due their low cost has made beet pulp a choice of feed in animals. However, the year-round availability of beet pulp in comparison to short-seasoned hybrids of corn is a limiting factor for its usage as feedstock for animals. Improvement in storage of beet pulps and the development of new hybrids of beet sugar can help in the development of beet pulp in animal feed markets.

The digestive system of a horse is considered a wonder in animal physiology, due to ammonia imbalances, ruminants to compaction and other several digestive maladies. Horses use secum, unlike rumen in other animals, for the digestion of cellulose. Reports suggest that beet pulp can be best suited for horses, due to compatibility to their unique digestive system. A trial of horses fed on beet pulp has demonstrated that the utilization of beet pulp as feed has slightly reduced the digestion of crude fats and the pulp has also been found to facilitate the formation of urea from ammonia (Olsman *et al.*, 2004). One study trial on beet-pulp fed horses, which were normal in exercise and body development, has been found to have higher insulin levels, making more energy available to these horses (Crandell *et al.*, 1999). Sugar beet pulp can also be used as a substitute for oats in equine diets, showing the same effects on their performance or metabolism (Palmgren-Karlsson *et al.*, 2002). Based on these studies, and the fact that horse farming or petting has mostly been performed by the more affluent communities, the prospect of beet pulp for the utilization as feed for horses is very promising.

The beet pulp bases substitute for hormones, and the bovine somatotropin is considered as the market niche in Europe. The hormones and bovine somatotropin stimulate milk and beef production in animals. The dairy and beef industry is required to certify that no bovine somatotropin or other hormones have been utilized for the cows. The beet pulp has been found to induce similar effects to that of bovine somatotropin. The increasing demands for beef and milk give the choice to the livestock industry to utilize the beet pulp as feed or as an additive to the feed (Watts and Associates, 2007).

The pulp molasses pellets containing total protein (5–6%) are traditionally used as feed for ruminants. Its protein enrichment is desirable for use as an alternative for good feed in poultry and pigs. The molasses are rich in fermentable sugar, which can be used by microbes for microbial protein production and such protein-rich materials could subsequently be used for animal feed. The sugar beet pulp protein could be upgraded to 31% protein concentration by solid state fermentation with the help of microbes (Nigam, 1994).

21.5.2 Animals Feed from Cane Sugar Industries

Sugar cane is one of the principal agricultural crops in tropical countries. Each year, it generates about 279 million metric tons of biomass in terms of bagasse and leaf residues. These biomass residues have been explored for their use in different applications, including biotechnological and non-biotechnological (Chandel *et al.*, 2012). Meat and dairy industries can flourish well in those regions where easy and smooth availability of low-cost feed is in abundance (Suresh and Reddy, 2011). Sugar cane top, and green and dry leaves are applicable as animals feed. Sugar cane bagasse and leaf trash contains a reasonable amount of cellulosic, hemicellulosic and lignin compounds, which can be converted into simple usable sugars such as glucose, mannose, galactose and arabinose etc. with the help of physical, chemical and enzymatic treatments (Krishnan *et al.*, 2010). Beside sugar cane bagasse, its leaf trash can be produced in large amounts, equal to 6–8 tonnes from 1 hectare of sugar cane crop (Singh *et al.*, 2008).

The biomass of sugar cane bagasse as feed for animals is limited due to its complex structure based on cellulose and hemicelluloses. The bagasse is therefore mostly used for biotechnological purposes, but its easy and low-cost availability are still making it attractive for use as an animal feed in combination with other feed materials and after physical or chemical treatments (Chandel *et al.*, 2012). Ruminants can digest cellulose but cannot digest lignin, which is the major constraint in overall digestibility of bagasse. Thus, chemical and physical (thermal) treatment of bagasse before its use as feed can improve its digestibility. Bagasse is treated with sodium hydroxide followed by blending and mixing with molasses. Urea can improve the quality and digestibility of bagasse and pith feeds. The use of sodium hydroxide should be optimized as its excessive use is harmful for animals.

Thermal treatment is the best alternative for increasing the digestibility of bagasse to avoid costly chemical treatment and its contraindications in animals. Under high temperatures, bagasse expands rapidly and produces acetic acid by acetyl radicals generated in hemicelluloses, helping in acid hydrolysis of cellulose. Two types of thermal processing, low temperature (170°C for 15–30 min) and high temperature (210–220°C for 3–5 minute) at high pressure (2.1–1.8 Mpa) can be applied for thermal treatments, which result in 55–60% increased digestibility (Delgado and Casanova, 2001).

The most widely-used by-product of sugar cane for animal feed is the final molasses. Molasses as an animal feed is a good source of sugar and phosphorus for the animal body. The urea molasses diet supplemented with protein and other forage has been successfully fed to lactating animals and growing calves. Molasses at concentrations of 10–15% mixed in the diet can also increase the edibility (Verma, 1997; Ranjhan, 1993). Sugar cane tops mixed in with leguminous feeds can be used for cattle diets (Verma, 1997). Digestibility of the unprocessed bagasse is low, which can be improved by high pressure and thermal treatments (Sindhu *et al.*, 2002).

Yeast biomass after the fermentation process for the production of alcohol, citric acid, monosodium glutamate etc. is a rich source of microbial protein for animal feed (Cheeke, 1991). Torula yeast is a common animal feed containing more than 40% protein content, with a high rate of growth. It can utilize molasses for its growth and production of different valuable chemicals. During the fermentation process, in the presence of oxygen, it produces a large quantity of biomass, which is then separated mechanically from the final stage and spray dried for use as animal feed and single cell proteins, as it contains approximately 45% and 20–30% carbohydrate content (Delgado and Casanova, 2001).

Sugar cane press residues, commonly known as sugar cane mud, filter press cake, fiber cake and scum, are the by-product of the sugar industry, with a potential for use in other valuable processing. This dark brown, soft and spongy material is composed of sugars, fibers, wax, organic salts and inorganic matters (nitrogen, phosphorous, calcium, iron and manganese). Currently, it is used for soil improvement, but its excessive use can pose environmental complications. The depithing residues containing bagasse and some fraction of pith are also used as animal feed after hydrolyzing (Suresh and Reddy, 2011). It has been reported that birds can tolerate low levels (3–15%) of sugar cane press residues (Suma *et al.*, 2007; Budeppa *et al.*, 2008), yet its inclusion as a regular poultry diet needs substantial research.

21.6 Acids

Ferulic acid has several industrial applications. Due to being a powerful antioxidant, it is utilized in many health and cosmetic industries. In several reports, ferulic acid has shown promising anti-cancerous, anti-inflammatory, neuro-protective, anti-aging, anti-apoptotic and anti-diabetic properties (Watts and Associates, 2007). Ferulic acid has commonly been isolated from Chinese herbs used for traditional health benefits and can also be applied as a food additive. Ferulic acid can easily be packaged with vitamin D or can be included in sports and energy drinks, where it can add anti-microbial and anti-mutagenic nutraceutical properties to that product. Ferulic acid can easily be metabolized by humans and has shown very low levels of toxicity (Ou and Kwok, 2004). The use of ferulic acid in cosmetics from sugar beet has been well known for many years (Graf, 1992). Many products, such as antioxidant creams, UV protection materials and serums, are available as patents and commercially-available products (Watts and Associates, 2007).

Ferulic acid has readily been found in common plants, but recent developments in sugar by-products have suggested that sugar beet can be a good source of ferulic acid. The extraction process of ferulic acid from sugar beet is similar to that of other plants (Watts and Associates, 2007). Researchers from the University of Alcala, Spain have reported many bacterial strains that are capable of extracting approximately 80% *in vitro* of the ferulic acid in free form from the sugar beet pulp (Ferris and Mayne, 2006). It has also demonstrated anti-radiative potential. In one trial study, the liver cells were divided into two groups. In one group, ferulic acid was added to the cells. Both groups were exposed to radiation. The group containing ferulic acid showed resistance against the destructive and mutagenic effects of radiation (Srinivasan *et al.*, 2006).

The beet pulp has high levels of oxalic acid, which is one of the major causes of oxalate stones in the kidney. However, in a study at Oklahoma State University, the participants consuming the sugar beet pulp were found to have less oxalic acid constituents than spinach or the ones taking pills of oxalic acid (Hanson *et al.*, 1989). This shows that the pulp from sugar beet either does not participate in stone formation or can be useful in stone prevention. D-galacturonic, commonly known as galacturanic acid or poly-galacturonic acid, is derived from pectin. The acids from sugar beets can be extracted by the process of fractionation. There is still ongoing research on the cost-effectiveness of the commercial-scale production of these acids from sugar industries (Spagnuolo *et al.*, 1999).

21.7 Pectins

Pectins are the biopolymers which occur naturally in many fruits. They are particularly present in large quantities in the soft fruits, providing shape and body to fruits. The physiological functions of pectins are similar to those of gelatins, making semi-solid gel states by binding liquids to the solid components. The pectins obtained from sugar industries, particularly from beet pulp, are found to be more complex in structure than the pectins occurring in other plants. This property of beet pulp pectins adds more value to them, as they can be more resistant to digestion in the human body, until it reaches the large intestine and colon, the lower gastrointestinal tract (Liu *et al.*, 2003). It has been found preferable to oral intake of anti-inflammatory pills, coated with pectin obtained from sugar beet, to treat the colon region, especially for suppository application (Yang *et al.*, 2002). There has been an enormous amount of continuous research on the extractions and applications of beet pectin, due to its complex chemical structure. Researchers have reported that the pectin obtained from beet is able to bind with cellulose, which makes it resistant to digestion in the digestive tract of humans (Zykwinska *et al.*, 2005). The side chains of beet pectin have also been found to have special applications in biotechnology (Ralet *et al.*, 2005). Cross-linkages found in beet pectin further highlight its slow digestible properties (Oosterveld *et al.*, 2000).

21.8 Functional Foods and Nutraceuticals

Non-pharmaceutical products, which are consumed in tablets, powders or other medicinal forms for different health benefits, are termed as nutraceuticals. Naturally-occurring attributes of food with health benefits are defined as functional foods. The intake of raw prunes for their dietary fiber and the tablets of St John's Wort, for enhancing the immune system, are examples of functional foods and nutraceuticals, respectively (Watts and Associates, 2007).

The fiber obtained from sugar beet has been suggested as a good source of dietary fiber by the researchers McKee and Latner (2000). Danisco, a Denmark-based firm, is one of the leading food supplement producers. White fiber obtained from highly-refined sugar beet pulp is the leading non-animal feed product of Danisco. The fractionating process of sugar beet pulp for obtaining the beet components has been patented by them. There is also an increasing demand for utilizing beet fiber for the reduction of cholesterol levels in humans. Trials at Billings Clinic have suggested that sugar beet fiber has greater potential for cholesterol reduction than oat fiber, which is a leading functional food used in breakfast and cereal products or as a nutraceutical for the control in cholesterol levels. Further research can enhance the functional food and nutraceutical properties of beet sugar pulp (Watts and Associates, 2007).

Microorganisms utilize fermentable sugars from pulp and molasses for their growth and produce single cell proteins. Different fungal cultures could be utilized as a single cell protein for animal feeds (Nigam and Vogel, 1991). Pandey *et al.* (1988) reported the bioconversion of sugar beet pulp into protein rich materials by scarification and fermentation processes using enzymes and microorganisms.

21.9 Anti-Desiccants

The term anti-desiccant is applied to the substances that prevent loss of moisture or drying. The term has also been extensively use for the applications or products designed for the extension of the storage life of materials, which can be sensitive to humidity or are perishable. The beet pulp has been found capable of water retention of more than four to six times its volume. This makes it a useful emulsifier to be applied in storage of food products, especially for blended meat products and sausages. Currently, mustard powder and mustard seeds are extensively used in meat processing industries for flavoring, water retention and emulsifying characteristics. However, the strong flavor induced by the mustard can be undesirable for consumers. Alternatively, there are reports testifying the applications of beet pulp for meat preservation rather than mustard (Javidipour *et al.*, 2005).

The use of ferulic acid has already been described in detail in the section on acids portion. Due to its antioxidant potentials, it has also been tested as a preservative in extending shelf life, when used coupled with rosemary leaves and an altered gas environment in meat packaging for retail. The technology used for the altered gas environment is used to inject inert gases into the meat packaging rather than using the air. The results inferred by the color of aging meat, have shown a dramatic slower oxidation in the retail meat. The use of altered gas technology may not be recommended for the preservation of natural products; however, the ferulic acids qualify for the same. Trials have shown promising results in the use of ferulic acid by adding shelf life to meat from 8 to 14 days (Martinez *et al.*, 2006). The trials of using beet pulp on improving shelf life of beef have also shown good effects. The antioxidant capacity of beet pulp can be used in organic meats, where the uses of altered atmosphere technologies are deterring the interests of consumers (Watts and Associates, 2007).

Carboxy-methyl-cellulose (CMC), obtained from beet pulp, has also been found to be a good preservative of fruit, in a study conducted in France. CMC is obtained from the beet pulp by doping it through the process of etherification and soybean oil, and then some other natural products are added to form a plasticized film on drying. The Mandarin, a citrus fruit closely resembling an orange, has been found to have an extended shelf life of 6 to 14 days, when sprayed by CMC obtained from beet pulp (Togrul and Arslan, 2004).

21.10 Biodegradable Plastics and Biopolymers

Cellulose is the basic component of plants cell walls. It provides rigidity, shape and structure for the entire plant. In nature, the cellulose is most abundantly found as biological fiber, making a substantial portion contributing to the total mass of the plant. The pure form of cellulose can be found in cotton lint. Linters are strands that are too fine or tangled, which are discarded in the milling process of cotton. Linters are a pure form cellulose produced at a very low cost. The linters are found to have tremendous uses in cellulose-based applications in industries such as food casings, plastics, lacquers, films, plastics and rayon.

Cellulose from cotton, having low-cost productive, enjoys more applications than the cellulose from sugar industries. Cellulose from sugar industries has to undergo separation from lignins, pectins, biological acids, hemicelluloses, proteins and other components of the mash or pulp, before it can be used in any application. The low cellulose composition of fibers from sugar industries makes them a poor choice for obtaining pure cellulose (Bhattacharya and Chatterjee, 2004). In comparison to mash fiber from sugar cane, the beet fiber can still have better prospects, even though the composition of cellulose is approximately 20%, as compared to 80% of hemicelluloses, sucrose, pectin, lignin, fat and proteins. However, there are many applications of fiber obtained from beet, which does not require high levels of purification. Researchers have reported that the characteristics of fibers and films obtained from beet pulp, having the interaction of cellulose and pectin, may actually be beneficial in strengthening the biodegradable films obtained from the beet pulp (Dufresne, 1997). One study has shown that if the beet fiber is added with pure cellulose when making the plastic film, it degrades the quality of the film (Chauvelon *et al.*, 1998).

Agricultural research has mostly focused on utilizing the potentials of starch from the beet pulp for applications in, for example, packaging of natural products. Some reports suggest that polymerization of starch, through the fermentation process of lactic acid, provides the material with similarity in strength and several other characteristics similar to that of petroleum-based plastics. The materials made by beet pulp are found to have high degradation when exposed to moisture, sunlight or microbes, degrading into the components that possess very little or almost no environmental threats. However, there is still a need for extensive research to minimize the cost of production of these bioplastics from beet pulp (Finkenstadt *et al.*, 2007). According to definition, beet pectin can be considered as a biopolymer. Its applications, particularly in the food and pharmaceutical industries, have already been discussed in detail. It can also be applied in bioplastics. The researchers are doing extensive studies to compare pectin-based biopolymers and polymers, which are derived from the cellulose, fibers, hemicelluloses and other components of plants (Zykwinska *et al.*, 2005). Bacterial production of poly-3-hydroxybutyrate, a well-known bio-plastic, has been reported from waste effluent of molasses using *Staphylococcus epidermidis*. Bacterial growth was supported by micronutrients such as calcium, phosphates and ascorbic acid (vitamin C) present in the effluent (Sarkar *et al.*, 2014).

21.11 Food Products, Flavorings and Aromas

The sugar in practical applications from sugar industries is applied in many food industries, such as the confectionary industry. Cane juice can be used in several other food products. The lack of distinctive flavor, antioxidant potential and water absorption characteristics make beet pulp applicable in a number of food products.

The pulp, which is a by-product of beet, is not considered appropriate for direct consumption by humans, due to its inappropriate smell and taste. However, the ferulic acid obtained from beet pulp can be converted into vanillin, which is a common and popular food flavoring. Ferulic acid, which is obtained from the lignin, is further treated in a series of reactions by using bacteria and enzymes to provide vanillin (Gasson *et al.*, 1998). There is a strong similarity between the molecules of vanillin and ferulic acid. This process was earlier carried out synthetically, by breaking the bonds carrying the hydroxyl and oxygen molecules, which are double bonds, to provide the carbon arm on the aromatic ring (Amend *et al.*, 1993). The use of biological metabolism of ferulic acid to vanillin has been found to be cost-effective (Muheim and Lerch, 1999). Vanillin is not only produced for flavoring purposes in food products, but approximately 13% of vanillin produced can also be utilized in the synthetic conversion of other chemicals in several chemical industries.

The aromatic compounds in chemistry are characterized by those having the ring structure present in their basic chemical structure, such as benzene. They can react with the olfactory receptors in the brain and can produce the scent response in the humans. As discussed earlier, vanillin and ferulic acid both obtained from beet pulp are composed of the six carbon, aromatic ring. The consumers relate the scent of vanillin with a sense of well-being, which makes it well applicable in the perfume industry (Watts and Associates, 2007). Vanillin is also utilized in aroma detergents, and in winter clothing, such as sweaters. The vanillin has the characteristic smell like that of vanilla. The use of vanillin does not only influence the flavoring in the food products but also adds the vanilla-like smell in the products of baking industry and home bakeries. The tobacco industry has also been known to add the smell of vanillin to cigarettes to attract a certain group of consumers (Cook *et al.*, 2003).

Simple alcohols, such as methanol and ethanol, lack the aromatic ring but are still used extensively in the cologne and perfume industries, because of being inexpensive, volatile, evaporate readily and spread the aroma into the surrounding environment, blend easily and are inert with the aromatic compounds. Methanol, which was previously designated as toxic, particularly dangerous for vision (Erdener and Ilhan, 1997), is now so commonly used in the perfume sector, that it is considered as the standard solvent into which the compounds having aroma can be dissolved (Zhong *et al.*, 1996). Ethanol is also commonly used in perfume industries. However, its properties as an alcoholic beverage are easily denatured so that it cannot be consumed by the user.

21.12 Char and Biofertilizers

Sugar cane biomass contains different components such as lignin, fiber, pith and pentosans. All these components have different applications in bioprocessing and in agricultural fields (Yadav and Solomon, 2006). Sugar press mud, composed from lingocellulosic by-products of sugar industries, is used as a rich source of fertilizer for plant growth and soil maintenance in terms of nutritional components for the plant, which also help the plant to protect itself from various soil-borne diseases (Rolz *et al.*, 2010). Each year, approximately 500 000 tonnes of press mud is generated by the sugar industries (Cifuentes *et al.*, 2013). Organic fertilizer for organic farming and due to its environmental-friendly nature, is gaining more appraisals as compared to chemical fertilizers. Bio-compost fertilizers are efficient in providing both organic and inorganic nutrients, which result in good-quality and healthy crops (Zhu *et al.*, 2007). It also improves the microbial quality of the soil, which is directly proportional to soil fertility (Bulluck *et al.*, 2002). Sugar press mud from organic waste of sugar industries are rich in nitrogen and phosphorous, enhancing nitrogen fixation bacterial activities, and also contain high amount of moistures (50–70%) favoring earthworm growth (Domınguez and Edwards, 1997).

Bagasse is carbon0rich source derived from sugar cane, and is highly suitable for biochar and biofuel production (Mothe and de Miranda, 2009). Its anaerobic digestion with the help of microbes could also be a source of biofuel in term of biogases (Osman *et al.*, 2006). It have a complex lignicellulosic substance, containing cellulose (50%), lignin (25%) and hemicelluloses (25%), whereas it is also a rich source of inorganic compounds (Pandey *et al.*, 2000). The maximum digestibility of bagasse for biofuel is almost 75% by weight, using different pre-treatment methods for anaerobic digestion. The remaining 25% of the bagasse residue (sludge) after the digestion process could be use directly as compost for land fertility or as a treated biochar. It has been reported that pyrolysis of anaerobic digestion residues of bagasse for the production of biochar to enhance yields could be environmental-friendly and economically beneficial (Inyanga *et al.*, 2010). Bagasse conversion into biochar for soil amendment could be another valuable utilization of by-products from sugar industries. Conversion of filtercake into biochar could help in soil improvement and eutrophication, and reduce waste from sugar industries (Eykelbosh *et al.*, 2014).

Biochar has adsorption abilities for dissolved nitrates, which increases the opportunity for plant roots to absorb the nitrates, can result in reduced leaching of nitrates from the soil and so result in better growth of the plants. It also increases the water retention time and porosity of soil. High efficiency in terms of increase soil fertility and structure has been reported by applying bagasse-derived biochar (Kameyama, *et al.*, 2012). Aerobic composting of sugar industry by-products into organic manure and the biodegradation of its lignocellulosic wastes with the help of integrated composting systems with microbes and vermicomposting, can result in a nutrition-rich matter and humas for plant growth that is also ideal for the breeding of earthworms, thus enhancing indirect soil fertility. Such integrated approaches also reduce the composting time (Kumar *et al.*, 2010).

21.13 Waste Water Treatment and Environmental Bioremediation

Dyes and pigments are frequently used for industrial processing in textile, carpet, paper and printing industries for different purposes. These dyes and pigments have a long life and are common industrial pollutants contaminating the environment due to their long stability. Intoxication due to dyes and pigments are disturbing aquatic life and environment. Degradation of dyes is relatively hard due to their structure and molecular size (Sun and Yang, 2003). Bagasse from sugar industries can be helpful for the removal of dyes by adsorption mechanisms. In a study, bagasse pretreated with formaldehyde and sulphuric acid was successfully used for the removal of methyl red and azo dye from simulated wastewater (Abdullah *et al.*, 2005). The use of adsorption materials for dye removal has an edge over other methods, due to its ability of complete removal of dyes, even present in diluted form. Bagasse could be efficiently and cost-effectively utilized for dye and color removal, replacing activated carbon commonly used as absorbents (Vijayakumar *et al.*, 2014).

Activated carbon from sugar industrial waste can be a valuable adsorbent for the removal of pollutants, particularly heavy metals and its re-use as a valuable by-product, resulting from wastes and manufacturing processes (Fahim *et al.*, 2006). Heavy metal removal from waste water using conservative technologies is one of the most efficient ways. Perez-Armendariz *et al.* (2004) reported the successful use of microorganisms isolated from bagasse pith for the re-habilitation of hydrocarbon-contaminated tropical soils. Wastewater treatment using cost-effective methods are required nowadays. Biomass from agriculture has been used as a carbon source for wastewater treatment, which could include molasses from sugar industries as a good source of nutrition for microbial growth and degradation of waste materials (Kumar and Sahu, 2013). Bagasse fly ash of the sugar industry as a low-cost treatment for wastewater has been used for the treatment of their wastewater; it was able to reduce the BOD significantly (Lakdawala and Oza, 2011).

21.14 Energy and Biogas from Sugar Industries

Sugar cane industries are considered as self-sustaining in energy production due to energy supplies from burning bagasse (Brooks *et al.*, 2008). Dry molasses from sugar industries contain a huge amount of calcium, potassium and sodium, two to four times higher than conventional molasses. Sugar beet pulp is also reported as a valuable source for biogas production (Brooks *et al.*, 2008). Sugar cane and beet plants alone and in co-digestion with other materials have been reported for biogas production, with both environmental and economical benefits (Fang *et al.*, 2011). The uses of sugar cane bagasse and other by-products are helpful in reducing the risk of climate change and global warming challenges. In Brazil, 572 million tonnes of sugar cane were used to produce 23 billion liter of ethanol in the year 2011/2012. Ethanol as an environmental-friendly fuel is the major component of the energy matrix. In addition, sugar cane bagasse is directly used for the generation of power (electricity) and steam energy. The electricity generated from bagasse burning is the second largest source in Brazil (Sbrana, *et al.*, 2013).

From the 200 million tonnes of dry bagasse per year, 19 GJ/t specific energy can be produced, which could supply 2 009 106 MWh electricity per year, if converted at an efficiency of 20%. The major constraint in bagasse conversion to energy is the drying cost (Stanmore, 2010). Burning the bagasse has an advantage over fossil fuels, as it poses comparatively lower environmental impacts in terms of no poisonous gases being released into the environment and less risk to the carbon balance. There are no reports of the release of SO_2 and SO_3 from bagasse burning and the carbon emissions from it are not a problem, as they are already captured by the sugar cane during its growth and development, so it is considered as a renewable source of energy production (Sbrana, *et al.*, 2013).

21.15 Sprays and Colors

Sugar processing results in the formation of different color compounds, mainly divided in three major groups, melanins, melanoidins and caramels. Chemical reactions between amino and carbohydrate compounds are mainly responsible for their formation. Despite these compounds being considered as pollutants, they can still be useful as coloring agents (Mudoga *et al.*, 2008). Amongst them, melanoidins have significant commercial and nutritional value. Melanoidin can be used to add color and flavor to different food products, whereas some hazardous health effects of the same compound have also been reported (Kanawade, 2014). From sugar industries, the beet pulp in particular has a number of applications in home furnishing industries, such as providing the sprays and color products. Similarly, the refined sugar and sucrose, obtained from the sugar industries, is utilized for the formation of polyurethane sprays (Watts and Associates, 2007).

21.16 Solvents

As discussed earlier in more detail, sugar industries harbor different kinds of alcohols, such as ethanol, methanol and butanol. These three alcohols, particularly butanol, are widely utilized as commercial or laboratory solvents. There are well established markets for these products. Researchers are efficiently working to minimize the cost of these alcohols obtained from the sugar industries (Watts and Associates, 2007).

21.17 Bio-Filters

The fiber from sugar industries, when dried, can be applied for bio-filtration processes (Lehrfeld, 1996). There is considerable evidence that the beet pulp has promising prospects in bio-filtration. Research indicates that the lack or absence of harmful residues and slow degradation of beet pulp makes it applicable in the filtration process. The beet pulp, epichloropphydrin and base can be doped to improve bio-filter characteristics as a cation exchanger in the bio-filtration process in industry.

21.18 Microbial Substrates

In biological research, microbes require a medium for their growth, called a substrate. Microbes can be grown on solid media, usually an agar, or in liquid media called broth. The agar is a biopolymer obtained from the sugars or their by-products from sugar industries. Dried fiber baggase or the beet pulp can be used directly or in a refined form as suitable media for the growth of certain microorganisms. Beet pulp in particular is rich in certain sugars, pectins and acids, which are beneficial or required for the growth of many microorganisms. The beet pulp or bagasse, however, must go through several refining processes and modifications to be compatible with agar, which is a uniform media.

21.19 Summary and Future Prospects

The by-products from the sugar industries can have numerous uses and applications in other processes and industries. The sugar processing plants can harbor many by-products, which can further produce its components. The challenging prospects for the utilization of these by-products and their components depend on the cost of production, refining and processing. For example, if the fiber from sugar industries can be extracted and purified at a cost comparable to mustard seed, then it can replace it in the market. Similarly, if pectin and ferulic acids can be produced cost-effectively from the by-products of sugar industries, then they can provide numerous further prospects in other industries.

References

Abdullah, A.G.L., Salleh, M.A.M., Mazlina, M.K.S., Noor, M.J.M.M., Osman, M.R., *et al.* (2005) Azo dye removal by adsorption using waste biomass: Sugarcane bagasse. *International Journal of Engineering and Technology*, **2**(1): 8–13.

Akbar, N.M and Khwaja, M.A. (2006) *Study on Effluents from Selected Sugar Mills in Pakistan: Potential Environmental, Health, and Economic Consequences of an Excessive Pollution Load.* Sustainable Development Policy Institute (SDPI) Islamabad, Pakistan.

Ali, I., Akbar, A., Punnapayak, H., Prasongsuk, S. and Yanwisetpakdee, B. (2014a) Seven big challenges for Pakistan – and the lessons they could teach. *The Futurist*, **48**(5): 22–26.

Ali, I., Kanhayuwa, L., Rachdawong, S. and Rakshit, S.K. (2013) Identification, phylogenetic analysis and characterization of obligate halophilic fungi isolated from a man-made solar saltern in Phetchaburi province, Thailand. *Annals of Microbiology*, **63**: 887–895.

Ali, I., Rakshit, S.K. and Kanhayuwa, L. (2011) Biohydrogen production from microalgae of *Chlorella* sp. *The International Conference on Sustainable Community Development, Khonkaen Thailand*, pp. 74–77.

Ali, I., Siwarungson, N., Punnapayak, H., Lotrakul, P., Prasongsuk, S., *et al.* (2014b) Screening of potential biotechnological applications from obligate halophilic fungi, isolated from a man-made solar saltern located in Phetchaburi province, Thailand. *Pakistan Journal of Botany*, **46**: 983–988.

Amend, J.R., Mundy, B.P. and Armold, M.T. (1993) *General, Organic, and Biological Chemistry*, 2nd Edition. Saunders Publishing, New York.

Bhattacharya, D. and Chatterjee, A. (2004) *Ecocomposites Reinforced with Cellulose Nanoparticles: An Alternative to Existing Petroleum Based Polymer Composites.* Powerpoint Presentation to State University of New York.

Brooks, L., Parravicini, V., Svardal, K., Kroiss, H. and Prendl, L. (2008) Biogas from sugar beet press pulp as substitute of fossil fuel in sugar beet factories. *Water Science and Technology*, **58**(7): 1491–1503.

Bruscino, M. (2009) Biorefineries: Fact or Fiction? *Hydrocarbon Processing,* **August**: 65–68.

Budeppa, H.B., Reddy, B.S.V., Singh, K.C. and Doss, R.G. (2008) Influence of sugarcane press mud on serum and plasma inorganic phosphorus in broilers. *Indian Journal of Animal Nutrition*, **25**(1): 93–96.

Bulluck, L.R., Brosius, M., Evanylo, G.K. and Ristaino, J.B. (2002) Organic and synthetic fertility amendments influence soil microbial, physical and chemical properties on organic and conventional farms. *Applied Soil Ecology*, **19**(2): 147–160.

Chandel, A.K., Da Silva, S.S., Carvalhoa, W. and Singh, O.V. (2012) Sugarcane bagasse and leaves: foreseeable biomass of biofuel and bio-products. *Journal of Chemical Technology and Biotechnology*, **87**(1): 11–20.

Chauvelon, G., Renard, C.M., Saulnier, L., Buleon, A. and Thibault, J.F. (1998) Preliminary study of formation of films from cellulose-enriched agricultural by-products. *Journal of Applied Polymer Science*, **68**(2): 331–337.

Cheeke, P.R. (1991) *Applied Animal Nutrition, Feeds and Feeding.* Macmillan Publishing Company, New York.

Cifuentes, R., Leon, R. de., Porres, C. and Rolz, C. (2013) Windrow composting of waste sugar cane and press mud mixtures. *Sugar Technology*, **15**(4): 406–411.

Cook, B.L., Wayne, G.F., Keithly, L. and Connolly, G. (2003) One size does not fit all: How the tobacco industry has altered cigarette design to target consumer groups with specific psychological and psychosocial needs. *Addiction*, **98**(11): 1547–1561.

Crandell, K.G., Pagan, J.D., Harris, P. and Duren, S.E. (1999) A comparison of grain, oil and beet pulp as energy sources for the exercised horse. *Equine Veterinary Journal Supplement*, **30**: 485–489.

Delgado, A.V. and Casanova, C.A. (2001) *Sugar Processing and By-products of the Sugar Industry*. Food and Agriculture Organization, Agricultural Services Bulletins.

Domınguez, J. and Edwards, C.A. (1997) Effects of stocking rate and moisture content on the growth and maturation of *Eisenia Andrei* (Oligochaeta) in pig manure. *Soil Biology and Biochemistry*, **29**: 743–746.

Dufresne, A., Cavaille, J.Y. and Vignon, M.R. (1997) Mechanical behavior of sheets prepared from sugar beet cellulose microfibrils. *Journal of Applied Polymer Science*, **64**(6): 1185–1194.

Erdener, U. and Ilhan, B. (1997) Cologne-water induced visual loss. *Ophthalmic Epidemiology*, **4**: 177–178.

Eykelbosh, A.J., Johnson, M.S., de Queiroz, E.S., Dalmagro, H.J. and Couto, E.G. (2014) Biochar from sugarcane filtercake reduces soil CO_2 emissions relative to raw residue and improves water retention and nutrient availability in a highly weathered tropical soil. *PLOS ONE*, **9**(6): e98523.

Fahim, N.F., Barsoum, B.N., Eid, A.E. and Khali, M.S. (2006) Removal of chromium (III) from tannery wastewater using activated carbon from sugar industrial waste. *Journal of Hazardous Materials*, **136**(2): 303–309.

Fang, C., Boe, K. and Angelidaki, I. (2011) Anaerobic co-digestion of desugared molasses with cow manure; focusing on sodium and potassium inhibition. *Bioresource Technology*, **102**: 1005–1011.

Ferris, C.P. and Mayne, C.S. (2006) Effects on milk production of feeding silage and sugar-beet pulp as separate components, in a mixed ration or as an ensiled blend. *Grass and Forage Science*, **49**: 229–240.

Finkenstadt, V.L., Liu, L.S. and Willett, J.L. (2007) Preparation of poly(lactic acid) and sugar beet pulp green composites. *Polymers and the Environment* **15**: 1–6.

Galves, T. (1988) *Handbook of Sugar Cane Derivatives*. Geplacea, Mexico.

Gasson, M.J., Kitamura, Y., McLauchlan, W.R., Narbad, A., Parr, A.J., *et al.* (1998) Metabolism of ferulic acid to vanillin. A bacterial gene of the Enoyl-SCoA hydratase/isomerase super family encodes an enzyme for the hydration and cleavage of a hydroxycinnamic acid SCoA thioester. *Journal of Biological Chemistry*, **273**(7): 4163–4170.

Graf, E. (1992) Antioxidant potential of ferulic acid. *Free Radical Biological Medicine*, **13**(4): 435–448.

Hanson, C.F., Frankos, V.H. and Thompson, W.O. (1989) Bioavailability of oxalic acid from spinach, sugar beet fibre and a solution of sodium oxalate consumed by female volunteers. *Food Chemistry and Toxicology*, **27**(3): 181–184.

Haynes, W.M. (ed.) (2011) *CRC Handbook of Chemistry and Physics*, 91st edition. CRC Press Inc., Boca Raton, FL. pp. 3–232.

US Environmental Protection Agency (2014) Information on sugar production process. Office of Air Quality Planning and Standards (1997) Sugarcane processing, Chapter 9. In: *Food and Agricultural Industries In AP 42: Compilation of Air Pollutant Emission*

Factors, vol. **1**, 5th edition. Research Triangle Park, NC: US EPA.: http://www.epa.gov/ttn/chief/ap42/ch09/bgdocs/b9s10-1a.pdf Accessed 28 December.

Inyanga, M., Gao, B., Pullammanappallil, P., Ding, W. and Zimmerman, A.R. (2010) Biochar from anaerobically digested sugarcane bagasse. *Bioresource Technology*, **101**: 8868–8872.

Jai, S.P.H.P., Ananth, N.B., Girish, K. and Puttaiah, E.T. (2012) Biodegradation of solid waste molasses from Aranthangi sugar industry, Tamilnadu by using *Aspergillus niger*. *International Journal of Environmental Sciences*, **1**(4): 289–293.

Javidipour, I., Vural, H., Ozen, O. and Tekin, A. (2005) Effects of inter-esterified vegetable oils and sugar beet fibre on the quality of Turkish-type salami. *International Journal of Food Science and Technology*, **40**: 177–182.

Kameyama, K., Miyamoto, T., Shiono, T. and Shinogi, Y. (2012) Influence of sugarcane bagasse-derived biochar application on nitrate leaching in calcaric dark red soil. *Journal of Environmental Quality*, **(S.S)**: 1131–1137.

Kanawade, S.M. (2014) Reuse of waste in sugar plants. *International Journal of Chemistry and Material Science*, **2**(3): 54–58.

Krishnan, C., Sousa, L.C., Jin, M., Chang, L., Dale, B.E., Balan, V. and Alkalibased, A.F.E.X. (2010) Pretreatment for the conversion of sugarcane bagasse and cane leaf residues to ethanol. *Biotechnology Bioengineering*, **107**: 441–450.

Kumar, R., Verma, D., Singh, B.L., Kumar, U. and Shweta (2010) Composting of sugar-cane waste by-products through treatment with microorganisms and subsequent vermicomposting. *Bioresource Technology*, **101**: 6707–6711.

Kumar, S. and Sahu, O. (2013) Sugar industry waste as wastewater treatment. *Journal of Environment Pollution and Human Health*, **1**(2): 21–25.

Lakdawala, M.M. and Oza, B.N. (2011) Removal of BOD contributing components from sugar industry waste water using bagasse fly ash-waste material of sugar industry. *Der Chemica Sinica*, **2**(4): 244–251.

Lehrfeld, J. (1996) Conversion of agricultural residues into cation exchange materials. *Journal of Applied Polymer Science*, **61**: 2099–2105.

Leite, R.C.C., Leal, M.R.L.V., Cortez, L.A.B., Griffin, W.M. and Scandiffio, M.I.G. (2009) Can Brazil replace 5% of the 2025 gasoline world demand with ethanol? *Energy*, **34**: 655–661.

Liu, L., Fishman, M.L., Kost, J. and Hicks, K.B. (2003) Pectin-based systems for colon-specific drug delivery via oral route. *Biomaterials*, **24**(19): 3333–3343.

Mahalaxmi, S. and Williford, C. (2012) Biochemical conversion of biomass to fuels. In: *Handbook of Climate Change Mitigation* (eds W-Y. Chen, J. Seiner, T. Suzuki and M. Lackner). Springer Science & Business Media, LLC, New York. pp. 965–999.

Martinez, L., Cilla, I., Beltran, J.A. and Roncales, P. (2006) Combined effect of modified atmosphere packaging and addition of rosemary (*Rosmarinus officinalis*), ascorbic acid, beet root (*Beta vulgaris*), and sodium lactate and their mixtures on the stability of fresh pork sausages. *Journal of Agricultural Food Chemistry*, **54**(13): 4674–4680.

McKee, L.H. and Latner, T.A. (2000) Underutilized sources of dietary fiber: A review. *Journal Plant Foods for Human Nutrition*, **55**(4): 285–304.

Mothe, C.G. and de Miranda, I.C. (2009) Characterization of sugarcane and coconut fibers by thermal analysis and FTIR. *Journal of Thermal Analysis and Calorimetry*, **97**: 661–665.

Mudoga, H.L., Yucel, H. and Kincal, N.S. (2008) Decolorization of sugar syrups using commercial and sugar beet pulp based activated carbons. *Bioresourse Technology*, **99**: 3528–3533.

Muheim, A. and Lerch, K. (1999) Towards a high-yield bioconversion of ferulic acid to vanillin. *Journal of Applied Microbiology and Biotechnology*, **51**: 456–461.

Nigam, P. (1994) Process selection for protein-enrichment: Fermentation of the sugar industry by-products molasses and sugar beet pulp. *Process Biochemistry*, **29**: 337–342.

Nigam, P. and Vogel, M. (1991) Bioconversion of sugar industry by-products-molasses and sugar beet pulp for single cell protein production by yeasts. *Biomass and Bioenergy*, **1**(6): 339–345.

Olsman, F.S., Huurdeman, C.M, Jansen, W.L., Haaksma, J., Sloet van Oldruitenborgh Oosterbaan, M.M. and Beynen, A.C. (2004) Macronutrient digestibility, nitrogen balance, plasma indicators of protein metabolism and mineral absorption in horses fed a ration rich in sugar beet pulp. *Journal of Animal Physiology and Animal Nutrition*, **88**: 321–328.

Oosterveld, A., Beldman, G. and Voragen, A.G. (2000) Oxidative cross-linking of pectic polysaccharides from sugar beet pulp. *Carbohydrate Research*, **328**(2): 199–207.

Osman, G.A., El-Tinay, A.H. and Mohamed, E.F. (2006) Biogas production from agricultural wastes. *Journal of Food Technology*, **4**: 37–39.

Ou, S. and Kwok, K-C. (2004) Ferulic Acid: pharmaceutical functions, preparation and applications in foods. *Journal of the Science of Food and Agriculture*, **84**(11): 1261–1269.

Palmgren-Karlsson, C., Jansson, A., Essen-Gustavsson, B. and Lindberg, J.E. (2002) Effect of molassed sugar beet pulp on nutrient utilisation and metabolic parameters during exercise. *Equine Veterinary Journal Supplement*, **34**: 44–49.

Pandey, A., Nigam, P. and Vogel, M. (1988) Process selection for bioconversion of sugar beet pulp into microbial protein. *Biological Wastes*, **26**: 71–75.

Pandey, A., Soccol, C.R., Nigam, P. and Soccol, V.T. (2000) Biotechnological potential of agro-industrial residues. I: sugarcane bagasse. *Bioresource Technology*, **74**: 69–80.

Perez-Armendariz, B., Loera-Corral, O., Fernandez-Linares, L., Esparza-Garcıa, F. and Rodrıguez-Vazquez, R. (2004) Biostimulation of micro-organisms from sugarcane bagasse pith for the removal of weathered hydrocarbon from soil. *Letters in Applied Microbiology*, **38**: 337–377.

Poopak, S. and Reza A.R. (2012) Environmental Benefit of using bagasse in paper production – a case study of LCA in Iran. in global warming – impacts and future perspectives. InTech Publishing, UK, pp. 205–222.

Raghukumar, C. and Rivonkar, G. (2001) Decolorization of molasses spent wash by the white-rot fungus *Flavodon flavus*, isolated from a marine habitat. *Applied Microbiology and Biotechnology*, **55**(4): 510–514.

Rainey, T.J. (2012) A comparison between highly depithed and conventionally depithed bagasse pulp. *Appita Journal (Technology, Innovation, Manufacturing, Environment)*, **65**(2): 178–183.

Rainey, T.J. and Clark, N.B. (2004) An overview of bagasse as a resource for the Australian paper industry. *International Sugar Journal*, **1271**: 608–611.

Ralet, M.C., Cabrera, J.C., Bonnin, E., Quemener, B., Hellin, P. and Thibault, J.F. (2005) Mapping sugar beet pectin acetylation pattern. *Phytochemistry*, **66**(15): 1832–1843.

Ranjhan, S.K. (1993) *Animal Nutrition in Tropics*, 3rd edition. New Delhi.

Rolz C., León, R. de, Cifuentes, R. and Porres, C. (2010) Windrow composting of sugarcane and coffee byproducts. *Sugar Technology*, **12**(1): 15–20.

Sarkar, K., Ray, B., Banerjee, R., Saha, S., Roy, S. and Chatterjee, S. (2014) Poly-β-hydroxybutyrate (bio-plastic) production utilizing waste effluent of a sugar industry. *IOSR Journal of Environmental Science, Toxicology and Food Technology*, **8**(4)**V.II**: 26–31.

Sbrana, C., Fernandes, E.A.N., Marson, J.C., Bacchi, M.A., Sarries, G.A. and Reyes, A.E.L. (2013) Clean energy from sugarcane bagasse: Quality evaluation by neutron activation analysis. *Journal of Radioanalytical and Nuclear Chemistry*, **296**: 939–942.

Sindhu, A.A., Khan, M.A., Nisa, M.U. and Sarwar, M. (2002) Agro-industrial by-products as a potential source of livestock feed. *International Journal of Agriculture and Biology*, **4**(2): 307–310.

Singh, P, Suman, A, Tiwari, P, Arya, N, Gaur, A. and Shrivastava, A.K. (2008) Biological pretreatment of sugarcane trash for its conversion to fermentable sugars. *World Journal of Microbiology and Biotechnology*, **24**: 667–673.

Smith, J.L. and Workman, J.P. (2004) *Alcohol for Motor Fuels Colorado State University Extension Publication* 5.010. http://www.ext.colostate.edu/PUBS/FARMMGT/05010.html. Last accessed 13 December 2014.

Spagnuolo, M., Crecchio, C., Pizzigallo, M.D. and Ruggiero, P. (1999) Fractionation of sugar beet pulp into pectin, cellulose, and arabinose by arabinases combined with ultrafiltration. *Biotechnology and Bioengineering*, **64**(6): 685–691.

Srinivasan, M., Sudheer, A.R., Pillai, K.R., Kumar, P.R., Sudhakaran, P.R. and Menon, V.P. (2006) Influence of ferulic acid on gamma-radiation induced DNA damage, lipid peroxidation and antioxidant status in primary culture of isolated rat hepatocytes. *Toxicology*, **228**: 249–258.

Stanmore, B.R. (2010) Generation of energy from sugarcane bagasse by thermal treatment. *Waste Biomass Valorization*, **1**: 77–89.

Suma, N., Reddy, B.S.V., Gloridoss, R.G., Rao, N.R., Singh, K.C., *et al.* (2007) Egg quality traits of layers influenced by supplementation of different levels of sugarcane press residue. *International Journal of Poultry Science*, **6**(2): 102–106.

Sun, Q. and Yang, L. (2003) The adsorption of basic dyes from aqueous solution on modified peat-resin particle. *Water Research*, **37**: 1535–1544.

Suresh, B.N. and Reddy, B.S.V. (2011) Dried sugarcane press residue as a potential feed ingredient source of nutrients for poultry. *Asian-Australian Journal of Animal Science*, **24**(11): 1595–1600.

Togrul, H. and Arslan, N. (2004) Carboxymethyl cellulose from sugar beet pulp cellulose as a hydrophilic polymer in coating of mandarin. *Journal of Food Engineering*, **62**(3): 271–279.

Verma, D.N. (1997) *A Text Book of Animal Nutrition*, 1st edition, R. 814. New Rajinder Nagar, New Delhi.

Vijayakumar, S., Arunprakash, C., Gobinath, R., Poornimasree, S.R. and Shidhardhan, C.K. (2014) Preliminary adsorption studies on textile industry waste water using sugarcane bagasse as an adsorbant. *The International Journal of Science and Technology*, **2**(3): 55–58.

Watts and Associates, Inc. (2007) Alternative sugar beet by-product uses research and feasibility analysis, http://www.ams.usda.gov/AMSv1.0/getfile?dDocName=STELPRDC5058621

Yadav, R.L. and Solomon, S. (2006) Potential of developing sugarcane by-product based industries in India. *Sugar Technology*, **8**(2-3): 104–111.

Yang, L., Chu, J.S. and Fix, J.A. (2002) Colon-specific drug delivery: New approaches and *in vitro/in vivo* evaluation. *International Journal of Pharmacology*, **235**(1–2): 1–15.

Zhong, C., Hui-Gai, L., Bin-Feng, W., Xu, D. and Yu, R.Q. (1996) A perfume odour-sensing system using an array of piezoelectric crystal sensors with plasticized PVC coatings fresenius. *Journal of Analytical Chemistry*, **355**: 194–199.

Zhu, P., Ren, J., Wang, L., Zhang, X. and Yang, X. (2007) Long-term fertilization impacts on corn yields and soil organic matter on a clay-loam soil in Northeast China. *Journal of Plant Nutrition and Soil Sciences*, **170**: 219–223.

Zykwinska, A.W., Ralet, M.C., Garnier, C.D. and Thibault, J.F. (2005) Evidence for *in vitro* binding of pectin side chains to cellulose. *Plant Physiology*, **139**(1): 397–340.

22
Regulatory and Legislative Issues for Food Waste Utilization

Lavaraj Devkota[1], Didier Montet[3], and Anil Kumar Anal[2]

[1]*Department of Chemical Engineering, Monash University, Clayton, Australia*
[2]*Asian Institute of Technology, Klong Luang, Thailand*
[3]*UMR Qualisud, CIRAD, Montpellier, France*

22.1 Introduction

It is estimated that about 795 million people of the 7.3 billion people in the world, or one in nine, were suffering from chronic under-nourishment in 2014–2016. Almost all these hungry people, 780 million, live in developing countries, representing 12.9%, or one in eight, of the population of developing counties. Also, there are 11 million people under-nourished in developed countries (IFAD, 2015). On the other hand, global food waste is 1.3 billion tonnes per year, which is one-third of total food production. In wealthy countries, every year, 222 million tonnes of food is wasted, which is almost equal to the entire net food production of sub-Saharan Africa (230 million tonnes) (FAO, 2013). These two contrary statistics prove the urgency and necessity of addressing the issue of food waste utilization.

Food waste does not have a uni-dimensional impact but comes with social, cultural, economic and environmental adversities. The environmental and social impacts are the most pronounced. increasing global hunger, food insecurity, greenhouse gas emission and global warming, which have a greater impact in the long term. Squandering of resources means that the economic consequences will have a major impact. Waste management becomes a big headache for processing industries, as it involves additional costs. This has focused intense policy activity on sustainable production, processing and manufacturing of food products. Food processing and consumption has a most important role to play in determining the environmental impacts of resource use, as identified by key policy and research reports (Martindale *et al.*, 2014). So the legislation and regulations for valorization of food waste is of utmost importance.

Food waste often contains many valuable components, among which phenolic compounds, antioxidants, pectin, proteins, flavonoids, dietary fibers, anthocyanin and enzymes are commonly of higher interest to food scientists and researchers

Food Processing By-Products and their Utilization, First Edition. Edited by Anil Kumar Anal.

(Galanakis *et al.*, 2015). But recovery of them and use in food may be a challenging job to accomplish, mainly for two reasons, i.e. stability of these components in processing parameters and technological difficulty in large-scale production. Other prominent factors may be dealing with contamination and safety of these components.

Waste in the food industry is highly complex and characterized by a high ratio of product-specific waste. These wastes mainly comprise of organic residues and can rarely be altered or avoided. Both utilization and disposal of product-specific waste is difficult, due to its biological instability, potentially pathogenic nature, high water content, potential for rapid auto-oxidation, and high levels of enzymatic activity. The complexities of valorization of food waste are not just limited to the complex nature of the food itself, but are also dependent on lack of effective policy. So framing and harmonization of legal and regulatory standards should be of interest to the stakeholders. In this chapter we deal with the prevailing rules and regulations in valorization of food waste, the challenges and opportunities in utilization of food waste as value-added products in food, and the scope for further improvements in statutory rules, codes of conduct and guidelines.

22.2 Possible Mitigation Measures for Food Processing Wastes

Management of industrial waste has now become a benchmark for food industries with enforcement of ISO 14001 (Environmental Management System). As introduced earlier, food processing waste is not merely trash, but instead is complex and at the same time contains highly valuable food components. To meet the demand of the growing population, urbanization and consumer needs, food industries have grown tremendously in recent years, so the production of waste is unambiguous. This had led to serious economic loss and is of much concern to environmentalists and environmental engineers. Food waste, being complex and organic in nature, under aerobic fermentation and putrefaction, results in spoilage odor. Solid waste may not convert completely into compost; liquid waste may flow into water sources and contaminate aquatic life. Broadly, mitigation of food processing waste may be carried out by the following.

22.2.1 Composting and Land Spreading of Food Processing Waste

Composting and land spreading allows naturally occurring bacteria and fungi to decompose organic matter present in food waste, converting it into useful soil additives. Food processing waste from fruit and vegetable industries can be composted to manure or land spread. Liquid waste from these industries may be directly added to the ground by spreading or injection. However, waste from meat and dairy industries may contain fat, grease and parts of carcasses, which may be unsuitable for composting or spreading, but could be used to feed livestock.

Although variation exists in laws and regulation among countries for land spreading and composting, much research is done at the individual level (Hershaft, 1972). Small- and medium-sized waste producers may use on-site composting options, which has to be done in a controlled environment and requires active management. Close monitoring is needed to ensure favorable condition for bacteria and fungi to function. Similarly, land spreading of organic food waste is helpful as the biologically breakdown provides necessary nutrients to plants.

22.2.2 Feeding Food Processing Waste to Livestock

Livestock and poultry feed needs to be formulated to meet the specific nutrient requirements of each species, consistently for their regular growth and development. Two key concerns to be considered in using food waste as partial ingredients or major components are efficacy and safety of these ingredients. The debate remains as to whether proper use of food waste can offer good value to livestock and poultry farmers and other potential users such as aquaculture and companion animal food industries. Food wastes have been used in animal feed from ancient times. These wastes may be used as such or are passed through simple processes such as drying, extruding or pelleting. Cattle performance was acceptable in terms of digestibility and palatability when processed food waste was incorporated into their diet (McClure *et al.*, 1970).

Feeding to livestock is a better option to land spreading and composting in many ways, as it may be fed as such, may be in solid or liquid form and used in any season. Food waste that can be included as partial or major ingredients may include solid wastes from meat industries, vegetable and fruit processing industries, slaughter-houses, spent off grains in alcohol industries, sugar industries, eateries etc. Often, blood, offal, beef carcass and poultry industries, together with grain supplements, is used in livestock feed.

Despite being ample nutrients, these wastes also contain potential pathogens, high moisture, and physical and chemical contaminants, which may pose a threat to health. There have been cases of outbreaks of diseases associated with uncooked waste. The main concerns are from foreign animal diseases such Hog cholera, foot and mouth diseases, African swine fever, and swine vesicular disease, which are significantly due to domestic pathogens like *Salmonella, Campylobacter, Trichinella* and *Toxoplasm* (Westendorf, 2000). So any feed producer who uses food waste in feed formulation should abide by the prevailing rules and regulations. This urges strong enforcement of prevailing regulations and framing new ones to bridge food safety and waste management.

22.2.3 Utilization of Food Processing Waste as Feed/Food Supplement through Value Addition or Modification in Processing Method

Today, a large number of scientist and researchers are working in the sector of food waste recovery, value addition and innovations in food sustainably. Food waste consists of valuable compounds such as antioxidants, dietary fibers, flavonoids, anthocyanin, proteins and enzymes, and valuable chemicals (Jayathilakan *et al.*, 2012; Schieber *et al.*, 2001; Waldron, 2007). Researchers also have proved that the recovery of these target compounds and their re-utilization is possible to use in the human food chain as additives, supplements, fortification and other minor ingredients (Baiano, 2014; Roselló-Soto *et al.*, 2015; Smithers, 2008). Even though, there has been much research, many patents and a plethora of indulgence in this field, not much have been achieved, mainly due to five major obstacles:

1. Low energy efficiency of conventional extraction processes and higher cost involvement in extraction of compounds, which exist in lower concentrations than in whole food.
2. Conventional extraction processes often are carried out in non-food grade solvents, which restrict further use of the extract in the food system.
3. Conventional processes may involve processes like spray drying and thermal processing, which may lead to instability and cause deterioration in the concentration and quality of target compounds.
4. High moisture content, the organic nature, potential source of pathogens, and presence of physical and chemical contaminants make food waste difficult to deal with.
5. Consumer's perception towards the safety of products from waste has been and will be skeptical.

In an FAO study, 54% of the world's food wastage occurs 'upstream' during production, post-harvest handling and storage, while 46% of it happens 'downstream' at the processing, distribution and consumption stages. So we need significant changes in both downstream and upstream processing strategies, which will be successful through innovation and creative process designs, well-planned systems and effective execution. There prevail huge opportunities and challenges in utilization of valuable components of food waste in the human food system. Regulatory and legislative issues in utilization of food waste will be discussed in a later part of this chapter.

22.2.4 Food Processing Source Reduction and Waste Management

Although unavoidable in most cases, food waste can be minimized by effective management of processing parameters and efficient handling. This is the most effective way of decreasing the volume of waste, which in turn is directly related to cost reduction. This may be achieved by simple modification of processing technology, cleaning procedures and maximum utilization of prevailing resources. For example, dedicating mixing lines to a specific product reduces product loss and clean-up requirements over product change, pre-cleaning using dry procedures, use of high-pressure cleaning, and careful handling to minimize spills, leaks etc.

Since valorization as feed and food supplementation is highly limited to a few products, reduction of waste is the most appropriate available mitigation strategy. Recent years have shown much interest from all stakeholders in reduction of food processing waste (Godfray *et al.*, 2010; Hodges *et al.*, 2011). In a work by Parfitt *et al.* (2010), to quantify the food waste in the supply chain and explain potential for change until 2050, the authors clearly put forward the way to tackle increasing food waste. Use of recent innovations to increase shelf life through packaging, maintaining the ideal environmental conditions throughout the supply chain, and consumer awareness towards handling and storage, are some of the methods proposed in their work.

22.3 Impact of Waste Disposal on Environment and Human Health

A significant aspect of the ecosystem is its sustainability in utilization of natural resources and re-use of waste and by-products, which minimizes over-exploitation of natural resources and depletion of environmental resources. But the magnitude of waste production is so alarming that it has disturbed the balance of the ecosystem and the fundamental requirements for survival of human life such as potable water, fresh air, unadulterated food, shelter and constant and continuous climate change effects.

Every hour, about 150 000 tonnes of solid and liquid waste from food industries are generated, whose proper segregation and management will not only have a significant economic loss but also deteriorate the environment we live in and finally enter the food system. Food waste by nature contains a large segment of bio-degradable components, which on anaerobic fermentation result in putrid odors and noxious gases that pollute the air. This creates a favorable environment for various pathogenic microorganisms and vectors to cause and transmit communicable disease. Environment degradation, especially water resources contamination, is furthered by leaching out of toxic nitrates, inorganic salts, pesticide residues and other chemical or biological pollutants. A descriptive analysis has been done in a report by the FAO, describing impacts of food loss on environment, economy and ecosystem (FAO, 2013).

The food chain follows a unique and complex supply chain from farm to fork; various activities occur along this supply chain before the food reaches our plates. Thailand is one of the top producers and exporter of food products, including processed food. More than 10 000 Thai food industries account for 23% of the national GDP and the country has gained a reputation of proudly being called the 'kitchen of the world'. Of course, these large industries and food production will produce enormous food waste and processing waste, which finally have impacts on various facets of human life and the ecosystem (Thailand Board of Investment, 2015).

Segregation of food waste from other solid waste is an important aspect of food waste management, and the inability to do so will make municipal waste non-degradable, while food waste undergoes biodegradation via anaerobic fermentation, which in turn results in favorable conditions for spoilage and pathogenic bacteria and fungi to proliferate, thus producing putrid foul odors. Such waste again cannot be land spread or composted, so food waste and other waste should be separately collected and managed. So reduction of waste, proper disposal or use in feed should be done to manage the food waste, whatever its type. Despite huge economic loss and social and cultural impact, waste degrades the condition of the planet we all share, which in turn affects the health and livelihood of all creatures on Earth, so this has to be dealt with concern and care, as we cannot afford the consequences thereof.

22.4 Need of Legislative and Regulatory Guidelines

Food consumers and stakeholders are highly concerned about the quality and safety of the food they consume. Regulations are promulgated under enacted law by a community towards safeguarding themselves and for maintenance of peace and sustainable utilization of available resources. Legislations, codes of conduct, guidelines, laws and regulation are all intended towards safety and guarantee of the rights of people. In the case of food waste utilization or management, stakeholders should also abide by the same rules and guidelines, so it does not intrude on the rights of others, such as their right to eat safe food, to breathe clean air, to enjoy good health etc.

Today's world is modernizing very rapidly in terms of urbanization, industrialization and technological advancements. These achievements are not coming for free but in exchange for the long-term consequences of climate change, global warming and extinction of many plant and animal species. Enormous waste, mostly solid, produced every day, has led to serious consequences for both land and aquatic life. Developed and developing countries thus needed regulations for effective management of these wastes, especially where the production of waste is particularly high. The Environmental Protection Act has been framed, which encourages all stakeholders to manage and think of their environment through effective waste treatment.

Stronger and more effective regulations are needed when food waste has to be turned into feed for animals or food for human consumption, with the safety of consumers as the center point. Extensive laboratory research, transfer to pilot plant, scale-up production, together with strong policy, safety standards and reliability, will only make food utilization policy stronger and more achievable.

22.5 Concept of Policies, Legislations, Code of Conduct and Regulations for Food Waste Utilization

Policy, as defined by the *Cambridge English Dictionary*, is a set of ideas or a plan of what to do in particular situations, which has been officially agreed by a group of people, an organization, a government or a political party. It involves three fundamental principles of planning, setting goals and developing concrete regulations. According to Chandrasekaran (2012b), policy development may be carried out on three levels:

1. *Supra-national level policies*: a document of agreement developed through conventions, discussions or programs towards resolving challenges at the international level.
2. *National level policies*: these are generally government documents of a country based on its needs and interests to address their social, economic or environmental issues. Their alignment is towards global policy and is often ambitious.
3. *Sub-national level policies*: these are policies developed in a specific region, state or at a municipal level. Although they are framed to address the issues of that specific region, it follows the national policy.

On the basis of policies decided, laws and acts are passed by the legislative body. The enforcing body then prepares a set of regulations to implement the law passed. Regulations are broader and wider than the law, as it comprise of detailed explanation of the law and much more practical issues. Regulation for food safety for example covers standard of building design for the food industry, equipment design, production standard, employee qualifications and so on.

Codes of conduct are a set of rules outlining the social norms and rules and responsibilities of, or proper practices for, an individual, party or organization. Guidelines are a set of standards devised for smooth function of routine activities and are not abided by any rules, but are intended to be followed on ethical and voluntary principles.

22.6 Prevailing Legislation and Regulatory Guidelines for Food Waste Utilization

The production, processing, distribution, retail, packaging and labeling of food stuffs are governed by a mass of laws, regulations, codes of practice and guidelines. Food waste utilization policies are more centered towards reduction of waste, management of waste, and utilization in animal feed. Although many patented technologies for food waste utilization for human food and animal exist and few are in operation as well (Galanakis *et al.*, 2015), policy on food waste utilization as a source of valuable compounds for human consumption are highly limited.

The following section tries to briefly overview the prevailing supra national and national policy on food waste utilization.

22.6.1 European Union

The EU is a supranational body that promulgates policy and standards that are followed by most of its member countries. In the EU, food waste is expected to rise to about 126 million tonnes a year by 2020, from a baseline of 89 million tonnes in 2006, unless action is taken to halt this trend. Although regarded no longer sustainable, food waste management policies in its member countries are followed as the European Commission in 2008 formulated 'The Landfill Directive' and 'The Waste Directive legislations'. The waste management legislations in Europe, Asian and the USA are well explained by Chandasekaran (2012a) in his book on valorization of food processing by-products.

The Waste Framework Directive highlights the relative priority of the different methods of managing waste, a concept that will affect future legislation and policies. In particular, the waste management hierarchy includes, in descending order of priority, the following actions: waste prevention; re-use; recycling; other recovery (e.g. energy recovery) Figure 22.1; and lastly, disposal. The bio-economy strategy is one of the major policies among food waste prevention policies in EU 9. According to the European Commission, the European nations are obliged to change their approach to production, consumption, processing, storage, recycling and disposal of biological resources. Recently, the EU and Member States have made commitments towards meeting the Sustainable Development Goals (SDG) adopted in September 2015,

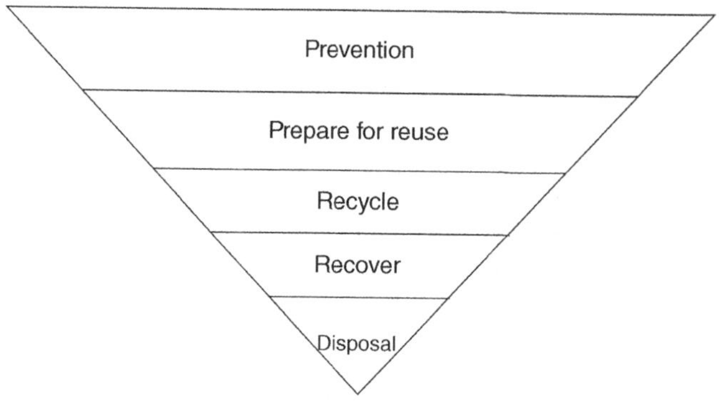

Figure 22.1 The waste hierarchy (adapted from the European Parliament Council, 2008).

including a target to halve per capita food waste at the retail and consumer level by 2030, and reduce food losses along the food production and supply chains.

Legal requirements are the baseline of the possibility of use of food waste as components of food. In order to assess the permissibility of a product used as an input substance (i.e. a substance used as raw material for food processing), this substance has to be qualified and the intended use of the output substance must be known, since they are determined by the legislation applied (Heeres and Waldron, 2009). If the output is a food/food ingredient, the input must be a food/food ingredient in agreement with Article 2 of EC Regulation No, 178/2002. According to this article, food is 'any substance or product, whether processed, partially processed or unprocessed, intended to be, or reasonably expected to be ingested by humans' and 'includes drink, chewing gum and any substance, including water, intentionally incorporated into the food during its manufacture, preparation or treatment.' However, the definition of 'food' does not include substances such as feed, live animals unless they are prepared for placing on the market for human consumption, plants prior to harvesting, medicinal products, cosmetics, tobacco and tobacco products, narcotic or psychotropic substances, residues and contaminants. The qualification of a certain substance as food has great implications, since it means that generic and specific food legislation must be applied. EC Regulation No, 258/97 (1997) also allows input and output material to be called 'novel'.

Novel foods are foods and food ingredients that were not used before 15 May 1997 within the European Community for human consumption to a significant degree. In order to ensure the protection of human health, novel foods must undergo a safety assessment before their placement on the EU market. In particular, whoever is aimed to place a substance as a novel food has to submit a request to the Member State in which the product will be sold for the first time and to wait for their authorization. The request must be submitted according to EC Recommendation No, 97/618.

Food waste utilization is also governed by a set of legislation and statutory directives related to food safety. The EU sets legal standards and requirements for contaminants of food, which may be pesticides, drug residues, or other physical and biological agents. The basic principles of EU legislation on contaminants of food, laid down in Council Regulation 315/93/EEC of 8 February 1993, require that:

1. Food containing contaminants to an amount unacceptable for human health and in particular at a toxicological level, shall not be placed on the market;
2. Contaminant levels shall be kept as low as reasonably achievable following recommended good working practices (ALARA-principle); and
3. The Commission may establish the maximum tolerance for certain contaminants that can threaten public health. Some of the perceived legislative obstacles and proposals for overcoming these barriers are presented in a study carried out by the FAO, which may also be applied in reduction of food waste and utilization.

Although some efforts have been made in recent years, food waste utilization policy in EU and member countries focuses on reduction or prevention of food waste and proper waste management, rather than on utilization of waste as a source of valuable components for food and feed. Food waste in the EU is broadly subjected to non-hazardous waste and same management principles and controls are applied as for other types of non-hazardous waste. Notable effort however has been made in the case of animal by-product wastes, which are subject to uniform and stringent controls with respect to their storage, transport, treatment and disposal, principally through Regulation (EC) No. 1069/2009 (which replaces Regulation (EC) No. 1774/2002 and others), so as to prevent associated risks to animal and public health (except when these wastes are intended for incineration, land-filling or use in biogas or composting plants).

Certainly some technical obstacles are present and their solutions are also on the way to development, with much effort being made by governments and stakeholders in framing strong legal policies to encourage food waste utilization. One great issue under discussion recently is the use of food waste as livestock feed, e.g. for swine feeding. IN 2002, The EU banned the use of food waste as feed after the UK foot and mouth disease epidemic, which was believed to be started by illegal feeding of uncooked food waste to pigs. But swill feeding is efficiently in practice in many nations such Japan, South Korea, Thailand and Taiwan (zu Ermgassen *et al.*, 2016). If well-regulated, this food waste can be used to convert to livestock feed and reduce economic loss and environmental depletion. The EU waste directive set a target that no biodegradable waste (including food wastes) be land-filled by 2025 (EC, 2014). This may also affect food waste utilization policy and regulations in days to come.

22.6.2 The USA

Documents revealing food waste utilization as components of food is limited in the USA, like elsewhere in the developed countries, most probably owing to consumer health and safety. The FDA has a primary responsibility in the federal government to ensure food safety and have enforced the Federal Food, Drug and Cosmetic Act (FFDCA) and the related food safety aspect of Public Health Service Act (PHSA 42 US C). This generic food safety law may be applicable for any input into food processing. The HACCP concept for example was developed by the USA, which has been proven as best in identifying and preventing physical, biological or chemical hazards that may otherwise pose risks to human health.

Food waste can be used in animal and poultry food, which is well-regulated by the FFDCA. The FFDCA (Section 201(f)) defines food as ‘articles used for food or drink for man or other animals and articles used for components of any such article.’ Therefore, any waste product, garbage, by-product, or co-product, regardless of source, that is intended to be used as a feed ingredient or to become part of an ingredient or feed, is considered a ‘food’ under the FFDCA and thus subject to regulation by the FDA. Section 402 (a) furthermore regulates adulteration of food and has provisions for several types of adulterants. The most appropriate one for food waste, by-products and co-products are (a) 1 and (a) 3. There are contradictions among various States in the use of food waste as feed for animals.

The legal status for safe (GRAS) is one measure to ensure safety of any food product. This safety status may be acquired by ensuring sufficient safety data are available and being judged by an unbiased panel of experts. If such data exist and the panel reach an agreement, the product will be allowed a GRAS standard and this can be further used in food or feed under approval from governing bodies. The GRAS certified product may also be assessed and monitored by the FDA on a timely basis and based on the information obtained may be or may no longer be regarded as safe.

22.6.3 Asian Region

The Asian region is now the fastest growing economy on Earth. With the tremendous rate of urbanization industrialization, the amount of waste produced will become much higher. This region is also of higher significance in terms of food waste, as almost half a billion of its population suffer from hunger and about half of the food produced will be wasted. Food waste as animal feed is mostly common in Asian countries with diverse cultural, traditional and geographical diversity. Japan, South Korea and Taiwan have regulated food waste feeding to pigs and have benefited mostly in terms of economy and sustainability. While the feeding of uncooked meat wastes to pigs can transmit diseases, including foot-and-mouth and the classic swine fever, appropriate heat treatment de-activates these viruses and renders food waste safe for animal feed (García *et al.*, 2005). In these countries, the industry is tightly regulated: the heat treatment of food waste is carried out by registered 'Ecofeed' manufacturers.

Japan has been making unprecedented efforts to reduce pressure on landfills by reviving its national policy on waste disposal. The Promotion of Utilization of Recyclable Food Waste Act (Food Recycling Law) of 2001 requires food industries to reduce their food waste and to promote recycling. The target food industries or food waste emitters that waste more than a 100 tonnes of food per year are food manufacturing, wholesale, retail and food services. The law was revised in 2007, which allows recycling facilities, which also collect food waste, to transport food waste across municipal borders. In addition, the law requires food waste emitters to report the amount of recycled food waste, and to purchase farm products that use food waste-derived products, such as compost and animal feed. The cyclical use of food waste is called a 'recycling loop', and this loop is completed when the farm products are purchased by food waste emitters (Takata *et al.*, 2012).

22.7 Possible Amendments and Scope for the Development of New Regulations on Food Waste Utilization

Governments, scientists, regulators, industries and other stakeholders continuously strive to close the ambiguities and inadequacies in certain sections of regulatory guidelines. Continuous improvements and amendments are carried out to ensure the regulations and guidelines are compatible with the latest technology trend and developments with time.

Very few regulations and guidelines exist in the sector of utilization of food waste as raw material, so there is still is a long way to go. The generic food safety law can be applied to food waste also, but this should be by mutual harmonization of these standards worldwide. The way food waste has been defined to date is confusing and unclear, so food waste should be segregated from other solid waste or municipal waste and treated accordingly. Significant amounts of research have been done and are still being done for recovery of various essential components from so-called food waste, but due to lack of regulatory and legal provision, most of them are limited to scientific papers and patents.

To list the possible changes and future regulatory provisions, the following points can be summarized as follows:

(a) There is a need for uniformity in the way food waste has been and shall be defined. Food waste has to be distinguished from other municipal waste and regulation has to be made accordingly for its valorization. Food waste should be categorized according to its source so that waste from these food sources can be treated differently.
(b) A strong and new regulation regarding food waste utilization as a component of food should be framed. International harmonization has to occur, based on the strong fundamentals of technical and safety data available. This will change consumer perception and guarantee safety aspects of such food.
(c) Industries and stakeholders should make their own standards in reducing waste and recovering essentials from waste.

Any regulation and legal documents should be dynamic and open, as technological advancements occur every day. Sustainability will be achieved if well-regulated recovery strategies for food waste are to be implemented, both in terms of economy and environment.

22.8 Use of Recent Advancements in Food Waste Utilization

Biotechnology has gained a larger degree of attention than it has ever received earlier. The advancement in biotechnology has definitely affected the way stakeholders see waste, especially food waste. The surging and fast-moving developments in genetic engineering applications, high-pressure technology and other inter-disciplinary technological advancements have provided lots of input to the scientists and researchers working in the food waste valorization field. Ample research has been done on application of these technologies in recovery of essential and valuable components from food waste (Barba *et al.*, 2015; Galanakis, 2013; Galanakis *et al.*, 2013, 2015a.b; Roselló-Soto *et al.*, 2015).

Genetic engineering in particular has been under debate since its first marketed application in food in 1994. There still remain differences in the regulations and guidelines among different countries regarding genetically-modified food, which have to be finalised. More genetically-modified foods may enter the food system; under such circumstances the stakeholders, particularly the food processers, should have a clear vision to trace the probable consequences along the supply chain. Harmonization of the existing regulation will be of great importance to regulate the GM food in the food system.

Food waste utilization policies should be based on strong technological baselines. Use of high pressure technology, pulse electric field and ultrasound assisted extraction can be of great importance, as these avoid the use of chemicals which may be toxic to use in food. The amount of food products on the market with application of high-pressure processing technology suggests the potential for its use in food waste recovery. Ultra sonication on the other hand has been extensively used in the extraction process of proteins and enzymes from food waste. In recent work at the Asian Institute of Technology, the effects of ultra-sonication on enzymatic hydrolysate of egg shell membrane proteins were studied and found to be of superior quality (Jain and Anal, 2016). Pulsed electric field treatment for extraction of valuable components from food waste can be a good option, as it has been proved successful in many instances and patents have been submitted. An electric field generated by applying voltage causes more electroporation and makes extraction easier and more efficient.

One more alternative is the use of sub-critical water, like the described by Mazza and Cacace (2011), for the extraction of phytochemicals from plant materials. The technology uses a water supply in a high-pressure, temperature-controlled extraction vessel with some basic accessories. Lastly, there have been recent discussions on use of so-called green solvents for extraction of valuable components. Use of edible solvent of various mixing ratios, temperatures and pressures were used by Todd (2000) in extraction and concentration of carotenoid pigment from paprika, red pepper and plants of the genus *Capsicum*.

22.9 Conclusion

Food waste should not be defined in the way it has been to date and cannot be categorized as the same as other municipal waste. To maintain sustainability, efforts need to be centered towards zero waste or exploitation of material before it actually becomes waste. Nevertheless, despite tremendous interest and on-going research, the re-use of components of co-products is still in its infancy, due to the necessity to invest in research, new recovery technologies, and/or new production lines, whose costs are still higher than landfill taxes currently in force and composting costs. The consumer's perception towards food recovered from waste, and the safety issues of the prevailing extraction techniques among other hidden regulatory obstacles, persist.

In order to utilize food waste as a component of food, again much work has to be done. Rigorous discussions on policy level and regulation framing bodies have to be conducted. Exploitation of the co-products can take advantage of greater collaboration between research institutes and industry, adaptation and use of already existing technologies and plants for the recovery of biomolecules from co-products, and the application of economies of scale needed for the treatment of large volumes of biomass characterized by low specific economic values. Regulation should be based on strong fundamentals of technical knowledge, research and data and they require a mobile document that can be updated and amended as per technological demand.

References

Baiano, A. (2014) Recovery of biomolecules from food wastes – A review. *Molecules*, **19**(9): 14821–14842.

Barba, F.J., Galanakis, C.M., Esteve, M.J., Frigola, A. and Vorobiev, E. (2015) Potential use of pulsed electric technologies and ultrasounds to improve the recovery of high-added value compounds from blackberries. *Journal of Food Engineering*, **167**(A): 38–44.

Chandrasekaran, M. (2012a) Regulatory issues and concerns of valorization of food processing by-products. In: *Valorization of Food Processing By-Products*. CRC Press, New York.

Chandrasekaran, M. (2012b) *Valorization of Food Processing By-products*. CRC Press, New York.

Regulation (EC) No 258/97 of the European Parliament and of the Council of 27 January 1997 concerning novel foods and novel food ingredients.

Directive 2008/1/EC of the European Parliament and of the Council of 15 January 2008 Concerning Integrated Pollution Prevention and Control.

FAO (2013) Food wastage footprint-Impacts on natural resources: Summary report: Food and Agriculture Organization of the United Nations, Rome.

Galanakis, C.M. (2013) Emerging technologies for the production of nutraceuticals from agricultural by-products: A viewpoint of opportunities and challenges. *Food and Bioproducts Processing*, **91**(4): 575–579.

Galanakis, C.M., Markouli, E. and Gekas, V. (2013) Recovery and fractionation of different phenolic classes from winery sludge using ultrafiltration. *Separation and Purification Technology*, **107**: 245–251.

Galanakis, C.M., Barba, F.J. and Prasad, K.N. (2015a) Cost and safety issues of emerging technologies against conventional techniques. In: *Food Waste Recovery*. Academic Press, San Diego, pp. 321–336.

Galanakis, C.M., Martinez-Saez, N., del Castillo, M.D., Barba, F.J. and Mitropoulou, V.S. (2015b) Patented and commercialized applications, Chapter 15. In: *Food Waste Recovery, Processing Technologies and Industrial Techniques*. Academic Press, San Diego.

García, A.J., Esteban, M.B., Márquez, M.C. and Ramos, P. (2005) Biodegradable municipal solid waste: Characterization and potential use as animal feedstuffs. *Waste Management*, **25**(8): 780–787.

Godfray, H.C.J., Beddington, J.R., Crute, I.R., Haddad, L., Lawrence, D., *et al.* (2010). Food security: the challenge of feeding 9 billion people. *Science*, **327**(5967): 812–818.

Heeres, H. and Waldron, K. (2009) Regulatory requirements for valorisation of food-chain co-products in the European Union. *Handbook of Waste Management and Co-Product Recovery in Food Processing*, **2**: 33–55.

Hershaft, A. (1972) Solid waste treatment technology. *Environmental Science and Technology*, **6**(5): 412–421.

Hodges, R.J., Buzby, J.C. and Bennett, B. (2011) Postharvest losses and waste in developed and less developed countries: Opportunities to improve resource use. *The Journal of Agricultural Science*, **149**(S1): 37–45.

IFAD (2015) *The State of Food Insecurity in the World*. WFP, Rome.

Jain, S. and Anal, A.K. (2016) Optimization of extraction of functional protein hydrolysates from chicken egg shell membrane (ESM) by ultrasonic assisted extraction (UAE) and enzymatic hydrolysis. *LWT – Food Science and Technology*. doi: https://dx.doi.org/10.1016/j.lwt.2016.01.057

Jayathilakan, K., Sultana, K., Radhakrishna, K. and Bawa, A. (2012) Utilization of by-products and waste materials from meat, poultry and fish processing industries: A review. *Journal of Food Science and Technology*, **49**(3): 278–293.

Martindale, W., Finnigan, T. and Needham, L. (2014) Current concepts and applied research in sustainable food processing. InL *Sustainable Food Processing*, John Wiley & Sons, Ltd, pp. 9–38.

Mazza, G. and Cacace, J.E. (2011) Extraction of phytochemicals: Google Patents.

McClure, K., Klosterman, E. and Johnson, R. (1970) Feeding garbage to cattle and sheep. *Ohio Report on Research and Developopment in Agriculture Home Economics and Natural Resources*, vol. **55**. Ohio Agriculture Research Development Center, Wooster, US, pp. 78–79.

Parfitt, J., Barthel, M. and Macnaughton, S. (2010) Food waste within food supply chains: quantification and potential for change to 2050. *Philosophical Transactions of the Royal Society of London B: Biological Sciences*, **365**(1554): 3065–3081.

Roselló-Soto, E., Galanakis, C. M., Brnčić, M., Orlien, V., Trujillo, F.J., *et al.* (2015) Clean recovery of antioxidant compounds from plant foods, by-products and algae assisted by ultrasounds processing. Modeling approaches to optimize processing conditions. *Trends in Food Science and Technology*, **42**(2): 134–149.

Schieber, A., Stintzing, F. and Carle, R. (2001) By-products of plant food processing as a source of functional compounds—recent developments. *Trends in Food Science and Technology*, **12**(11): 401–413.

Smithers, G.W. (2008) Whey and whey proteins – From 'gutter-to-gold'. *International Dairy Journal*, **18**(7): 695–704.

Takata, M., Fukushima, K., Kino-Kimata, N., Nagao, N., Niwa, C. and Toda, T. (2012) The effects of recycling loops in food waste management in Japan: Based on the environmental and economic evaluation of food recycling. *Science of the Total Environment*, **432**: 309–317.

Thailand Board of Investment. (2015) Inustry Focus, Food Industry. Thailand Ivestment Review, 25 August 2015. Retrieved from http://www.boi.go.th/tir/issue_content.php?issueid=126;page=42 website

Todd, G.N. (2000) High temperature countercurrent solvent extraction of capsicum solids: Google Patents.

Waldron, K. (2007) *Handbook of Waste Management and Co-product Recovery*. vol. **1**, Woodhead Publishing Ltd, Cambridge.

Westendorf, M.L. (2000) Food waste as animal feed: An introduction. In: *Food Waste to Animal Feed*, Chapter 1. Iowa State University Press, Ames, IO.

zu Ermgassen, E.K.H.J., Phalan, B., Green, R.E. and Balmford, A. (2016) Reducing the land use of EU pork production: Where there's swill, there's a way. *Food Policy*, **58**: 35–48.

Index

Note: Page numbers in *italic* denote figures, those in **bold** denote tables.

Food Processing By-Products and their Utilization, First Edition. Edited by Anil Kumar Anal.

Printed and bound by CPI Group (UK) Ltd, Croydon, CR0 4YY
17/09/2026
14964607-0003